TABLE A-6
Student's *t* distribution

Example For 15 degrees of freedom, the *t* value that corresponds to an area of 0.05 in both tails combined is 2.131.

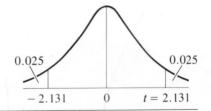

0.025   0.025

−2.131   0   *t* = 2.131

Degrees of Freedom	Area in Both Tails Combined			
	0.10	0.05	0.02	0.01
1	6.314	12.706	31.821	63.657
2	2.920	4.303	6.965	9.925
3	2.353	3.182	4.541	5.841
4	2.132	2.776	3.747	4.604
5	2.015	2.571	3.365	4.032
6	1.943	2.447	3.143	3.707
7	1.895	2.365	2.998	3.499
8	1.860	2.306	2.896	3.355
9	1.833	2.262	2.821	3.250
10	1.812	2.228	2.764	3.169
11	1.796	2.201	2.718	3.106
12	1.782	2.179	2.681	3.055
13	1.771	2.160	2.650	3.012
14	1.761	2.145	2.624	2.977
15	1.753	2.131	2.602	2.947
16	1.746	2.120	2.583	2.921
17	1.740	2.110	2.567	2.898
18	1.734	2.101	2.552	2.878
19	1.729	2.093	2.539	2.861
20	1.725	2.086	2.528	2.845
21	1.721	2.080	2.518	2.831
22	1.717	2.074	2.508	2.819
23	1.714	2.069	2.500	2.807
24	1.711	2.064	2.492	2.797
25	1.708	2.060	2.485	2.787
26	1.706	2.056	2.479	2.779
27	1.703	2.052	2.473	2.771
28	1.701	2.048	2.467	2.763
29	1.699	2.045	2.462	2.756
30	1.697	2.042	2.457	2.750
40	1.684	2.021	2.423	2.704
60	1.671	2.000	2.390	2.660
120	1.658	1.980	2.358	2.617
Normal Distribution	1.645	1.960	2.326	2.576

Source: From Table III of Fisher and Yates: *Statistical Tables for Biological, Agricultural and Medical Research*, published by Longman Group, Ltd., London (1974) 6th edition (previously published by Oliver and Boyd, Ltd., Edinburgh), and by permission of the authors and publishers.

Make No Marks
except

Yellow Highlighting

Textbook Department
Central Missouri State University
Union Room 121
816-429-4171

PROPERTY OF
TEXTBOOK DEPARTMENT
CMS UNIVERSITY STORE

Union Room 121
Warrensburg, MO 64093
816-429-4171

Copy No. _____ 216 _____

This book is rented to you for a small sum as
a favor, and it should be used with care.

You are responsible for the return of this book
in good condition. Damage to the book may
result in the purchase of the book.

A late fee of $5/book will be assessed if this
book is returned later than 5:00 p.m. the last day
of finals of the current semester.

Write your name below on the first blank space.
It will help if the book is lost or stolen.

Name _____

_____

_____

_____

_____

CMSU 973 - 89

STATISTICAL ANALYSIS FOR DECISION MAKING

FIFTH EDITION

STATISTICAL ANALYSIS FOR DECISION MAKING

FIFTH EDITION

 MORRIS HAMBURG
THE WHARTON SCHOOL
UNIVERSITY OF PENNSYLVANIA

HBJ

HARCOURT BRACE JOVANOVICH, PUBLISHERS
AND ITS SUBSIDIARY, *ACADEMIC PRESS*
SAN DIEGO   NEW YORK   CHICAGO   AUSTIN   WASHINGTON, D.C.
LONDON   SYDNEY   TOKYO   TORONTO

PREFACE

As in previous editions, the objectives of this fifth edition of *Statistical Analysis for Decision Making* are to present the fundamental concepts and methods of statistics in a clear, straightforward manner and to help students develop critical judgment and decision-making abilities through the use of quantitative methods. To achieve these objectives more effectively, I have incorporated many suggestions and comments from my colleagues at The Wharton School and from instructors at other universities who have taught from the fourth edition. From my own students I have gained considerable insight into making the material easier to understand while retaining the precision of the theory and concepts involved. This book is designed for a first course in statistics for students in business and economics, although the topical presentation and development of the methodology are appropriate for students in public administration, social sciences, and liberal arts as well. I hope to convey the idea that statistics is an exciting field that deals with a scientific method for acquiring, analyzing, and using numerical data for decision making and inference and at the same time focus on the practical applications of statistics. Many of the applications, examples, and exercises pertain to the analysis and solution of problems in managerial decision making, but there are also problems in such fields as finance, accounting, market research, quality control, consumer research, economics, psychology, and education.

The structure of the book gives the instructor considerable flexibility in designing a course. With appropriate selection of chapters, this book may be used for either a one- or a two-semester course. Because this book presents the *power* of modern statistical reasoning and the *versatility* of methods, without bogging down the discussion in mathematical formalities, the philosophy, concepts, and methods of modern statistical analysis can be understood by any diligent student with a modest mathematical background. Mathematical derivations have not been included in the body of the text, but some are given in footnotes and in Appendix C. Points for which an explanation in the language of calculus is particularly helpful also appear in footnotes.

In keeping with today's use of computers to perform many of the tedious and time-consuming calculations involved in statistical methods, there are available with this edition computer diskettes designed for use with the Minitab and EasyStat software packages. The Minitab diskette contains exercises that appear at the end of most chapters in the text, data sets that correspond to the exercises, and solutions to the

Minitab exercises. The EasyStat diskette also contains problems and data sets and has an accompanying manual. Therefore, instructors using the textbook have the flexibility of using Minitab, a powerful commercial software package with the capability of carrying out many statistical procedures both within and beyond the scope of this book, or EasyStat, a menu-driven educational statistical software package, less powerful and comprehensive than Minitab, but user-friendly and simple to apply.

Although the advantages of integrating the use of computers with a statistics book are evident, please note that most of the exercises in this book may be solved with pencil, paper, and hand calculators. Thus, the fifth edition should appeal to instructors who favor heavy use of computers in their courses, as well as those instructors who share my conviction that one better understands the fundamental theory and methodology of statistical analysis by thinking through and carrying out calculations than by placing sole reliance on the computer.

Many combinations of chapters in the book may be used in developing a basic statistics course. The fundamentals of classical statistics—descriptive statistics, probability and random variables, sampling, statistical inference, and regression and correlation analysis are presented in Chapters 1–10. Chapters 11, 12, and 13 are self-contained discussions of time series, index numbers, and nonparametric statistics, respectively. Chapters 14, 15, and 16 deal with statistical decision analysis, and Chapter 17 compares classical statistics with Bayesian decision analysis.

Within this edition numerous changes have been made in organization, emphasis, and coverage, and a number of sections were rewritten to achieve greater clarity. Most of the exercises are new to this edition, and many of the examples have been changed, revised, or updated. New review exercises, which appear at the end of Chapters 5, 8, and 10, generally require the student to use material from more than one section or chapter. Such review exercises are also provided for the Minitab material.

In Chapter 1, illustrative computer outputs for tabular and graphic presentation and box plot displays have been added, as well as a new optional section on the geometric mean. In Chapter 3, expected values and variances of discrete probability distributions have been introduced and summarized to provide a more integrated discussion of the probability distribution material. Also, a section has been added on computer displays of probability distributions from Minitab and EasyStat software.

In Chapter 5, on sampling distributions, the sequence of topics has been substantially reorganized based on perceptive comments of professors who have used the fourth edition. An optional section has been added to Chapter 6 on the "jackknife" technique, a very versatile procedure for setting up confidence intervals for any parameter from data observed in a single sample. Chapter 7, which deals with hypothesis testing, places greater emphasis on the use of the p value as an alternative to significance levels in evaluating sample evidence. Also, a section on the use of Minitab for hypothesis testing has been added and includes the appropriate commands shown for each test. In Chapter 8, two-factor analysis of variance with interaction has been added along with a discussion of how such an analysis is carried out in Minitab. A brief subsection has also been added on the use of Minitab for chi-square tests.

A section has been added to Chapter 9 on the use of Minitab for two variable regression analysis. Also, a chapter-long example of regression analysis has been revised and updated to include Minitab commands and output. In Chapter 10, multiple regression and correlation analysis, the Minitab computer application is used to represent a detailed, step-by-step example of how a multiple regression model is built. Minitab commands, graphs, and the resulting output are shown for each step in the analysis. In Chapter 11 on time series analysis, Minitab examples illustrate how easily the necessary calculations are carried out with this software.

For the first time with this fifth edition, Key Terms and Key Formulas appear at the end of most chapters. The endpapers contain four of the most useful tables from the text and Appendix A, repeated for easy reference. A glossary of symbols follows Appendix E, with each symbol keyed to the section in which that symbol is introduced.

Several supplementary pedagogical aids for this fifth edition are available in computer diskettes or separate publications.

- A computer diskette for use with Minitab software contains computer exercises, solutions, and data sets. Many exercises are also shown in the text at the end of appropriate chapters; and as an aid to students, instructions for Minitab use immediately follow the Preface. Some of the data sets were derived from the Citibase Series and Slater-Hall Information Products and are larger and more realistic than the data in typical textbook exercises.

- A diskette for the EasyStat computer program operates within the framework of a spreadsheet and includes a data set for 1,000 urban families. Students can select random samples for carrying out exercises. The Minitab and EasyStat programs are available for students who have access to an IBM-PC. An EasyStat manual that is available to students accompanies the EasyStat diskette.

- The Study Guide provides supplementary exercises, as well as worked-out problems and explanations.

- Solutions to all of the text exercises and Minitab exercises appear in the Solutions Manual. (Only the even-numbered text solutions appear in the book.)

- Additional exercises (with solutions) suitable for homework or testing are available in a Testbook, which also has a set of multiple-choice and true–false questions for examinations. The exercises and questions have doubled in number from those available with the fourth edition.

My thanks again are extended to the many organizations and individuals that have made contributions to this book. My colleagues in the Statistics Department of The Wharton School of the University of Pennsylvania deserve special mention for constructive comments and recommendations stemming from their experience in teaching from previous editions and from interacting with students. I also express grateful appreciation to Richard W. Gideon, president of Dick Gideon Enterprises, for the television and other data and estimates that he continues to make available for us in the chapter on time series.

I am very grateful to those who assisted in the development of earlier editions and special thanks go to Thomas Johnson (North Carolina State University), Brenda Masters (Oklahoma State University), and Byung T. Cho (University of Notre Dame) whose excellent constructive comments and suggestions were critical to the preparation of the fifth edition. This fifth edition has also benefited greatly from the candid comments and criticisms obtained from the following professors: Joe D. Berry (Marion Military Institute), James M. Cannon (Florida Institute of Technology), Carl Gordon (California State University, Sacramento), Rudy Moore (Rosemont College), Manuel G. Russon (University of South Alabama), William L. Steglitz (Nichols College), Kishor Thanawala (Villanova University), Robert J. Thornton (Lehigh University), and Ebenge Usip (Youngstown State University).

The comments of teachers and students who have used previous editions have continued to be very helpful. My appreciation is especially extended to the following instructors: Richard W. Andrews (University of Michigan), David Ashley (University of Missouri), Eileen C. Boardman (Colorado State University), William D. Coffey (St. Edward's University), F. Damanpour (La Salle College), Maynard M. Dolecheck (Northeast Louisiana University), Fred H. Dorner (Trinity University), Linda W. Dudycha (University of Wisconsin), David Eichelsdorfer (Gannon University), David Frew (Gannon University), James C. Goodwin (University of Richmond), Charles R. Gorman (La Salle College), I. Greenberg (George Mason University), John B. Guerard, Jr. (Lehigh University), Jack A. Holt (University of Virginia), Yutaka Horiba (Tulane University), Gary Kern (University of Virginia), Burton J. Kleinman (Widener University), Ming-Te Lu (St. Cloud State University), William F. Matlack (University of Pittsburgh), Patrick McKeown (University of Georgia), Elias Alphonse Parent (George Mason University), Chander T. Rajaratnam (Rutgers University), S. R. Ruth (George Mason University), James R. Schaefer (University of Wisconsin), Stanley R. Schultz (Cleveland State University), Marion G. Sobol (Southern Methodist University), William R. Stewart, Jr. (College of William and Mary), Chris A. Theodore (Boston University), Charles F. Warnock (Colorado State University), Roger L. Wright (University of Michigan), and Thomas A. Yancey (University of Illinois).

An especially warm expression of gratitude goes to Bryan Sayer and Zhongquan Zhou for their expert work in the preparation of the diskettes and related material for the EasyStat and Minitab computer programs, respectively. A special debt is also owed to Delores Johnson for cheerful, extremely competent assistance in word processing and related secretarial assistance. My thanks are again extended to the fine professional staff of Harcourt Brace Jovanovich: Scott Isenberg, Acquisitions Editor; Dee W. Salisbury, Production Editor; Suzanne Montazer, Designer; and Mary Kay Yearin, Production Manager.

I am grateful to the Literary Executor of the Late Sir Ronald A. Fisher, F.R.S., to Dr. Frank Yates, F.R.S., and to Longman Group, Ltd., London, for permission to reprint Tables III and IV from their book *Statistical Tables for Biological, Agricultural, and Medical Research* (6th edition, 1974). My gratitude also goes to the other authors and publishers whose generous permission to reprint tables or excerpts from tables has been acknowledged at appropriate places.

As in previous editions, I dedicate this book to my wife, June, and my children, Barbara and Neil.

Morris Hamburg

INSTRUCTIONS FOR MINITAB USAGE

If you are using a personal computer and have Minitab software installed in drive C, you can enter Minitab by typing **cd Minitab** and then **Minitab**.

Once you are in Minitab, you can familiarize yourself with the program by using the HELP command. When working problems, you will need to retrieve data sets from the disk that accompanies the Solutions Manual by typing:

MTB > retrieve 'a:xxxx'

where **xxxx** is the name of the data set specified in the problem.

To output your own solution, use Minitab command

MTB > outfile 'yyyy.zzz'

where **yyyy.zzz** is a filename of your choice. When you have finished a problem, type:

MTB > nooutfile

to close your own solution file. To modify and print your solution file, you have to exit Minitab. To exit Minitab, simply type:

MTB > stop

You can print and edit your solution file in many ways depending on the computer software with which you are working. We suggest that you use a word processing software, for example, WordPerfect, to edit your solution file first. You can also use the software to print the solution file. On the other hand, to print in DOS, simply type:

C:\> print 'yyyy.zzz'

Listed below are filenames for solutions and data sets. The chapter numbers in the solutions correspond to the same chapter numbers in the text.

Chapter Numbers and Filenames of Solutions

Chapter	Filename (Solutions)
1	**abort.ans**
1	**mstat.ans**
1	**bankroa.ans**

Chapter Numbers and Filenames of Solutions *(continued)*

Chapter	Filename (Solutions)
3	discrete.ans
5	random.ans
Review	rev1-5.ans
6	roa.ans
7	testing.ans
8	table.ans
Review	rev6-8.ans
9	regress.ans
10	multiple.ans
11	series.ans
13	nonpar.ans

A List of Filenames (Data)

ability.mtw	abort.mtw	bankroa.mtw
butter.mtw	customer.mtw	databank.mtw
dmotor.mtw	dow.mtw	dummy.mtw
gnp.mtw	invest.mtw	lab.mtw
lifetime.mtw	mstat.mtw	profit.mtw
project.mtw	rating.mtw	roa.mtw
salary.mtw	sales.mtw	vote.mtw

Remarks:

- These instructions, problems, and solutions were developed with PC Minitab, Release 6.1.1. If you use an earlier or later release, there will be some very minor differences in the appearance of some output.

- Some sets of the data were taken from the Citibase Series and Slater-Hall Information Products. Citibase Series is a databank containing a large number of mainly U.S. macroeconomic time-series data. Slater-Hall Information Products records both economic and demographic statistics at the following four levels: national, state, county, and city.

- Although some instructions for the use of many Minitab commands are given in the computer problems, a Minitab handbook may be useful for problem solving. Also, you can always type **HELP** in Minitab to get information when you are uncertain. Two Minitab handbooks are listed below for your reference:

    1. Ryan, Barbara F., Brian L. Joiner, and Thomas A. Ryan. *Minitab Handbook*, 2nd ed. Boston: Duxbury Press, 1985.

    2. Miller, Robert B., *Minitab Handbook for Business and Economics*. 2nd ed. Boston: PWS-Kent Publishing Company, 1988.

CONTENTS

2

INTRODUCTION TO PROBABILITY

69

3

DISCRETE RANDOM VARIABLES AND PROBABILITY DISTRIBUTIONS

111

4

STATISTICAL INVESTIGATIONS AND SAMPLING   191

5

SAMPLING DISTRIBUTIONS   217

6

ESTIMATION   273

7

HYPOTHESIS TESTING

8

CHI-SQUARE TESTS AND
ANALYSIS OF VARIANCE

9

REGRESSION ANALYSIS AND CORRELATION ANALYSIS

10

MULTIPLE REGRESSION AND CORRELATION ANALYSIS

14

DECISION MAKING USING
PRIOR INFORMATION

15

DECISION MAKING WITH
POSTERIOR PROBABILITIES

16

DEVISING OPTIMAL STRATEGIES
PRIOR TO SAMPLING

17

COMPARISON OF CLASSICAL AND BAYESIAN STATISTICS

INTRODUCTION

What do the following questions have in common? Such questions occur in every aspect of our lives. Personal questions: Whom should I marry? What field should I choose as my life's work? Which automobile should I buy? What clothes should I wear today?

A private business corporation's questions: Should the corporation buy or lease this building? Should this new product be placed on the market? In which of these capital investment proposals should the corporation participate? Should the corporation invest now or wait a year? Or two?

A national government's questions: Should the Federal Reserve Bank place more emphasis on controlling inflation or fighting the possibility of a forthcoming recession? Should a proposed new educational approach to the drug problem be instituted? Should a proposed national long-term nursing home care program be adopted? If so, which one of three competing programs should be selected?

All of these questions require that decisions be made under conditions of uncertainty. In most cases, the costs and benefits associated with these decisions can be estimated only roughly, and the outcomes that will affect these costs and benefits are uncertain. The required choices must be based on incomplete information. Nevertheless, choices must be made. Even a failure to make a decision constitutes a choice—one that may have net benefits far less or far more desirable than those that flow from an explicit decision.

In its modern interpretation, *statistics* is a body of theory and methodology for drawing inferences and making decisions under conditions of uncertainty. From this interpretation, it would seem that the field of statistics has much to contribute toward answering some of the questions posed earlier, and indeed it does. The raw material of statistics is *statistical data* or numbers that represent counts and measurements of events or objects. The theory and methodology of statistics aid in determining what data should be compiled and how they should be collected, analyzed, interpreted, and presented to make the best inferences and decisions.

However, you may protest, "The field of statistics is not going to help me answer some of the earlier questions, such as whom should I marry or what clothes should I wear today?"

For some decisions, a careful scientific approach based on quantitative data does not seem very appropriate. Many decisions require a substantial component of intuitive judgment. Nonetheless, in many areas of human activity, statistical analysis provides a solid foundation for decision making. Statistical concepts and methods bring a logical, objective, and systematic approach to decision making in social, governmental, business, and scientific problems. They are not meant to replace intuition and common-sense judgments. On the contrary, they assist in structuring a problem and in bringing the application of judgment to it.

As an example, let us return to the question that probably seems least likely to be subject to a meaningful solution by statistical reasoning: "Whom should I marry?" In structuring this problem, let us consider the following line of thought. If you marry a (the?) "right" person, you have made no error. Also, if you do not marry someone you should not have married, you have again made a correct decision. On the other hand, if you fail to marry someone with whom you would have been happy, you have made an error. Or, if you marry someone with whom your future life falls far short of a state of connubial bliss, you have also made an error. After studying the subject of hypothesis testing, you will recognize the difference between the two types of errors. Which type of error is more serious? If you are considering the possibility of marrying someone, which is the best course of action, to marry or not to marry? With which action is the higher expected "payoff" associated? Although the problem and questions here are raised tongue in cheek and without the expectation of a serious answer, they illustrate a way in which you may begin to structure any problem and apply judgment to it. We cannot dispute that someone's romantic sensibilities may be offended by any attempt to use the suggested method of analysis to solve this problem. Our purpose is merely to indicate that even so unlikely a problem for quantitative analysis may be structured in a framework that can assist decision making. It is currently recognized that the field of statistics aids in providing the basis for arriving at rational decisions regarding a tremendous variety of matters related to business, public affairs, science, and other fields. In many of these matters, such quantitative analysis has been found to be not only applicable but also extremely helpful.

The most widely known statistical methods are those that summarize numerical data in terms of **averages** and other descriptive measures.

If we are interested in the incomes of a group of 1,000 families chosen at random in a particular city, important characteristics of these incomes may be described by calculating an average income and a measure of the spread, or dispersion, of these incomes around the average.

The essence of modern statistics, however, is the theory and the methodology of **drawing inferences** that extend beyond the particular set of data examined and of **making decisions** based on appropriate analyses of such inferential data.

We are probably not so much interested in the incomes of the particular 1,000 families included in the sample as we are interested in an inference about the incomes of all families in the city from which the sample was drawn.

- Such an inference might be in the form of a **test of a hypothesis** that the average income of all families in the city is $35,000 or less.
- The inference could also be in the form of a single figure, an **estimate** of the average income of all families in the city based on the average income observed in the sample of 1,000 families.
- Or the marketing department of a company may want the information in order to decide among different types of advertising programs based on the identification of the city as a low-, medium-, or high-income area.

The mathematical theory of probability provides the logical framework for the mental leap from the sample of data studied to the inference about all families in the city and for decisions such as the type of advertising program to be used.

The preceding example presents three points:

- We may have wanted an inference about the incomes of *all* families in the city. The totality of families in the city (or, more generally, the totality of the elements about which the inference is desired) is referred to in statistics as the **universe** or **population**.
- Because it would have been too expensive and too time-consuming to obtain the income data for every family in the city, only the sample of 1,000 families was observed. The 1,000 families, which represent a collection of only some elements of the universe, are referred to as a **sample**.
- In statistics, sample data are collected in order to make **inferences** or **decisions** concerning the populations from which samples are drawn.

Note that the sample was drawn "at random" from the population. A **random sample** is one drawn in such a way that the probability, or likelihood, of inclusion of every element in the population is known. However, even though these probabilities of inclusion may be known, the average income that would be observed for a random sample of 1,000 families would vary from sample to sample. These sample-to-sample variations are known as **chance sampling fluctuations**.

Although we cannot predict with certainty what the average income will be for any particular sample, the **theory of probability** enables us to compute how often these different sample results occur in the long run. It is an intriguing and remarkable fact that even though there is *uncertainty* concerning which particular sample may have been drawn, probability theory provides a rational basis for inference and decision making about the population from which the sample was taken. This textbook deals with the theory and methods by which inferences and decisions are made.

The Role of Statistics

Statistical concepts and methods are applied in many areas of human activity. They are used in the physical, natural, and social sciences, in business and public administration, and in many other fields.

In the sciences, the applications extend from the design and analysis of experiments to the testing of new and competing hypotheses. In industry, statistics makes its contributions in both short- and long-range planning and decision making. Many firms use statistical methods to analyze patterns of change and to forecast economic trends for the firm, the industry, and the economy as a whole. Such forecasts provide the foundation for corporate planning and control; purchasing, production, and inventory control depend on short-range forecasts, while capital investment and long-term development decisions depend on long-range forecasts. Statistical methods are employed in production control, inventory control, and quality control. To control the quality of manufactured products, for example, statistical methods are used to differentiate between variation attributable to chance causes and variation too great to be considered a result of chance; the latter type of variation can be analyzed and remedied. Application of statistical quality control methods results in substantial improvement in the quality of products and in lower costs because of reduction in rework and spoilage. Such statistical quality control methods have been a major factor in the improvement of the quality of Japanese-manufactured products since World War II.*

Over the past four decades, a body of **quantitative techniques and procedures** has been developed in the fields of business and government to aid and improve managerial decision making. The field of statistics has provided many fruitful ideas and techniques in this development. Applications of statistics are evident in most activities of business firms, including production, financial analysis, distribution analysis, market research, research and development, and

---

* The American statistician Dr. W. Edwards Deming was a major contributor to the introduction of these methods in Japan. An annual award in Deming's name is made to a Japanese firm singled out for distinguished achievements in quality control.

accounting. Statistical methods are used often and with increasing sophistication in every field in which they have been introduced, and they are an integral part of the development of rational and quantitative approaches to the solution of business problems. One characteristic of this development has been the increased adoption of scientific decision-making approaches using **mathematical models**. These models are mathematical formulas or equations that state the relationship among the important factors or variables in a problem or system. (For example, an equation may be developed to represent the relationship between a company's sales and the economic and other variables that influence sales.) Chapters 9 and 10 of this text discuss the methods for deriving one such mathematical model.

Extensive statistical activities are conducted by federal, state, and local governments. There are many applications of statistical ideas and methods in governmental administration; governments collect and disseminate a great deal of statistical data. The most highly organized and extensive **statistical information systems** are those of the federal government. Such information systems—which include national income and product accounts, input–output accounts, flow of funds accounts, balance of payments accounts, and national balance sheets—depend on massive statistical collection and distribution systems. Statistical methods are applied to the resulting data to assess past trends and current status and to project future economic activity. These methods provide measures of human and physical resources and of economic growth, well-being, and potential. They are essential tools for appraising the performance and for analyzing the structure and behavior of an economy.

Data collection and dissemination activities are also carried out by government and private agencies in fields such as population, vital statistics, education, labor force, employment and earnings, business and trade, prices, housing, medical care, public health, agriculture, natural resources, welfare services, law enforcement, area and industrial development, construction, manufacturing, transportation, and communications.

Statistical analysis constitutes a body of theory and methods that plays an important role in this wide variety of human activities. It is extremely useful for communicating information, for drawing conclusions from data, and for the guidance of planning and decision making.

1

FREQUENCY DISTRIBUTIONS AND SUMMARY MEASURES

Variation is a basic fact of life. As individuals, we differ in age, sex, height, weight, and intelligence as well as in the quantities of the world's goods we possess, the amount of our good or bad luck, and countless other characteristics. In the business world, variations are observed in the articles produced by manufacturing processes, in the yields of the economic factors of production, in production costs, financial costs, marketing costs, and so forth.

The methods discussed in this chapter are useful for describing patterns of variation in data. Such variations occur in data observed at a particular time and in data occurring over a period of time.

Summarizing Variations in Data

We begin our discussion by asking how you might go about summarizing the variation in a large body of numerical data. Suppose that data were available in a computer file on the ages of all individuals in the United States and that you wanted to describe these approximately 250 million figures in some generally useful manner. How might you go about it, assuming that adequate resources for processing the data were available? Since it would be difficult to see important characteristics of the data by merely listing the data, you probably would group the figures into classes. For example, you might set up age classes of under five years, at least five but under 10 years, and so forth. You could have the computer tabulate the number of persons in each class. If you divide these

Group by class and record frequency of occurrence

numbers by the total population, you have the proportion of the population in each class. You would then find it relatively easy to summarize the general characteristics of the age distribution of the population. If you compared similar distributions, say, for the years 1920 and 1990, a number of important features would be observable without any further statistical analysis. For example, the range of ages in both distributions would be clear at a glance. The higher proportions of persons under the age of 20 and of persons over the age of 65 in 1990 than in 1920 would stand out. Also smaller percentages of persons in the age categories from 50 to 60 years would be observed for 1990, reflecting the decline in births during the 1930–1940 decade. Thus, the simple device of grouping the age figures into classes and recording the frequencies of occurrence in these classes shows us some of the underlying characteristics of the nation's age composition for each year. Generalizations about age patterns thus become easier to make.

Calculate an average and a measure of spread

You might also want to continue your description of the age distributions by calculating one or more types of average. For example, you might be interested in computing an average age for the population in 1920 and in 1990 to determine whether this average had increased or decreased. Furthermore, if you wished to give a more exact description of the fact that in the later period there were heavier concentrations of persons in the younger and older age groups, you might attempt to construct a measure of how the ages were spread around an average age. Because of these concentrations, the measure of spread of dispersion around the average would tend to be larger in 1990 than in the earlier period.

Statistical terms

The types of statistical techniques that might be used to summarize and describe the characteristics of the age data constitute the subject of this chapter.

- The table into which the data are grouped is referred to as a **frequency distribution**.
- The average or averages that can be computed are measures of **central tendency** or **central location** of the data.
- The measure of spread around the average is a measure of **dispersion**.

These and other techniques that group, summarize, and describe data are referred to as **descriptive statistics**. If the data treated by descriptive statistics represent a sample from a larger group or population, as noted in the Introduction, inferences may be desired about this larger group. Ways of making such statistical inferences are discussed in subsequent chapters.

Significance of order of occurrence of data

The order or occurrence of data is frequently significant in data analysis. **Cross-sectional data** refers to data observed at a point in time, whereas **time series data** are sets of figures that vary over a period of time.

Frequency distribution analysis is concerned with cross-sectional data; in particular, such analysis deals with data in which the order of recording observations is of no importance (for example, the ages of the present members of the labor force in the United States, the present wage distribution of employees in the automobile industry, or the distribution of U.S. corporations by net worth on a given date).

Times series analysis would be used to record quality control data for a manufactured product, as we would be very much concerned with the order in which the articles were produced. If a run of defective articles was produced, we would want to know when this run occurred and what the general time pattern of production of defective and good articles was. Similarly, in the study of economic growth, we might be interested in the variation over time of such data as real income per person or real gross national product per person. General methods of time series analysis are treated in Chapter 11.

1.1
FREQUENCY DISTRIBUTIONS

When we are confronted with masses of ungrouped data (that is, listings of individual figures), it is difficult to generalize about the information the masses contain. However, if a frequency distribution of the figures is formed, many features become readily discernible.

> A **frequency distribution** or **frequency table** records the number of cases that fall in each class of the data.

The numbers in each class are referred to as **frequencies**; hence the term "frequency distribution." When the numbers of items are expressed by their proportion in each class, the table is usually referred to as a **relative frequency distribution** or a **percentage distribution**.

How the classes of a frequency distribution are described depends on the nature of the data. In all cases, data are obtained either by counting or by measuring. For example, individuals have characteristics such as race, nationality, sex, and religion, and counts can be made of the number of persons who fall in each of the relevant categories. If a classification of employees in a manufacturing firm by union membership is used, the frequency distribution may be shown as in Table 1-1.

Describing the classes of a frequency distribution

TABLE 1-1
Distribution of the number of employees in a manufacturing firm classified by union membership

Membership Classification	Number of Employees
Union Members	16,365
Nonunion Members	12,472
Total	28,837

FIGURE 1-1

Number of employees in a manufacturing firm classified by union membership

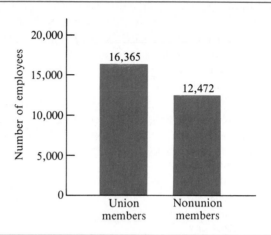

Counting discrete variables

Characteristics such as nativity, color, sex, and religion that can be expressed in qualitative classifications or categories are often referred to as **attributes** or **discrete variables**. It is always possible to encode the attribute classifications to make them numerical. Thus, in our illustration, "union members" could have been denoted 0 and "nonunion members" 1. In certain cases the data seem to fall naturally into simple numerical classifications. For example, families may be grouped according to number of children; the classes could be labeled 0, 1, 2, and so forth.

Data for qualitative characteristics or discrete variables can be presented graphically in terms of simple bar charts. Figure 1-1 gives a bar chart representation of the data given in Table 1-1.

Measuring continuous variables

To obtain **continuous variables**, or **continuous data** (that is, data that can assume any value in a given range), we perform numerical measurements rather than counts. When large numbers of measurements are made, it is convenient to use intervals or groupings of values and to list the number of cases in each class. With this procedure, a few problems have to be resolved concerning the number of class intervals, the size of these intervals, and the manner in which class limits should be stated.

1.2
CONSTRUCTION OF A FREQUENCY DISTRIBUTION

The decisions about the number and size of the classes in a frequency distribution are essentially arbitrary. However, these two choices are clearly interrelated.

Selecting size and number of class intervals

> The smaller the intervals chosen, the more intervals will be needed to cover the range of the scores.

Frequency distributions generally are constructed with from 5 to 20 classes. When class intervals are of equal size, comparisons of classes are easier and subsequent calculations from the distribution are simplified. However, this is not always a practical procedure. For example, with data on the annual incomes of families, in order to show the detail for the portion of the frequency distribution where the majority of incomes lie, class intervals of $4,000 may be used up to about $20,000; then intervals of $5,000 may be used up to $45,000 and a final class of $45,000 and over. Clearly, maintaining equal-sized classes of, say, $1,000 throughout the entire range of income would result in too many classes. On the other hand, if much larger class intervals were used, too many families would be lumped together in the first one or two classes, and we would lose the information concerning how these incomes were distributed. The use of unequal class intervals and an open-ended interval for the highest class provides a simple way out of the dilemma.

An **open-ended class interval** is one that contains only one specific limit and an "open" or unspecified value at the other end, as for example *$35,000 and over* or *110 pounds and under*. The use of **unequal class sizes** and open-ended intervals generally becomes necessary when most of the data are concentrated within a certain range, when gaps appear in which relatively few items are observed, and when there are a very few extremely large or extremely small values. Open-ended intervals are sometimes also used to retain confidentiality of information. For example, the identity of the small number of individuals or companies in the highest class may be general knowledge, and stating an upper limit for the class might be considered excessively revealing.

Open-ended class interval

Unequal class sizes

We illustrate the construction of a frequency distribution by considering the figures in Table 1-2, which represents the scores obtained on the commercial banking examination given in Urbanbank's training program during the past six months. Although the data have been arrayed from lowest score to highest, it is difficult to discern patterns in the ungrouped figures. However, when a frequency distribution is constructed, the nature of the data clearly emerges. There is no single perfect frequency distribution for a given set of data. Several

TABLE 1-2
Scores obtained on the commercial banking examination given in Urbanbank's training program during the past six months

57	70	75	78	82	84	86	88	89	94
62	70	75	78	82	84	86	88	90	94
62	71	75	79	82	84	87	88	90	95
64	72	76	79	83	84	87	88	90	95
65	72	76	79	83	85	87	88	91	96
67	72	76	80	83	85	87	88	92	96
67	73	77	80	84	85	87	88	92	98
68	73	78	81	84	86	87	89	92	98
68	74	78	81	84	86	87	89	93	99
69	74	78	81	84	86	88	89	93	99

alternative distributions with different class interval sizes and different highest and lowest values may be equally appropriate.

Let us assume that we would like to set up a frequency distribution with 8 classes for the list of figures shown in Table 1-2 and that we want the classes *Formula for class* to be of equal size. A simple formula to obtain an estimate of the appropriate *interval size* interval size is

$$i = \frac{H - L}{k}$$

where i = the size of the class intervals
   H = the value of the highest item
   L = the value of the lowest item
   k = the number of classes

This formula for class interval size simply divides the total range of the data (that is, the difference between the values of the highest and lowest observations) by the number of classes. The resulting figure indicates how large the class intervals would have to be in order to cover the entire range of the data in the desired number of classes. Other considerations involved in determining an appropriate number of classes are discussed in section 1.4.

In the case of the examination scores data,

$$i = \frac{(99 - 57)}{8} = 5.25$$

Rounding off for Since it is desirable to have convenient sizes for class intervals, the 5.25 figure *convenient class* may be rounded to 5, and the distribution may be tentatively set up on that *intervals* basis. The frequency distribution shown in Table 1-3 results from a tally of the number of items that fall in each 5 point class interval.

TABLE 1-3
Frequency distribution of the scores obtained on the commercial banking examination given in Urbanbank's training program during the past six months

Examination Score	Frequency
55 and under 60	1
60 and under 65	3
65 and under 70	6
70 and under 75	10
75 and under 80	15
80 and under 85	19
85 and under 90	27
90 and under 95	11
95 and under 100	8
Total	100

Some important features of these data are immediately seen from the frequency distribution. The approximate value of the range, or the difference between the values of the highest and lowest items, is revealed. (Of course, since the identity of the individual items is lost in the grouping process, we cannot tell from the frequency table alone what the exact values of the highest and lowest items are). Also, the frequency distribution gives at a glance some notion of how the elements are clustered. For example, more of the examination scores fall in the interval from 85 to 90 than in any other single class. When the frequencies in the classes immediately preceding and following the 85 to 90 grouping are added to the 27 in that interval, a total of 57 of the scores are accounted for. Furthermore, the distribution shows how the data are spread or dispersed throughout the range from the lowest to the highest value. We can quickly determine whether the items are bunched near the center of the distribution or spread rather evenly throughout. Also, we can see whether the frequencies fall away symmetrically on both sides of the center of the distribution or whether they tend to fall mostly to one side of the center. We now consider various statistical measures for describing these characteristics more precisely, but much information can be gained by simply studying the distribution itself.

1.3
CLASS LIMITS

The way in which class limits of a frequency distribution are described depends on the nature of the data. Figures on ages provide a good illustration of this point. Suppose that ages were recorded as of the *last* birthday. A clear and unambiguous way of stating the class limits is as follows: 15 and under 20, 20 and under 25 and so on. (Of course, there are other ways of wording the limits, such as *at least 15 but under 20* or *from 15 up to but not including 20*.)

Describing class limits

Consider the first class interval, "15 and under 20". Since ages have been recorded as of the last birthday, this class encompasses individuals who have reached at least their fifteenth birthday but not their twentieth birthday. If you are 19.999 years of age, that is, a fraction of a day away from your twentieth birthday, you fall in the first class. However, upon attaining your twentieth birthday, you fall in the second class, "20 and under 25." Thus, these class intervals are five years in size. The **midpoints** of the classes—that is, the values located halfway between the class limits—are 17.5, 22.5, 27.5, 32.5, 37.5, and so on. These values are used in computations of statistical measures for the distribution. Note that with class limits established and stated this way, the **stated limits** are in fact the true boundaries, or **real limits**, of the classes.

Stated limits and real limits

Suppose, on the other hand, that age data were rounded to the *nearest* birthday. We could follow a widely used convention and state the class limits as follows: 15-19, 20-24, and so on.

> Even though the stated limits in each class are only 4 years apart, the size of these class intervals is still 5 years.

For example, since the ages are given as of the nearest birthday, everyone between 14.5 and 19.5 years of age falls in the class 15-19. Thus, when data recorded to the nearest unit are grouped into frequency distribution classes, the lower real limit or lower boundary of any given class lies one-half unit below the lower stated limit and the upper real limit or upper boundary lies one-half unit above the upper stated limit. The midpoints of the class intervals may be obtained by averaging the lower and upper real limits or the lower and upper stated limits. For example, the midpoint of the class 15-19 is 17, which is the same figure obtained by averaging 14.5 and 19.5.

Summary
- When raw data are rounded to the *last* unit, the stated class limits and real class limits are identical.
- When raw data are rounded to the *nearest* unit, the real limits are one-half unit removed from the stated limits.
- With both types of data, the midpoints of classes are halfway between the stated limits or, equivalently, halfway between the real limits.

Class intervals should always be mutually exclusive, and the class each item falls into should always be clear. If class limits are stated as 30–40, 40–50, and so on, for example, it is not clear whether 40 belongs to the first class or the second.

Unfortunately, conventions are not universally observed. Often, one must use a frequency distribution constructed by others, and the nature of the raw data may not be clearly indicated. The producer of a frequency distribution should always indicate the nature of the underlying data.

1.4
OTHER CONSIDERATIONS IN CONSTRUCTING FREQUENCY DISTRIBUTIONS

Location of midpoint

A number of other points should be taken into account in the construction of a frequency distribution. If there are concentrations of particular values, it is desirable that these values be the midpoints of the class intervals. For example, assume that data are collected on the amounts of the lunch checks in a student cafeteria. Suppose these checks predominantly occur in multiples of five cents, although not exclusively so. If class intervals are set up as $3.70–$3.74, $3.75–$3.79, and so on, a preponderance of items would be concentrated at the lower limits.

> In calculating certain statistical measures from the frequency distribution, the assumption is made that the midpoints of classes are average (arithmetic mean) values of the items in these classes.

If, in fact, most of the items lie at the lower limits of the respective classes, a systematic error will be introduced by this assumption, because the actual averages within classes will typically fall below the midpoints.

Another factor to be considered in constructing a frequency distribution is the desirability of having a relatively smooth progression of frequencies. In many frequency distributions of business and economic data, one class contains more items than any other single class and the frequencies drop off more or less gradually on either side of this class. Table 1-3 is an example of such a distribution. (As indicated in section 1.2, the distribution may not be at all symmetrical). However, erratic increases and decreases of frequencies from class to class tend to obscure the overall pattern, and such erratic variations often arise from the use of class intervals that are too small. Increasing the size of class intervals usually results in a smoother progression of frequencies, but wider classes reveal less detail than narrower classes. A compromise must be made in the construction of every frequency distribution.

Smooth progression of frequencies

- If we use class interval sizes of one unit each, every item of raw data is assigned to a separate class.
- If we use only one class interval as wide as the range of the data, all items fall in the single class.

Within the limits of these considerations, some freedom exists for the choice of an appropriate class interval size.

1.5
GRAPHIC PRESENTATION OF FREQUENCY DISTRIBUTIONS

The use of graphs for displaying frequency distributions will be illustrated for the data on examination scores shown in Table 1-3. One method is to represent the frequency of each class by a rectangle or bar. Such a chart is generally referred to as a **histogram**. A histogram for the frequency table given in Table 1-3 is shown in Figure 1-2. In agreement with the usual convention, values of the variable are depicted on the horizontal axis and frequencies of occurrence are shown on the vertical axis.

Histogram

Many computer software packages are available that compute and/or display the various statistical measures and procedures discussed in this book. Virtually all of the computer displays, examples, and exercises discussed in

Minitab Commands

FIGURE 1-2

Histogram of frequency distribution of the scores obtained on the commercial banking examination given in Urbanbank's training program during the past six months

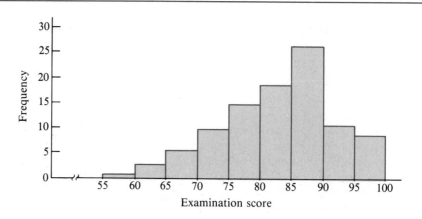

the book are in terms of PC Minitab, Release 6.1.1. It is assumed that for the carrying out of computer exercises, the reader is generally familiar with the workings of an IBM PC or an IBM compatible PC and with MS-DOS. When displays and examples are shown here, they are accompanied by the appropriate Minitab commands or command sequences. It may be noted that, for consistency, computer commands and output have generally been shown in this book in capital letters, even though in Minitab the output sometimes appears in lowercase letters.

In Figure 1-3, two possible frequency distributions are shown as Minitab output for the examination score data given in Table 1-2. The graphic form of output given in Figure 1-3 is referred to both as a frequency table or histogram. Note that in the Minitab command, the data were read into column 1—"hist" is an abbreviation for histogram—and the analyst must specify the midpoint of the first class and size of the class interval. The first histogram in Figure 1-3 assumes a midpoint of 55 for the first class and a class interval size of 5. The second histogram assumes a midpoint of 57.5 for the first class and a class interval size of 5. Of course, the second histogram displays the same frequencies as are shown in Table 1-3, but the Minitab output gives midpoints of classes rather than class limits.

A histogram that shows class limits as in Table 1-3 is given in Figure 1-4. This represents the computer output from EASYSTAT software.[1] Because EASYSTAT is menu driven, no commands are shown. In EASYSTAT, the computer prompt asks you to specify the interval width for the table and the lower limit of the first class.

Frequency polygon   An alternative method for the graphic presentation of a frequency distribution is the **frequency polygon**. In this type of graph, the frequency of each

---

[1] See B. Sayer, and M. Hamburg, EASYSTAT: *A Guide to Accompany Hamburg's* **Statistical Analysis for Decision Making, 5th edition**.

```
MTB > HIST C1 55 5


HISTOGRAM OF C1      N = 100


MIDPOINT    COUNT
   55.00       1    *
   60.00       2    **
   65.00       4    ****
   70.00       9    ********
   75.00      11    **********
   80.00      16    ****************
   85.00      26    **************************
   90.00      19    *******************
   95.00       8    ********
  100.00       4    ****


MTB > HISTOGRAM OF C1 57.5 5


HISTOGRAM OF C1      N = 100


MIDPOINT    COUNT
   57.50       1    *
   62.50       3    ***
   67.50       6    ******
   72.50      10    **********
   77.50      15    ***************
   82.50      19    *******************
   87.50      27    ***************************
   92.50      11    ***********
   97.50       8    ********
```

FIGURE 1-3

Two possible frequency distributions for the examination scores data shown in Table 1-2 (Minitab output)

class is represented by a dot have the midpoint of each class at a height corresponding to the frequency of the class. The dots are joined by line segments to form a many-sided figure, or polygon. A frequency polygon can also be thought of as the line graph obtained by joining the midpoints of the tops of the bars in a histogram. By convention, the polygon is connected to the horizontal axis by line segments drawn from the dot representing the frequency in the lowest class to a point on the horizontal axis one half a class interval below the lower limit of the first class, and from the dot representing the frequency in the highest class to a point one half a class interval above the upper limit of the last class. A frequency polygon for the distribution given in Table 1-3 is shown in Figure 1-5. It is important to realize that the line segments are

---

FIGURE 1-4

A possible frequency distribution for the examination scores data shown in Table 1-2 (EASYSTAT output)

URBAN.DAT

```
INTERVAL                 FREQUENCY

                         0    10    20    30    40    50    60    70    80    90

 55.  UNDER  60.      1  :*
 60.  UNDER  65.      3  :***
 65.  UNDER  70.      6  :******
 70.  UNDER  75.     10  :**********
 75.  UNDER  80.     15  :***************
 80.  UNDER  85.     19  :*******************
 85.  UNDER  90.     27  :***************************
 90.  UNDER  95.     11  :***********
 95.  UNDER 100.      8  :********
100.  UNDER 105.      0  :
```

---

drawn only for convenience in reading the graph and that the only significant points are the plotted frequencies for the given midpoints. Interpolation for intermediate values between such points would be meaningless. Often, midpoints of classes are shown on the horizontal axis rather than class limits as are shown in Figure 1-5.

If the class sizes in a frequency distribution were gradually reduced and the number of items increased, the frequency polygon would approach a smooth curve more and more closely. Thus, as a limiting case, the variable of interest may be viewed as continuous rather than discrete, and the polygon would assume the shape of a smooth curve. The frequency curve approached by the polygon for the examination scores would appear as shown in Figure 1-6.

FIGURE 1-5

Frequency polygon of distribution of the scores obtained on the commercial banking examination given in Urbanbank's training program during the past six months

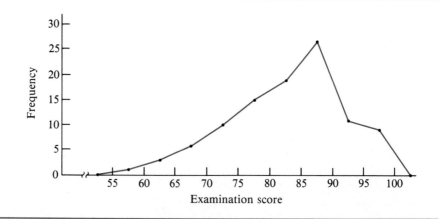

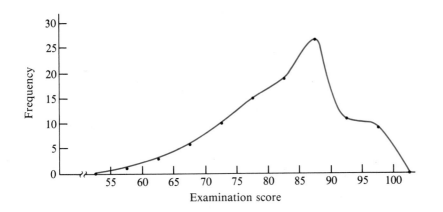

FIGURE 1-6

Frequency curve for distribution of the scores obtained on the commercial banking examination given in Urbanbank's training program during the past six months

1.6
CUMULATIVE FREQUENCY DISTRIBUTIONS

When interest centers on the number of cases that lie below or above specified values rather than within intervals, it is convenient to use a **cumulative frequency distribution** instead of the usual frequency distribution. Table 1-4 shows a so-called "less than" cumulative distribution for the examination scores data shown in Table 1-2 and summarized in Table 1-3. The cumulative numbers of examination scores less than the lower class limits of 55, 60, and so on are given. Thus, there were no scores less than 55, one score less than 60, four scores less than 65, and so on.

The graph of a cumulative frequency distribution is referred to as an **ogive** (pronounced "ōjive"). The ogive for the cumulative distribution shown in Table

Ogive

TABLE 1-4

Cumulative frequency distribution of the scores obtained on the commercial banking examination given in Urbanbank's training program during the past six months

Examination Score	Cumulative Frequency
Less than 55	0
Less than 60	1
Less than 65	4
Less than 70	10
Less than 75	20
Less than 80	35
Less than 85	54
Less than 90	81
Less than 95	92
Less than 100	100

FIGURE 1-7

Ogive for the distribution of the scores obtained on the commercial banking examination given in Urbanbank's training program during the past six months

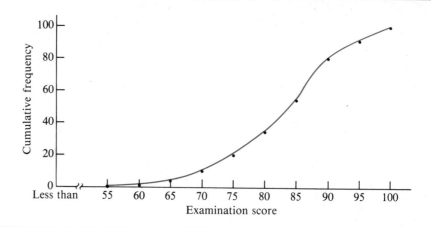

1-4 is given in Figure 1-7. The plotted points represent the number of scores less than the figure shown on the horizontal axis. The vertical coordinate of the last point represents the sum of the frequencies (100 in this case). The **S**-shaped configuration depicted in Figure 1-7 is quite typical of the appearance of a "less than" ogive. A "more than" ogive for the examination scores distribution would have class limits reading "more than 55" and so on. In this case, a reverse **S**-shaped figure would have been obtained, sloping downward from the upper left to the lower right.

EXERCISES 1.6

1. Expo-Classic, an exhibitor for recreational games' manufacturers and hobbyists, drew the following daily audiences at 30 performances during one of its tours:

976	1,069	1,336
1,027	1,919	1,159
1,411	1,583	1,432
1,327	1,101	1,810
917	1,646	1,391
1,291	1,680	1,630
1,048	1,257	1,988
942	1,190	1,457
1,036	1,745	669
711	934	1,517

a. Construct a frequency distribution using five or six classes.
b. Show the distribution in part (a) by sketching a frequency polygon and frequency curve.

2. Union-Path, a national charity, has compiled the following list of the donations received for the current year:

Donations received for the current year at
Union-Path Offices (in thousands of dollars)

253.0	173.4	117.0	191.2	151.4
182.0	132.0	162.0	212.9	155.9
221.0	158.0	135.0	124.4	68.9
89.7	95.6	84.1	135.1	123.2
101.0	126.5	142.8	20.2	119.0

Assuming the data are recorded to the last hundred dollars:
a. Determine a suitable frequency distribution. (Hint: Use between five and seven classes.) Specify class limits and midpoints.
b. Construct a histogram and a "less than" ogive on the basis of part (a).

3. Compute the cumulative frequency distribution for the following distribution:

Analysis of Ordinary Life Insurance

Policy Size	Number of Ordinary Life Insurance Policies in Force per 1,000 Policies in a Certain Month
Under $1,000	31
$1,000–2,499	181
2,500–4,999	108
5,000–9,999	213
10,000 or more	467

4. Comment critically on the following systems of designating class intervals:
a. 83-102   b. 33 and under 102
    102-121   103 and under 121
    and so on   122 and under 141
                       and so on

5. Kayley Hall and Company, an East Coast management consulting firm, compiled data on labor absenteeism in Ultra Corporation's plants. After

selecting plants with approximately the same size work force, Kayley Hall derived the following figures:

Plant Location	Number of Worker-days Lost Last Year due to Absenteeism			
Midwest	254.0	105.2	80.5	133.9
(20 plants)	138.1	100.1	92.6	178.9
	76.5	83.2	158.0	144.7
	115.8	107.3	109.1	211.5
	78.7	82.0	118.5	78.9
South	210.1	100.0	191.0	129.1
(16 plants)	198.5	112.3	122.4	162.0
	154.5	184.7	105.4	169.3
	142.1	162.8	138.7	178.5

a. Using five or six classes, sketch a histogram for the midwestern and southern regions combined.
b. Using the same class intervals, repeat part (a) for the two regions individually.
c. Which region, in your opinion, has the greater problem? Why? What information do you get from part (a)? From part (b)?

6. The Tolkien Associates, a Philadelphia consulting firm, analyzed the market demand for Frosted Freakies, a breakfast cereal sold at many Philadelphia–Camden area retail food markets. The firm producing the cereal used this analysis to make its advertising and distribution planning decisions. Twenty-two Philadelphia food stores and 14 Camden food stores were surveyed for a one-week period, and the number of boxes sold in each store during that period was recorded as follows:

Philadelphia

2,511	3,180	1,600	2,752	2,849	2,466	753
3,375	2,754	3,200	2,115	2,961	3,162	2,632
1,447	3,414	2,648	2,970	3,334	2,387	1,789
2,893						

Camden

1,874	1,280	1,659	2,062	1,866	1,731	832
2,633	1,509	1,167	2,374	514	1,901	1,326

a. Using six classes and "and under" upper class limits, construct and graph the frequency distribution for the total number of boxes sold in the two cities combined during the one-week period.
b. Using the same classes, construct and graph the frequency distribution for each city separately.
c. Which answer, that for part (a) or that for part (b), gives a better picture of the distribution of demand for Frosted Freakies? Explain your answer.

7. The vice-president in charge of corporate credit for a large bank is reviewing the outstanding credit positions of the bank's largest corporate clients. The following list of the 25 largest clients and their credit positions has been compiled.

Company	Credit (in $millions)	Company	Credit (in $millions)
A	25.0	N	23.7
B	15.8	O	3.5
C	33.3	P	26.2
D	6.3	Q	13.9
E	13.5	R	5.1
F	9.2	S	11.0
G	18.6	T	14.4
H	2.0	U	29.3
I	10.4	V	22.0
J	15.7	W	19.9
K	8.0	X	12.5
L	23.0	Y	17.1
M	26.8		

a. Set up a frequency table having seven classes and specify the midpoint of each class.
b. Compute the cumulative frequency distribution.

1.7
DESCRIPTIVE MEASURES FOR FREQUENCY DISTRIBUTIONS

As indicated in section 1.2, once a frequency distribution is constructed from a set of figures, certain features of the data become readily apparent. For most purposes, however, it is necessary to have a more quantitative description of these characteristics than can be ascertained by a casual glance at the distribution. Analytical measures are usually computed to describe such characteristics as the central tendency, dispersion, and skewness of the data.

Averages

Averages are the measures used to describe the characteristic of **central tendency** or **central location** of data. Averages convey with a single number the notion of "central location" or the "middle property" of a set of data. The most familiar average is the **arithmetic mean**; in fact, it is often referred to as "the average." In ordinary conversation or in print, we encounter such terms as "average income," "average growth rate," "average profit rate," and "average person." Actually, several different types of averages, or measures of central tendency, are implied in these terms. The type of average to be employed depends on the purpose of the application and the nature of the data being summarized. In this section, we consider the most commonly employed and most generally useful averages.

Dispersion   **Dispersion** refers to the spread, or variability, in a set of data. One method of measuring this variability is in terms of the difference between the values of selected items in a distribution, such as the difference between the values of the highest and lowest items. Another more comprehensive method is in terms of some average of the deviations of all the items from an average. Dispersion is an important characteristic of data because we are frequently interested as much in the variability of a set of data as in its central frequency.

Skewness   **Skewness** refers to the symmetry or lack of symmetry in the shape of a frequency distribution. This characteristic is useful in judging the typicality of certain measures of central tendency.

We begin the discussion of averages or measures of location by considering the most familiar one, the arithmetic mean. We assume throughout this chapter that the term **set of data** means a set of observations on a single numerical variable.

1.8
THE ARITHMETIC MEAN

Probably the most widely used and most generally understood way of describing the central tendency, or central location, of a set of data is the average known as the arithmetic mean.

> The **arithmetic mean**, or simply the **mean**, is the total of the values of a set of observations divided by the number of observations.

For example, if $X_1, X_2, \ldots, X_n$ represent the values of n items or observations, the arithmetic mean of these items, denoted $\bar{X}$, is defined as

$$\bar{X} = \frac{X_1 + X_2 + \cdots + X_n}{n} = \frac{\sum\limits_{i=1}^{n} X_i}{n}$$

For simplicity, subscript notation such as that given above will usually not be used in this book. (However, before continuing, you should turn to Appendix B and work out the examples given there.) Thus, when the subscripts are dropped, the formula becomes

(1.1) $$\bar{X} = \frac{\sum X}{n}$$

(Some of the x's)   *(x₁+x₂+x₃+...)*   *ungrouped data formula*

Σ = *The sum of*   where the capital Greek letter Σ (sigma) means "the sum of."

For example, suppose that an accounting department established accounts receivable in the following amounts during a one-hour period: $600,

$350, $275, $430, and $520. The arithmetic mean of the amounts of the accounts receivable is

$$\bar{X} = \frac{\$600 + \$350 + \$275 + \$430 + 520}{5} = \frac{\$2,175}{5} = \$435$$

The mean, $435, may be thought of as the size of each account receivable that would have been set up if the total of the five accounts was the same ($2,175) but all the accounts were the same size. That is, the mean is the value each item would have if they were all identical and the total value and number of items remained unchanged.

Symbolism

In keeping with the standard statistical practice of denoting sample statistics by Roman letters, we use the symbol $\bar{X}$ to denote the mean of a sample of observations. The number of observations in the sample is represented by the lower-case letter n. A value such as $\bar{X}$ computed from sample data is referred to as a **statistic**. A statistic may be used as an estimate of an analogous population measure known as a **parameter**. Thus, the statistic $\bar{X}$ (the sample mean) may be thought of as an estimate of a parameter, the mean of the population from which the sample was drawn. The number of observations in the population is represented by the capital letter N.

Just as Roman letters represent sample statistics, Greek letters represent population parameters. Accordingly, we will denote a population mean by the lower-case Greek letter μ (mu).

If the population mean were calculated directly from the data collected from the entire population of N members, then

(1.2)
$$\mu = \frac{X_1 + X_2 + \ldots + X_N}{N} = \frac{\Sigma X}{N}$$

In practice, the population mean μ is often not calculated, because it may be infeasible or inadvisable to accomplish a complete enumeration of the population.

Grouped Data

When data have been grouped into a frequency distribution, the arithmetic mean can be computed by a generalization of the definition for the mean of ungrouped data. As given in equation 1.1, the formula for the mean of a set of ungrouped data is $\bar{X} = (\Sigma X)/n$. However, with grouped data, since the identity of the individual items has been lost, an estimate must be made of the total of the values of the observations, ΣX. This estimate is obtained by multiplying the midpoint of each class in the distribution by the frequency of that class and summing over all classes. In symbols, if m denotes the midpoint of a class and f denotes the frequency, the arithmetic mean of a frequency distribution may be estimated from the following formula, known as the **direct method**:

Direct method

(1.3)
$$\bar{X} = \frac{\Sigma fm}{n}$$

m = class mark same as X_i

The computation of $\bar{X}$ for the frequency distribution of the examination scores data shown in Table 1-3 is given in Table 1-5. The mean score figure of 82.65, calculated from the frequency distribution, is very close to the corresponding mean of 82.38 for the ungrouped data given in Table 1-2. The small difference in these two figures illustrates the slight loss of accuracy involved in calculating statistical measures from frequency distributions rather than from ungrouped data.

Shortcut formulas are often useful for calculating the arithmetic mean and other measures for frequency distributions of secondary data, that is, data for which someone else has constructed the frequency distribution. One such shortcut formula, known as the the **step-deviation method**, is explained in Appendix D.

Step-deviation method

In many instances, for a variety of reasons, frequency distributions are shown with open intervals, usually in the first or last class. For example, such classes may appear as "Losses of $5,000 or more" or "Sales of $1,000,000 and over." In these cases, assumptions must be made concerning the midpoints of the classes in order to calculate statistical measures such as the arithmetic mean.

TABLE 1-5
Calculation of the arithmetic mean for grouped data by the direct method: examination scores data

Examination Score	Frequency f	Midpoints m	fm
55 and under 60	1	57.5	57.5
60 and under 65	3	62.5	187.5
65 and under 70	6	67.5	405.0
70 and under 75	10	72.5	725.0
75 and under 80	15	77.5	1,162.5
80 and under 85	19	82.5	1,567.5
85 and under 90	27	87.5	2,362.5
90 and under 95	11	92.5	1,017.5
95 and under 100	8	97.5	780.0
$n = \Sigma f = 100$			8,265.0

$$\bar{X} = \frac{\Sigma fm}{n} = \frac{8,265.0}{100} = 82.65$$

In averaging a set of observations, it is often necessary to compute a **weighted** *Weighted average*
average in order to arrive at the desired measure of central location. For example, suppose a company consists of three divisions, all selling different lines of
products. The ratios of net profit to sales (expressed as percentages) for these
divisions for the year 1990 were 5% for Division A, 6% for Division B, and 7%
for Division C. Assume that we want to find the net profit to sales percentage
for the three divisions combined, or equivalently, *for the company as a whole.*
This ratio is the figure that results from dividing total net profits by total sales
for the three divisions combined. Clearly, if we have only the profit *percentages*
for the three divisions, we do not have enough information to compute the required figure. However, if we are given the dollar sales for each of the three
divisions (that is, the denominators of the three ratios of net profit to sales from
which the percentages were computed), then these figures can be used as
"weights" in calculating the desired figure for the entire company. Specifically,
the **weighted arithmetic mean** would be calculated as shown in Table 1-6 by *Weighted arithmetic mean*
carrying out the following steps:

1. "Weight" (that is, multiply) the net profit to sales percentage for each division by the sales of that division. As indicated in Table 1-6, the resulting
   figures are the net profits of the divisions.

2. Sum the net profits obtained in step 1 to obtain total net profit for the
   three divisions combined.

3. Sum the sales figures (the weights) to obtain total sales for the three divisions
   combined.

4. Calculate the desired average by dividing the total net profits figure (obtained in step 2) by the total sales figure (found in step 3).

TABLE 1-6
Calculation of a weighted arithmetic mean: net profit to sales percentage for the three
divisions of a company combined, 1990

Division	Net Profit to Sales Percentage X	Sales w	Net Profit wX
A	5	$10,000,000	$ 500,000
B	6	10,000,000	600,000
C	7	30,000,000	2,100,000
		$50,000,000	$3,200,000

$$\text{Weighted mean} = \bar{X}_w = \frac{\Sigma wX}{\Sigma w} = \frac{\$3,200,000}{\$50,000,000} = 6.4\%$$

Symbolically, the weighted arithmetic mean is given by the formula

(1.4)
$$\bar{X}_w = \frac{\Sigma w X}{\Sigma w}$$

where $\bar{X}_w$ = the weighted arithmetic mean
X = the values of the observations to be averaged (net profits to sales percentages in this case)
w = the weights applied to the X values (sales, in this case)

Note that the weighted average of 6.4% computed in Table 1-6 is interpreted as the net profit to sales percentage for the three divisions combined (that is, for the company as a whole). Since dollar sales are in the denominator of the $\Sigma w X / \Sigma w$ ratio, the answer of 6.4% may be interpreted in terms of dollars of profit per dollars of sales, that is, an average of $0.064 profit per dollar of sales.

On the other hand, suppose we had asked instead, "What is the arithmetic mean net profit to sales ratio *per division*, without regard to the sales size of these divisions?" The answer is given by

$$\bar{X} = \frac{\Sigma X}{n} = \frac{5\% + 6\% + 7\%}{3} = \frac{18\%}{3} = 6\% \text{ per division}$$

where X = the percentages for the three divisions
n = the number of the divisions

Unweighted arithmetic mean   The result, $\bar{X}$, may be referred to as an **unweighted arithmetic mean** of the three ratios. In this computation, the net profit to sales percentages were totaled and the result divided by the number of the company's divisions. The result is therefore stated as 6% *per division*, because of the appearance of the number of divisions (3) in the denominator. Clearly, this calculation disregards differences that may exist in the amount of sales of the three divisions; however, this does not make the average meaningless. If we are interested in obtaining a "representative" or "typical" profit ratio and there are no extreme values to distort the representativeness of the unweighted mean, then this type of computation is a valid one. Of course, if one were seeking a typical or representative figure, it would be desirable to have more than the three observations present in this illustration, and other averages, such as the median or mode, may be preferable to the arithmetic mean in the determination of a typical or representative value. However, the unweighted arithmetic mean is not a meaningless figure; indeed, it is the correct answer to the question just posed.

Returning to weighted mean calculation, the weights that were applied to the three profit to sales percentages in this problem were the actual dollar amounts of sales for the three divisions. That is, the weights used were the values of the denominators of the original ratios. An alternative procedure would be to weight the ratios by a percentage breakdown of the denominators (in this case, a percentage breakdown of total sales). For example, in the computation shown in Table 1-6, weights of 20%, 20%, and 60% could have been applied instead of weights of $10 million, $10 million, and $30 million, and the same answer of 6.4% would have resulted. Indeed, any figures in the same proportions as $10 million, $10 million, and $30 million would have led to the same numerical answer. Note that the reason the weighted arithmetic mean of 6.4% exceeded the unweighted mean of 6.0% was that in the weighted mean calculation, greater weight was applied to the 7% profit figure for Division C than to the corresponding 5% figure for Division A. This had the effect of pulling the weighted average up toward the 7% figure.

EXERCISES 1.9

1. The Center City Bank reports bad debt ratios (dollar losses to total dollar credit extended) of 0.04 for personal loans and 0.02 for industrial loans in 1990. For the same year, the Neighborhood Bank reports bad debt ratios of 0.05 for personal loans and 0.03 for industrial loans. Can one conclude from this that Center City's overall bad debt ratio is less than Neighborhood's? Justify your response.

2. The frequency distribution for the percentage return on sales of 150 U.S. companies for one year is as follows:

Percentage Return on Sales	Number of Companies
0% and under 5%	34
5% and under 10%	46
10% and under 15%	38
15% and under 20%	22
20% and under 25%	10

   a. Compute the arithmetic mean of this distribution.
   b. If one totaled the profit figures for the 150 firms and divided by the total sales of all the companies, would the resulting average equal the answer to part (a)?

3. Consider the following data for the percentage of the civilian labor force unemployed in a certain area:

County	Percentage Unemployed	Civilian Labor Force
A	3.6	114,395
B	3.8	214,758
C	2.5	206,324
D	6.5	843,160

   a. What is the unweighted average of the percentage of the labor force
       unemployed per county?
   b. What is the weighted average of the percentage of the labor force unem-
       ployed for the four counties combined?
   c. Explain the reason for the difference in the figures obtained in parts
       (a) and (b).

4. Assume that the FDA recently discovered a new carcinogen, Oxy-toxin, in
   the ingredients generally found in sandwich spread. Brandex, a sandwich
   spread manufacturer, has compared Oxy-toxin levels in its ingredients with
   Oxy-toxin levels in the ingredients of the leading spread and has published
   its results as follows:

	Oxy-toxin Level (%)	
Ingredient	Brandex	Leading Spread
Moonflower oil	1.2	1.8
Processed butter-glop	0.6	0.8
Flavoring	0.3	0.3
Coloring	0.5	0.6

   Is this information sufficient to decide that Brandex is a safer sandwich
   spread (in terms of Oxy-toxin content)? Why or why not? If the information
   is insufficient, what additional data would you wish to have?

5. The Peerless Gauge Company manufactures its own parts for assembling
   into Model M-40/B gauges. One of these parts, serial number X-007, is
   made in runs of 1,000 each. Each run is inspected for defective items, and
   the percentage of defective items is recorded. The last 50 runs produced
   the following data:

Percentage of Defective Items	Number of Runs
0.0 and under 0.5	3
0.5 and under 1.0	12
1.0 and under 1.5	24
1.5 and under 2.0	9
2.0 and under 2.5	2
Total	50

   a. Compute the overall percentage of defective items for the last 50 runs.
   b. Suppose a record were kept of the total *number* of defective items rather
       than the percentage. If you calculated the overall percentage of de-
       fective items by dividing the total number of defective items by 50,000
       (50 runs × 1,000 per run), would this give the same answer as in part
       (a)? Why or why not?

6. Pacific-Poorfield, a manufacturer of high-quality lubricants, suspected that
    some of its base-oil storage tanks contained a chemical pollutant. Upon
    checking Tanks 11, 15, and 16, company inspectors discovered the following
    levels of the pollutant:

Tank Number	Tank Contents (in liters)	Pollutant Level (in parts per 10,000)
11	261,432	4.5
15	118,300	15.0
16	287,456	21.3

a. What is the weighted averaged of pollution level in the three tanks?
b. What is the unweighted average?
c. Are you surprised by the answers to parts (a) and (b)? Why or why not?

7. Mortimer Hutton, a wealthy investor who spends a large amount of time
    watching the quote machine in his broker's office, is weighing the relative
    merits of two periodic investment strategies. The first method is to buy
    the same number of shares of stock each investment period. The second
    method requires an investment of a constant dollar amount each period
    regardless of the stock price.

    His broker demonstrates the result of using each strategy on two
    stocks, United Aerodynamics and Mitton Industries, for a 5-year period.
    The purchases in each case are made at midyear at the prevailing stock
    price. The constant dollar amount for the second strategy is $1,000 (that
    is, the amount invested in the stock is as close to $1,000 as possible).

a. Calculate the average cost per share for each stock in each strategy.
b. Which strategy achieved the lower average cost for United Aerody-
    namics? For Mitton Industries?
c. Explain these differences in terms of the weights used in calculating
    the average cost per share of each stock and strategy.

Comparison of Two Investment Strategies

	First Strategy						Second Strategy					
	United Aerodynamics			Mitton Industries			United Aerodynamics			Mitton Industries		
Year	Price per Share	Number of Shares	Total Cost	Price per Share	Number of Shares	Total Cost	Price per Share	Number of Shares	Total Cost	Price per Share	Number of Shares	Total Cost
1	$54	25	$1,350	$25	25	$625	$54	19	$1,026	$25	40	$1,000
2	50	25	1,250	30	25	750	50	20	1,000	30	33	990
3	42	25	1,050	43	25	1,075	42	24	1,008	43	23	989
4	35	25	875	40	25	1,000	35	29	1,015	40	25	1,000
5	30	25	750	49	25	1,225	30	33	990	49	20	980

1.10
THE MEDIAN

The median is a well known and widely used average. It has the connotation of the "middlemost" or "most central" value of a set of numbers.

Ungrouped data

> For ungrouped data, the **median** is defined simply as the value of the central item when the data are arrayed by size.

If there is an odd number of observations, the median is directly ascertainable. If there is an even number of items, there are two central values, and by convention, the value halfway between these two central observations is designated as the median.

For example, suppose that a test of a brand of gasoline in five new small economy cars yielded the following numbers of miles per gallon: 27, 29, 30, 32, and 33. Then the median number of miles per gallon would be 30. If another car were tested and the number of miles per gallon obtained was 34, the array would now read 27, 29, 30, 32, 33, 34. The median would be designated as 31, the number halfway between 30 and 32.

> Another way of viewing the median is as a value below and above which lie an equal number of items.

Thus, in the preceding illustration involving five observations, two lie above the median and two below. In the example involving six observations, three fall above and three fall below the median. Of course, in the case of an array with an even number of items, any value lying between the two central items may, strictly speaking, be referred to as a median. However, as indicated earlier, the convention is to use the midpoint between the two central items. In the case of tied values at the center of a set of observations, there may be no value such that equal numbers of items lie above and below it. Nevertheless, the central value, as defined in the preceding paragraph, is still designated as the median. For example, in the array 52, 60, 60, 60, 60, 61, 62, the number 60 is the median, although unequal numbers of items lie above and below this value.

Frequency distribution

In a frequency distribution, the median is necessarily an estimated value, since the identity of the original observations is not retained. Because in a frequency distribution the data are arranged in order of magnitude, frequencies can be cumulated to determine the class in which the median observation falls. It is then necessary to make some assumption about how observations are distributed in that class. Conventionally, the assumption is made that observations are equally spaced, or evenly distributed, throughout the class containing the median. The value of the median is then established by a linear interpolation.

TABLE 1-7
Calculation of the median for a frequency distribution of personal loan data

Class Interval	Number of Loans f
$ 0 and under $ 200	6
200 and under 400	18
400 and under 600	25
	$\Sigma f_p = 49$
600 and under 800	20
800 and under 1,000	17
1,000 and under 1,200	14
	100

The procedure is illustrated for the distribution of personal loans given in Table 1-7. First, the calculation of the median is explained without the use of symbols. Then the procedure is generalized by stating it as a formula.

There are 100 loans represented in the distribution shown in Table 1-7, so the median lies between the fiftieth and fifty-first loans. Since 49 loans occur prior to the class "$600 and under $800," the median must be in that class. Assuming that the 20 loans are evenly distributed between $600 and $800, we can determine the median observation by interpolation $\frac{1}{20}$ of the distance through this $200 class. The median is calculated by adding $\frac{1}{20}$ of $200 to the $600 lower limit of the class containing the median. That is,

$$Md = \$600 + \left(\frac{50 - 49}{20}\right)\$200 = \$600 + \left(\frac{1}{20}\right)\$200 = \$610$$

Thus, the formula for calculating the median of a frequency distribution is

(1.5)
$$Md = L_{Md} + \left(\frac{n/2 - \Sigma f_p}{f_{Md}}\right)i$$

do need to know

where Md = the median
L_{Md} = the (real) lower limit of the class containing the median
n = the total number of observations in the distribution
Σf_p = the sum of the frequencies of the classes preceding the one containing the median
f_{Md} = the frequency of the class containing the median
i = the size of the class interval

It may seem that we have located the value of the fiftieth observation rather than one falling midway between the fiftieth and fifty-first. However, the value determined is indeed one lying halfway between the fiftieth and fifty-first observations, if we use the assumption that items are evenly distributed within the class in which the median falls.

FIGURE 1-8
Diagram depicting the
meaning of the assumption
concerning an even distribution
of observations

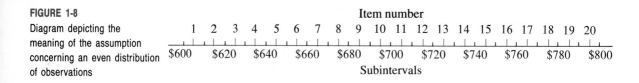

If there are 20 observations in the $200 class from $600 to $800, we may divide the class into 20 equal subinterval of $10 each, as depicted in Figure 1-8. Since the items are assumed to be evenly distributed in the class from $600 to $800, they must be located at the midpoints of these subintervals. An interpolation of $\frac{1}{20}$ through the class interval brings us to the end of the first subinterval, $610, which is the value halfway between the first and second items. Since 49 frequencies preceded this class, the median of $610 is a value lying midway between the fiftieth and fifty-first observations.

1.11
CHARACTERISTICS AND USES OF THE ARITHMETIC MEAN AND MEDIAN

The preceding sections have concentrated on the mechanics of calculating means and medians for ungrouped and grouped data. We now turn to a few of the characteristics and uses of these averages.

Arithmetic mean

> The **arithmetic mean**—the most familiar measure of central tendency—is defined as the total of the values of a set of observations divided by the number of these observations.

It thus has the advantage of being a rigidly defined mathematical value that can be manipulated algebraically. For example, the means of two related distributions can be combined by suitable weighting. If the arithmetic mean income in 1990 of 10 marketing executives of a company was $85,000 and the corresponding mean income of 19 marketing executives of another company was $95,000, then the arithmetic mean income for the 29 executives combined was, by equation 1.4,

$$\bar{X}_w = \frac{10(\$85,000) + 19(95,000)}{10 + 19} = \$91,552$$

On the other hand, if we knew the median income of the 10 executives from the first company and the median income of the 19 executives from the second company, there would be no way of averaging those two numbers to obtain the median income for the 29 executives combined. Because of such mathematical properties, the arithmetic mean is used more often than any other average in advanced statistical techniques.

A disadvantage of the mean is its tendency to be distorted by extreme values at either end of a distribution. In general, it is pulled in the direction of these extremes. Thus, the arithmetic mean of the five figures $110, $126, $132, $157, and $1,000 is

$$\frac{\$110 + \$126 + \$132 + \$157 + \$1,000}{5} = \$305$$

a value that is greater than four of the five items averaged. In such situations, the arithmetic mean may not be a typical or representative figure.

The **median** is also a useful measure of central tendency. Its relative freedom from distortion by skewness in a distribution makes it particularly useful for conveying the idea of a typical observation. It is primarily affected by the number of observations rather than their size. Consider an array in which the median has been determined; if the largest item is multiplied by 100 (or any large number), the median remains unchanged. The arithmetic mean would, of course, be pulled toward the large extreme item.

Median

The major disadvantage of the median is that it is an average of position and hence is more difficult to deal with than is the arithmetic mean. Thus, as indicated earlier, if one knows the medians of each of two distributions, there is no algebraic way of averaging the two figures to obtain the median of the combined distribution.

1.12
THE MODE

Another measure of central tendency that is sometimes useful but often not explicitly calculated is the mode. In French, to be "a la mode" is to be in fashion. The **mode** is the observation that occurs with the greatest frequency and thus is the most "fashionable" value. The mode is usually not determined for ungrouped data. The reason is that even when most of the data items are clustered toward the center of the array of observations, the item that occurs more often than any other may lie at the lower or upper end of the array and thus be an unrepresentative figure. Therefore, determination of the mode is generally attempted only for grouped data.

Ungrouped data

Grouped data

When data are grouped into a frequency distribution, it is not possible to specify the single observation that occurs most frequently, since the identity of the individual items is lost. However, we can determine the **modal class**, the class that contains more observations than any other. Of course, class intervals should be of the same size when this determination is made. When the location of the modal class is considered along with the arithmetic mean and median, much useful information is generally conveyed not only about central tendency but also about the skewness of a frequency distribution.

FIGURE 1-9
The location of the mode

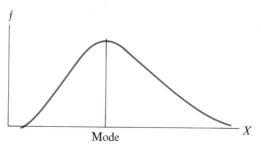

In this and subsequent graphs, the following convention is used:
The horizontal axis label X = values of the observations.
The vertical axis label f = frequency of occurrence.

Location of mode   Several formulas have been developed for determining the location of the mode within the modal class. These usually involve the use of frequencies in the classes preceding and following the modal class as weighting factors that tend to pull the mode up or down from the midpoint of the modal class. We shall not present any of these formulas here. For our purposes, the midpoint of the modal class may be taken as an estimate of the mode. To understand the meaning of the mode clearly, let us visualize the frequency polygon of a distribution and the frequency curve approached as a limiting case when class size is gradually reduced. In the limiting situation, the variable under study may be thought of as continuous rather than as discrete. The mode may then be thought of as the value of the horizontal axis lying below the maximum point on the frequency curve (see Figure 1-9).

        The mode of a frequency distribution has the connotation of a typical or representative value, a location in the distribution at which there is maximum clustering. In this sense, it serves as a standard against which to judge the representativeness or typicality of other averages.

> If a frequency distribution is symmetrical, the mode, median, and mean coincide.

As noted earlier, extreme values in a distribution pull the arithmetic mean in the direction of these extremes. Stated somewhat differently, in a skewed distribution, the mean is pulled away from the mode toward the extreme values. The median also tends to be pulled away from the mode in the direction of skewness but is not affected as much as the mean. If the mean exceeds the
Positive skewness   median,[2] a distribution is said to have **positive skewness** or to be **skewed to the**

[2] Sometimes the mode, rather than the median, is used for this comparison.

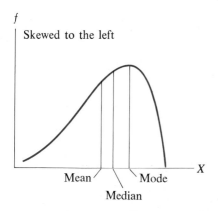
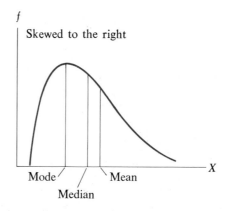

FIGURE 1-10

Skewed distributions depicting typical positions of averages

right; if the mean is less than the median, the terms **negative skewness** and **skewed to the left** are used. The order in which averages tend to fall in skewed distributions is shown in Figure 1-10.

Many distributions of economic data in the United States are skewed to the right. Examples include the distributions of incomes of individuals, savings of individuals, corporate assets, sizes of farms, and company sales within many industries. In many of these instances, the arithmetic mean is pulled so far from the median and mode as to be a very unrepresentative figure.

Multimodal Distributions

If more than one mode appears, the frequency distribution is referred to as **multimodal**; if there are two modes, it is referred to as **bimodal**. Extreme care must be exercised in analyzing such distributions. For example, consider a situation in which you want to compare the mean wages of workers in two different companies. Assume that the mean calculated for Company A exceeds that of Company B. If you conclude from this finding that workers in Company A earn higher wages, on the average, than those in Company B, without taking into account the fact that the wage distribution for each of these companies is bimodal, you may make serious errors of inference. To illustrate the principle involved, let us assume that the mean annual wage is $15,000 for unskilled workers and $25,000 for skilled workers at each of these companies. Let us also assume that the individual distributions of wages of unskilled and skilled workers are symmetrical and that there are the same total number of workers in each company. Further, let us assume that these companies have workers only in the aforementioned two skill classifications. However, suppose 75% of the Company A workers are skilled, whereas only 50% of the Company

FIGURE 1-11

Bimodal frequency
distributions: Annual
wages of workers in
two companies

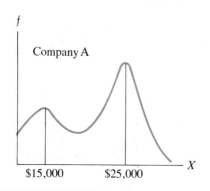

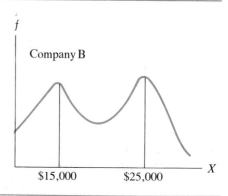

B workers are skilled. Figure 1-11 shows the frequency curves of the distributions of annual wages at the two companies. Clearly, the mean annual wage of workers in Company A exceeds that in Company B, simply because there is a higher percentage of skilled workers at Company A. However, if you were ignorant of this fact, you might be tempted to infer that workers at Company A earn more than those at Company B. The fact of the matter is that unskilled workers at both companies earn the same wages, on the average, and the same

Separate into
two distributions   holds true for the skilled workers. What is required here is to separate two wage distributions at each company, one for skilled workers and one for unskilled. A comparison of the mean wages of unskilled workers (and of skilled workers) at the two companies would reveal their equality.

The principle involved here is one of **homogeneity** of the basic data. The fact that a wage distribution is bimodal (that is, has two values around which frequencies are clustered) suggests that two different "cause systems" are present and that two distinct distributions should be recognized. The data on wages

Nonhomogeneous data   may be said to be **nonhomogeneous** with respect to skill level. Bimodal distributions would also result, for example, from the merging of height data for men and for women or weight data for men and for women.

> If a basis for separating a bimodal distribution into two distributions cannot be found, then extreme care must be used in describing the data.

In cases such as that shown for Company B in Figure 1-11, where the heights of the two modes are about equal, the arithmetic mean and median will probably fall between the modes and will not be representative of the large concentrations of values lying at the modes below and above these averages.

EXAMPLE 1-1

The following distribution gives the dollar cost per unit of output for 200 plants in the same industry:

Dollar Cost per Unit of Output	Number of Plants
$1.00 and under $1.02	6
1.02 and under 1.04	26
1.04 and under 1.06	52
1.06 and under 1.08	58
1.08 and under 1.10	39
1.10 and under 1.12	15
1.12 and under 1.14	5
1.14 and under 1.16	1
Total	200

a. Calculate the arithmetic mean of the distribution.
b. Calculate the median of the distribution.
c. Is the answer in part (a) the same as the one you would have obtained by computing the following ratio?

$$\frac{\text{Total dollar cost for the 200 plants}}{\text{Total number of units of output of the 200 plants}}$$

d. Can you say that 50% of the units produced cost less than the median calculated in part (b)? Explain.
e. Suppose that the last class had been "$1.14 and over." What effect, if any, would this have had on your calculation of the arithmetic mean and of the median? Briefly justify your answers.

SOLUTIONS

a. $\bar{X} = \dfrac{\$213.20}{200} = \$1.066$

b. Median class = $1.06 and under $1.08

$$\text{Median} = \$1.06 + \left(\frac{100 - 84}{58}\right)(\$0.02) = \$1.066$$

c. No. Conceptually, the ratio in part (c) is a weighted mean of 200 cost per unit output ratios. The mean calculated in part (a) is unweighted. It is an estimate of the figure that we would have obtained by adding up the original 200 cost per unit output ratios and dividing by 200 plants.
d. No. We have no way of telling how many units each plant produces. For example, the six plants with costs under $1.02 per unit might produce 60% of the total number of units in the industry. However, we can say that the grouped data procedure indicates that 50% of the plants had a cost per unit less than the median figure of $1.066.
e. The open-ended interval would have no effect on the median, which is concerned only with the order of the frequencies. However, the mean, which is concerned with the actual values, would have to be recalculated. A mean value for the items in the open interval would have to be assumed and multiplied by the frequency to obtain an estimate of the fm value for that class.

EXAMPLE 1-2

What is the proper average(s), if any, to use in the following situations? Briefly justify your answers.

a. Determining which members are in the upper half of a class with respect to their overall grade average.

b. Determining the average death rate for six cities combined, given the death rate of each.

c. Finding the average amount each worker is to receive in a profit-sharing plan in which a firm wishes to ensure equal distribution.

d. Determining how high to make a bridge (not a drawbridge). The distribution of the heights of boats expected to pass under the bridge is known and is skewed to the left.

e. Determining a typical wage figure for use in later arbitration for a company that employs 100 workers, several of whom are highly paid specialists.

f. Determining the average annual percentage rate of net profit to sales of a company over a 10-year period.

SOLUTIONS

a. The median, since (generally) one-half of the grade averages will fall above and below this figure.

b. The weighted arithmetic mean. The death rates of the six cities would be weighted by the population figures of these cities.

c. The arithmetic mean. Divide the total profits to be shared (ΣX) by the number of workers (n).

d. None of the averages would be appropriate, since they would result in a large number of boats being unable to pass under the bridge. The bridge should be high enough to allow all expected traffic to pass under it.

e. The median. Because of the tendency of the arithmetic mean to be distorted by a few extreme items, it would tend to be a less typical figure.

f. The weighted arithmetic mean. The profit rates would be weighted by the sales figures for each year.

EXERCISES 1.12

1. Nu-Coat, an outerwear manufacturer, had the following closing stock prices last year on 30 random days (rounded to the nearest dollar):

$30	$45	$30	$29	$48	$39
15	49	41	28	33	50
40	49	33	52	32	45
26	51	47	45	51	46
42	33	48	30	41	21

Calculate the arithmetic mean, the mode, and the median.

2. Explain or criticize the following statement: The frequency distributions of family income, size of business, and wages of skilled employees all tend to be skewed to the right.

3. The following table presents a frequency distribution of the bank deposits in the First Metropolis Bank and Trust Company at the end of 1990:

Deposit Size	Number of Customers
$0–$499	7178
500– 999	8296
1000–1499	8342
1500–1999	9347
2000–2499	9546
2500–2999	6147
3000–3499	4239
3500–3999	2103
4000–4499	1445
4500 and over	676

a. Why do you think open-ended intervals are used in this distribution?
b. Which is the modal class?
c. Which is the median class?

4. The following data represent chainstore prices paid by farmers for two products during a certain month:

Composition Roofing (price range for a 90-pound roll)	Number of Purchases Reported	Douglas and Inland Firs, 2′ × 4′ s Standard Grade or Better (price range per thousand pound-foot)	Number of Purchases Reported
$2.35 and under $2.75	1	$ 76–85	2
2.75 and under 3.15	6	86–95	4
3.15 and under 3.55	33	96–105	8
3.55 and under 3.95	51	106–115	7
3.95 and under 4.35	121	116–125	28
4.35 and under 4.75	50	126–135	48
4.75 and under 5.15	44	136–145	57
5.15 and under 5.55	13	146–155	71
5.55 and under 5.95	5	156–165	45
		166–175	20
		176–185	1

a. Graph the frequency distribution for each product.
b. Indicate the median and modal class for each distribution.
c. Calculate and graph the cumulative distribution for each frequency distribution.

    d. Compare the skewness of the two distributions. (No calculations are required.)

5. Criticize or explain the following statement:

      In a bimodal frequency distribution, the median gives a more typical value than the arithmetic mean and therefore should be used for comparison with other distributions.

6. The following are the earnings for all salespersons of the Watson Corporation for a certain week:

Earnings for Week	Number of Salespersons with Given Earnings
$600 and under $650	13
650 and under 700	24
700 and under 750	29
750 and under 800	36
800 and under 850	34
850 and under 900	21
900 and under 950	11
	168

    a. Compute the arithmetic mean of the distribution.

    b. Is the answer to (a) the same as the one you would have obtained had you calculated the following ratio?

$$\frac{\text{Total earnings of all salespersons for the week}}{\text{Total number of salespersons for the week}}$$

    Why or why not?

    c. Compute the median of the distribution.

    d. In which direction are these data skewed?

    e. The comptroller of the company stated that the total payroll for the week was $129,832. Do you have any reason to doubt this statement? Support your position very briefly.

    f. Do you think the arithmetic mean computed in (a) provides a satisfactory description of the typical earnings of these 168 employees for the given week? Why or why not?

1.13
THE GEOMETRIC MEAN

~ SKIP ~

In business and economic problems, questions frequently arise concerning average percentage rates of change over time. Neither the mean, median, nor mode is the appropriate average to use in these instances. For example, consider

the following time series for annual sales of the Gallison Equipment Company
from 1987 through 1990:

Year	Sales
1987	$10,000,000
1988	8,000,000
1989	12,000,000
1990	15,000,000

What was the average percentage rate of change per year in sales? To answer
this question, we must specify what we mean by the "average percentage rate
of change per year." The most generally useful interpretation of this term is
the constant percentage rate of change which if applied each year would take
us from the first to the last figure. Hence, in the above illustration we would
be interested in that constant yearly percentage rate of change which would be
required to move from $10 million of sales in 1987 to $15 million in 1990.
Clearly, none of the previously discussed averages provides the correct answer
to this question. The correct answer can be obtained through the use of the geometric mean, or what amounts to the same thing, through the use of the familiar
compound interest formula. In the discussion which follows, the geometric mean
is defined, and the relationship between this average and compound interest calculations is indicated. Then these concepts are applied to the average percentage
rate of change per year in sales of the Gallison Company.

The geometric mean, (G), of n numbers is defined as the nth root of the
product of n numbers.

(1.6)
$$G = \sqrt[n]{X_1 \cdot X_2 \cdots X_n}$$

This definition can be contrasted to that of the arithmetic mean, $(\bar{X})$,
which is the sum of the n numbers divided by n,

$$\bar{X} = \frac{X_1 + X_2 + \cdots + X_n}{n}$$

In the calculation of the geometric mean, the n numbers are multiplied
together, then the nth root is extracted, whereas in the case of the arithmetic
mean, the n numbers are added together, and then this total is divided by n.
It can be shown that the geometric mean of n numbers is always equal to or
less than the arithmetic mean. When all of the numbers are identical, the two
averages are equal. For example, consider the numbers 9 and 4. The arithmetic
mean is $(9 + 4)/2 = 6\frac{1}{2}$. The geometric mean is $\sqrt{9 \times 4} = 6$. On the other
hand, if the two numbers are identical, such as 6 and 6, the arithmetic mean is
$(6 + 6)/2 = 6$ and the geometric mean is $\sqrt{6 \times 6}$ or also 6.

In the preceding simple examples, because the geometric mean involved a
number whose square root was an integer, its calculation was trivial. In more
realistic problems, logarithms must be used in the calculation. Thus, to determine the geometric mean of n numbers, we first obtain the logarithm of the
geometric mean by adding together the logs of the n numbers and dividing

this total by n. Then the antilogarithm is taken to yield the geometric mean. Symbolically, we have

(1.7)
$$\log G = \frac{\Sigma \log X}{n}$$

and

(1.8)
$$G = \text{antilog} \frac{\Sigma \log X}{n}$$

For example, suppose the geometric mean of the three numbers 10, 12, and 16 is required. That is,

$$G = \sqrt[3]{10 \times 12 \times 16}$$

Taking logarithms gives

$$\log G = (1/3)(\log 10 + \log 12 + \log 16)$$

From Appendix Table A-4, we have

$$\log 10 = 1.000$$
$$\log 12 = 1.0792$$
$$\underline{\log 16 = 1.2041}$$
$$3.2833$$

$$\log G = (1/3)(3.2833) = 1.0944$$

$$G = \text{antilog } 1.0944 = 12.43$$

Now we turn to the compound interest formula as an application of the geometric mean. Suppose \$100 was invested at 5% compound interest for three years. Let

$$P_0 = \text{original principal (\$100)}$$

$$P_1 = \text{amount accumulated at the end of the first year}$$

$$P_2 = \text{amount accumulated at the end of the second year}$$

$$P_3 = \text{amount accumulated at the end of the third year}$$

Then

$$P_1 = \$100(1.05) = \$105.0$$

$$P_2 = \$100(1.05)(1.05) = 100(1.05)^2 = \$110.25$$

$$P_3 = \$100(1.05)(1.05)(1.05) = \$100(1.05)^3 = \$115.76$$

In general, if

$$P_0 = \text{original principal}$$

$$r = \text{rate of interest expressed as a decimal}$$

$$n = \text{number of compounding periods}$$

$$P_n = \text{amount accumulated at the end of } n \text{ periods}$$

then,

(1.9)
$$P_n = P_0(1 + r)^n$$

If interest is compounded at different rates in each time period, and if these successive rates are denoted $r_1, r_2, \ldots, r_n$, then the amount accumulated at the end of n periods with an original principal of P_0(dollars) is

(1.10)
$$P_n = P_0(1 + r_1)(1 + r_2) \cdots (1 + r_n)$$

Thus, if interest were earned on \$100 at 3%, 5%, and 6%, respectively, in three periods, the amount accumulated at the end of the third period would be

$$P_3 = \$100(1.03)(1.05)(1.06) = \$114.64$$

The relationship between compound interest and the geometric mean can now be seen by a joint consideration of formulas (1.9) and (1.10). If the expression for P_n in (1.9) is substituted on the left-hand side of (1.10), we have, after cancellation of P_0,

$$(1 + r)^n = (1 + r_1)(1 + r_2) \cdots (1 + r_n)$$

Taking the nth root of both sides of this equation gives,

(1.11)
$$(1 + r) = \sqrt[n]{(1 + r_1)(1 + r_2) \cdots (1 + r_n)}$$

Thus, $1 + r$ is seen to be the geometric mean of the $(1 + r_1), (1 + r_2), \ldots, (1 + r_n)$ values, since it is the nth root of the product of these n quantities. Let us interpret this result. The $r_1, r_2, \ldots, r_n$ figures may be thought of as percentage rates of change between the successive $P_0, P_1, \ldots, P_n$ values, whereas the $(1 + r_1), (1 + r_2), \ldots, (1 + r_n)$ figures represent the relative relationships between each P value and the preceding one. For example, in the preceding illustration, where $r_1 = 3\%$,

$$1 + r_1 = \frac{P_1}{P_0} = \frac{\$103}{\$100} = 1.03$$

Therefore, the \$103 total at the end of the first period was 3% greater than \$100, and expressed as a relative rate of change, 103% of the \$100. The $1 + r$ value in equation (1.11) is thus the *average relative* and r is the *average percentage rate* of *change* per period in the time series of P values. If the $r_1, r_2, \ldots, r_n$ are not all equal, then the calculated r value can be interpreted as the rate that would have prevailed had there been a constant percentage rate of change from P_0 to P_n. Actually, this average percentage rate of change can be computed by using only the first and last figures, P_0 and P_n, respectively. An explicit formula for this average is given by solving equation (1.9) for r,

(1.12)
$$r = \sqrt[n]{\frac{P_n}{P_0}} - 1$$

Another way of showing that the average percentage rate of change over time depends only on the first and last figures in a time series is to substitute P_1/P_0

for $1 + r_1$, P_2/P_1 for $1 + r_2$, etc., in equation (1.11). All P values other than P_0 and P_n cancel as follows:

$$1 + r = \sqrt[n]{\frac{P_1}{P_0} \cdot \frac{P_2}{P_1} \cdot \frac{P_3}{P_2} \cdots \frac{P_{n-1}}{P_{n-2}} \cdot \frac{P_n}{P_{n-1}}} = \sqrt[n]{\frac{P_n}{P_0}}$$

These ideas will be illustrated by the case of the Gallison Equipment Company referred to earlier. The calculation is carried out for illustrative purposes in two different ways, first using equation (1.11) and then equation (1.12). The computation by equation (1.11) is shown in Table 1-8. Sales for each year relative to the preceding year are obtained first. Then the geometric mean sales relative, $(1 + r)$, is derived by obtaining the logarithms of the sales relatives, the mean of these logarithms (0.0587) and the antilogarithm of that figure (1.145). Subtracting 1 gives an average rate of change of 14.5% per year.

The same figure is obtained by equation (1.12), where $P_0 = \$10,000,000$, $P_n = \$15,000,000$, and $n = 3$.

$$r = \sqrt[3]{\frac{\$15,000,000}{\$10,000,000}} - 1$$

$$= \sqrt[3]{1.5} - 1 = 1.145 - 1 = .145$$

$$= 14.5\% \text{ per year}$$

Ordinarily, the first method, equation (1.11) is not used, since the calculation by the simple form of the compound interest formula (1.12) is easier. However, it is important to keep in mind that the typicalness of the computed

---

TABLE 1-8

Calculation of average annual percentage rate of change in sales of the Gallison Equipment Company, 1987–1990

Year	(1) Sales	(2) $(1 + r_i)$ Sales Relatives	(3) $(\log(1 + r_i))$ Logs of Sales Relatives
1987	$10,000,000		
1988	8,000,000	0.8000	9.9031 − 10
1989	12,000,000	1.5000	0.1761
1990	15,000,000	1.2500	0.0969
			10.1761 − 10

$$\log(1 + r) = \frac{\Sigma \log(1 + r_i)}{n} = \frac{0.1761}{3} = 0.0587$$

$$1 + r = 1.145$$

$$r = .145 = 14.5\% \text{ per year}$$

average can only be judged in terms of the individual percentage rates of change. For example, in the case of the sales data presented in Table 1-8, the annual rates of change were -20%, $+50\%$, and $+25\%$. Clearly, the 14.5% average rate of change is not typical of such widely fluctuating figures. On the other hand, if sales had gone from $10 million to $15 million at relatively constant annual rates of change, the computed average would be a typical figure.

In interpreting an average, such as the 14.5% rate of change in the preceding example, it is important to note the effect of using only the terminal points, P_0 and P_n, of the time series. In dealing with economic data in which cycles appear, if the measurement of an average percentage rate of change is made by taking the first figure, P_0, at the trough of a business cycle, while the last point, P_n, is taken at the peak of a cycle, clearly a greater average rate will tend to be obtained than if the two selected terminal points were both midway between troughs and peaks of cycles. This effect is often well illustrated in national political campaigns when candidates refer to economic growth rates under earlier administrations. Often, grossly inconsistent statements seem to be made by opposing candidates about average growth rates, even when measured in terms of similar data such as real gross national product per capita. A closer examination generally reveals substantially different selections of terminal points over which rates of change are measured. In more careful analytical work, trend lines are often fitted to time series data, and average growth or decline rates may be determined from these trend lines. This certainly is a fairer procedure than measuring rates of change between arbitrarily selected terminal points. Trend lines are discussed in Chapter 11.

EXERCISES 1.13

1. Sales of Tarn Engineering Inc. for the year ending June 30, 1980 were $845,600. Sales for the year ending June 30, 1990 were $1,255,100. What was the yearly average percentage rate of increase?

2. Observe the following time series:

Sales of Product Produced by
the ABC Company

Year	Number of Units Sold
1960	4,000,000
1990	2,000,000

Compute the average percentage rate of change in number of units sold per five-year period.

3. The following table shows the market value of production of the J.M. Trister Company from 1965 to 1990 at five-year intervals:

Year	Production (in millions of dollars)
1965	$20,000
1970	15,114
1975	28,267
1980	32,467
1985	36,572
1990	25,000

a. Compute the *mean amount* of change *per five*-year *period* in value of production.

b. Compute the *average percentage rate* of change *per five-year* period in value of production.

c. Is the average computed in part (b) typical? Why or why not?

4. a. Mr. Compound wants to earn 9% on his investment each year. He wishes to invest $4,000 and reinvest interest received each year. How many years must he follow his plan to double his money? Give your answer as a whole number.

b. If he had followed his plan half as long, how much would he have?

c. If he had followed his plan twice as long (as your answer to part (a)) how much would he have? Comment on the relationship among your answers.

1.14
DISPERSION: DISTANCE MEASURES

Central tendency, as measured by the various averages already discussed, is an important descriptive characteristic of statistical data. However, although two sets of data may have similar averages, they may differ considerably with respect to the spread, or dispersion, of the individual observations. Measures of dispersion describe this variation in numerical observations.

There are two types of measures of dispersion: distance measures and average deviation measures.

> Distance measures describe the spread of data in terms of the distance between the values of selected observations.

The simplest distance measure is the **range**, or the difference between the values   *Range*
of the highest and lowest items. For example, if the loans extended by a bank
to five corporate customers are $82,000, $125,000, $140,000 $212,000, and
$245,000, the range of these loans is $245,000 − $82,000 = $163,000. Such a
measure of dispersion may be useful for obtaining a rough notion of the spread
in a set of data, but it is certainly inadequate for most analytical purposes.

A disadvantage of the range is that it describes dispersion in terms of only
two selected values in a set of observations. Thus, it ignores the nature of the
variation among all other observations, which may be either tightly clustered
in one small interval or spread out rather evenly between the extreme values.
Furthermore, the two numbers used, the highest and the lowest, are extreme
rather than typical values.

Other distance measures of dispersion, like the interquartile range, employ   *Quartiles*
more typical values. **Quartiles** are special cases of general measures known as
fractiles, which refer to values that exceed specified fractions of the data—the
ninth decile exceeds $\frac{9}{10}$ of the items, the ninety-ninth percentile exceeds $\frac{99}{100}$ of the
items in a distribution, and so on. Clearly, many arbitrary distance measures
of dispersion could be developed, but they are infrequently used in practical
applications.

The **interquartile range** is the difference between the third-quartile and the   *Interquartile range*
first-quartile values. The **third quartile** is a figure such that three quarters of
the observations lie below it; the **first quartile** is a figure such that one quarter
of the observations lie below it. Thus, the distance between these two numbers
measures the spread between the values that bound the middle 50% of the
values in a distribution. A main disadvantage of such a measure is that it does
not describe the variation *among* the middle 50% of the items (nor among the
lower and upper fourths of the values).

The method of calculating quartile values will not be explicitly discussed
here, but for frequency distribution data, its calculation proceeds in a manner   *Calculating*
completely analogous to that of the median, which is itself the second-quartile   *quartile values*
value (that is, two quarters of the observations in a distribution lie below the
median and two quarters lie above it).

1.15
DISPERSION: AVERAGE DEVIATION METHODS

The most comprehensive descriptions of dispersion are stated in terms of the
average deviation from some measure of central tendency. The most important
average deviation measures are the variance and the standard deviation.

The **variance** of the observations in a population—denoted by the Greek   *Variance*
lower-case letter σ^2 (sigma, squared)—is the arithmetic mean of the squared
deviations from the population mean. In symbols, if $X_1, X_2, \ldots, X_N$ represent
the values of the N observations in a population and if μ is the arithmetic mean

of these values, the population variance is defined by

(1.13)
$$\sigma^2 = \frac{(X_1 - \mu)^2 + (X_2 - \mu)^2 + \cdots + (X_N - \mu)^2}{N}$$
$$= \frac{\Sigma(X - \mu)^2}{N}$$

As usual, subscripts have been dropped in the simplified form on the right side of equation 1.13.

Although the variance measures the extent of variation in the values of a set of observations, it is in units of squared deviations or squares of the original numbers. To obtain a measure of dispersion in terms of the units of the original data, the square root of the variance is taken. The resulting measure is known as the standard deviation. Thus, the standard deviation of a population is given by

(1.14)
$$\sigma = \sqrt{\frac{\Sigma(X - \mu)^2}{N}}$$

By convention, the positive square root is used.

Standard deviation | The **standard deviation** is a measure of the spread in a set of observations.

If all the values in a population were identical, each deviation from the mean would be 0, and the standard deviation would thus be equal to 0, its minimum value. On the other hand, as items are dispersed more and more widely from the mean, the standard deviation becomes larger and larger.

If we now consider the corresponding measures for a sample of n observations, it would seem logical to substitute the sample mean $\bar{X}$ for the population mean μ and the sample number of observations n for the population number N in equations 1.13 and 1.14. However, it can be shown that when the sample variance and standard deviation are defined with $n - 1$ in the divisors, better estimates are obtained of the corresponding population parameters.[3] Hence, in keeping with modern usage, we defined the sample variance and sample standard deviation, respectively, as

Sample variance   (1.15)
$$s^2 = \frac{\Sigma(X - \bar{X})^2}{n - 1}$$

---

[3] A brief justification for the division by $n - 1$: It can be shown mathematically that for an infinite population, when the sample variance is defined with divisor $n - 1$ as in equation 1.15, it is a so-called "unbiased estimator" of the population parameter σ^2. This means that if all possible samples of size n were drawn from a given population and the variances of these samples were averaged (using the arithmetic mean), this average would be equal to the population variance σ^2. Thus, when the sample variance is defined with divisor $n - 1$, on the average, it correctly estimates the population variance. Of course, for large samples, the difference in results obtained by using n rather than $n - 1$ as the divisor would tend to be very slight, but for small samples the difference can be rather substantial.

and

$$(1.16) \qquad s = \sqrt{\frac{\Sigma(X - \bar{X})^2}{n - 1}}$$

The term s^2 is usually referred to simply as the "sample variance" and s as the "sample standard deviation."

We will now concentrate on methods of computing the standard deviation, both for ungrouped and for grouped data. Then we discuss how this measure of dispersion is used. However, the major uses of the standard deviation are in connection with sampling theory and statistical inference, which are discussed in subsequent chapters.

Assume that the following figures represent the amounts of five accounts receivable. The arithmetic mean is $435. Regarding these data as a sample, the calculation of the standard deviation using equation 1.16 is shown in Table 1-9.

The resulting standard deviation of $129.71 is an absolute measure of dispersion, which means that it is stated in the units of the original data. Whether this is a great deal or only a small amount of dispersion cannot be immediately determined. This sort of judgment is based on the particular type of data analyzed (in this case, accounts receivable data). Furthermore, as we shall see in Section 1.16, relative measures of dispersion are preferable to absolute measures for comparative purposes.

In calculating the standard deviation for data grouped into frequency distributions, it is merely necessary to adjust the foregoing formulas to take account of this grouping. The defining equation 1.16 generalizes to

$$(1.17) \qquad s = \sqrt{\frac{\Sigma f(m - \bar{X})^2}{n - 1}}$$

TABLE 1-9
Calculation of the standard deviation for ungrouped data by the direct method: accounts receivable data

Amount of Accounts Receivable X	Deviation from Mean $X - \bar{X}$	Squared Deviations $(X - \bar{X})^2$
$600	$600 - $435 = \quad $165	27,225
350	350 - \quad 435 = \quad -85	7,225
275	275 - \quad 435 = -160	25,600
430	430 - \quad 435 = \quad -5	25
520	520 - \quad 435 = \quad 85	7,225
	$0	67,300

$$s = \sqrt{\frac{\Sigma(X - \bar{X})^2}{n - 1}} = \sqrt{\frac{67,300}{4}} = \sqrt{16,825} = \$129.71$$

TABLE 1-10
Calculation of the standard deviation for group data by the direct method:
examination scores data

Examination Scores	Frequency f	Midpoints m	$(m - \bar{X})$	$(m - \bar{X})^2$	$f(m - \bar{X})^2$
55 and under 60	1	57.5	-25.2	635.04	635.04
60 and under 65	3	62.5	-20.2	408.04	1,224.12
65 and under 70	6	67.5	-15.2	231.04	1,386.24
70 and under 75	10	72.5	-10.2	104.04	1,040.40
75 and under 80	15	77.5	-5.2	27.04	405.60
80 and under 85	19	82.5	-0.2	0.04	0.76
85 and under 90	27	87.5	4.8	23.04	622.08
90 and under 95	11	92.5	9.8	96.04	1,056.44
95 and under 100	8	97.5	14.8	219.04	1,752.32
	100				8,123.00

$\bar{X} = 82.7$ (rounded)

$$s = \sqrt{\frac{\Sigma f(X - \bar{X})^2}{n - 1}} = \sqrt{\frac{8123.00}{99}} = \sqrt{82.0505}$$

$$= 9.06$$

where, as usual for grouped data, m represents the midpoint of a class, f is the frequency in a class, $\bar{X}$ is the arithmetic mean, and n is the total number of observations. This calculation is illustrated in Table 1-10 for the frequency distribution of examination scores data previously given in Tables 1-3 and 1-5. As shown in Table 1-10, the standard deviation is equal to 9.06 points. This figure for the standard deviation is quite close to the corresponding figure of 9.076 for the ungrouped data shown in Table 1-2.

Calculation of the sample standard deviation by the defining formula can be tedious, particularly if the class midpoints and frequencies contain several digits and the arithmetic mean is not a round number. A shortcut method of calculation often useful in such cases is given in Appendix D. Of course, if the original data are available, calculations can be carried out on a PC or even a hand calculator. However, if the frequency distribution represents secondary data, such shortcut formulas can be helpful.

Uses of the Standard Deviation

The standard deviation of a frequency distribution is useful in describing the general characteristics of the data. For example, in the so-called normal distribution (a bell-shaped curve), which is discussed extensively in Chapters 5, 6, and 7, the standard deviation is used in conjunction with the mean to indicate the percentage of items that fall within specified ranges. Hence, if a population is in the form of a normal distribution, the following relationships apply:

$\mu \pm \sigma$ includes 68.3% of all of the items

$\mu \pm 2\sigma$ includes 95.5% of all of the items

$\mu \pm 3\sigma$ includes 99.7% of all of the items

For example, if a production process is known to produce items that are normally distributed with a mean length of $\mu = 10$ inches and a standard deviation of 1 inch, then we can infer that 68.3% of the items have lengths between $10 - 1 = 9$ inches and $10 + 1 = 11$ inches. About 95.5% have lengths between $10 - 2 = 8$ and $10 + 2 = 12$ inches, and 99.7% have lengths between $10 - 3 = 7$ and $10 + 3 = 13$ inches. Thus, a range of $\mu \pm 3\sigma$ includes virtually all the items in a normal distribution. We shall see in Chapter 5 that the normal distribution is perfectly symmetrical. If the departure from a normal distribution is not too great, the rough generalization that virtually all the items are included within a range from 3σ below the mean to 3σ above the means still holds. Another even more general idea for providing summary statistics is **Tchebycheff's rule**,   *Tchebycheff's rule* which states the following:

In *any* distribution of data, the proportion of observations within k standard deviations from the mean is at least $1 - 1/k^2$, where $k > 1$.

Some examples of this statement follow:

If $k = 2$, then $1 - 1/k^2 = 1 - 1/2^2 = 3/4$

If $k = 3$, then $1 - 1/k^2 = 1 - 1/3^2 = 8/9$

Therefore, if μ and σ are the mean and standard deviation of a population, we can say the following:

the range of $\mu \pm 2\sigma$ includes at least 3/4 of the observations

the range of $\mu \pm 3\sigma$ includes at least 8/9 of the observations

Although these statements are framed in terms of population values, exactly the same types of statements pertain to samples as well. As an example, in the examination scores shown in Table 1-2, for which the mean and standard deviation (rounded to one decimal place are 82.4 and 9.1 respectively), we can make the following statements:

The range of $82.4 \pm 2(9.1)$, i.e., 64.2 to 100.6, includes at least 3/4 of the observations

The range of $82.4 \pm 3(9.1)$, i.e., 55.1 to 109.7, includes at least 8/9 of the observations

As noted above, Tchebycheff's rule gives meaningful statements only for values of k greater than 1. However, it is a useful tool because it enables us to make general statements about distributions of data (and probability distributions as discussed in Chapter 3) when we do not know the shapes of the data distributions (or probability distributions) involved. Of course, when we know the form of the distribution, we can make more specific statements as was indicated above for the normal distribution.

Other uses   The standard deviation is also useful in describing how far individual items in a distribution depart from the mean of the distribution. Suppose the population of students who took a certain aptitude test displayed a mean score of $\mu = 100$ with a standard deviation of $\sigma = 20$. Then a score of 80 on the examination can be described as lying one standard deviation below the mean. The terminology usually employed is that the standard score is -1; that is, if the examination score is denoted X, then

$$\frac{X - \mu}{\sigma} = \frac{80 - 100}{20} = -1$$

The **standard score** of an observation is the number of standard deviations the observation lies below or above the mean of the distribution.

Hence, the score of 80 deviates from the mean by -20 units, which is equal to -1 in terms of units of standard deviations away from the mean. If standard scores are computed from sample rather than universe data, the formula $(X - \bar{X})/s$ would be used instead.

Comparisons can thus be made for items in distributions that differ in order of magnitude or in the units employed. For example, if a student scored 120 on an examination in which the mean was $\mu = 150$ and $\sigma = 30$, the standard score would be $(120 - 150)/30 = -1$. Thus, the score of 120 is the same number of standard deviations below the mean as the 80 in the preceding example. We could also compare standard scores in a distribution of wages with comparable figures in a distribution of length of employment service, and so on.

The standard deviation is doubtless the most widely used measure of dispersion, and considerable use is made of it in later chapters of this text.

1.16
RELATIVE DISPERSION: COEFFICIENT OF VARIATION

Although the standard score discussed earlier is useful for determining how far an *item* lies from the mean of a set of data, we often are interested in comparing the dispersion of *an entire set of data* with the dispersion of another set. As observed earlier, the standard deviation is an absolute measure of dispersion, whereas a relative measure is required for purposes of comparison. A relative measure is essential when the sets of data to be compared are expressed in different units or when the data are in the same units but are of different orders of magnitude. Such a relative measure is obtained by expressing the standard deviation as a percentage of the arithmetic mean. The resulting figure is referred to as the **coefficient of variation** (CV) and is defined symbolically as

(1.18)
$$CV = \frac{s}{\bar{X}}$$

Thus, for the frequency distribution of examination scores, the standard deviation is 9.06 with a mean of 82.65 ounces. The coefficient of variation is

$$CV = \frac{9.06}{82.65} = 11.0\%$$

Let us assume that the corresponding figures for a second set of examination scores revealed a standard deviation of 9.2 with an arithmetic mean of 93.1. The coefficient of variation for this set of data is $CV = \frac{9.2}{93.1} = 9.9\%$. Therefore, the scores in the second examination were relatively more uniform—that is, they displayed relatively less variation than did the scores of the first examination. Note that the scores of the second examination had the larger standard deviation, but because of the higher average score, relative dispersion was less.

Both absolute and relative measures of dispersion are widely used in practical sampling problems. For example, a question may arise about the sample size required to yield an estimate of a universe parameter with a specified degree of precision. Specifically, a finance company may want to know how large a random sample of its loans it must study in order to estimate the average dollar size of its delinquent loans. If the company wants this estimate accurate within a specified number of *dollars*, an absolute measure of dispersion is appropriate. On the other hand, if the company wants the estimate to be within a specified *percentage* of the true average figure, a relative measure of dispersion would then be used.

1.17
ERRORS IN PREDICTION

In this chapter, we have discussed descriptive measures for empirical frequency distributions, with emphasis on measures of central tendency and dispersion. Some interesting relationships between these two types of measure are observable when certain problems of prediction are considered. Suppose we want to guess or "predict" the value of an observation picked at random from a frequency distribution. Let us refer to the penalty of an incorrect prediction as the "cost of error." If there were a **fixed cost error** on each prediction, no matter what the size of the error, we should guess the **mode** as the value of the random

Fixed cost error

observation. This would give us the highest probability of guessing the *exact value* of the unknown observation. Assuming repeated trials of this prediction experiment, we would thus minimize the average (arithmetic mean) cost of error.

Suppose, on the other hand, that the **cost of error varies directly with the size of error regardless of its sign**, that is, regardless of whether the actual observation is above or below the predicted value. In this case, we would want

Absolute error   a prediction that minimizes the average **absolute error**. The **median** would be the "best guess," since it minimizes average absolute deviations. The mean deviation about the median would be a measure of this minimum cost of error.

Finally, suppose **the cost of error varies according to the square of the error** (for example, an error of two units costs four times as much as an error of one unit). In this situation, the **mean** should be the predicted value, since it can be demonstrated mathematically that the average of the squared deviations about it is less than around any other figure. Here the variance, which may be inter-preted as the average cost of error per observation, would represent a measure

Minimum error   of this minimum error. Another point previously observed for the mean is that the average amount of error, taking account of sign, would be 0.

A practical business application of these ideas is in the determination of the optimum size of inventory to be maintained. Let us assume a situation in which the cost of overstocking a unit (cost of overage) is equal to the cost of being short one unit (cost of underage). Further, it may be assumed that the cost of error varies directly with the absolute amount of error. For example, having two units in excess of demand costs twice as much as one unit. In this situation, the optimum stocking level is the median of the frequency distribution of numbers of units demanded.

1.18
PROBLEMS OF INTERPRETATION

Many of the most common misinterpretations and misuses of statistics involve measures and concepts such as those discussed in this chapter: averages, disper-sion, and skewness.

> Sometimes misleading interpretations are drawn from the use of averages that are not "typical" or "representative."

Reference was made in section 1.11 to the possible distortion of the typicality of the arithmetic mean because of the presence of extreme items at one end of a distribution. An interesting example of this distortion occurred in the case of a survey conducted by a popular periodical. One of the purposes of the survey was to determine the current status of persons who had graduated from college during the early Depression years. Among those included were the graduates of Princeton University for three successive years during the early 1930s. The results of the Princeton survey indicated that the arithmetic mean income of the respondents in the class that graduated in the second year was far higher than the corresponding mean income for the first- and third-year classes. The analysts attempted to rationalize this result in various ways. However, a re-examination of the data yielded a simple explanation, which precluded potential misinterpretations. It turned out that one of the graduates of the second-year class was a member of one of the wealthiest families in the United States and was an heir to an immense fortune. His very large income exerted an obvious upward pull on the mean income of his class, making it an unrepresentative average.

> Misinterpretations of averages often arise because dispersion is not taken into account.

Prospective college students are sometimes discouraged when they observe the mean scholastic aptitude test scores of classes admitted to colleges or universities in which they are interested. Admissions officers have commented that students sometimes erroneously assume they will not be admitted to a school because their test scores are somewhat below the published mean scores for that school. Of course, such students fail to take into account dispersion around this average. Assuming a roughly symmetrical distribution, about one half of the admitted students on whom the published means were based had test scores that fell below that average.

> Because of the shape of the underlying frequency distribution, sometimes no average will be typical.

In section 1.12, reference was made to bimodal distributions of wages of workers. Arithmetic means or medians for such distributions tend to fall somewhere between the two modes; they are not typical of the groups characterized by either of the modes. As indicated in section 1.12, the solution when nonhomogeneous data are present is to separate the distinct distributions. However, sometimes U-shaped frequency distributions are encountered in which the separation into different distributions is not warranted. In such distributions, frequencies are concentrated at both low and high values of the variable under consideration. For example, suppose the test scores of a mathematics class yield grades that are either very high (in the 90s) or very low (in the 60s). Means or medians, which might be about 75, would clearly be unrepresentative or the concentrations at either end of the distribution. When averages are presented for such distributions, without some indication of the nature of the underlying data, misinterpretations can occur quite easily.

EXERCISES 1.18

Note: In calculating the variance and standard deviation in the following exercises, use formulas with an $n - 1$ divisor.

1. The closing prices of two common stocks traded on the New York Stock Exchange for a week in December 1990 were

	Highfly	Stabil
Monday	$28	$28
Tuesday	34	26
Wednesday	18	22
Thursday	20	24
Friday	25	25

    a. Compare the two stocks simply on the basis of measures of central tendency.
    b. Compute the standard deviation for each of the two stocks. What information do the standard deviations give concerning the price movements of the two stocks?

2. For the period 1981–1990, the annual earnings per share for the Tiny Tot Toy Company and the Gigantic Game Corporation are as follows:

	Tiny Tot	Gigantic Game
1981	$ .50	$6.40
1982	.80	7.00
1983	.90	6.80
1984	1.20	7.60
1985	1.00	8.00
1986	.80	8.30
1987	1.20	7.90
1988	1.40	8.50
1989	1.50	8.60
1990	1.70	8.90

    a. Compute the arithmetic mean and standard deviation of the earnings per share for each firm. Which firm showed the greater absolute variation in earnings per share?
    b. Compute the coefficient of variation for each firm. Which firm showed relatively greater variation in earnings per share?

3. The vice president of a Central Motors plant is interested in investigating whether production is lower on Mondays than other days. The production figures on six successive Mondays were 10, 15, 17, 21, 15, and 12.

a. What is the mean of the six production figures? What is the median?
b. Production for the same time period on the other days of the week was summarized in 5-unit groupings. Is the mean of the other days' production greater than the mean production on Mondays?
c. Do the Monday production figures have greater relative dispersion than the other days?

Production	Number of Days
10 through 14	9
15 through 19	9
20 through 24	3
25 through 29	3
	24

4. There are five members in a family. The father's age is 45 and the mother's age is 42. The children are 21, 17, and 10 years old. At present, the mean age of the five people is 27 years and the standard deviation is 15.6 years.
a. Calculate the mean and standard deviation of the ages of this family 6 years later.
b. What generalizations concerning the mean and standard deviation are suggested by your results to part (a)?

5. An applicant for a position received a score of 84 on aptitude test A and a score of 70 on aptitude test B. On aptitude test A, the mean and the standard deviation were 80 and 4, respectively. On aptitude test B, the mean and the standard deviation were 60 and 7, respectively. On which aptitude test was the applicant's performance relatively better?

1.19
EXPLORATORY DATA ANALYSIS*

A number of useful tools for data description and preliminary data analysis—developed primarily by John W. Tukey of Princeton University—are referred to as *exploratory data analysis*.[4] We comment briefly on a couple of them here, namely, *stem and leaf displays* and *box plots*. We include also a brief discussion of the output of the Minitab command DESCRIBE, which produces a number of descriptive measures referred to in this chapter.

* Optional material is indicated with this symbol (*); such material may be omitted without interfering with the continuity of the discussion.
[4] See John W. Tukey, *Exploratory Data Analysis* (Philippines copyright: Addison-Wesley Publishing Company, Inc., 1977).

Stem and leaf displays combine some of the characteristics of histograms and portray the general trend of a batch of values as a whole. Some aspects of stem and leaf displays are illustrated in the following examples. The STEM-AND-LEAF command in Minitab gives the type of display shown in the examples plus a couple of other details that will not be discussed here.

EXAMPLE 1-3

An automobile manufacturer, about to introduce a new car model measured the mileage obtained by 24 test versions of the model and rounded the results to the nearest mile. Present the data as a simple stem and leaf display.

30	33	18	27	32	40	26	28	21	28	35	20
27	19	32	29	36	29	30	22	25	16	17	30

SOLUTION

The first line on the left of the display represents a tabulation of the four mileages from 10 to 19, namely 16, 17, 18, and 19. The asterisk (*) is a place holder, or place filler, indicating that we have a two-digit number, in this case, a 1 followed by 6, 7, 8, and 9, respectively. Each line is a **stem**, and each item of information on the stem is a **leaf**. Here, the label for the stem is the first part of a number, which is followed in turn by each leaf. The label 1* is called the **starting part**.

	Gas Mileage per Gallon	(#)
1*	6789	(4)
2	01256778899	(11)
3	00022356	(8)
4*	0	(1)
		(24, ✓)

The efficiency of this type of notion is apparent. For example, 1*|6789 uses seven characters to represent 16, 17, 18, 19. The actual form requires eight characters, or 12 if the commas and period are counted. A tally notation for these four numbers, such as 10–19|xxxx, requires 10 characters, while the identity of the individual items is lost. Note that the leaves in this example have been arranged in ascending order, as in 1*|6789. This is not a necessary procedure, and for many purposes the observations might very well be recorded in the order in which they occur.

The figures on the right side of the display represent tallies or counts of the leaves. Taken together with the stems, they constitute a frequency distribution of the original data. By convention we use the symbol (#) to denote "count" or "frequency of occurrence." At the bottom of the "count" column is the "check count" (24, ✓), signifying that 24 mileages were recorded. The check mark indicates that we have counted the number of leaves and that this figure agrees with the number of original observations. The need for checking is especially evident when we deal with large bodies of data.

EXAMPLE 1-4

Disturbed because of increasing complaints from retailers who have not been receiving promised shipments of radios, a manufacturer of mobile radios, decides to run a check on the current radio distribution network. Each of the 40 warehouses owned by the manufacturer throughout the United States is instructed to maintain at least 300 mobile radios in stock at all times. The inventory levels of the 25 warehouses that were checked are listed.

a.   Present the data as a two-digit stem and leaf display.
b.   Are the warehouses keeping the required inventories?

150	280	60	10	85	160	305	70
253	180	0	150	90	0	50	110
300	25	400	610	100	320	200	330
210							

a.   Inventory of mobile radios:

SOLUTION

	Unit = 1 Radio	(#)
0**	00, 00, 10, 25, 50, 60, 85, 90	(8)
1	00, 10, 50, 50, 60, 70, 80	(7)
2	00, 10, 53, 83	(4)
3**	00, 05, 20, 30	(4)
4	00	(1)
5		
6**	10	(1)
		(25, ✓)

b.   No, because only 6 out of 25 warehouses have inventories of at least 300 units.

   A few points may be noted in this solution. First, all displays of data are in terms of some unit. In this example, the unit is a radio. In another situation, perhaps in stating prices, $10 may be an appropriate unit. Then $360 becomes 36 and is stated as 3*|6.

   In the present example, a two-digit leaf has been used. Hence, 3**|20 represents 320, 6**|10 is 610, and so forth. Note that 0**| is used to record values between zero and 99. Had negative numbers been possible, −0**| would have been used for tabulating values from zero to −100.

1.20
MINITAB DESCRIPTION OF UNGROUPED DATA

We turn now to the summary of ungrouped data provided by the Minitab command DESCRIBE. The computer output given by this command represents a summary of a number of statistical measures discussed in this chapter, and is useful in discussing the next topic, *box plots*.

TABLE 1-11

Minitab output for the examination scores data shown in Table 1-2

```
MTB > DESCRIBE C1

               N          MEAN        MEDIAN       TRMEAN        STDEV       SEMEAN
C1           100        82.380        84.000       82.644        9.076        0.908

             MIN           MAX            Q1           Q3
C1        57.000        99.000        76.000       88.000
```

The examination scores data shown in Table 1-2 were originally SET into column C1 of Minitab's worksheet. If we respond to the MTB prompt with the command DESCRIBE C1, we obtain the output shown in Table 1-11, which is interpreted as follows.

There are 100 examination scores with a mean of 82.380 and a median of 84.000. TRMEAN is an abbreviation for the "trimmed mean." In calculating this type of mean, Minitab deletes or "trims" the smallest 5% of the observations, and the largest 5% of the observations, rounded to the nearest integer. The arithmetic mean is calculated for the remaining 90%. In this case the smallest five and largest five examination scores are deleted and the remaining 90% are averaged to obtain a trimmed mean of 82.644. The standard deviation is equal to 9.076, as we have observed previously. SEMEAN is an abbreviation for the standard error of the mean, which is equal to the standard deviation divided by the square root of the number of observations. Thus, we have $0.908 = 9.076/\sqrt{100}$. MIN and MAX are respectively the smallest and largest observations and Q_1 and Q_3 are respectively the first quartile and third quartile values.

Box Plot

The data plot shown in Figure 1-12 is obtained through the command BOXPLOT and represents another useful way of displaying data. In terms of the examination scores, the edges of the box are shown at the first quartile

FIGURE 1-12

Minitab boxplot for the examination scores data shown in Table 1-2

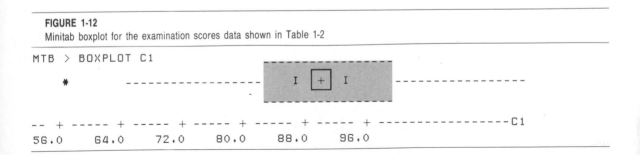

value ($Q_1 = 76$) and the third quartile value ($Q_3 = 88$). Thus the box encloses the interquartile range (IQR) or the middle one-half of the data. The vertical line of the plus symbol in the box is the median (Md = 84). Potential outliers are designated by a * symbol, while serious outliers are designated by a 0 symbol. Tukey defines potential outliers (*) as values falling more than $1\frac{1}{2}$ IQRs below Q_1 or above Q_3 and serious outliers as values falling more than 3 IQRs below Q_1 or above Q_3. In Figure 1-12 the lowest value of 57 has been designated as a potential outlier because it lies more than 18 units below 76. That is, $1\frac{1}{2}$ IQRs $= 1\frac{1}{2}(88 - 76) = 18$. The lines (sometimes called "whiskers") emanate from the edges of the box to the smallest and largest non-outliers.

In summary, the boxplot displays the main features of the data in terms of the center, the middle half, the range of the data and unusual observations.

MINITAB EXERCISES

Problems 1 through 4 are based on the data stored in the file named 'abort.mtw'. If you are in Minitab, you can get the data by using the following command:

```
MTB > retrieve 'abort'
MTB > information
```

Column	Name	Count
C1	ABORT81	51
C2	ABORT85	51

In C1, the data correspond to the number of abortions per 1,000 live births in the 50 states and in the District of Columbia in 1981. C2 is the same sort of data for 1985.
(Source of data: *Statistical Abstract of the United States*, 1989.)

1. Obtain the means, medians, minimums, maximums, and standard deviations for C1 and C2. You can, of course, use separate Minitab commands, but it is simpler to use:

   ```
   MTB > describe C1 and C2
   ```

   Interpret the computer output. Is there a clear indication that the number of abortions decreased from 1981 to 1985?

2. (*optional*) Find the number of states in which the number of abortions per 1,000 live births increased from 1981 to 1985. Do the following commands give the desired result? Why or why not?

   ```
   MTB > let k1 = (sum(signs(C2 − C1)) + 50)/2
   MTB > print k1
   ```

3. Type:

   ```
   MTB > print C1 and C2
   ```

   and see if there is a close relationship between C1 and C2. To visualize, practice the following command:

   ```
   MTB > plot C2 against C1
   ```

   and state the relationship between C1 and C2 that is shown on your computer screen.

4. Make boxplots for C1 and C2 via:

   ```
   MTB > boxplot C1
   MTB > boxplot C2
   ```

   and label the elements of these plots. Are there outliers? If yes, do outliers occur in the same state?

Problems 5 through 8 are all related to the following data: Minitab file 'mstat.mtw' contains data about three types of money aggregates: M1, M2, and M3.

Column	Name	Description
C1	YEAR	From 1985 to 1987
C2	MONTH	From January, Coded 1, to December, Coded 12
C3	M1	Currency Plus Checking Accounts
C4	M2	M1 Plus Savings Deposits & Small-Time Deposits
C5	M3	M2 Plus Time Deposits.

(Source: *Slater-Hall Information Products & Business Statistics*, 1988.)

5. Retrieve the data from the file labeled 'mstat.mtw'. Because the money aggregates data were observed over time, make time-series plots for M1, M2, and M3 by using, for example:

   MTB > tsplot C3

   Read the plots to determine the pattern of increase of each series.

6. Note that in Minitab, the commands

   MTB > let C10 = 'YEAR' + 'MONTH'/12
   MTB > name C10 'TIME'

   will create equally spaced time intervals for plotting. Note also that one can plot M1, M2, and M3 on the same graph against time with the following command:

   MTB > mplot C3 C10 C4 C10 and C5 C10

   It should be apparent that these paths do not intersect. Besides that, what else can you say about the plot?

7. Economists are interested not only in the size of money aggregates, but in their month-to-month percentage changes as well. Find the month-to-month percentage changes for M1. Hint:

   MTB > let C13 — (C3 — lag(C3))/lag(C3)

   Do a time-series plot for the percentage changes. What does the plot imply about the volatility of money supply?

8. Make a stem-and-leaf plot for C13 by letting

   MTB > stem-and-leaf C13

   and guess the skewness of the data from the display. Calculate the mean and median of C13 using Minitab, and then justify your guess.

Problems 9 through 11 refer to the data as described here: A survey of 193 state banks with standard industrial code 6022 was carried out to obtain information about the returns on assets. The data have been stored in the Minitab file labeled 'bankroa.mtw'. A description of the data is as follows:

Column	Name	Count	Description
C1	BANKROA	193	Returns on Assets (%)

9. Retrieve the data from the disk. Use Minitab to calculate the mean, median, and standard deviation. Then calculate the coefficient of variation.

10. It is known that the median minimizes the average absolute error. Therefore, k1 < k2, where k1 and k2 are defined below:

    MTB > let k1 = sum(absolute(C1 − median(C1)))
    MTB > let k2 = sum(absolute(C1 − mean(C1)))
    MTB > print k1 and k2

    Calculate k1 and k2. Is it true that k1 < k2?

11. A simple way to look at the shape of the data is to use:

    MTB > dotplot C1

    or

    MTB > histogram C1

    From the dotplot and histogram of C1, what can you say about the skewness of the data?

KEY TERMS

Arithmetic Mean (μ or $\bar{X}$)   the total of the values of a set of observations divided by the number of observations.

Attributes (discrete variables)   characteristics that can be expressed in qualitative classifications.

Average   a measure of central tendency or central location of data.

Bimodal Distribution   a frequency distribution with two modes.

Coefficient of Variation (CV)   the standard deviation divided by the arithmetic mean.

Continuous Data   data that can assume any value in a given range.

Cross-Sectional Data   data observed at a point of time.

Cumulative Frequency Distribution   a table that presents the number of cases that lie above or below specified values.

Dispersion   the spread, or variability, in a set of data.

Distance Measure   a measure that describes dispersion in terms of the distance between the values of selected observations.

Exploratory Data Analysis   a set of methods for describing numerical data.

Frequency Distribution   a table that records the numbers of items that fall in classes of data.

Frequency Polygon   a line graph that represents a frequency distribution.

Histogram   a bar chart that represents a frequency distribution.

Interquartile Range   the difference between the third-quartile and the first-quartile values.

Median (Md)   the value of the central item when data are arranged by size.

Midpoints   values located halfway between class limits of a frequency distribution.

Modal Class   the class in a frequency distribution that contains more observations than any other.

Mode   the value that occurs with the greatest frequency.

Ogive   the graph of a cumulative frequency distribution.

Parameter   a population value.

Quartiles   for example, the third quartile is a figure such that three quarters of the observations lie below it.

Sample Variance (s^2)   the total of the squares of the deviations of sample values from the sample mean divided by the number of sample values minus one.

Skewness   the symmetry or lack of symmetry in the shape of a frequency distribution.

Standard Deviation (σ or s)   the square root of the variance.

Standard Score   the number of standard deviations an observation lies below or above the mean of a distribution.

Statistics   a value computed from sample data.

Stem and Leaf Display   a method for displaying the data of a frequency distribution.

Time-series Data   a set of data arranged in chronological order.

Variance of a Population (σ^2)   the arithmetic mean of the squared deviations from the population mean.

KEY FORMULAS

Class Interval Size

$$i = \frac{H - L}{k}$$

Median of a Frequency Distribution

$$Md = L_{Md} + \left(\frac{\frac{n}{2} - \Sigma f_P}{f_{Md}}\right)i$$

Arithmetic Mean of a Frequency Distribution

$$\bar{X} = \frac{\Sigma fm}{n}$$

Arithmetic Mean of a Sample

$$\bar{X} = \frac{\Sigma X}{n}$$

Weighted Arithmetic Mean

$$\bar{X}w = \frac{\Sigma wx}{\Sigma w}$$

Standard Deviation of a Population

$$\sigma = \sqrt{\frac{\Sigma(X - \mu)^2}{N}}$$

Variance of a Population

$$\sigma^2 = \frac{\Sigma(X - \mu)^2}{N}$$

Coefficient of Variation

$$CV = \frac{s}{\bar{X}}$$

Standard Deviation of a Sample

$$s = \sqrt{\frac{\Sigma(X - \bar{X})^2}{n - 1}}$$

Standard Deviation of a Frequency Distribution

$$s = \sqrt{\frac{\Sigma f(m - \bar{X})^2}{n - 1}}$$

2
INTRODUCTION TO PROBABILITY

In Chapter 1, we discussed the frequency distribution as a device for summarizing the variation in sets of numerical observations in convenient tabular form. It is often very useful to describe and draw generalizations about patterns of variation using the concept of probability. The discussions of probability in this chapter and in Chapter 3 lay the foundation for our treatment of statistical analysis for decision making.

2.1
THE MEANING OF PROBABILITY

The development of a mathematical theory of probability began during the seventeenth century when the French nobleman Antoine Gombauld, known as the Chevalier de Méré, raised certain questions about games of chance. Specifically, he was puzzled about the probability of obtaining two 6s at least once in 24 rolls of a pair of dice. (This is a problem you should have little difficulty solving after reading this chapter.) De Méré posed the question to Blaise Pascal, a young French mathematician, who solved the problem. Subsequently, Pascal discussed this and other puzzlers raised by de Méré with the famous mathematician Pierre de Fermat. In the course of their correspondence, the mathematical theory of probability was born.

Theory of probability

    The several different methods of measuring probabilities represent different conceptual approaches and reveal some of the current controversy concerning the foundations of probability theory. In this chapter we discuss three

Methods of measuring probabilities

conceptual approaches: **classical probability**, **relative frequency of occurrence**, and **subjective probability**. Regardless of the definition of probability used, the same mathematical rules apply in performing the calculations (that is, measures of probability are always added or multiplied under the same general circumstances).

Classical Probability

Since probability theory had its origin in games of chance, it is not surprising that the first method developed for measuring probabilities was particularly appropriate for gambling situations. According to the so-called **classical** concept of probability, the probability of an event A is defined as follows: If there are a possible outcomes favorable to the occurrence of the event A and b possible outcomes unfavorable to the occurrence of A, and if all outcomes are equally likely and mutually exclusive, then the probability that A will occur, denoted $P(A)$, is

$$P(A) = \frac{a}{a+b} = \frac{\text{Number of outcomes favorable to occurrence of } A}{\text{Total number of possible outcomes}}$$

Thus, if a fair coin with two faces, denoted head and tail, is tossed into the air, the probability that it will fall with the head uppermost is $P(\text{Head}) = 1/(1 + 1) = \frac{1}{2}$. In this case, there is one outcome favorable to the occurrence of the event "head" and one unfavorable outcome. (The extremely unlikely situation that the coin will stand on end is defined out of the problem; that is, it is not classified as an outcome for the purpose of the probability calculation.)

The preceding equation can also be used to determine the probability that a certain face will show when a true die is rolled. (A die is a small cube with 1, 2, 3, 4, 5, or 6 dots on each of its faces.) A **true die** is one that is equally likely to show any of the six numbers on its uppermost face when rolled. The probability of obtaining a 1 if such a die is rolled is $P(1) = \frac{1}{1+5} = \frac{1}{6}$. Here, there is one outcome favorable to the event "1" and five unfavorable outcomes.

Terminology   Some of the terms used in classical probability require further explanation. The **event** whose probability is sought consists of one or more possible outcomes of the given activity (tossing a coin, rolling a die, or drawing a card). These activities are known in modern terminology as **experiments**, a term referring to processes that result in different possible outcomes or observations. The term **equally likely** in referring to possible outcomes is intuitively clear. Two or more outcomes are said to be **mutually exclusive** if when one of the outcomes occurs, the others cannot. Thus, the appearances of a 1 and a 2 on a die are mutually exclusive events, since if a 1 results, a 2 cannot. All possible results of an experiment are conceived of as a complete, or **exhaustive**, set of mutually exclusive outcomes.

Classical probability measures have two very interesting characteristics.

Characteristics of classical probability   • The objects referred to as *fair* coins, *true* dice, or *fair* decks of cards are abstractions in the sense that no real-world object possesses exactly the features postulated. For example, in order to be a **fair coin** (equally likely to fall "head" or "tail") the object would have to be a perfectly flat, homogeneous disk—an unlikely object.

● In order to determine the probabilities in the above examples, no coins had to be tossed, no dice rolled, nor cards shuffled. That is, no experimental data had to be collected; the probability calculations were based entirely on logic.

In the context of this definition of probability, if it is *impossible* for an event A to occur, the probability of that event is said to be zero. For example, if the event A is the appearance of a 7 when a single die is rolled, then $P(A) = 0$. A probability of 1 is assigned to an event that is *certain* to occur. Thus, if the event A is the appearance of any one of the numbers 1, 2, 3, 4, 5, or 6 on a single roll of a die, then $P(A) = 1$. According to the classical definition, as well as all others,

> the probability of an event is a number between 0 and 1, and the sum of the probabilities that the event will occur and that it will not occur is 1.

Relative Frequency of Occurrence

Although the classical concept of probability is useful for solving problems involving games of chance, serious difficulties occur with a wide range of other types of problems. For example, it is inadequate for determining the probabilities that (a) a black male American, age 30, will die within the next year, (b) a consumer in a certain metropolitan area will purchase a company's product during the next month, or (c) a production process used by a particular firm will produce a defective item. In none of these situations is it feasible to establish a complete set of mutually exclusive outcomes, each equally likely to occur. For example, in (a), only two occurrences are possible: The individual will either live or die during the ensuing year. The likelihood that he will die is, of course, much smaller than the likelihood that he will live, but how much smaller? The probability that a 30-year-old black male American will live through the next year is greater than the corresponding probability for a 30-year-old black male inhabitant of India. However, how much greater is it and precisely what do these probabilities mean? Questions of this type require reference to data.

We know that the life insurance industry establishes mortality rates by observing how many of a sample of, say, 100,000 black American males, age 30, die within a one-year period. In this instance, the number of deaths divided by 100,000 is the **relative frequency of occurrence** of death for the 100,000 individuals studied. It may also be viewed as an estimate of the **probability** of death for Americans in the given color-sex-age group. This relative frequency of occurrence concept can also be illustrated by a simple coin-tossing example.

Suppose you are asked to toss a coin known to be biased (that is, not a fair coin). You are not told whether the coin is more likely to produce a head or a tail, but you are asked to determine the probability of the appearance of a head by means of many tosses of the coin. Assume that 10,000 tosses of the coin result in 7,000 heads and 3,000 tails. Another way of stating this is that the relative frequency of occurrence of heads is $\frac{7,000}{10,000}$, or 0.70. It certainly seems reasonable to assign a probability of 0.70 to the appearance of a head with this particular coin. On the other hand, if the coin had been tossed only three times and one head resulted, you would have little confidence in assigning a probability of $\frac{1}{3}$ to the occurrence of a head.

> The **relative frequency concept of probability** may be interpreted as the proportion of times an event occurs in the long run under uniform or stable conditions.

In practice, past relative frequencies of occurrence are often used as probabilities. Hence, if 800 of the 100,000 30-year-old black male Americans died during the year, the relative frequency of death or probability of death is said to be $\frac{800}{100,000}$ for individuals in this group.

Subjective Probability

The **subjective** or **personalistic** concept of probability is a relatively recent development.[1] Its application to statistical problems has occurred almost entirely in the post–World War II period.

> According to the **subjective** concept, the probability of an event is the degree of belief, or degree of confidence, placed in the occurrence of an event by a particular individual based on the evidence available.

This evidence may consist of data on relative frequency of occurrence and any other quantitative or nonquantitative information. The individual who considers it unlikely that an event will occur assigns a probability close to 0 to it; if an individual believes it is very likely an event will occur, he or she assigns it a probability close to 1. Thus, for example, in a consumer survey, an individual may assign a probability of $\frac{1}{2}$ to the event of purchasing an automobile during the next year. An industrial purchaser may assert a probability of $\frac{4}{5}$ that a future incoming shipment will have 2% or fewer defective items.

[1] The concept was first introduced in 1926 by Frank Ramsey, who presented a formal theory of personal probability in F.P. Ramsey, *The Foundation of Mathematics and Other Logical Essays* (London: Kegan Paul; New York: Harcourt Brace Jovanovich, 1931). The theory was developed primarily by B. de Finetti, B.O. Koopman, I.J. Good, and L.J. Savage.

Subjective probabilities should be assigned on the basis of all objective and subjective evidence currently available and should reflect the decision maker's current degree of belief. Reasonable persons might arrive at different probability assessments because of differences in experience, attitudes, values, and so on. Furthermore, these probability assignments may be made for events that will occur only once, in situations where neither classical probabilities nor relative frequencies appear to be appropriate.

This approach is a broad and flexible one, permitting probability assignments to events for which there are no objective data or for which there is a combination of objective and subjective data. However, the assignments of the probabilities must be consistent. For example, if the purchaser assigns a probability of $\frac{4}{5}$ to the event that a shipment will have no more than 2% defective items, then a probability of $\frac{1}{5}$ must be assigned to the event that a shipment will have more than 2% defective items. The concept of subjective, or personal, probability is a useful one, particularly in the context of business decision making.

Sample Spaces and Experiments

The concept of an experiment is a central one in probability and statistics. In this connection,

> an **experiment** is any process of measurement or observation of different **outcomes** (results).

The experiment may be real or conceptual. The collection or totality of the possible outcomes of an experiment is referred to as its **sample space**. Thus, the collection of outcomes of the experiment of tossing a coin once (or twice, or any number of times) is a sample space. The objects that constitute the sample space are referred to as its **elements**. The elements are usually enclosed within braces, and the symbol S is conventionally used to denote a sample space.

Terms

On a single toss of a coin, there are two possible outcomes, tail (T) and head (H). Therefore,

$$S = \{T, H\}$$

If the coin is tossed twice, there are four possibilities:

$$S = \{(T, T), (T, H), (H, T), (H, H)\}$$

In the first case, the experiment consists of **one trial**, a single toss of the coin; in the second case, the experiment contains **two trials**, the two tosses of the coin. In these examples, a physical experiment may actually be performed, or we may easily conceive of the possible outcomes of such an experiment.

In other situations, although no sequence of repetitive trials is involved, we may conceive of a set of outcomes as an experiment. These outcomes may simply be the result of an observational process and need not bear any resemblance to a laboratory experiment, as long as they are well defined. Thus, we may think of each of the following two-way classifications as constituting sample spaces:

0	1
Customer was granted credit.	Customer was not granted credit.
Employee elected a stock purchase plan.	Employee did not elect a stock purchase plan.
The merger will take place.	The merger will not take place.
The company uses direct mail advertising.	The company does not use direct mail advertising.

The elements in these two-element sample spaces may be designated 0 and 1, as indicated by the column headings. Therefore, each of the four illustrative sample spaces may be conveniently symbolized as

$$S = \{0, 1\}$$

Two methods of graphically depicting sample spaces are: (1) graphs using the conventional rectangular coordinate system and (2) tree diagrams. These methods are most useful for sample spaces with relatively small numbers of sample points (or elements). Tree diagrams are more manageable because of the obvious graphic difficulties encountered by the coordinate system method beyond three dimensions. These methods are illustrated in Example 2-1.

EXAMPLE 2-1

In a certain city, a market-research firm studied the differences among consumers classified according to (1) income groups "low," "middle," or "high" and (2) whether or not they purchased a given product during a one-month period. Consumers were classified as those who (1) did not purchase or (2) did purchase the product at least once. Graph this situation.

SOLUTION

Let 0 represent "low" income, 1, "middle" income, and 2, "high" income. For purchasing activity, let 0 stand for "did not purchase" and 1 stand for "did purchase at least once." The sample space may be expressed

$$S = \{(0, 0), (1, 0), (2, 0), (0, 1), (1, 1), (2, 1)\}$$

The graph is given in Figure 2-1(a), and the tree diagram is presented in Figure 2-1(b). To illustrate how a tree diagram is read, let us consider the element (0, 1). This element is depicted in the tree by starting at the left-hand side, following the uppermost branch to 0, and then continuing from this fork down the branch leading to a 1. This (0, 1) element denotes a consumer classified as "low" for income and as "did purchase at least once" for buying activity.

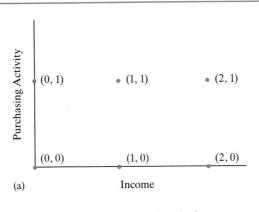

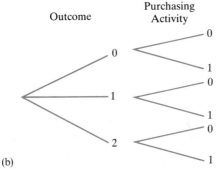

FIGURE 2-1
Example 2-1 solution: Graph
and tree diagram

Events

The meaning of the term "event" as used in ordinary conversation is usually clear. However, because the concept of an event is fundamental to probability theory, it requires an explicit definition.

> Once a sample space S has been specified, an **event** is defined as a collection of elements each of which is also an element of S.

An **elementary event** is a single possible outcome of an experiment. It is an *Elementary event* event that cannot be further divided into other events.

For example, a single roll of a die constitutes an experiment. The sample space generated is

$$S = \{1, 2, 3, 4, 5, 6\}$$

We may define an event A, say, as the "appearance of a 2 or a 3," which may be expressed as

$$A = \{2, 3\}$$

If either a 2 or a 3 appears on the upper face of the die when it is rolled, the event A is said to have occurred. Note that A is not an elementary event, since it can be divided into the two elementary events, {2} and {3}.

The **complement** of an event A in the sample space S is the collection of elements that are not in A. We will use the symbol $\bar{A}$ (pronounced "A bar") for the complement of A. For example, in the experiment of rolling a die once, the complement of the event that a 1 appears on the uppermost face is the event that a 2, 3, 4, 5, or 6 appears.

Certain (sure) event   When a sample space S has been defined, S itself is referred to as the **certain event**, or the **sure event**, since in any single trial of the experiment that generates S, one or another of its elements must occur. Hence, in the experiment of rolling a die once, the certain event is that either a 1, 2, 3, 4, 5, or 6 appears.

Mutually exclusive event   Two events A_1 and A_2 are said to be **mutually exclusive events** if when one of these events occurs, the other cannot. On a single roll of a die, the event that a 1 appears and the event that a 2 appears are mutually exclusive. On the other hand, the appearance of a 1 and the appearance of an odd number (1, 3, or 5) are not mutually exclusive events, since 1 is an odd number.

Probability and Sample Spaces

Probabilities may be thought of as numbers assigned to points in a sample space.

Probabilities must have the following characteristics:

1. They are numbers greater than or equal to 0.
2. They must add up to 1.

EXAMPLE 2-2

A real estate analyst conducted a study of transfers of ownership of 550 parcels of real property in a large city over a 20-year period. The table gives the number of parcels that were transferred 0, 1, 2, etc. times. The relative frequencies in the table have been derived by dividing the number of parcels in each category by the total number of categories.

Number of Transfers	Number of Parcels	Relative Frequencies
0	101	0.209
1	182	0.377
2	73	0.151
3	57	0.118
4	42	0.087
5 or more	28	0.058
	483	1.000

The analyst decided to use the relative frequencies as probabilities for numbers of transfers of similar real estate parcels in this city for the next 20-year period. In the classification of numbers of transfers, there would thus be six events, (0, 1, 2, 3, 4, 5 or more).

The two characteristics of probabilities are satisfied. Each probability is greater than or equal to 0, and the probabilities add up to 1. Note that the events in a sample space need not be elementary events. For example, the event "5 or more" consists of the elementary events 5, 6, etc. Of course, the events defining the sample space must be a complete set of mutually exclusive events, and the sum of the probabilities of the occurrence of an event A and its complement must be equal to 1, that is, $P(A) + P(\bar{A}) = 1$.

EXAMPLE 2-3

A corporation economist was concerned about the surging of credit demand and the effect of inflationary pressures in the economy on the company's activities during the next six-month period. The economist established the following subjective probability distribution concerning the possible action of the Federal Reserve System in the area of monetary policy during the next half year:

Events E_i	Probability of Events E_i
E_1: Tighten monetary policy	0.6
E_2: Leave current monetary policy unchanged	0.3
E_3: Relax monetary policy	0.1
	1.0

Thus, this economist assigned the value 0.9 to the probability that the Federal Reserve System will either tighten monetary policy or leave the current monetary situation unchanged.

Odds Ratios

Regardless of the definition used, we sometimes prefer to express probabilities in terms of **odds**. Thus, if the probability that a 6 will appear on the roll of a die is $\frac{1}{6}$, then the odds that it will appear are one to five, written 1:5. If an economist assesses the probability that the Federal Reserve System will adopt a tightened monetary policy as $\frac{1}{2}$, then this probability expressed in terms of odds is 1:1.

If the probability that an event A will occur is

$$P(A) = \frac{a}{n}$$

then the odds in favor of the occurrence of A are

$$\text{Odds in favor of } A = \frac{a}{n-a} = a:(n-a)$$

The odds against A are

$$\text{Odds against } A = \frac{n-a}{a} = (n-a):a$$

Another way of viewing the odds ratio is as the ratio of the probability of the occurrence of an event A to the probability of its complement, $\bar{A}$.

Thus

$$\text{Odds in favor of } A = \frac{P(A)}{P(\bar{A})} = \frac{a/n}{(n-a)/n} = \frac{a}{n-a} = a:(n-a)$$

$$\text{Odds against } A = \frac{P(\bar{A})}{P(A)} = \frac{(n-a)/n}{a/n} = \frac{n-a}{a} = (n-a):a$$

EXERCISES 2.1

1. Let 0 represent the default of a loan, and let 1 represent the repayment of a loan. Depict the set of possible outcomes that represent the repayment of at least 3 out of 4 loans made.

2. An investor tries to predict whether interest rates will fall, remain the same, or rise. Let the vertical axis represent her predictions, and let the horizontal axis represent the actual outcome. Graph the possible combinations of predictions and actual outcomes.

3. A major stockholder is seeking bids for his shares in his company. He decides to accept the first bid that exceeds a certain price. Let R represent his rejection, and let A represent his acceptance of a bid. Write the sample space of possible outcomes of the bidding.

4. Hercules Company manufactures 3 different products. Let 0 denote the event that sales of a product exceed forecasts, let 1 denote the event that sales of a product meet forecasts, and let 2 denote the event that sales of a product fall below forecasts. What set of outcomes represents the event that sales of no more than one product meets or exceeds forecasts?

5. Lucky Stores plans to open 3 news offices, one in each of 3 regions. The possible candidates in the Western Region are Los Angeles and Seattle; the possible candidates in the Central Region are Detroit and Chicago; and the possible candidates in the Eastern Region are Philadelphia, Baltimore, and Atlanta. Specify and count the possible ways of choosing the 3 cities for the new offices.

6. Refer to exercise 5. Draw a tree diagram to represent the possible ways of choosing the 3 cities.

7. A bucket contains 2 red balls and 1 black ball. Draw a tree diagram depicting the possible outcomes of choosing one ball at a time out of the bucket, without replacement.

8. Are the following mutually exclusive events?
    a. In a batch of 3 articles:
        (i)   All are defective
        (ii) Some are defective.
    b. The price of gold:
        (i)   rises one day, and falls the next.
        (ii) falls on both days.
    c. In an analysis of 4 companies (A, B, C, and D), exactly 1 is found to be profitable.
        (i)   Company A is profitable.
        (ii) Company B is profitable.
    d. In an analysis of 4 companies (A, B, C, and D), exactly 2 are found to be profitable.
        (i)   Companies A and B are profitable.
        (ii) Companies A and C are profitable.
    e. (i)   On a certain day, London stocks close higher.
        (ii) On the same day, Tokyo stocks close higher.

2.2
ELEMENTARY PROBABILITY RULES

In most applications of probability theory, we are interested in combining probabilities of events that are related in some important way. In this section, we discuss two fundamental ways of combining probabilities: *addition* and *multiplication*.

Before considering the combining of probabilities by addition, we will define two new terms.

- The symbol $P(A_1$ or $A_2)$ refers to the probability that *either* event A_1 or event A_2 occurs—*or* that they both occur. For example, if A_1 refers to the event that an individual is a male and A_2 refers to the event that the individual is a college graduate, then the symbol $P(A_1$ or $A_2)$ denotes the probability that the individual is *either* a male *or* a college graduate. The term "or" is used inclusively; that is, it includes the case of a person who is both a male and a college graduate. If events A_1 and A_2 cannot both occur, $P(A_1$ or $A_2)$ refers to the probability that *either* A_1 or A_2 occurs. For example, if A_1 and A_2 refer respectively to the obtaining of a head and a tail on a toss of a coin, then $P(A_1$ or $A_2)$ denotes the probability of obtaining *either* a head *or* a tail.

- The symbol $P(A_1$ and $A_2)$ is used to denote the probability that both events A_1 *and* A_2 will occur. $P(A_1$ and $A_2)$ is called the **joint probability** of the events A_1 and A_2. Continuing the example in the preceding paragraph,

$P(A_1$ and $A_2)$ is the probability that the individual is both a male *and* a college graduate.

Addition rule for any two events A_1 and A_2

We now state the general addition rule for any two events A_1 and A_2 in a sample space S.

(2.1) $$P(A_1 \text{ or } A_2) = P(A_1) + P(A_2) - P(A_1 \text{ and } A_2)$$

Addition rule for two mutually exclusive events A_1 and A_2

If A_1 and A_2 are **mutually exclusive events**, that is, if they cannot both occur, then

$$P(A_1 \text{ and } A_2) = 0$$

This leads to a special case of the general addition rule.

(2.2) $$P(A_1 \text{ or } A_2) = P(A_1) + P(A_2)$$

Applying equation 2.2 to the rolling of a die, we note that the probability that either a 1 or a 2 will occur on a single roll is $P(A_1 \text{ or } A_2) = P(A_1) + P(A_2) = \frac{1}{6} + \frac{1}{6} = \frac{1}{3}$. In Example 2-4 we consider an application of equation 2.1 for two events that are not mutually exclusive.

EXAMPLE 2-4

What is the probability of obtaining a 6 on the first or second roll of a die or on both? Another way of wording this question is, "What is the probability of obtaining a 6 at least once in two rolls of a die?"

SOLUTION

Let A_1 denote the appearance of a 6 on the first roll and A_2 the appearance of a 6 on the second roll. We want to find the value of $P(A_1 \text{ or } A_2)$. (As explained earlier, because of the inclusive meaning of "or," the symbol $P(A_1 \text{ or } A_2)$ means the probability that a 6 appears *either* on the first *or* on the second roll *or* on both rolls.) Consider the sample space of 36 equally likely elements listed below. These are all possible outcomes of two rolls of the die; the numbers in each element represent the outcomes on the first and second rolls, respectively.

$$
\begin{array}{cccccc}
1,1 & 2,1 & 3,1 & 4,1 & 5,1 & 6,1 \\
1,2 & 2,2 & 3,2 & 4,2 & 5,2 & 6,2 \\
1,3 & 2,3 & 3,3 & 4,3 & 5,3 & 6,3 \\
1,4 & 2,4 & 3,4 & 4,4 & 5,4 & 6,4 \\
1,5 & 2,5 & 3,5 & 4,5 & 5,5 & 6,5 \\
1,6 & 2,6 & 3,6 & 4,6 & 5,6 & 6,6 \\
\end{array}
$$

The probability that a 6 will appear on both the first and second rolls is $\frac{1}{36}$, that is, $P(A_1 \text{ and } A_2) = \frac{1}{36}$. The probability that a 6 will appear on the first roll is $P(A_1) = \frac{1}{6}$, and on the second roll, $P(A_2) = \frac{1}{6}$. Hence, applying the addition rule, we have

$$P(A_1 \text{ or } A_2) = P(A_1) + P(A_2) - P(A_1 \text{ and } A_2)$$

$$= \frac{1}{6} + \frac{1}{6} - \frac{1}{36}$$

$$= \frac{11}{36}$$

The term $P(A_1 \text{ and } A_2)$ must be subtracted in this calculation in order to avoid double counting. That is, if we incorrectly solved the problem by using the addition theorem for mutually exclusive events, computing $P(A_1 \text{ or } A_2) = P(A_1) + P(A_2) = \frac{1}{6} + \frac{1}{6} = \frac{12}{36}$, we would have counted the event $(6, 6)$ twice, because $(6, 6)$ is an elementary event both of A_1 (6 on the first roll) and of A_2 (6 on the second roll). Note that 11 elements in the sample space listed above represent the event "6 at least once in two trials." Therefore, the same result could have been obtained by using the classical definition of probability. The ratio of outcomes favorable to the event "6 at least once" to the total number of outcomes is $\frac{11}{36}$.

The ideas involved in the use of the addition rule are portrayed in Figure 2-2. The interiors of the rectangles represent the sample space, and the two events A_1 and A_2 are displayed as circles. The event "A_1 or A_2," whose probability is to be found, is shown as the tinted region in Figure 2-2(a). If we added all the outcomes in A_1 to those in A_2, which would be implied when we add $P(A_1)$ and $P(A_2)$, we would count the outcomes associated with the event "A_1 and A_2" twice; the latter event is shown in Figure 2-2(b) as the tinted region where the circles overlap. That is why we must subtract the term $P(A_1 \text{ and } A_2)$ from the sum of $P(A_1)$ and $P(A_2)$ to obtain the desired probability, $P(A_1 \text{ or } A_2)$.

On the other hand, if A_1 and A_2 are mutually exclusive events, they cannot occur jointly. This is indicated in Figure 2-2(c), which depicts these events as circles that do not overlap. The event "A_1 or A_2" is represented by the tinted area of the sample space. Since the circles do not intersect, there is no double

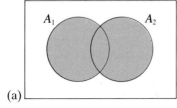

(a)

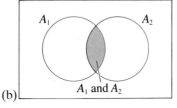

$A_1 \text{ and } A_2$

(b)

where A_1 and A_2 are not mutually exclusive,
$$P(A_1 \text{ or } A_2) = P(A_1) + P(A_2) - P(A_1 \text{ and } A_2)$$

FIGURE 2-2
Symbolic portrayal of the addition rules

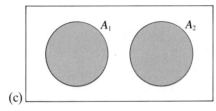

(c)

where A_1 and A_2 are mutually exclusive,
$$P(A_1 \text{ or } A_2) = P(A_1) + P(A_2)$$

counting when $P(A_1)$ is added to $P(A_2)$ to obtain $P(A_1$ or $A_2)$. (Note again that since A_1 and A_2 cannot occur together, $P(A_1$ and $A_2) = 0$.)

The addition rule can, of course, be extended to more than two events. The generalization for n mutually nonexclusive events will not be given here *Addition rule for n* because of its complexity. If the n events $A_1, A_2, \ldots, A_n$ are mutually exclusive, *mutually exclusive events* then we have the following rule:

(2.3) $$P(A_1 \text{ or } A_2 \text{ or } \ldots \text{ or } A_n) = P(A_1) + P(A_2) + \cdots + P(A_n)$$

This addition rule is applicable whenever we are interested in the probability that any one of several mutually exclusive events will occur. For example, returning to Example 2-3, the computation of the probability that the Federal Reserve System will either tighten monetary policy or leave the current monetary situation unchanged illustrates this rule. In probability notation, we have

$$P(E_1 \text{ or } E_2) = P(E_1) + P(E_2)$$

Joint Probability Tables

In many applications, we are interested in the probability of the **joint occurrence** of two or more events. To illustrate joint probabilities, consider the data in Table 2-1. These figures represent the results of a market research survey in which 1,000 persons were asked which of two competitive products they preferred, product ABC or product XYZ. To simplify the discussion, as shown in Table 2-1, we designate A_1, A_2, B_1, and B_2, respectively, as "male," "female," "prefers product ABC," and "prefers product XYZ." Hence, the joint outcome that an individual is male and prefers product ABC is denoted as "A_1 and B_1," and the joint probability that a randomly selected individual is male and prefers product ABC is $P(A_1$ and $B_1)$. Analogous notation is used for the other possible joint outcomes and joint probabilities.

Joint probabilities may be illustrated in the following manner. If a person is selected at random from this group of 1,000, the joint probability that the individual is a male and prefers product ABC is

$$P(A_1 \text{ and } B_1) = \frac{200}{1{,}000} = 0.20$$

TABLE 2-1

1,000 persons classified by sex and by product preference

Sex	Prefers Product ABC B_1	Prefers Product XYZ B_2	Total
A_1: Male	200	300	500
A_2: Female	100	400	500
	300	700	1,000

TABLE 2-2
Joint probability table for 1,000 persons, classified by sex and product preference

Sex	Prefers Product ABC B_1	Prefers Product XYZ B_2	Marginal Probabilities
A_1: Male	0.20	0.30	0.50
A_2: Female	0.10	0.40	0.50
Marginal probabilities	0.30	0.70	1.00

Similarly, the probability that a randomly selected person is a female and prefers product XYZ is

$$P(A_2 \text{ and } B_2) = \frac{400}{1,000} = 0.40$$

It is useful to construct a so-called **joint probability table** by dividing all entries in Table 2-1 by the total number of individuals (1,000). The resulting joint probability table is shown as Table 2-2. The figures in the table are examples of probabilities calculated as relative frequencies of occurrence, as discussed in section 2.1.

Marginal Probabilities

In addition to the joint probabilities mentioned earlier, we can also separately obtain from Table 2-2 probabilities for each of the two classifications "sex" and "product preference." These probabilities, which are shown in the margins of the joint probability table, are referred to as **marginal probabilities**, or **unconditional probabilities**. For example, the marginal probability that a randomly chosen individual is a male is $P(A_1) = 0.50$, and the marginal probability that a person prefers product ABC is $P(B_1) = 0.30$.

The marginal probabilities for each classification are obtained by summing the appropriate joint probabilities. For example, the marginal probability that an individual prefers product ABC is 0.30. The event B_1, "prefers product ABC," consists of two mutually exclusive parts, "A_1 and B_1" ("male and prefers product ABC") and "A_2 and B_1" ("female and prefers product ABC"). Hence, we have

$$P(B_1) = P[(A_1 \text{ and } B_1) \text{ or } (A_2 \text{ and } B_1)]$$
$$= P(A_1 \text{ and } B_1) + P(A_2 \text{ and } B_1)$$
$$= 0.20 + 0.10 = 0.30$$

Thus, to obtain marginal probabilities for product preference, we add the appropriate joint probabilities over the other classification (in this case, sex). This is an application of the addition rule for mutually exclusive events. All other marginal probabilities in Table 2-2 may be similarly calculated.

Conditional Probabilities

Often we are interested in how certain events are related to the occurrence of other events. In particular, we may be interested in the probability of the occurrence of an event given that another related event has occurred. Such probabilities are referred to as **conditional probabilities**. For instance, returning to the events discussed in Table 2-2, we may be interested in the probability that an individual prefers product ABC given that the individual is a male. This conditional probability is denoted $P(B_1|A_1)$ and is read "the probability of B_1 given A_1." The vertical line is read "given," and the event following the line, in this case A_1, is the one known to have occurred. We can now formally define a conditional probability.

Conditional probability of B_1 given A_1

Let A_1 and B_1 be two events in a sample space S. The conditional probability of B_1 given A_1, denoted $P(B_1|A_1)$, is

(2.4)
$$P(B_1|A_1) = \frac{P(A_1 \text{ and } B_1)}{P(A_1)} \quad \text{where } P(A_1) > 0$$

The statement that $P(A_1) > 0$ is included in order to rule out the possibility of dividing by 0.

This definition may be illustrated by the example given in Table 2-2. The conditional probability that a person prefers product ABC given that the person is male is

$$P(B_1|A_1) = \frac{P(A_1 \text{ and } B_1)}{P(A_1)} = \frac{0.20}{0.50} = 0.40$$

We note that the conditional probability of event B_1 given A_1 is found by dividing the joint probability of A_1 and B_1 by the marginal probability of A_1. The rationale of this procedure becomes clear by returning to Table 2-1, where we see that the proportion of males who prefer product ABC is $\frac{200}{500} = 0.40$. Thus, the conditional probability of B_1 given A_1 is simply the proportion of times that B_1 occurs out of the total number of times that A_1 occurs.

Note that a chronological order is not necessarily implied in conditional probability. That is, in $P(B_1|A_1)$, event A_1 does not necessarily precede B_1 in time. In fact, using the same method of definition given in equation 2.4, we would have

(2.5)
$$P(A_1|B_1) = \frac{P(A_1 \text{ and } B_1)}{P(B_1)} \quad \text{where } P(B_1) > 0$$

Thus, in Table 2-2 the conditional probability that an individual is a male given that the person prefers product ABC is

$$P(A_1|B_1) = \frac{P(A_1 \text{ and } B_1)}{P(B_1)} = \frac{0.20}{0.30} = 0.67$$

Conditional probabilities are important concepts in our everyday affairs and in managerial decision making. For instance, we may be interested in the probability that a friend will arrive at an appointment punctually given that he has said that he will appear at a certain time. A university is concerned about the probability that an applicant for admission will have a satisfactory academic performance at the university given that the applicant has certain aptitude test scores. A marketing executive would be interested in the probability that the sales volume for one of her company's products will increase given that she has made a commitment for an expensive sales promotion campaign for that product. In the illustration given in Table 2-2, the users of the market research survey data would be interested in the relationship between sex and product preference. Indeed, the following conditional probabilities in that illustration are revealing:

$$P(B_1|A_1) = \frac{0.20}{0.50} = 0.40 \quad \text{and} \quad P(B_1|A_2) = \frac{0.10}{0.50} = 0.20.$$

That is, the conditional probability of preference for product ABC given that the person is a male is 40%, whereas the corresponding conditional probability for a female is only 20%. Such sex differences in product preferences may be very important, for example, in choosing types of promotional effort to be used in selling the products.

Multiplication Rule

The multiplication rule for two events A_1 and B_1 follows immediately from the definition of conditional probability given in equation 2.4. Multiplying both sides of equation 2.4 by $P(A_1)$, we obtain the multiplication rule.

(2.6) $$P(A_1 \text{ and } B_1) = P(A_1)P(B_1|A_1)$$

Multiplication rule for any two events A_1 and A_2

Equivalently, from equation 2.5, the multiplication rule may be stated as

(2.7) $$P(A_1 \text{ and } B_1) = P(B_1)P(A_1|B_1)$$

Hence, $P(A_1 \text{ and } B_1)$ can be computed by either equation 2.6 or 2.7. As an example, let us apply these equations to the data of Table 2-2 to find the joint probability that a randomly selected person would be male and would prefer product ABC. By equation 2.6, we have

$$P(A_1 \text{ and } B_1) = P(A_1)P(B_1|A_1)$$
$$= (0.50)(0.40) = 0.20$$

and by equation 2.7, we obtain

$$P(A_1 \text{ and } B_1) = P(B_1)P(A_1|B_1)$$
$$= (0.30)(0.67) = 0.20$$

Of course, these calculations are given only to illustrate the principles involved in the multiplication rule. We would ordinarily not compute joint probabilities this way if the joint probabilities were already available as in Table 2-2.

The generalization of the multiplication rule to three or more events is straightforward.

Multiplication rule for any three events A_1, A_2, and A_3

(2.8) $$P(A_1 \text{ and } A_2 \text{ and } A_3) = P(A_1)P(A_2|A_1)P(A_3|A_2 \text{ and } A_1)$$

Multiplication rule for any n events A_1, A_2, ..., A_n

(2.9) $$P(A_1 \text{ and } A_2 \text{ and } \ldots \text{ and } A_n)$$
$$= P(A_1)P(A_2|A_1)P(A_3|A_2 \text{ and } A_1)\ldots P(A_n|A_{n-1} \text{ and } \ldots \text{ and } A_1)$$

This notation means that the joint probability of the n events is given by the product of the probability that the first event A_1 has occurred, the conditional probability of the second event A_2 given that A_1 has occurred, the conditional probability of the third event A_3 given that both A_2 and A_1 have occurred, and so on. Of course, the n events can be numbered arbitrarily; any one of them may be the first event, any of the remaining $n - 1$ may be the second event, and so forth.

Let us consider a couple of examples of these ideas. Suppose the probability that a sales representative following up on a lead will make the sale is 0.2. Past experience indicates that 40% of such sales are for amounts in excess of $100. What is the probability that the representative will make a sale in excess of $100?

To solve this problem, let us use the following symbols:

$P(A_1)$:   Probability that a sale is made

$P(A_2|A_1)$:   Probability that the sale is in excess of $100 given that a sale is made

The required probability is given by

$$P(A_1 \text{ and } A_2) = P(A_1)P(A_2|A_1)$$
$$= (0.2)(0.4) = 0.08$$

It should be noted that $P(A_2|A_1)$ is a conditional probability because the 0.40 probability that a sale will exceed $100 depends on the sale actually being made.

Our second example is typical of many situations involving the sampling of human populations. Suppose we had a list of 10 individuals, 5 of whom reside in New York, 3 in Pennsylvania, and 2 in New Jersey. Suppose we select 3 names at random from the list, one at a time, so that at each drawing all remaining names have an equal chance of being selected. (The actual techniques involved in such sampling procedures are discussed in Chapter 4.) What is the probability that all 3 names are those of New York residents?

We designate the events of the first-drawn, second-drawn, and third-drawn names of New York residents as A_1, A_2, and A_3, respectively. The required joint probability is

$$P(A_1 \text{ and } A_2 \text{ and } A_3) = P(A_1)P(A_2|A_1)P(A_3|A_2 \text{ and } A_1)$$

$$= \left(\frac{5}{10}\right)\left(\frac{4}{9}\right)\left(\frac{3}{8}\right) = 0.083$$

Let us consider the factors on the right side of equation. Because 5 of the 10 names are those of New York residents, $P(A_1) = \frac{5}{10}$. Given that a New York resident's name is selected, 4 of the 9 remaining names are of New York residents, so $P(A_2|A_1) = \frac{4}{9}$. Similarly, after the second New York name is chosen, there are 8 names remaining, 3 of which are of New York residents.

An important point concerning the use of the multiplication rule may be noted from this example. The probability calculation pertains to 3 names being drawn successively. What is the probability of obtaining 3 New York names if the 3 names are drawn *simultaneously* from the list? The answer to this question is exactly the same as in the preceding calculation. We can think of one of the 3 names as the "first," another as the "second," and another as the "third," although they have been drawn together. A_1 is again used to denote the event "the first name is of a resident of New York," and so on.

Because the *joint* probability of the *simultaneous* occurrence of events $A_1, A_2,$ and A_3 is the same as their successive occurrence, we can state the following generalization about the use of the multiplication rule.

> The multiplication rule may be used to obtain the joint probability of the successive or simultaneous occurrence of two or more events.

The example of sampling names from a list has interesting implications. In the example, we drew a sample of 3 names at random *without replacement* of the sampled elements. In human populations, sampling is usually carried out without replacement; that is, after the necessary data are obtained, an individual drawn into the sample is usually not replaced prior to the choosing of another individual. We have seen how the partial exhaustion of the population because of sampling had to be taken into account in the calculation of conditional probabilities. If the sampling had been with replacement, then the basic probabilities of selection remain unchanged after each item is replaced in the population. For example, if the sampling from the list had been with replacement, then the respective probabilities for New York residents would have been $P(A_1) = \frac{5}{10}$, $P(A_2) = \frac{5}{10}$, and $P(A_3) = \frac{5}{10}$. Hence the joint probability of the occurrence of $A_1, A_2,$ and A_3 would have been

Sampling with replacement

$$P(A_1 \text{ and } A_2 \text{ and } A_3) = \left(\frac{5}{10}\right)\left(\frac{5}{10}\right)\left(\frac{5}{10}\right) = 0.125$$

> In sampling *without replacement*, we are dealing with *dependent* events; in sampling *with replacement*, we are dealing with *independent* events.

The meaning of these concepts and the multiplication rules for independent events are discussed next.

Statistical Independence

In our discussion of Table 2-2, we saw that there was a relationship between product preference and sex. For example, we found that the probability that a male preferred product ABC was 0.40, whereas the corresponding probability for a female was 0.20. In other words, product preference *depends* on sex. Hence, the corresponding events involved (for example, "male" and "prefers product ABC") are said to be **dependent**. Since it is important in analysis and decision making to detect such relationships, it is correspondingly important to know when we are dealing with **independent** events.

We now turn to the concept of independence, usually referred to as **statistical independence**. If two events, say A_1 and B_1, are statistically independent, then knowing that one of them has occurred does not affect the probability that the other will occur; in such a case, $P(B_1|A_1) = P(B_1)$. For example, in tossing a fair coin twice, suppose we use the following notation for events:

A_1: Head on first toss
B_1: Head on second toss

The marginal, or unconditional, probability of obtaining a head on the second toss is $P(B_1) = \frac{1}{2}$. The conditional probability of obtaining a head on the second toss given that a head was obtained on the first toss is $P(B_1|A_1) = \frac{1}{2}$. Of course, the probability of obtaining a head on the second toss does not depend on whether a head or tail was obtained on the first toss. Thus, we note

$$P(B_1|A_1) = P(B_1) = \frac{1}{2}$$

We can now define statistical independence of two events. Two events A_1 and B_1 are *statistically independent* if

$$P(B_1|A_1) = P(B_1)$$

When this equality holds, it is also true[2] that

$$P(A_1|B_1) = P(A_1)$$

---

[2] If several events are *collectively independent* (or *mutually independent*), then every possible conditional probability for every combination of events must be equal to the corresponding unconditional probability. For example, for 3 events A_1, A_2, and A_3 to be collectively independent, the necessary and sufficient conditions are the following:

$$P(A_1) = P(A_1|A_2) \qquad P(A_1) = P(A_1|A_2 \text{ and } A_3)$$
$$P(A_2) = P(A_2|A_3) \qquad P(A_2) = P(A_2|A_1 \text{ and } A_3)$$
$$P(A_3) = P(A_3|A_1) \qquad P(A_3) = P(A_3|A_1 \text{ and } A_2)$$

When two events A_1 and B_1 are statistically independent, the multiplication rule given in equations 2.6 and 2.7 can be simplified.

(2.10) $$P(A_1 \text{ and } B_1) = P(A_1)P(B_1)$$

Multiplication rule for two statistically independent events A_1 and B_1

Note that if two events A_1 and B_1 are statistically independent, equation 2.10 holds. The converse is also true. That is if two events A_1 and B_1 are related according to equation 2.10, the two events are statistically independent; furthermore, $P(B_1|A_1) = P(B_1)$ and $P(A_1|B_1) = P(A_1)$.

The generalization of the multiplication rule for n collectively independent events (any one event is independent of any combination of the others) is given next.

(2.11) $$P(A_1 \text{ and } A_2 \text{ and } \ldots \text{ and } A_n) = P(A_1)P(A_2)P(A_3) \ldots P(A_n)$$

Multiplication rule for n collectively independent events $A_1, A_2, \ldots, A_n$

As Examples 2-5 and 2-6 show, it is possible to solve a variety of probability problems using only the addition and multiplication rules.

EXAMPLE 2-5

A national franchising firm is interviewing prospective buyers in the Morganville area. The probability that a prospective buyer will offer to buy the franchise is 0.1. Assuming statistical independence, what is the probability that in interviewing five prospective buyers, the firm will receive at least one offer to buy the franchise?

Let O and F represent, respectively, an offer and a failure to offer to buy the franchise. Then, for any prospective buyer,

SOLUTION

$$P(O) = 0.1 \text{ and } P(F) = 0.9$$

Let E denote the event "at least one offer to buy the franchise in interviewing five prospective buyers." Then $\bar{E}$, the complement of E, denotes the event "no offers to buy the franchise in interviewing five prospective buyers." Since the event $\bar{E}$ represents the successive occurrence of five failures to offer to buy the franchise and since we have assumed independence of events, we obtain by the multiplication rule for independent events

$$P(\bar{E}) = P(F \text{ and } F \text{ and } F \text{ and } F \text{ and } F) = P(F)P(F)P(F)P(F)P(F)$$
$$= (0.9)(0.9)(0.9)(0.9)(0.9) = 0.59$$

Therefore,

$$P(E) = 1 - P(\bar{E}) = 1 - 0.59 = 0.41$$

EXAMPLE 2-6

The following table refers to the 2500 employees of the Johnson Company, classified by sex and by opinion on a proposal to emphasize fringe benefits rather than wage increases in an impending contract discussion.

| | Opinion | | | |
Sex	In Favor	Neutral	Opposed	Total
Male	900	200	400	1500
Female	300	100	600	1000
Total	1200	300	1000	2500

a. Calculate the probability that an employee selected from this group will be
   (1) a female opposed to the proposal
   (2) neutral
   (3) opposed to the proposal, given that the employee selected is a female
   (4) either a male or opposed to the proposal
b. Are opinion and sex independent for these employees?

SOLUTION

We use the following representation of events

$$A_1: \text{Male} \quad B_1: \text{In favor}$$
$$A_2: \text{Female} \quad B_2: \text{Neutral}$$
$$B_3: \text{Opposed}$$

a. In (a), we have the following probabilities:
   (1) $P(A_2 \text{ and } B_3) = \frac{600}{2500} = 0.24$

   (2) $P(B_2) = \frac{300}{2500} = 0.12$

   (3) $P(B_3 | A_2) = \dfrac{P(B_3 \text{ and } A_2)}{P(A_2)} = \dfrac{600/2500}{1000/2500} = 0.60$

   (4) $P(A_1 \text{ or } B_3) = P(A_1) + P(B_3) - P(A_1 \text{ and } B_3)$

   $$= \frac{1500}{2500} + \frac{1000}{2500} - \frac{400}{2500}$$

   $$= \frac{2100}{2500} = 0.84$$

b. In (b), in order for opinion and sex to be statistically independent, the joint probability of each pair of A events and B events would have to be equal to the product of the respective unconditional probabilities. That is, the following equalities would have to hold:

$$P(A_1 \text{ and } B_1) = P(A_1)P(B_1) \quad P(A_2 \text{ and } B_1) = P(A_2)P(B_1)$$

$$P(A_1 \text{ and } B_2) = P(A_1)P(B_2) \quad P(A_2 \text{ and } B_2) = P(A_2)P(B_2)$$

$$P(A_1 \text{ and } B_3) = P(A_1)P(B_3) \quad P(A_2 \text{ and } B_3) = P(A_2)P(B_3)$$

Clearly, these equalities do not hold; for example,

$$P(A_1 \text{ and } B_1) \neq P(A_1)P(B_1)$$

$$\frac{900}{2500} \neq \frac{1500}{2500} \times \frac{1200}{2500}$$

Another way of viewing the problem is that each conditional probability would have to be equal to the corresponding unconditional probability. Thus, the following equalities would have to hold:

$$P(A_1|B_1) = P(A_1) \quad P(A_1|B_2) = P(A_1) \quad P(A_1|B_3) = P(A_1)$$

$$P(A_2|B_1) = P(A_2) \quad P(A_2|B_2) = P(A_2) \quad P(A_2|B_3) = P(A_2)$$

These equalities do not hold; for example,

$$P(A_1|B_1) \neq P(A_1)$$

$$\frac{900}{1200} \neq \frac{1500}{2500}$$

The nature of the dependence (lack of independence) can be summarized briefly as follows: The proportion of males declines as we move from favorable to opposed opinions. This type of relationship is sometimes described by saying that there is a *direct* relationship between the proportion of males and favorableness of opinion. Correspondingly, there is an *inverse* relationship between the proportion of females and favorableness of opinion. The discovery and interpretation of such dependence lays the groundwork for improved decision making.

EXERCISES 2.2

1. Given the following probabilities of bonds and stocks prices falling:

$$P(\text{stock prices fall}) = 0.4 \quad = 40\%$$

$$P(\text{bond prices fall}) = 0.3 \quad = 30\%$$

$$P(\text{stock and bond prices both fall}) = 0.2 \quad = 20\%$$

   a. Are the events "bond prices fall" and "stock prices fall" mutually exclusive?
   b. Are the events "bond prices fall" and "stock prices fall" statistically independent? NO, THEY ARE BOTH DEPENDENT ON OTHER EVENTS

2. Of the articles produced by a certain machine, 4% are too big and 2% too small. What is the probability that an article chosen at random is of the right size? 94%   HAS TO ADD UP TO 100%

3. Two randomly chosen articles, made by the same machine as in exercise 2, are inspected. What is the probability that both are of the wrong size? 6% · 6% = 36/ 36 · 100 .0036 %

4. There is a 20% chance that a sales person will not meet her weekly sales quota. What are the odds that she will meet her sales quota?

5. The marketing department of a firm forecasts that each of the following sales levels is equally likely:

   $50,000 $100,000 $150,000 $200,000 $250,000 $300,000

a.   Assuming that the forecasts are accurate, what is the probability that sales will be $150,000?

b.   What is the probability that sales will be no more than $150,000?

c.   Assuming that sales must be $200,000 or more in order for the company to break even, what is the probability that the company will break even?

6.   Three members of a labor union about to vote on whether to go on strike are interviewed. Assume that each member is equally likely to vote either way and that their votes are statistically independent. Let S represent a "strike" vote, and let N represent a "no-strike" vote.

a.   List the elementary events in the sample space.

b.   What are the probabilities of the following ordered events:
   (i) SSN
   (ii) NNN
   (iii) SSS

c.   Find the probability that the 3 members do not vote unanimously?

d.   Find the probability that at most 1 of the 3 members casts a strike vote.

7.   There is a 40% chance that a stock will give a positive return on investment. If you own 5 different stocks, what is the probability that none of them will give a positive return? Assume that all stocks' returns are statistically independent. Is this assumption reasonable?

8.   A research department has come up with 3 new product ideas. The manager estimates that the probabilities of the company's directors approving these product ideas are 0.8, 0.6, and 0.5 respectively.

a.   What is the probability that all 3 ideas will be accepted?

b.   What is the probability that none will be accepted?

c.   What statistical assumption did you make?

9.   A new chairman wants the 5 labor unions representing her company's workers to take a wage cut. Assume that there is a 0.8 probability that a union will reject her proposal and that the unions' decisions are independent of each other. What is the probability that:

a.   All 5 will accept.

b.   All 5 will reject.

c.   4 or more unions will reject the proposal.

10. An accounting firm surveys 500 corporations to investigate their opinion of a proposed tax reform. Of the 200 "small-size" corporations, 150 find the measure favorable, while a total of 250 corporations found it unfavorable.

a.   Construct a table with 2 rows and 2 columns for "Small-size corporations," "Large-size corporations," and "favor," "not in favor" classifications.

b.   Find the probability that a randomly selected corporation is small-size and does not favor the tax reform.

c. Find the probability that a randomly chosen corporation does not favor the tax reform.

d. Find the probability that a small-size corporation does not favor the tax reform.

e. Are size of corporation and opinion of tax reform statistically independent? Justify your answer.

11. Data obtained from a survey of 200 families are classified under the following criteria:

(1) Age of the head of family

(2) Income level

Income	Age in years		
	Less than 35	Between 35 and 45	More than 45
Low	50	15	5
Medium	25	30	15
High	20	15	25

a. Find the probability that a family earns a medium level of income.

b. Find the probability that the head of a family is over 45 years old.

c. Given that a family has a high income level, find the probability that the head of the family is over 45 years of age.

d. Given that the head of a family is over 45 years old, find the probability that the family has a low income level.

e. Are age and income level independent?

2.3
BAYES' THEOREM

The Reverend Thomas Bayes (1702–1761), an English Presbyterian minister and mathematician, considered the question of how to make inferences from observed sample data about the larger groups from which the data were drawn. His motivation was his desire to prove the existence of God by examining the sample evidence of the world about him. Mathematicians had previously concentrated on the problem of deducing the consequences of specified hypotheses. Bayes was interested in the inverse problem of drawing conclusions about hypotheses from observations of consequences. He derived a theorem that calculated probabilities of "causes" based on the observed "effects." The theorem may also be considered a means of revising probabilities of events based on additional information. In the period since World War II, a body of knowledge known as **Bayesian decision theory** has been developed to solve problems involving decision making under uncertainty.

Bayes' theorem is really nothing more than a statement of conditional probabilities. The following problem illustrates the nature of the theorem. Assume that 1% of the inhabitants of a country suffer from a certain disease. Let A_1 represent the event "has the disease" and A_2 denote "does not have the disease." Now suppose a person is selected at random from this population. What is the probability that this individual has the disease? Since 1% of the population has the disease and it is equally likely that any individual would be selected, we assign a probability of 0.01 to the event "has the disease." This probability, $P(A_1) = P(\text{has the disease}) = 0.01$, is referred to as a **prior probability** in the sense that it is assigned prior to the observation of any empirical information. Of course, $P(A_2) = 0.99$ is the corresponding prior probability that the individual does not have the disease.

Prior probability

Now let us assume that a new but imperfect diagnostic test has been developed, and let B denote the event "test indicates the disease is present." Suppose that through past experience it has been determined that the conditional probability that the test indicates the disease is present, given that the person has the disease, is

$$P(B\,|\,A_1) = 0.97$$

and the corresponding probability, given that the person does *not* have the disease, is

$$P(B\,|\,A_2) = 0.05$$

Suppose a person is selected at random and given the test and the test indicates that the disease is present. What is the probability that this person actually has the disease? In symbols, we want the conditional probability $P(A_1\,|\,B)$. It is important to note the nature of this question. The probability $P(\text{has the disease}\,|\,\text{test indicates the disease is present}) = P(A_1\,|\,B)$ is referred to as a **posterior probability**, or a **revised probability**, because it is assigned after the observation of empirical or additional information. The posterior probability is the type of probability computed with the help of Bayes' theorem. We will derive the value of $P(A_1\,|\,B)$ from probability principles developed in this chapter.

Posterior probability

By the definition of conditional probability given in equations 2.4 and 2.5, we have

(2.12)
$$P(A_1\,|\,B) = \frac{P(A_1 \text{ and } B)}{P(B)}$$

We compute the joint probability $P(A_1 \text{ and } B)$ in the numerator of equation 2.12 by the multiplication rule given in equation 2.6.

(2.13)
$$P(A_1 \text{ and } B) = P(A_1)P(B\,|\,A_1)$$

To compute the denominator of equation 2.12, we observe that

$$P(B) = P[(A_1 \text{ and } B) \text{ or } (A_2 \text{ and } B)]$$

Hence,

(2.14) $$P(B) = P(A_1 \text{ and } B) + P(A_2 \text{ and } B)$$

since the two joint events $(A_1 \text{ and } B)$ and $(A_2 \text{ and } B)$ are mutually exclusive.

If we express the joint probabilities $P(A_1 \text{ and } B)$ and $P(A_2 \text{ and } B)$ according to the multiplication rule, equation 2.14 becomes

(2.15) $$P(B) = P(A_1)P(B|A_1) + P(A_2)P(B|A_2)$$

Now, substituting into the numerator of equation 2.12 the expression for the joint probability $P(A_1 \text{ and } B)$ given in equation 2.13, and substituting into the denominator the marginal probability $P(B)$ given in equation 2.15, we obtain the result known as Bayes' theorem.

(2.16) $$P(A_1|B) = \frac{P(A_1)P(B|A_1)}{P(A_1)P(B|A_1) + P(A_2)P(B|A_2)}$$

Bayes' theorem for two basic events, A_1 and A_2

In terms of the disease example, by substituting the known values into the Bayes' theorem formula in equation 2.16, we find that

$$P(A_1|B) = \frac{(0.01)(0.97)}{(0.01)(0.97) + (0.99)(0.05)} = \frac{0.0097}{0.0592} = 0.16$$

Hence, the posterior probability that the individual has the disease given that the test indicated the presence of the disease is 0.16. In summary, if a person were selected at random from the population of this country, the prior probability that the individual has the disease is 0.01. On the other hand, after we have the empirical information that the test indicated the disease is present, we revise the probability that the individual has the disease upward to 0.16.

Although the posterior probability (0.16) is 16 times as large as the prior probability (0.01), 0.16 is still a surprisingly low probability that the individual has the disease given that the test indicated the disease was present. This results from the fact that there are a large number of persons in this population who do not have the disease but for whom the test would (falsely) indicate that the disease is present.

As we have seen, Bayes' theorem weights prior information with empirical evidence. The manner in which it does this may be seen by laying out the calculations in a form such as Table 2-3. The first column of Table 2-3 gives the basic events of interest, "has the disease" and "does not have the disease." The second column shows the prior probability assignments to these basic events. The third column shows the conditional probabilities of the additional information given the basic events. As noted in the column heading, such conditional probabilities are referred to as "likelihoods." In our illustration these are, respectively, $P(B|A_1) = P(\text{test indicates disease is present}|\text{has the disease})$ and $P(B|A_2) = P(\text{test indicates disease is present}|\text{does not have the disease})$. The fourth column gives the joint probabilities of the basic events and the additional information. Note that as indicated in equation 2.15, the sum of these joint probabilities is the marginal probability $P(B)$. When the joint probabilities

TABLE 2-3
Bayes' theorem calculations for illustrative problem

Events A_i	Prior Probabilities $P(A_i)$	Likelihoods $P(B\|A_i)$	Joint Probabilities $P(A_i)P(B\|A_i)$	Revised Probabilites $P(A\|B)$
A_1: Has the disease	0.01	0.97	0.0097	$\dfrac{0.0097}{0.0592} = 0.16$
A_2: Does not have the disease	0.99	0.05	0.0495	$\dfrac{0.0495}{0.0592} = 0.84$
	1.00		$P(B) = 0.0592$	$\dfrac{0.0592}{0.0592} = \dfrac{1.00}{1.00}$

are divided by their total, $P(B)$ (in this case, 0.0592), the results are the revised probabilities shown in the last column. The first probability shown in the last column ($0.0097/0.0592 = 0.16$) is the Bayes' theorem calculation required in the illustrative problem.

In the preceding problem, there were 2 basic events (A_i). If there are more than 2 basic events, then correspondingly, additional terms appear in the denominator of equation 2.16. Hence, we can make the following formal statement of Bayes' theorem for n basic events. Assume a set of complete and mutually exclusive events $A_1, A_2, \ldots, A_n$. The appearance of one of the A_i events is a necessary condition for the occurrence of another event B, which is observed. The probabilities $P(A_i)$ and $P(B|A_i)$ are known. The posterior probability of event A_1 given that B has occurred is given by Bayes' theorem.

Bayes' theorem for (2.17)
n basic events,
$A_1, A_2, \ldots, A_n$

$$P(A_1|B) = \frac{P(A_1)P(B|A_1)}{P(A_1)P(B|A_1) + P(A_2)P(B|A_2) + \cdots + P(A_n)P(B|A_n)}$$

Although the theorem was stated in equations 2.16 and 2.17 for a particular A_i, namely A_1, it is a general statement, since any of the n events A_i can be designated A_1. In most applications of the theorem to decision problems, A_i represents events that precede the occurrence of the observed event B. In this connection, we can think of the theorem as answering the question "Given that event B has occurred, what is the probability that it was preceded by event A_1?" Or, as indicated earlier, $P(A_1|B)$ is the revised probability assigned to event A_1 after event B is observed.

In modern Bayesian decision theory, subjective prior probability assignments are made in many applications. It is argued that it is meaningful to assign prior probabilities concerning hypotheses based on degree of belief. Bayes' theorem is then a means of revising these probability assignments. In business applications, this means that executives' intuitions, subjective judgments, and current quantitative knowledge are captured in the form of prior probabilities; these figures undergo revision as relevant empirical data are collected. This procedure seems sensible and fruitful for a wide variety of applications. Examples 2-7 through 2-9 suggest some of the many possible types of applications of this interesting theorem. Example 2-9 continues the foregoing disease illustration.

EXAMPLE 2-7

The probability that a correct tactical decision is made in a particular type of military operation has been found to be 0.80. Due to the length of time required to complete an operation of this type, the correctness of the decision cannot be immediately ascertained. However, by means of military intelligence, information can be quickly obtained as to whether or not the decision is correct and this intelligence is accurate 75% of the time. A tactical decision has recently been made and intelligence information has just been received indicating that the decision is correct. What probability would you assign for the correctness of the decision given this intelligence information?

$$
\begin{array}{c|cc}
 & P.B. & P(B/A) \\
A_1 & .80 & 75 = 60 \\
A_2 & .20 & .25 = .05 \\
 & 1.00 & 65
\end{array}
$$

$$\frac{60}{65} = .92\%$$

Let A_1 be the event "a correct tactical decision is made," and let A_2 be "an incorrect tactical decision is made."

Let B represent the event "intelligence information indicates that a correct tactical decision has been made."

$$P(A_1 | B) = \frac{(0.8)(0.75)}{(0.8)(0.75) + (0.2)(0.25)} = 0.92$$

EXAMPLE 2-8

A man regularly plays darts in the recreation room of his home, observed by his eight-year-old son. The father has a history of making bull's-eyes 30% of the time. The son, who habitually sits injudiciously close to the dart board, reports to his father whether or not bull's-eyes have been made. However, as is often the case with eight year-old boys, the son is an imperfect observer. He reports correctly 90% of the time. On a particular occasion, the father throws a dart at the board. The son reports that a bull's-eye was made. What is the probability that a bulls-eye was indeed scored?

Let A be the event "bull's-eye scored" and $\bar{A}$ "bull's-eye not scored." Let B represent "son reports that a bull's-eye has been scored." Then,

$$P(A | B) = \frac{P(A)P(B|A)}{P(A)P(B|A) + P(\bar{A})P(B|\bar{A})} = \frac{(0.3)(0.9)}{(0.3)(0.9) + (0.7)(0.1)} = 0.79$$

This is illustrative of a class of problems involving an information system which transmits uncertain knowledge. That is, the son may be thought of as an information system of 90% "reliability." A more colorful way of expressing it is that the son has "error in him" or is a "noisy" information system (where the term "noise" refers to random error).

EXAMPLE 2-9

Assume that 1% of the inhabitants of a country suffer from a certain disease. A new diagnostic test is discovered which gives a positive indication 97% of the time when an individual has this disease and gives a negative indication 95% of the time when the disease is absent. An individual is selected at random, is given the test, and reacts positively. What is the probability that he has the disease?

SOLUTION

Let D represent "has the disease" and $\bar{D}$ "does not have the disease." Let $+$ and $-$ represent positive and negative indications, respectively. Then, by Bayes' theorem,

$$P(D\,|\,+) = \frac{(0.01)(0.97)}{(0.01)(0.97) + (0.99)(0.05)} = 0.16$$

This is a startlingly low probability which doubtless runs counter to intuitive feelings based on the 97% and 95% figures given above. If the "cost" incurred is high when a person is informed that he has a disease (e.g., such as cancer) on the basis of such a test when in fact he does not, it would appear that action based solely on conditional probabilities such as the above 97% and 95% might be seriously misleading and costly.

The implications of this application are quite subtle. Actually, correct decisions would be made a very high proportion of the time on the basis of this test. The probability of correct decisions is

$$P\left(\begin{array}{c}\text{correct} \\ \text{decisions}\end{array}\right) = P(D \text{ and } +) + P(\bar{D} \text{ and } -)$$
$$= P(D)P(+\,|\,D) + P(\bar{D})P(-\,|\,\bar{D})$$
$$= (0.01)(0.97) + (0.99)(0.95) = 0.9502$$

However, from the $P(D\,|\,+) = 0.16$ calculation given above it follows that if a person has a positive indication on this test and is informed he has the disease, the probability of error is 0.84. On the other hand, a corresponding Bayes' theorem calculation shows that the probability is only 0.00032 that a person would have the disease, given a negative indication. Thus, only very rarely would an error be committed if a negative indication were obtained on the test.

In summary, the Bayes' theorem computations indicate the following:

$$P(D\,|\,+) = 0.16 \quad P(D\,|\,-) = 0.00032$$
$$P(\bar{D}\,|\,+) = 0.84 \quad P(\bar{D}\,|\,-) = 0.99968$$

The method by which optimal decisions are made, taking into account all of these probabilities in addition to the seriousness of the two types of error implied by $(\bar{D}\,|\,+)$ and $(D\,|\,-)$, is a basic topic of statistical decision theory.

EXERCISES 2.3

$(.7)(.9)$

$(.7)(.9)+(.3)(.1)$

1. A team of oil company planners estimates that there is a 70% probability of finding oil in Bombay High. Seismic tests have have favorable. If there is really oil, the probability that the tests will be favorable is 0.9, and if there is no oil, the probability that the tests will be favorable is 0.2. What is the revised probability of finding oil?

2. Of the female students in an MBA program, 4% were awarded distinctions compared with 7% for the males. If 25% of the students are female, what is the probability that a student selected randomly from those who scored a distinction is a male?

3. The percentages of American employees, Japanese employees, and employees of other nationalities in a selected list of international companies are 30%, 10%, and 60%, respectively. If for these nationality categories the percentages of white-collar workers are 80%, 75%, and 5%, respectively, what is the probability that a non-white-collar worker selected randomly is Japanese?

4. Management's prior probability assessment of the demand for a newly developed product is 0.55 for high demand, 0.25 for average demand, and 0.20 for low demand. A survey taken to help determine the true demand for the product indicates that demand is high. The reliability of the survey is such that it will indicate high demand 80% of the time when demand is actually high, 60% of the time when demand is actually average, and 15% of the time when demand is actually low. In the light of this information, reassess the probabilities of these three states of demand.

2.4
COUNTING PRINCIPLES AND TECHNIQUES

In the problems we have encountered so far, the pertinent sample spaces have been comparatively simple. However, in many situations the numbers of points in the appropriate sample spaces are so great that efficient methods are needed to count these points in order to arrive at required probabilities or answer other questions of interest. In this connection, it is useful to return to the concept of sequences of experimental trials to specify a simple but important fundamental principle.

The Multiplication Principle

In an experiment can result in n_1 distinct outcomes on the first trial, n_2 distinct outcomes on the second trial, and so forth for k sequential trials, then the total number of different sequences of outcomes in the k trials is $(n_1)(n_2) \cdots (n_k)$.

It is sometimes helpful to think in terms of the sequential performance of tasks rather than trials of an experiment. Thus, using somewhat different language than was employed in the context of experimental trials, if the first of a sequence of tasks can be performed in n_1 ways, the second in n_2 ways, and so forth for k tasks, then the sequence of k tasks can be carried out in $(n_1)(n_2) \cdots (n_k)$ ways.

Hence, if a coin is tossed and then a card is drawn at random from a standard deck of cards, there are $2 \times 52 = 104$ possible different sequences. For example, one such sequence of outcomes might be head, king of spades.

FIGURE 2-3

Tree diagram for trip
from Philadelphia to
Washington

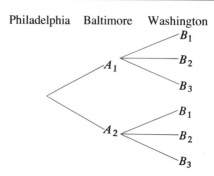

If a die is rolled 3 times, there are $6 \times 6 \times 6 = 216$ different sequences.

If it is possible to go from Philadelphia to Baltimore in 2 different ways and from Baltimore to Washington in 3 different ways, then there are $2 \times 3 = 6$ ways of going from Philadelphia to Washington via Baltimore.

A tree diagram is often helpful in demonstrating the total possible number of sequences. For example, in the case of the trip from Philadelphia to Washington, if A_1 and A_2 denote the two ways of going to Baltimore and B_1, B_2, and B_3 the 3 ways of proceeding from Baltimore to Washington, then the total number of possible sequences is indicated by the total number of different paths through the tree from left to right (see Figure 2-3).

In the following sections, the multiplication principle is used in many different types of problems.

Permutations

To handle the problem of counting points in complicated sample spaces, the counting techniques of *combination* and *permutation* are used. In this connection, it is helpful to think in terms of objects that occur in groups. These groups may be characterized by type of object, the number belonging to each type, and the way in which the objects are arranged. For example, consider the letters a, b, c, d, and e. There are 5 objects, one of each type. If we have the letters a, b, b, c, and c, there are 5 objects, one of type a, 2 of b, and 2 of c. Returning to the first group of objects, $a\,b\,c\,d\,e$, $b\,a\,c\,d\,e$, and $c\,d\,e\,a\,b$ differ in the *order* in which the 5 objects are arranged, but each of these groups contains the same number of objects of each type.

Suppose we have a group of n different objects. In how many ways can these n objects be arranged in order in a line? Applying the multiplication principle, we see that any one of the n objects can occupy the first position, any of the $n - 1$ remaining objects can occupy the second position, and so forth until we have only one possible object to occupy the nth position. Thus, the number of different possible arrangements of the n objects in a line consisting of n positions is

$$n! = (n)(n-1)\cdots(2)(1)$$

The symbol $n!$ is read "n factorial."

We shall be concerned only with cases for which n is a nonnegative integer. By definition, $0! = 1$.

Some examples of **factorials** are   *Factorials*

$$1! = 1$$

$$2! = 2 \times 1 = 2$$

$$3! = 3 \times 2 \times 1 = 6$$

$$\vdots$$

$$10! = 10 \times 9 \times \cdots \times 2 \times 1 = 3{,}628{,}800$$

It is useful to note that $n! = n(n-1)!$. Thus, $10! = 10 \times 9!$, and so on. We see from this relation that it makes sense to define $0! = 1$, since if we let $n = 1$, we have

$$1! = 1 \times 0!$$

and $0! = 1$. This enables us to maintain a consistent definition of factorials for all nonnegative integers.

Factorials obviously increase in size very rapidly. For example, how many different arrangements can be made of a deck of 52 cards if the cards are placed in a line? The answer, $52!$, is a number that contains 68 digits.[3]

Frequently we are interested in choosing and arranging in order some subgroup of n different objects. If x of the objects ($x \leqslant n$) are to be selected and arranged in order, as in a line, then each such *ordered arrangement* is said to be a **permutation** of the n objects taken x at a time. The number of such   *Permutation $_nP_x$* permutations is denoted $_nP_x$. For example, suppose there are 50 persons competing in a contest for 3 rankings—first, second, and third. How many permutations of the 50 people taken 3 at a time are possible, that is, how many different rankings are possible? The answer is

$$_{50}P_3 = 50 \times 49 \times 48 = 117{,}600$$

This is true because any one of the 50 persons can occupy the first position, any of the remaining 49 the second position, and any of 48 could fill the third place. By the multiplication principle, the number of different sequences of first, second, and third rankings is obtained by the indicated product.

---

[3] Warren Weaver points out in *Lady Luck* (Garden City, NY: Doubleday, 1963), p. 88, about the number of possible arrangements in $52!$, that, "If every human being on earth counted a million of these arrangements per second for twenty-four hours a day for lifetimes of eighty years each, they would have made only a negligible start in the job of counting all these arrangements—not a billionth of a billionth of one percent of them!"

We can now generalize this procedure to obtain a convenient formula for the number of permutations of n different objects taken x at a time.

$$_nP_x = (n)(n-1)\cdots[n-(x-1)]$$
$$= (n)(n-1)\cdots(n-x+1)$$
$$= \frac{(n)(n-1)\cdots(n-x+1)(n-x)!}{(n-x)!}$$

(2.18) $$_nP_x = \frac{n!}{(n-x)!}$$

It can be seen, in general, that if there are x positions to be filled, the first position can be filled in n ways; after one object has been placed in the first position, $x-1$ positions remain. The second can be filled in $n-1$ ways, the third in $n-2$ ways, and so forth down to the xth or last position, which can be filled in $n-(x-1)$ ways. In writing down the factors that must be multiplied together, 0 is subtracted from n in the first position, 1 is subtracted from n in the second position, and so forth down to $(x-1)$ subtracted from n in the xth position. The formula $n!/(n-x)!$ follows from the definition of a factorial, since $(n-x)!$ cancels out all factors after $(n-x+1)$ in the numerator. Thus, in the contest problem, where $n=50$ and $x=3$, we have

$$_{50}P_3 = \frac{50!}{(50-3)!} = \frac{50!}{47!} = \frac{50 \times 49 \times 48 \times 47!}{47!} = 50 \times 49 \times 48$$

A special case of the formula for permutations occurs when all n objects are considered together. In this situation, we are concerned with the number of permutations of n different objects taken n at a time, which is

Permutation $_nP_n$

(2.19) $$_nP_n = \frac{n!}{(n-n)!} = \frac{n!}{0!} = n!$$

For example, if a consumer were given one cup of coffee of each of five brands and asked to rank these according to preference, the total number of possible rankings (excluding the possibility of ties) would be

$$_5P_5 = 5! = 120$$

Combinations

In the case of permutations of objects, the order in which the objects are arranged is of importance.

> When order is not important, we are concerned with *combinations* of objects rather than permutations.

A simple example will illustrate the difference. Suppose the president of a company is interested in setting up a finance committee of two people and plans to select them from a group of 3 executives named Brown, Jones, and Smith. How many possible committees could be formed? Obviously, order is of no importance in this situation. That is a committee consisting of Brown and Jones is no different from a committee of Jones and Brown. Using first letters to symbolize the 3 names, we can list 3 possible committees: BJ, BS, and JS.

This is an example of combinations of 3 objects taken 2 at a time. The terminology is similar to that used for permutations, so in general, we refer to the number of combinations that can be made of n different objects taken x at a time.

Using the same group of letters and treating them merely as symbols, if order of arrangement were important, the number of permutations of the 3 objects taken 2 at a time would exceed the number of combinations of 3 objects taken 2 at a time. The following 6 permutations can be made in this case:

BJ   JB
BS   SB
JS   SJ

To develop a formula for combinations, we need merely consider the relationship between numbers of combinations and numbers of permutations for the same group of n objects taken x at a time. Fixing attention for the moment on any particular combination, there are x objects filling x positions. How many permutations can be made of these x objects in the x positions? Clearly, any one of the x objects may fill the first position, $x - 1$ the second, and so forth down to one object for the xth position. Thus, $x!$ distinct permutations can be formed of the x objects in x positions. Therefore, the number of permutations that can be formed of n different objects taken x at a time is $x!$ times the number of combinations of these n objects taken x at a time.

> The symbol for the number of combinations of n different objects taken x at a time is $\binom{n}{x}$.

Thus,

$$\binom{n}{x} x! = {}_nP_x$$

Solving for $\binom{n}{x}$ yields the following formula:

(2.20)
$$\binom{n}{x} = \frac{{}_nP_x}{x!} = \frac{n!}{x!(n-x)!}$$

Returning to the committee illustration, the number of combinations of the 3 people taken 2 at a time is

$$\binom{3}{2} = \frac{3!}{2!1!} = 3$$

which was the number previously listed. Similarly, the number of permutations of 3 objects taken 2 at a time is

$$_3P_2 = \frac{3!}{1!} = 6$$

which was the number of ordered arrangements listed earlier.

EXAMPLE 2-10

A brief market research questionnaire requires the respondent to answer each of 10 successive questions with either a "yes" or "no." The sequence of 10 yes-no responses is defined as the respondent's profile. How many different possible profiles are there?

SOLUTION

There are 2 possible responses for each question. Therefore, by the multiplication principle, there are $2 \times 2 \times \cdots \times 2 = 2^{10} = 1024$ different profiles.

EXAMPLE 2-11

A sales manager wishes to place an advertisement in 2 journals. There are 5 feasible journals in which to advertise. In how many journals can the advertisement be placed?

SOLUTION

$$\binom{5}{2} = \frac{5!}{2!3!} = 10$$

EXAMPLE 2-12

A shopkeeper wishes to place each brand of detergent that she sells on a shelf.
a. If she sells 4 brands of detergent, in how many ways can she arrange these brands on the shelf?
b. If there was space available for displaying only 2 of the 4 brands, how many arrangements are possible?

SOLUTION

a. $_4P_4 = \dfrac{4!}{(4-4)!} = 24$

b. $_4P_2 = \dfrac{4!}{(4-2)!} = 12$

EXAMPLE 2-13

Consider a group of 5 persons, consisting of 3 men and 2 women, all of whom belong to an organization.
a. How many committees of 3 persons can be formed from the group?
b. In how many ways can the two positions, president and vice-president be formed?

c. What is the probability that a committee of 2 persons chosen at random will consist of one man and one woman?

a. $\dbinom{5}{3} = \dfrac{5!}{3!2!} = 10$

b. $_5P_2 = \dfrac{5!}{(5-2)!} = 20$

c. $\dfrac{\dbinom{3}{1}\dbinom{2}{1}}{\dbinom{5}{2}} = \dfrac{6}{10}$

EXAMPLE 2-14

A firm desires to build 6 new factories: 2 factories in the 13 Southern states, one factory in the 6 Middle Atlantic states, one factory in the Far West states, and 2 factories in the 8 Midwest states. If the firm wants to study the desirability of each possible combination of locations, how many combinations would the firm have to consider?

$$\binom{13}{2}\binom{6}{1}\binom{4}{1}\binom{8}{2} = 52{,}416$$

EXERCISES 2.4

1. A newspaper carrier has to make 10 deliveries in a given area. In how many sequences can he make his stops? $10! \approx$ $3{,}628{,}000$

2. Rittleman Furs, Inc. has just purchased data cards to handle its accounts receivable. Each data card contains 70 columns in which a number from zero to nine or a letter may be punched to represent information about an account. It is decided that each account will be assigned 3 identification symbols, which will be punched in the first 3 columns of the data card to identify each card with a particular account.
    a. If only numbers are used, how many accounts can be handled by this method?
    b. If the first column is to contain a letter and the next 2 are to contain numbers, how many accounts can be handled?
    c. If either a letter or a number may be punched in each column, how many accounts can be handled?

3. A motivational researcher shows a woman 12 projected colors for new spring clothes and asks her to pick out her 4 favorites.
    a. Give a specific possible outcome of the experiment. $\dbinom{12}{4}$

COMBINATION **b.** How many such outcomes are there? $\frac{12!}{(n-4!)\,4!}$

  c. If one of the color choices is azure, how many possible outcomes will contain that color?

  d. What is the probability that the woman will choose azure as one of her 4 favorites?

  e. What is the probability that she will not choose either azure or carmine (another color available) in her selection of favorites?

4. In a determination of preference of package design, a panel of consumers was given 4 different packaging designs and asked to rank them. How many different possible rankings could the panel have given (excluding ties)?

5. Professor Tom Robbins anticipates teaching the same course for the next few years. In order not to become bored with his own jokes, he decides to tell a set of exactly 3 jokes each year. He may repeat one or two jokes from year to year, but he vows never to repeat the same set of 3 jokes. How many years can he last with a repertoire of 7 jokes?

6. A committee consists of 8 union and 6 nonunion workers. In how many ways can a subcommittee of 6 workers, 3 union and 3 nonunion, be formed? What would the answer be if all the subcommittee members to be chosen had to be

  a. union workers?

  b. nonunion workers?

7. Lemon Motors orders 7 different upholstering colors for its cars and 12 different colors of body paint.

  a. How many different color combinations of body and upholstering are available to the customer?

  b. If Lemon Motors allows the customer to order a roof color different from the basic body color, how many additional different color coordinations of body, roof, and upholstering are available to the customer?

8. Vivian Chan, the financial vice-president of the Petro Chemical Company, is considering 5 similar investment proposals for the upcoming fiscal period. After analyzing the financial condition of the firm and estimating the firm's cash flow for the period, she decides that all portfolios consisting of 3 proposals are equally feasible and desirable in the long run.

  a. Assuming that the 3 investments in the adopted portfolio are made simultaneously, how many different portfolios are there from which to choose?

  b. If the 3 adopted proposals are implemented sequentially and it is determined that, due to differing cash payback periods, the order of implementation is a distinguishing factor in the comparison of otherwise identical portfolios, how many portfolio arrangements are there?

  c. Under the assumptions of (b) and assuming random selection, what is the probability that the portfolio selected will include proposals A, B, and C?

9. A committee consists of 10 people. It is decided to appoint a chairman, a vice-chairman, and a secretary-treasurer. How many different ways can this be done?

10. A certain organization consists of 8 men and 8 women, from whom a committee of 7 people is to be formed.

    a. What is the probability that there will be exactly one woman on the committee?

    b. Given that Mr. Greene is to be chairman of the committee, what is the probability that of the 6 remaining to be selected, exactly one will be a woman?

    c. Given that Mr. Greene is to be chairman of the committee, what is the probability that all of the 6 remaining to be selected will be women?

    d. Recalculate the probability for part (a) given that the organization consists of 9 men and 9 women.

KEY TERMS

Classical Probability   the ratio of (1) the number of outcomes favorable to an event to (2) the total number of possible outcomes.

Combination $\begin{pmatrix} n \\ x \end{pmatrix}$ an arrangement of n objects taken x at a time, in which the order of arrangement is not important.

Elements   the objects that constitute a sample space.

Event   a collection of elements of a sample space.

Independence of Events A_1 and A_2   knowing that one of the two events has occurred does not affect the probability that the other will occur.

Mutually Exclusive Events A_1 and A_2   if A_1 occurs, A_2 cannot occur and vice versa.

Odds Ratio   the ratio of the probability that an event will occur to the probability that the event will not occur.

$P(A_1$ and $A_2)$ the joint probability that events A_1 and A_2 occur.

$P(A_1$ or $A_2)$ the probability that either event A_1 or event A_2 occurs, or that they both occur.

$P(B_1|A_1)$ the conditional probability that event B_1 occurs given that event A_1 occurs.

Permutation ($_nP_x$)   an ordered arrangement of n objects taken x at a time.

Posterior Probability   a probability assigned after the observation of empirical information.

Prior Probability   a probability assigned prior to the observation of empirical information.

Probability of an Event   a number between zero and one. The sum of the probabilities that the event will occur and that it will not occur is 1.

Relative Frequency Probability   the proportion of times an event occurs in the long run under uniform or stable conditions.

Sample Space   the collection of the possible outcomes of an experiment.

Subjective Probability   the degree of belief or confidence placed in the occurrence of an event.

KEY FORMULAS

Addition Rule for Any Two Events

$$P(A_1 \text{ or } A_2) = P(A_1) + P(A_2) - P(A_1 \text{ and } A_2)$$

Addition Rule for Two Mutually Exclusive Events

$$P(A_1 \text{ or } A_2) = P(A_1) + P(A_2)$$

Conditional Probability of B_1 Given A_1

$$P(B_1 \mid A_1) = \frac{P(A_1 \text{ and } B_1)}{P(A_1)}$$

Multiplication Rule for Any Two Events

$$P(A_1 \text{ and } B_1) = P(A_1)P(B_1 \mid A_1)$$

Multiplication Rule for Any Two Independent Events

$$P(A_1 \text{ and } B_1) = P(A_1)P(B_1)$$

Bayes' Theorem for Two Basic Events

$$P(A_1 \mid B) = \frac{P(A_1)P(B \mid A_1)}{P(A_1)P(B \mid A_1) + P(A_2)P(B \mid A_2)}$$

Permutations of n Objects Taken x at a Time

$$_nP_x = \frac{n!}{(n-x)!}$$

Combinations of n Objects Taken x at a Time

$$\binom{n}{x} = \frac{n!}{x!(n-x)!}$$

3

DISCRETE RANDOM VARIABLES
AND PROBABILITY DISTRIBUTIONS

Managerial decisions are ordinarily made under conditions of uncertainty.

- A corporation treasurer makes investment decisions in the face of uncertainties concerning future movements of interest rates and stock market prices.

- A corporate executive committee may make a decision concerning expansion of manufacturing facilities despite uncertainty about future levels of demands for the company's products.

- An advertising manager makes decisions on advertising expenditures in various media without being certain of the sales that will be generated by these outlays.

In each of these cases, the outcomes of concern—such as interest rates, stock market prices, levels of demand, and sales that result from advertising—may assume a variety of values; we refer to the outcomes as **variables**. In statistical analysis, such variables are usually called random variables.

3.1
RANDOM VARIABLES

Probability distributions

A **random variable** may be defined roughly as a variable that takes on different numerical values because of chance.[1]

In this chapter, we will be concerned primarily with **probability distributions** of random variables. This concept is central to all of statistics, and although we introduce it here in the context of the business decision problem, it is used in every field in which statistical methods are applied. Examples 3-1, 3-2, and 3-3 introduce the idea of a probability distribution.

EXAMPLE 3-1

Harnett Industries, a large corporation, was interested in diversifying its product line. In that connection, Harnett was negotiating the acquisition of Chase Products, a smaller company. Mary Prescott, manager of mergers and acquisitions at Harnett Industries, contemplated the purchase of Chase Products in the very near future, but she was uncertain about the price Harnett would have to pay per share for Chase Products common stock. Ms. Prescott set up the probability distribution shown in Table 3-1 for the stock price.

The two columns shown in Table 3-1 constitute the probability distribution for the random variable "price of Chase Products common stock." The following type of symbolism is conventionally used: If we let the symbol X stand for the random variable (in this example price of Chase Products common stock), then we represent the values the random variable can assume by x. The probability that the random variable X will take on the value x is symbolized as $P(X = x)$ or simply $f(x)$. In Table 3-1, the values of the random variable are listed under the column headed x and the probabilities of

TABLE 3-1
Probability distribution for the price of Chase Products common stock

Price of Chase Products Common Stock (to the nearest dollar) x	Probability $f(x)$
$33	0.10
34	0.25
35	0.50
36	0.10
37	0.05

[1] From a mathematical viewpoint, a random variable is a function consisting of the elements of a sample space and the numbers assigned to these sample elements. Ordinarily, a shortcut method of referring to a random variable is used. For example, we may refer to the random variables "the price of XYZ stock at some specified future time," "annual volume of sales of product ABC," or, in a coin-tossing example, "number of heads obtained in two tosses of a coin."

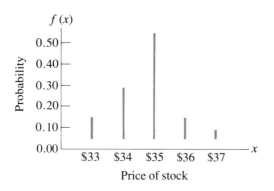

FIGURE 3-1

Graph of probability
distribution of price of Chase
Products common stock

these values are shown under $f(x)$. Thus,

$$P(X = \$33) = f(\$33) = 0.10$$

$$P(X = \$34) = f(\$34) = 0.25$$

$$P(X = \$35) = f(\$35) = 0.50$$

$$P(X = \$36) = f(\$36) = 0.10$$

$$P(X = \$37) = f(\$37) = 0.05$$

Note that these probabilities sum to 1.

This is an example of a subjective probability distribution. We observe that by assigning a probability of 0.50 to a price of $35, Ms. Prescott indicated that she felt the odds were 50:50 that the stock would be priced at that figure as opposed to any other figure. She felt that a price of $35 was twice as likely to occur as a price of $34, to which she assigned a probability of 0.25. She felt that a price of $33, $36, or $37 was rather unlikely.

When a probability distribution is graphed, it is conventional to display the values of the random variables on the horizontal axis and their probabilities on the vertical scale. The graph of the probability distribution for the common stock price example is shown in Figure 3-1.

EXAMPLE 3-2

A department store has classified its list of charge customers into two mutually exclusive categories: (1) high-volume purchasers, and (2) nonhigh-volume purchasers. Twenty percent of the customers are high-volume purchasers. Assume that a sample of four customers is drawn at random from the list. What is the probability distribution of the random variable "number of high-volume purchasers"?

SOLUTION

It may be assumed that the list of charge customers is so large that even though the sample is drawn without replacement, there is only a negligible loss of accuracy if the computations are performed as though the sampling were carried out with replacement. That is, the partial exhaustion of the list because of drawing of the items in the

TABLE 3-2
Elements of the sample space for the experiment of drawing a random sample of four customers

$\bar{A}\,\bar{A}\,\bar{A}\,\bar{A}$	$A\,A\,\bar{A}\,\bar{A}$	$A\,A\,A\,\bar{A}$
	$A\,\bar{A}\,A\,\bar{A}$	$A\,A\,\bar{A}\,A$
$A\,\bar{A}\,\bar{A}\,\bar{A}$	$A\,\bar{A}\,\bar{A}\,A$	$A\,\bar{A}\,A\,A$
$\bar{A}\,A\,\bar{A}\,\bar{A}$	$\bar{A}\,A\,A\,\bar{A}$	$\bar{A}\,A\,A\,A$
$\bar{A}\,\bar{A}\,A\,\bar{A}$	$\bar{A}\,A\,\bar{A}\,A$	
$\bar{A}\,\bar{A}\,\bar{A}\,A$	$\bar{A}\,\bar{A}\,A\,A$	$A\,A\,A\,A$

sample is so small that for practical purposes, the probabilities of obtaining the two types of purchasers remain unchanged.

Let A represent the occurrence of a high-volume purchaser and $\bar{A}$ the occurrence of a nonhigh-volume purchaser. The elements of the sample space for the experiment of drawing the sample of four customers are listed in Table 3-2.

We denote by X the random variable "number of high-volume purchasers." X can take on the values 0, 1, 2, 3, 4. As can be seen from Table 3-2, one sample element corresponds to the occurrence of no high-volume purchasers; 4 elements to one high-volume purchaser; 6 elements to two high-volume purchasers; 4 elements to three high-volume purchasers; and 1 element to four high-volume purchasers. However, the sample elements are not equally likely. The probability of a high-volume purchaser is 0.2, and that of a nonhigh-volume purchaser is 0.8. Considering one sample element each for zero, one, two, three, and four high-volume purchasers, we have the following probabilities:

$$P(\bar{A}\,\bar{A}\,\bar{A}\,\bar{A}) = (0.8)(0.8)(0.8)(0.8) = (0.8)^4$$

$$P(A\,\bar{A}\,\bar{A}\,\bar{A}) = (0.2)(0.8)(0.8)(0.8) = (0.2)(0.8)^3$$

$$P(A\,A\,\bar{A}\,\bar{A}) = (0.2)(0.2)(0.8)(0.8) = (0.2)^2(0.8)^2$$

$$P(A\,A\,A\,\bar{A}) = (0.2)(0.2)(0.2)(0.8) = (0.2)^3(0.8)$$

$$P(A\,A\,A\,A) = (0.2)(0.2)(0.2)(0.2) = (0.2)^4$$

For any given value of the random variable "number of high-volume purchasers," each elementary event has the same probability. For example, in the case of one high-volume purchaser, each of the four elementary events (sample elements) has a probability of $(0.2)(0.8)^3$. If we multiply the specified probabilities for elementary events by the number of such elements in the composite events "no high-volume purchasers," "one high-volume purchaser," and so forth, we have

$$P(X = 0) = f(0) = 1(0.8)^4 \qquad = 0.4096$$

$$P(X = 1) = f(1) = 4(0.8)^3(0.2) = 0.4096$$

$$P(X = 2) = f(2) = 6(0.8)^2(0.2)^2 = 0.1536$$

$$P(X = 3) = f(3) = 4(0.8)(0.2)^3 = 0.0256$$

$$P(X = 4) = f(4) = 1(0.2)^4 \qquad = \frac{0.0016}{1.0000}$$

TABLE 3-3
Probability distribution of number of high-volume purchasers

x	$f(x)$
0	0.4096
1	0.4096
2	0.1536
3	0.0256
4	0.0016

These values are summarized in Table 3-3 in the form of a probability distribution. Note that the probabilities add up to 1.

Actually, it would not have been necessary to list all the elements in the sample space in order to derive this probability distribution. A much briefer method for computing the required probabilities is explained in section 3.4, where the binomial distribution is discussed.

A graph of this probability distribution is given in Figure 3-2.

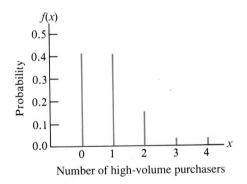

Number of high-volume purchasers

FIGURE 3-2

Graph of probability distribution of number of high-volume purchasers

EXAMPLE 3-3

Raymond Taylor, a corporation economist, develops a subjective probability distribution for the change that will take place in GNP during the following year. He establishes the following five categories of change and defines a random variable by letting the indicated numbers correspond to each category.

Change in GNP	Number Assigned
Down more than 5%	−2
Down 5% or less	−1
Unchanged	0
Up 5% or less	+1
Up more than 5%	+2

TABLE 3-4
Subjective probability distribution of change in GNP

x	$f(x)$
-2	0.1
-1	0.1
0	0.2
$+1$	0.4
$+2$	0.2

On the basis of all information available to him, Mr. Taylor assigns probabilities to each of these possible events as indicated in Table 3-4.

Types of Random Variables

Discrete random variables   Random variables are classified as either discrete or continuous. A **discrete random variable** is one that can take on only a finite or countable number of distinct values. Examples 3-1, 3-2, and 3-3 illustrate probability distributions of discrete random variables.

Continuous random variables   A random variable is said to be continuous in a given range if the variable can assume any value in that range. The term **continuous random variable** implies that variation takes place along a continuum. Examples of continuous variables include weight, length, velocity, rate of production, dosage of a drug, and the length of life of a given product. While discrete variables can be *counted*, continuous variables can be *measured* with some degree of accuracy.

> Although measured data are essentially discrete in the real world, the variable under measurement is often continuous.

It may be argued that, in the real world, all data are discrete. For example, if we measure weight with a measuring instrument that permits a determination only to the nearest thousandth of a pound, then the resulting data will be discrete in units of thousandths of a pound. Despite this discreteness of data caused by limitations of measuring instruments, it is nevertheless useful in many instances to use mathematical models that treat certain variables as continuous. We may conceive of a continuous mathematical model of heights of individuals—where the underlying data are measured and discrete—as a model of reality that is more accurate than the discrete data from which the model was derived rather than conceiving of such a model as merely a convenient approximation.

On the other hand, we often find it convenient to convert a variable that is conceptually continuous into a discrete one. Thus, in the case of heights of individuals, we may set up classifications such as tall, medium, and short rather than using measurements along a continuous scale of inches.

It is sometimes said that one indication of progress in science is the extent to which discrete variables can be converted into continuous variables.

- The physicist treats color in terms of the *continuous variable* of wavelengths rather than the *discrete* classification of names of colors.

- Measurement of temperature by means of a thermometer treats human body temperature as varying along a *continuous* scale (although the resulting measurements in degrees are discrete) rather than as *discrete* (as when temperature is judged to be "normal" or "high" by a hand placed on the forehead).

In applied problems, where a probability model is used to represent a real-world situation, we may work in terms of either discrete or continuous random variables, whichever system is most appropriate for the problem or decision-making situation in question. Only probability distributions of discrete random variables will be discussed in the remainder of this chapter.

Characteristics of Probability Distribution

In Examples 3-1, 3-2, and 3-3, we saw that the sum of the probabilities in each probability distribution was equal to one. It is possible to summarize the characteristics of probability distributions somewhat more formally.

A probability distribution of a discrete random variable X whose value at x is $f(x)$ possesses the following properties:[2]

1. $f(x) \geq 0$ for all real value of X.
2. $\sum_{x} f(x) = 1$

Property 1 simply states that probabilities are greater than or equal to zero. The second property states that the sum of the probabilities in a probability

---

[2] A somewhat simplified notation is used here. A mathematically more elegant notation would represent the values that the random variable X could assume as $x_1, x_2, \ldots, x_n$ with associated probabilities $f(x_1), f(x_2), \ldots, f(x_n)$. Then the two properties would appear as

1. $f(x_i) \geq 0$ for all i

2. $\sum_{i=1}^{n} f(x_i) = 1$

If X takes on an infinite number of values, then the second property would appear as

$$\sum_{i=1}^{\infty} f(x_i) = 1$$

distribution is equal to one. The notation

$$\sum_x f(x)$$

means "sum of the values of $f(x)$ for all values that x takes on."

While we will ordinarily use the term **probability distribution** to refer to both discrete and continuous variables, other terms are sometimes used to refer to probability distributions (also called probability functions).

- Probability distributions of discrete random variables are often referred to as **probability mass functions** or simply **mass functions** because the probabilities are *massed* at distinct points, for example, along the x axis.
- Probability distributions of continuous random variables are referred to as **probability density functions** or **density functions**.

Cumulative Distribution Functions

Frequently, we are interested in the probability that a random variable is less than, equal to, or greater than a given value. The **cumulative distribution function** is particularly useful in this connection. We may define this function as follows.

> Given a random variable X, the value of the cumulative distribution function at x, denoted $F(x)$, is the probability that X takes on values less than or equal to x.

Hence,

(3.1) $$F(x) = P(X \leqslant x)$$

In the case of a discrete random variable, it is clear that

(3.2) $$F(c) = \sum_{x \leqslant c} f(x)$$

The symbol

$$\sum_{x \leqslant c} f(x)$$

means "sum of the values of $f(x)$ for all values of x less than or equal to c."

EXAMPLE 3-4

We return to Example 3-1, involving the probability distribution for the price of Chase Products common stock. The probability that the price would be $33 or less is $P(X \leqslant \$33) = F(\$33) = 0.10$; $34 or less, $P(X \leqslant \$34) = F(\$34) = 0.35$; and so on. In Table 3-5, the probability distribution and cumulative distribution function for the random variable "price of Chase Products common stock" are shown.

TABLE 3-5
Probability distribution and cumulative distribution function for the price of
Chase Products common stock

Price of Stock x	Probability $f(x)$	Cumulative Probability $F(x)$
$33	0.10	0.10
34	0.25	0.35
35	0.50	0.85
36	0.10	0.95
37	0.05	1.00

A graph of the cumulative distribution function is given in Figure 3-3. This graph
is a *step function*; that is, the values change in discrete "steps" at the indicated integral
values of the random variable, X. Thus, $F(x)$ takes the value 0 to the left of the point
$x = \$33$, steps up to $F(x) = 0.10$ at $x = \$33$, and so on. The dot shown at the left of
each horizontal line segment indicates the probability for the integral values of x. At
these points, the values of the cumulative distribution function are read from the *upper*
line segments.

We note the following relations in this problem, which follow from the definition
of a cumulative distribution function:

$$F(\$33) = f(\$33) = 0.10$$

$$F(\$34) = f(\$33) + f(\$34) = 0.10 + 0.25 = 0.35$$

$$F(\$35) = f(\$33) + f(\$34) + f(\$35) = 0.10 + 0.25 + 0.50 = 0.85$$
and so on.

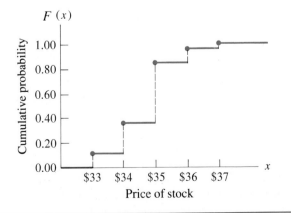

FIGURE 3-3
Graph of cumulative
distribution function of the
price of Chase Products
common stock

The probabilities that the price would be more than \$33, \$34, and \$35 are, respectively,

$$1 - F(\$33) = 0.90$$

$$1 - F(\$34) = 0.65$$

$$1 - F(\$35) = 0.15$$

EXAMPLE 3-5

Let us return to Example 3-3, which discussed an economist's subjective probability distribution of change in GNP. A few questions will illustrate some uses of the cumulative distribution function.

What was the probability assigned by Mr. Taylor to

a. the event that the change in GNP will not exceed an increase of 5%?
b. the event that GNP will not decline?
c. a change in GNP of 5% or less?

SOLUTIONS

a.
$$F(1) = f(-2) + f(-1) + f(0) + f(1) = 0.8$$

or

$$F(1) = 1 - f(2) = 1 - 0.2 = 0.8$$

b. The event "GNP will not decline" is the event "GNP will remain unchanged or will increase" and is thus the complement of the event "GNP will decrease." Thus, it is given by

$$1 - F(-1) = 1 - [f(-2) + f(-1)] = 1 - 0.2 = 0.8$$

or

$$1 - F(-1) = f(0) + f(1) + f(2) = 0.8$$

c.
$$f(-1) + f(0) + f(1) = 0.7$$

or

$$F(1) - F(-2) = 0.8 - 0.1 = 0.7$$

EXERCISES 3.1

1. State whether the following random variables are discrete or continuous
   C a. Strength of a steel beam in pounds per inch
   C b. The actual weight of a supposedly 16-ounce box of breakfast cereal
   D c. X equals 0 if the weight of a supposedly 16-ounce box of cereal is less than 16 ounces and 1 if the weight is 16 ounces or more
   D d. The number of defective batteries in a lot of 1000

2. Distinguish between the probability distribution and cumulative distribution of a discrete random variable X. Graph both functions for the following examples:

   a. $f(x) = \dfrac{1}{14} x^2 \qquad x = 1, 2, 3,$

   b. $f(x) = \dfrac{(x + 2)}{10} \qquad x = -1, 2, 3$

   c. Show that these functions satisfy the two properties of probability distributions.

3. Find k such that the following are probability functions.

   a. $\dfrac{k}{x + 1} \qquad x = 1, 2, 3$

   b. $\dfrac{x^2 - x}{k} \qquad x = 0, 1, 2$

   c. $kx^2 \qquad x = 1, 2, 3, 4$

4. As part of its diversification strategy, Food Products, Inc. is developing a new beverage. To determine interest in this new product, its market research department surveyed consumers in a large metropolitan area. The consumers in the survey were drawn from names in a telephone book serving the area (that is, we can assume the population sampled is large enough to treat sampling as having been done with replacement). Of the consumers listed in the telephone book, 35% use Kool Kola, a similar product sold by the firm. Assume that a sample of three names is drawn.

   a. Show the elements of the sample space for the drawing. Let K represent the occurrence of a Kool Kola customer and $\bar{K}$ represent a noncustomer.

   b. Determine the probability distributions of the random variable, "number of Kool Kola customers."

   c. Graph the probability distribution obtained in part (b).

5. An ice cream plant manufacturing chocolate and vanilla flavors estimates that it takes about X minutes to drain the ice cream filler of one flavor when changing to the other flavor. The probability distribution of X can be described as follows:

$$f(x) = \dfrac{x}{15} \qquad x = 1, 2, 3, 4, 5$$

   a. Prove that $f(x)$ of a probability function.

   b. What is the probability that it will take exactly three minutes to drain the filler?

   c. What is the probability that it will take at least two minutes but not more than four minutes?

   d. Find the cumulative distribution (in tabular form).

6. Consider the following random variable for the lifetime of a particular machine:

$X = 1$ if the machine lasts less than two years before wearing out
$X = 2$ if the machine lasts at least two but less than three years
$X = 3$ if the machine lasts at least three but less than four years
$X = 4$ if the machine lasts at least four but less than five years
$X = 5$ if the machine lasts more than five years.

Let

$$F(x) = \frac{x^2}{25}$$

Find $f(x)$ in tabular form.

3.2
PROBABILITY DISTRIBUTIONS OF DISCRETE RANDOM VARIABLES

In many situations, it is useful to represent the probability distribution of a random variable by a general algebraic expression. Probability calculations can then be conveniently made by substituting appropriate values into the algebraic model. The mathematical expression is a compact summary of the process that has generated the probability distribution. Thus, the statement that a particular probability distribution is appropriate in a given situation contains a considerable amount of information concerning the nature of the underlying process. In the following sections, we discuss the uniform, binomial, multinomial, hypergeometric, and Poisson probability distributions of discrete random variables.

3.3
THE UNIFORM DISTRIBUTION

Sometimes, equal probabilities are assigned to all the possible values a random variable may assume. Such a probability distribution is referred to as a **uniform distribution**. For example, suppose a fair die is rolled once. The probability is $\frac{1}{6}$ that the die will show any given number on its uppermost face. The probability mass function in this case may be written as

$$f(x) = \tfrac{1}{6} \quad \text{for } x = 1, 2, \dots, 6$$

A graph of this distribution is given in Figure 3-4.

As another illustration, let us consider the case of the Amgar Power Company, which produces electrical energy from geothermal steam fields. It takes about two years to build a production facility. In planning its production

3�7ㅗ   4ㄱㄱ+

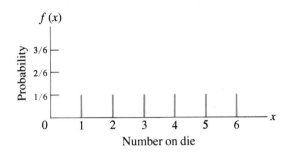

FIGURE 3-4
Graph of probability mass function of numbers obtained in a roll of a fair die

strategies, the company concludes that it is equally likely that demand two years hence will be 80,000, 90,000, 100,000, 110,000, or 120,000 kilowatts. Therefore, the probability distribution established by Amgar Power Company for this future demand is

$$f(x) = 0.20 \quad \text{for } x = 80,000, 90,000, \ldots, 120,000$$

Later, we will examine how such information is used in decision-making procedures.

EXERCISES 3.3

1. A roulette wheel has 38 equally spaced openings with numbers 00, 0, 1, 2, ..., 36. Write the probability function of X, where X is any number appearing after a spin of the wheel.

2. An international airline flight is scheduled to arrive at Nairobi Airport at 7:30 A.M. A study has shown that the actual arrival time is uniformly distributed by minutes in the range 7:05 to 8:40 A.M. Let $X = 1$ represent arrival at 7:05 A.M., $X = 2$, arrival at 7:06 A.M., and so on.
   a. Write the mathematical expression for $f(x)$. $f(x) = \frac{1}{96}$
   b. What is the probability that the plane will be late? $96 - 26 = 70 = \frac{70}{96}$
   c. What is the probability that the plane will arrive after 8:00 A.M.? $\longrightarrow \frac{40}{96}$
   d. What is the probability that the plane will arrive at or after 8:00 A.M.? $\longrightarrow \frac{41}{96}$
   e. What is the probability that the plane will arrive before 7:30 A.M.? $\longrightarrow \frac{25}{96}$

3. Let X denote the number of bricks a mason will lay in one hour, and assume that X is uniformly distributed in the range 150 to 200. If a certain project is 170 bricks short of completion and a second project is waiting to be started as soon as this one is finished,
   a. What is the probability that the mason will start the second project in the next hour?

b.   What is the probability that more than 25 bricks will have been laid on the second project at the end of the next hour?

c.   What is the probability that the first project will be more than ten bricks short of completion at the end of the hour?

d.   What is the probability that the mason will lay exactly 175 bricks during the next hour?

3.4
THE BINOMIAL DISTRIBUTION

The **binomial distribution**, in which there are two possible outcomes on each experimental trial, is undoubtedly the most widely applied probability distribution of a discrete random variable. It has been used to describe a large variety of processes in business and the social sciences as well as other areas. The process that gives rise to the binomial distribution is usually referred to as a *Bernoulli process* **Bernoulli trial** or a **Bernoulli process**.[3] The mathematical model for a Bernoulli process is developed from a specific set of assumptions involving the concept of a series of experimental trials.

Let us envision a process or experiment characterized by repeated trials taking place under the following conditions or assumptions:

Assumptions 1.   On each trial, there are two mutually exclusive possible outcomes, which are referred to as "success" and "failure." In somewhat different language, the sample space of possible outcomes on each experimental trial is $S = \{$failure, success$\}$.

2.   The probability of a success, denoted p, remains constant from trial to trial. The probability of a failure, denoted q, is equal to $1 - p$.

3.   The trials are independent. That is, the outcomes on any given trial or sequence of trials do not affect the outcomes on subsequent trials.

Random process The outcome on any specific trial is determined by chance. Processes having this characteristic are referred to as **random processes**, or **stochastic processes**, and Bernoulli trials are one example of such processes.

Our aim is to develop a formula for the probability of x successes in n trials of a Bernoulli process. We start with a simple, specific case of a series of Bernoulli trials, 5 tosses of a coin. We calculate the probability of obtaining exactly 2 heads in 5 tosses; the resulting expression can then be generalized.

If we are tossing a fair coin 5 times, we may treat each toss as one Bernoulli trial. The possible outcomes on any particular trial are a head and a tail. Assume that the appearance of a head is a success. (Of course, the classification

---

[3] Named after James Bernoulli (1654–1705), a member of a family of Swiss mathematicians and scientists, who did some of the early significant work on the binomial distribution.

of one of the 2 possible outcomes as a "success" is completely arbitrary, and there is no necessary implication of desirability or goodness involved. For example, we may choose to refer to the appearance of a defective item in a production process as a success; or, if a series of births is treated as a Bernoulli process, the appearance of a female (male) may be classified as a success.) Suppose that the sequence of outcomes is

$$H \, T \, H \, T \, T$$

where H and T denote head and tail, as usual. We now introduce a convenient coding device for outcomes on Bernoulli trials. Let

$$x_i = 0 \quad \text{if the outcome on the } i\text{th trial is a failure}$$

$$x_i = 1 \quad \text{if the outcome on the } i\text{th trial is a success}$$

Then the outcomes of the previous sequence of tosses may be written as

$$1 \, 0 \, 1 \, 0 \, 0 \quad \text{(representing } H \, T \, H \, T \, T)$$

Since the probability of a success and a failure on a given trial are, respectively, p and q, the probability of this particular sequence of outcomes is, by the multiplication rule,

$$P(1, 0, 1, 0, 0) = pqpqq = q^3 p^2$$

In this notation, for simplicity, commas have been used to separate the outcomes of the successive trials. Actually, though, this is the joint probability of the events that occurred on the 5 trials, that is, the probability of obtaining the specific sequences of successes and failures in the order in which they occurred. However, we are interested not in any specific order of results, but rather in the probability of obtaining a given number of successes in n trials. What then is the probability of obtaining exactly 2 successes in 5 Bernoulli trials? There are 9 other sequences that satisfy the condition of exactly 2 successes in 5 trials.

$$
\begin{array}{lll}
1\,1\,0\,0\,0 & 0\,1\,1\,0\,0 & 0\,0\,1\,1\,0 \\
1\,0\,0\,1\,0 & 0\,1\,0\,1\,0 & 0\,0\,1\,0\,1 \\
1\,0\,0\,0\,1 & 0\,1\,0\,0\,1 & 0\,0\,0\,1\,1
\end{array}
$$

By the same reasoning used earlier, each of these sequences has the same probability, $q^3 p^2$. We can obtain the number of such sequences from the formula for the number of combinations of n objects taken x at a time given in equation 2.20. Thus, the number of possible sequences in which two ones can occur is $\binom{5}{2}$. We indicated in equation 2.20 that

$$\binom{n}{x} = \frac{n!}{x!(n-x)!}$$

Then,

$$\binom{5}{2} = \frac{5!}{2!3!} = 10$$

and we may write

$$P(\text{exactly 2 successes}) = \binom{5}{2}q^3p^2$$

In the case of the fair coin example, we assign a probability of $\frac{1}{2}$ to p and $\frac{1}{2}$ to q. Hence,

$$P(\text{exactly 2 heads}) = \binom{5}{2}\left(\frac{1}{2}\right)^3\left(\frac{1}{2}\right)^2 = \frac{10}{32} = \frac{5}{16}$$

This result may be generalized to obtain the probability of (exactly) x successes in n trials of a Bernoulli process. Let us assume $n - x$ failures occurred followed by x successes, in that order. We then represent this sequence as

$$\underbrace{0\ 0\ 0 \cdots 0}_{n - x \text{ failures}}\ \underbrace{1\ 1\ 1 \cdots 1}_{x \text{ successes}}$$

The probability of this particular sequence is $q^{n-x}p^x$. The number of possible sequences of n trials resulting in exactly x successes is $\binom{n}{x}$.[4] Therefore, the probability of obtaining x successes in n trials of a Bernoulli process is given by[5]

(3.3)

$$f(x) = \binom{n}{x}q^{n-x}p^x \quad \text{for } x = 0, 1, 2, \ldots, n$$

If we denote by X the random variable "number of successes in these n trials," then

$$f(x) = P(X = x)$$

---

[4] Because the combination notation is universally used in connection with the binomial probability distribution, that convention is followed here. However, conceptually, we have here the number of distinct permutations that can be formed of n objects, $n - x$ of which are of one type and x of the other. Since the number of such permutations turns out to be equal to $\binom{n}{x}$, the combination notation may be used instead of that for permutations.

[5] The following method of writing the mathematical expression for such a probability distribution is often used:

$$f(x) = \binom{n}{x}q^{n-x}p^x \quad \text{for } x = 0, 1, 2, \ldots, n$$

$$= 0, \text{ elsewhere}$$

In this and other places where it is clear that $f(x)$ is equal to 0 for values of the random variable other than the specified ones, the notation on the last line will not be included.

The fact that this is a probability distribution is verified by noting the following conditions.

1. $f(x) \geqslant 0$ for all real values of x
2. $\sum_x f(x) = 1$

The first condition is verified by noting that since p and n are nonnegative numbers, $f(x)$ cannot be negative. The second condition is true because (as shown mathematically)

$$\sum_x \binom{n}{x} q^{n-x} p^x = (q + p)^n = 1^n = 1$$

Therefore, the term **binomial probability distribution**, or simply **binomial distribution**, is usually used to refer to the probability distribution resulting from a Bernoulli process.

> In problems where the assumptions of a Bernoulli process are met, we can obtain the probabilities of zero, one, or more successes in n trials from the respective terms of the binomial expansion of $(q + p)^n$, where q and p denote the probabilities of failure and success on a single trial and n is the number of trials.

The binomial distribution has two parameters, n and p.[6] Each pair of values for these parameters establishes a different distribution. Thus, the binomial distribution is actually a family of probability distributions. Since computations become laborious for large values of n, it is advisable to use special tables. Selected values of the binomial cumulative distribution function are given in Table A-1 of Appendix A. The values of

$$F(c) = P(X \leqslant c) = \sum_{x \leqslant c} f(x) \quad \text{for } x = 0, 1, 2, \ldots, n$$

are shown in that table for $n = 2$ to $n = 20$ and $p = 0.05$ to $p = 0.50$ in multiples of 0.05. Values of $f(c)$, cumulative probabilities for p values greater than 0.50, and probabilities that x is greater than a given value or lies between two values can be obtained by appropriate manipulation of these tabulated values. Some of the examples that follow illustrate the use of the table. More extensive tables have been published in other places, but even these tables usually do not go beyond $n = 50$ or $n = 100$. For large values of n, approximations are available for the binomial distribution, and the exact values generally need not be determined.

---

[6] In this context, the term "parameters" refers to numerical quantities that are sufficient to specify a probability distribution. When particular values are assigned to the parameters of a probability function, a specific distribution in the family of possible distributions is defined. For example, $n = 10$, $p = \frac{1}{2}$ specifies a particular binomial distribution; $n = 20$, $p = \frac{1}{2}$ specifies another.

For convenience in looking up individual terms of the binomial probability distribution, selected values of the binomial probability distribution are given in Table A-2 of Appendix A. The values of

$$f(x) = P(X = x) \quad \text{for } x = 0, 1, 2, \ldots, n$$

are shown in that table for $n = 1$ to $n = 20$ and $p = 0.05$ to $p = 0.50$ in multiples of 0.05. Of course, Table A-2 is particularly useful when values of individual terms of the binomial distribution are desired rather than sums of terms.

> In the case of the binomial distribution, as with any other mathematical model, the correspondence between the real-world situation and the model must be carefully established.

In many cases, the underlying assumptions of a Bernoulli process are not met. For example, suppose that in a production process, items produced by a certain machine tool are tested as to whether they meet specifications. If the items are tested in the order in which they are produced, then the assumption of independence would doubtless be violated. That is, whether an item meets specifications would not be independent of whether the preceding item(s) did. If the machine tool had become subject to wear, it is quite likely that if it produced an item that did not meet specifications, the next item would fail to conform to specifications in a similar way. Thus, whether or not an item is defective would *depend* on the characteristics of preceding items. In the coin-tossing illustration, on the other hand, we imagined an experiment in which a head or tail on a particular toss did not affect the outcome on the next toss.

We can see from the assumptions underlying a Bernoulli process that the following is true.

> The binomial distribution is applicable to situations of *sampling from a finite population with replacement or sampling from an infinite population with or without replacement.*

In either of these cases, the probability of success may be viewed as remaining constant from trial to trial. If the population size is large relative to sample size—that is, if the sample constitutes only a small fraction of the population—and if p is not very close in value to zero or to one, the binomial distribution is often sufficiently accurate, even though sampling may be carried out from a finite population without replacement. It is difficult to give universal rules of thumb on appropriate ratios of population size to sample size for this purpose. Some practitioners suggest a population size at least 10 times the sample size. However, the purpose of the calculations must determine the required degree of accuracy.

An important property of the binomial distribution is that when $p = 0.50$, the distribution is symmetrical. For example, see Figure 3.5(a), where $p = 0.50$ and $n = 10$. When $p \neq 0.50$, the distribution is asymmetrical (skewed). This property is illustrated in Figure 3.5(b) and (c), where the binomial distributions for $p = 0.10$, $n = 10$ and for $p = 0.90$, $n = 10$ are plotted.

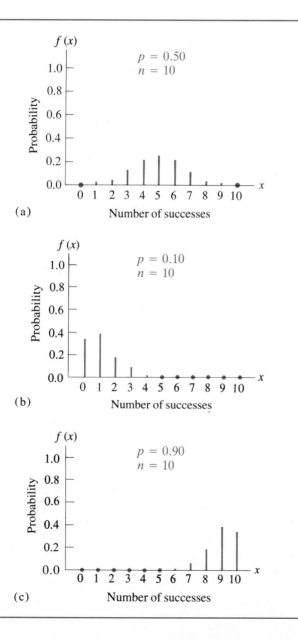

(a)

(b)

(c)

FIGURE 3-5

Graphs of binomial distributions. When $p = 0.50$, the distribution is symmetrical, as in (a); when $p \neq 0.50$, as shown in (b) and (c), the distribution is asymmetrical (skewed)

EXAMPLE 3-6

The tossing of a fair coin 5 times was used earlier as an example of a Bernoulli process; the probability of obtaining 2 heads (successes) was calculated. Compute the probabilities of all possible numbers of heads and thus establish the particular binomial distribution that is appropriate in this case.

SOLUTION

This problem is an application of the binomial distribution for $p = \frac{1}{2}$ and $n = 5$. Letting X represent the random variable "number of heads," the probability distribution is as follows:

x	$f(x)$
0	$\binom{5}{0}\left(\frac{1}{2}\right)^5\left(\frac{1}{2}\right)^0 = \dfrac{1}{32}$
1	$\binom{5}{1}\left(\frac{1}{2}\right)^4\left(\frac{1}{2}\right)^1 = \dfrac{5}{32}$
2	$\binom{5}{2}\left(\frac{1}{2}\right)^3\left(\frac{1}{2}\right)^2 = \dfrac{10}{32}$
3	$\binom{5}{3}\left(\frac{1}{2}\right)^2\left(\frac{1}{2}\right)^3 = \dfrac{10}{32}$
4	$\binom{5}{4}\left(\frac{1}{2}\right)^1\left(\frac{1}{2}\right)^4 = \dfrac{5}{32}$
5	$\binom{5}{5}\left(\frac{1}{2}\right)^0\left(\frac{1}{2}\right)^5 = \dfrac{1}{32}$
	$\overline{1}$

EXAMPLE 3-7

Calculate the probability of obtaining at least one 6 in two rolls of a die (or in one roll of two dice) using the binomial distribution.

SOLUTION

We view the two rolls of the die as Bernoulli trials. If we define the appearance of a 6 as a success, $p = \frac{1}{6}$, $q = \frac{5}{6}$, and $n = 2$. It is instructive to examine the entire probability distribution

x	$f(x)$
0	$\binom{2}{0}\left(\frac{5}{6}\right)^2\left(\frac{1}{6}\right)^0 = \left(\dfrac{5}{6}\right)^2$
1	$\binom{2}{1}\left(\frac{5}{6}\right)^1\left(\frac{1}{6}\right)^1 = 2\left(\dfrac{5}{6}\right)\left(\dfrac{1}{6}\right)$
2	$\binom{2}{2}\left(\frac{5}{6}\right)^0\left(\frac{1}{6}\right)^2 = \left(\dfrac{1}{6}\right)^2$
	$\overline{1}$

The expressions at the right side of the $f(x)$ column are in the form with which the student is probably most familiar for the terms in the expansion of $(\frac{5}{6} + \frac{1}{6})^2$.

The required probability is

$$P(\text{at least one } 6) = f(1) + f(2) = 2\left(\frac{5}{6}\right)\left(\frac{1}{6}\right) + \left(\frac{1}{6}\right)^2 = \frac{11}{36}$$

EXAMPLE 3-8

An interesting correspondence took place in 1693 between Samuel Pepys (author of the famous *Diary*) and Isaac Newton, in which Pepys posed a probability problem to the eminent mathematician. The question as originally stated by Pepys was:

A has six dice in a box, with which he is to fling a 6.
B has in another box twelve dice, with which he is to fling two 6s.
C has in another box eighteen dice, with which he is to fling three 6s.
(Question)—Whether B and C have not as easy a task as A at even luck?[7]

In rather flowery seventeenth century English, Newton replied and said, essentially, "Sam, I do not understand your question." Newton asked whether individuals A, B, and C were to throw independently and whether the question pertained to obtaining *exactly* one, two, or three 6s or *at least* one, two, or three 6s.

After an exchange of letters, in which Pepys supplied little help in answering these queries, Newton decided to frame the question himself. In modern language, Newton's wording would appear somewhat as follows:

If A, B, and C toss dice independently, what are the probabilities that:
A will obtain at least one 6 in a roll of six dice?
B will obtain at least two 6s in a roll of twelve dice?
C will obtain at least three 6s in a roll of eighteen dice?

SOLUTION

Newton's reply to these questions involved some rather tortuous arithmetic. His work doubtless represented a respectable intellectual feat, considering the infantile state of probability theory at that time. Today, almost any beginning student of probability theory, standing on the shoulders of the giants who came before, would immediately see the application of the binomial distribution to the problem. Let us denote by $P(A)$, $P(B)$, and $P(C)$ the probabilities that A, B, and C would obtain the specified events. Then

$$P(A) = 1 - \binom{6}{0}\left(\frac{5}{6}\right)^6\left(\frac{1}{6}\right)^0 \approx 0.67$$

$$P(B) = 1 - \binom{12}{0}\left(\frac{5}{6}\right)^{12}\left(\frac{1}{6}\right)^0 - \binom{12}{1}\left(\frac{5}{6}\right)^{11}\left(\frac{1}{6}\right)^1 \approx 0.62$$

$$P(C) = 1 - \binom{18}{0}\left(\frac{5}{6}\right)^{18}\left(\frac{1}{6}\right)^0 - \binom{18}{1}\left(\frac{5}{6}\right)^{17}\left(\frac{1}{6}\right)^1 - \binom{18}{2}\left(\frac{5}{6}\right)^{16}\left(\frac{1}{6}\right)^2 \approx 0.60$$

Thus, $P(A) > P(B) > P(C)$.

[7] Schell, Emil D., "Samuel Pepys, Isaac Newton and Probability," *The American Statistician*, October 1960, pp. 27–30.

Pepys admitted frankly that he did not understand Newton's calculations and furthermore that he did not believe the answer. He argued that since B throws twice as many dice as A, why can't he simply be considered two A's? Thus, he would have at least as great a probability of success as A. Of course, Pepys' question indicated that he was rather confused. There is no reason why the probability of at least two 6s in a roll of twelve dice should be twice the probability of at least one 6 in a roll of six dice, and as seen by the above calculations, indeed it is not.

EXAMPLE 3-9

The customer service manager of Courier Express is responsible for expediting late mail delivery. From past experience, she knows that prompt deliveries occur about 90% of the time. View this situation as a Bernoulli process and determine from Table A-1 of Appendix A the probabilities that in 10 deliveries

a. 3 or fewer deliveries will be late.
b. between 3 and 5 deliveries (inclusive) will be late.
c. 3 or more deliveries will be prompt.
d. at most, 8 deliveries will be prompt.
e. exactly 2 deliveries will be late.
f. 7 or more deliveries will be late.

SOLUTION

Let $p = 0.10$ stand for the probability that a delivery will be late. (We define the probability of a success this way because Table A-1 gives p values only up to $p = 0.50$.) Therefore, $q = 0.90$ and $n = 10$. Let X represent the number of deliveries that do not arrive on time. Note that a failure to deliver mail on time is considered a "success" in this problem despite the undesirability of this outcome.

a. From Table A-1 of Appendix A, $F(3) = 0.9872$.

$$P(X \leqslant 3) = F(3) = \sum_{x=0}^{3} \binom{10}{x}(0.90)^{10-x}(0.10)^{x}$$

b. The probability of obtaining 3, 4, or 5 successes is the difference between "5 or fewer successes" and "2 or fewer successes." Thus,

$$P(3 \leqslant X \leqslant 5) = F(5) - F(2) = 0.9999 - 0.9298 = 0.0701$$

c. The event "3 or more failures" is the same as the event "7 or fewer successes." Hence,

$$P(X \leqslant 7) = F(7) = 1.0000$$

d. The event "at most, 8 failures" is the same as "8 or fewer failures" or "2 or more successes."

$$P(2 \leqslant X \leqslant 10) = F(10) - F(1) = 1.0000 - 0.7361 = 0.2639$$

e. The probability of "exactly 2 successes" is the difference between the probabilities "2 or fewer successes" and "one or fewer successes."

$$P(X = 2) = F(2) - F(1) = 0.9298 - 0.7361 = 0.1937$$

This probability may also be obtained directly by reference to Table A-2 where for $n = 10$, $x = 2$, we again find

$$f(x) = P(X = 2) = 0.1937$$

Of course, it is simpler to look up probabilities of exact numbers of successes in Table A-2 than in Table A-1, but both solutions are shown here.

f. The event "7 or more successes" is the complement of the event "6 or fewer successes." Therefore,

$$P(X \geqslant 7) = 1 - P(X \leqslant 6) = 1 - F(6) = 1 - 1 = 0$$

EXERCISES 3.4

In some of these exercises, assumptions of (1) constant probability of success from trial to trial and (2) independence among trials seem reasonable and are necessary in order to obtain solutions.

1. Mr. Morrison, an investor in industrial and mining stocks, prepares a portfolio consisting of eight stocks. He feels that the probability of any one stock's price going down is 0.60 and that the price movements of the stocks are independent.
   a. What is the probability that exactly five of his stocks will decline?
   b. What is the probability that five or more will decline?
   c. Does the assumption of independence here seem logical? If not, is the binomial distribution the appropriate probability distribution for the problem?

2. Builders, Inc. submits bids for construction projects together with five other companies. In the past, Builders, Inc. was awarded contracts 15% of the time. What is the probability that this company will be awarded contracts for exactly three out of the next four projects? Explain what it means to assume independent trials in this case.

3. Political analysts estimate that 40% of a particular candidate's supporters are inclined to vote for the other candidate in a municipal election. Verify that in a random sample of two voters selected from among the first candidate's supporters, the probabilities that 0, 1, and 2 people are inclined to vote for the other candidate are 0.36, 0.48, and 0.16, respectively.

4. After returning from Atlantic City, Mr. Hugh Stakes announces that he has accomplished the impossible. He has won 10 consecutive times in betting on the color red in the game of roulette. (There are 16 red slots, 16 black slots, and two green slots on a roulette wheel.) How likely is this feat?

5. Assume that on any given day, the price of your Georesource common stock can advance, decline, or remain unchanged with identical chances.
   a. What is the probability of five successive advances?
   b. What is the probability of at least two advances in three days?
   c. A stock broker advises you to purchase another stock, which has advanced for 10 consecutive days. What are the chances of the stock

repeating this experience, assuming that the basic probability of a price advance is the same as for Georesource?

d. What assumptions have you made in your answers to part (a), (b), and (c)?

6. The quality control unit of Forte Iron Corporation estimates that 5% of the iron poles that the company produces are defective. Forte Iron has a prospective sale to a builder who uses 10 poles in the construction of a bungalow. The builder has determined that if two or more poles are defective, the bungalow will not be properly balanced. The builder will close the sale only if the likelihood of an unbalanced bungalow because of defective poles is less than 10%. Should she contract with Forte Iron Corporation?

7. The probability that Mr. Arkay, a shrewd trader, makes a profit on any business deal is 0.80. What is the probability that he will make a profit exactly seven times in 10 successive independent deals?

8. A fighter bomber will hit its target $\frac{2}{3}$ of the time. Suppose military tacticians assign four planes to strike a crucial target area one time each. Assuming independence among strikes by different planes, determine the probability that the target will be hit two or three times.

9. To stimulate sales, Smokey Toes, a well-known pub, decides to give its customers coupons, 25% of which are for a free beer and 75% of which are for a free cola drink. Three customers enter and each draws one coupon. The number of beer coupons that will be drawn in defined as X.
   a. Find the probability distribution of X and tabulate its cumulative distribution.
   b. Find the probability that exactly two beer coupons will be drawn from among the three.

10. Five fair dice are tossed. Would you agree that the probability of obtaining no ones is the same as that of obtaining fives ones? If not, what are the exact probabilities of these two events?

11. The sales manager of a growing cosmetics company has just hired a new sales representative for the firm's line of pressed powder. From experience, the manager knows that an above-average sales representative will make one sale for every two attempts; an average representative will make two sales for every five attempts; and a below-average representative will have a sales ratio of 0.10. In 10 attempts, the new employee makes one sale. Assuming independence, what is the probability of this event if
   a. the employee is below average?
   b. the employee is above average?
   c. the employee is average?

12. Sketch or display on a computer screen the binomial distribution for $n = 5$ and $p = 0.30, 0.50,$ and 0.70. What can be said about the skewness of the binomial distribution as the value of p departs from 0.50?

13. A certain delicate manufacturing process produces 25% defective items. In testing a new process, a sample of 50 items is produced. The new process will be installed if the sample yields ten or fewer defective items. Assume that the production of 50 items corresponds to 50 independent trials of the process. Write a symbolic expression, specifying all numerical values involved, for the exact probability of ten or fewer defective items, if the new process also produces 25% defective items on the average. Do not carry out the arithmetic.

14. The sales manager of a small electronics firm has just hired a new sales representative for the firm's line of calculators. From experience, the manager knows that an average sales representative will make one sale for every five customers approached, an above average one will make one sale in four attempts, and a below average one will have a sales ratio of 1:10. In 16 attempts, the new employee makes one sale. Assuming independence, what is the probability of this event if
    a. the employee is below average?
    b. the employee is average?
    c. the employee is above average?

15. An electronics engineer estimates that 25% of the components in a certain batch are defective and will fail when tested. Assuming independence, what is the probability that a device requiring three of these components will work properly when tested if
    a. all three are required for operation of the device?
    b. two of the three are backup units (that is, only one component need work properly for the device to operate)?

3.5
THE MULTINOMIAL DISTRIBUTION

In the case of the binomial distribution, there were two possible outcomes on each experimental trial. The **multinomial distribution** represents a straightforward generalization of the binomial distribution for the situation where there are more than two possible outcomes on each trial.

The assumptions underlying the multinomial distribution are completely analogous to those of the binomial distribution.

1. On each trial, there are k mutually exclusive possible outcomes, which may be referred to as $E_1, E_2, \ldots, E_k$. Therefore, the sample space of possible outcomes on each trial is $S = \{E_1, E_2, \ldots, E_k\}$. *Assumptions*

2. The probabilities of outcomes $E_1, E_2, \ldots, E_k$, denoted $p_1, p_2, \ldots, p_k$, remain constant from trial to trial.

3. The trials are independent.

Under these assumptions, the probability that there will be x_1 occurrences of E_1, x_2 occurrences of $E_2, \ldots$, and x_k occurrences of E_k in n is given by

(3.4)
$$f(x_1, x_2, \ldots, x_k) = \frac{n!}{x_1! x_2! \cdots x_k!} p_1^{x_1} p_2^{x_2} \cdots p_k^{x_k}$$

where $x_1 + x_2 + \cdots + x_k = n$ and $p_1 + p_2 + \cdots + p_k = 1$.

The expression $f(x_1, x_2, \ldots, x_n)$ is the general term of the multinomial distribution

$$(p_1 + p_2 + \cdots + p_k)^n$$

Analogously, in the binomial distribution, the probability of x successes in n trials is given by

$$f(x) = \frac{n!}{x!(n-x)!} q^{n-x} p^x$$

which is the general term of $(q + p)^n$. In terminology similar to that of the multinomial distribution, $(q + p)^n$ may be written

$$(p_1 + p_2)^n$$

where $p_1 + p_2 = 1$.

EXAMPLE 3-10

The Bargain Center store is featuring a unique sale of ladies' stockings. For only 50¢, any customer may reach into a closed box containing a large number of pairs of stockings and remove a pair. The stockings have been classified into three boxes according to small, medium, or large sizes. Barbara Kane, a bargain hunter, draws 10 pairs from the "medium" box. Because of an error in size classification, about 15% of the pairs in the "medium" box are large, 5% are small, and the rest are medium sizes.

a. What is the probability that Ms. Kane will get exactly 5 pairs of medium size, one pair of large size, and 4 pairs of small size stockings?
b. What is the probability that she will get 7 medium size, 2 large size, and one small size?

SOLUTION

a. Applying the multinomial distribution, we find the probability of exactly 5 pairs of medium size, one pair of large size, and 4 pairs of small size stockings is

$$f(5, 1, 4) = \frac{10!}{5!1!4!} (0.80)^5 (0.15)(0.05)^4$$

$$= 0.0004$$

b.
$$f(7, 2, 1) = \frac{10!}{7!2!1!} (0.80)^7 (0.15)^2 (0.05)$$

$$= 0.0849$$

TABLE 3-6
Multinomial distribution for numbers of DS, HP, and P grades: $p_1 = 0.10$, $p_2 = 0.20$, $p_3 = 0.70$, and $n = 3$

(x_1, x_2, x_3)	$f(x_1, x_2, x_3)$
3, 0, 0	0.001
0, 3, 0	0.008
0, 0, 3	0.343
2, 1, 0	0.006
2, 0, 1	0.021
1, 2, 0	0.012
1, 1, 1	0.084
1, 0, 2	0.147
0, 2, 1	0.084
0, 1, 2	0.294
	1.000

EXAMPLE 3-11

The distribution of grades among MBAs in a well-known eastern school is as follows: 10% Distinguished (DS), 20% High Pass (HP), and 70% Pass (P). Suppose you select a sample of three student records at random from the records for the entire school. Give the probability distribution of the numbers of each grade level in the sample, assuming that the multinomial distribution is applicable.

SOLUTION

If X_1, X_2, and X_3 stand for the numbers of DS, HP, and P grades, then the appropriate multinomial distribution is

$$f(x_1, x_2, x_3) = \frac{3!}{x_1! x_2! x_3!} (0.10)^{x_1}(0.20)^{x_2}(0.70)^{x_3}$$

$$x_1 + x_2 + x_3 = 3 \quad \text{and} \quad x_i = 0, 1, 2, 3$$

This probability distribution is given in Table 3-6.

EXERCISES 3.5

1. Charles Nolson, vice-president of sales in a large finance firm, regularly reviews the monthly sales reports from 30 regional offices in the United States. In one such report, Jacob Donnelly writes that of the 12 new credit customers in Texas, 2 have defaulted and 4 have asked to renegotiate their loans. From experience, Nolson realizes that any new credit customer has a 1% probability of default and a 5% chance to renegotiate terms. Assuming independence, does Nolson have reason to doubt Donnelly's report?

2. Winter Wear Company specializes in selling "factory reject" coats at one-half the regular price. The four different types of flaws in the coats and the probabilities of these errors occurring among the coats are as follows:

Type of Error	Probability
e_1: Imperfect darts	0.40
e_2: Uneven seam	0.30
e_3: Uneven collar	0.20
e_4: Uneven sleeve length	0.10

In the last shipment of 15 coats, 3 had imperfect darting, 4 had uneven seams, 6 had uneven collars, and 2 had uneven sleeve length. The prices of defective coats vary according to the type of error. For example, those with imperfect darts are priced higher than those with uneven sleeve length, so the distribution of errors is critical. Assuming independence, what is the probability of this combination of errors?

3. A prestigious school on the East Coast claims that for every 100 applicants to its graduate school, only 30 are accepted. Of those who are accepted, 5% graduate with distinction, 30% drop out of school, and the rest graduate with passing marks.

   a. What is the probability that of 20 new students accepted, 10 will pass the graduate program, 8 will drop out, and 2 will achieve distinction?

   b. What is the probability that from among 20 applicants, at most 10 will be accepted for graduate studies?

3.6
THE HYPERGEOMETRIC DISTRIBUTION

In section 3.4, the binomial distribution was discussed as the appropriate probability distribution for situations in which the assumptions of a Bernoulli process were met. A major application of the binomial distribution was the computation of probabilities for the sampling of finite populations *with replacement*. In most practical situations, sampling is carried out without replacement. In this section, we discuss the **hypergeometric distribution** as the appropriate model for sampling *without replacement*.

For example, if a sample of families is selected in a city in order to estimate the average income of all families in the city, sampling units are ordinarily not replaced prior to the selection of subsequent ones. That is, families are not replaced in the original population and thus given an opportunity to appear more than once in the sample. In fact, such samples are usually drawn in a single operation, without any possibility of drawing the same family twice. Moreover, in a sample drawn from a production process, articles are generally not replaced and given an opportunity to reappear in the sample. Thus, for both human populations and universes of physical objects, sampling is ordinarily carried out without replacement.

Suppose we have a list of 1,000 persons, 950 of whom are adults and 50 of whom are children. Numbers from 1 to 1,000 are assigned to these individuals. These numbers are printed on 1,000 identical chips, which are placed in a large bowl. We draw a sample of 5 chips at random from the bowl without replacement. These 5 chips may be drawn simultaneously or successively. For simplicity, we shall refer to this situation as the drawing of a random sample of 5 persons from the group of 1,000, although, of course, chips rather than persons are sampled. The process of numbering chips to correspond to persons and sampling the chips is simply a device to ensure randomness in sampling the population of 1,000 persons.

What is the probability that none of the 5 persons in the sample is a child? An alternative way of wording the question is, "What is the probability that all 5 persons in the sample are adults?" The sample space of possible outcomes in this experiment is the total number of samples of 5 persons that can be drawn from the population of 1,000 persons. This is the number of combinations that can be formed of 1,000 objects taken 5 at a time $\binom{1,000}{5}$. We can compute the required probability by obtaining the ratio of the number of sample points favorable to the event "none of the 5 persons is a child" to the total number of points in the sample space. The number of ways 5 adults can be drawn from the 950 adults is $\binom{950}{5}$. The number of ways no children can be selected from the 50 children is $\binom{50}{0}$. Therefore, the total number of ways of selecting 5 adults and no children from the population of 950 adults and 50 children is $\binom{950}{5}\binom{50}{0}$, by the multiplication principle.

Hence, the probability of obtaining no children (and 5 adults) in this sample of 5 persons is

$$\frac{\binom{950}{5}\binom{50}{0}}{\binom{1,000}{5}}$$

If we carry out the arithmetic, we see a very interesting fact.

$$\frac{\binom{950}{5}\binom{50}{0}}{\binom{1,000}{5}} = \frac{\dfrac{950!}{5!945!} \times \dfrac{50!}{0!50!}}{\dfrac{1,000!}{5!995!}} = \frac{950 \times 949 \times 948 \times 947 \times 946}{1,000 \times 999 \times 998 \times 997 \times 996} = 0.7734$$

Grouping the product obtained as a multiplication of 5 factors, we have

$$P(\text{no children}) = \left(\frac{950}{1,000}\right)\left(\frac{949}{999}\right)\left(\frac{948}{998}\right)\left(\frac{947}{997}\right)\left(\frac{946}{996}\right)$$

which is the result we would have arrived at if we had simply solved the original problem in terms of conditional probabilities. That is, the probability of obtaining an adult on the first draw is $\frac{950}{1,000}$; the probability of obtaining an adult on the second draw given that an adult was obtained on the first draw is $\frac{949}{999}$, and so on. Therefore, the joint probability of obtaining no children (5 adults) if the sampling is carried out without replacement is given by the multiplication of the 5 factors shown.

We can now state the general nature of this type of problem and the hypergeometric distribution as a solution to it. Suppose there is a population containing N elements, X of which are termed "successes," $N - X$ of which are denoted "failures." The corresponding terminology for a random sample of n elements drawn without replacement is that we require x successes and therefore $n - x$ failures. The data of this general problem are tabulated below:

Population	Required Sample
X = Number of successes[8]	x = Number of successes
$N - X$ = Number of failures	$n - x$ = Number of failures
N = Total number in population	n = Total number in sample

The **hypergeometric distribution**, which gives the probability of x successes in a random sample of n elements drawn *without replacement*, is

(3.5)
$$f(x) = \frac{\binom{N - X}{n - x}\binom{X}{x}}{\binom{N}{n}} \quad \text{for } x = 0, 1, 2, \ldots, [n, X]$$

where the symbol $[n, X]$ means the smaller of n and X. For example, in the preceding illustration, if there had been only 10 children in the population (X) and the sample size had been 50 (n), the largest value that the number of children in the sample (x) could take on would be 10 (X). On the other hand, if X exceeded n, clearly x could be as large as n.

> The hypergeometric distribution bears an interesting relationship to the binomial distribution.

Suppose, in the case of the population containing 950 adults and 50 children, we had been interested in the same probability (of obtaining no children in a random sample of 5 persons) but the sample was randomly drawn *with replacement*. Then, letting $q = 0.95$, $p = 0.05$, and $n = 5$ in the binomial distribution,

---

[8] Note that in order to maintain parallel notation for the population and sample in this case, the symbol X does *not* denote the *random variable* for number of successes in the sample, but is instead the total number of successes in the population being sampled.

we have

$$f(0) = \binom{5}{0}(0.95)^5(0.05)^0 = (0.95)^5 = 0.7738$$

Just as in the case of the hypergeometric distribution, where the required probability could have been computed by using the multiplication rule for *dependent* events, in the case of the binomial distribution, the probability could have been computed by simply using the multiplication rule for *independent* events

$$P(\text{no children}) = \left(\frac{950}{1,000}\right)\left(\frac{950}{1,000}\right)\left(\frac{950}{1,000}\right)\left(\frac{950}{1,000}\right)\left(\frac{950}{1,000}\right).$$

Note that the hypergeometric and binomial probability values are extremely close in this illustration, agreeing exactly in the first three decimal places. We can show that when N increases without limit, the hypergeometric distribution approaches the binomial distribution.

> The binomial probabilities may be used as approximations to hypergeometric probabilities when n/N is small.

A frequently used rule of thumb is that the population size should be at least 10 times the sample size $(N > 10n)$ for the approximations to be used. However, the governing considerations, as usual, include such matters as the purpose of the calculations, whether a sum of terms rather than a single term is being approximated, and whether terms near the center or the extremes of the distribution are involved.

> Just as the multinomial distribution represents the generalization of the binomial distribution when there are more than two possible classifications of outcomes, the hypergeometric distribution can be similarly extended.

No special name is given to the more general distribution; it also is referred to as the hypergeometric distribution. Assume a population that contains N elements, X_1 of type one, X_2 of type two, ..., and X_k of type k. Suppose we require, in a sample of n elements drawn without replacement, that there be x_1 elements of type one, x_2 of type two, ..., and x_k of type k. Tabulating the data in an analogous fashion to the two-outcome case, we have

Population (number of elements)	Required Sample (number of elements)
X_1 of type one	x_1 of type one
X_2 of type two	x_2 of type two
$\vdots$	$\vdots$
X_k of type k	x_k of type k

The **hypergeometric distribution**, which gives the probability of obtaining x_1 occurrences of type one, x_2 occurrences of type two, ..., and x_k occurrences of type k in a random sample of n elements drawn without replacement, is

(3.6)
$$f(x_1, x_2, \ldots, x_k) = \frac{\binom{X_1}{x_1}\binom{X_2}{x_2}\cdots\binom{X_k}{x_k}}{\binom{N}{n}}$$

$$\text{for } x_i = 0, 1, 2, \ldots, [n, X_i]$$

where

$$\sum_{i=1}^{k} X_i = N \quad \text{and} \quad \sum_{i=1}^{k} x_i = n$$

EXAMPLE 3-12

The Carleton Oil Corporation has 100 service stations in a certain community; it has classified them according to merit of geographic location as follows:

Merit of Location	Number of Stations
Excellent	22
Good	38
Fair	27
Poor	10
Disastrous	3
	100

The corporation has a computer program for drawing random samples (without replacement) of its service stations. In a random sample of 20 of these stations, what is the joint probability of obtaining 6 excellent, 6 good, 4 fair, 3 poor, and one disastrous station?

SOLUTION

$$f(6, 6, 4, 3, 1) = \frac{\binom{22}{6}\binom{38}{6}\binom{27}{4}\binom{10}{3}\binom{3}{1}}{\binom{100}{20}}$$

EXAMPLE 3-13

Sales representative Nina Waterton receives 100 leads a day, 5 of which are excellent sales prospects. She draws 5 at random to visit on a particular day from the 100 leads received that day. What is the exact probability that she does not draw any of the 5 excellent prospects? Calculate an approximate probability.

Exact probability:

SOLUTIONS

$$\frac{\binom{95}{5}\binom{5}{0}}{\binom{100}{5}} = 0.7699$$

Approximate probability:

$$\binom{5}{0}(0.95)^5(0.05)^0 = 0.7738$$

EXAMPLE 3-14

An investment banking firm plans to hire 5 summer corporate research trainees from a pool of 18 first-year MBA students. Nine of the 18 students attend the Cambridge Graduate School of Business, 5 attend the Philadelphia School of Finance, and the other 4 are students at the Palo Alto Business School. If the firm decides that all 18 applicants are equally qualified and therefore will select the potential trainees randomly from the pool, what is the probability that

a.   all 5 trainees selected are Cambridge students?
b.   no Palo Alto business students are selected?
c.   2 trainees are Cambridge students, 2 are Philadelphia students, and one is a Palo Alto student?
d.   3 trainees are Philadelphia students and the other 2 attend Palo Alto Business School?

a. $$\frac{\binom{9}{5}\binom{9}{0}}{\binom{18}{5}} = \frac{126}{8568} = 0.0147$$

SOLUTIONS

b. $$\frac{\binom{4}{0}\binom{14}{5}}{\binom{18}{5}} = \frac{2002}{8568} = 0.2337$$

c. $$\frac{\binom{9}{2}\binom{5}{2}\binom{4}{1}}{\binom{18}{5}} = \frac{1440}{8568} = 0.1681$$

d. $$\frac{\binom{9}{0}\binom{5}{3}\binom{4}{2}}{\binom{18}{5}} = \frac{60}{8568} = 0.0070$$

EXERCISES 3.6

1. The Bostonian Consulting Group (BCG) plans to hire 6 summer corporate research trainees from a pool of 20 first-year MBA students. Ten of the 20 students attend business schools in the East, 5 attend midwestern schools, and the other 5 students attend schools on the West Coast. If all 20 applicants are equally qualified, what is the probability that in a random selection of 6 students, BCG will select
   a. 5 students from eastern schools?
   b. no students from eastern schools?
   c. 2 eastern, one midwestern, and 3 West Coast trainees?

2. Write an expression for calculating the probability that a bridge hand (13 cards) will contain exactly 3 diamonds, 4 spades, one club, and 5 hearts.

3. In a lot formed from a production run of calculator parts, 4 out of 12 units are found to be defective. A sample of 3 units is taken randomly from this lot of 12. What is the probability of selecting *at least one* defective part if the samples are drawn
   a. without replacement?
   b. with replacement?

4. The bylaws of the Thrifty Bank and Trust Company stipulate that membership on committees of the board of directors shall be determined by lottery and that committees may make recommendations only when there exists unanimous agreement. A 3-person committee on managerial appointments is to be selected from the full board of 12 directors to recommend a new vice-president to head the foreign department. Seven members of the full board favor selection of Mrs. Werth, while 5 members favor Mr. Hedge. Assuming that the directors do not change their minds, what is the probability that the managerial appointments committee will unanimously recommend the appointment of Mr. Hedge? What would be the probability of a committee recommendation of Mr. Hedge if only a majority vote of the committee were required?

5. As a result of a crisis in the Middle East total oil production of one of the OPEC countries has declined by 30%. This country decides to allocate current oil production among 3 Japanese, 2 European, and 5 American tankers. The country decides to select 6 tankers randomly from these 10 tankers. What is the probability that the chosen tankers will include:
   a. no Japanese tankers?
   b. 3 American, 2 Japanese, and one European tanker?
   c. one American tanker?

6. Mrs. Betsy Coleman is picked as the lucky contestant in a famous California television program called "The Price Is Almost Right." As a contestant, she can win any 3 of the following 6 prizes: a trip around the world, 2 new Toyota Celicas, or 3 living room sets. Suppose Mrs. Coleman's lucky streak

is at its height. Determine the probability that in her 3 random choices she wins the following:

a.   A car, a living room set, and a trip around the world.
b.   A trip around the world.
c.   2 cars.
d.   If you were the contestant, what 3 choices would you prefer? Evaluate the probability of getting these 3 choices.

7.   Tax returns submitted annually can be classified into three categories: (a) tax returns with refund, (b) tax returns with penalty, and (c) balanced tax returns (no refund or penalty). In a given week, 20 tax returns are submitted of which 5 require refunds, 6 require penalties, and the remainder are balanced. What is the probability that a tax inspector will receive 2 tax returns requiring refunds and 3 requiring penalties from a sample of 6 returns selected randomly?

3.7
THE POISSON DISTRIBUTION

Another useful probability function is the Poisson distribution, named for the Frenchman who developed it during the first half of the nineteenth century.[9] We will discuss the Poisson distribution first as a distribution in its own right, which is by far the most important use and which has many fruitful applications in a wide variety of fields. Then we will discuss the Poisson distribution as an approximation to the binomial distribution.

The Poisson Distribution Considered in Its Own Right

The Poisson distribution has been usefully employed to describe the probability functions of such phenomena as

● product demand
● demands for service
● numbers of telephone calls that come through a switchboard
● numbers of accidents
● numbers of traffic arrivals (such as trucks at terminals, airplanes at airports, ships at docks, and passenger cars at toll stations)
● numbers of defects observed in various types of lengths, surfaces, or objects.

   All of the preceding illustrations have two elements in common.

[9] Siméon Denis Poisson (1781–1840), was particularly noted for his applications of mathematics to the fields of electrostatics and magnetism. He wrote treatises in probability, calculus of variations, Fourier's series, and other areas.

WHOLE #'s

Elements of a Poisson distribution

- The given occurrences can be described in terms of a discrete random variable, which takes on values 0, 1, 2, and so forth.

- There is some rate that characterizes the process producing the outcome. That **rate** is the number of occurrences per interval of *time* or *space*.

For example, product demand can be characterized by the number of units purchased in a specified period; the number of defects in a specified length of electrical cable can be counted. Product demand may be viewed as a process that produces random occurrences in continuous time; the observance of defects is a process that produces random occurrences in a continuum of space. In cases such as the defects example, the continuum may be one of *area* or *volume* as well as *length*. Thus, there may be a count of the number of blemishes in areas of sheetmetal used for aircraft or a count of the number of a certain type of microscopic particle in a unit of volume such as a cubic centimeter of a solution. We can indicate the general nature of the process that produces a Poisson probability distribution by examining the occurrence of defects in a length of electrical cable. The length of cable has some rate of defects per interval: say, two defects per meter. If the entire length of cable is divided into sub-intervals of one millimeter each, then we might make the following assumptions:

Assumptions

1. The probability that exactly one defect occurs in each subinterval is a small number that is constant for each such subinterval.

2. The probability of two or more defects in a millimeter is so small that it may be considered to be zero.

3. The number of defects that occur in a millimeter does not depend on where that subinterval is located.

4. The number of defects that occur in a subinterval does not depend on the number of defects in any other nonoverlapping subinterval.

Although the subinterval was a unit of *length* in the preceding example, analogous sets of assumptions would characterize examples in which the sub-interval is a unit of *area*, *volume*, or *time*.

The Nature of the Poisson Distribution

As indicated previously,

the Poisson distribution results from occurrences that can be described by a discrete random variable.

This random variable, denoted X, can take on values $x = 0, 1, 2, \ldots$ (where the three dots mean *ad infinitum*). That is, X can take on the values of all non-negative integers. The probabillity of exactly x occurrences in the Poisson distribution is

(3.7)
$$f(x) = \frac{\mu^x e^{-\mu}}{x!} \quad \text{for } x = 0, 1, 2, \ldots$$

where μ is the mean number of occurrences per interval and $e = 2.71828\ldots$ (the base of the Naperian or natural logarithm system).

We can see from equation 3.7 that the Poisson distribution has a single parameter symbolized by the Greek lower-case letter μ (mu). If we know the value of μ, we can write out the entire probability distribution.

> The parameter μ can be interpreted as the average number of occurrences per interval of time or space that characterizes the process producing the Poisson distribution.

(The average referred to here is the arithmetic mean.) Thus, μ may represent an average of three units of demand per day, 5.3 demands for service per hour, 1.2 aircraft arrivals per five minutes, 1.5 defects per 10 feet of electrical cable, and so on.

In order to illustrate how probabilities are calculated in the Poisson distribution, we consider the following example. A study revealed that the number of telephone calls per minute coming through a certain switchboard between 10:00 A.M. and 11:00 A.M. on business days is distributed according to the Poisson probability function with an average μ of 0.4 calls per minute. What is the probability distribution of the number of telephone calls per minute during the specified time period?

Let X represent the random variable "number of telephone calls per minute" during the given time period. Then $\mu = 0.4$ calls per minute is the parameter of the Poisson probability distribution of this random variable. The probability that no calls will occur (come through the switchboard) in a given minute is obtained by substituting $x = 0$ in the Poisson probability function, equation 3.7. Hence,

(3.8)
$$P(X = 0) = f(0) = \frac{(0.4)^0 e^{-0.4}}{0!}$$

Since $(0.4)^0 = 1$ and $0! = 1$, equation 3.8 becomes simply

(3.9)
$$f(0) = e^{-0.4} = 0.670$$

The value 0.670 for $f(0)$ can be found in Table A-10 of Appendix A, where exponential functions of the form e^x and e^{-x} are tabulated for values of x from 0.00 to 6.00 at intervals of 0.10.

Continuing with the calculation of the Poisson probability distribution, we find the probability of exactly one call in a given minute by substituting $x = 1$ in equation 3.7. Hence, $f(1)$ is given by

(3.10)
$$P(X = 1) = f(1) = \frac{(0.4)^1 e^{-0.4}}{1!} = (0.4)(0.670) = 0.268$$

To find the other values of $f(x)$, we can use Table A-3 of Appendix A, which lists values of the cumulative distribution function for the Poisson distribution. That is, values of

$$F(c) = P(X \leqslant c) = \sum_{x=0}^{c} f(x)$$

or the probabilities of c or fewer occurrences, are provided for selected values of the parameter μ. As in Table A-1 for the binomial cumulative distribution, probabilities such as $1 - F(c)$ or $a \leqslant f(x) \leqslant b$ can be obtained by appropriate manipulation of the tabulated values.

The use of Table A-3 will be illustrated in terms of our phone call example. To obtain the probability of no calls in a given minute, using $c = 0$ and $\mu = 0.4$ in Table A-3, we find the value of $F(0) = 0.670$. Of course, this is also the value of $f(0)$, since the probability of zero or fewer occurrences equals the probability of zero occurrences. Therefore, as before, $f(0) = 0.670$.

We find the probability of exactly one telephone call per minute by subtracting the probability of no calls from the probability of one or fewer calls, that is,

$$f(1) = F(1) - F(0) = 0.938 - 0.670 = 0.268$$

Similarly, using Table A-3, we find the values of $f(2)$, $f(3)$, and $f(4)$:

$$f(2) = F(2) - F(1) = 0.992 - 0.938 = 0.054$$

$$f(3) = F(3) - F(2) = 0.999 - 0.992 = 0.007$$

$$f(4) = F(4) - F(3) = 1.000 - 0.999 = 0.001$$

Although, as indicated earlier, the random variable X in the Poisson distribution takes on the values $0, 1, 2, \ldots, F(4) = 1.00$ in this problem. This means that the probabilities of $5, 6, \ldots$, occurrences are so small that they would appear as zero when rounded to three decimal places.

The required probability distribution for this problem is given in Table 3-7. Several other illustrations of the use of the Poisson distribution are given in Examples 3-15, 3-16, and 3-17.

TABLE 3-7
Poisson probability distribution of the number of telephone calls per minute coming through a certain switchboard between 10:00 A.M. and 11:00 A.M. on business days

Number of Calls x	Probability $f(x)$
0	0.670
1	0.268
2	0.054
3	0.007
4	0.001
	1.000

EXAMPLE 3-15

On weekdays at a certain small airport, airplanes arrive at an average rate of three for the one-hour period 1:00 P.M. to 2:00 P.M. If these arrivals are distributed according to the Poisson probability distribution, what are the probabilities that
a. exactly zero airplanes will arrive between 1:00 P.M. and 2:00 P.M. next Monday?
b. either one or two airplanes will arrive between 1:00 P.M. and 2:00 P.M. next Monday?
c. a total of exactly two airplanes will arrive between 1:00 P.M. and 2:00 P.M. during the next three weekdays?

In this problem, we may use the parameter $\mu = 3$ arrivals per day for the period 1:00 P.M. to 2:00 P.M. Let X represent the random variable "number of arrivals during the specified time period." The mathematical solutions are given for parts (a), (b), and (c) to illustrate the theory involved. However, the answers may also be determined by looking up values in Table A-3 of Appendix A as indicated.
a. The random variable X follows the Poisson distribution with the parameter $\mu = 3$. Thus,

$$P(X = 0) = f(0) = \frac{3^0 e^{-3}}{0!} = 0.050$$

This value may be obtained from Table A-3 of Appendix A for $\mu = 3, c = 0$. We note that $f(0) = F(0)$.
b. Since exactly one arrival and exactly two arrivals are mutually exclusive events, we have, by the addition rule,

$$P(X = 1 \text{ or } X = 2) = f(1) + f(2) = \frac{3^1 e^{-3}}{1!} + \frac{3^2 e^{-3}}{2!} = 0.373$$

This value can be obtained from Table A-3 of Appendix A for $\mu = 3$. The required probability is $F(2) - F(0) = 0.423 - 0.050 = 0.373$.
c. A total of exactly two arrivals in three weekdays during the period 1:00 P.M.–2:00 P.M. can be obtained, for example, by having two arrivals on the first day, none on the second day, and none on the third day during the specified one-hour period. The total number of ways in which the event in question can occur is shown in Table 3-8.

TABLE 3-8
Possible ways of obtaining a total of exactly 2 arrivals in 3 weekdays

	Number of Arrivals	
Day 1	Day 2	Day 3
2	0	0
0	2	0
0	0	2
1	1	0
1	0	1
0	1	1

Let P_2 represent the required probability. Using the multiplication and addition rules, and again using the parameter $\mu = 3$ arrivals per day during the period 1:00 P.M.– 2:00 P.M., we have

$$P_2 = 3[f(2)][f(0)]^2 + 3[f(1)]^2[f(0)]$$

$$= 3\left(\frac{3^2 e^{-3}}{2!}\right)\left(\frac{3^0 e^{-3}}{0!}\right)^2 + 3\left(\frac{3^1 e^{-3}}{1!}\right)^2\left(\frac{3^0 e^{-3}}{0!}\right)$$

$$= \frac{81}{2} e^{-9} = 0.005$$

The solution is greatly simplified if we change the time interval for which the parameter μ is stated. This has the effect of changing the random variable in the problem. Thus, if μ = three arrivals *per day* during the period 1:00 P.M.–2:00 P.M., then μ = nine arrivals *per three days* during the same time period. The probability of exactly two arrivals in three weekdays during the given one-hour period can then be obtained by computing $P(X = 2)$, where X is a Poisson-distributed random variable denoting the number of arrivals *per three days*. The required probability is, therefore, obtained by simply computing $f(2)$ in a Poisson distribution with the parameter $\mu = 9$.

$$P_2 = f(2) = \frac{9^2 e^{-9}}{2!} = \frac{81}{2} e^{-9} = 0.005$$

This value can be obtained from Table A-3 of Appendix A for $\mu = 9$. The probability is given by $F(2) - F(1) = 0.006 - 0.001 = 0.005$.

This problem illustrates the point that considerable simplification of computations for Poisson processes can often be accomplished by convenient choice of parameters.

Appropriateness   We should note a few points concerning the appropriateness of the Poisson distribution in Example 3-15. We stated at the beginning of the problem that the airplane arrivals were distributed according to the Poisson distribution. Whether it is appropriate to consider the past arrival distribution as a Poisson distribution during the specified time periods depends on the nature of the past data. Actual relative arrival frequencies can be tabulated and compared with the theoretical probabilities given by a Poisson distribution. Tests of "goodness of fit" for judging the closeness of actual and theoretical frequencies are discussed in Chapter 8.

In practice, the question often arises whether a given mathematical model is applicable in a certain situation. This requires careful examination of whether the underlying assumptions of the model are likely to be fulfilled by the real-world phenomena. For example, suppose certain cargo deliveries are made either on Mondays or Tuesdays between 1:00 P.M. and 2:00 P.M. Assuming that if a delivery is made on Monday, it will not be made on Tuesday, the fourth assumption of a Poisson process (given on page 146) is clearly violated. That is, the number of arrivals during the one-hour period on Tuesday *depends* on the number of arrivals during the corresponding period on Monday, and vice versa. Furthermore, if the nature of the aircraft arrivals is such that Monday and Tuesday always have more arrivals between 1:00 P.M. and 2:00 P.M.

than do other weekdays, then the third assumption is violated. That is, if we were to count arrivals for the one-hour period for a given day (or two days, and so forth), then the number of occurrences obtained would depend on the day on which the count was begun.

> The assumptions of a probability distribution rarely are met perfectly by a real-world process. Actual comparison of *the data generated by a process* with *the probabilities of the theoretical distribution* is the best way of determining the appropriateness of the distribution.

Determining appropriateness of a distribution

Of course, even if a given mathematical model (or other type of model) has provided a good description of past data, there is no guarantee that this state of affairs will continue. The analyst must be alert to changes in the environment that would make the model inapplicable. Experience in a given field aids considerably in judging whether so great a departure from assumptions has occurred that a model may no longer be applicable.[10]

EXAMPLE 3-16

A department store has determined in connection with its inventory control system that the demand for a certain brand of portable radio was Poisson-distributed with the parameter $\mu = 4$ per day.
a. Determine the probability distribution of the daily demand for this item.
b. If the store stocks 5 of these items on a particular day, what is the probability that the demand will be greater than the supply?

SOLUTION

Let X represent the random variable "number of portable radios of this brand sold per day."
a. The probability distribution of X is

x	$f(x)$
0	0.018
1	0.074
2	0.146
3	0.195
4	0.196
5	0.156
6	0.104
7	0.060
8	0.030
9	0.013
10	0.005
11	0.002
⋮	⋮

---

[10] An appropriate thought here is perhaps contained in the anonymous bit of advice. "Good judgment comes from experience, and experience comes from poor judgment."

The sum of the probabilities for demand from zero through 11 units is 0.999. Therefore, the sum of the probabilities for 12 or more units is only 0.001.

The probabilities can be obtained from Table A-3 of Appendix A using the relationship $f(x) = F(x) - F(x - 1)$.

b. The probability that demand will be greater than 5 units is the complement of the probability that it will be 5 units or less. Thus, from Table A-3 of Appendix A we have

$$P(x > 5) = 1 - F(5) = 1 - 0.785 = 0.215$$

EXAMPLE 3-17

A check of a hundred 50-pound crates of oranges from Florida reveals the following data on the number of spoiled oranges in each crate:

Spoiled Oranges in Crate	Number of Crates
0	38
1	35
2	18
3	7
4	2
	100

It is hypothesized that the number of spoiled oranges per crate has a Poisson distribution with a mean of 1.0. Calculate the probabilities of finding zero, one, two, three, and four spoiled oranges in a randomly selected crate and compare these results with the above data. Is the hypothesis valid?

SOLUTION

x	$f(x)$	Hypothetical Value in 100	Actual
0	0.368	37	38
1	0.368	37	35
2	0.184	18	18
3	0.061	6	7
4	0.015	2	2

Because the actual frequencies are quite close to the theoretical ones, we can accept the hypothesis as valid.

The Poisson Distribution as an Approximation to the Binomial Distribution

The foregoing discussion concerned the use of the Poisson probability function as a distribution in its own right. We turn now to a consideration of the Poisson distribution as an approximation to the binomial distribution.

We saw that the Bernoulli process gives rise to a two-parameter probability function, the binomial distribution. Since computations involving the binomial distribution become quite tedious when n is large, it is useful to have a simple method of approximation.

> The Poisson distribution is particularly suitable as an approximation to the binomial distribution when n is large and p is small.

Assume in the expression for $f(x)$ of the binomial distribution that, as n is permitted to increase without bound, p approaches zero in such a way that np remains constant. Let us denote this constant value for np as μ (which denotes the mean number of successes in n trials). Under these assumptions, it can be shown that the binomial expression for $f(x)$ approaches the value

$$f(x) = \frac{\mu^x e^{-\mu}}{x!}$$

where $\mu = np$ and e is the base of the natural logarithm system. Thus, we can see from equation 3.7 that the value approached by the binomial distribution under the given conditions is the value of the Poisson distribution. Hence, the Poisson distribution can be used as an approximation to the binomial probability function. In this context, the Poisson distribution is similar to the binomial distribution, because it gives the probability of observing x successes in n trials of an experiment, where p is the probability of success on a single trial. That is x, n, and p are interpreted in the same way as in the binomial distribution.

Because of the assumptions underlying the derivation of the Poisson distribution from the binomial distribution, the approximations to binomial probabilities are best when n is large and p is small.

> A frequently used rule of thumb is that the Poisson approximation to the binomial distribution is appropriate when $p \leqslant 0.05$ and $n \geqslant 20$.

However, the Poisson distribution sometimes provides surprisingly close approximations even in cases when n is not large nor p very small. As an illustration of how these approximations may be carried out, we return to the problem of sampling the population of 950 adults and 50 children. The probability of observing no children in a random sample of 5 persons drawn with replacement was previously computed from the binomial distribution. We now compute the same probability using the Poisson distributiion. Since n is only 5 in this problem, this is not an ideal situation for the use of the Poisson distribution for approximating binomial probabilities. Rather, it is an example of the surprisingly small errors observed in certain cases, even though n is small, and it is used here simply to carry out the arithmetic for a familiar illustration.

The binomial parameters in this problem were $p = 0.05$ and $n = 5$. Therefore,

$$\mu = np = 5 \times 0.05 = 0.25$$

Thus, in the Poisson distribution, the probability of no successes (children) is

$$f(0) = \frac{(0.25)^0 e^{-0.25}}{0!} = e^{-0.25}$$

From Table A-3 of Appendix A with $c = 0$ and $\mu = 0.25$, we find the value of $F(0)$, which in this case is equal to $f(0)$ (since the probability of zero or fewer successes equals the probability of zero successes). Therefore,

$$f(0) = 0.779$$

This figure is the same in the first two decimal places as the corresponding number (0.7738) obtained from the binomial probability (see section 3.6).

However, the percentage errors would be much larger for the other terms of the binomial distribution, representing probabilities of one, two, or more, successes. It is recommended, therefore, that the Poisson approximations not be used unless the conditions for n and p meet the limits outlined in the rule of thumb.

The parameter $\mu = np$ can be interpreted as the average number of successes per sample of size n.

Since p is the probability of success per trial and n is the number of trials, multiplication of n by p gives the average number of successes per n trials. In terms of the foregoing problem, we can interpret μ as a long-run relative frequency. The proportion of children in the population is $p = 0.05$, and a random sample of $n = 5$ persons was drawn with replacement from this population. Suppose samples of size $n = 5$ were repeatedly drawn with replacement from the same population and the number of children was recorded for each sample. It can be proven mathematically, and it seems intuitively reasonable, that the *average proportion* of children per sample of 5 persons is equal to $p = 0.05$. Furthermore, it follows that the *average number* of children per sample of 5 persons is equal to $np = 5(0.05) = 0.25$ children. The average referred to here is the arithmetic mean, obtained by totaling the proportions or numbers of children for all samples and dividing by the number of samples.

Example 3-18 represents a more justifiable use of the Poisson approximation to binomial probabilities than the preceding illustration, which involved a small sample size.

EXAMPLE 3-18

An oil exploration firm is formed with enough capital to finance 20 ventures. The probability of any exploration being successful is 0.10. What are the firm's chances of
a.   exactly one successful exploration?
b.   at least one successful exploration?
c.   2 or fewer successful explorations?
d.   3 or more successful explorations?
Assume that the population of possible explorations is sufficiently large to warrant binomial probability calculations.

a.   Let $p = 0.10$ stand for the probability that an exploration will be successful and X *SOLUTIONS*
for the number of successful explorations in 20 trials. Using the binomial distribution with parameters $p = 0.10$ and $n = 20$, we find that the probability of exactly one successful exploration is

$$P(X = 1) = f(1) = \binom{20}{1}(0.9)^{19}(0.1)^1$$

This probability may be determined from Table A-1 of Appendix A as

$$P(X = 1) = F(1) - F(0) = 0.3917 - 0.1216 = 0.2701$$

An approximation to this probability is given by the Poisson distribution with parameter

$$\mu = np = 20(0.10) = 2$$

The Poisson probability of exactly one success is

$$f(1) = \frac{2^1 e^{-2}}{1!}$$

which may be determined from Table A-3 of Appendix A as

$$P(X = 1) = F(1) - F(0) = 0.406 - 0.135 = 0.271$$

Thus, the percentage error is about 1 in 270, or about 0.4%.
Using Tables A-1 and A-3 of Appendix A for parts (b), (c), and (d), we have the following:

	Binomial	Poisson
b.	$P(X \geqslant 1) = 1 - F(0)$	$P(X \geqslant 1) = 1 - F(0)$
	$= 1 - 0.1216 = 0.8784$	$= 1 - 0.135 = 0.865$
c.	$P(X \leqslant 2) = F(2) = 0.6769$	$P(X \leqslant 2) = F(2) = 0.677$
d.	$P(X \geqslant 3) = 1 - F(2)$	$P(X \geqslant 3) = 1 - F(2)$
	$= 1 - 0.6769 = 0.3231$	$= 1 - 0.677 = 0.323$

EXERCISES 3.7

1. Airplanes arrive at a certain airport at the rate of 4 per hour according to a Poisson distribution.
   a. What is the probability that there will be exactly 5 arrivals in a given one-hour period?
   b. What is the probability that there will be exactly 10 arrivals in a 2 hour period?
   c. Explain why it is less probable to have 10 arrivals in 2 hours than 5 arrivals in one hour.

2. In a textile manufacturing process, an average of one flaw per 100 square feet of linen material has appeared. What is the probability that a roll measuring 60 feet by 10 feet will have no flaws?

3. An average of 2.8 telephone calls per minute is made through a central switchboard according to the Poisson distribution. What is the probability that during a given minute
   a. exactly 2 calls will be made?
   b. at least 2 calls will be made?
   c. at most 2 calls will be made?

4. A barbershop has on the average 12 customers during the lunch break from 12:00–1:00 P.M. Customers arrive according to the Poisson distribution.
   a. What is the probability that the barbershop will have exactly 12 customers during this period?
   b. What is the probability that the barbershop will have more than 10 customers?
   c. What is the probability that the barbershop will have fewer than 6 customers?

5. The probability that a child from one to 10 years old will die from a certain rare disease during a period of one year has been estimated to be 0.0001. If a life insurance company has 10,000 policyholders in this age group, find the approximate probability that the company must pay at least 4 claims in one year because of death from this disease. (Use the Poisson distribution.)

6. The Science Computer Center uses five computers to generate scientific forecasts. The probability that one computer will not work on a given day is 0.05.
   a. Assume that the binomial distribution is the appropriate model.
      (1) What is the probability that all computers will be working on a given day?
      (2) What is the probability that at least three computers will not be working on a given day?
   b. Use the Poisson approximation to solve part a and compare your results.

Although the calculations and table lookups for the binomial, Poisson and hypergeometric probability distributions have been explained in the preceding sections in order to emphasize basic principles, these distributions are easily available from many statistical software packages.

Minitab commands and subcommands for printing out the probability density function (pdf) and cumulative distribution function (cdf) for a binomial distribution with $n = 5$ and $p = 0.50$ are shown in Table 3-9. Probabilities can be obtained for single terms by typing the desired number of successes after pdf or cdf.

The Minitab command and subcommand for printing out the pdf and cdf of a Poisson distribution with a mean equal to 3 are shown in Table 3-10.

TABLE 3-9

Minitab output for the PDF and CDF of a binomial distribution with $n = 5$ and $p = 0.50$

```
MTB > PDF;
SUBC> BINOMIAL N=5, P=0.5,

    BINOMIAL WITH N = 5 P = 0.500000
       K            P( X = K )
       0               0.0313
       1               0.1562
       2               0.3125
       3               0.3125
       4               0.1562
       5               0.0313


MTB > CDF;
SUBC> BINOMIAL N=5, P=0.5,

    BINOMIAL WITH N = 5 P = 0.500000
       K  P( X LESS OR = K )
       0               0.0313
       1               0.1875
       2               0.5000
       3               0.8125
       4               0.9688
       5               1.0000
```

TABLE 3-10
Minitab output for the PDF and CDF of a Poisson distribution with a mean equal to 3

```
MTB > PDF;
SUBC> POISSON 3.

   POISSON WITH MEAN = 3.000
      K          P( X = K )
      0            0.0498
      1            0.1494
      2            0.2240
      3            0.2240
      4            0.1680
      5            0.1008
      6            0.0504
      7            0.0216
      8            0.0081
      9            0.0027
     10            0.0008
     11            0.0002
     12            0.0001
     13            0.0000


MTB > CDF;
SUBC> POISSON 3.

   POISSON WITH MEAN = 3.000
      K     P( X LESS OR = K )
      0            0.0498
      1            0.1991
      2            0.4232
      3            0.6472
      4            0.8153
      5            0.9161
      6            0.9665
      7            0.9881
      8            0.9962
      9            0.9989
     10            0.9997
     11            0.9999
     12            1.0000
```

TABLE 3-11
EASYSTAT output for the PDF and CDF of a hypergeometric distribution with $S = 5$, $N = 10$, and $n = 5$

HYPERGEOMETRIC PROBABILITIES

N = 10,	S = 5,	n = 5,
r	PROBABILITY	CUMUL. PROB.
0.	0.00396825397	0.00396825397
1.	0.09920634921	0.10317460317
2.	0.39682539683	0.5
3.	0.39682539683	0.89682539683
4.	0.09920634921	0.99603174603
5.	0.00396825397	1.

Although Minitab does not have such calculations for the hypergeometric distribution, EASYSTAT does provide this output. The pdf and cdf of a hypergeometric distribution with $X = 5$, $N = 10$, and $n = 5$ are shown in Table 3-11. The EASYSTAT symbols that are equivalent to the corresponding symbols used for the hypergeometric distribution in section 3.6 are $S = X$, $r = x$, $N = N$, and $n = n$.

3.9
SUMMARY MEASURES FOR PROBABILITY DISTRIBUTIONS

In Chapter 1 we discussed summary, or descriptive, measures for empirical frequency distributions. We now turn to the corresponding measures for theoretical frequency distributions, that is, for probability distributions. These **summary measures for probability distributions** are essential components of modern quantitative techniques employed as aids for decision making under conditions of uncertainty.

Earlier in this chapter, we considered certain probability distributions for discrete random variables as appropriate mathematical models for real-world situations under specific sets of assumptions. Sometimes, from the nature of a problem, it is relatively easy to specify a suitable probability model. In other situations, the appropriate model is suggested only after substantial numbers of observations have been taken and empirical frequency distributions have been constructed. Whatever the method by which we arrive at probability distributions, we must be able to capture their salient properties in a few summary measures. These measures are the subject of the next two sections.

3.10
EXPECTED VALUE OF A RANDOM VARIABLE

Suppose the following game of chance were proposed to you: A fair coin is tossed. If it lands "heads," you win $10; if it lands "tails," you lose $5. What is the average amount that you would win per toss?

On any particular toss, you will either win $10 or lose $5. However, let us think in terms of a repeated experiment in which we toss the coin and play the game many times. Since the probability assigned to the event "heads" is $\frac{1}{2}$ and to the event "tails" $\frac{1}{2}$, in the long run, you would win $10 on half of the tosses and lose $5 on half. Therefore, the average (arithmetic mean) winnings per toss would be obtained by weighting both of the outcomes—$10 and $-$5—by $\frac{1}{2}$ to yield a weighted mean of $2.50 per toss. In terms of equation 1.4, the weighted mean is

$$\bar{X}_w = \frac{\Sigma wX}{\Sigma w} = \frac{\$10(\frac{1}{2}) + (-\$5)(\frac{1}{2})}{\frac{1}{2} + \frac{1}{2}} = \$2.50 \text{ per toss}$$

Of course, when the weights are probabilities in a probability distribution, as in this case, the formula can be written without showing the division by the sum of the weights.

The average of $2.50 per toss is referred to as the **expected value** of the winnings. Note that on a single toss, only two outcomes are possible, namely win $10 or lose $5. If these two possible outcomes are viewed as the possible values of a random variable, which occur with probabilities of $\frac{1}{2}$ each, then the expected value of the random variable is the mean of its probability distribution. More formally, if X is a discrete random variable that takes on the value x with probability $f(x)$, then the expected value of X, denoted $E(X)$, is

(3.11)
$$E(X) = \sum_x xf(x)$$

The expected value of a discrete random variable is obtained by multiplying (1) each value that the random variable can assume by (2) the probability of occurrence of that value; then all of these products are totaled.

Since the expected value is used frequently as a measure of central tendency for probability distributions, $E(X)$ is given as a simpler symbol, μ.

Hence, equation 3.11 can be rewritten as

$$\mu = E(X) = \sum_x xf(x)$$

The calculation of the expected value for the problem of tossing a coin is shown in Table 3-12. This calculation is similar to that for the mean of an empirical frequency distribution.

TABLE 3-12
Calculation of the expected value for the coin-tossing problem

Value of Winnings x	Probability $f(x)$	Weighted Winnings $xf(x)$
$10	$\frac{1}{2}$	$5.00
−5	$\frac{1}{2}$	−2.50
	1	$2.50

$$\mu = E(X) = \sum_x xf(x) = \$2.50$$

The expected value has a wide variety of applications in situations involving uncertain outcomes. Example 3-19 gives a simple illustration, in which the uncertainty is summarized in terms of a probability distribution of death for a particular type of insurance policyholder. Such probability distributions are ordinarily based on a large sample of observed experience, that is, on past relative frequencies of mortality. On the other hand, business problems often involve "one-time" decisions; with such problems, uncertainties can be summarized in terms of relevant subjective probability distributions. Example 3-20 illustrates such a situation.

EXAMPLE 3-19

Suppose an insurance company offers a 45-year-old man a $1000 one-year term insurance policy for an annual premium of $12. Assume that the number of deaths per 1000 is 5 for persons in this age group. What is the expected gain for the insurance company on a policy of this type?

We may think of this problem as representing a chance situation in which there are two possible outcomes, (1) the policy purchaser lives or (2) he dies during the year. Let X be a random variable denoting the dollar gain to the insurance company for these 2 outcomes. The probability that the man will live through the year is 0.995. In this case, the insurance company collects the premium of $12. The probability that the policy purchaser will die during the year is 0.005. In this case, the company has collected a premium of $12 but must pay the claim of $1000, for a net gain of −$988. Thus, X takes on the values $12 and −$988 with respective probabilities 0.995 and 0.005. The calculation of expected gain for the insurance company is displayed in Table 3-13.

A couple of points may be noted. First, in insuring only one person, the company would not realize the expected gain but would have either the gain of $12 or the loss of $988. Hence, in order to realize expected gains, insurance companies "play the averages" by insuring large numbers of individuals. Second, in setting a premiun for this policy, the insurance company would have to take into account usual expenses of doing business as well as the expected gain calculation.

TABLE 3-13
Calculation of expected gain for an insurance company on a one-year term policy

Outcome	x	$f(x)$	$xf(x)$
Policy holder lives	$12	0.995	$11.94
Policy holder dies	−$988	0.005	−$4.94
		1.000	$7.00

$$E(X) = \sum_x xf(x) = \$7.00$$

EXAMPLE 3-20

Lawrence Carlton, a financial manager for a corporation, is considering two competing investment proposals. For each of these proposals, he has carried out an analysis in which he has determined various net profit figures and has assigned subjective probabilities to the realization of these returns. For proposal A, his analysis shows net profits of $20,000, $30,000, or $50,000 with respective probabilities 0.2, 0.4, and 0.4. For proposal B, he concludes that there is a 50% chance of a successful investment, estimated as producing net profits of $100,000, and of an unsuccessful investment, estimated as a break-even situation involving $0 of net profit. Assuming that each proposal requires the same dollar investment, which is preferable solely from the standpoint of expected monetary return?

Denoting the expected net profit on these proposals as $E(A)$ and $E(B)$, we have

$$E(A) = (\$20,000)(0.2) + (\$30,000)(0.4) + (\$50,000)(0.4) = \$36,000$$

$$E(B) = (\$0)(0.5) + (\$100,000)(0.5) = \$50,000$$

Mr. Carlton would maximize expected net profit by accepting proposal B.

3.11
VARIANCE OF A RANDOM VARIABLE

As we have seen, the expected value of a random variable is analogous to the arithmetic mean of a frequency distribution of data. Similarly, the variability of a random variable may be measured in the same general way as the variability of a frequency distribution. In section 1.15 the variability of the observations in a population of data was referred to as the *variance*; the square root of the variance was called the *standard deviation*. The same terms are used for random variables. The **variance of a random variable** is the average (expected value) of the squared deviations from the expected value. The **standard deviation of a random variable** is the square root of the variance. The variance of a random variable denoted σ_X^2, VAR(X), or simply σ^2 may be written as $E(X - \mu)^2$ and may be calculated for discrete random variables by the following expression:

(3.12)
$$\sigma^2 = \text{VAR}(X) = \sum_x (x - \mu)^2 f(x)$$

A mathematically equivalent formula that is easier for purposes of calculation is

(3.13) $$\sigma^2 = \text{VAR}(X) = E(X^2) - [E(X)]^2$$

where

$$E(X^2) = \sum_x x^2 f(x)$$

and

$$E(X) = \sum_x x f(x)$$

In Examples 3-21 and 3-22, we illustrate the calculation of the variance and standard deviation for a few random variables. The calculation of the expected value as an intermediate step is also shown in Example 3-22.

EXAMPLE 3-21

Compute the mean and variance for the total obtained on the uppermost faces in a roll of two unbiased dice.

SOLUTION

Let X denote the specified total on the two dice. Then

$$\mu = \sum_{x=2}^{12} x f(x)$$

$$= 2(\tfrac{1}{36}) + 3(\tfrac{2}{36}) + 4(\tfrac{3}{36}) + 5(\tfrac{4}{36}) + 6(\tfrac{5}{36}) + 7(\tfrac{6}{36}) + 8(\tfrac{5}{36}) + 9(\tfrac{4}{36}) + 10(\tfrac{3}{36})$$
$$\quad + 11(\tfrac{2}{36}) + 12(\tfrac{1}{36})$$

$$= 7$$

$$\sigma^2 = \sum_{x=2}^{12} x^2 f(x) - \mu^2$$

$$= 4(\tfrac{1}{36}) + 9(\tfrac{2}{36}) + 16(\tfrac{3}{36}) + 25(\tfrac{4}{36}) + 36(\tfrac{5}{36})$$
$$\quad + 49(\tfrac{6}{36}) + 64(\tfrac{5}{36}) + 81(\tfrac{4}{36}) + 100(\tfrac{3}{36}) + 121(\tfrac{2}{36}) + 144(\tfrac{1}{36}) - (7)^2$$

$$= 54\tfrac{5}{6} - 49 = 5\tfrac{5}{6}$$

EXAMPLE 3-22

On the basis of past experience, a store manager assesses the probability distribution for the number of units of a particular product sold daily. The distribution is as shown in the first two columns of Table 3.14. What are the expected value and standard deviation of the number of units sold daily?

SOLUTION

The calculation of the expected value and standard deviation from the definitional formulas equations 3.11 and 3.12, is given in Table 3-14. Also shown is the alternative calculation for the standard deviation, following from (3.13).

TABLE 3-14
Expected value and standard deviation of number of units of a product sold

Number of Units Sold x	Probability $f(x)$	$xf(x)$	$x - \mu$	$(x - \mu)^2$	$(x - \mu)^2 f(x)$
0	0.15	0.00	-1.85	3.4225	0.513375
1	0.25	0.25	-0.85	0.7225	0.180625
2	0.30	0.60	0.15	0.0225	0.006750
3	0.20	0.60	1.15	1.3225	0.264500
4	0.10	0.40	2.15	4.6225	0.462250
	1.00	1.85			1.427500

$$\mu = E(X) = \sum_{x=0}^{4} xf(x) = 1.85 \text{ units sold}$$

$$\sigma^2 = \text{VAR}(X) = E(X - \mu)^2 = \sum_{x=0}^{4} (x - \mu)^2 f(x) = 1.427500$$

$$\sigma = \sqrt{\sigma^2} = \sqrt{1.427500} = 1.19 \text{ units sold}$$

Alternative Calculation of the Standard Deviation

$$E(X^2) = \sum_{x=0}^{4} x^2 f(x) = 0(0.15) + 1(0.25) + 4(0.30) + 9(0.20) + 16(0.10) = 4.85$$

$$\sigma^2 = E(X^2) - [E(X)]^2 = 4.85 - (1.85)^2 = 1.4275$$

$$\sigma^2 = \sqrt{1.4275} = 1.19 \text{ units sold}$$

EXAMPLE 3-23

Ventures Limited, Inc., is considering a proposal to develop a new calculator. The initial cash outlay would be $1 million, and development time would be one year. If successful, the firm anticipates that revenues over the 5-year life cycle of the product will be $1.5 million. If moderately successful, revenues will reach $1.2 million. If unsuccessful, the firm anticipates zero cash inflows. The firm assigns the following probabilities to the 5-year prospects for this product: successful, 0.60; moderately successful, 0.30; and unsuccessful, 0.10. What are the expected net profit and standard deviation of revenues? We will ignore the time value of money in this calculation.

SOLUTION

Let X denote profit for the 5-year period, equal to total revenues for the 5-year period less development costs.

$$E(X) = \$500,000(0.6) + (\$200,000)(0.3) + (-\$1,000,000)(0.1)$$
$$= \$260,000$$

$$\sigma^2 = (\$500,000 - \$260,000)^2(0.6) + (\$200,000 - \$260,000)^2(0.3)$$
$$+ (-\$1,000,000 - \$260,000)^2(0.1)$$
$$= 19.44 \times 10^{10}$$

$$\sigma = \sqrt{19.44 \times 10^{10}} = \$440,908$$

We now discuss some of the important properties of expected values and variances of random variables. It can be shown that

(3.14) $$E(X_1 + X_2 + \cdots + X_N) = E(X_1) + E(X_2) + \cdots + E(X_N)$$

> The expected value of a sum of N random variables is equal to the sum of the expected values of these random variables.

A somewhat analogous relationship holds for variances of *independent* random variables. If $X_1, X_2, \ldots, X_N$ are N *independent* random variables, then

(3.15) $$\begin{aligned} \mathrm{VAR}(X_1 + X_2 + \cdots + X_N) = {} & \mathrm{VAR}(X_1) + \mathrm{VAR}(X_2) \\ & + \cdots + \mathrm{VAR}(X_N) \end{aligned}$$

> The variance of the sum of N *independent* random variables is equal to the sum of their variances.

Note that there is no restriction of independence in the case of equation 3.14. That is, equation 3.14 holds whether or not the variables are independent. Other properties of expected values and variances are summarized in Appendix C.

Example 3-24 (in the following section) illustrates the properties of expected values and variances discussed in this section.

3.13
TCHEBYCHEFF'S RULE

In this chapter we have discussed probability distributions of discrete random variables. In Chapter 5, we discuss probability distributions of continuous random variables, focusing on the normal distribution, probably the best-known and most widely applied probability function. In some situations, the underlying processes that produce data may not be well understood. Therefore, the appropriate form of probability distribution may be unknown.

In section 1.15 we discussed Tchebycheff's rule as a useful tool for making statements about sets of observed data when we do not know the shape of their frequency distributions. Tchebycheff's rule is also useful in dealing with random variables because it sets bounds on probability statements regardless of the form of the underlying probability distribution and because it holds true for both discrete and continuous variables.

For probability distributions, Tchebycheff's rule takes the following form:

> If μ and σ are the mean and standard deviation of a probability distribution, then for any $k > 1$, at least $1 - (1/k^2)$ of the distribution is included within k standard deviations from the mean, that is, within the interval $\mu \pm k\sigma$.

Therefore, we can make the following statement. Regardless of the form of the probability distribution,

the range of $\mu \pm 2\sigma$ includes at least $\frac{3}{4}$ (i.e. $1 - (\frac{1}{2})^2$) of the total probability

the range of $\mu \pm 3\sigma$ includes at least $\frac{8}{9}$ (i.e. $1 - (\frac{1}{3})^2$) of the total probability

As in section 1.15, we note that if we know the form of a probability distribution, we can make more specific statements about the distribution. We again state that in Chapter 5, we will see that in a normal distribution with mean μ and standard deviation σ,

the range of $\mu \pm 2\sigma$ includes 95.5% of the total probability

the range of $\mu \pm 3\sigma$ includes 99.7% of the total probability

An example of the use of Tchebycheff's rule is given in Example 3-24.

EXAMPLE 3-24

Nancy Schultz is in charge of a project to develop a scale model of an assembly to be placed in production at a later time. She has scheduled the following activities to complete this project: (1) design of equipment, (2) training of personnel, (3) assembly of equipment. The times required to complete these activities are independent of one another. Based on her experience with similar projects in the past, she estimates the means (expected times) and standard deviations of the completion times in weeks for each activity as follows:

Activity	Expected Time (in weeks)	Standard Deviation (in weeks)
1. Design of equipment	10	4
2. Training of personnel	5	1
3. Assembly of equipment	6	3

Estimate the expected value and standard deviation of completion time for the entire project.

Using Tchebycheff's rule, give two probability statements concerning completion times for the entire project.

Let μ_1, μ_2, and μ_3 denote the expected values and σ_1, σ_2, and σ_3 the standard deviations of the completion times of the three activities. Then the expected value and standard deviation of completion time for the entire project are:

$$\mu = \mu_1 + \mu_2 + \mu_3 = 10 + 5 + 6 = 21 \text{ weeks}$$
$$\sigma = \sqrt{\sigma_1^2 + \sigma_2^2 + \sigma_3^2} = \sqrt{4^2 + 1^2 + 3^2} = 5.1 \text{ weeks}$$

The probability is at least 75% that the completion time of the entire project will lie in the range of

$$\mu \pm 2\sigma = 21 \pm 2(5.1) \text{ weeks}$$
$$= 10.8 \text{ to } 31.2 \text{ weeks}$$

The probability is at least 89% that the completion time of the entire project will lie in the range of

$$\mu \pm 3\sigma = 21 \pm 3(5.1) \text{ weeks}$$
$$= 5.7 \text{ to } 36.3 \text{ weeks}$$

EXERCISES 3.13

1. In the Fair Gain Bazaar, a booth offers you the chance to toss a fair coin until the first tail appears. If a tail appears on the first toss, you will be paid $2 for participating. If the first tail appears on the second toss, you will be paid $4 for participating; if the first tail appears on the third toss, you will receive $8. Thus, your payment is increased by a factor of two for each head that appears, and the game ends with the appearance of a tail. How much should you be willing to pay to participate in this game if you intend to quit on the third try, whether you win or lose?

2. The probability is 0.97 that a 50-year-old miner will survive one year. An insurance company offers to sell this man a $10,000 one-year term life insurance policy at a premium of $315. What is the company's expected gain? Does this mean that the company will earn that much money on *this particular policy?*

3. The corporate planning group of BT & T is considering two investment alternatives for expanding its telephone service. Investment *A* limits its geographic expansion to three big coast cities, while investment *B* covers five major cities on the East Coast. The net profits for identical periods

and probabilities of success for investments A and B are given in the following table:

Net Profits (in thousands of dollars)	Probability of Return	
	Investment A	Investment B
8,000	0.0	0.1
9,000	0.3	0.2
10,000	0.4	0.4
11,000	0.3	0.2
12,000	0.0	0.1

Handwritten annotations on the table:
- Investment A: 8,000 → 0; 9,000 → 2700; 10,000 → 4000; 11,000 → 3300; 12,000 → 0; sum 10,000
- Investment B: 8,000 → 800; 9,000 → 1800; 10,000 → 4000; 11,000 → 2200; 12,000 → 1200; sum 10,000

a. Which investment yields a higher expected net profit? B

b. Compute the variance of each investment alternative. Can you make a decision on which investment alternative is better, given this additional information?

4. A contractor is bidding on an oil-drilling job in the Middle East. If he gets the contract, he will need new heavy earth-moving equipment. Unfortunately, he does not know exactly how much equipment he will need until the feasibility of commercial drilling is tested. These tests will not be completed until after the equipment is shipped to the Middle East. He has promised to purchase as many as four of these pieces of equipment from Heavy Machineries, Inc. If the equipment is unavailable, he will purchase additional equipment from Heavy Machineries' competitor, which has a warehouse in the Middle East. Heavy Machineries, Inc., estimates a net gain of $5,000 on each piece of equipment sold and a net loss of $1,000 on each piece of equipment sent to the Middle East but not sold. Consulting engineers at Heavy Machineries estimate the following probability distribution for X, the number of pieces of equipment purchased by the contractor:

x	$f(x)$
0	0.50
1	0.10
2	0.20
3	0.15
4	0.05

a. Calculate the net return to Heavy Machineries, Inc., if it sends three units and the contractor requires zero, one, two, three, or four pieces of equipment.

b. Evaluate the expected net return of the decision to send three pieces of equipment. Compute the standard deviation associated with this decision.

c. Is there a way to determine the optimal number of pieces of equipment that Heavy Machineries, Inc., should send? Explain.

Handwritten marginal notes (left side):

$(8000 - 10,000)^2.1 +$
$(9000 - 10,000)^2.2 +$
$(10,000 - 10,000)^2.4 +$
$(11,000 - 10000)^2.2 +$
$(12,000 - 10000)^2.1 =$
$1,200,000$

A.

B.

LESS RISK

820   1200

$(9000 - 10000)^2.3 +$
$(10,000 - 10,000)^2.4 +$
$(11,000 - 10,000)^2.3 =$
$600,000$

5. A used car sales representative has the opportunity to work for either dealer A or dealer B. The salesperson assesses the sales prospects at these dealerships as follows:

	Dealer A A = Number of cars that can be sold per week		Dealer B B = Number of cars that can be sold per week	
	A	P(A)	B	P(B)
	0	0.4	0	0.2
	1	0.3	1	0.6
	2	0.2	2	0.2
	3	0.1	3	0.0
		1.0		1.0

(handwritten annotations:)
$(0-1)^2 .4 +$
$(1-1)^2 .3 +$
$(2-1)^2 .2 +$
$(3-1)^2 .1 =$

0.4 = 0
0.3 = .3
0.2 = .4
0.1 = .3

0.2 = 0
0.6 = .6
0.2 = .4
0.0 = 0

a. Compare the expected number of cars that the salesperson would sell with dealer A and dealer B. Would the "weekly sales record" be any more consistent with dealer A than with B?

b. Dealer B offers the sales representative $100 per week plus $100 per car sold. Evaluate her expected weekly earnings. What is the standard deviation of weekly earnings? (Hint: See rules 3 and 7 of Appendix C.)

6. The King Mart Retail Store operates four major department stores in which net profit per month, as represented by expected value and variance, is given in the following table:

Store	Expected Profit per Month	Variance
1	$6,500	10,000
2	6,200	6,800
3	6,000	8,000
4	5,800	7,500

King Mart has an opportunity to buy another high-volume outlet with net profit averaging $8,000 per month, but with a higher variance of 15,000. Assuming independence, compare King Mart's expected total net profit with and without the fifth outlet, and comment on the variance in sales that will result if the fifth outlet is acquired.

7. You are the only economist on the Isle of Famboanga, whose basic monetary unit is the "pound silver." Given the isle's productive activities, you estimate next year's national income (Y) to be as follows:

Y (in millions of pounds)	P(Y)
8	0.3
10	0.5
11.5	0.1
12	0.1
	1.0

a. Find $E(Y)$ and $VAR(Y)$.

b. Oil is discovered off the coast of Famboanga. A foreign oil conglomerate offers the government 5 million pounds per year for drilling rights. How does this affect your forecast of Y? Assume that the drilling rights fee is exogenous to the economy, that is, let $Y' = Y + 5$. (Hint: See rules 3 and 7 of Appendix C.)

8. The checkout counter of Druggists Unlimited has display space for only one weekly magazine. The owner is deciding between two possibilities: *TV Times* (T) or *Sports Weekly* (S). *TV Times* sells for 40¢ per copy and costs 15¢ per copy; unsold copies are discarded. *Sports Weekly* sells for 60¢ and costs 30¢; a refund of 5¢ per copy is received for unsold copies. Based on a market survey, the demand distribution can be defined as follows:

Potential Demand (D) (copies per week)	TV Times Probability P(T)	Sports Weekly Probability P(S)
0	0.02	0.10
1	0.04	0.10
2	0.12	0.10
3	0.20	0.15
4	0.30	0.20
5	0.20	0.15
6	0.12	0.20

a. Calculate $E(T)$, $VAR(T)$, $E(S)$, and $VAR(S)$ and compare the results, where $E(T)$ = expected demand for *TV Times*, and so on.

b. Compute the weekly profit from *TV Times* and *Sports Weekly*, using expected demands $E(S)$ and $E(T)$. Assume that six issues are stocked each week. {Hint: Weekly profit = [(price)(demand)] − [(cost per copy)(number stocked)] + [(refund per copy)(number stocked − number demanded)].}

c. Based on your analysis, which weekly publication should the owner place at the checkout counter?

9. The probability distribution for the number of clients telephoning the senior partner in charge of management services for a Big Eight public accounting firm each day is as follows:

x	$f(x)$	x	$f(x)$
7	0.08	11	0.13
8	0.13	12	0.10
9	0.21	13	0.06
10	0.24	14	0.05

What is the expected number of clients telephoning the partner on a given day?

10. The Discount King department store chain has 12 stores in the Detroit metropolitan area. The net profit per month from each has an expected value of $6500 and a variance of 30,000. What is the expected total net profit per month and the standard deviation of total net profit for all 12 stores combined? Assume independence.

11. To relieve the strain of decision-making pressures on its executives, the Barker Brothers Game and Novelty Company has set up a roulette wheel in the executive dining room. A roulette wheel has 38 slots, numbered 1 through 36, 0, and 00. On a spin of the wheel, it is equally probable that a ball will rest on any of the 38 numbers. During lunch hour, any executive can place a $5 bet on any number on the wheel. If the ball rests on the chosen number, the payoff is $125. Twelve executives from the financial and marketing divisions have formed a group that bets on numbers 1 to 12 inclusive each day. Each person contributes $5 for every spin of the wheel and shares equally in the winnings, if any. What is the expected profit (or loss) for the group on each spin? What is the expected profit (or loss) for the group on each spin? What is the expected profit (or loss) for 500 spins of the wheel if the group places its usual bet combination each time?

12. After completing a study on the possible side effects of a new pain reliever for headaches, the research and marketing divisions of the Aspiring Medical Drug Company are trying to decide whether to test the drug further or place it on the market now. They have determined the following probabilities for the percentage of users who will experience an unpleasant side effect of the new drug:

Percentage of Users Affected	Probability	Percentage of Users Affected	Probability
0	0.03	6	0.14
1	0.06	7	0.10
2	0.07	8	0.09
3	0.11	9	0.07
4	0.17	10	0.02
5	0.14		

Find the expected value and standard deviation of percentage of users affected.

13. A certain machinist produces five to eight finished pieces during an eight-hour shift. An efficiency expert wants to assess the value of this machinist, where value is defined as value added minus the machinist's labor cost. The

value added for the work the machinist does is estimated at $9 per item, and the machinist earns $5 per hour. From past records, the machinist's output per eight-hour day is known to have the following probability distribution:

Number of Pieces Produced	Probability
5	0.2
6	0.4
7	0.3
8	0.1

What is the expected monetary value of the machinist to the company per eight-hour day?

14. The Amstar Motor Company is in the process of planning its new model line for 1992. The company has defined four stages in the development of a new automobile model. First, the research and marketing people decide what market they are trying to enter or expand their sales in and, hence, what general type of automobile should be built. Then the designers and engineers, with the goals of the marketing division in mind, develop the new product from drawing board to full-scale model. Next, the production division retools the plant where the new car will actually be built. Finally, the quality control people test the first automobiles off the line and recommend improvements and adjustments. After completion of this last step, the company is ready to begin production of the new model. Each stage must be completed before the next step is started. In an attempt to get an estimate of when the new automobile will be ready for introduction, senior management has obtained the following information from the various division heads in charge of the four stages:

Stage	Expected Completion Time of Stage (in weeks)	Standard Deviation (in weeks)
I	5	2
II	14	5
III	8	3
IV	3	1

What are the expected value and standard deviation of the total time for the new model's complete development process? Assume that the times for completion of the stages are independent.

LOOK → **3.14**

EXPECTED VALUES AND VARIANCES
OF WELL-KNOWN DISCRETE
PROBABILITY DISTRIBUTIONS

In sections 3.10 and 3.11, we discussed expected values and variances of random variables. It is useful to derive the mathematical expressions for these parameters of the probability distributions of well known random variables. For example, the expected value and variance of the binomial distribution are of particular interest in statistical inference. Examples 3-25, 3-26, and, 3-27 illustrate the idea of the expected value, variance and standard deviation of probability distributions of random variables. Examples 3-25, and 3-26 pertain to the binomial distribution, in which the random variables are "number of successes" and "proportion of successes", respectively. Example 3-27 illustrates the *use of the Poisson distribution for the random variable "number of successes."*

EXAMPLE 3-25

Assume that a restaurant has determined that there is a probability of 20% that a customer will order Blitz beer. If at a particular time, there are five customers in the restaurant, what is the expected value, variance, and standard deviation of the number of customers who will order Blitz beer?

SOLUTION

General expressions for the expected value and standard deviation of the binomial distribution for the random variable "number of successes" can be derived by substituting $\binom{n}{x} q^{n-x} p^x$ for $f(x)$ in the definitional formulas in equations 3.14 and 3.15 and performing appropriate manipulations. The following formulas are obtained:

Let x = number of successes in n trials of a Bernoulli process
p = probability of success on a single trial
q = probability of failure on a single trial
n = number of trials
It can be shown that

(3.16)
$$E(X) = np$$

and

(3.17)
$$\text{VAR}(X) = npq$$

Therefore, in the Blitz beer example,

$$E(X) = np = 5(0.20) = 1 \text{ customer}$$

$$\text{VAR}(X) = npq = 5(0.20)(0.80) = 0.80$$

$$\sigma = \sqrt{0.80} = 0.89 \text{ customers}$$

It may be noted that the use of these formulas is much simpler than obtaining the specific probabilities for $x = 0, 1, 2, 3, 4, 5$, and then using the definitional formulas given in equations 3.14 and 3.15.

EXAMPLE 3-26

In Example 3-25, let us now word the question in terms of proportion of successes, rather than number of successes. That is, if at a particular time, there are five customers in the restaurant, what are the expected value, variance and standard deviation of the *proportion* of customers who will order Blitz beer?

SOLUTION

We can convert the numbers of successes to proportions of successes by dividing by the sample size. The sample proportion, denoted p (pronounced "p-bar") may be calculated from

(3.18)
$$\bar{p} = \frac{x}{n}$$

In the present example $\bar{p}$ takes on the values $\frac{0}{5} = 0.00$, $\frac{1}{5} = 0.20$, ..., $\frac{5}{5} = 1.00$ with the same probabilities as the corresponding numbers of customers $0, 1, \ldots, 5$. (Note that the number of customers in a sample of size n is given by $x = n\bar{p}$.
It can be shown that

(3.19)
$$E(\bar{p}) = p$$

(3.20)
$$VAR(\bar{p}) = \frac{pq}{n}$$

Therefore, in this example

$$E(\bar{p}) = 0.20 = 20\%$$

$$VAR(\bar{p}) = (0.20)(0.80)/5 = 0.0320$$

$$\sigma_{\bar{p}} = \sqrt{0.0320} = 0.1789 = 17.89\%$$

Hence, in this example, if the sample statistic or random variable is proportion of customers who will purchase Blitz beer, the expected value is $E(\bar{p}) = p = 0.20$ or 20% purchasers. The standard deviation $\sigma_{\bar{p}} = \sqrt{pq/n} = 0.1789$ or 17.89%, is a measure of the chance effects of sampling, expressed in terms of *porportion* of customers who will purchase Blitz beer.

EXAMPLE 3-27

It has been determined in a certain factory that the number of accidents per month is distributed according to the Poisson distribution with an average (arithmetic mean) of $\mu = 4$ accidents per month. What is the expected value, variance, and standard deviation of the number of accidents per month at this factory?

It can be shown that the expected value and variance of the random variable "number of successes" (number of occurrences) in a Poisson distribution are

(3.21)
$$E(X) = \mu$$

(3.22)
$$VAR(X) = \mu$$

Therefore, we obtain

$$E(X) = 4 \text{ accidents per month}$$

$$VAR(X) = 4$$

$$\sigma = \sqrt{4} = 2 \text{ accidents per month}$$

Again, note the simplicity of these calculation as compared to obtaining specific probabilities for numbers of occurrences and performing calculations according to equations 3.14 and 3.15.

Table 3-15 summarizes the parameters of the probability distributions discussed in this chapter. Although no example was given for the hypergeometric

TABLE 3-15
Parameters of probability distributions

	Expected Value	Variance
Random Variable **Binomial Distribution**		
$x = n\bar{p}$	np	npq
(number of successes)		
$\dfrac{x}{n} = \bar{p}$	p	$\dfrac{pq}{n}$
(proportion of successes)		
Poisson Distribution		
x	μ	μ
(number of successes)		
Hypergeometric		
x	$\dfrac{n}{N} X$	$\dfrac{nX(N-X)(N-n)}{N^2(N-1)}$
(number of successes)		

distribution, the parameters for that distribution are included in Table 3-15 for completeness.

3.15
JOINT PROBABILITY DISTRIBUTIONS

The discussion to this point has been concerned with probability distributions of discrete random variables considered one at a time. These are often referred to as *univariate probability distributions*, or *univariate probability mass functions*. In such distributions, probabilities are assigned to events pertaining to a single random variable.

In most realistic decision-making situations, however, more than one factor at a time must be taken into account. Frequently, the joint effects of several variables, some or all of which are interdependent, must be analyzed in terms of their impact on some objective the decision maker wishes to achieve.

In this section, we consider the joint probability distributions of discrete random variables, or, stated differently, probability distributions of two or more discrete random variables. Such functions are frequently referred to as **multivariate probability distributions**, the term **bivariate probability distribution** being used for the two-variable case.

We return to Table 2-2, the joint probability table for 1,000 persons classified by sex and product preference, for a simple example of a bivariate probability distribution. Table 2-2 is reproduced here for convenience as Table 3-16.

We now introduce some symbolism for the discussion of joint bivariate probability distributions. Instead of using the terminology of events, as we did earlier in discussing Table 2-2, we now use the language of random variables. Let X represent the random variable "product preference" and Y the random variable "sex". We let X and Y take on the following values:

$$X = 1 \text{ for "prefers product ABC"}$$

$$X = 2 \text{ for "prefers product XYZ"}$$

$$Y = 1 \text{ for "male"}$$

$$Y = 2 \text{ for "female"}$$

Then $P(X = x \text{ and } Y = y)$ denotes the joint probability that X takes on the value x and Y takes on the value y.[11] For example, $P(X = 1 \text{ and } Y = 1)$ is the joint probability $P(\text{prefers product ABC and male}) = 0.20$, and so on. The bivariate probability distribution of X and Y is given in Table 3-17.

---

[11] The notation $P(X = x \text{ and } Y = y)$ should really read $P\{(X = x) \text{ and } (Y = y)\}$. The simplified symbolism is in common use and will be employed in this book.

TABLE 3-16
Joint probability distribution for 1,000 persons classified by sex and product preference

| | Product Preference | | |
Sex	Prefers Product ABC	Prefers Product XYZ	Marginal Probabilities
Male	0.20	0.30	0.50
Female	0.10	0.40	0.50
Marginal probabilities	0.30	0.70	1.00

TABLE 3-17
Bivariate probability distribution corresponding to Table 3-16

| | x | | |
y	1	2	$P(Y = y)$
1	0.20	0.30	0.50
2	0.10	0.40	0.50
$P(X = x)$	0.30	0.70	1.00

Marginal Probability Distributions

The values in the cells of Table 3-17 are the joint probabilities of the respective outcomes denoted by the column and row headings for X and Y. Also displayed in the table are the separate univariate probability distributions of X and Y. Earlier, we referred to the probabilities in the margins of the table as marginal probabilities. These probabilities form **marginal probability distributions**. The marginal distribution of X consists of the values of X shown in the column headings and the column totals at the bottom; the marginal distribution of Y consists of the values of Y shown in the row headings and the row totals at the right side of the table. These distributions are shown in Table 3-18. As indicated, the symbols $P(X = x)$ and $P(Y = y)$ are used to denote the probabilities for the marginal probability distributions of X and Y.

TABLE 3-18
Marginal probability distributions of X and Y

x	Product Preference $P(X = x)$	y	Sex $P(Y = y)$
1	0.30	1	0.50
2	0.70	2	0.50
	1.00		1.00

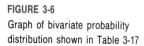

FIGURE 3-6

Graph of bivariate probability
distribution shown in Table 3-17

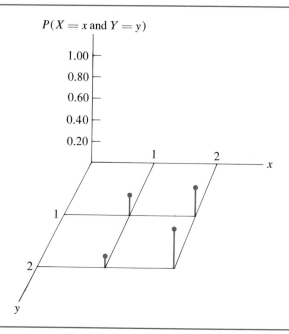

$P(X = x \text{ and } Y = y)$

Graph of Bivariate Probability Distribution

The probability distribution of a single discrete random variable is graphed by displaying the values of the random variable along the horizontal axis and the corresponding probabilities along the vertical axis. In the case of a bivariate distribution, two axes are required for the values of the random variables and a third for the measurement of probabilities. Usually, the joint values of the two variables are depicted on a plane (the x-y plane) and the associated probabilities are read along an axis perpendicular to the plane. A graph of the joint probability distribution of Table 3-17 is shown in Figure 3-6.

Conditional Probability Distributions

Another important type of probability distribution obtainable from a joint probability distribution is the **conditional probability distribution**. Using equations 2.4 and 2.5, we may calculate the conditional probability that X takes on a particular value (say, $X = 2$) given that Y takes on a particular value (say, $Y = 1$) by dividing the joint probability that $X = 2$ and $Y = 1$ by the marginal probability that $Y = 1$. For example, the conditional probability $P(\text{prefers product XYZ} | \text{male}) = P(X = 2 | Y = 1)$ is computed as follows:

$$P(X = 2 | Y = 1) = \frac{P(X = 2 \text{ and } Y = 1)}{P(Y = 1)} = \frac{0.30}{0.50} = 0.60$$

TABLE 3-19
Conditional probability distribution for product preference (X) given that the individual is male (Y = 1)

x	$P(X = x \mid Y = 1)$
1	$\dfrac{0.20}{0.50} = 0.40$
2	$\dfrac{0.30}{0.50} = 0.60$
	$\overline{1.00}$

We can also find the conditional probability $P(X = 1 \mid Y = 1)$, thus forming a conditional probability distribution for X given $Y = 1$. This distribution is shown in Table 3-19.

Similarly, we can obtain the conditional probability distribution for X given $Y = 2$ by dividing the joint probabilities in the row $Y = 2$ by the row total, $P(Y = 2) = 0.50$.

Corresponding calculations give us the conditional probability distributions for Y given particular values of X. For example, the conditional probability distribution of sex (Y) given that the individual prefers product XYZ $(X = 2)$ is shown in Table 3-20.

Independence

Returning to the terminology of events, we saw in section 2.2 that if two events A_1 and B_1 are statistically independent, $P(B_1 \mid A_1) = P(B_1)$ and $P(A_1 \mid B_1) = P(A_1)$. Using random variable notation, the analogous statement is that if X and Y are two *independent* random variables, then

$$P(X = x \mid Y = y) = P(X = x)$$

TABLE 3-20
Conditional probability distribution for sex (Y) given that the individual prefers product XYZ (X = 2)

y	$P(Y = y \mid X = 2)$
1	$\dfrac{0.30}{0.70} = 0.43$
2	$\dfrac{0.40}{0.70} = 0.57$
	$\overline{1.00}$

and

$$P(Y = y \,|\, X = x) = P(Y = y)$$

for all pairs of outcomes (x, y). For example, the first of these statements means that the conditional probability that X takes on a particular value x, given that Y takes on a particular value y, is equal to the marginal, or unconditional, probability that X takes on the particular value x.

Again returning to the terminology of events, we observed in section 2.2 that by the multiplication rule, if two events A_1 and B_1 are statistically independent, $P(A_1 \text{ and } B_1) = P(A_1)P(B_1)$. The analogous statement for random variables follows.

If X and Y are two independent random variables, then

$$P(X = x \text{ and } Y = y) = P(X = x)P(Y = y)$$

for all pairs of outcomes (x, y).

We illustrate this definition of independence by returning to Table 3-17. Suppose we consider the outcome pair $(1, 1)$, that is, $X = 1$ and $Y = 1$. In this case,

$$P(X = 1 \,|\, Y = 1) = \frac{0.20}{0.50} = 0.40$$

and

$$P(X = 1) = 0.30$$

Since $P(X = 1 \,|\, Y = 1)$ is not equal to $P(X = 1)$, X and Y are not independent random variables. Alternatively, we observe

$$P(X = 1 \text{ and } Y = 1) = 0.20$$
$$P(X = 1) = 0.30$$
$$P(Y = 1) = 0.50$$

We note that

$$P(X = 1 \text{ and } Y = 1) = 0.20 \quad \neq \quad P(X = 1)P(Y = 1) = (0.30)(0.50) = 0.15$$

Again, we conclude that X and Y are not independent random variables. Our reasoning is based on the fact that for X and Y to be independent random variables, the independence conditions stated earlier must hold for *all* pairs of outcomes (x, y). Since the conditions do not hold for the pair $(1, 1)$, that is, for $(X = 1, Y = 1)$, then X and Y are not independent.

If we retain the same marginal probability distributions shown in Table 3-17, what would the values of the joint probabilities have been if product preference (X) and sex (Y) had been *independent* random variables? To answer

TABLE 3-21
Joint probability distribution for 2 independent random variables X and Y

y	1	2	$P(Y = y)$
	x		
1	0.15	0.35	0.50
2	0.15	0.35	0.50
$P(X = x)$	0.30	0.70	1.00

this question, we consider the marginal probability distributions in Table 3-17:

y	1	2	$P(Y = y)$
	x		
1			0.50
2			0.50
$P(X = x)$	0.30	0.70	1.00

We compute the required joint probabilities as follows:

$$P(X = 1 \text{ and } Y = 1) = P(X = 1)P(Y = 1) = (0.30)(0.50) = 0.15$$

$$P(X = 2 \text{ and } Y = 1) = P(X = 2)P(Y = 1) = (0.70)(0.50) = 0.35$$

and so on.

In Table 3-21, we show the resulting joint probability distribution if product preference and sex had been independent random variables. We now observe that $P(X = x | Y = y) = P(X = x)$ and $P(Y = y | X = x) = P(Y = y)$ for all pairs of outcomes (x, y). For example, for the outcome (1, 1), $P(X = 1 | Y = 1) = \frac{0.15}{0.50} = 0.30 = P(X = 1)$. Also, we observe that $P(X = x \text{ and } Y = y) = P(X = x)P(Y = y)$ for all pairs of outcomes (x, y). For example, for the outcome (1, 1), $P(X = 1 \text{ and } Y = 1) = 0.15 = P(X = 1)P(Y = 1) = (0.30)(0.50) = 0.15$.

Note in Table 3-21 that 30% ($\frac{0.15}{0.50} = 0.30$) of the males prefer product ABC and 30% of the females prefer product ABC. Correspondingly, 70% of the males prefer product XYZ, and the same percentage applies for females as well. Hence, we can say that product preference is *independent* of sex.

EXAMPLE 3-28

Table 3-22 presents the results of a sample survey of the employment status of the labor force by age in a certain community:
a. If a person is drawn at random from this labor force, what is
   (1) the joint probability that the individual is under 25 years of age and un-employed?
   (2) the marginal probability that the individual is under 25 years of age?
   (3) the conditional probability of being unemployed, given that the individual is under 25 years of age?

TABLE 3-22

Sample of the labor force in a certain community classified by age group and employment status

Employment Status	Age Group (in years)			All Ages
	Under 25	25 and Under 45	45 and Over	
Unemployed	120	250	130	500
Employed	1880	4750	2870	9500
Labor force	2000	5000	3000	10,000

b. What are the marginal probability distributions of age and employment status?
c. What is the conditional probability distribution of age given that an individual in this labor force is unemployed?
d. Is there evidence of dependence between employment status and age in this labor force?
e. If the same marginal totals are retained, what would the number of persons in each cell of the table have to be for independence to exist between age and employment status?

SOLUTION

In order to answer these questions, we convert the table of numbers of occurrences (absolute frequencies) to probabilities (relative frequencies), by dividing each number in Table 3-22 by 10,000, the total number of persons in the sample. These relative frequencies are given in Table 3-23.

Let X and Y be random variables representing age group and employment status, respectively, and taking on values as follows:

Age Group	x	Employment Status	y
Under 25	0	Unemployed	0
25 and under 45	1	Employed	1
45 and over	2		

The joint probability distribution now appears as given in Table 3-24. The answers to the questions asked may now be given.

TABLE 3-23

Relative frequencies of occurrence for a sample of the labor force of a certain community classified by age group and employment status

Employment Status	Age Group (in years)			All Ages
	Under 25	25 and Under 45	45 and Over	
Unemployed	0.012	0.025	0.013	0.050
Employed	0.188	0.475	0.287	0.950
Labor force	0.200	0.500	0.300	1.000

TABLE 3-24
Joint probability distribution derived from Tables 3-22 and 3-23

y \ x	0	1	2	$P(Y = y)$
0	0.012	0.025	0.013	0.050
1	0.188	0.475	0.287	0.950
$P(X = x)$	0.200	0.500	0.300	1.000

a. (1) P(under 25 and unemployed) $= P(X = 0$ and $Y = 0) = 0.012$
    (2) P(under 25) $= P(X = 0) = 0.200$
    (3) P(unemployed | under 25) $= P(Y = 0 | X = 0) = \frac{0.012}{0.200} = 0.06$
b. The marginal probability distributions given in the margins of the probability table are

x	$P(X = x)$		y	$P(Y = y)$
0	0.200		0	0.050
1	0.500		1	0.950
2	0.300			1.000
	1.000			

c. This conditional probability distribution is given by

$$P(X = x | Y = 0) = \frac{P(X = x | Y = 0)}{P(Y = 0)}, \quad \text{where } x = 0, 1, 2$$

Thus, each of the joint probabilities in the first row of Table 3-23 or 3-24 is divided by the marginal probability, or total, of that row. That is, the joint probabilities of being unemployed and in the specified age groups are divided by the marginal probability of being unemployed. Therefore, the required conditional probability distribution is

x	$P(X = x | Y = 0)$
0	$0.012/0.050 = 0.24$
1	$0.025/0.050 = 0.50$
2	$0.013/0.050 = 0.26$
	1.00

d. Employment status and age are *not independent* random variables. That is, it is not true that $P(X = x | Y = y) = P(Y = y)P(X = x)$ for *all* values of x and y. For example,

$$P(X = 0 \text{ and } Y = 0) \neq P(X = 0)P(Y = 0)$$

Numerically,

$$(0.012) \neq (0.200)(0.050)$$

It may be noted that for age group "25 and under 45" ($X = 1$), the joint probabilities do factor into the product of the respective marginal probabilities. For example,

$$P(X = 1 \text{ and } Y = 0) = P(X = 1)P(Y = 0)$$

or

$$0.025 = (0.500)(0.050)$$

However, this is not sufficient. The equality must hold for *all* values of X and Y for these variables to be considered independent.

Another way of indicating that age and employment status are not independent random variables is to note that the marginal distributions are not equal to the corresponding conditional distributions. Thus, for age, the conditional and marginal probability distributions are

x	$P(X = x \mid Y = 0)$	$P(X = x \mid Y = 1)$	$P(X = x)$
0	0.240	0.198	0.200
1	0.500	0.500	0.500
2	0.260	0.302	0.300
	1.000	1.000	1.000

The specific nature of the dependence between age and employment status is that the percentage of unemployed individuals decreases with age. The unemployment rates are: under 25, 6.0%; 25 and under 45, 5.0%; 45 and over, 4.3%.

It is important to observe that the relationship between age and employment status depends to a certain extent on the arbitrary classifications used for the two variables. For example, if narrower age classifications had been used, it might have been found that the heaviest unemployment rates in the "under 25" group were among the teenagers and that although the unemployment rate was low for the "45 and over" group, the rate was quite high for persons 60 years and older. Similarly, if a narrower breakdown for employment had been used (including, for instance, part-time employment and multiple-job-holding classifications), greater insights might have been gained into the underlying relationships between age and employment status.

e. As indicated in (d), if age and employment status were independent, the marginal probability distributions would be equal to the corresponding conditional probability distributions. One way of interpreting this in terms of the problem is to observe that since 0.05 of persons in all age groups were unemployed ($P(Y = 0) = 0.050$), then under the assumption of independence, the same proportion of individuals in *each* age group would be unemployed. Thus, 5% of 2000, 5000, and 3000 persons would be unemployed in the three respective age groups. In terms of marginal

TABLE 3-25
Classification of persons under the assumption of independence between age and employment status

Employment Status	Age Group (in years)			All Ages
	Under 25	25 and Under 45	45 and Over	
Unemployed	100	250	150	500
Employed	1900	4750	2850	9500
Labor force	2000	5000	3000	10,000

totals, the arithmetic for each age group would be:

$$\text{Under 25:} \qquad \frac{500}{10,000} \times 2000 = 100 \text{ unemployed}$$

$$\text{25 and under 45:} \qquad \frac{500}{10,000} \times 5000 = 250 \text{ unemployed}$$

$$\text{45 and over:} \qquad \frac{500}{10,000} \times 3000 = 150 \text{ unemployed}$$

Therefore, the numbers of persons in each cell of the table under the assumption of independence are as given in Table 3-25.

EXERCISES 3.15

1. The following table gives the results of a sample survey of sales of different types of packaging in 200 test communities:

Sales	Type of Packaging				Total
	Ordinary Bottles	Ordinary Cans	Flip-top Cans	Screw-top Bottles	
Under 2,000 cases	45	40	30	25	140
At least 2,000 cases	15	20	10	15	60
Total	60	60	40	40	200

a. If a test community is selected at random for further study, what is
   (1) the probability that it had sales of at least 2,000 cases and was tested with flip-top cans?
   (2) the marginal probability that it was tested with flip-top cans?

   (3) the conditional probability that it had sales of at least 2,000 cases given that it was tested with flip-top cans?

  b. What are the marginal distributions of sales and type of packaging?

  c. What is the conditional distribution of sales given that ordinary bottles are used?

  d. Is there evidence of dependence between packaging and sales?

  e. If we retain the same marginal totals, what would the numbers in the cells of the table have to be for independence to exist between type of packaging and sales?

2. A poll of the employees of Bajo, Inc., on a particular labor proposal gives the following results:

	Type of Position		
Opinion	Skilled	Unskilled	Total
For	275	225	500
Against	200	500	700
No opinion	125	75	200
Total	600	800	1,400

  Does a person's type of position affect his or her opinion on the labor proposal? Interpret your answer.

3. The table below gives the results of a market survey conducted to determine consumers' use of cranberry sauce, classified according to their interest in cooking. Four segments are identified with the following characteristics:

  (1) "Convenience-oriented" segment prefers natural foods that are easy to cook.

  (2) "Enthusiastic" segment enjoys taking time to prepare meals and likes fancy dishes and colorful salads.

  (3) "Decorator" segment likes festive meals appropriate for formal occasions or holidays.

  (4) "Disinterested" segment does not like cooking.

	Type of Consumer				
Type of Sauce	Convenience-oriented	Enthusiastic	Decorator	Disinterested	Total
Jellied cranberry	35	29	19	17	100
Whole cranberry	30	40	16	14	100
Total	65	69	35	31	200

  a. If a consumer is selected at random, what is the

   (1) probability that he or she is a decorator and uses jellied cranberry?

      (2) probability that he or she is an enthusiastic cook?

      (3) conditional probability that he or she is a convenience-oriented cook given the use of jellied cranberry?

b. What is the marginal distribution of types of cranberry sauce? Of types of consumers?

c. Is there evidence of dependence between types of consumers and types of sauce used?

4. An operations research analyst developed the following mathematical expression for the joint distribution of two discrete economic variables:

$$f(x, y) = \frac{1}{56}(2x + y^2) \quad x = 0, 1, 2, 3, \quad y = 0, 1, 2$$

a. Display the joint bivariate distribution of X and Y.

b. State the marginal probability distributions of X and Y.

c. State the conditional probability distributions of the Y variable.

d. Are X and Y independent?

5. An analyst who was studying the operations of two dealerships owned by Barkley Holding, Inc., derived a bivariate probability distribution for sales per day of the two dealerships.

		X	
Y	0	1	2
0	0.1	0.1	0
1	0.1	0.5	0
2	0	0	0.2

X = the number of sales per day for Exotic Motors, Inc.

Y = the number of sales per day for Phlegmatic Motors, Inc.

a. Calculate $E(X + Y)$ and interpret your result.

b. Given that the sales of Exotic Motors, Inc., are one per day, what are the expected value and standard deviation of the number of sales per day for Phlegmatic Motors, Inc.?

c. What is the probability that the daily profit for the two dealerships combined would exceed $4,000 if the sales prices were $14,500 per unit at Exotic Motors and $14,200 per unit at Phlegmatic Motors and cost per unit at both dealerships was $13,000?

MINITAB EXERCISES

The following two exercises will familiarize you with the binomial and Poisson distributions. The techniques used can be employed to solve problems relating to other discrete distributions as well.

1. Minitab commands

   MTB > set C1
   DATA > 0:20
   DATA > end

   create a column consisting of integers from 0 to 20. Find the probability density function for the binomial distribution with $n = 20$, $p = 0.4$, by using:

   MTB > pdf C1 C2;
   SUBC > binomial 20 0.4.

   Print the density function on your computer screen to find the mode of the density. Graph the density function. Find the median of the density by using the cumulative distribution function defined below:

   MTB > cdf C1 C3;
   SUBC > binomial 20 0.4.

   Finally obtain the mean of the distribution with the command:

   MTB > let k1 = sum(C1 ∗ C2)
   MTB > print k1

   Is $k1 = np = 8$, as it should be theoretically? If not, why?

2. For values from 0 to 25, produce the density function for the Poisson distribution with mean = 6 (*Hint:* first generate c1, a column of integers from 0 to 25, and then use:

   MTB > pdf C1 C2;
   SUBC > poisson 6:

   to get the pdf values). Do the same thing for the Poisson distribution with mean = 10, and denote the column by C3. Plot C3 and C2 on the same graph with C1 as the horizontal axis. Comment on the impact of a change in mean values on the density functions. Obtain the mode, median, mean, and variance for the Poisson distribution with mean = 6. Is mean = variance, as it should be in theory?

KEY TERMS

Binomial Distribution   the probability distribution of a Bernoulli process, i.e. (1) one of two outcomes occurs on each trial, (2) the probability of a success remains constant from trial to trial, and (3) the trials are independent.

Conditional Distribution   a probability distribution of a random variable given a particular value of another random variable.

Continuous Random Variable   a variable that can assume any value in a given range.

Discrete Random Variable   a random variable that can take on only a finite or countable number of distinct values.

Expected Value   the long run average (arithmetic mean).

Hypergeometric Distribution   a probability distribution appropriate for sampling without replacement.

Joint Probability Distribution   a probability distribution of two or more random variables.

Marginal Distribution   a probability distribution of a single random variable.

Multinomial Distribution   the generalization of the binomial distribution for the situation of more than two possible outcomes on each trial.

Poisson Distribution   a probability distribution for random occurrences in a continuum or for approximating the binomial distribution.

Probability Density Function   a probability distribution of a continuous random variable.

Probability Mass Function   probability distribution of a discrete random variable.

Random (stochastic) Process   a process in which the outcome on any specific trial is determined by chance.

Random Variable   a variable that takes on different values because of chance.

Standard Deviation of a Random Variable   the square root of the variance.

Tchebycheff's Rule   a method for setting bounds on probability statements.

Uniform Distribution   a probability distribution in which equal probabilities are assigned to all values of a random variable.

Variance of a Random Variable   the average (expected value) of the squared deviations from the expected value.

KEY FORMULAS

Cumulative Probability for Discrete and Continuous Random Variables

$$F(x) = P(X \leq x)$$

Probability for a Discrete Random Variable

$$f(x) = P(X = x)$$

Binomial Distribution—Probability of x Successes in n Trials

$$f(x) = \binom{n}{x} q^{n-x} p^x$$

Hypergeometric Distribution—Probability of *x* Successes in a Sample of Size *n* Drawn Without Replacement

$$f(x) = \frac{\binom{N-X}{n-x}\binom{X}{x}}{\binom{N}{n}}$$

Poisson Distribution—Probability of *x* Successes; μ is the Average Number of Occurrences Per Interval of Time or Space

$$f(x) = \frac{\mu^x e^{-\mu}}{x!}$$

Expected Value of a Discrete Random Variable *X*

$$E(X) = \sum_x xf(x)$$

The Variance of a Discrete Random Variable *X*

$$\text{VAR}(X) = \sigma^2 = \sum_x (x-\mu)^2 f(x)$$

Expected Value of the Sum of Any *N* Random Variables

$$E(X_1 + X_2 + \cdots + X_N) = E(X_1) + E(X_2) + \cdots + E(X_N)$$

Variance of the Sum of *N* *Independent* Random Variables

$$\text{VAR}(X_1 + X_2 + \cdots + X_N) = \text{VAR}(X_1) + \text{VAR}(X_2) + \cdots + \text{VAR}(X_N)$$

Tchebycheff's Rule

At least $1 - \left(\dfrac{1}{k^2}\right)$ of the distribution lies within $\mu \pm k\sigma$

4

STATISTICAL INVESTIGATIONS AND SAMPLING

Throughout our lives, we are involved in answering questions and solving problems. For many of these questions and problems, careful, detailed investigations are simply inappropriate. For example, to answer the question, "What clothes should I wear today?" or the question, "What type of transportation should I take to get to a friend's house?" may not require painstaking, objective, scientific investigation.

This book is concerned with the investigation of problems that do require careful planning and an objective, scientific approach to arrive at meaningful solutions.

4.1
FORMULATION OF THE PROBLEM

A statistical investigation arises out of the need to solve some sort of a problem. Problems may be classified in a variety of ways.

> Choosing among alternative courses of action is one sort of problem.

The problems of managerial decision making typically fall under this category. For example, an industrial corporation wishes to choose a particular plant site from several alternatives; a financial vice-president wishes to decide among alternative methods of financing planned increases in productive capacity; or an advertising manager must select advertising media from many possible choices.

> Reporting information is another sort of problem.

Many illustrations can be given of statistical data collected for reporting purposes. For example, a trade association may report to its member firms on the characteristics of these companies. A research organization may publish data on the relationship between achievement of school children and the socioeconomic characteristics of their parents. An economist may report data on the frequency distribution of family incomes in a particular city. Even in the case of informational reporting, the data collected should have an ultimate decision-making purpose for someone.

Many problems originally arise as rather vague questions. These questions must be translated into a series of more specific questions, which then form the basis of the investigation. In most carefully planned investigations, the problem will be defined and redefined many times. The purposes and importance of an investigation will determine the type of study to be conducted.

There are instances when the objectives are particularly hard to define because of the many uses that will be made of the data and the large number of research consumers who will utilize the results of the study. For example, the U.S. Bureau of the Census publishes a wide variety of data on population, housing, manufacturers, and retail trade. It cannot specify in advance the many uses that will be made of these data.

In all studies it is critical to spell out as meticulously as possible the purposes and objectives of the investigation. All subsequent analysis and interpretation depends on these objectives, and only by stating very carefully what these objectives are can we know what questions have been answered by the inquiry.

4.2
DESIGN OF THE INVESTIGATION

Some investigations may be referred to as "controlled inquiries." We are all familiar with the scientist who controls variables in the laboratory. The scientist exercises control by manipulating the things and the events being investigated. For example, a chemist may hold the temperature of a gas constant while varying pressure and observing changes in volume. We begin by discussing observational or comparative studies, in which manipulation as in a laboratory is not possible.

Observational or Comparative Studies

In most statistical investigations in business and economics, it is not possible to manipulate people and events as directly as a physical scientist manipulates experimental materials. For example, if we want to investigate the effect of income on a person's expenditure pattern, it would not be feasible for us to vary this individual's income. On the other hand, we can observe the different expenditure patterns of people who fall in different income groups, and therefore we can make statistical generalizations about how expenditures vary with differences in income. This would be an example of a so-called **observational** study. In this type of study, the analyst essentially examines historical relationships among variables of interest. If one observes the important and relevant properties of the group under investigation, the study can be carried out in a controlled manner. For example, if we are interested in how family expenditures vary with family income and race, we can record data on family expenditures, family incomes, and race and then tabulate data on expenditures by income and racial classifications, such as white or black. Moreover, if we observe the differences in family expenditures for white and black families within the same income group, we have, in effect, "controlled" for the factor of income. That is, since the families observed are in the same income group, income cannot account for any differences in the expenditures observed.

If observational data represent historical relationships, it may be particularly difficult to ferret out causes and effects. For example, suppose we observe past data on the advertising expenses and sales of a particular company. Let us also assume that both of these series have been increasing over time. It may be quite incorrect to assume that the changes in advertising expenditures have caused sales to increase. If a company's practice in the past had been to budget 3% of last year's sales for advertising expense, one may state that advertising expenses depend on sales with a one-year lag. However, in this situation sales might be increasing quite independently of changes in advertising expenses. Therefore, one certainly would not be justified in concluding that changes in advertising expenses cause changes in sales. The point should also be made that many factors other than advertising may have influenced changes in sales. If data were not available on these other factors, it would not be possible to infer cause-and-effect relationships from these past observational data. The specific difficulty in attempting to derive cause-and-effect relationships in mathematical terms from historical data is that the various pertinent environmental factors will not ordinarily have been controlled or have remained stable.

Direct Experimentation or Controlled Studies

Direct experimentation studies are being increasingly used in fields other than the physical sciences, where they have traditionally been employed. In **direct experimentation** studies, the investigator directly controls or manipulates factors that affect a variable of interest. For example, a marketing experimenter may vary the amounts of direct mail exposure to a particular consumer audience. It is also possible to use different types of periodical advertising and observe the effects upon some experimental group. Various combinations of these direct mail exposures and periodical advertising may be used, as well as other types of promotional expenditures, such as a sales force. Thus, the investigator may be able to observe from the experiment that high levels of periodical advertising produce high levels of sales only if there is a high concentration of sales force activity. Such scientifically controlled experiments for generating statistical data, to which only brief reference is being made here, can be efficiently utilized to reduce the effect of uncontrolled variations. The real importance of this type of planning or design is that it gives greater assurance that the statistical investigation will yield valid and useful results.

Ideal Research Design

An important concept of a statistical inquiry is the **ideal research design**—meaning the investigators should think through at the design stage what the ideal research experiment would be, without reference either to the limitations of data available or to what data can feasibly be collected. Then if compromises must be made because of the practicalities of the real-world situation, the investigator will at least be completely aware of the specific compromises and expedients that have been employed. As an example, suppose we want to answer the question whether women or men are better automobile drivers. Clearly, it would be incorrect simply to obtain past data on the accident rates of men versus women. First of all, men may drive under quite different conditions than do women. For example, driving may constitute a large proportion of the work that many men do, whereas a larger proportion of women than men may drive primarily in connection with activities associated with the care of a home and family. The conditions of such driving differ considerably with respect to exposure to accident hazards. Many other reasons may be indicated for differences in accident rates between men and women apart from the essential driving ability of these two groups. Thus, as a first approximation to the ideal research design, perhaps we would like to have data for quite homogeneous groups of men and women, for example, women and men of essentially the same age, driving under essentially the same conditions, using the same types of automobiles. It may not be within the resources of a particular statistical investigation to gather data of this sort. However, once the ideal data required for a meaningful answer to the question have been thought through, the limitations of other, somewhat more practical sets of data become apparent.

EXERCISES 4.2

1. What is the difference between an observational study and a controlled study?

2. Why is it ordinarily very difficult to determine cause-and-effect relationships from historical data?

3. Why is it desirable to think through the ideal research design at the beginning stage of a statistical investigation?

An important phase of a statistical investigation is the construction of the conceptual or mathematical model to be used. A **model** is simply a representation of some aspect of the real world. Mechanical models are profitably used in industry as well as other fields of endeavor. For example, airplane models may be tested in a wind tunnel, or ship models may be tested in experimental water basins. Experiments may be carried out by varying certain factors and observing the effect of these variations on the mechanical models employed. Thus, we can manipulate and experiment with the models and draw corresponding inferences about their real-world counterparts. The advantages of this procedure are obvious compared with attempting to manipulate an experiment using the real-world counterparts, such as actual airplanes or ships, after they are constructed. In statistical investigations, mathematical models are often used to state in mathematical terms the relationships among the relevant variables.

Mechanical models

Mathematical models

> **Mathematical models** are conceptual abstractions that attempt to describe, to predict, and often to control real-world phenomena.

For example, the law of gravity describes and predicts the relationship between the distance an object falls and the time elapsed. Such models can be tested by physical experimentation.

> In well-designed statistical investigations, the nature of the model or models to be employed should be carefully thought through in the planning phases of the study.

In fact, the nature of these models provides the conceptual framework that dictates the type of statistical data to be collected. Let us consider a few simple examples. Suppose a market research group wants to investigate the relationship between expenditures for a particular product and income and several other socioeconomic variables. The investigators may want to use a mathematical model such as a **regression equation** (discussed later in this book), which states in mathematical form the relationship among the variables. When the investigators determine the variables that are most logically related to the expenditures for the product, they also determine the types of data that will have to be collected in order to construct their model.

Nature of models

Even in the case of relatively simple informational reporting, there is a conceptual model involved. For example, suppose an agency wishes to determine the unemployment rate in a given community. Also, assume that the agency must gather the data by means of a sample survey of the labor force in this community. The ratio "proportion unemployed" is itself a model. It states

Informational reporting models

a mathematical relationship between the numerator (number of persons unemployed) and the denominator (total number of persons in the labor force). The agency may wish to go further and state the range within which it is highly confident that the true unemployment rate falls. In such a situation—as we shall see later when we study estimation of population values—there is an implicit model, namely the probability distribution of a sample proportion.

Suppose a company wishes to establish a systematic procedure for accepting or rejecting shipments from a particular supplier. Various types of models have been used to solve this sort of problem. The company may decide to accept or reject shipments on the basis of testing some hypothesis concerning the percentage of defective items observed. On the other hand, it may decide to base its acceptance procedure on the arithmetic mean value of some characteristic that is considered important. Other procedures are possible; for example, *Formal decision model*   a **formal decision model** may be constructed. For these types of models, the probability distribution of the percentage of defective items produced by this company in the past may be required as well as data on the percentage of defective articles observed in a sample drawn from the particular incoming shipment in question. Obviously, the nature of the data to be observed and the nature of the analysis to be carried out will flow from the type of conceptual model used in the investigation.

4.4
SOME FUNDAMENTAL CONCEPTS

Statistical Universe

In the problem formulation stage, we must define very carefully the relevant **statistical universe** of observations. The universe, or **population**, consists of the total collection of items or elements that fall within the scope of a statistical investigation.

> The purpose of defining a statistical population is to provide explicit limits for the data collection process and for the inferences and conclusions that may be drawn from the study.

The items or elements that form the population may be individuals, families, employees, schools, corporations, and so on. Time and space limitations must be specified, and it should be clear whether or not any particular element falls within or outside the universe.

In survey work, a listing of all the elements in the population is referred to as the **frame**, or **sampling frame**. A **census** is a survey that attempts to include every element in the universe. The word "attempts" is used here because complete coverage may not be effected in surveys of very large populations, despite every effort to do so. Thus, for example, the Bureau of the Census readily admits that its national "censuses" of population invariably result in under-enumerations. Strictly speaking, any partial enumeration of a population constitutes a **sample**, but the term "census" is used as indicated here. In most practical applications, it is not even feasible to attempt complete enumerations of populations, and therefore, typically, only samples of items are drawn. If the population is well defined in space and time, the problem of selecting a sample of elements from it is considerably simplified.

Let us illustrate some of these ideas by means of a simple example. Suppose we draw a *sample* of 1,000 families in a large city to estimate the arithmetic mean family income of all families in the city. The aggregate of all families in the city constitutes the *universe*, and each family is an *element* of the universe. The income of the family is a *characteristic* of the unit. A listing of all families in the city would comprise a *frame*. If, instead of drawing the sample of 1,000 families, an attempt had been made to include all families in the city, a *census* would have been conducted. The definition of the universe would have to be specific as to the geographic boundaries that constitute the city and also the period for which income would be observed. The terms "family" and "income" would also have to be rigidly defined. Of course, the precise definitions of all of these concepts would depend on the underlying purposes of the investigation.

Summary of terms

The terms *universe* and *sample* are relative. An aggregate of elements that constitutes a population for one purpose may merely be a sample for another. Thus, if we want to determine the average weight of students in a particular classroom, the students in that room would represent the population. However, if we were to use the average weight of these students as an estimate of the corresponding average for all students in the school, then the students in the one room would be a sample of the larger population. The sample might not be a good one from a variety of viewpoints (such as representativeness), but nevertheless, it is a sample.

Relative terms

If the number of elements in the population is fixed, that is, if it is possible to count them and come to an end, the population is said to be **finite**. Such universes may range from a small to a large number of elements. For example, a small population might consist of the three vice-presidents of a corporation; a large population might be the retail transactions occurring in a large city during a one-year period. A point of interest concerning these two examples is that the vice-presidents represent a fixed and unchanging population, whereas the retail transactions illustrate a dynamic population that might differ considerably over time and space.

Finite populations

Infinite Populations

An **infinite population** is composed of an uncountable number of items. Usually such populations are conceptual constructs in which data are generated by processes that may be thought of as repeating indefinitely, such as the rolling of dice and the repeated measurement of the weight of an object. Sometimes, the population sampled is finite but so large that it makes little practical difference if it is considered to be infinite. For example, suppose that a population consisting of 1,000,000 manufactured articles contains 10,000 defectives. Thus, 1% of the articles is defective. If two articles are randomly drawn from the lot in succession, without replacing the first article after it is drawn, the probability of obtaining two defectives is

$$\left(\frac{10,000}{1,000,000}\right)\left(\frac{9,999}{999,999}\right)$$

For practical purposes, this product is equal to $(0.01)(0.01) = 0.0001$. If the population is considered infinite with 1% defective articles, the probability of obtaining two defectives is exactly 0.0001. Frequently, in situations when a finite population is extremely large relative to sample size, it is simpler to treat this population as infinite. Since a finite population is depleted by sampling without replacement whereas an infinite population is inexhaustible, and if the depletion causes the population to change only slightly, it may be simpler for computational purposes to consider the population infinite.

> Sometimes, an infinite population may be a *process* that produces finite populations.

For example, a company may draw a sample from a lot from a particular supplier in order to decide whether to purchase from this supplier in the future. Thus, the purchaser makes a decision concerning the *manufacturing process* that produces future lots. The particular lot sampled for test purposes is a finite population. The process that produces the particular lot may be viewed as an infinite population. Care must be exercised in such situations to ensure that the manufacturing process is indeed a stable one and may validly be viewed as a single universe. Future testing may in fact reveal differences of such a magnitude that the conceptual universe should be viewed as having changed.

Target Populations

Another useful concept is the **target population**, or the universe about which inferences are desired. Sometimes in statistical work, it is impractical or perhaps impossible to draw a sample directly from this target population, but it is possible to obtain a sample from a closely related population. The list of elements that constitutes this sampled **frame** may be related to—but is definitely different from—the list of elements contained in the target population.

Sampling frame

For example, suppose we wish to predict the winner in a forthcoming municipal election by means of a polling technique. The target population is the collection of individuals who will cast votes on election day. However, it is not possible to draw a sample directly from this population, since the specific individuals who will show up at the polls on election day are unknown. It may be possible to draw a sample from a closely related population, such as the eligible voting population. In this case, the list of eligible voters constitutes the sampling frame. The percentage of eligible voters who would vote for a given candidate may differ from the corresponding figure for the election day population. Furthermore, the percentage of the eligible voter population who would vote for a given candidate will probably change as the election date approaches. Thus, we have a situation in which the population that can be sampled changes over time and is different from that about which inferences are to be made. In the case of election polling, the situation is further complicated by the fact that at the time the sample is taken, many individuals may not have made up their minds concerning the candidate for whom they will vote. Therefore, some assumption must be made about how these "undecideds" will break down as to voting preferences. It is common to use in-depth interviews, in which the undecided voters are questioned about the issues and individuals in the campaign, to help determine for whom the respondents will probably vote. In carefully run election polls, numerous sample surveys are taken, spaced through time, including some investigations near the electron date. Then, trends can be determined in voting composition, and inferences can be made from populations that are defined close to election day. Some instances of incorrect predictions in national elections have resulted from failures to deal properly with the problem of undecided voters and from cessation of sampling too long before election day. There is no easy answer to the question of how to adjust for the fact that *Adjustments* the populations sampled are different from the election day population. For example, one approach to the problem of nonvoters is to conduct postelection surveys to determine the composition of the nonvoting group and to estimate their probable voting pattern had they shown up at the polls. Historical information of this sort could conceivably be used to adjust future polls of eligible voters. However, this is an expensive procedure, and the appropriate method of adjustment is fraught with problems.

In many statistical investigations, the target population coincides with a population that can be sampled. However, in any situation when one must sample a past statistical universe and yet make estimates for a future universe, the problem of inference about the target universe is present.

Control Groups

Probably the most familiar setting involving the concept of a control group is the situation in which an experimental group is given some type of treatment. In order to determine the effect of the treatment, another group is included in the experiment but is not given the treatment. These "no-treatment" cases are known as the **control group**. The effect of the treatment can then be determined by comparing the relevant measures for the "treatment" and "control" groups. For example, in testing the effectiveness of an inoculation against a particular disease, the inoculation may be administered to a group of school children (the treatment group) and not given to another group of school children (the control group).[1] The effectiveness of the inoculation can then be determined by comparing the incidences of the disease in the two groups. The experiment should be designed so that there is no systematic difference between the two groups at the outset that would make one group more susceptible to the disease than the other. Therefore, such experiments are sometimes designed with so-called "matched pairs," in which pairs of persons having similar characteristics, one from the treatment group and one from the control group, are drawn into the experiment. For example, if age and health are thought to have some effect on incidence of the disease, the experiment may require pairs of school children who are similar with respect to these characteristics so that the treatment can be given to one child of each pair and not to the other. Since the children are of similar ages and have the same health backgrounds, these factors cannot explain the fact that one child contracts the disease whereas the other does not. In the language of experimental design, age and general health conditions are said to have been **designed out** of the experiment. Numerous other techniques are employed in experimental design to ensure that treatment effects can be properly measured.

The concept of a control group is important in many statistical investigations in business and economics.

Forethought in selecting control groups

> The results of an investigation may be uninterpretable unless one or more suitable control groups have been included in the study.

Sadly, it is often *after* statistical investigations are completed at considerable expense when it is found that because of faulty design and inadequate planning, the results cannot be meaningfully interpreted or the data collected are inappropriate for testing the hypotheses in question.

> It is of paramount importance that during the planning stage, the investigators think ahead to the completion of the study.

They should ask, "If the collected data show thus-and-so, what conclusions can we reach?" This simple yet critical procedure will often highlight difficulties connected with the study design.

Example of use of control groups

The following example illustrates the use of control groups in statistical studies. Suppose a mail-order firm decides to conduct a study to determine the characteristics of its high-volume purchasers. Its purpose is to determine the distinguishing characteristics of these customers in order to direct future campaigns to noncustomers who have similar attributes. Assume that the firm decides to do this by studying all its high-volume customers. At the conclusion

[1] Difficult ethical questions arise in cases of this sort involving human experimentation. If the inoculation is indeed effective, its use clearly should not be withheld from anyone who wants it. In cases where the effectiveness of a new treatment is highly questionable, yet human experimentation appears necessary, the treatment group often is composed entirely of volunteers.

of the investigation, it will be able to make statements such as, "The mean income of high-volume customers is so-many dollars." Or it may calculate that $X\%$ of these purchasers have a certain characteristic. Such population figures will be of virtually no use unless the company has an appropriate comparison group against which to assess them. The company wants to be able to isolate the distinguishing characteristics of high-volume purchasers. Thus, in studying its customers, the company should have separated them into two groups, "high-volume" and "non-high-volume." If it studied both groups, it would be in a position to determine those properties that are different between the two groups. Thus, if the company found that the high-volume and non-high-volume customers had the *same* mean incomes and that in *both* groups $X\%$ possessed a certain characteristic, it could not use these properties to distinguish between the two groups. The properties that differed most between the two groups would obviously be the ones most useful for spelling out the distinguishing characteristics of high-volume purchasers. In summary, the firm could have used the non-high-volume customers as a control group against which to compare the properties of the high-volume groups, which in the terminology used earlier would represent the "treatment group."

Care must be used in the selection of the properties of the two groups to be observed. These properties should bear some logical relationships to the characteristic of high-volume versus non-high-volume purchasers. Otherwise, the properties may be spurious indicators of the distinguishing characteristics of the two groups. For example, income level would be logically related to purchasing volume; if the high-volume purchaser group had a substantially higher income than the non-high-volume group, then income would evidently be a reasonable distinguishing characteristic. On the other hand, suppose the high-volume purchaser group happened to have a higher percentage of persons who wore black shoes at the time of the survey than did the non-high-volume group. This characteristic of shoe color would *not* seem to be logically related to volume of purchases. Hence, we would not be surprised if the relationship between shoe color and volume of purchases disappeared in subsequent investigations or even reversed itself.

Selection of properties of control groups

A couple of comments may be made on the construction of control groups. If the treatment group is symbolized as A and the control group as B, then an alternative control group to the one used would have been the treatment and control groups combined, or $A + B$. Thus, in the above example, if the relevant data had been available for the high-volume and non-high-volume customers combined (that is, for all customers), this group could have constituted the control. For example, let us assume for simplicity that there were equal numbers of high-volume and non-high-volume customers. Suppose that 90% of high-volume customers possessed characteristic X, whereas only 50% of non-high-volume customers had this characteristic. The same information would be given by stating that 90% of high-volume customers possessed characteristic X, whereas 70% of *all* customers possessed this characteristic. (The 70% figure, of course, is the weighted mean of 90% and 50%.) With the knowledge of equal numbers of persons in the high-volume and non-high-volume groups, we can infer that

The construction of control groups

50% of non-high-volume customers had the property in question. This point is important, because sometimes historical data may be available for an entire group $A + B$, whereas available resources may permit a study only of the treatment group A or (the more usual case) only a sample of this group. However, drawing conclusions about the present from a historical control group is dangerous, because systematic changes may have taken place in the treatment and control groups over time or in the surrounding conditions of the experiment. Therefore, the more scientifically desirable procedure is to design the treatment and control groups for the specific investigation in question.

Objectives determine nature of control groups

The general objectives of an investigation determine the control groups to be used. Thus, individuals who are not customers of a firm could constitute an appropriate control group for an experiment to determine the distinguishing characteristics of the firm's customers; or customers who have not purchased a specific product could constitute the control group in an experiment to find the particular traits of the customers who purchase the product. Note that time considerations and available resources usually permit only drawing of samples from the treatment and control populations rather than complete enumerations of these populations.

Examples of control groups

Some other brief examples of the use of control groups will be given here. A national commission wanted to investigate insurance conditions in cities in which civil disturbances in the form of riots had occurred. Specifically, the commission wished to study cancellation rates for burglary, fire, and theft policies in sections of these cities ("riot areas") primarily affected by the riots. The main purpose was to determine whether individuals and businesses in these areas were having difficulty retaining such policies because of cancellations by insurance companies. It became clear in the planning stages of the study that it would not be sufficient merely to measure cancellation rates in the cities in which riots had occurred, because it would not be possible in the absence of other information to judge whether these rates were low, average, or high. Therefore, a sample of individuals and businesses in cities that had not experienced riots was used as a control group. Another control group was established consisting of individuals and businesses in the "nonriot areas" of the cities that had experienced riots. Thus, the data on cancellation rates could be meaningfully interpreted. Comparisons were made between cities that had experienced riots and those that had not. Further comparisons were made between cancellation rates in riot areas and nonriot areas in cities where these disturbances had been present. The data disclosed that burglary, fire, and theft insurance cancellation rates were higher in cities in which riots had occurred. Furthermore, within cities in which these civil disturbances had been present, cancellation rates were higher in riot sections that in nonriot sections. It may be noted, parenthetically, that if the company policies on cancellations were known and data were available in suitable form, the same information could have been obtained from the company records. However, such information was not available, so the sample survey was required to obtain the indicated data.

Another illustration is the case of a company that wished to determine whether its labor costs were very different from such costs throughout the company's industry group. It obtained the ratio of labor costs to total operating costs for the company and compared these with a published distribution of such ratios for all firms in the industry for the same period. In this situation, all firms in the industry constituted the control group. This is an illustration of a rather obvious need for and choice of a control group. In many situations, the need and choice are somewhat more subtle.

Types of Measurement Errors

The concept of error is central throughout all statistical work. Wherever we have measurement, inference, or decision making, the possibility of error is present. In this section, we deal with **errors of measurement**. Errors of inference and decision making are treated in subsequent chapters.

Systematic errors

It is useful to distinguish the two types of errors that may be present in statistical measurements, namely, systematic errors and random errors. **Systematic errors** cause a measurement to be incorrect in some systematic way. Such errors are built into the procedures and they **bias** the results of a statistical investigation. These errors may occur in the planning stages or during or after the collection process. Examples of causes of systematic error are faulty design of a questionnaire (such as misleading or ambiguous questions), systematic mistakes in planning and carrying out the collection and processing of the data, nonresponse and refusals by respondents to provide information, and a great discrepancy between the sampling frame and the target universe. If observations have arisen from a sample drawn from a statistical universe, systematic errors are those that persist even when the sample size is increased. As a generalization, these errors may be viewed as arising primarily from inaccuracies or deficiencies in the measuring instrument.

Random errors

On the other hand, **random errors** or **sampling errors** or **experimental errors** may be viewed as arising from the operation of a large number of uncontrolled factors, conveniently described by the term "chance." As an example of this type of error, if repeated random samples of the same size are drawn from a statistical universe (with replacement of each sample after it is drawn), then a particular statistic, such as an arithmetic mean, will differ from sample to sample. These sample means tend to distribute themselves below and above the "true" population parameter (arithmetic mean), with small deviations between the statistic and the parameter occurring relatively frequently and large deviations occurring relatively infrequently. The word "true" has quotation marks around it because it refers to the figure that would have been obtained through complete coverage of the universe, that is, a complete census using the same definitions and procedures used in the samples. The difference between the mean of a particular sample and the population mean is said to be a *random error* (or a *sampling error*, as it is termed in later chapters). The complete collection of factors that could explain why the sample mean differed from the population mean is unknown, but we can conveniently lump them all together and refer to the difference as a random or chance error.

> **Random errors** are those that arise from differences found between the outcomes of trials (or samples) and the corresponding universe value using the same measurement procedures and instruments.

The sizes of the differences indicate reliability or precision.

> Random errors decrease on the average as sample size increases.

It is precisely for this reason that we prefer a large sample of observations to a small one, all other things being equal; that is, since sampling errors are on the average smaller for large samples, the results are more reliable or more precise.

Systematic and random errors may occur in experiments in which the variables are manipulated by the investigator or in survey work where observations are made on the elements of a population without any explicit attempt to manipulate directly the variables involved. A few examples will be given to show how bias, or systematic error, may be present in a statistical investigation. The problems of *how random errors are measured* and *what constitutes suitable models for the description of such errors* represent central topics of statistical methods and are discussed extensively later in this chapter and in Chapters 5 through 8.

Systematic Error: Biased Measurements

The possible presence of biased measurements in an experimental situation may be illustrated by a simple example. Suppose that a group of individuals measured the length of a 36-inch table top using the same yardstick. Let us further assume that the yardstick, although calibrated as though it were 36 inches long, was in fact 35 inches long and this fact was unknown to the individuals making the measurements. There would then be a systematic error of one inch present in each of the measurements, and a statistic such as an arithmetic mean of the readings would reflect this bias. In this situation, the systematic error could be detected if a correctly calibrated yardstick were used as a standard against which to test the incorrect yardstick. This is an important methodological point.

> Systematic error often can be discovered through the use of an independent measuring instrument.

Even if the independent instrument is inaccurate, a comparison of the two measuring instruments may give clues about where the search for sources of bias should be made. The variation among the individual measurements made with the incorrect yardstick would be a measure of random error because the differences are not attributable to specific causes of variation. Note that the observations may have been very precise (although inaccurate), in the sense that each person's measurement was close to that of every other person. Thus, the random errors would be small, and there would be good *repeatability*, because in repeating the experiment, each measurement would be close to preceding measurements. These random or chance errors may be assumed to be compensating, in the sense that some observations would tend to be too large and some too small. Since the table top is 36 inches long, the measurements would tend to cluster around a value about one inch greater than the true length of the table top. In summary, we have a model in which each individual measurement may be viewed as the sum of three components: (1) the true value, (2) systematic error, and (3) random error. This relationship can be stated in equation form.

(4.1)
$$\frac{\text{Individual}}{\text{measurement}} = \frac{\text{True}}{\text{value}} + \frac{\text{Systematic}}{\text{error}} + \frac{\text{Random}}{\text{error}}$$

Systematic Error: Literary Digest *Poll*

A classic case of systematic error in survey sampling procedure is that of the *Literary Digest* prediction of the presidential election of 1936. During the election campaign between Franklin D. Roosevelt and Alf Landon, the *Literary Digest* magazine sent questionnaire ballots to a large list of persons whose names appeared in telephone directories and automobile registration lists. Over two million ballots—about one-fifth of the total number sent out—were returned by the respondents. On the basis of these replies, the *Literary Digest* erroneously predicted that Landon would be the next president of the United States.

The reasons that the results of this survey were so severely biased are rather clear. In 1936, during the Great Depression, the presidential vote was cast largely along economic lines. The group of the electorate that did not own telephones or automobiles did not have an opportunity to be included in the sample. This group, which represented a lower economic level than owners of telephones and automobiles, voted predominantly for Roosevelt, the Democratic candidate.

A second reason stemmed from the nonresponse group, which represented about four-fifths of those polled. Typically, individuals of high educational and high economic status are more apt to respond to voluntary questionnaires than those with low economic and educational status. Therefore, the nonresponse group doubtless contained a higher percentage of this low-status group than did the group that responded to the questionnaire. Again, this factor added a bias due to underrepresentation of Democratic votes.

In summary, the sample used for prediction purposes contained a greater proportion of persons of high socioeconomic status than were present in the target population, namely, those who cast votes on election day. Since this factor of socioeconomic status was related to the way people voted, a systematic overstatement of the Republican vote was present in the sample data.

Two methodological   Two methodological lessons can be derived from this example.
lessons

1. It is a dangerous procedure to sample a frame that differs considerably from the target population.
2. Procedures must be established to deal with the problem of nonresponse in statistical surveys.

Clearly, even if the proper target universe had been sampled in the previous case, the problem of nonresponse would still have to be properly handled.

Systematic Error: Method of Data Collection

Another example of bias will illustrate that the direction of systematic error may be associated with the nature of the agency that collects the data as well as the method by which the data are collected.

The alumni society of a large eastern university decided to gather information from the graduates of that institution to determine a number of characteristics, including their current economic status. One of the questions of interest was the amount of last year's gross income, suitably defined. A questionnaire was mailed to a random sample of graduates, and the results were tabulated from the returns. When frequency distributions were made and averages were calculated by year of graduation, it became clear that the income figures were unusually high compared with virtually any existing external data that could be examined; in other words, the income figures were clearly biased in an upward direction. It is fairly easy in this case to speculate on the causes of this upward systematic error. In this type of mail questionnaire, a higher nonresponse rate could be expected from those graduates whose incomes were relatively low than from those with higher incomes. That is, it appears reasonable that those with higher incomes would have a greater propensity to respond than others. Furthermore, if there were instances of misreporting of incomes, these probably tended to be overstatements rather than understatements, because of the desire to appear relatively economically successful.

On the other hand, let us consider the same sort of data as reported to the Internal Revenue Service on annual income tax returns. Doubtless, it is safe to say that relatively little overstatement of gross incomes occurs. Indeed, it seems reasonable to suppose that a downward bias exists in these data in the aggregate. Note that since responses to the Internal Revenue Service are mandatory, the effect of nonresponse may be considered negligible. Thus, the interesting situation is presented here of the same type of data being gathered by two different agencies, one set biased in an upward direction, the other in the opposite direction.

> In using secondary statistical information, we must exercise informed critical judgment to extract meaningful inferences.

This judgment must include practical considerations such as methods of data collection and auspices under which studies are conducted. Of course, false reporting is not easily overcome, particularly in situations in which no independent objective data are available against which the reported information may be checked.

EXERCISES 4.4

1. Explain the difference between a sample and a census.

2. Describe a situation in which the population of interest would be
   a. an infinite population.
   b. considered an infinite population, yet in reality is a finite population.
   c. considered an infinite population, because it is generated by a process that produces repeated observations.
   d. a target population.

3. Explain the way a *control group* should be used by criticizing the following situations:
   a. A study of the effectiveness of aspirin in relieving headaches showed that 90% of the adults in the study stated that they had obtained relief from pain 2 hours after taking 10 grains of aspirin.
   b. A stockbroker claims that because 85% of his recommendations last year performed better than market expectations (appropriately defined), everyone should invest with him.
   c. A fertilizer company conducted a study that showed that farmers who use Moregro fertilizer last year experienced a 10% increase in crop yield over the preceding year. This company is now advertising that Moregro will increase crop yields by 10%.

4. Determine an appropriate control group for each of the following studies, and briefly state why you feel it is appropriate.
   a. The American Athletic Association wants to determine whether a new form of heat treatment is effective in relieving its members of post-exercise cramps.
   b. A research center wishes to determine the effect of visual aids on children's recall capabilities (that is, can children recall the subject matter in the aids better if they are shown the aids first than if they are not shown the aid first).
   c. A medical investigator wishes to test the assertion that marathon runners do not suffer heart attacks.

5. Explain why the following control group was not adequate for the experiment conducted, and identify a more appropriate control group.

   A market research firm wanted to determine the effectiveness of a new advertising campaign on current users of a product. The campaign was presented to a group of individuals who already used the product and to another group of non-users. The non-users represented the control group. After the campaign, the users increased their purchases but the non-users showed no significant tendency to buy the product. The firm concluded that the campaign would be effective in increasing the consumption of the product by current users.

6. The personnel director of a large manufacturing concern wishes to determine employee opinion on the annual Christmas party. In the past, the party has been for employees only, but the director thought employees' spouses might be included for the next party. Of a random sample of 100 persons who attended the party the previous year, 80 want the party for employees only, and 20 request that spouses be included. Based on the results, the director decides to continue the current practice. Briefly evaluate the procedure employed by the director, indicating any possible sources of bias and nonhomogeneity.

7. Explain how the concept of a control group should be used by criticizing the following situations:
    a. Of the persons in the United States who contracted polio in a certain year, only 6% were from families with annual incomes over $10,000. Therefore, the study concluded that families in that income group had less to fear from polio than those in lower income groups.
    b. In a certain university, only 20% of those students failing to graduate were women. This statistic suggests that in this university women are better students than men.
    c. A manufacturer of a preparation for treating the common cold claimed that 95% of cold sufferers who used his product were free of their colds within a one-week period.

8. A magazine with 200,000 subscribers mailed a questionnaire to a random sample of 10,000 of these persons, inquiring about their attitudes toward the magazine's editorial policy. About 1,500 respondents mailed back completed questionnaires. Do you think that both sampling and systematic errors are present in the results of this study? Explain.

9. State whether each of the following errors should be considered random, systematic, or both, and why.
    a. In a study that attempted to estimate the percentage of students who smoke, the first 100 students who entered the student lounge—the only area in the building where students are permitted to smoke—were asked if they smoked. The study resulted in an overestimate of the true percentage.
    b. In a study to estimate the average life of a certain type of vacuum tube, five tubes were purchased, each one from a different store in a different wholesale sales region. The average life of the five tubes tested was shorter than the "true" average life.
    c. In a study to determine the effectiveness of a process that fills one-pound cans, 50 cans were selected randomly and weighed on a scale that mea-

sured 0.1 ounce too heavy. The process actually filled the cans, on the average, with 1 pound of material, but the 50 cans averaged 1.06 pounds.
d. The mayor of a community sent 1,000 questionnaires asking citizens to rate community services (police, fire, garbage collection, and so on) as bad, adequate, or good. Of the 87 responses, a majority of the people in the community considered the services adequate.

4.5
FUNDAMENTALS OF SAMPLING

Purposes

Sampling is important in most applications of quantitative methods to managerial and other business problems for a number of reasons. In certain instances, sampling may represent the only possible or practicable method of obtaining the desired information. For example, in the case of processes, such as manufacturing, in which the universe is conceptually infinite (including all future as well as current production), a complete enumeration of the population is not possible. On the other hand, if sampling is a destructive process, a complete enumeration of the universe may be possible, but it would not be practical to do so. For example, if a military procurement agency wanted to test a shipment of bombs, it could detonate all of the bombs in a testing procedure and obtain complete information concerning the quality of the shipment. However, since there would be no usable product remaining, a sampling procedure is clearly the only practical way to assess the quality of the shipment.

Sampling procedures are often employed for overall effectiveness, cost, timeliness, and other reasons. A complete census, although it does not have sampling error introduced by a partial enumeration of the universe, nevertheless often contains greater total error than does a sample survey, because greater care can usually be exercised in a sample survey than in carrying out censuses. Errors in collection, classification, and processing of information may be considerably smaller in sample surveys, which can be carried out under far more carefully controlled conditions than large-scale censuses. For example, it may be possible to reduce response errors arising from lack of information, misunderstood questions, faulty recall, and other reasons only by intensive and expensive interviewing and measurement methods, which may be feasible in the case of a sample but prohibitively costly for a complete enumeration.

The employment of sampling rather than censuses for purposes of timeliness occurs in a variety of areas. A notable example is the wide array of government data on economic matters such as income, employment, and prices, which are collected on a sample basis at periodic intervals. Timeliness of publication of these results is of considerable importance. The more rapid collection and processing of data afforded by sampling procedures represents an important advantage over corresponding census methods.

Random and Nonrandom Selection

Judgment sampling

Items can be selected from statistical universes in a variety of ways. It is useful to distinguish random from nonrandom methods of selection. In this book, attention is focused on **random**, or **probability**, **sampling**, that is, sampling in which the probability of inclusion of every element in the universe is *known*. **Nonrandom sampling** methods are referred to as **judgment sampling** because judgment is exercised in deciding which elements of a universe to include in the sample. Such judgment samples may be drawn by choosing "typical" elements or groups of elements to represent the population. They may even involve random selection at one stage but allow the exercise of judgment in another. For example, areas may be selected at random in a given city, and interviewers may be instructed to obtain specified numbers of persons of given types within these areas but may be permitted to make their own decisions as to which individuals are brought into the sample.

This book deals only with random, or probability, sampling methods (rather than judgment sampling) because of the clear-cut superiority of probability selection techniques.

> Random sampling is preferable to judgment sampling because of the precision with which estimates of population values can be made from the sample itself. In judgment selection there is no objective method of measuring the precision or reliability of estimates made from the sample.

This is an important advantage, since random sampling techniques provide an objective basis for measuring errors due to the sampling process and for stating the degree of confidence placed on estimates of population values.

Judgment samples can sometimes be usefully employed in the planning and design of probability samples. For example, when expert judgment is available, a pilot sample may be selected on a judgment basis in order to obtain information that will aid in the development of an appropriate sampling frame for a probability sample.

Simple Random Sampling

We have seen that a *random sample* or *probability sample* is a sample drawn in such a way that the probability of inclusion of every element in the population *is known*. A wide variety of types of such probability samples exist, particularly in the area of sample surveys. Experts in survey sampling have developed a large body of theory and practice aimed toward the optimal design of probability samples. This highly specialized area is often allocated an entire course or two in a graduate program in statistics. We will concentrate on the simplest and most fundamental probability method, namely, *simple random sampling*. The major body of statistical theory is based on this method of sampling.

We first define a simple random sample for the case of a finite population of N elements.

Finite population sampling without replacement

> A **simple random sample** of n elements is a sample drawn in such a way that *every* combination of n elements has an *equal* chance of being the sample selected.

Since most practical sampling situations involve sampling *without replacement*, it is useful to think of this type of sample as one in which each of the N population elements has an equal probability, $1/N$, of being the one selected on the first draw, each of the remaining $N - 1$ elements has an equal probability, $1/(N - 1)$, of being selected on the second draw, and so on until the nth sample item has been drawn. Since there are $\binom{N}{n}$ possible samples of n items, the probability that any sample of size n will be the one drawn is $1 \left/ \binom{N}{n} \right.$.

This concept of a simple random sample may be illustrated by the following example. Let the population consist of three elements, A, B, and C. Thus, $N = 3$. Suppose we wish to draw a simple random sample of two elements; then $n = 2$. Using equation 2.20 for the number of combinations that can be formed of N objects taken n at a time, we find that the number of possible samples is

$$\binom{3}{2} = \frac{3!}{2!1!} = 3$$

These three possible samples contain the following pairs of elements: (A, B), (A, C), and (B, C). The probability that any one of these three samples will be the one selected is $\frac{1}{3}$.

It is a property of simple random sampling that every element in the population has an equal probability of being included in the sample. However, many other sample designs possess this property as well, as for example, certain stratified sample and cluster sample procedures.

Simple random samples were defined above for the case of sampling a finite population without replacement. If a finite population is sampled *with replacement*, the same element could appear more than once in the sample. Since, for practical purposes, this type of sampling is virtually never employed, it will not be discussed any further here.

Finite population sampling with replacement

On the other hand, simple random sampling of *infinite populations* is important, particularly in the context of sampling of processes. The following definition corresponds to the one given for finite populations.

Infinite population sampling

> For an infinite population, a simple random sample is one in which on *every* selection, each element of the population has an *equal* probability of being the one drawn.

This is difficult to visualize in terms of actual sampling from a physical population. Therefore it is helpful to take a more formal approach and to use the language of random variables. Thus, we view the drawing of the sample as an experiment in which observations of values of a random variable are generated, and the successive sample observations or elements are the outcomes of trials of the experiment. Then a simple random sample of n observations is defined by the presence of two conditions: (1) the n successive trials of the experiment are independent, and (2) the probability of a particular outcome of a trial defined as a success remains constant from trial to trial. In terms of sampling a physical population, we may interpret these as meaning that (1) the n successive sample observations are independent and (2) the population composition remains constant from trial to trial.

To aid in the interpretation of the above definition, let us consider the case of drawing a simple random sample of n observations from a Bernoulli process. Assume a situation in which a fair coin is tossed. A simple random sample of n observations would be the sample consisting of the outcomes on n *independent* tosses of the coin. Thus, if the number 0 denotes the appearance of a tail and 1 the appearance of a head, the following notation might designate a particular simple random sample of five observations, in which two tails and three heads were obtained in the indicated order (0, 1, 0, 1, 1). The random variable in this illustration may be designated as "number of heads (or tails) obtained in five tosses of a fair coin." In summary, we note that (1) the tosses of the coin were statistically independent and (2) the probability of obtaining a head (or tail) remained constant from trial to trial. It is conventional to use an abbreviated method of referring to such a sample as "a sample of five independent observations from a Bernoulli process," or "a sample of five independent observations from a binomial distribution."

The term "random sample," although it properly refers to a sample drawn with known probabilities, is often used to mean "simple random sample." The student should be aware of this alternative usage.

Methods of Simple Random Sampling

Although it is easy to state the definition of a simple random sample, it is not always obvious how such a sample is to be drawn from an actual population. The following are two useful methods:

Drawing chips from a bowl   We first restrict our attention to the most straightforward situation, in which the population is finite and the elements are easily identified and can be numbered. For example, suppose there are 100 students in a college freshman class and we wish to draw a simple random sample of 10 of these students without replacement. We could assign numbers from 1 to 100 to each of the students and place these numbers on physically similar disks (or balls, slips of paper, and so on), which could then be placed in a bowl. We shake the bowl to mix the disks thoroughly and then proceed to draw the sample. The first disk is drawn, and we record the number written on it. We then shake the bowl again, draw the second disk, and record the result. The process is repeated until we have drawn 10 numbers. The students corresponding to these 10 numbers constitute the required simple random sample.

If the population size is very large, the above procedure can become quite unwieldy and time-consuming. Furthermore, it may introduce biases if the disks are not thoroughly mixed. Therefore, tables of random digits are often used for the purpose of drawing such samples. These tables are useful for the selection of other types of probability samples, as well.

Tables of random numbers

A **table of random digits** is simply a table of digits that has been generated by a random process. For ease of use, the digits are usually combined into groups, for example, of five digits each. Thus, a table of random digits could be generated by the process of drawing chips from a bowl similar to the one just described. The digits 0, 1, 2, . . . , 9 could be written on disks and the disks placed in a bowl and then drawn, one at a time, *replacing the selected disk after each drawing.* Thus, on each selection, the population would consist of the 10 digits. The recorded digits would constitute a particular sequence of random digits. These tables are now usually produced by a computer that has been programmed to generate random sequences of digits.

We now illustrate the use of random digits using Table 4-1. Suppose there were 9,241 undergraduates at a large university and we wished to draw a simple random sample of 300 of these students. Each of the 9,241 students could be assigned a four-digit number, say, from 0001 to 9241. This list of names and numbers would constitute the sampling frame. We now turn to a table of random digits in order to select a simple random sample of 300 such four-digit numbers. We may begin on any page in the table and proceed in any systematic manner to draw the sample. Assume we decide to use the first four columns of each group of five digits, beginning at the upper left and reading downward. Starting with the first group of digits, we find the sequence 98389. Since we are using the first four digits, we have the number 9838. This exceeds the largest number in our population, 9241, so we ignore this number and read down to pick up the next four-digit number, 1724. This is the number of the first student in the sample. Reading down consecutively, we find 0128,9818 (which we ignore), 5926, and so on until 300 four-digit numbers between 0001 and 9241 have been specified. If any previously selected number is repeated, we simply ignore the repeated appearance and continue. In this illustration, we read downward on the page, but we could have read laterally, diagonally, or in any other systematic fashion. The important point is that each four-digit number has an equal probability of selection, regardless of what systematic method of drawing is used, and regardless of what numbers have already preceded.

Using random digits

Methods are available for drawing types of samples other than simple random samples and even for situations in which the elements have not been listed beforehand. Many of the tables include instructions for their use, but we will not pursue the subject further here.

Computers are often used these days to generate sequences of random numbers. However, even when a computer is used, it is useful to refer back to

Computer-generated sequences

TABLE 4-1
Random digits

98389	95130	36323	33381	98930	60278	33338	45778	86643	78214
17245	58145	89635	19473	61690	33549	70476	35153	41736	96170
01289	68740	70432	43824	98577	50959	36855	79112	01047	33005
98182	43535	79938	72575	13602	44115	11316	55879	78224	96740
59266	39490	21582	09389	93679	26320	51754	42930	93809	06815
42162	43375	78976	89654	71446	77779	95460	41250	01551	42552
50357	15046	27813	34984	32297	57063	65418	79579	23870	00982
11326	67204	56708	28022	80243	51848	06119	59285	86325	02877
55636	06783	60962	12436	75218	38374	43797	65961	52366	83357
31149	06588	27838	17511	02935	69747	88322	70380	77368	04222
25055	23402	60275	81173	21950	63463	09389	83095	90744	44178
35150	34706	08126	35809	57489	51799	01665	13834	97714	55167
61486	33467	28352	58951	70174	21360	99318	69504	65556	02724
44444	86623	28371	23287	36548	30503	76550	24593	27517	63304
14825	81523	62729	36417	67047	16506	76410	42372	55040	27431
59079	46755	72348	69595	53408	92708	67110	68260	79820	91123
48391	76486	60421	69414	37271	89276	07577	43880	08133	09898
67072	33693	81976	68018	89363	39340	93294	82290	95922	96329
86050	07331	89994	36265	62934	47361	25352	61467	51683	43833
84426	40439	57595	37715	16639	06343	00144	98294	64512	19201
41048	26126	02664	23909	50517	65201	07369	79308	79981	40286
30335	84930	99485	68202	79272	91220	76515	23902	29430	42049
33524	27659	20526	52412	86213	60767	70235	36975	28660	90993
26764	20591	20308	75604	49285	46100	13120	18694	63017	85112
85741	22843	16202	48470	97412	65416	36996	52391	81122	95157

Source: The Rand Corporation, *A Million Random Digits with 100,000 Normal Deviates* (New York: Free Press, 1955), excerpt from page 387. Copyright 1955 by The Rand Corporation. Used by permission.

the idea of tables of random numbers to understand how a universe of elements is mixed in the abstract for sampling purposes.

The commands for producing a table of four-digit random numbers by Minitab, and 99 such numbers, are given in Table 4-2.

EXERCISES 4.5

1. Indicate whether you think a sample or census would be advisable in each of the following situations. Give a brief reason for your choice in each case.
    a. A professor wishes to determine whether the students in one of his classes would prefer an open- or closed-book final examination.
    b. The publisher of a nationally distributed magazine wishes to determine the socioeconomic characteristics of the magazine's subscribers.

TABLE 4-2
Four-digit random numbers (99) produced by Minitab

```
MTB > RANDOM 99 C1;
SUBC> INTEGERS 0 9999.
MTB > PRINT C1

C1
   7915    9413    6698    7351    8947    8437    4652    1507    8414    1803    5499
   7498    7351    8989    3745    8219    7414    6778    7331    6453    6742    2776
   7099    7407    5981    7698    2344    3036    9583    7880    5082    5339    7461
   2671    3989    8634    9277    9677    9633    4206    5831    8921    5132    1536
   2005    0626    8365    5649    6209    6126    5767    0987    3441    0187    3478
   4839    4903    2915    4497    2176    2023    2974    1787    3456    2074    9289
   1195    9399    4989    3724    5604    0520    5005    5695    1920    0040    5086
   5795    4402    0303    7980    7509    8673    4225    8149    8711    8919    4938
   7340    7535    1914    9294    1823    7966    5785    3147    3394    4328    1123

MTB > STOP
```

   c. A manufacturer wants to determine the level of customer satisfaction among purchasers of its toaster ovens.

   d. A trade association of 25 firms wishes to obtain current data on numbers of employees in these firms.

2. The school board of Lower Fenwick wished to determine voter opinion concerning a special assessment that would permit the expansion of school services. Lower Fenwick, an industrial community on the fringe of a metropolitan area, has a population of 25,000. There are 5,000 students enrolled in the public schools of the community. The board selected a random sample of these students and sent questionnaires to their parents.

   a. Identify the statistical universe from which this sample was drawn.

   b. Is the sample chosen a simple random sample of parents in Lower Fenwick? Of parents of public school children in Lower Fenwick? Why or why not?

   c. If you had been asked to assist the school board, would you have approved the universe it studied? Defend your position.

3. State whether each of the following statements is true or false and explain your answer:

   a. Judgment sampling is good, because we can get an objective measure of the random error.

   b. When costs permit a census to be taken, a census is usually preferable to sampling.

   c. Systematic errors can often be reduced by better procedures, while random errors can only be reduced by larger sample sizes.

KEY TERMS

Census   a survey that attempts to include every element in the population.

Control Group   a group that is not given the treatment of interest in an experiment.

Direct Experimentation (controlled) Study   a study in which the investigator controls or manipulates factors that affect one or more variables.

Finite Population   a population in which the number of elements is fixed or countable.

Ideal Research Design   a design for research without reference to the availability of data or feasibility of their collection.

Infinite Population   a population in which the number of items is uncountable.

Judgment Sampling   nonrandom (nonprobability) sampling.

Mathematical Models   mathematical formulations that attempt to describe, to predict or to control real-world phenomena.

Observational Study   a study in which the investigator observes historical or present relationships among variables.

Random (sampling or experimental) Errors   errors attributable to chance.

Random (probability) Sample   a sample in which the probability of inclusion of every element in the population is known.

Sample   any partial enumeration of a population.

Sampling Frame   a listing of all of the elements in a population.

Simple Random Sample   a sample of n elements drawn in a way that every combination of n elements has the same probability of being the sample selected.

Statistical Universe (population)   the total collection of items that fall within the scope of a statistical investigation.

Systematic Error (bias)   a type of error that causes statistical measurements to be incorrect in a systematic way.

Table of Random Digits   a table of digits that has been generated by a random process.

Target Population   population about which inferences are desired.

5

SAMPLING DISTRIBUTIONS

In Chapter 1, we examined how to compute the arithmetic mean and standard deviation from the data contained in a sample. We now consider how such statistics differ from sample to sample if repeated simple random samples of the same size are drawn from a statistical population. The probability distribution of such a statistic is referred to as its **sampling distribution**. Thus, we may have a sampling distribution of a proportion, a sampling distribution of a mean, and so on.

> Sampling distributions are the foundations of statistical inference and are of considerable importance in modern statistical decision theory as well.

We begin our discussion by examining the concept of the samping distribution of a statistic using as an example the sampling distribution of the mean.

In many, perhaps most, business and managerial applications in which data are collected as an aid to drawing an inference or making a decision, a single sample of n observations is drawn from the relevant population of individuals or physical entities. Assuming that we draw a simple random sample from the population, we cannot tell beforehand what value we would obtain for a statistic, such as the arithmetic mean of the sample. Even if we knew the population mean and other population parameters, we could not predict exactly the value of the mean of the sample. The sample mean is a random variable and is subject to chance sampling variations. Similar to other random variables, the sample mean has a probability distribution. We refer to this probability distribution of a sample statistic as a *sampling distribution*. Since the sampling distribution is a probability distribution, it consists of two elements, the possible values of the statistic and the probabilities of those values (or interval of values) for the given sample size. We illustrate in Example 5-1 the concept of the sampling distribution of a mean in a simple sampling context.

EXAMPLE 5-1

A population consists of the six values (3, 4, 5, 6, 7, 8). If we draw a simple random sample of two of these items, we can calculate a statistic such as the arithmetic mean for the sample. Find the sampling distribution of the mean in this situation.

SOLUTION

The population can be represented as a discrete, uniform distribution as shown in Figure 5-1. That is, the six population values 3, 4, 5, 6, 7, 8 are equally likely and are therefore each assigned a probability of 1/6. The mean population value is

$$\mu = \frac{\Sigma X}{N} = \frac{3 + 4 + 5 + 6 + 7 + 8}{6} = 5.5$$

Suppose we now draw a simple random sample of two items without replacement from this population. There are $\binom{6}{2} = 15$ such possible samples. The 15 possible samples are listed in column (1) of Table 5-1 and the mean for each sample is given in column (2). For example, the first sample might consist of a 3 and 4 and its mean is equal to 3.5, and so forth. Since the number of possible samples is 15, the probability that a sample of size 2 would have a mean of 3.5 is 1/15. The sampling distribution of the mean consists of the sample means in column (2) and their probabilities in column (4). The sampling distribution is graphed in Figure 5-2.

FIGURE 5-1

Graph of the population distribution in Example 5-1

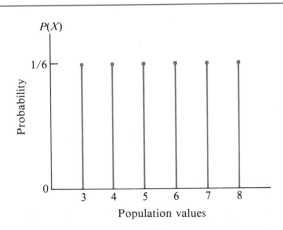

TABLE 5-1
The sampling distribution of the mean for simple random samples of 2 items from
a population of 6 items ($n = 2$, $N = 6$)

(1) Possible Samples	(2) Sample Mean ($\bar{X}$)	Number of Samples	Probability $P(\bar{X})$
(3, 4)	3.5	1	1/15
(3, 5)	4.0	1	1/15
(3, 6)(4, 5)	4.5	2	2/15
(3, 7)(4, 6)	5.0	2	2/15
(3, 8)(5, 6)(4, 7)	5.5	3	3/15
(4, 8)(5, 7)	6.0	2	2/15
(5, 8)(6, 7)	6.5	2	2/15
(6, 8)	7.0	1	1/15
(7, 8)	7.5	1	1/15
		15	1.00

Interpretation of the Sampling Distribution

A few properties of the sampling distribution given in Table 5-1 and Figure 5-2 may be noted. The mean of the six population values is $\mu = 5.5$. The mean of the sample means ($\bar{X}$ values) is also equal to 5.5. Another way of stating this is that $E(\bar{X}) = \mu$. When the expected value of the sample mean ($\bar{X}$) is equal to the population mean (μ), then $\bar{X}$ is said to be an unbiased estimator of μ. This property pertains under simple random sampling. Of course, ordinarily only one sample is drawn from a population and a sample statistic such as the sample mean, $\bar{X}$, is used as an estimate of the corresponding population parameter, μ. Since the sampling distribution gives us information about how sample means might vary from the population mean value, the sampling distribution enables us to determine the accuracy of using the sample mean as an estimate of the population mean. For example, in the present illustration, we can say that the probability that the sample mean will differ from the population mean by no more than one unit is the probability that the sample mean will be between 4.5 and 6.5, or $2/15 + 2/15 + 3/15 + 2/15 + 2/15 = 11/15$.

It is important to realize that a sampling distribution is a theoretical distribution. It is not ordinarily empirically derived, except in occasional computer simulation experiments. It is useful to interpret the sampling distribution in a relative frequency context as follows. One takes repeated simple random samples (theoretically an infinite number of samples) of the same size from a population and records a statistic, such as the mean, for each sample. The frequency distribution of the sample statistic approaches the theoretical sampling distribution of the statistic.

FIGURE 5-2

Graph of the sampling
distribution of the mean for
Example 5-1

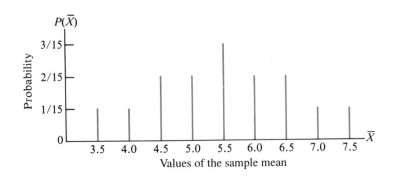

EXERCISES 5.1

1. Consider a population consisting of the following five elements: (4, 5, 6, 7, 8). Assume these values represent the number of sales per day made by a sales representative.

   a. Obtain the sampling distribution of the mean for all possible simple random samples of size 2, drawn without replacement.

   b. Compute the mean of the sampling distribution and compare it with the mean of the population values.

   c. What proportion of elements in the population lie between 5 and 7 sales? What proportion of means lie between 5 and 7 sales?

2. For the same population of 5 elements given in exercise 1, now assume sampling with replacement.

   a. Obtain the sampling distribution of the mean for all possible simple random samples of size 2.

   b. Compute the mean of the sampling distribution and compare it with the mean of the population values. What generalization is suggested by comparing this result with the corresponding result for exercise 1, part (b)?

5.2
CONTINUOUS DISTRIBUTIONS

Thus far, we have dealt solely with probability distributions of *discrete* random variables. Probability distributions of continuous random variables are also of considerable importance in statistical theory and practice. We turn now to an examination of such distributions. You may want to review the definitions of discrete and continuous variables given in section 3.1.

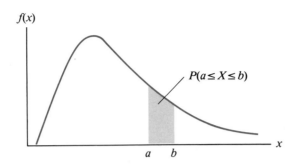

FIGURE 5-3

Graph of a continuous
distribution: the shaded area
represents the probability that
the random variable X lies
between a and b

The probability distribution of a continuous random variable is represented by a smooth curve as in Figure 5-3. The curve is usually referred to as
a *probability density function* or *probability distribution*. The areas under the
curve represent probabilities for the random variable X.

Specifically, the area under the curve lying between the two vertical lines
erected at points a and b on the x axis represents the probability that X lies
between a and b.[1] The total area under a continuous curve representing the
probability distribution is equal to *one*. Note that $P(a \leqslant X \leqslant b)$ and $P(a <
X < b)$ are equal because the probability is zero that $X = a$ or $X = b$. That is,
for continuous probability distributions, probabilities are only assigned to
intervals, not points.

In the continuous case, since there are an infinite number of points between a and b, the probability that X lies between a and b may be viewed as
the sum of an infinite number of ordinates erected from a to b. Intuitively, we
see this sum as identical with the area bounded by the curve, the horizontal
axis, and the ordinates at a and b.

In the discrete case, $f(x)$ denotes the probability that a random variable
X takes on the value x. In the continuous case, $f(x)$ cannot be interpreted as
the probability of a event x, since there are an infinite number of x values and
the probability of any one of them must be considered zero.[2]

---

[1] Let the value of the probability distribution of a random variable X at x be denoted $f(x)$.
If X is discrete, the probability that X lies between a and b inclusive [in the closed interval (a, b)] is

$$P(a \leqslant X \leqslant b) = \sum_{x=a}^{b} f(x)$$

If X is continuous, the probability that X lies between a and b is

$$P(a \leqslant X \leqslant b) = \int_{a}^{b} f(x)\, dx$$

The reader acquainted with integral calculus can see that this definition in the continuous
case is the counterpart of the summation in the discrete case. Also, it can be seen that the graphic
interpretation of the probability in the continuous case is the area bounded by the curve whose
value at x is $f(x)$, the x axis, and the ordinates at a and b. If the probability distribution is
continuous at a and b, it makes no difference whether we consider $P(a \leqslant X \leqslant b)$ or $P(a < X < b)$,
because the probability is zero that X is exactly equal to a or exactly equal to b.

[2] For example, $P(X = a) = \int_{a}^{a} f(x)\, dx = 0$.

FIGURE 5-4

Graph of the normal distribution

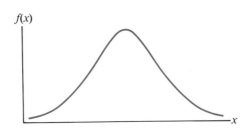

Probability density functions can have a variety of shapes. Perhaps the most generally familiar one is the bell-shaped curve, known as the normal curve or normal distribution, depicted in Figure 5-4.

We turn now to a discussion of the normal distribution, followed by its use as a model for the sampling distributions of the mean. Then we discuss sampling distributions of a number of occurrences and proportions of occurrences. For each of these sampling distributions, the normal curve is used as an approximation to the binomial distribution.

5.3
THE NORMAL DISTRIBUTION

The normal distribution plays a central role in statistical theory and practice, particularly in the area of statistical inference. Before we consider such applications, let us examine the basic properties of the normal distribution.

Properties of the Normal Curve

Probability distributions of continuous random variables can be described by the same types of measures (such as means, medians, and standard deviations) as are used for discrete random variables.

> An important characteristic of the normal curve is that we need know only the mean and standard deviation to compute the entire distribution.

The normal probability distribution is defined by the equation

(5.1)
$$f(x) = \frac{1}{\sqrt{2\pi}\sigma} e^{-(1/2)[(x-\mu)/\sigma]^2}$$

In this equation, the mean μ and the standard deviation σ, which determine the location and spread of the distribution, are said to be the two parameters

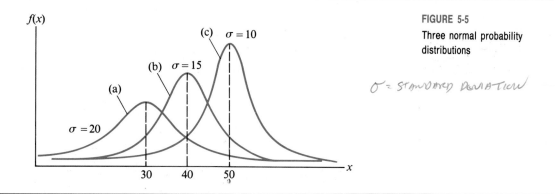

FIGURE 5-5

Three normal probability distributions

$\sigma = $ STANDARD DEVIATION

of the normal distribution.[3] Thus, for given values of μ and σ, if we substitute a value of x into equation 5.1, we can compute the corresponding value of $f(x)$. Following the usual convention, the values x of the random variable of interest are plotted along the horizontal axis and the corresponding ordinates $f(x)$ along the vertical axis. Figure 5-5 shows three normal probability distributions that differ in their locations and spreads. Distribution (c) has the largest mean (50), and distribution (a) has the smallest mean (30). On the other hand, the standard deviation of (c), which is 10, is the smallest of the three, while the standard deviation of (a), which is 20, is the largest. Thus, the normal distribution defined by equation 5.1 represents a family of distributions, with each specific member of that family determined by particular values of the parameters μ and σ.

Graphically, a normal curve is bell-shaped and symmetrical around the ordinate erected at the mean, which lies at the center of the distribution. Recall that the total area under the graph of a continuous probability distribution is equal to one; since one-half of the area (representing probability) lies to the left (right) of the mean, the probability is 0.5 that a value of x will fall below (above) the mean. The values of x range from minus infinity to plus infinity. As we move farther away from the mean, either to the right or to the left, the ordinates $f(x)$ get smaller. Thus, moving in either direction from the mean, the curve is **asymptotic** to the x axis; that is, the curve gets closer to the horizontal axis but never reaches it. However, for practical purposes, we rarely need to consider x values lying beyond three or four standard deviations from the mean, since nearly the entire area is included within this range. Stated differently, there is virtually no area in the tails of a normal distribution beyond three or four standard deviations from the mean.

---

[3] The numbers π and e are simply constants that arise in the mathematical derivation; their approximate values are $\pi = 3.1416$ and $e = 2.7183$. The number π is the familiar quantity that appears in numerous mathematical formulas, such as the expression for the area of a circle, $A = \pi r^2$, where A denotes the area and r the radius. The constant e is the base of the natural logarithm system, as indicated in the discussion of the Poisson distribution in section 3.7.

Areas under the Normal Curve

We now turn to the use of the areas under the normal curve. Although it was important to define the distribution as in equation 5.1 in order to observe the relationship between x values and $f(x)$ values, in most applications in statistical inference we are not interested in the ordinates of the curve. Rather, since the normal curve is a continuous distribution and since it is a useful probability distribution, we are interested in the areas under the curve.

Normally distributed variables   It is convenient to use the term **normally distributed** for variables that have normal probability distributions, and we shall do so here. The term "normal" merely refers to probability distributions described by equation 5.1; it does not imply that other distributions are in some sense "abnormal." Normally distributed variables occur in a variety of units, such as dollars, pounds, inches, and hours. Any normally distributed variable can be transformed into a form applicable to a single table of areas under the normal curve, regardless of the units of the original data. The transformation used for this purpose is that of the standard unit or, as it is often called in the case of a normal distribution, the **standard score**. As we noted earlier, to express an observation of a variable in standard units, we obtain the deviation of this observation from the mean of the distribution and then state this deviation in multiples of the standard deviation. For example, suppose a variable is normally distributed with mean 100 pounds and standard deviation 10 pounds. If one observed value of this variable is 120 pounds, what is this number in standardized units? The deviation of 120 pounds from 100 pounds is $+20$ pounds, in units of the original data. Dividing $+20$ pounds by 10 pounds, we obtain $+2$. Thus, if one standard deviation equals 10 pounds, a deviation of $+20$ pounds from the mean lies *two standard deviations* above the mean.

Standard score   Let us state this notion in general form. The standard score, that is, the number of standard units z for an observation x from a probability distribution is defined by

(5.2)
$$z = \frac{\text{Value} - \text{Mean}}{\text{Standard deviation}} = \frac{x - \mu}{\sigma}$$

where x = the value of the observation
μ = the mean of the distribution
σ = the standard deviation of the distribution

In the illustration, $z = \frac{120-100}{10} = \frac{20}{10} = +2$. The "$+2$" indicates a value lying two standard deviations *above* the mean. If the observation had been 80, we would have $z = \frac{80-100}{10} = \frac{-20}{10} = -2$. The "$-2$" denotes a value lying two standard deviations *below* the mean.

We now turn to an example to illustrate the use of a table of areas under the normal curve to compute probabilities. The Watts Renewal Corporation has a manufacturing process that produces light bulbs whose lifetimes are normally distributed with an arithmetic mean of 1,000 hours and a standard

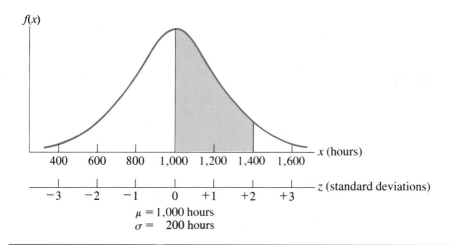

FIGURE 5-6

Relationship between
x values and z values

deviation of 200 hours. Figure 5-6 shows the relationship between values of
the original variable (*x* values) and values in standard units (*z* values).

Suppose we wish to determine the proportion of light bulbs produced by
this process with lifetimes between 1,000 and 1,400 hours, indicated by the
shaded area in Figure 5-6. We can obtain this value from Table A-5 of Appendix A, which gives areas under the normal curve lying between vertical lines
erected at the mean and at specified points above the mean stated in multiples
of standard deviations (*z* values).

● The left column of Table A-5 gives *z* values to one decimal place.

● The column headings give the second decimal place of the *z* value.

● The entries in the body of Table A-5 represent the area included between
  the vertical line at the mean and the line at the specified *z* value.

Thus, returning to our example, the *z* value for 1,400 hours is $z = (1,400 -
1,000)/200 = +2$. In Table A-5, we find the value 0.4772; hence 47.72% of the
area in a normal distribution lies between the mean and a value two standard
deviations above the mean. We conclude that 0.4772 is the proportion of light
bulbs produced by this process with lifetimes between 1,000 and 1,400 hours.

We now note a general point about the distribution of *z* values. Comparing the *x* scales and *z* scales in Figure 5-6, we see that for a value at the
mean in the distribution of *x*, $z = 0$. If an *x* value is at $\mu + \sigma$ (that is, one
standard deviation above the mean), $z = +1$, and so on. Therefore the probability distribution of *z* values, referred to as the **standard normal distribution**,
is simply a normal distribution with mean zero and standard deviation one.[4]

---

[4] We note that Table A-5 gives values of the integral

$$\int_0^{z_0} f(z)\, dz \quad \text{where} \quad z_0 = \frac{x_0 - \mu}{\sigma} \quad \text{and} \quad f(z) = \frac{1}{\sqrt{2\pi}} e^{-z^2/2}$$

EXAMPLE 5-2

What is the proportion of light bulbs produced by the Watts Renewal Corporation with lifetimes between 800 and 1,200 hours?

SOLUTION

First, we transform these values to deviations from the mean in units of the standard deviation.

$$\text{If } x = 1,200, \quad z = \frac{1,200 - 1,000}{200} = +1$$

$$\text{If } x = 800, \quad z = \frac{800 - 1,000}{200} = -1$$

Thus, we want to determine the area in a normal distribution that lies within one standard deviation of the mean. Table A-5 of Appendix A gives entries only for positive z values. However, since the normal distribution is symmetrical, the area between the mean and a value one standard deviation *below* the mean is the same as the area between the mean and a value one standard deviation *above* the mean. From Table A-5 we find that 34.13% of the area lies between the mean and a value one standard deviation above the mean. Hence, we double this area to find that about 68.3% of the light bulbs produced by this process have lifetimes between 800 and 1,200 hours. The required area is shown in Figure 5-7(a). Note that in general, about 68.3% of the area in a normal distribution lies within one standard deviation of the mean.

FIGURE 5-7

Areas corresponding to
Examples 5-1 through 5-4

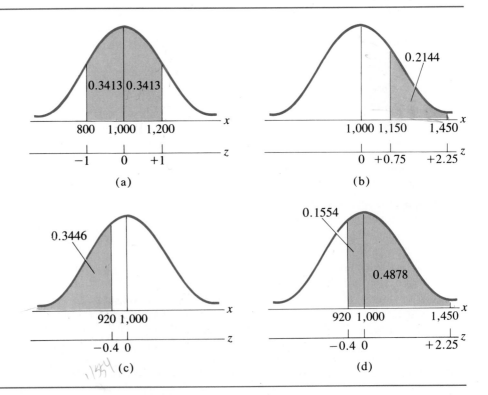

EXAMPLE 5-3

What is the proportion of light bulbs produced by the Watts Renewal Corporation with lifetimes between 1,150 and 1,450 hours?

Both 1,150 and 1,450 lie above the mean of 1,000 hours. We can determine the required probability by obtaining (1) the area between the mean and 1,450 and (2) the area between the mean and 1,150 and then subtracting (2) from (1).

$$\text{If } x = 1,450, \quad z = \frac{1,450 - 1,000}{200} = \frac{450}{200} = 2.25$$

$$\text{If } x = 1,150, \quad z = \frac{1,150 - 1,000}{200} = \frac{150}{200} = 0.75$$

Table A-5 gives 0.4878 as the area corresponding to $z = 2.25$ and 0.2734 for $z = 0.75$. Subtracting 0.2734 from 0.4878 yields 0.2144, or 21.44%, as the result. This area is shown in Figure 5-7(b).

EXAMPLE 5-4

What is the proportion of light bulbs produced by the Watts Renewal Corporation with lifetimes less than 920 hours?

The observation "920 hours" lies below the mean. We solve this problem by determining the area between the mean and 920 and subtracting this value from 0.500, which is the entire area to the left of the mean.

$$\text{If } x = 920, \quad z = \frac{920 - 1,000}{200} = -0.40$$

Only positive z values are shown in Table A-5, but we look up $z = 0.40$ and find 0.1554, which is also the area between the mean and $z = -0.40$. Subtracting 0.1554 from 0.5000 gives the desired result, 0.3446. The area corresponding to this probability is shown in Figure 5-7(c).

EXAMPLE 5-5

What is the proportion of light bulbs produced by this process with lifetimes between 920 and 1,450 hours?

Since 920 lies below the mean and 1,450 lies above the mean, we determine (1) the area lying between 920 and the mean and (2) the area lying between 1,450 and the mean, and we add (1) to (2). The respective z values for 920 and 1,450 were previously determined as -0.40 and $+2.25$, with corresponding areas of 0.1554 and 0.4878. Adding these two figures, we obtain 0.6432 as the proportion of light bulbs with lifetimes between 920 and 1,450 hours. The corresponding area is shown in Figure 5-7(d).

TABLE 5-2

Percentages of area that lie within specified intervals around the mean in a normal distribution

Interval	Percentage of Area
$\mu \pm \sigma$	68.3
$\mu \pm 2\sigma$	95.5
$\mu \pm 3\sigma$	99.7

We stated that in the normal distribution, the range of the x variable extends from minus infinity to plus infinity. Yet, in Examples 5-2, 5-3, 5-4, and 5-5, negative lifetimes are not possible. This illustrates the point that a variable may be said to be normally distributed provided that the normal curve constitutes a good fit to its empirical frequency distribution within a range of about three standard deviations from the mean. Since virtually all the area is included in this range, the situation in the tails of the distribution is considered negligible.

It is useful to note the percentages of area that lie within integral numbers of standard deviations from the mean of a normal distribution. These values have been tabulated in Table 5-2. Hence, as we observed in Example 5-2, about 68.3% of the area in a normal distribution lies within plus or minus one standard deviation from the mean. The reader should verify the other figures from Table A-5. Let us restate these probability figures in terms of rough statements of odds.

> Since about two-thirds of the area in a normal distribution lies within one standard deviation, the odds are about two to one that an observation will fall within that range.

Correspondingly, the odds are about 95:5, or 19:1, for the two standard deviation range and 997:3, or about 332:1, for three standard deviations.

EXERCISES 5.3

 1. A regional survey group tabulated the starting salaries of recent graduates from two eastern universities and found that the salaries were normally distributed. Roughly sketch the probability distributions for the two universities (A and B) on the same graph for each of the following cases:

a. A has a mean of $35,000 and a standard deviation of $1,000; B has a mean of $35,000 and a standard deviation of $2,000.

b.   *A* has a mean of $35,000 and a standard deviation of $2,000, *B* has a mean of $40,000 and a standard deviation of $2,000.

2. The weights of male students at a university are normally distributed with a mean of 175 pounds and a standard deviation of 10 pounds. For each of the following questions, indicate the corresponding area under the normal curve. If a male student is selected at random from this university, what is the probability that he will weigh
a.   under 175 pounds?
b.   more than 200 pounds?
c.   between 160 and 190 pounds?
d.   either more than 180 pounds or less than 170?
e.   at least 195 pounds?
f.   at most 185 pounds?

3. Assume that *X*, the weight of a crate of grapefruit, is normally distributed with a mean of 20 pounds and a standard deviation of one pound. A crate is selected at random, and its weight is noted. Change each of the following probability statements made in terms of *X* into statements about the standardized variable *z*. What is the probability that the weight will be
a.   at least 21 pounds?
b.   less than 18.5 pounds?
c.   between 18 and 22 pounds?
d.   between 18.75 and 21.25 pounds?
e.   either more than 21.50 or less than 18.5 pounds?
f.   at most 22 pounds?

4. The scores on an achievement test given to 544,302 high school seniors are normally distributed with a mean of 450. The distribution has a standard deviation of 75.
a.   What is the probability that an achievement score is less than 450? 50%
b.   What is the probability that an achievement score is between 300 and 600? $\pm 2\sigma$ 95.5%
c.   The probability is 0.85 that a score is more than what value? 20's $75 \times 1.04 = 78$

5. One form of the random walk hypothesis for stock market prices says that successive price changes in individual securities are independent and normally distributed. Thus, if the price of a stock at time *t* is P_t, this characteristic of price changes can be expressed as $P_t = P_{t-1} + \varepsilon_t$ where ε_t has a normal distribution and is independent of ε_{t+k} where $k \neq 0$. Suppose this distribution has a mean of 0 and a standard deviation of $1 for a particular stock.
a.   What is the probability that the stock price will increase from time $t - 1$ to t?
b.   What is the probability that the stock price will change by at least $2?
c.   What is the probability that P_t will be at most $P_{t-1} + \$1.50$?

6. While studying a group of children, you determine two measures of the aggression manifested by any child in an aggression-inducing situation.

You theorize that the first measure of aggression X is a normally distributed random variable with a mean of three and a variance of one. The second measure Y is also thought to be normally distributed, but with a mean of six and a variance of four. Suppose that your theories are correct and that X and Y are independent random variables.

a. Find $P(2.0 < X < 4.0)$.

b. Find $P(X < 2.8$ and $4.5 < Y < 6.5)$.

c. Find $P(X < 2.8$ and $Y > 3.0)$.

d. Find $P(X < 2.8$ or $Y > 3.0)$.

e. Between what two numbers do the middle 51% of the scores lie as measured by X, the first measure of aggression?

7. The ages of subscribers to a weekly news magazine are normally distributed with a mean of 36.8 years and a standard deviation of 5.4 years. If a subscriber is chosen at random, what is the probability that the age of the subscriber is

a. more than 36.8 years?

b. more than 40 years?

c. between 30 and 40 years?

8. The Klondike Manufacturing Company hires an efficiency expert to monitor the time required by the transistor radio chassis assembly station to complete its assigned task. The results of the study indicate that the time required for the station to produce the chassis and send it along to the next station is normally distributed with a mean of 18 minutes and a standard deviation of three minutes.

a. Calculate the probability that the station will complete its task in less than 21.5 minutes.

b. What is the probability that the station does as well as the optimal time of 16.5 minutes?

c. The probability is 0.0668 that the required time will be under a certain value; find this value.

5.4
SAMPLING DISTRIBUTION OF THE MEAN

We now turn to the use of the normal distribution as the sampling distribution of the arithmetic mean. For brevity, we will use the term **the sampling distribution of the mean**, or simply **the sampling distribution of** $\bar{x}$.[5] To illustrate the

---

[5] Although we used the symbol $\bar{X}$ in Chapter 3 to denote a sample mean, we will henceforth use instead the symbol $\bar{x}$ in sampling theory and statistical inference. The lower-case notation is more convenient because of the use of $\bar{x}$ as a subscript.

nature of this distribution, let us return to the Watts Renewal Corporation manufacturing process that produces light bulbs whose lifetimes are normally distributed with an arithmetic mean of 1,000 hours and a standard deviation of 200 hours. We now interpret this distribution as an infinite population from which simple random samples can be drawn. It is possible for us to draw a large number of such samples of a given size, say $n = 5$, and compute the arithmetic mean lifetime of the five light bulbs in each sample. In accordance with our usual terminology, each such sample mean is referred to as a statistic. Since these statistics will usually differ from one another, we can consider them values of a random variable for which we can construct a frequency distribution. The universe mean of 1,000 hours is the parameter around which these sample statistics will be distributed, with some sample means lying below 1,000 and some lying above it. If we draw any finite number of samples, the sampling distribution is referred to as an **empirical sampling distribution**. On the other hand, if we conceive of drawing all possible samples of the given size, the resulting sampling distribution is a **theoretical sampling distribution**. Statistical inference is based on these theoretical sampling distributions, which are nothing more than probability distributions of the relevant statistics.

> In most practical situations, only one sample is drawn from a statistical population in order to test a hypothesis or to estimate the value of a parameter.

As noted in section 5.1, the work implied in generating a sampling distribution by drawing repeated samples of the same size is virtually never carried out, except perhaps as a learning experience. However, the reader must realize that the sampling distribution provides the underlying theoretical structure for decisions based on single samples.

Sampling from Normal Populations

What are the salient characteristics of the sampling distribution of the mean, if samples of the same size are drawn from a population in which values are normally distributed? To answer this question, we begin by assuming that the sample size is 5. In terms of our problem, this means that a random sample of 5 light bulbs is drawn from the above-mentioned population, and the mean lifetime of these 5 bulbs, denoted $\bar{x}_1$, is determined. Then, another sample of 5 bulbs is drawn, and the mean $\bar{x}_2$ is determined. Let us assume that the first mean is 990 hours, which falls below the population mean, and that the second mean is 1,022 hours, which lies above the population mean. The theoretical frequency distribution of $\bar{x}$ values of all such simple random samples of 5 bulbs would constitute the sampling distribution of the mean for samples of size 5.

FIGURE 5-8

Relationship between a
normal population
distribution and normal
sampling distributions of
the mean for $n = 5$ and
$n = 50$

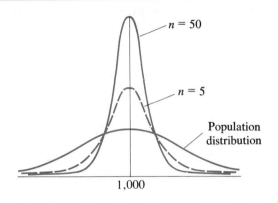

Characteristics
of the sampling
distribution

Intuitively, we can see what some of the characteristics of such a distribution might be. A sample mean would be just as likely to lie above the population mean of 1,000 hours as below it. Small deviations from 1,000 hours would occur more frequently than large deviations. Furthermore, because of the effect of averaging, we would expect less dispersion or spread among these sample means than among the values of the individual items in the original population; that is, the standard deviation of the sampling distribution of the mean should be less than the standard deviation of the values of individual items in the population.

Other characteristics of sampling distributions of the mean might be noted. If samples of sizes 50 rather than 5 had been drawn, another sampling distribution of the mean would be generated. Again, we would expect the means of these samples to cluster around the population mean of 1,000 hours. However, we would expect to find even less dispersion among these sample means than in the case of samples of size 5, because the larger the sample, the closer the sample mean is likely to be to the population mean. Thus, the standard deviation of the sampling distribution, which measures chance error inherent in the process of using samples to approximate population values, would decrease with increasing sample size. Another characteristic of these sampling distributions, which is not at all intuitively obvious but can be proved mathematically, is that if the original population distribution is normal, sampling distributions of the mean will also be normal. Figure 5-8 displays the relationships we have just discussed for the case of a normal population. For the population distribution, the horizontal axis represents values of individual items (x values). For the sampling distributions, the horizontal axis represents the means of samples of size 5 and 50. Since all three of the distributions are probability distributions of continuous random variables, the vertical axis pertains to probability densities.

The foregoing material introduces the following theorem:

Theorem 5.1

If a random variable X is normally distributed with mean μ and standard deviation σ, then the random variable "the mean $\bar{x}$ of a simple random sample of size n"[6] is also normally distributed with mean $\mu_{\bar{x}} = \mu$ and standard deviation

$$\sigma_{\bar{x}} = \frac{\sigma}{\sqrt{n}}$$

In this statement of the theorem, we have used somewhat more formal language than in the preceding discussion. Instead of saying that the values of individual items in a population and sample means are normally distributed, we refer to normal distributions of the random variables X and $\bar{x}$.

An interesting aspect of the theorem is that the expected value (mean) of the sampling distribution of the mean, symbolized $\mu_{\bar{x}}$, is equal to the original population mean μ. Although we noted this relationship informally in section 5.1, it is proved in rule 13 of Appendix C for the more general case of simple random samples of size n from *any* infinite population. The standard deviation of the sampling distribution of the mean—usually referred to as the **standard error of the mean** and denoted $\sigma_{\bar{x}}$—is given by

(5.3)
$$\sigma_{\bar{x}} = \frac{\sigma}{\sqrt{n}}$$

This relationship is proved in rule 12 of Appendix C, again for the more general case of sampling from any infinite population.

Important implications

Important implications follow from equation 5.3. We can think of any sample mean $\bar{x}$ as an estimate of the population mean μ. The difference between the statistic $\bar{x}$ and the parameter μ, $\bar{x} - \mu$, is referred to as a **sampling error**. (For example, if $\bar{x}$ were exactly equal to μ and were used as an estimate of μ, there would be no sampling error.) Therefore $\sigma_{\bar{x}}$, which is a measure of the spread of the $\bar{x}$ values around μ, is a measure of **average sampling error;** that is, it measures the amount by which $\bar{x}$ can be expected to vary from sample to sample. Another interpretation is that $\sigma_{\bar{x}}$ is a measure of the *precision* with which μ can be estimated using $\bar{x}$. Referring to equation 5.3, we see that $\sigma_{\bar{x}}$ varies directly with the dispersion in the original population σ and inversely with the square root of the sample size n. Thus, the greater the dispersion among the items in the original population, the greater the expected sampling error in using $\bar{x}$ as an estimate of μ; similarly, the smaller the population dispersion, the smaller the expected sampling error. In the limiting case in which every item in the population has the same value, the population standard deviation is zero; therefore, the standard error of the mean is also zero. This indicates that the mean of a sample from such a population would be a perfect estimate

---

[6] See footnote 4.

of the corresponding population mean, since there could be no sampling error. For example, if every item in the population weighed 100 pounds, the population mean weight would be 100 pounds. All the items in any sample would weigh 100 pounds, and the sample mean would be 100 pounds; thus, the sample mean would estimate the population mean with perfect precision. As the population dispersion increases, estimation precision decreases.

> The fact that the standard error of the mean varies inversely with the square root of sample size means that there is a diminishing return in sampling effort.

Quadrupling sample size only halves the standard error of the mean; multiplying sample size by nine cuts the standard error only to one-third its previous value.

Sampling from Nonnormal Populations

We mentioned in the foregoing discussion that if a population is normally distributed, the sampling distribution of $\bar{x}$ is also normal. However, many population distributions of business and economic data are not normally distributed. What then is the nature of the sampling distribution of $\bar{x}$?

> It is a remarkable fact that for almost all types of population distributions, the sampling distribution of $\bar{x}$ is approximately normal for sufficiently large samples.

Central Limit Theorem

This relationship between the shapes of the population distribution and of the sampling distribution of the mean has been summarized in what could be the most important theorem of statistical inference, the **Central Limit Theorem**. The theorem is stated in terms of the z variable for the sampling distribution of the mean and the approach of the distribution of this variable to the standard normal distribution. The formal statment of the theorem follows.

Central Limit Theorem

Theorem 5.2

> If a random variable X, either discrete or continuous, has a mean μ and a finite standard deviation σ, then the probability distribution of $z = (\bar{x} - \mu)/\sigma_{\bar{x}}$ approaches the standard normal distribution as n increases without limit.

This theorem is a general one, since it makes no restrictions on the shape of the original population distribution. The requirement of a finite standard deviation is not a practical restriction at all, since virtually all distributions involved in real-world problems satisfy this condition. Note also that the z variable in this theorem is a transformation in terms of deviations of the sample mean $\bar{x}$ from the mean of the sampling distribution of means μ stated in multiples of the standard deviation of that distribution $\sigma_{\bar{x}}$. Thus, it is exactly the same type of transformation as was used in the illustrative examples given earlier in this chapter.

$$z = \frac{\text{Value} - \text{Mean}}{\text{Standard deviation}}$$

Here, however, all the values pertain to the sampling distribution of the mean.

It is useful at this point to restate Theorems 5.1 and 5.2 in the less formal context of sampling applications and to summarize the importance of the concepts involved.

If a population distribution is normal, the sampling distribution of $\bar{x}$ is also normal for samples of all sizes.

Theorem 5.1 restated

If a population distribution is nonnormal, the sampling distribution of $\bar{x}$ may be considered approximately normal for a large sample.

Theorem 5.2 restated

The first of these theorems states that the sampling distribution of $\bar{x}$ will be *exactly* normal if the population is normal. The second, the Central Limit Theorem, assures us that no matter what the shape of the population distribution, the sampling distribution of $\bar{x}$ approaches normality as the sample size increases. For a wide variety of population types, samples do not even have to be very large for the sampling distribution of $\bar{x}$ to be approximately normal. For example, only in the case of highly skewed populations would the sampling distribution of $\bar{x}$ be appreciably skewed for samples larger than about 20. For most types of populations, the approach to normality is quite rapid as n increases.

EXAMPLE 5-6

Management of a wholesale outlet is considering a new sales campaign. It is known that the average purchase per customer is $200 and the standard deviation is $15. If 36 customers are randomly chosen, calculate the probability that the average (mean) purchase is more than $204.

FIGURE 5-9
Sampling distribution of the
mean and standard normal
distribution for the average
purchase example
(Example 5-6)

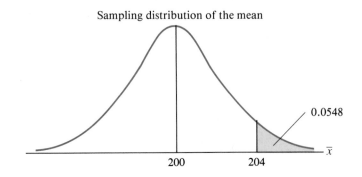

Sampling distribution of the mean

0.0548

200   204

$\bar{x}$

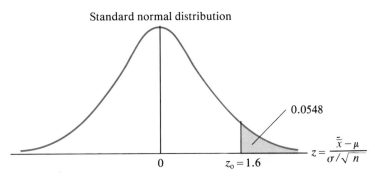

Standard normal distribution

0.0548

0            $z_0 = 1.6$

$z = \dfrac{\bar{x} - \mu}{\sigma/\sqrt{n}}$

SOLUTION

See Figure 5-9.

$$z_0 = (204 - 200)/(15/\sqrt{36}) = 1.6$$

$$P(\bar{x} > \$204) = P(z > 1.6) = 0.0548$$

EXAMPLE 5-7

A production manager is considering the purchase of new card-printing machines. On average, these machines can print 1,900 cards per day and the standard deviation is 200. Assume that the number of cards printed per day is normally distributed.

a. What is the probability that a machine can print more than 2,100 cards per day?

b. If 30 machines are randomly chosen from the manufacturing company, what is the probability that the mean number of cards printed per day is fewer than 1,850?

SOLUTION

See Figure 5-10.

a. $z_0 = (2,100 - 1,900)/200 = 1$ b. $z_0 = (1,850 - 1,900)/(200/\sqrt{30}) = -1.37$

$P(x > 2,100) = P(z > 1) = 0.1587$ $P(\bar{x} < 1,850) = P(z < -1.37) = 0.0853$

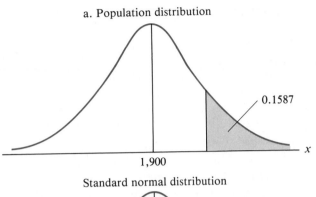

a. Population distribution

Standard normal distribution

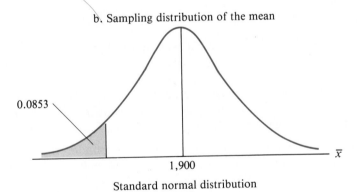

b. Sampling distribution of the mean

Standard normal distribution

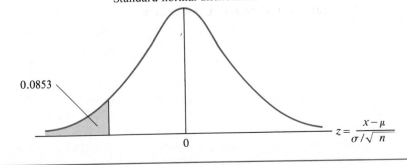

FIGURE 5-10

Population distribution, sampling distribution of the mean, and corresponding standard normal distributions for the card printing example (Example 5-7)

EXERCISES 5.4

1. The mean score of high school seniors taking a certain test 10 years ago was 450 with a standard deviation of 50. In a random sample of 900 high school seniors taking the test this year, the mean score is 445.
   a. If the true mean score is 450, what is the probability that the mean score in a random sample of this size would be less than 445?
   b. Is it correct to state that the mean score of this year's high school seniors is less than 450?

2. The life of a Rollmore tire is normally distributed with a mean of 32,000 miles and a standard deviation of 1,500 miles.
   a. What is the probability that a tire will last at least 30,000 miles?
   b. What is the probability that a tire will last more than 35,000 miles?
   c. If four tires are purchased, what is the probability that the average life of the four tires exceeds 30,000 miles? Assume independence; that is, assume that the tires are used on four different vehicles.

3. The number of arrests made per week in an urban area is normally distributed with a mean of 164 and a standard deviation of 19. Calculate the range in which 95.5% of the weekly arrest figures fall. If arrest figures are averaged every month (that is, the arithmetic mean is calculated for a four-week period), calculate the range in which 95.5% of these average weekly arrest figures will fall.

4. Do you agree or disagree with the following statements? Explain your answer.
   a. According to the Central Limit Theorem, if the mean and variance of a variable are known, we can use the normal distribution to approximate the probability that the variable will exceed some number.
   b. If X is normally distributed, we need to know only its mean and standard deviation to answer probability statements about it.
   c. The mean of a sample is always exactly normally distributed.

5. What is the probability of observing a mean of 40 or more when drawing a simple random sample of size 100 from a population with a mean of 38? The population variance is 64. Do we have to assume that the population is normal? No   as long as n is large

   $n = 100$ $40 - 38 = 2$

   $\mu = 38$ $\sigma = \sqrt{64} = 8$

   $\dfrac{\sigma}{\sqrt{n}} = \dfrac{8}{\sqrt{100}} = \dfrac{8}{10}$ $\dfrac{2}{.8} = 2.50$

6. What is the probability of drawing a simple random sample of size 100 with a mean of 30 or more from a population with a mean of 28? The population variance is 81. Do we have to assume that the population is normal?

   $z = \dfrac{x - \mu}{\sigma}$ $\dfrac{34,000 - 36,000}{2500} = .8$

7. The life of an Ever-Steady spark plug is a normally distributed variable with a mean of 36,000 miles and a standard deviation of 2,500 miles under normal driving conditions.
   a. What is the probability that a spark plug will last at least 34,000 miles?

   $= .2881$

   34,000   36,000

36000 40,000

40,000 - 36,000
————————
2500

$\mu = 36,000$

b. What is the probability that a spark plug will last more than 40,000 miles?

c. If you buy eight spark plugs for your car, what is the probability that the average life of the eight plugs exceeds 34,000 miles?

8. The weekly demand for Snoozeburgers at a local Hamburger Queen franchise is normally distributed with a mean of 4,500 and a standard deviation of 450. Calculate the range in which 95.5% of the weekly demand figures will fall. If demand figures are averaged every month (that is, the arithmetic mean is calculated for four weekly demands), calculate the range in which 95.5% of these average weekly demand figures would fall.

9. It is known from national statistics that the mean family income in a certain suburban area is $5,160 per month with a standard deviation of $800 per month. In a random sample of 50 families drawn from this suburban area, what is the probability that the mean family income will be

5000 - 5160
———————— = .20
800

= .028

a. greater than $5,000?
b. less than $5,000?
c. between $5,200 and $5,300?

10. The family income distribution in a certain city is characterized by skewness to the right. A census reveals that the mean family income is $39,000 and the standard deviation is $2,000. If a simple random sample of 100 families is drawn, what is the probability that the sample mean family income will differ from the city mean income of $39,000 by more than $200?

46 56.0

5.5
SAMPLING DISTRIBUTION OF A
NUMBER OF OCCURRENCES

There are many sampling distributions that are useful in statistical inference. In section 5.4, we dealt with the sampling distribution of the mean. Often, the statistic of interest is either in the form of a number of occurrences or a proportion of occurrences. In this section we discuss the sampling distribution of a number of occurrences and in section 5.6 we deal with the sampling distribution of a proportion of occurrences.

We can illustrate the meaning and properties of the sampling distribution of a number of occurrences by means of a typical example. Let us assume a 10% probability that any given customer who enters Johnson's Supermarket will purchase ice cream. We conceive of the purchasing process as an infinite population. Thus, we may view the successive purchases and failures to purchase as outcomes of a series of Bernoulli trials. That is, in terms of this problem, the three requirements of a Bernoulli process are as follows.

Bernoulli trials

1. There are two possible outcomes on each trial: customer purchases ice cream or customer fails to purchase ice cream.

2. The probability of a purchase of ice cream remains constant from trial to trial.

3. The trials are independent.

Suppose we draw a simple random sample of five customers who enter the supermarket on a particular day and note the number of purchasers of ice cream in the sample. This number is a random variable that can take on the values 0, 1, 2, 3, 4, or 5. Since we are dealing with a Bernoulli process, the probabilities of obtaining these numbers of purchasers may be computed by means of a binomial distribution with $p = 0.10$, $q = 0.90$, and $n = 5$. Therefore, the respective probabilities are given by the expansion of the binomial $(0.9 + 0.1)^5$. This probability distribution is shown in Table 5-3, using the same notation as in Chapter 3.

We can also interpret the probability distribution given in Table 5-3 as a sampling distribution. Since the number of ice cream purchasers observed in a sample of five customers is a sample statistic, Table 5-3 displays the probability distribution of this sample statistic. If we took repeated simple random samples of five customers each and if the probability that any customer would purchase ice cream was 10%, then we would observe no ice cream purchasers

TABLE 5-3
Probability distribution of the number of ice cream purchasers in a simple random sample of 5 customers: $p = P(\text{ice cream purchaser}) = 0.10$

Number of Ice Cream Purchasers x	Probability $f(x)$
0	$\binom{5}{0}(0.9)^5(0.1)^0 = 0.59$
1	$\binom{5}{1}(0.9)^4(0.1)^1 = 0.33$
2	$\binom{5}{2}(0.9)^3(0.1)^2 = 0.07$
3	$\binom{5}{3}(0.9)^2(0.1)^3 = 0.01$
4	$\binom{5}{4}(0.9)^1(0.1)^4 \approx 0.00$
5	$\binom{5}{5}(0.9)^0(0.1)^5 \approx 0.00$
	$\overline{1.00}$

in 59% of these samples, one ice cream purchaser in 33% of the samples, and so forth.

> The probability distribution may now be called a *sampling distribution of number of occurrences.*

5.6
SAMPLING DISTRIBUTION OF A PROPORTION

Frequently, it is convenient to consider proportion of occurrences rather than number of occurrences. We can convert the numbers of occurrences to proportions by dividing by the sample size. The sample proportion, denoted $\bar{p}$ (pronounced "p-bar"), may be calculated from

(5.4)
$$\bar{p} = \frac{x}{n}$$

where x is the number of occurrences of interest and n is the sample size. In the above example, $\bar{p}$ takes on the possible values $\frac{0}{5} = 0.00$, $\frac{1}{5} = 0.20, \ldots,$ $\frac{5}{5} = 1.00$ with the same probabilities as the corresponding numbers of ice cream purchasers. (Note that the number of occurrences in a sample of size n is given by $x = n\bar{p}$. This may be seen by multiplying both sides of equation 5.4 by n.) The sampling distribution of $\bar{p}$ is given in Table 5-4. In keeping with the usual convention, the probabilities are denoted $f(\bar{p})$.

TABLE 5-4
Sampling distribution of the proportion of ice cream purchasers in a simple random sample of 5 customers: $p = P(\text{ice cream purchaser}) = 0.10$

Proportion of Ice Cream Purchasers $\bar{p}$	Probability $f(\bar{p})$
0.00	0.59
0.20	0.33
0.40	0.07
0.60	0.01
0.80	0.00
1.00	0.00
	1.00

Properties of the Sampling Distributions of $\bar{p}$ and $n\bar{p}$

We turn now to the properties of the sampling distributions of number of occurrences $n\bar{p}$ and proportion of occurrences $\bar{p}$. The means and standard deviations of these distributions are of particular interest in statistical inference. The calculation of these two measures is given in Table 5-5 for number of ice cream purchasers and in Table 5-6 for proportion of ice cream purchasers.

TABLE 5-5
Calculation of the mean and standard deviation of the sampling distribution of the number of ice cream purchasers in a simple random sample of 5 customers: $p = P$(ice cream purchaser) $= 0.10$

$x = n\bar{p}$	$f(x)$	$xf(x)$	$x - \mu_x$	$(x - \mu_x)^2$	$(x - \mu_x)^2 f(x)$
0	0.59	0.00	-0.5	0.25	0.1475
1	0.33	0.33	$+0.5$	0.25	0.0825
2	0.07	0.14	$+1.5$	2.25	0.1575
3	0.01	0.03	$+2.5$	6.25	0.0625
4	0.00	0.00	$+3.5$	12.25	0.0000
5	0.00	0.00	$+4.5$	20.25	0.0000
	$\overline{1.00}$	$\overline{0.50}$			$\overline{0.4500}$

$\mu_{n\bar{p}} = \mu_x = \Sigma xf(x) = 0.50$ ice cream purchasers

$\sigma_{n\bar{p}} = \sigma_x = \sqrt{\Sigma(x - \mu)^2 f(x)} = \sqrt{0.4500} = 0.67$ ice cream purchasers

---

TABLE 5-6
Calculation of the mean and standard deviation of the sampling distribution of the proportion of ice cream purchasers in a simple random sample of 5 customers: $p = P$(ice cream purchaser) $= 0.10$

$\dfrac{x}{n} = \bar{p}$	$f(\bar{p})$	$\bar{p}f(\bar{p})$	$\bar{p} - \mu_{\bar{p}}$	$(\bar{p} - \mu_{\bar{p}})^2$	$(\bar{p} - \mu_{\bar{p}})^2 f(\bar{p})$
0.00	0.59	0.000	-0.10	0.01	0.0059
0.20	0.33	0.066	$+0.10$	0.01	0.0033
0.40	0.07	0.028	$+0.30$	0.09	0.0063
0.60	0.01	0.006	$+0.50$	0.25	0.0025
0.80	0.00	0.000	$+0.70$	0.49	0.0000
1.00	0.00	0.000	$+0.90$	0.81	0.0000
		$\overline{0.100}$			$\overline{0.0180}$

$\mu_{\bar{p}} = \mu_{x/n} = \Sigma \bar{p}f(\bar{p}) = 0.10 = 10\%$ ice cream purchasers

$\sigma_{\bar{p}} = \sigma_{x/n} = \sqrt{\Sigma(\bar{p} - \mu_{\bar{p}})^2 f(\bar{p})} = \sqrt{0.0180} = 0.134 = 13.4\%$ ice cream purchasers

---

TABLE 5-7
Formulas for the mean and standard deviation of a binomial distribution

Random Variable	Mean	Standard Deviations
Number of occurrences ($n\bar{p}$)	$\mu_{n\bar{p}} = np$	$\sigma_{n\bar{p}} = \sqrt{npq}$
Proportion of occurrences ($\bar{p}$)	$\mu_{\bar{p}} = p$	$\sigma_{\bar{p}} = \sqrt{\dfrac{pq}{n}}$

Subscripts are used to indicate the random variable for which these measures are computed. For example, $\mu_{n\bar{p}}$ denotes the mean of the random variable $n\bar{p}$. The definitional formulas (3.11) and (3.12) were used for these computations. In actual applications, calculations such as those in Tables 5-5 and 5-6 are never made to obtain the mean and standard deviation of a binomial distribution because convenient computational formulas are available. The calculations are given here only to aid in understanding the meaning of sampling distributions. The general formulas for the mean and standard deviation of a binomial distribution given earlier in Table 3-12 (variances are shown in the table) are summarized here for convenience in Table 5-7.

Let us illustrate the use of the formulas given in Table 5-7 for the preceding distributions of number and proportion of ice cream purchasers. Substituting $p = 0.10$, $q = 0.90$, and $n = 5$, we obtain

Number of Ice Cream Purchasers

Mean $\quad \mu_{n\bar{p}} = np = 5 \times 0.1 = 0.50$ ice cream purchasers

Standard deviation $\quad \sigma_{n\bar{p}} = \sqrt{npq} = \sqrt{5 \times 0.1 \times 0.9} = \sqrt{0.45}$

$$= 0.67 \text{ ice cream purchasers}$$

Proportion of Ice Cream Purchasers

Mean $\quad \mu_{\bar{p}} = p = 0.10 = 10\%$ ice cream purchasers

Standard deviation $\quad \sigma_{\bar{p}} = \sqrt{\dfrac{pq}{n}} = \sqrt{\dfrac{0.10 \times 0.90}{5}} = \sqrt{0.0180} = 0.134$

$$= 13.4\% \text{ ice cream purchases}$$

Of course, these are the same results obtained in the longer calculations shown in Tables 5-5 and 5-6. Let us interpret these results in terms of the appropriate sampling distributions. In this example, where the sample statistic is

Number   *number* of ice cream purchasers $n\bar{p}$, the mean of the binomial distribution is $\mu_{n\bar{p}} = np = (5)(0.10) = 0.50$ ice cream purchasers. This says that if we take repeated simple random samples of five customers each and if the probability is 10% that any customer would purchase ice cream, then there will be, on the average, one-half an ice cream purchasers per sample. (If the sample size were $n = 200$ with $p = 0.10$, then we would expect to obtain, on the average, $(200)(0.10) = 20$ ice cream purchasers per sample.) The standard deviation, $\sigma_{n\bar{p}} = \sqrt{npq} = 0.67$, is a measure of the variation in *number* of ice cream purchasers attributable to the chance effects of random sampling.

Proportion   When the sample statistic is *proportion* of ice cream purchasers $\bar{p}$, the mean $\mu_{\bar{p}} = p = 0.10$, or 10% purchasers. This means that if we draw simple random samples of five customers each, we will observe, on the average, 10% ice cream purchasers in the samples. The standard deviation, $\sigma_{\bar{p}} = \sqrt{pq/n} = 0.134$ or 13.4%, is a measure of variation attributable to the chance effects of sampling, expressed in terms of *proportion* of ice cream purchasers.

The binomial distribution in this example is skewed, as shown in Figure 5-11. Two horizontal scales are shown in this graph to depict corresponding values of $n\bar{p}$ and $\bar{p}$.

As we saw in section 3.4, if p is less than 0.5, the distribution tails off to the right as in Figure 5-11. On the other hand, if p exceeds 0.5, the skewness is to the left. If p is held fixed and if the sample size n is made larger, the sampling distributions of $n\bar{p}$ and $\bar{p}$ become more symmetrical. This is an important property of the binomial distribution from the standpoint of sampling theory and practice, and we examine it further in section 5.7. An illustration of this property is given in Figure 5-12, in which p is held fixed at 0.20. The sampling distribution of $\bar{p}$ is shown for sample sizes of $n = 5$, 10, and 100.

FIGURE 5-11
Graph of sampling
distribution of $n\bar{p}$ and $\bar{p}$
for $p = 0.1$, $q = 0.9$,
and $n = 5$

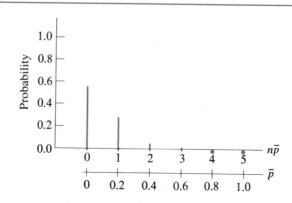

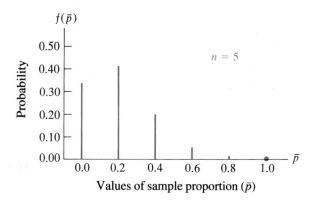

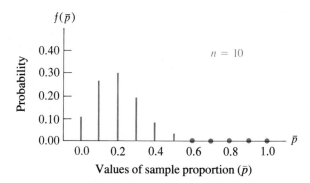

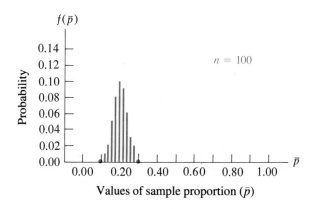

FIGURE 5-12

Binomial distributions for different values of *n*, with $p = 0.20$

EXERCISES 5.6

1. A manufacturer asked a random sample of 100 people if they had ever purchased one of the firm's refrigerators. Assume that 20% of the consumers had, at some time, purchased one of the firm's refrigerators. Assuming Bernoulli trials, calculate the mean and standard deviation for the number of affirmative replies and the proportion of affirmative replies.

2. An econometric forecasting service makes the following claim to Fashionable Dresses, Inc.: "We can predict with 90% accuracy whether U.S. consumer demand for dresses in each of six regions will increase or decrease next month compared with this month." The forecasting service makes demand predictions for dresses for each of the six regions at the end of a particular month. Assume independence regarding the accuracy of these six predictions.
   a. List all possible proportions of correct predictions the service can have in the one-month period.
   b. Assign probabilities to each possible outcome, assuming the service's claim is correct.
   c. Assign probabilities to each possible outcome under the assumption that the service makes random guesses (and thus that the probability of any correct answer is 0.5).

3. In a large city, 20% of the residents have watched a particular TV program. Assuming independence, for a random sample of 10, what are the mean and standard deviation of the sampling distribution of the number watching the program? For a sample of 20 people? For a sample of 40 people?

4. At a large eastern university, the ratio of men to women is 3:1. Calculate the mean and the standard deviation of the sampling distribution of the proportion of women in a random sample of 20 students. Perform the same calculation for a random sample of 40 students. Assume independence.

5. Five percent of the companies headquartered in a certain industrialized nation lost money from normal operations. If a random sample of 20 companies is drawn from this large population, what is the mean and standard deviation of the sampling distribution of the proportion of money losers in the random sample? Perform the same calculation for a random sample of 40 companies. Assume independence.

5.7
THE NORMAL CURVE AS AN APPROXIMATION TO THE BINOMIAL DISTRIBUTION

The normal curve can be used as an approximation to the binomial distribution in calculating probabilities for which the binomial distribution is the theoretically correct distribution. This approximation is particularly useful for large sample sizes (n values) for which binomial tables may not have entries. For example, Tables A-1 and A-2 of this text give tabulations for the binomial distribution only up to $n = 20$.

The binomial distribution that we have been discussing in this chapter is an example of a probability distribution of a *discrete* random variable. We have graphed such distributions by erecting *ordinates* (vertical lines) at distinct values along the horizontal axis. To gain some insight into how the normal distribution, which is the probability distribution of a continuous random variable, can be used as an approximation to the binomial distribution, which is the probability distribution of a discrete random variable, let us graph a binomial distribution as a *histogram* (bar graph). We assume a situation in which a fair coin is tossed twice and the random variable of interest is the number of heads obtained. The probabilities of zero, one, and two heads are, respectively, $\frac{1}{4}, \frac{1}{2},$ and $\frac{1}{4}$. This is an illustration of a binomial distribution in which $p = \frac{1}{2}$ and $n = 2$. If the coin is tossed four times, the probabilities of zero, one, two, three, and four heads are respectively, $\frac{1}{16}, \frac{4}{16}, \frac{6}{16}, \frac{4}{16},$ and $\frac{1}{16}$. This is a binomial distribution in which $p = \frac{1}{2}$ and $n = 4$. Graphs of these distributions in the form of histograms are given in Figure 5-13. Using these histograms, let us now interpret zero, one, two, three, and four heads not as discrete values, but rather as midpoints of classes whose respective limits are $-\frac{1}{2}$ to $\frac{1}{2}, \frac{1}{2}$ to $1\frac{1}{2},$ $1\frac{1}{2}$ to $2\frac{1}{2}$, and so on. The probabilities or relative frequencies associated with these classes are represented on the graphs by the areas of the rectangles or bars. Thus, in the graph for $n = 4$, since the rectangle for the class interval $2\frac{1}{2}$ to $3\frac{1}{2}$ has four times the area of that from $3\frac{1}{2}$ to $4\frac{1}{2}$, it represents four times the probability. If we were to represent the histogram for $n = 4$ with a smooth continuous curve, the curve would pass through the rectangle for three heads as shown in Figure 5-14(a). In Figure 5-14(b), the curve is simplified to a straight line, and it is clear that the shaded area under the curve for the class interval $2\frac{1}{2}$ to $3\frac{1}{2}$ is approximately equal to the area of the rectangle representing the probability of three heads, because the included area ABC is about equal to the excluded area CDE.

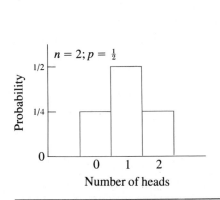

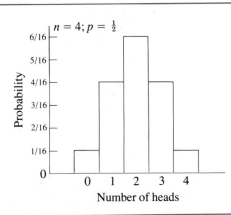

FIGURE 5-13

Histograms of the binomial distribution for $n = 2$, $p = \frac{1}{2}$, and $n = 4$, $p = \frac{1}{2}$

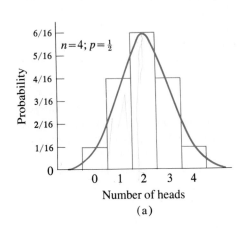

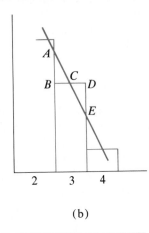

Number of heads
(a)

(b)

Summary

> In the approximation of a histogram by a smooth curve, the area under the curve bounded by the class limits for any given class represents the probability of occurrence of that class.

In the foregoing illustration, if we had increased n greatly, say to 50 or 100, and decreased the width of the rectangles, the corresponding shape of the histogram would approach that of a continuous curve more closely.

> Since the total area of the rectangles in a histogram representing a probability distribution of a discrete random variable is equal to *one*, the total area under a continuous curve representing the probability distribution of a continuous random variable is correspondingly equal to *one*.

Furthermore, the area under the curve lying between the two vertical lines erected at points a and b on the x axis represents the probability that the ran-

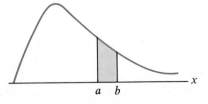

dom variable X takes on values in the interval a to b.[7] This is depicted in Figure 5-15.

By mathematics beyond the scope of this text, it can be shown that if p is held fixed while n is increased without limit, in the binomial distribution, the distribution approaches a particular continuous distribution referred to as the **normal distribution**, **normal curve**, or **Gaussian distribution**.[8] Although our illustration has been for the case $p = \frac{1}{2}$, this is not a necessary condition for the proof. Even if the binomial distribution is not symmetrical (that is, $p \neq \frac{1}{2}$), it still approaches the normal distribution as n increases. The shape of the normal curve is shown in Figure 5-4. Actually, in the early mathematical derivations, the binomial variable x was expressed in **standard units**, that is, as

$$\frac{x - \mu_{n\bar{p}}}{\sigma_{n\bar{p}}} = \frac{x - np}{\sqrt{npq}}$$

and n was assumed to increase without limit. Modern proofs use other approaches to arrive at the same result.

A brief comment on standard units is useful at this point, because such units are widely employed, particularly in sampling theory and statistical inference. Standard units are merely an example of the previously mentioned standard score (see section 1.15).[9]

> The **standard score** or **standard unit** is the deviation of a value from the mean stated in units of the standard deviation.

In general, it is of the form $(x - \mu)/\sigma$, where x denotes the value of the item and μ and σ are the mean and standard deviation of the distribution. In the case of a binomially distributed random variable, as indicated in Table 5-6, the mean and standard deviation of X (the number of successes in n trials) are,

---

[7] Let the value of the probability distribution of a random variable X at x be denoted $f(x)$. If X is discrete, the probability that X lies between a and b inclusive [in the closed interval (a, b)] is

$$P(a \leqslant X \leqslant b) = \sum_{x=a}^{b} f(x)$$

If X is continuous, the probability that X lies between a and b is

$$P(a \leqslant X \leqslant b) = \int_{a}^{b} f(x)\, dx$$

The reader acquainted with integral calculus can see that this definition in the continuous case is the counterpart of the summation in the discrete case. Also, it can be seen that the graphic interpretation of the probability in the continuous case is the area bounded by the curve whose value at x is $f(x)$, the x axis, and the ordinates at a and b. If the probability distribution is continuous at a and b, it makes no difference whether we consider $P(a \leqslant X \leqslant b)$ or $P(a < X < b)$ because the probability is zero that X is exactly equal to a or exactly equal to b.

[8] After the German mathematician and astronomer Karl Friedrick Gauss.

[9] Other terms used to refer to standard scores or standard units include *standardized unit*, *standardized form*, and *standard form*.

respectively, np and $\sqrt{npq}$. Hence, the standard score or standard unit is $(x - np)/\sqrt{npq}$.

In summary, the normal curve can be used an an approximation to the binomial distribution for the calculation of probabilities for which the binomial distribution is the theoretically correct distribution. This approximation is possible because the binomial distribution approaches the normal distribution when n becomes large. In general, the approximations are better when the value of p in the binomial distribution is close to $\frac{1}{2}$ than when p is close to zero or one, because for $p = \frac{1}{2}$ the binomial distribution is symmetrical, and as we have seen, the normal curve is a symmetrical distribution. However, the normal distribution often provides surprisingly good approximations even when $p \neq \frac{1}{2}$ and even when n is not very large.

> A popular rule states that the normal distribution is an appropriate approximation to the binomial distribution when both $np \geq 5$ and $n(1 - p) \geq 5$.

Under these conditions, the binomial distribution can be closely approximated by a normal curve with the same mean and standard deviation. We illustrate the use of the normal curve as an approximation to the binomial distribution by two examples. Example 5-8 illustrates the approximation of the probability of a single term in the binomial distribution by a normal curve calculation. Example 5-9 illustrates a corresponding calculation for a sum of terms in the binomial distribution.

EXAMPLE 5-8

Assume that 20% of a large population smoke at least one pack of cigarettes a day. What is the probability that a randomly drawn sample of 20 individuals will contain exactly 4 who smoke at least one pack a day?

SOLUTION

Using equation 3.3 for the binomial distribution with $n = 20$, $p = 0.20$, and $q = 0.80$, we have

$$P(X = 4) = f(4) = \binom{20}{4}(0.80)^{16}(0.20)^4$$

This probability is evaluated from Table A-1 of Appendix A as

$$P(X = 4) = F(4) - F(3) = 0.6296 - 0.4114 = 0.2182$$

In order to obtain the normal curve approximation to this probability of exactly 4 such smokers, we set up a normal curve with the same mean and standard deviation as the given binomial distribution and find the area between 3.5 and 4.5, as shown in Figure 5-16. We obtain the area between 3.5 and 4.5 because the random variable in the binomial distribution is discrete, whereas in the case of the normal curve, it is continuous.

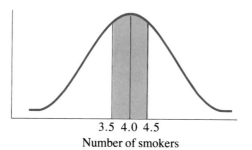

FIGURE 5-16

Area under the normal curve for the probability of obtaining exactly 4 individuals who smoke at least one pack of cigarettes a day in a randomly drawn sample of 20 individuals: $p = 0.20$

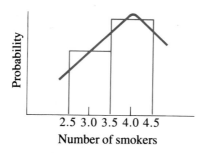

FIGURE 5-17

Representation of a binomial distribution as a histogram and the corresponding normal curve approximation

Hence, as shown in Figure 5-17, if the binomial probabilities are depicted graphically as a histogram, the true probability of 4 occurrences is given by the area of the rectangles centered at 4.0. To approximate this area by a corresponding area under the normal curve, we treat 4 smokers as the value at the midpoint of a class whose limits are 3.5 and 4.5. The mean and standard deviation of the binomial distribution in this problem are

$$\mu = np = (20)(0.20) = 4$$

$$\sigma = \sqrt{npq} = \sqrt{(20)(0.20)(0.80)} = 1.79$$

Using these numbers as the mean and standard deviation of the approximating normal curve, we calculate the z values for 3.5 and 4.5 as follows:

$$z_1 = \frac{3.5 - 4}{1.79} = -0.28 \qquad z_2 = \frac{4.5 - 4}{1.79} = 0.28$$

The area corresponding to a z value of 0.28 is 0.1103, and doubling this yields the desired approximation, $2(0.1103) = 0.2206$. Hence, 0.2206 is the normal curve approximation to the true binomial probability, which is 0.2182.

EXAMPLE 5-9

Referring to Example 5-8, what is the probability that a randomly drawn sample of 20 individuals will contain 3 or more persons who smoke at least one pack of cigarettes a day?

FIGURE 5-18

Area under the normal
curve for the probability of
obtaining 3 or more
persons who smoke at least
on pack of cigarettes a day:
$p = 0.20$

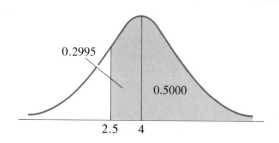

0.2995

0.5000

2.5   4

SOLUTION

Summing the appropriate terms in equation 3.3, we find

$$P(X \geqslant 3) = \sum_{x=3}^{20} f(x) = \sum_{x=3}^{20} \binom{20}{x}(0.80)^{20-x}(0.20)^x$$

This probability is evaluated from Table A-1 of Appendix A as

$$P(X \geqslant 3) = 1 - F(2) = 1 - 0.2061 = 0.7939$$

The corresponding normal curve approximation is shown graphically in Figure 5-18. The z value for 2.5 is calculated as follows:

$$z = \frac{2.5 - 4}{1.79} = -0.84$$

Therefore, the desired area is $0.2995 + 0.5000 = 0.7995$. The closeness of this approximation to the true binomial probability of 0.7939 illustrates that normal curve approximations involving sums of terms usually are closer to the true probabilities than are approximations for individual terms in the binomial distribution.

When the probability of a *single term* in the binomial distribution is desired, as in Example 5-8, it is necessary to use an interval *one unit wide* centered on that term, because the binomial distribution is discrete whereas the normal curve is continuous. On the other hand, this correction is often dispensed with when *sums of terms* in the binomial distribution are desired and the *sample size is large*. Thus, in Example 5-9, if the sample size had been, say, 100, we could have used the z value for 3 rather than for 2.5 in calculating the probability of 3 or more persons who smoke at least one pack of cigarettes a day.

EXAMPLE 5-10

Goyle's Rebound Theory for the stock market, simply stated, is: If in one week the market as a whole has declined, then in the following week, 70% of the stocks will show an increase in price. Assume that Goyle's Theory treats each stock price change as an independent event.

This week a random sample of 60 stocks is observed. According to Goyle, what is the probability that 36 or more of these will show an increase in price?

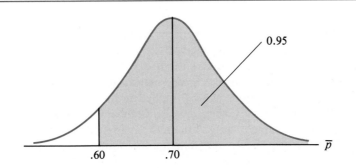

FIGURE 5-19
Area under the normal curve
for Example 5-10 in terms of
proportions of successes

We give the solution first in terms of proportion of successes. Then we show the equivalent solution in terms of numbers of successes. The corresponding sampling distributions are shown in Figures 5-19 and 5-20, respectively.

SOLUTION

Proportion of successes

$$\mu = p = 0.7$$

$$\sigma = \sigma_{\bar{p}} = \sqrt{\frac{pq}{n}} = \sqrt{\frac{(0.7)(0.3)}{60}} = 0.05916$$

$$z = (0.60 - 0.70)/0.05916 = -1.69$$

$$P(\bar{p} \geqslant 0.60) = P(z \geqslant -1.69) = 0.95$$

Numbers of successes

$$\mu = np = (60)(0.7) = 42 \text{ increases}$$

$$\sigma = \sqrt{npq} = \sqrt{(60)(0.7)(0.3)} = 3.55 \text{ increases}$$

$$z = (36 - 42)/3.55 = 0.05916$$

$$P(x \geqslant 36) = P(z \geqslant -1.69) = 0.95$$

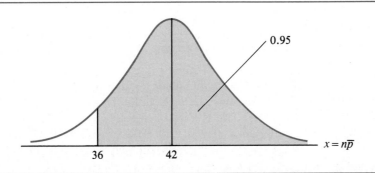

FIGURE 5-20
Area under the normal curve
for Example 5-10 in terms of
numbers of successes

Finite Population Multiplier

In our discussion of sampling distributions, we have dealt with infinite populations. However, many of the populations in practical problems are finite, as for example, the employees in a given industry, the households in a city, and the counties in the United States. As a practical matter, formulas already obtained for infinite populations can be applied in most cases to finite populations as well. In those cases when the results for infinite populations are not directly applicable, a simple correction factor is applied to the formula for the standard deviation of the relevant sampling distribution.

In simple random sampling from an *infinite population*, we have seen that the sampling distribution of $\bar{x}$ has a mean $\mu_{\bar{x}}$ equal to the population mean μ and a standard deviation $\sigma_{\bar{x}}$ equal to $\sigma/\sqrt{n}$.

Standard error of the mean for finite populations   In sampling from a *finite population*, the mean $\mu_{\bar{x}}$ of the sampling distribution of $\bar{x}$ again is equal to the population mean μ but the standard deviation (standard error of the mean) is given by the following formula.

(5.5)   5% or less, not needed →

$$\sigma_{\bar{x}} = \sqrt{\frac{N - n}{N - 1}} \frac{\sigma}{\sqrt{n}}$$

Here N is the number of elements in the population and n is the number in the sample. The quantity $\sqrt{(N - n)/(N - 1)}$ is usually referred to as the **finite population correction** or the **finite correction factor**. Thus, we see that in the case of a finite population, the standard error of the mean is equal to the finite population correction multiplied by $\sigma/\sqrt{n}$, which is the standard error of the mean in the infinite case. The finite population correction is approximately equal to one when the population size N is large relative to the sample size n; therefore, when we are choosing samples of size n from a much larger (but finite) population, the standard error of the mean $\sigma_{\bar{x}}$ is for practical purposes equal to $\sigma/\sqrt{n}$, as was the case when we sampled an infinite population.

To see why the finite correction factor is approximately one when population size is large relative to sample size, note that the factor $\sqrt{(N - n)/(N - 1)}$ is approximately equal to $\sqrt{(N - n)/N}$ for large populations, since the subtraction of one in the denominator is negligible. We can now write

$$\sqrt{\frac{N - n}{N}} = \sqrt{1 - \frac{n}{N}} = \sqrt{1 - f}$$

where $f = n/N$ is referred to as the **sampling fraction** because it measures the fraction of the population contained in the sample. Thus, if the population size is $N = 1,000$ and the sample size is $n = 10$, then $f = \frac{10}{1,000} = \frac{1}{100}$. In such a case, the finite population correction is very close to one. Here, for example, $\sqrt{1 - \frac{1}{100}} \approx 1$ (the symbol $\approx$ means "is approximately equal to"). In summary, in this case, $\sigma_{\bar{x}}$ is practically equal to $\sigma/\sqrt{n}$

> A generally employed rule of thumb is that the formula $\sigma_{\bar{x}} = \sigma/\sqrt{n}$ may be used whenever the size of the population is at least 20 times that of the sample, or in other words, whenever the sample represents 5% or less of the population.

A striking implication of equation 5.5 is that as long as the population is large relative to the sample, sampling precision becomes a function of sample size alone and does not depend on the proportion of the population sampled. Of course, we assume in this statement that the population standard deviation is constant. For example, let us assume a situation in which we draw a simple random sample of size $n = 100$ from each of two populations. Each population has a standard deviation equal to 200 units ($\sigma = 200$). In order to observe the effect of increasing the number of elements in the population, we further assume the population are of different sizes, namely, $N = 10,000$ and $N = 1,000,000$. The standard error of the mean for the population of 10,000 elements is, by equation 5.5,

$$\sigma_{\bar{x}} = \sqrt{\frac{10,000 - 100}{10,000 - 1}} \left(\frac{200}{\sqrt{100}} \right) \approx \sqrt{1} \left(\frac{200}{10} \right) \approx 20$$

For the population of 1,000,000 elements, we have

$$\sigma_{\bar{x}} = \sqrt{\frac{1,000,000 - 100}{1,000,000 - 1}} \left(\frac{200}{\sqrt{100}} \right) \approx \sqrt{1} \left(\frac{200}{10} \right) \approx 20$$

Thus, increasing the population size from 10,000 to 1,000,000 has virtually no effect on the standard error of the mean, since the finite population correction is approximately equal to one in both instances. Indeed, if the population size were increased to infinity, the same result would again be obtained for the standard error.

The finding that it is the absolute size of the sample, and not the proportion of the population sampled, that basically determines sampling precision is difficult for many people to accept intuitively. In fact, prior to the introduction of statistical quality control procedures in American industry, arbitrary methods such as sampling 10% of the items of incoming shipments, regardless of shipment size, were quite common. Managers had vague feelings in these cases that approximately the same sampling precision was obtained by maintaining a constant sampling fraction. However, widely different standard errors resulted from large variations in the absolute sizes of the samples. The interesting principle that emerges from this discussion is this:

> For cases in which the populations are large relative to the samples, the absolute amount of work done (sample size), not the amount of work that might conceivably have been done (population size), is important in determining sampling precision.

We leave it to the reader's judgment whether this finding can be applied to other areas of human activity as well.

In our subsequent discussion of statistical inference, we will be concerned with measures of sampling error for proportions as well as for means. Therefore, we note the corresponding formula for the standard error of a proportion in sampling a finite population. In section 5.6, we indicated that the standard deviation of the sampling distribution of a proportion, which we now refer to as the **standard error of a proportion**, is given by $\sigma_{\bar{p}} = \sqrt{pq/n}$. Since our discussion referred to sampling as a **Bernoulli process**, it pertained to the sampling of an infinite population. The corresponding formula for the standard error of a proportion for a simple random sample of size n from a finite population is as follows.

Standard error of a proportion for finite populations

(5.6)

$$\sigma_{\bar{p}} = \sqrt{\frac{N-n}{N-1}} \sqrt{\frac{pq}{n}}$$

The same sorts of approximation considerations discussed in the case of the mean are pertinent here as well. Hence, if the size of the population is at least 20 times that of the sample, the formula for infinite populations may be used.

Other Sampling Distributions

In this chapter, we have discussed sampling distributions of numbers and proportions of occurrences and sampling distributions of the mean. Just as we were able to use the binomial distribution as a sampling distribution of numbers of occurrences under the appropriate conditions, other distributions may similarly be used under other sets of conditions. However, it is frequently far simpler to use normal curve methods based on the operation of the Central Limit Theorem. Two other continuous sampling distributions, which we have not yet examined, are important in elementary statistical methods: the student t distribution and the chi-square distribution. They will be discussed at the appropriate places in connection with statistical inference.

EXERCISES 5.7

1. According to a manufacturer of canned soups, 10% of all consumers purchase its product regularly. Assume independence, and suppose that 20 consumers are randomly chosen and questioned about their buying behavior.
   a. Find the exact probability that two or more of the consumers questioned are regular purchasers of the firm's soups.

b. Use the normal approximation to the binomial distribution with the continuity correction factor to obtain an approximation to the exact probability.

2. A major city's smog level is classified as unacceptable approximately 15% of the time. If 40 days are selected at random, assuming independence, what is the probability that
   a. exactly six days are unacceptable?
   b. more than six days are unacceptable?
   c. fewer than six days are unacceptable?

3. According to a certain study, the probability that a reader will read any particular advertisement in *Scanners Digest* is 0.2. Multiroyal, Inc., in a large advertising campaign, places 25 advertisements in *Scanners Digest* in three months. Assuming independence, use the normal approximation to the binomial distribution with the continuity correction factor to find the approximate probability that a person reads
   a. exactly one Multiroyal advertisement
   b. exactly two Multiroyal advertisements
   c. exactly three Multiroyal advertisements
   d. exactly five Multiroyal advertisements

4. The Crude Oil Company has two main oil fields, one in the Alaskan tundra region and one in the Texas Panhandle. The company has enjoyed 25% success in finding retrievable oil deposits with its drillings in the Alaskan field, whereas its successful drillings in Texas have been only 20% of the total drillings. In 1991, the company plans 400 new drillings in each field. If the success percentages stay the same, what is the probability of the company establishing 90 or more new successful oil wells in Alaska? In Texas? Assume independence.

5. a. Bank A has 5,024 savings account depositors, with the average amount in each account equal to $512. Assume that the standard deviation is $150. Find the mean and standard deviation of the sampling distribution of the average balance per account for random samples of 1,000 depositors.
   b. Bank B has 10,244 savings account depositors, with the average amount in each account equal to $564. Assume that the standard deviation is also $150. Find the mean and standard deviation of the sampling distribution of the average balance per account for random samples of 1,000 depositors.
   c. Compare the standard deviations of the sampling distributions in parts a and b. Why are they different?

6. The following information is known about the persons seeking employment through a certain employment service: (1) fifteen percent of the applicants have high scores on a standard intelligence test given by the agency and (2) the scores of the applicants are normally distributed with a mean of 68 and a standard deviation of 12.

a. If 10 applicants are randomly selected, what is the probability of obtaining exactly one high scorer?

b. If one applicant is selected, what is the probability that his or her score will be between 56 and 92?

c. In a random sample of 49 test scores, what is the probability that the sample mean will be between 70 and 74?

d. If one applicant's test score is randomly selected, the probability is 0.1357 that it will exceed a certain number. What is this number?

7. The 20 top executives of a certain company spend an average of 12 hours per week in meetings. The standard deviation is three hours. Six executives are selected at random and asked various questions. The answers to these questions will be used for the orientation of newly hired management aspirants. One of the questions concerns the amount of time per week spent in meetings. The mean for the six executives questioned was 11.5 hours with a standard deviation of two hours.

a. Determine if the 11.5 hours and two hours are the mean and standard deviation of the sampling distributions of samples of six executives selected from the population of 20.

b. Define the sampling distribution of the mean time spent per week in meetings for samples of six executives selected from the 20. Calculate the mean and standard deviation of the sampling distribution.

8. The personnel department of a manufacturing company found that 80% of the employees want to join a trade union. If a random sample of 100 employees is drawn, what is the probability that the sample will contain at least 75% who want to join a trade union?

9. Do you agree or disagree with the following statements? Explain your answer.

a. The probability that there are 10 defective items in a sample of 200 is the same as the probability that 95% of the items in a sample of 200 are not defective.

b. You could be reasonably certain that you would get exactly 5,000 billion "heads" if you flipped a fair coin 10,000 billion times—because the probability that the proportion of "heads" equals $\frac{1}{2}$ converges to 1 as the number of tosses increases to infinity.

c. According to the Central Limit Theorem, if the mean and variance of a variable are known, we can use the normal distribution to approximate the probability that the variable will exceed some number.

d. If X is normally distributed, the only information we need know about X to answer probability statements about it is its mean and standard deviation.

e. The mean of a sample is always exactly normally distributed.

10. The mean salary of the presidents of 100 small electronic controls companies is $98,900 with a standard deviation of $4,210. A certain business magazine decides to make a study of the presidents of these 100 firms.

a. Suppose 25 presidents are selected at random. What are the mean and standard deviation of the distribution of average salaries for all possible samples of size 25?

b. Suppose 50 presidents are selected at random. What are the mean and standard deviation of the distribution of average salaries for all possible samples of size 50?

c. Suppose 100 presidents are selected at random. What are the mean and standard deviation of the distribution of average salaries for all possible samples of size 100?

MINITAB EXERCISES

Problems 1 and 2 deal with some simple simulation problems. Random number generation is not only important in helping us understand a complex statistical model, but also useful for random sampling purposes.

1. Generate a sample of 60 observations of numbers of successes from a binomial distribution with $n = 20$ and $p = 0.6$. Since both np and $n(1 - p)$ are larger than 5, the normal distribution should be a good approximation to the binomial distribution. In fact,

   MTB > random 60 C1;
   SUBC > binomial 20 0.6.

   will produce such a sample. To check the normality of C1, one needs to see if the following plot

   MTB > nscore C1 C2;
   MTB > plot C2 C1

   looks like a straight line. Check the normality of a sample from the binomial distribution with $n = 70$ and $p = 0.3$.

2. The lognormal distribution is a market pricing model. It states that the log of the ratio of tomorrow's price to today's price is normally distributed. Assuming that the log of the ratio is distributed according to the standard normal distribution and today's price is 1, generate and graph a random sample of prices for the next 30 days. *Hint:* First try

   MTB > random 30 C1;
   SUBC > normal 0 1.

   to get a sample of 30 observations from the standard normal distribution, then

   MTB > parsum C1 C2

   will give you partial sums in column C2. For example, if C1 contains 1, 2, 3, 4, then "parsum C1 C2" puts 1, 1 + 2 = 3, 1 + 2 + 3 = 6, 1 + 2 + 3 + 4 = 10 into C2. And finally,

   MTB > let C3 = expo(C2)

   will produce a sample of the prices. Are you convinced?

KEY TERMS

Central Limit Theorem   as sample size increases, the sampling distribution of the mean approaches the normal distribution, regardless of the shape of the population distribution.

Finite Population Correction (finite correction factor) a correction factor applied to a standard error for the measurement of error in sampling a finite population.

Normal Distribution a continuous probability distribution characterized by two parameters, the mean and standard deviation. Graphically, the distribution is a bell-shaped curve.

Sampling Distribution a probability distribution of a statistic (for example, proportion or mean).

Sampling Distribution of the Mean ($\bar{x}$) the probability distribution of the sample mean.

Sampling Distribution of Number of Occurrences ($n\bar{p}$) the probability distribution of a number of occurrences.

Sampling Distribution of Proportion ($\bar{p}$) the probability distribution of a proportion.

Standard Error of the Mean the standard deviation of the sampling distribution of the mean.

Standard Error of a Proportion the standard deviation of the sampling distribution of a proportion.

Sampling Fraction the fraction of a population contained in a sample.

Standard Score the deviation of a value from the mean stated in units of the standard deviation.

KEY FORMULAS

A Sample Proportion

$$\bar{p} = \frac{x}{n} = \frac{\text{Number of occurrences}}{\text{Sample size}}$$

Standard Error of a Number of Occurrences

$$\sigma_{n\bar{p}} = \sqrt{npq}$$

Standard Error of a Proportion

$$\sigma_{\bar{p}} = \sqrt{\frac{pq}{n}}$$

Standard Score (in general)

$$z = \frac{\text{Value} - \text{Mean}}{\text{Standard deviation}} = \frac{x - \mu}{\sigma}$$

Standard Error of a Mean

$$\sigma_{\bar{x}} = \frac{\sigma}{\sqrt{n}}$$

Finite Population Correction

$$\sqrt{\frac{(N - n)}{(N - 1)}} \approx \sqrt{1 - \frac{n}{N}}$$

REVIEW EXERCISES
FOR CHAPTERS 1 THROUGH 5

1. The following is known about the clients of a large stockbrokerage firm:
   (1) Twenty percent of the clients are "heavy traders" and therefore generate rather large commissions.
   (2) The dollar sizes of the clients' accounts (defined in terms of annual value of transactions) are normally distributed with an arithmetic mean of $20,000 and a standard deviation of $1,500.
   a. If 50 accounts are randomly selected, what is the probability that 25% or more "heavy traders" would be included in the sample?
   b. If one account is selected, what is the probability that its size will be between $18,500 and $23,000?
   c. If one account is randomly selected, the probability is 0.209 that its size will exceed a certain dollar amount. What is this dollar amount?

2. The consumption of natural gas (units of hundreds of cubic feet per day) for cooking and water heating in a home had this frequency distribution for $n = 21$ months:

Consumption	Months
0.3	1
0.4	1
0.5	0
0.6	0
0.7	2
0.8	1
0.9	1
1.0	1
1.1	2
1.2	7
1.3	2
1.4	3

Mean = 1.06   Variance = 0.0986   Standard deviation = 0.3140

   a. Does a normal distribution appear to be a reasonable model for the average rates? Justify your answer by computing the probability that consumption is at least 1.2 using the normal distribution and compare this to the empirical results.

   b. What are the median and modal consumption rates?

3. Workers in a large company are given aptitude tests. Designers of the test have found that the scores follow a normal distribution with a mean of 120 points and a standard deviation of 20 points.

   a. What is the probability that a worker chosen at random in this company will score 147 or higher?

   b. A simple random sample of the scores of 36 workers is to be drawn. What is the probability that the sample mean will be within 5 points of the true mean?

   c. Those who score in the upper 10% on this aptitude test are classified as "high potential" workers. In a simple random sample of 100 workers in this company, what is the probability that 14 or more workers would obtain "high potential" scores?

4. In a "mini-lottery," an individual selects two numbers from 01 to 20 to be placed on a ticket. The State chooses three numbers from 01 to 20 without replacement. A ticket wins if both numbers on the ticket are among the three numbers chosen by the State.

   a. Verify that the probability that a ticket wins is approximately 0.016.

   b. If fifty people each buy one ticket by selecting numbers independently, what is the approximate probability that at least one person wins? You may use 0.016 as the probability that an individual ticket wins.

   c. A ticket costs $1. If you win, you receive $51 in return (i.e., the purchase price plus an additional $50). Let X = the gain from buying one ticket. Find the expected value of X.

5. Ten "old grads" convene in a hotel suite on homecoming weekend for a night of card playing. Five are managers, 3 are lawyers, and 2 are investment bankers.

   a. They decide to break up into 2 groups of 5 each. In how many ways can this be done?

   b. What is the probability, assuming random assignment, that the 2 investment bankers are in the same group?

   c. What is the probability that the first group consists of 3 managers and 2 lawyers?

6. The arrival times of scheduled air flights for a certain air carrier can be assumed to follow a normal distribution with a mean of +5 (minutes) and a standard deviation of 9. ('0' indicates on time, negative values indicate early arrivals.)

   a. Assuming random arrival times, what is the probability that there will be 3 early arrivals on the next 5 incoming flights? (approximate calculations.)

b. For the next 9 arrivals, what is the probability that the mean arrival time is at least $+4$ (minutes)?

7. Assume that the Department of Labor has estimated that 30% of the U.S. labor force is dissatisfied with their jobs. Twenty presently employed candidates are being interviewed for a position at Booker Industries, Inc. Assuming perfect randomness in choosing these 20 candidates from the labor force:

a. What is the probability that at least 4 are dissatisfied with their present jobs?

b. Using the Poisson approximation, find the probability in (a).

c. Using the Normal approximation, find the same probability.

d. Are the approximations in (b) and (c) justified? Why or why not?

8. The daily productivity of three employees A, B, and C is described by the following density functions:

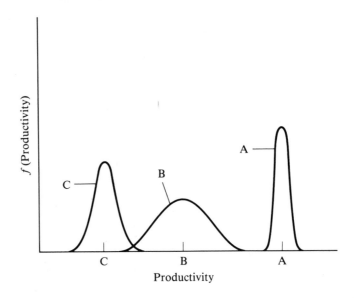

a. Which employee has the lowest mean productivity? Explain.

b. Which employee's productivity has the highest standard deviation? Explain.

c. Which employees's productivity has the lowest coefficient of variation? Explain.

9. Each of the 25 employees in the new training program referred to in question 3 is evaluated subsequently by his/her supervisor. The intent is to have the evaluations follow a normal distribution with $\mu = 70$ and $\sigma = 10$.

a. If the employees are evaluated according to this distribution, what would be the probability distribution for the *number of employees* with evaluations exceeding 90? Give the parameters of the probability distribution.

b.   One issue is whether the supervisors evaluate differently. The 25 employees are randomly assigned to each of five supervisors, that is, five
employees are assigned to each supervisor. Assuming that all supervisors evaluate in an identical manner, what is the probability that the
five employees with the highest scores are assigned to the same
supervisor?

10. A survey of the numbers of people arriving at a bank teller's window per
5-minute period shows the following:

$$x: 2, 3, 1, 4, 2, 5, 3, 0 \text{ (note: } n = 8)$$

a.   Find the standard deviation for the sample.
b.   A consultant suggests that a Poisson distribution is appropriate in
this case. If so, what is the probability of 4 or more arrivals per 5-
minute period?

11. A corporation has two subsidiary companies that are involved in the development of new products. In planning for next year, the corporate planning
group has set up the following models for the numbers of successful products to be forthcoming next year for the two companies:
    The predicted number of successful new products in the first company
(X) for a total of eight new products in the next year is distributed as a
binomial distribution with $n = 8$ and $p = 0.25$.
    The predicted number of successful new products in the second com-
pany (Y) for next year is distributed as

$$g(y) = \frac{y}{10}; y = 1, 2, 3, 4$$

a.   Find the mean of $X + Y$, the sums of the predicted numbers of new
products in the two companies combined.
b.   Assume that X and Y are independent random variables. What is the
joint probability that the first company will have 4 or more successful
products and the second company will also have four or more successful products next year?

12. Employees of a certain company were asked to complete a questionnaire
relating to their lifestyles and health knowledge as part of the company
health promotion program. One of the scales derived from the response
was termed a Well-Being Score. The scale ranged from 1 (poorest) to 6
(best). Two hundred individuals, representing 80 percent of the employees,
completed the questionnaire. The results were:

Well-Being Score:	1	2	3	4	5	6
Frequency:	6	22	30	35	62	45

a.   What was the median score?

b. Assume that the sample variance (s^2) is 2. What proportion of the responding employees fell in the interval: Mean $\pm$ one standard deviation?

c. Consider the group of responders who scored "1" or "2" on the scale. In a pilot study the company records of 5 members of this group were randomly drawn for further analysis. Write a mathematical expression for the probability that at least one of the 5 scored a "1" on the scale. You need not carry out the computation.

d. In a second pilot study, the records of 10 of the total 250 employees were randomly selected. What is the approximate probability that 8 or more of the 10 were responders?

13. A study of manufacturers of hand-assembled electronic components revealed the following joint distribution of productivity (X) and the number of employees (Y) at the company. The values in the following table represent $f(x, y) = P(X = x \text{ and } Y = y)$:

		Productivity (X, in units per employee)		
		10	20	30
Number of	5	0.10	0.10	0.15
Employees	10	0.05	0.30	0
(Y)	15	0.20	0.10	0

a. What is the expected level of productivity for a firm with 5 employees?

b. If a manufacturer whose characteristics obey this joint distribution is randomly selected, what is the probability that the firm will produce less than 100 items in a day?

14. A department store obtained the following frequency distribution for daily demand for a certain home furnishing item based on a random sample of 200 business days.

Number of Units Demanded	Number of Days
0	73
1	74
2	37
3	12
4	4
	200

a. Find the mean and variance of this sample.

b. What well known probability distribution approximates this empirical frequency distribution? Give your reasoning including a statement about the parameter(s) of the approximating distribution.

c. Using the probability distribution that you referred to in part (b), what is your best estimate of the number of days on which 2 to 4 units inclusive would be demanded in the next 250 business days?

15. In reviewing its accounts receivable, a firm determined the probability that an account should be classified as "overdue" to be 0.15. Assume this probability is constant from account to account and assume independence.

   a. In a random sample of 10 such accounts, what is the probability that more than 3 would be overdue?

   b. In five such groups of 10 accounts, what is the probability that more than 3 overdue accounts would be observed exactly twice?

   c. Assume that a group of 1000 accounts receivable were being worked on in which 100 were overdue. If a sample of 20 accounts were drawn from the 1000 without replacement what is the probability that there would be fewer than two overdue accounts? Set up an expression which when evaluated would yield the answer. You need not carry out the arithmetic.

16. The following is the joint probability distribution of the net profits for two proposed investments of a corporation. X and Y denote the net profits on investment proposals 1 and 2, respectively, in millions of dollars.

			X	
		0	1	2
	0	0.10	0.05	0.30
Y	2	0.10	0.05	0
	4	0.30	0.10	0

   a. Calculate the expected value of net profit for the two investment proposals *combined*.

   b. Calculate $E(Y|X = 1)$ and explain the meaning of your result.

17. A study conducted at Hanover Products Inc. during the first week of December determined that the following numbers of defective articles were produced per day;

$$x: 18, 16, 11, 25, 10$$

Assume that about the same number of articles is produced per day.

   a. A study throughout the year at this company yielded a mean of 20 defectives per day and a standard deviation of 2 defectives per day. On the basis of these results, what can be said about the *relative variability* of the number of defectives produced per day during the first week of December and the entire-year period?

   b. Suppose in fact that the true mean number of defectives is 20 per day with a standard deviation of 2 per day. What is the probability that there will be between 15 and 21 defective articles produced on a given day? Assume normality.

c. For a sample of 5 days, what is the probability of obtaining a sample mean of less than 18 defectives per day?

d. Suppose the true mean number of defectives per day (μ) is unknown but the standard deviation (σ) is still 2. Again a sample of 5 days is drawn, and the sample mean $\bar{X}$ is calculated. How likely is it that a difference between $\bar{X}$ and μ of more than 2 defectives per day will be obtained?

18. A study of the number of incoming telephone calls at a real estate office revealed the following data for weekday afternoons. The calls appear to occur randomly and independently.

Time Period	Mean Rate per Hour
1:00–1:59 p.m.	2
2:00–2:59 p.m.	2
3:00–3:59 p.m.	4

a. What is the probability of the following pattern of the number of calls on a given day for the first 3 hours in the afternoon (that is, starting at 1:00 p.m.)?

$$x = 2, 2, 3$$

b. What is the probability of receiving between 4 and 6 calls (inclusively) during the period, 2:00–3:59 p.m. on a given day (If you feel that you need to make any assumptions, do so).

c. On 5 weekday afternoons, between 3:00–3:59 p.m., what is the probability that there will be at least 3 calls during this hour on exactly 4 of the 5 days? (Approximate calculations).

19. A market analyst has used past data to obtain the following joint probability distribution of monthly supply and demand for a certain product. Assume the data are in appropriate units.

$$X = \text{quantity demanded}$$

$$Y = \text{quantity supplied}$$

		Y		
		1	2	3
	0	.1	.1	0
X	1	.2	0	0
	2	.4	.1	.1

a. Compute $P\left(\dfrac{X}{Y} > 0.6\right)$. Interpret your result.

b. Let $Z = Y - X$. Compute the expected value and standard deviation of Z, i.e., $E(Z)$ and σ_z. Give a brief interpretation of these results in the context of the problem.

MINITAB REVIEW EXERCISES FOR CHAPTERS 1 THROUGH 5

1. In a simplified model for stock price movements, the price of a stock (in terms of dollars) either increases or decreases by $1 every following day. Assume that one day, the stock price is at $40, and that with probability 0.6, the price will move up by $1, and with probability 0.4, the price will go down by $1. Find the probability that 30 days later, the price would be in the range of $42 to $50. *Hint:* Equivalently, find the probability that during the 30 days, there were 16 to 20 price increases. One way to find the probability is to try:

    ```
    MTB > set C1
    DATA > 16:20
    DATA > end
    MTB > pdf C1 C2;
    SUBC > binomial 30 0.6.
    MTB > let k1 = sum (C2)
    MTB > print k1
    ```

2. (*continuing Problem* 20) Generate a random sample for the path of price movement in the 30 days, and then make a time-series plot for the sample. Find the number of days in which the stock prices were up for the random sample. *Hint:* Minitab commands

    ```
    MTB > random 30 C1;
    SUBC > Bernoulli 0.6.
    MTB > let C1 = 2 * (C1 − 0.5)
    ```

    produce a random sample of 30 observations of 1's and −1's. For each observation, the commands generate 1 with probability 0.6. Since the initial price was $40, and later prices were cumulative in the model, the following commands

    ```
    MTB > let C2 = 40 + parsums(C1)
    MTB > tsplot C2
    ```

    will generate and plot a random sample of prices for the 30 days.

Problems 3–5 are based on the following data. An investor is interested in one of the two projects: Project A and Project B. He has conducted a survey of some firms engaged in the same business and the results were saved in 'project . mtw':

Column	Name	Description
C1	PROA	Profit ($) per Dollar Invested: 30 Firms in 1990
C2	PROB	Profit ($) per Dollar Invested: 25 Firms in 1990

3. Assume that the investor is risk-neutral, that is, (s)he only cares about expected profit per dollar invested. Thus, if the mean of PROA is greater

than the mean of PROB, the investor will invest in Project A, and vice versa. In which project will the investor invest?

4. Suppose the investor thinks that the median is a better estimate than the mean profit per dollar invested. Thus, if the median of PROA is greater than the median of PROB, the investor will invest in Project A, and vice versa. In which project will the investor invest?

5. If the investor is a risk-preferrer, he prefers both high expected value and high variance of profit per dollar invested. Assume he tries to maximize:

mean + sqrt(variance)

Which project will the investor prefer?

6. Use Minitab to obtain 16 random samples of size 4 without replacement from a population consisting of the data in C1 in the file labeled 'project . mtw'. Calculate the 16 sample means and make a histogram for them. Compare the mean of the sample means to the mean of the population, and compare the standard deviation of the sample means with that of the population. *Hint*:

MTB > store 'sample'
STOR > noecho
STOR > sample 4 C1 Ck1
STOR > let k2 = mean(Ck1)
STOR > stack C20 k2 C20
STOR > let k1 = k1 + 1
STOR > end
MTB > let k1 = 2
MTB > stack 1 1 C20
MTB > execute 'sample' 16
MTB > copy C20 C20;
SUBC > omit 1 2.
MTB > describe C1 C20

7. Generate 20 random samples of size 10 from the standard normal distribution. Find the mean and standard deviation of the sample means, and compare them to those of the population. *Hint*:

MTB > random 20 C1–C10;
SUBC > normal 0 1.
MTB > rmeans C1–C10 C20
MTB > describe C20

Problems 8–10 refer to the following data. It is often assumed that the number of customers arriving at a service point during a fixed amount of time is a random variable that is distributed according to a Poisson distribution. The numbers of customers arriving at two service points during 80 one-hour periods have been observed and the data were entered into 'customer . mtw'. From the file, you will find

Column	Name	Description
C1	SERV A	Number of Customers Arriving at SERV A During Each of the 80 hours
C2	SERV B	Number of Customers Arriving at SERV B During Each of the 80 hours

8. Find the mean, median, and variance of the data in C1. Is it roughly true that the mean is larger than the median, and that the variance is equal to the mean, as suggested by the properties of a Poisson distribution?

9. Generate a random sample of 80 observations from the Poisson distribution with the mean equal to the mean of C1, and enter the sample into C3. Make a dotplot for C1 and a dotplot for C3. Do the two dotplots look alike?

10. Based on the information in C1 and C2, which service point attracts more customers on average? What can be said about the relative variability in the number of customers arriving per hour at the two service points? Note that for a Poisson distribution, the mean is equal to the variance.

6

ESTIMATION

We said at the beginning of Chapter 3 that decisions must often be made when only incomplete information is available and there is uncertainty concerning the outcomes that must be considered by the decision maker. The remainder of this book deals with methods by which rational decisions can be made under such circumstances. In our brief introduction to probability theory, we have begun to see how probability concepts can be used to cope with problems of uncertainty. **Statistical inference** uses this theory as a basis for making reasonable decisions from incomplete data. Statistical inference treats two different classes of problems: (1) estimation, which is discussed in this chapter; and (2) hypothesis testing, which is examined in Chapters 7 and 8. In both cases, inferences are made about population characteristics from information contained in samples.

6.1
POINT AND INTERVAL ESTIMATION

The need to estimate population parameters from sample data stems from the fact that it is usually too expensive or not feasible to enumerate complete populations to obtain the required information. The cost of complete censuses of finite populations may be prohibitive; complete enumerations of infinite populations are impossible.

> **Statistical estimation** procedures provide us with the means to estimate population parameters with desired degrees of precision.

Numerous examples can be given of the need to estimate pertinent population parameters in business and economics. A marketing organization may be interested in estimates of average income and other socioeconomic characteristics of the consumers in a metropolitan area; a retail chain may want an estimate of the average number of pedestrians per day who pass a certain corner; or a bank may want an estimate of average interest rates on mortgages in a certain section of the country. Undoubtedly, in all of these cases, exact accuracy is not required, and estimates derived from sample data would probably provide appropriate information to meet the demands of the practical situation.

Two different types of estimates of population parameters are of interest: point estimates and interval estimates.

Point estimate

> A **point estimate** is a single number used as an estimate of the unknown population parameter.

For example, the arithmetic mean income of a sample of families in a metropolitan area may be used as a point estimate of the corresponding population mean for all families in that metropolitan area; or the percentage of a sample of eligible voters in a political opinion poll who state that they would vote for a particular candidate may be used as an estimate of the corresponding unknown percentage in the relevant population.

A distinction can be made between an *estimate* and an *estimator*. Consider the illustration of estimating the population figure for arithmetic mean income of all families in a metropolitan area from the corresponding sample mean.

Estimate
- The numerical value of the sample mean is said to be an **estimate** of the population mean figure.

Estimator
- The statistical measure used (that is, the *method* of estimation) is referred to as an **estimator**.

For example, the sample mean $\bar{x}$ is an *estimator* of the population mean. When a specific number is calculated for the sample mean, say $8,000, that number is an *estimate* of the population mean figure.

Choosing an estimate
Whether we use a point estimate rather than an interval estimate depends on the purpose of the investigation. For example, for planning purposes, a marketing department may estimate a single figure for annual sales of one of its company's products and may then break that figure down into monthly sales estimates. These figures may be passed on to the production department for the planning of production requirements. The production department may in turn convert its production requirements into materials purchasing plans for the purchasing department. If the marketing department estimated annual sales

as a range, say from $10 million to $12 million, rather than as a single figure, this could unduly complicate the subsequent steps of obtaining monthly break-downs and planning production and purchasing requirements. However, for many practical purposes, having only a single point estimate of a population parameter is not sufficient. Any single point estimate will be either right or wrong. It would certainly seem useful, and perhaps even necessary, to have in addition to a point estimate, some notion of the degree of error that might be involved in using this estimate. Interval estimation is useful in this connection.

> An **interval estimate** of a population parameter is a statement of two values between which we have some confidence that the parameter lies.

Interval estimate

Thus, an interval estimate in the example of the population arithmetic mean income of families in a metropolitan area might be $34,100 to $35,900. An interval estimate for the percentage of defectives in a shipment might be 3% to 5%.

We may have a great deal of confidence or very little confidence that the population parameter is included in the range of the interval estimate, so we must attach some sort of probabilistic statement to the interval. The procedure used to create such a probabilistic statement is **confidence interval estimation**.

Confidence interval estimation

The confidence interval is an interval estimate of the population para-meter. A confidence coefficient such as 90% or 95% is attached to this interval to indicate the degree of confidence or credibility placed on the estimated interval.

6.2
CRITERIA OF GOODNESS OF ESTIMATION

Numerous criteria have been developed by which to judge the goodness of point estimators of population parameters. A rigorous discussion of these criteria requires some complex mathematics that falls outside the scope of this text. However, it is possible to gain an appreciation of the nature of these criteria in an intuitive, nonrigorous way.

Let us return to our illustration of estimating the arithmetic mean income of families in a metropolitan area. This arithmetic mean—assuming suitable definitions of income, family, and metropolitan area—is an unknown popula-tion parameter that we designate as μ. Suppose we took a simple random sample of families from this population and calculated the arithmetic mean $\bar{x}$, the median Md, and the mid-range $(x_{max} + x_{min})/2$, where x_{max} and x_{min} are the largest and smallest sample observations. Which method would be the best estimator of the population mean? Probably you would answer that the sample mean $\bar{x}$ is the best estimator. In fact, if this question had not been raised, it might not even have occurred to you to use any statistic other than the sample mean as an estimator of the population mean. However, why do you think the sample mean represents the best estimator? It may not be easy to articulate your answer to that question. The sample mean is preferable to the other estimators by the generally utilized criteria of goodness of estimation of classical statistical inference. Let us briefly examine the nature of a few of these criteria: *unbiasedness, consistency,* and *efficiency.*

Unbiasedness   An estimator—such as a sample arithmetic mean—is a random variable, because it may take on different values, depending on which population elements are drawn into the sample. Rule 13 of Appendix C proves that the expected value of this random variable is the population mean. In symbols, we have

(6.1)
$$E(\bar{x}) = \mu$$

where $\bar{x}$ = the sample mean
μ = the population mean

> If the expected value of a sample statistic is equal to the population parameter for which the statistic is an estimator, the statistic or estimator is said to be **unbiased**.

If symbols, if θ is a parameter to be estimated and $\hat{\theta}$ is a sample statistic used to estimate θ, then $\hat{\theta}$ is said to be an *unbiased estimator* of θ if

(6.2)
$$E(\hat{\theta}) = \theta$$

If we say that a given estimator is unbiased, we are simply saying that this method of estimation is correct *on the average.* That is, if the method is employed repeatedly, the average of all estimates obtained from this estimator is equal to the value of the population parameter. Clearly, an unbiased estimator does not guarantee useful individual estimates. The differences of these individual estimates from the value of the population parameter may represent large errors. The simple fact that the bias, or long-run average of these errors, is zero may be of little practical importance. Furthermore, if we have two unbiased estimators, we require additional criteria in order to make a choice between them.

Just as the sample mean is an unbiased estimator of the population mean, the sample variance as defined in equation 1.15 is an unbiased estimator of the population variance.[1] In symbols, as proven in rule 14 of Appendix C, we have

(6.3)
$$E(s^2) = E\left[\frac{\Sigma(x - \bar{x})^2}{n - 1}\right] = \sigma^2$$

---

[1] We can see from the nature of the proof in rule 14 of Appendix C, $\Sigma(x - \bar{x})^2/(n - 1)$ is not an unbiased estimator of the variance of a *finite* population. Perhaps surprisingly, both $\sqrt{\Sigma(x - \bar{x})^2/n}$ and $\sqrt{\Sigma(x - \bar{x})^2/(n - 1)}$ are biased estimators of the population standard deviation.

Equation 6.3 is a mathematical restatement of the point made in section 1.15 that when the sample variance is defined with divisor $n - 1$, it is an unbiased estimator of the population variance.

It is clear from the preceding discussion that knowing only that an estimator is unbiased gives us insufficient information as to the goodness of that method of estimation. Closeness of the estimator to the parameter seems to be of primary importance. Both the concepts of *consistency* and *efficiency* deal with this property of closeness. Consider the sample mean $\bar{x}$ as an estimator of the population parameter μ for an infinite population. What happens to the possible values of $\bar{x}$ as the sample size n increases? On an intuitive basis, we would certainly expect $\bar{x}$ to lie closer to μ as n becomes larger. Generally, if an estimator, say $\hat{\theta}$, approaches closer to the parameter θ as the sample size n increases, $\hat{\theta}$ is said to be a **consistent estimator** of θ.

Consistency

In terms of sampling, this idea of consistency means that the sampling distribution of the estimator becomes more and more "tightly packed" around the population parameter as the sample size increases. Figure 6-1(a) illustrates this concept for the sample mean as an estimator of μ, the mean of an infinite population. The graph represents the respective sampling distributions of the

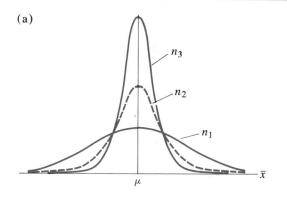

(a)

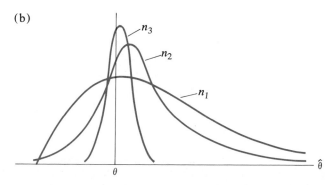

(b)

FIGURE 6-1

(a) Sampling distributions of $\bar{x}$ as the sample size increases; $n_3 > n_2 > n_1$

(b) Sampling distributions of a biased but consistent estimator $\hat{\theta}$; $n_3 > n_2 > n_1$

sample mean $\bar{x}$ for samples of size n_1, of size n_2, and of size n_3 drawn from the same population; n_3 is larger than n_2, which is larger than n_1. We know from the relationship $\sigma_{\bar{x}}^2 = \sigma^2/n$ that the sampling variance $\sigma_{\bar{x}}^2$ decreases as n increases. Note that all three sampling distributions center on the population parameter μ, since $\bar{x}$ is an unbiased estimator of μ.

Figure 6-1(b) illustrates the concept of consistency for $\hat{\theta}$, a **biased estimator** of a parameter θ. As in Figure 6-1(a), the graph represents the respective sampling distributions of an estimator, in this case $\hat{\theta}$, for samples of size n_1, of size n_2, and of size n_3 drawn from the same population, with $n_3 > n_2 > n_1$. None of the distributions is centered on the population parameter θ. Since $E(\hat{\theta})$ is not equal to θ, $\hat{\theta}$ is a biased estimator of θ. However, we can see from the graph that as the sample size increases, the sampling distribution becomes more "tightly packed" around θ.

Efficiency   The concept of efficiency refers to the sampling variability of an estimator. If two competing estimators are both unbiased, the one with the smaller variance (for a given sample size) is said to be relatively more efficient. More formally, if $\hat{\theta}_1$ and $\hat{\theta}_2$ are two unbiased estimators of θ, their relative efficiency is defined by the ratio

(6.4)
$$\frac{\sigma_{\hat{\theta}_2}^2}{\sigma_{\hat{\theta}_1}^2}$$

where $\sigma_{\hat{\theta}_1}^2$ is the smaller variance.

Let us consider as an example a simple random sample of size n drawn from a normal population with mean μ and variance σ^2. Suppose we want to consider the relative efficiency of the sample mean $\bar{x}$ and the sample median Md as estimators of the population mean μ. Both estimators are unbiased. We know that the variance of the sample mean $\bar{x}$ is $\sigma_{\bar{x}}^2 = \sigma^2/n$. It can be shown that the variance of the sample median Md is approximately $\sigma_{Md}^2 = 1.57\sigma^2/n$. Therefore, the relative efficiency of $\bar{x}$ with respect to Md is

$$\frac{\sigma_{Md}^2}{\sigma_{\bar{x}}^2} = \frac{1.57\sigma^2/n}{\sigma^2/n}$$

$$= 1.57$$

We can interpret this result in terms of sample sizes. If the sample median rather than the sample mean were used as an estimator of the mean μ of a normal population, then in order to obtain the same precision provided by the sample mean, a sample size 57% larger would be required. Stated differently, the required sample size for the sample median would be 157% of that for the sample mean. Figure 6-2 shows the sampling distributions of these two estimators.

Other criteria for goodness   Other criteria for goodness of estimation may be found in standard texts
of estimation   on mathematical statistics. As we have seen, sampling error decreases with increasing sample size, and biased but consistent estimators approach the population parameter as sample size increases. However, greater cost is incurred with larger sample sizes. Therefore, most practical estimation situations involve trade-offs between these considerations.

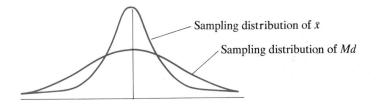

FIGURE 6-2

Sampling distributions of the
mean $\bar{x}$ and median Md for
the same sample size

Sampling distribution of $\bar{x}$

Sampling distribution of Md

Two problems of interest in this chapter are the estimation of a population proportion and a population mean. In the case of estimating a population proportion, if we assume a Bernoulli process generating n sample observations, it can be shown that the observed sample proportion of successes $\bar{p}$ = (number of successes)/(number of sample observations) is an unbiased, efficient, and consistent estimator. Similarly, the mean $\bar{x}$ of a simple random sample of n observations is an unbiased, efficient, and consistent estimator of the population mean. It is not surprising that in many applications of statistical methods, sample proportions or sample means are used as the "best" point estimators of the corresponding population parameters. Perhaps the reader has had occasion to calculate a sample proportion or mean and has found it intuitively appealing to use such a figure as an estimator of the corresponding population parameter. If so, this intuitive approach was supported by sound statistical theory.

EXERCISES 6.2

1. Determine whether each of the following statements describes an estimator or an estimate. If the statement concerns an estimate, is it a point estimate or an interval estimate?
    a. A meteorologist feels that the total accumulation of snow from a particular storm will be between 10 and 14 inches.
    b. In certain situations, the median of a class may be a better measure of central tendency than the mean. For example, in a research study on the savings habits of American families, the large savings held by the wealthy may distort the mean.
    c. After examining a sample of the bank's loan portfolio, a bank auditor decides that 5% of all loans in the portfolio will have to be written off as bad debts.
    d. A security analyst predicts that the Dow Jones average will be between 2,950 and 3,150 by the end of next year.

2. List three criteria used to judge the goodness of a point estimate, and discuss the meaning of each criterion.

3. State whether each of the following statements is true or false, and explain your answer:
   a. If a statistic is an unbiased estimator of a parameter, then it is the "best" estimator of the parameter.
   b. If an unbiased estimator of σ^2 is desired, it is best to calculate the sample variance as $\dfrac{\Sigma(x - \bar{x})^2}{n}$.
   c. If $\hat{\theta}_1$ is a biased estimator of θ and $\hat{\theta}_2$ is an unbiased estimator of θ, and the ratio $\sigma^2_{\hat{\theta}_2}/\sigma^2_{\hat{\theta}_1}$ is greater than one, then $\hat{\theta}_1$ is relatively more efficient than $\hat{\theta}_2$.

4. Determine which of the following estimates of sales for next year you would like to receive from the marketing department.
   a. Sales will be $1,000,000.
   b. Sales will fall between $500,000 and $2,500,000 with 100% probability.
   c. Sales will fall between $900,000 and $1,100,000 with 95% probability.

5. Sampling error decreases with increasing sample size, and biased but consistent estimators approach the population parameter as sample size increases. In view of the benefits of increasing sample sizes, why don't researchers try to take as large a sample as possible?

6.3
CONFIDENCE INTERVAL ESTIMATION (LARGE SAMPLES)

As indicated in section 6.1, for many practical purposes, it is not sufficient merely to have a single point estimate of a population parameter. It is usually necessary to have an estimation procedure that measures the degree of precision involved. In classical statistical inference, the standard procedure for this purpose is confidence interval estimation.

Confidence Interval Estimation

We will explain the rationale of confidence interval estimation in terms of an example in which a population mean is the parameter to be estimated. Suppose a manufacturer has a very large production run of a certain brand of tire and wants to obtain an estimate of their arithmetic mean lifetime by drawing a simple random sample of 100 tires and subjecting them to a forced life test. Let us assume that, from long experience in manufacturing this brand of tire, the manufacturer knows that the population standard deviation for a production run is $\sigma = 3,000$ miles. (Of course, ordinarily the standard deviation of a population is not known exactly and must be estimated from a sample, just as are the mean and other parameters. However, let us assume in this case that the population standard deviation is indeed known.) When the sample of 100 tires is drawn, a mean lifetime of 32,500 miles is observed. Thus, we denote $\bar{x} = 32,500$ miles. This sample mean is our best *point* estimate of the population mean lifetime, that is, of the mean lifetime of all tires in the production run. Additionally, we would like an *interval* estimate of the population mean lifetime. That is, we would like to be able to state that the population mean is between two limits, say $\bar{x} - 2\sigma_{\bar{x}}$ and $\bar{x} + 2\sigma_{\bar{x}}$ where $\bar{x} - 2\sigma_{\bar{x}}$ is the lower limit of the interval and $\bar{x} + 2\sigma_{\bar{x}}$ is the upper limit. Furthermore, we would like to have a high degree of confidence that the true population mean is included in this interval.

The procedure in confidence interval estimation is based on the concept of the sampling distribution. In this example, since we are dealing with the estimation of a mean, the appropriate distribution is the sampling distribution of the mean. We will review some fundamentals of this distribution to lay the foundation for confidence interval estimation. Figure 6-3 shows the sampling distribution of the mean for simple random samples of size $n = 100$ from a population with an unknown mean, denoted μ, and a standard deviation $\sigma = 3,000$. We assume that the sample is large enough so that by the Central Limit Theorem, stated in section 5.5, the sampling distribution may be assumed to be normal, even if the population is nonnormal. The standard error of the

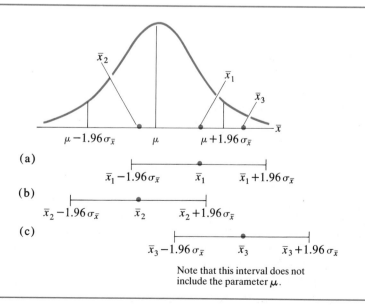

FIGURE 6-3

Sampling distribution of the mean and confidence interval estimates for three illustrative samples

mean,[2] which is the standard deviation of this sampling distribution, equals $\sigma_{\bar{x}} = \sigma/\sqrt{n}$.

In our work with the normal sampling distribution of the mean, we have learned how to make probability statements about sample means, given the value of the population mean. Thus, in terms of the data of this problem, we can state that if repeated simple random samples of 100 tires each were drawn from the production run, 95% of the sample means $\bar{x}$ would lie within 1.96 standard error units of the mean of the sampling distribution (the population mean), or between $\mu - 1.96\sigma_{\bar{x}}$ and $\mu + 1.96\sigma_{\bar{x}}$. This range is indicated on the horizontal axis of the sampling distribution in Figure 6-3. For emphasis, vertical lines show the endpoints of this range. As usual, we determine the 1.96 figure from Table A-5 of Appendix A, where we find that 47.5% of the area in a normal distribution is included between the mean of a normal distribution and a value 1.96 standard deviations to the right of the mean; thus, by symmetry, 95% of the area is included in a range of ± 1.96 standard deviations from the mean. In terms of relative frequency, 95% of the $\bar{x}$ values of samples of size 100 would lie in this range if repeated samples were drawn from the given population.

How then might we construct the desired interval estimate for the population parameter? Let us consider again the repeated simple random samples of size 100 from the population of the production run of tires. Suppose our first sample yields a mean that exceeds μ but falls between $\mu + 1.96\sigma_{\bar{x}}$. The position of this sample mean, denoted $\bar{x}_1$, is shown on the horizontal axis of Figure 6-3. Suppose now that we set up an interval from $\bar{x}_1 - 1.96\sigma_{\bar{x}}$ to $\bar{x}_1 + 1.96\sigma_{\bar{x}}$. This interval is shown immediately below the graph in Figure 6-3(a). As seen in the figure, this interval, which may be written as $\bar{x}_1 \pm 1.96\sigma_{\bar{x}}$, includes the population parameter μ. This follows from the fact that $\bar{x}_1$ fell less than $1.96\sigma_{\bar{x}}$ from the mean of the sampling distribution μ.

Now, let us assume our second sample from the same population yields the mean $\bar{x}_2$, which lies on the horizontal axis to the left of μ, but again at a distance less than $1.96\sigma_{\bar{x}}$ from μ. Again we set up an interval of the sample mean $\pm 1.96\sigma_{\bar{x}}$, or from $\bar{x}_2 - 1.96\sigma_{\bar{x}}$ to $\bar{x}_2 + 1.96\sigma_{\bar{x}}$. This interval, shown below the graph in Figure 6-3(b), includes the population mean μ.

Finally, suppose a third sample is drawn from the same population, with the mean $\bar{x}_3$ shown on the horizontal axis of Figure 6-3. This sample mean lies to the right of μ, but at a distance *greater than* $1.96\sigma_{\bar{x}}$ above μ. When we set up the range $\bar{x}_3 - 1.96\sigma_{\bar{x}}$ to $\bar{x}_3 + 1.96\sigma_{\bar{x}}$, this interval shown below the graph in Figure 6-3(c), does not include μ.

We can imagine a continuation of this sampling procedure. Since 95% of the sample means fall within $1.96\sigma_{\bar{x}}$ of μ, we can assert that 95% of the intervals of the type $\bar{x} \pm 1.96\sigma_{\bar{x}}$ include the population parameter μ. Now, we can get

---

[2] Strictly speaking, the finite population correction should be shown in this formula, but we will assume the population is so large relative to sample size that for practical purposes the correction factor is equal to one. The mean of the sampling distribution $\mu_{\bar{x}}$ is equal to the population mean μ.

to the crux of confidence interval estimation. In the problem originally posed, as in most practical applications, only one sample was drawn from the population, not repeated samples. On the basis of the single sample, we were required to estimate the population parameter. The procedure simply establishes the interval $\bar{x} \pm 1.96\sigma_{\bar{x}}$ and attaches a suitable statement to it. The interval itself is referred to as a **confidence interval**. Thus, in our original problem the required confidence interval is

Confidence interval

$$\bar{x} \pm 1.96\sigma_{\bar{x}} = \bar{x} \pm 1.96\frac{\sigma}{\sqrt{n}} = 32,500 \pm 1.96\frac{3,000}{\sqrt{100}}$$

$$= 32,500 \pm 588 = 31,912 \text{ to } 33,088 \text{ miles}$$

We must be very careful how we interpret this confidence interval. It is incorrect to make a probability statement about this *specific* interval. For example, it is incorrect to state that the probability is 95% that the mean lifetime μ of all tires falls in this interval. *The population mean is not a random variable;* hence, probability statements cannot be made about it. The unknown population mean μ either lies in the interval or it does not. We must return to the line of argument used in explaining the method and indicate that the values of the random variable are the intervals of $\bar{x} \pm 1.96\sigma_{\bar{x}}$, not μ. Thus, if repeated simple random samples of the same size were drawn from this population and the interval $\bar{x} \pm 1.96\sigma_{\bar{x}}$ were constructed from each of them, then 95% of the statements that the interval contains the population mean μ would be correct. Another way of putting it is that in 95 samples out of 100, the mean μ would lie within intervals constructed by this procedure. The 95% figure is referred to as a **confidence coefficient** to distinguish it from the type of probability calculated when deductive statements are made about sample values from known population parameters.[3]

Confidence coefficient

Despite the above interpretation of the meaning of a confidence interval, where the probability pertains to the estimation procedure rather than to the specific interval constructed from a single sample, the fact remains that we ordinarily must make an inference on the basis of the single sample drawn. We will not draw the repeated samples implied by the interpretational statement. For example, in the tire illustration, an inference was required about the production run based on the particular sample of 100 tires in hand, and an interval estimate of a mean lifetime of 31,912 to 33,088 miles was obtained. If the confidence coefficient attached to the interval estimate is high, then the investigator will assume that the interval estimate is correct. This interval may or may not encompass the actual value of the population parameter μ. However, since 95% of intervals so constructed would include the value of the mean lifetime μ

Interpretation and use of interval estimation

---

[3] This paragraph presents the standard interpretation of confidence intervals provided by classical statistics. Bayesian statistics (discussed in Chapters 14 through 17) disputes this interpretation, arguing that if μ is unknown, it may be treated as a random variable. Hence, in Bayesian statistics, "prior probability" statements may be made about μ based on subjective assessments, and "posterior probability" statements may be made that combine the prior probabilities and information obtained from sampling.

of all tires in the production run, we will behave as though this particular interval does include the actual value.

> It is desirable to obtain a relatively narrow interval with a high confidence coefficient associated with it. One without the other is not particularly useful.

Thus, for example, in estimating a proportion (say, the proportion of persons in the labor force who are unemployed), we can assert even without sample data that the percentage lies somewhere between zero and 100% with a confidence coefficient of 100%. Obviously, this statement is neither profound nor useful, because the interval is too wide. On the other hand, if the interval is narrow but has a low associated confidence coefficient, say 10%, the statement would again have little practical utility.

Confidence coefficients such as 0.90, 0.95, and 0.99 and limits of two or three standard errors (such as 0.955 or 0.997) are conventionally used. Limits of two or three standard errors are those obtained by making an estimate of a population parameter and adding and subtracting two or three standard errors to establish confidence intervals. For a fixed confidence coefficient and population standard deviation, the only way to narrow a confidence interval and increase the precision of the statement is to increase the sample size. This is readily apparent from the way the confidence interval was constructed in the tire example. We computed $\bar{x} \pm 1.96\sigma_{\bar{x}}$, where $\sigma_{\bar{x}} = \sigma/\sqrt{n}$. If the 1.96 figure and σ are fixed, we can decrease the width of the interval only by increasing the sample size n since $\sigma_{\bar{x}}$ is inversely related to $\sqrt{n}$. Thus, the marginal benefit of increased precision must be measured against the increased cost of sampling. Later in this section, we discuss a method of determining the sample size required for a specified degree of precision.

One final point may be made before turning to confidence interval estimation of different types of population parameters. Ordinarily, as indicated in the tire example, the standard deviation of the population σ is unknown. Therefore, it is not possible to calculate $\sigma_{\bar{x}}$, the standard error of the mean. However, we can estimate the standard deviation of the population from a sample and use this figure to calculate an estimated standard error of the mean. We use this estimation technique in the examples that follow.

Interval Estimation of a Mean (Large Samples)

We will use examples to discuss confidence interval estimation and will concentrate first on situations in which the sample size is large. Our discussion will then focus, in turn, on interval estimation of a mean, a proportion, the difference between means, and the difference between proportions. Finally, we will briefly treat corresponding estimation procedures for small samples.

Let us look at Example 6-1, an illustration of interval estimation of a mean from a large sample.

EXAMPLE 6-1

A group of students working on a summer project with a social agency took a simple random sample of 120 families in a well defined "poverty area" of a large city in order to determine the mean annual family income of this area. The sample results were $\bar{x} = \$6{,}810$, $s = \$780$, and $n = 120$. What would be the 99% confidence interval for the mean income of all families in this poverty area?

The only way this problem differs from the illustration of the mean lifetime of tires is that here the population standard deviation is unknown. The usual procedure for large samples ($n > 30$) is simply to use the sample standard deviation as an estimate of the corresponding population standard deviation. Using s as an estimator of σ, we can compute an estimated standard error of the mean, $s_{\bar{x}}$. We have

$$s_{\bar{x}} = \frac{s}{\sqrt{n}} = \frac{\$780}{\sqrt{120}} = \$71.20$$

Hence, we may use $s_{\bar{x}}$ as an estimator of $\sigma_{\bar{x}}$, and because n is large we invoke the Central Limit Theorem to argue that the sampling distribution of $\bar{x}$ is approximately normal. Again, we have assumed the finite population correction to be equal to 1. The confidence interval, in general, is given by

(6.5)
$$\bar{x} \pm z s_{\bar{x}}$$

where z is the multiple of standard errors and $s_{\bar{x}}$ now replaces $\sigma_{\bar{x}}$, which was used when the population standard deviation was known. For a 99% confidence coefficient, $z = 2.58$. Therefore, the required interval is

$$\$6{,}810 \pm 2.58(\$71.20) = \$6{,}810 \pm \$183.70$$

so the population mean is roughly between $6,626 and $6,994 with a 99% confidence coefficient. The same sort of interpretation given earlier for confidence intervals again applies here.

Interval Estimation of a Proportion (Large Samples)

In many situations, it is important to estimate a proportion of occurrences in a population from sample observations. For example, it may be of interest to estimate the proportion of unemployed persons in a certain city, the proportion of eligible voters who intend to vote for a particular political candidate, or the proportion of students at a university who favor changing the grading system. In all these cases, the corresponding proportions observed in simple random samples may be used to estimate the population proportions. Before turning to a description of how this estimation is accomplished, we will briefly establish the conceptual underpinnings of the procedure.

In Chapter 5, we saw that under certain conditions, the binomial distribution is the appropriate sampling distribution for the number of successes x in a simple random sample of size n. Furthermore, we noted that if p, the proportion to be estimated, is not too close to zero or one, the binomial distribution can be closely approximated by a normal curve with the same mean and standard deviation, that is, $\mu = np$ and $\sigma = \sqrt{npq}$.

Because we want to discuss interval estimation of proportion, we repeat some basic ideas concerning the sampling distribution of a proportion. It is a simple matter to convert from a sampling distribution of *number of successes* to the corresponding distribution of *proportion of successes*. If x is the number of successes in a sample of n observations, then the proportion of successes in the sample is $\bar{p} = x/n$. Hence, dividing the formulas in the preceding paragraph by n, we find that the mean and standard deviation of the sample proportion become

(6.6)
$$\mu_{\bar{p}} = p$$

and

(6.7)
$$\sigma_{\bar{p}} = \sqrt{\frac{pq}{n}}$$

Note that the subscript $\bar{p}$ has been used on the left sides of equations 6.6 and 6.7 in keeping with the symbolism used for the corresponding values for the sampling distribution of the mean, namely, $\mu_{\bar{x}}$ and $\sigma_{\bar{x}}$. Therefore, in summary, if p is not too close to zero or one, the sampling distribution of $\bar{p}$ can be closely approximated by a normal curve with the mean and standard deviation given in equations 6.6 and 6.7. It is important to observe in this connection that the Central Limit Theorem holds for sample proportions as well as for sample means.

Note that the above discussion pertains to cases in which the population size is large compared with the sample size. Otherwise, equation 6.7 should be multiplied by the finite population correction $\sqrt{(N - n)/(N - 1)}$. As we mentioned in Chapter 5, $\sigma_{\bar{p}}$ is referred to as the **standard error of a proportion**.

To illustrate confidence interval estimation for a proportion, we shall make assumptions similar to those in the preceding example. In Example 6-2, we assume a large simple random sample drawn from a population that is very large compared with the sample size.

EXAMPLE 6-2

In the town of Smallsville, a simple random sample of 800 automobile owners revealed that 480 would like to see the size of automobiles reduced. What are the 95.5% confidence limits for the proportion of all automobile owners in Smallsville who would like to see car size reduced?

In this problem, we want a confidence interval estimate for p, a population proportion. We have obtained the sample statistic $\bar{p} = 480/800 = 0.60$, which is the sample proportion who wish to see car size reduced. As noted earlier, for large sample sizes and for p values not too close to 0 or 1, the sampling distribution of $\bar{p}$ may be approximated by a normal distribution with mean $\mu_{\bar{p}} = p$ and $\sigma_{\bar{p}} = \sqrt{pq/n}$. Here we encounter the same type of problem as in interval estimation of the mean. The formula for the exact standard error of a proportion, $\sigma_{\bar{p}} = \sqrt{pq/n}$, requires the values of the unknown population parameters p and q. Hence, we use an estimation procedure similar to that used in the case of the mean. Just as we used s to approximate σ, we can substitute the corresponding sample statistics $\bar{p}$ and $\bar{q}$ for the parameters p and q in the formula for $\sigma_{\bar{p}}$ in order to calculate an estimated standard error of a proportion, $s_{\bar{p}} = \sqrt{\bar{p}\bar{q}/n}$. Using the same type of reasoning as that for interval estimation of the mean, we can state a two-sided confidence interval estimate for a population proportion as

(6.8)
$$\bar{p} \pm z s_{\bar{p}}$$

In this problem, $z = 2$, since the confidence coefficient is 95.5%. Hence, substituting into Equation 6.8, we obtain our interval estimate of the proportion of all automobile owners in Smallsville who would like to see car size reduced:

$$0.60 \pm 2 \sqrt{\frac{0.60 \times 0.40}{800}} = 0.60 \pm 0.0346$$

Thus, the population proportion is estimated to be included in the interval 0.5654 to 0.6346, or roughly between 56.5% and 63.5% with a 95.5% confidence coefficient.

Interval Estimation of the Difference between Two Means (Large Samples)

The foregoing examples of *estimation* of a population mean and proportions are based on single samples. We now examine interval estimation of the difference between means and the difference between proportions based on data obtained from two independent large samples. First, we examine an example of confidence interval estimation of the difference between two population means.

EXAMPLE 6-3

A large department store chain was interested in analyzing the difference between the average dollar amount of its delinquent charge accounts in the northeastern and western regions of the country for a certain year. The store took two independent simple random samples of these delinquent charge accounts, one from each region. The mean and standard deviation of the dollar amounts of these delinquent accounts were calculated to the nearest dollar. The northeastern region is denoted as 1 and the western region is denoted as 2.

The analysts decided to establish 99.7% confidence limits for $\mu_1 - \mu_2$, where μ_1 and μ_2 denote the respective population mean sizes of delinquent accounts. Of course, a point estimate of $\mu_1 - \mu_2$ is given by $\bar{x}_1 - \bar{x}_2$. The required theory for the interval estimate is based on the fact that the sampling distribution of $\bar{x}_1 - \bar{x}_2$ for two large independent samples is exactly normal, if the population of differences is normal, with

Sample 1	Sample 2
$\bar{x}_1 = \$76$	$\bar{x}_2 = \$65$
$s_1 = \$25$	$s_2 = \$22$
$n_1 = 100$	$n_2 = 100$

mean and standard deviation

(6.9)
$$\mu_{\bar{x}_1 - \bar{x}_2} = \mu_1 - \mu_2$$

and

(6.10)
$$\sigma_{\bar{x}_1 - \bar{x}_2} = \sqrt{\frac{\sigma_1^2}{n_1} + \frac{\sigma_2^2}{n_2}}$$

where σ_1 and σ_2 represent the respective population standard deviations of sizes of delinquent accounts.[4]

Since the population standard deviations σ_1 and σ_2 are unknown, and since the sample sizes are large, the sample standard deviations may be substituted into the formula for $\sigma_{\bar{x}_1 - \bar{x}_2}$ to give an estimated standard error of the difference between two means,

$$s_{\bar{x}_1 - \bar{x}_2} = \sqrt{\frac{s_1^2}{n_1} + \frac{s_2^2}{n_2}}$$

As usual with problems of this type, the population of differences may not be normal, and the population standard deviations are unknown. However, since the samples are large, we can use the Central Limit Theorem to assert that the sampling distribution of $\bar{x}_1 - \bar{x}_2$ is approximately normal. The required confidence limits are given by

(6.11)
$$(\bar{x}_1 - \bar{x}_2) \pm z s_{\bar{x}_1 - \bar{x}_2}$$

The calculation for $s_{\bar{x}_1 - \bar{x}_2}$ in this problem is

$$s_{\bar{x}_1 - \bar{x}_2} = \sqrt{\frac{(25)^2}{100} + \frac{(22)^2}{100}} = \$3.33$$

Since a 99.7% confidence interval is desired, the value of z is three. Therefore, substituting into equation 6.11 gives

$$(\$76 - \$65) \pm 3(\$3.33) = \$11 \pm \$9.99$$

Hence, to the nearest dollar, confidence limits for $\bar{x}_1 - \bar{x}_2$ are $1 and $21. It is a worthwhile exercise for the reader to attempt to express in words specifically what this confidence interval means.

Interval Estimation of the Difference between Two Proportions (Large Samples)

The procedure for constructing a confidence interval estimate for the difference between two proportions is analogous to the technique used in constructing a confidence interval estimate for means.

---

[4] Equation 6.9 follows from rule 4 of Appendix C. By rule 11 of Appendix C, the variance of the difference $\bar{x}_1 - \bar{x}_2$ is $\sigma_{\bar{x}_1 - \bar{x}_2}^2 = \sigma_{\bar{x}_1}^2 + \sigma_{\bar{x}_2}^2$. This is an application of the principle that the variance of the difference between two *independent* random variables is equal to the sum of the variances of these variables. Taking the square root of both sides of this equation, we obtain $\sigma_{\bar{x}_1 - \bar{x}_2} = \sqrt{\sigma_{\bar{x}_1}^2 + \sigma_{\bar{x}_2}^2}$, where $\sigma_{\bar{x}_1}^2$ and $\sigma_{\bar{x}_2}^2$ are simply the variances of the sampling distributions of $\bar{x}_1$ and $\bar{x}_2$. Substituting $\sigma_{\bar{x}_1}^2 = \sigma_1^2/n_1$ and $\sigma_{\bar{x}_2}^2 = \sigma_2^2/n_2$, we get equation 6.10.

EXAMPLE 6-4

A credit reference service investigated two simple random samples of customers who applied for charge accounts in two different department stores. The service was interested in the proportion of applicants in each store who had annual incomes exceeding \$30,000. Confidence limits of 90% were established for the difference $p_1 - p_2$, where p_1 and p_2 represent the population proportions of applicants in each store whose incomes exceeded \$30,000. The sample data were

Store 1	Store 2
$\bar{p}_1 = 0.50$	$\bar{p}_2 = 0.18$
$\bar{q}_1 = 0.50$	$\bar{q}_2 = 0.82$
$n_1 = 150$	$n_2 = 160$

where these symbols have their conventional meanings. As in the preceding example, we can start with a point estimate. The number $\bar{p}_1 - \bar{p}_2$ is the obvious point estimate of $p_1 - p_2$, and we can assume that the sampling distribution of $\bar{p}_1 - \bar{p}_2$ is approximately normal with mean

$$\mu_{\bar{p}_1 - \bar{p}_2} = p_1 - p_2$$

and standard deviation

$$\sigma_{\bar{p}_1 - \bar{p}_2} = \sqrt{\frac{p_1 q_1}{n_1} + \frac{p_2 q_2}{n_2}}$$

Since the population proportions p_1 and p_2 are unknown and the sample sizes are large, $\bar{p}_1$ and $\bar{p}_2$ may be substituted for p_1 and p_2 to obtain the estimated standard error of the difference between percentages.

(6.12)
$$s_{\bar{p}_1 - \bar{p}_2} = \sqrt{\frac{\bar{p}_1 \bar{q}_1}{n_1} + \frac{\bar{p}_2 \bar{q}_2}{n_2}}$$

As in the procedure for differences between means, we use the Central Limit Theorem to argue that the sampling distribution of $\bar{p}_1 - \bar{p}_2$ is approximately normal, and we establish confidence limits of

(6.13)
$$(\bar{p}_1 - \bar{p}_2) \pm z s_{\bar{p}_1 - \bar{p}_2}$$

In this problem, the value of $s_{\bar{p}_1 - \bar{p}_2}$ is

$$s_{\bar{p}_1 - \bar{p}_2} = \sqrt{\frac{(0.50)(0.50)}{150} + \frac{(0.18)(0.82)}{160}} = 0.051$$

and since a 90% confidence coefficient is desired, $z = 1.65$. Therefore, the required confidence interval for the difference in the proportion of the applicants in the two stores whose incomes exceeded \$30,000 is

$$(0.50 - 0.18) \pm 1.65(0.051) = 0.32 \pm 0.084$$

The confidence limits are 0.236 and 0.404.

EXERCISES 6.3

1. Explain the following statement in detail: The standard deviation of $\bar{x}$ ($\sigma_{\bar{x}}$) is both a measure of statistical error in point estimation and a measure of dispersion of a distribution.

2. Dairy Products Unlimited (DPU) is studying the characteristics of the ice cream industry in Brazil. DPU takes a random sample of 36 ice cream manufacturers in order to assess the average number of flavors currently marketed by Brazilian manufacturers. The arithmetic mean is 5.2 with a standard deviation of 0.9.
   a. Establish a 95.5% confidence interval for the population mean assuming a large number of ice cream manufacturers in Brazil.
   b. In general, what is the meaning of a 95.5% confidence interval?

3. C&G Cola Company completed a test market survey of the taste preferences of people in Bangkok. The company asked a random sample of 400 people to assess the C&G drink against a leading local beverage. The mean rating of the C&G drink was 105 with a standard deviation of 50. The rating for the local beverage was assumed to be 100. On the basis of the test market results, can C&G Cola Company conclude that the 95% confidence interval includes the rating of 100 for the local beverage?

4. A simple random sample of 400 air passengers arriving at the Hong Kong International Airport yielded a mean processing time of 50 minutes from disembarking until leaving the airport. The standard deviation was 15 minutes. If you wished to estimate the population mean with 97% confidence, what would your interval estimate be?

5. The mean current ratio of 100 firms selected at random from a total of 5,000 shoe manufacturing businesses is 1.2 with a standard deviation of 0.9. Determine the 95% confidence limits for the true mean current ratio of the 5,000 firms.

6. Based on a simple random sample of 100 departing air passenger check-in times, a 90% confidence interval of 0.5 to 3.5 minutes is established for the mean check-in time. Is it correct to conclude that the mean check-in time will be in this range with a probability of 90%?

7. In a simple random sample of 100 stockholders of Atlas Credit Corporation, 59 favored a new bond issue (with attached warrants for purchase of common stock), while 41 opposed the bond issue. Construct a 95.5% confidence interval for the actual proportion of all stockholders in favor of the new bond issue. Assume there is a very large number of stockholders.

8. An economist wishes to determine the mean population elasticity of supply of poultry farmers at their respective production levels. A random sample of 100 producers yields an average elasticity of supply of 1.9 and a standard deviation of 1.0. What would be the 90% confidence interval for the true elasticity of supply of all poultry farmers?

9. The prices of a stock on eight random dates in 1990 were $12.1, $11.9, $12.4, $12.3, $11.9, $12.1, $12.4, and $12.1. Find the 99% confidence limits for the population mean price of the stock assuming that the population variance is $.04.

10. As part of the planning of capital expenditures for expansion in the next 10 years, a large corporation's finance division asks the marketing research department to estimate future demand for one of the company's major products, homogenized milk. Marketing research chooses a random sample of 100 families from one large city to estimate the current average annual family demand for homogenized milk. The mean family demand in the sample is 150 gallons with a standard deviation of 40 gallons.
   a. Construct a 95.5% confidence interval for the mean annual demand for the product by all families in the city.
   b. If the range you obtained in part (a) is larger than you are willing to accept, in what ways can you narrow it?

11. In a random sample of 100 firms drawn from 10,000 lamp manufacturers, the average net profit-to-sales rate was 10.1% with a standard deviation of 2.5%.
   a. Construct a 99% confidence interval for the true mean.
   b. Explain your result in part (a).

12. In a simple random sample of 400 firms within a very large industry, the arithmetic mean number of employees was 232 with a standard deviation of 40.
   a. Establish a 95.5% confidence interval for the population mean.
   b. In general, what is the meaning of a 95.5% confidence interval?

13. Based on a simple random sample of 100 checking accounts at a small suburban commercial bank, a 90% confidence interval of 2% to 4% is established for the percentage of overdrawn accounts. Is it correct to conclude that the probability is 90% that between 2% and 4% of all of the checking accounts at this bank are overdrawn?

14. The following results were obtained from a random sample of 400 shoppers in the Philadelphia area. (1) Sixty percent preferred sales help when purchasing clothing. (2) Of the 100 shoppers from New Jersey, 30 thought store A had the lowest prices in town. (3) Of the 300 shoppers from Philadelphia, 80 thought store A had the lowest prices in town.
   a. What would be the 90% confidence interval for the true percentage of shoppers who prefer sales help when purchasing clothing?
   b. Construct a 95% confidence interval for the true difference in opinion of the New Jersey and Philadelphia shoppers about store A's prices. Interpret the interval in light of the problem.

15. A random sample of 100 invoices was drawn from some 10,000 invoices of the Ace Company. For the sample, the mean dollar sales per invoice was $23.50 with a standard deviation of $6.00.
   a. Construct a 99% confidence interval for the mean.
   b. Give the verbal meaning of your result in part (a).

6.4
CONFIDENCE INTERVAL ESTIMATION (SMALL SAMPLES)

The estimation methods discussed thus far are appropriate when the sample size is large. The distinction between large and small sample sizes is important when the population standard deviation is *unknown* and therefore must be estimated from sample observations. The main point is as follows. We have seen that the ratio $z = (\bar{x} - \mu)/\sigma_{\bar{x}}$ (where $\sigma_{\bar{x}} = \sigma/\sqrt{n}$) is normally distributed for all sample sizes if the population is normal and approximately normally distributed for large samples if the population is not normally distributed. In words, this ratio is $z = $ (sample mean − population mean)/*known* standard error. Furthermore, in section 6.3 we observed that for *large samples*, even if an *estimated* standard error is used in the denominator of this ratio, the sampling distribution may be assumed to be a standard normal distribution for practical purposes. However, the ratio (sample mean − population mean)/*estimated* standard error is not approximately normally distributed for *small* samples, so the theoretically correct distribution, known as the *t* distribution, must be used instead.[5] Although the underlying mathematics involved in the derivation of the *t* distribution is complex and beyond the scope of our book, we can get an intuitive understanding of the nature of that distribution and its relationship to the normal curve.

The ratio $(\bar{x} - \mu)/(s/\sqrt{n})$ is referred to as the *t* statistic. That is,

(6.14)
$$t = \frac{\bar{x} - \mu}{s/\sqrt{n}}$$

where, as previously defined in equation 1.9, the sample standard deviation $s = \sqrt{\Sigma(x - \bar{x})^2/(n - 1)}$ is an estimator of the unknown population standard deviation σ.

Let us examine the *t* statistic and its relationship to the standard normal statistic, $z = (\bar{x} - \mu)/\sigma_{\bar{x}}$. We noted that the denominator of the z ratio represents a *known* standard error, because it is based on a known population standard deviation. On the other hand, the denominator of the *t* statistic represents an *estimated* standard error, because s is an estimator of the population standard deviation.

The number $n - 1$ in the formula for s is referred to as the **number of degrees of freedom**, which we will denote by the Greek lower-case letter v

---

[5] Early work on the *t* **distribution** was carried out in the early 1900s by W. S. Gossett, an employee of Guinness Brewery in Dublin. Since the brewery did not permit publication of research findings by its employees under their own names, Gossett adopted "Student" as a pen name. Consequently, in addition to the term "*t* distribution" used here, the distribution has come to be known as "Student's distribution" or "Student's *t* distribution" and is so referred to in many books and journals.

(pronounced "nu"). It is not feasible to give a simple verbal explanation of this concept. From a purely mathematical point of view, the number of degrees of freedom v is simply a parameter that appears in the formula of the t distribution. However, in the present discussion, in which s is used as an estimator of the population standard deviation σ, the $n - 1$ may be interpreted as the number of independent deviations of the form $x - \bar{x}$ present in the calculation of s. Since the total of the deviations $\Sigma(x - \bar{x})$ for n observations equals zero, only $n - 1$ of them are independent. This means that if we were free to specify the deviations $x - \bar{x}$, we could designate only $n - 1$ of them independently. The nth one would be determined by the condition that the n deviations must add up to zero. Therefore, in the estimation of a population standard deviation or a population variance, if the divisor $n - 1$ is used in the estimator, then $n - 1$ degrees of freedom are present.

    The t distribution has been derived mathematically under the assumption of a normally distributed population.[6] As with the standard normal distribution, the t distribution is symmetrical and has a mean of zero. However, the standard deviation of the t distribution is greater than that of the normal distribution but approaches the latter figure as the number of degrees of freedom (and, therefore, the sample size) becomes large. It can be demonstrated mathematically that for an infinite number of degrees of freedom, the t distribution and normal distribution are exactly equal. The approach to this limit is quite rapid. Hence, a widely applied rule of thumb considers samples of size $n > 30$ "large," and for such samples, the standard normal distribution may appropriately be used as an approximation to the t distribution, even though the latter is the theoretically correct functional form. Figure 6-4 shows the graphs of several t curves for different numbers of degrees of freedom. We can see from these graphs that the t curves are lower at the mean and higher in the tails than the standard normal distribution. As the number of degrees of freedom increases, the t distribution rises at the mean and lowers at the tails until, for an infinite number of degrees of freedom, it coincides with the normal distribution. The use of tables of areas for the t distribution is explained in Example 6-5.

The t distribution

---

[6] The t distribution has the form

$$f(t) = c\left(1 + \frac{t^2}{v}\right)^{-(v+1)/2}$$

where $t = \dfrac{\bar{x} - \mu}{s_{\bar{x}}}$ (as previously defined)

       c = a constant required to make the area under the curve equal to one

       $v = n - 1$, the number of degrees of freedom

The variable t ranges from minus infinity to plus infinity. The constant c is a function of v, so that for a particular value of v, the distribution of $f(t)$ is completely specified. Thus, $f(t)$ is a family of functions, one for each value of v.

FIGURE 6-4

The t distributions for $v = 1$ and $v = 10$ compared with the normal distribution ($v = \infty$)

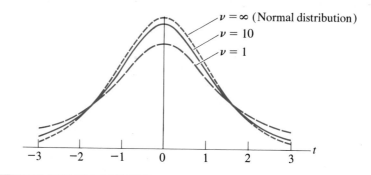

EXAMPLE 6-5

Assume that a simple random sample of 9 automobile tires was drawn from a large production run of a certain brand of tire. The mean lifetime of the tires in the sample was $\bar{x} = 32{,}010$ miles. This sample mean is the best single estimate of the corresponding population mean. The population standard deviation is unknown. Hence, an estimate of the population standard deviation is calculated by the formula $s = \sqrt{\Sigma(x - \bar{x})^2/(n - 1)}$. The result is $s = 2{,}520$ miles. What is the interval estimate for the population mean at a 95% level of confidence?

SOLUTION

Reasoning as we did in the case of the normal sampling distribution for means, we find that confidence limits for the population mean, using the t distribution, are given by

(6.15)
$$\bar{x} \pm t\,\frac{s}{\sqrt{n}}$$

where t is determined for $n - 1$ degrees of freedom. The number of degrees of freedom is one less than the sample size, that is, $v = n - 1 = 9 - 1 = 8$.

Just as the z values in Examples 6-1 and 6-2 represented multiples of standard errors, the t value in equation 6.15 represents a multiple of estimated standard errors. We find the t value in Table A-6 of Appendix A.

A brief explanation of Table A-6 is required. In the table of areas under the normal curve, areas lying between the mean and specified z values were given. However, in the case of the t distribution, since there is a different t curve for each sample size, no single table of areas can be given for all these distributions. Therefore, for compactness, a t table shows the relationship between areas and t values for only a few "percentage points" in different t distributions. Specifically, the entries in the body of the table are t values for areas of 0.01, 0.02, 0.05, and 0.10 in the two tails of the distribution combined.

In this problem, we refer to Table A-6 of Appendix A under column 0.05 for 8 degrees of freedom, and we find $t = 2.306$. This means that, as shown in Figure 6-5, for 8 degrees of freedom, a total of 0.05 of the area in the t distribution lies below $t = -2.306$ or above $t = 2.306$. Correspondingly, the probability is 0.95 that for 8 degrees of freedom, the t value lies between -2.306 and 2.306.

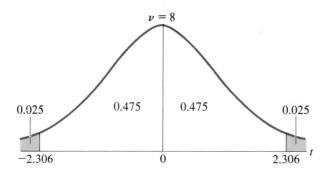

In this problem, substituting $t = 2.306$ into equation 6.15, we obtain the following 95% confidence limits:

$$32,010 \pm 2.306(840) = 32,010 \pm 1,937.04 \text{ miles}$$

Hence, to the nearest mile, the confidence limits for the estimate of the mean lifetime of all tires in the production run are 30,073 and 33,947 miles.

The interpretation of this interval and the associated confidence coefficient is the same as in the case of large samples and a normal distribution. Comparing this procedure with the corresponding method for large samples for 95% confidence limits, we note that the t value of 2.306 replaces the 1.96 figure that is appropriate for the normal curve. Thus, with small samples we have a wider confidence interval, leading to a vaguer result. This is to be expected, because σ is estimated by s using a small sample size n.

Note that since Table A-6 shows areas in the combined tails of the t distribution, we had to look under the column headed 0.05 for the t value corresponding to a 95% confidence interval. Correspondingly, we would find t values for 90%, and 98%, and 99% confidence intervals under the columns headed 0.10, 0.02, and 0.01, respectively.

EXERCISES 6.4

1. Are the following statements true or false? Explain your answers.
   a. For small samples, if you are estimating the population mean, you use $\bar{x} \pm ts_{\bar{x}}$.
   b. The more efficient of two unbiased estimators has the narrower confidence interval.

c.   If a 95% confidence interval for the average strength of a certain type of wood beam is 12–15 pounds per square inch, then you can conclude that in a sample of 100 wood beams, 95 will have strengths between 12 and 15 pounds per square inch.

2.   A high school gave a mathematics achievement test to nine of its seniors who were selected at random. The scores of these students were 78, 68, 80, 92, 64, 80, 58, 60, and 77.

a.   Construct a 99% confidence interval for the mean mathematics score.

b.   If the population scores followed a highly skewed distribution, would the range you set up in part (a) really be a 99% confidence interval?

3.   A large eastern university contemplated a change in the course requirements for its students. To determine faculty opinion of this move, the curriculum committee selected 15 faculty members at random and asked them to rank the course change proposal from zero to 10 (where 10 = Totally in favor). The results obtained from this sample were $\bar{x} = 3$ and $s = 1$.

a.   State a 95% confidence interval for μ, the population mean opinion of the course change proposal.

b.   Can the curriculum committee be sure that there is no more than a 5% chance that the population mean μ lies outside this interval? Explain.

4.   A security analyst selected a simple random sample of 20 stocks from those listed on the New York Stock Exchange in order to estimate the mean quarterly dividend payment of common stocks on the Exchange. The sample yielded a mean dividend payment of $.60 per share per quarter with a standard deviation of $.15.

a.   Construct a 99% confidence interval for the true mean quarterly dividend payment per share of all New York Stock Exchange common stocks.

b.   Repeat part (a) for a 95% confidence coefficient and for a 90% confidence coefficient.

5.   A sample of 16 high-technology firms selected at random yielded a mean annual expenditure on research and development of $600,000 with a standard deviation of $20,000.

a.   Construct a 98% confidence interval for the true mean annual research and development expenditures of all such firms.

b.   What would be the effect, if any, on the width of the interval in (a) if a higher level of confidence were used? If the interval had been based on a sample of 40 firms, assuming the same level of confidence and the same standard deviation?

6.   A company ran a test to determine the length of time required to complete service calls. The following times, in minutes, were obtained for a simple random sample of nine service calls: 48, 51, 28, 66, 81, 36, 40, 59, 50.

a.   Construct a 99% confidence interval for the mean time for completion of service calls.

b.   If times to complete service calls followed a highly skewed distribution, would the range you set up in part (a) really be a 99% confidence interval?

6.5
DETERMINATION OF SAMPLE SIZE

In all of the examples thus far, the sample size n was given. However, we could ask the following question: How large should a sample be in a specific situation? If a sample larger than necessary is used, resources are wasted; if the sample is too small, the objectives of the analysis may not be achieved.

Sample Size for Estimation of a Proportion

Statistical inference provides the following answer to the question of sample size. Let us assume an investigator desires to estimate a certain population parameter and wants to know how large a simple random sample is required. We assume that the population is very large relative to the prospective sample size.

> To specify the required sample size any investigator must answer two questions.
>
> 1. What degree of precision is desired?
> 2. What probability is attached to obtaining the desired precision?

Clearly, the greater the degree of desired precision, the larger will be the necessary sample size; similarly, the greater the probability specified for obtaining the desired precision, the larger will be the required sample size. We will use examples to indicate the technique of determining sample size for estimation of a population proportion and a population mean.

EXAMPLE 6-6

Suppose we would like to conduct a poll among eligible voters in a city in order to determine the percentage who intend to vote for the Democratic candidate in an upcoming election. We want a 95.5% probability that we will estimate the percentage that will vote Democratic within ± 1 percentage point. What is the required sample size?

SOLUTION

We answer the question by first indicating the rationale of the procedure and then condensing this rationale into a simple formula. The statement of the question gives a relationship between the sampling error that we are willing to tolerate and the probability of obtaining this level of precision. In this problem, $2\sigma_{\bar{p}}$ must equal 0.01. This means that we are willing to have a probability of 95.5% that our sample percentage $\bar{p}$ will fall within 0.01 of the true but unknown population proportion p (see Figure 6-6). We may now write

$$2\sigma_{\bar{p}} = 0.01$$

FIGURE 6-6

Sampling distribution of
a proportion showing
the relationship between the
sampling error and the
probability of obtaining this
degree of precision

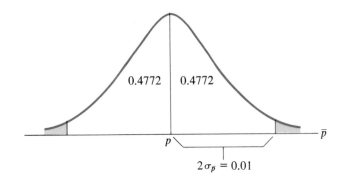

FIGURE 6-6

Sampling distribution of
a proportion showing
the relationship between the
sampling error and the
probability of obtaining this
degree of precision

or

$$2\sqrt{\frac{pq}{n}} = 0.01$$

and

$$\sqrt{\frac{pq}{n}} = 0.005$$

In all our previous problems, the sample size n was known, but here n is the un-known for which we must solve. However, it appears that there are too many un-knowns, namely, the population parameters p and q as well as n. Therefore, we must estimate values for p and q, and then we can solve for n. Suppose we wanted to make a conservative estimate for n. What should we guess as a value for p? In this context, a conservative estimate is an estimate made in such a way as to ensure that the sample size will be large enough to deliver the desired precision. In this problem, the "most conservative" estimate for n is given by assuming $p = 0.50$ and $q = 0.50$. This follows from the fact that the product $pq = 0.25$ is larger for $p = 0.50$ and $q = 0.50$ than for any other possible values of p and q where $p + q = 1$. For example, if $p = 0.70$ and $q = 0.30$, then $pq = 0.21$, which is less than 0.25. The relationships between possible values of p and the corresponding values of pq are shown in Figure 6-7. Thus, the largest, or "most conservative," value of n is determined by substituting $p = q = 0.50$ as follows:

$$0.005 = \sqrt{\frac{0.50 \times 0.50}{n}}$$

Squaring both sides gives

$$0.000025 = \frac{0.50 \times 0.50}{n}$$

and

$$n = \frac{0.50 \times 0.50}{0.000025} = 10,000$$

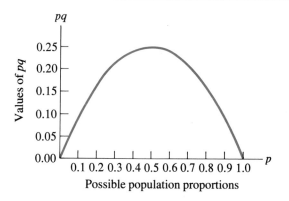

FIGURE 6-7

Relationship between possible values of p and the corresponding products pq (where $q = 1 - p$)

Hence, to achieve the desired degree of precision, a simple random sample of 10,000 eligible voters would be required. Of course, the large size of this sample is attributable to the high degree of precision specified. If $\sigma_{\bar{p}}$ were doubled from 0.005 to 0.01, the required sample size would be cut down to one-fourth of 10,000, or 2,500. This follows from the fact that the standard error varies inversely with the square root of sample size.

In this election problem, we assumed $p = q = 0.50$, although less conservative estimates are possible if we believe that $p \neq 0.50$. In problems involving proportions, we would use whatever past knowledge we have to estimate p. For example, suppose we wanted to determine the sample size to estimate an unemployment rate and we knew from past experience that for the community of interest the proportion of the labor force that was unemployed was somewhere between 0.05 and 0.10. We then would assume $p = 0.10$, since this would give us a more conservative estimate (larger sample size) than assuming $p = 0.05$ or any value between 0.05 and 0.10. (For example, if we assumed $p = 0.07$ and in fact the true value of p was 0.09, the sample size we determined from a calculation involving $p = 0.07$ would not be large enough to give us the specified precision.) Assuming $p = 0.10$ assures us of obtaining the desired degree of precision regardless of what the true value of p is, as long as it is in the range 0.05 to 0.10.

We can summarize this calculation for sample size by noting that we start with the following statement:

$$D = z\sigma_{\bar{p}}$$

where D is the tolerated deviation (in percentage points) and z is the multiple of standard errors corresponding to the specified probability of obtaining this precision. Then for infinite populations or large populations relative to sample size, we have

(6.16)

$$n = \frac{z^2(p)(1 - p)}{D^2}$$

Hence, in the preceding voting problem, application of equation 6.16 yields

$$n = \frac{(2)^2(0.50)(0.50)}{(0.01)^2} = 10,000$$

Sample Size for Estimation of a Mean

The required sample size for estimation of a mean can be determined by an analogous calculation. Suppose we wanted to estimate the arithmetic mean hourly wage for a group of skilled workers in a certain industry. Let us further assume that from prior studies we estimate that the population standard deviation of the hourly wages of these workers is about $.15. How large a sample size would be required to yield a probability of 99.7% that we will estimate the mean hourly wage of these workers within $\pm\$.03$?

Since the 99.7% probability corresponds to a level of three standard errors, we can write

$$3\sigma_{\bar{x}} = \$.03$$

or

$$\frac{3\sigma}{\sqrt{n}} = \$.03$$

and

$$\frac{\sigma}{\sqrt{n}} = \$.01$$

The population standard deviation σ is known from past experience. Hence, substituting $\sigma = \$.15$ gives

$$\frac{\$.15}{\sqrt{n}} = \$.01$$

and

$$\sqrt{n} = \frac{\$.15}{\$.01} = 15$$

Squaring both sides yields the solution

$$n = (15)^2 = 225$$

Therefore, a simple random sample of 225 of these workers would be required. In summary, if we calculate $\bar{x}$ for the hourly wages of a simple random sample of 225 of these workers, we can estimate the mean wage rate of all skilled workers in this industry within \$.03 with a probability of 99.7%.

In this problem, we assumed that an estimate of the population standard deviation was available from prior studies; this situation may exist for governmental agencies that conduct repeated surveys of wage rates, population, and the like. If the population standard deviations (or estimated population standard deviations) in these past studies are not erratic or excessively unstable, they provide useful bases for estimating σ values in the above procedures for computing sample size.

Of course, an estimate of the population standard deviation may not be available from past experience. It may be possible, however, to get a rough estimate of σ if there is at least some knowledge of the total range of the basic random variable in the population. For example, suppose we know that the difference between the wages of the highest- and lowest-paid workers is about \$1.20. In a normal distribution, a range of three standard deviations on either side of the mean includes virtually the entire distribution. Thus, a range of 6σ includes almost all the frequencies, and we may state

$$6\sigma \approx \$1.20$$

or

$$\sigma \approx \$.20$$

Of course, the population distribution is probably nonnormal and \$1.20 may not be exact. Consequently, the estimate of σ may be quite rough. Nevertheless, we may be able to obtain a reasonably good estimate of the required sample size in a situation where, in the absence of this "guestimating" procedure, we may be at a loss for any notion of a suitable sample size.

If we express in the form of an equation the technique for calculating a required sample size when estimating a population mean, we have

(6.17)
$$n = \frac{z^2 \sigma^2}{D^2}$$

where D is the tolerated deviation. For the problem involving hourly wages, this gives

$$n = \frac{(3)^2 (\$.15)^2}{(\$.03)^2} = 225$$

In equations 6.16 and 6.17, we assume that the n value determined is sufficiently large for the assumption of a normal sampling distribution as appropriate and that the populations are large relative to this sample size. If the populations are not large relative to sample size, appropriate formulas may be derived for n by taking into account the finite population correction. The formulas for the sample sizes required for estimation of a proportion and mean, respectively, are

(6.18)

$$n = \frac{p(1-p)}{\dfrac{D^2}{z^2} + \dfrac{p(1-p)}{N}}$$

(6.19)

$$n = \frac{\sigma^2}{\dfrac{D^2}{z^2} + \dfrac{\sigma^2}{N}}$$

In equations 6.18 and 6.19, N is the number of elements in the population (population size), and all other symbols are as previously defined. Equations 6.18 and 6.19 correspond, respectively, to equations 6.16 and 6.17, in which we assumed infinite populations.

For example, in the voting problem discussed earlier in this section, if the population size had been 50,000, then the required sample size computed from equation 6.18 would be

$$n = \frac{(0.50)(0.50)}{\dfrac{(0.01)^2}{(2)^2} + \dfrac{(0.50)(0.50)}{50,000}} = 8,333$$

This result may be compared with the sample size of 10,000 computed earlier from equation 6.16 assuming an infinite population size.

Analogously, in the problem involving the required sample size for estimating the population mean hourly wage for a group of skilled workers in a certain industry, let us assume a population size of 1,000 such skilled workers. Then the required sample size computed from equation 6.19 would be

$$n = \frac{(0.15)^2}{\dfrac{(\$.03)^2}{(3)^2} + \dfrac{(0.15)^2}{1,000}} = 184$$

This figure may be compared with the sample size of 225 computed previously from equation 6.17 assuming an infinite population size. It would be instructive for you to verify that if a population size of 1,000,000 had been assumed, the sample size computed from equation 6.19 would be 224.9, which rounds off to 225—the same figure obtained by assuming an infinite population size.

EXERCISES 6.5

1. A financial analyst wishes to estimate the proportion of multinational firms that recorded a net loss despite "hedging" against currency translation losses. The error of the estimate should be kept within four percentage points with a confidence interval of 96%. How large a simple random sample is required?

2. An economist makes 100 random observations of the profits recorded by monopolistic firms. In 56 of these observations, she finds that the profits are higher than the average profits under competitive conditions.

   a. Construct a 98% confidence interval for the proportion of monopolistic firms with profit levels higher than this average profit.

   b. The economist decided that the confidence interval in part (a) was too large. If a new sample were drawn, how large must it be in order to estimate the true proportion in part (a) within three percentage points with a confidence coefficient of 98%?

3. A consumer research group is surveying the consumer population of New England to estimate the proportion of consumers who use biodegradable laundry detergent.

   a. Assume that the proportion of consumers using this type of detergent is estimated from past experience to be 0.20 or less. How large a random sample should be drawn if the objective is to estimate the proportion of consumers using biodegradable detergent within three percentage points with 95% confidence?

   b. Assume, without regard to your answer in part (a), that the research group drew a random sample of 100 consumers and found that 15 used biodegradable laundry detergent. Construct a 99% confidence interval for the proportion of consumers using this type of detergent.

4. A promotion manager wishes to determine the average increase in sales in a given region following each advertising campaign. He examines company records and obtains point estimates of $150,045 for the mean and $55,840 for the standard deviation. When he takes his random sample, he wants to be 95.5% confident that the maximum error of estimate will not exceed $5,000.

   a. Discuss the sense in which he will have 95.5% confidence in his estimate.

   b. What size sample should he take?

5. A certified public accountant wishes to estimate the percentage of companies in the United States that use the LIFO method of pricing inventory. He intends to base his estimate on a random sample and wishes to be 95% confident that his estimate lies within three percentage points of the true percentage of companies using LIFO. He is quite certain that no more than 25% of the companies in the United States use this method of pricing inventories. How large a simple random sample should the accountant take?

6. You can always decrease the width of a confidence interval by increasing the sample size. Why then do you not always determine the desired width and then sample accordingly?

7. A radio talk show asked divorced women callers whether they would like to see their ex-husbands behind bars for stopping alimony payments. Of 100 random callers, 89 replied in the affirmative.

   a. Construct a 99% confidence interval for the proportion of women who would like to see their ex-husbands behind bars, assuming that the 100 callers constituted an appropriate random sample.

b.   What sample size is required to estimate the proportion in part (a) within five percentage points with 95% confidence?

8.   The Chicago Board of Trade wishes to select a random sample of all wheat futures contracts currently in force to estimate the average premium being paid for a one-month term. For example, in a normal market, the price paid per bushel for immediate delivery of the commodity will be less than the selling price per bushel in a futures contract with a delivery (and payment) date one month from now. The price of a bushel of wheat in a cash contract (immediate delivery) might be \$2.00, while the price of a bushel of wheat in a futures contract with a delivery date one month later might be \$2.25. The premium in this case would be \$2.25 − \$2.00 = \$.25. From past experience, the Board of Trade estimates that the standard deviation of this premium is about \$.05. How large a sample is needed to estimate the mean premium within \$.005 with a 95.5% confidence coefficient?

9.   A writer wishes to estimate the proportion of teenagers who obtain more sex information from television than from their parents. He interviewed 100 teenagers; 60 of them replied that they learned more about sex from television than from their parents.

a.   Construct a 90% confidence interval for the proportion in question.

b.   If the confidence coefficient is set at 99%, how large would the sample have to be in order to estimate this proportion within 0.05?

10.   A statistician wishes to determine the average hourly earnings for employees in a given occupation in a particular state. She runs a pilot study and obtains point estimates of \$6.20 for the mean and \$.50 for the standard deviation. She then specifies that when she takes her random sample, she wants to be 95.5% confident that the maximum error of estimate will not exceed \$.05.

a.   Discuss the sense in which she will have 95.5% confidence in her estimate.

b.   What size sample should she take?

11.   An estimate of p, the percentage of unemployed executives in the United States, is desired. How large a sample should be drawn if it is desired to estimate p within 0.005 with 98% confidence? It may be assumed that $0.01 \leqslant p \leqslant 0.05$.

6.6
THE JACKKNIFE*

In this chapter we have considered methods for setting up confidence intervals for the mean and for percentages. Suppose we wanted to establish a confidence interval for a median, variance, ratio or even more complex measure for which the theoretical standard error may not even have been derived. A computer simulation can be used, in which repeated random samples are drawn from the same population, and an estimate of the standard error can be calculated from the appropriate statistics observed in these samples. However, a more ingenious method known as the "the jackknife" has recently been developed that is useful for setting up confidence intervals for any parameter from data observed in a single sample. The term "jackknife" has been coined to suggest the broad scale usefulness of the tool, just as is true for the original namesake. We illustrate the use of this versatile technique by means of a simple example.

EXAMPLE 6-7

Suppose we were interested in estimating the weighted assessment-to-market value ratio of all homes sold in a city in a certain year. The "*weighted* assessment-to-market value ratio" is the ratio

$$\frac{\text{Total assessed values}}{\text{Total market values}}$$

for all homes sold in the city in the given year. Suppose that this figure is 30% but that we did not know it. How can we establish a confidence interval for this population parameter from a simple random sample of four home sales (sale price equals market, value) such as is shown in Table 6-1? Of course, this is an unrealistically small sample size but we use it to explain the methodology and keep the arithmetic simple.

SOLUTION

Table 6-1 shows the assessed values and market values for the sample of four homes. The ratio of the total assessed values to total market values of the four homes, $R = 31.7\%$ is an estimate of the population parameter. We would like to have, say, a 90% confidence interval around this estimate.

  Note that we cannot use the method for interval estimation of a proportion given in section 6.3. In that discussion the sample proportion (or percentage) pertains to a number of successes x in a simple random sample of size n. On the other hand, here our estimate R is a ratio of two random variables. The theory and methodology for constructing confidence intervals are very different for these two cases.

TABLE 6-1
Assessed values and market values (sales price) for a simple random sample of four home sales

Home	Assessed Value ($000)	Market Value ($000)
1	24	96
2	30	112
3	50	150
4	76	210
Total	180	568

$$R = \frac{180}{568} = 31.7\%$$

TABLE 6-2

Sample of four home sales split into two subsamples: Two subsample estimates
(X) of the population ratio of total assessments to total market values

Sample 1	Assessed Value ($000)	Market Value ($000)	Sample 2	Assessed Value ($000)	Market Value ($000)
Home 1	24	96	Home 3	50	150
Home 2	30	112	Home 4	76	210
Total	54	208		126	360

$$X = \frac{54}{208} = 26.0\% \qquad\qquad X = \frac{126}{360} = 35\%$$

We turn now to the general idea of the jackknife technique. The basis for confidence interval estimates in general is a sampling distribution in which we have repeated
observations of a statistic in many random samples. The standard deviation of this
sampling distribution is the standard error needed for setting up the confidence interval.
How can we obtain this repetition in order to see how the statistic varies from sample
to sample?

The Simplest Split into Subsamples   If we are restricted to using only the values in the
single sample of four home sales, probably the simplest way to obtain different estimates
of the population parameter would be to split the sample into two halves. Table 6-2
shows this split with the desired ratio calculated separately for the first two homes combined and the third and fourth homes combined.

Treating these two ratios, $X = 26\%$ and $X = 35\%$, as observed statistics in a
sampling distribution, we calculate their mean and standard deviation as in Table 6-3.
We can then set up the desired confidence interval as

$$\bar{X} \pm ts/\sqrt{n}$$

In this case, a 90% confidence interval for the population ratio of total assessed
values to total market value is:

$$30.5 \pm (6.31)\frac{6.4}{\sqrt{2}} = 30.5\% \pm 28.6\%$$

TABLE 6-3

Calculation of the mean and standard deviation of the two subsample estimates
(X) of the population ratio of total assessment values to total market values

Estimates (X)	$X - \bar{X}$	$(X - \bar{X})^2$
26.0	−4.5	20.25
35.0	4.5	20.25
61.0		40.50

$$\bar{X} = \frac{61.0}{2} = 30.5 \qquad s = \sqrt{\frac{40.50}{1}} = 6.4$$

where the number of degrees of freedom for determining the t value is $n - 1 = 2 - 1 = 1$. From Appendix Table A-6, under 0.10, we have $t = 6.314$.

Of course, in this case the confidence interval is ridiculously wide, and it certainly includes the actual population figure of 30%. If we had a sample of six homes, we could have split them into two samples of three homes each, or three samples of two homes each. However, can we split the sample up in somewhat more clever ways to obtain repeated estimates of the desired parameters?

A Typical Jackknife Split The reason for the large t value and wide confidence interval resulting from the splitting of the original sample into two subsamples was that the subsamples were too small, thus producing unreliable estimates. This would also have been true if our sample consisted of six or eight homes and we used similar types of splits. What is needed for improved reliability (smaller standard errors) is larger subsample sizes and more subsamples. The solution to this problem in the jackknife technique is illustrated in Table 6-4. The first subsample is created on the left-hand side of the table by deleting the observation for Home 1. The estimate of the desired ratio denoted X_{-1} is calculated from the observations for the remaining subsample of three items. The second subsample is constructed by deleting the observation for Home 2 as shown on the right-hand side of the table. The estimate for the resulting subsample is denoted X_{-2}. As shown in Table 6-4, the resulting estimates are $X_{-1} = 33.1\%$, $X_{-2} = 32.9\%$, $X_{-3} = 31.1\%$, and $X_{-4} = 29.1\%$.

We now have more subsamples than when we merely split the sample into two parts, and our subsamples are larger. However, observing these values of X_{-1}, X_{-2}, X_{-3}, and X_{-4}, we see that there is relatively little dispersion among them. This is not surprising, because each estimate used most of the values of the original sample. If our original sample size had been larger than four, we could have expected even more similarity among the subsample values. Furthermore, the subsample estimates are not independent, since they all use most of the items in the original sample. Independence is assumed in the setting up of confidence intervals as in the 90% confidence interval given for the simple split of the sample into two parts.

In an effort to mitigate the aforementioned problems of lack of dispersion and lack of independence in the subsample estimates, new estimates are calculated using the

---

TABLE 6-4
Examples of deletion of one item at a time: Four subsamples are obtained

Home	Assessed Value ($000)	Market Value ($000)	Home	Assessed Value ($000)	Market Value ($000)
~~1~~	~~24~~	~~96~~	1	24	96
2	30	112	~~2~~	~~30~~	~~112~~
3	50	150	3	50	150
4	76	210	4	76	210
Total	$\overline{156}$	$\overline{472}$		$\overline{150}$	$\overline{456}$

$$X_{-1} = \frac{156}{472} = 33.1\% \qquad X_{-2} = \frac{150}{456} = 32.9\%$$

$$X_{-3} = \frac{130}{418} = 31.1\% \qquad X_{-4} = \frac{104}{358} = 29.1\%$$

TABLE 6-5
Pseudovalues: Calculation of their mean and standard deviation

X_i	$X - \bar{X}$	$(X - \bar{X})^2$
27.5	−4.7	22.09
28.1	−4.1	16.81
33.5	1.3	1.69
39.5	7.3	53.29
128.6		93.88

$$\bar{X} = \frac{128.6}{4} = 32.2\% \qquad s = \sqrt{\frac{93.88}{3}} = 5.6\%$$

following formula illustrated for the first subsample:

$$X_{(1)} = X_{\text{all}} + (n - 1)(X_{\text{all}} - X_{-1})$$

(6.20)
$$X_{(1)} = 31.7 + (3)(31.7 - 33.1)$$

$$X_{(1)} = 27.5$$

Similarly

$$X_{(2)} = 31.7 + (3)(31.7 - 32.9) = 28.1$$

$$X_{(3)} = 31.7 + (3)(31.7 - 31.1) = 33.5$$

$$X_{(4)} = 31.7 + (3)(31.7 - 29.1) = 39.5$$

These new values, referred to as "pseudovalues" obtained from "pseudosamples" are much more spread out than the X_{-1}, X_{-2}, X_{-3}, and X_{-4} figures, and it can be shown that they behave like independent values. The calculation of the mean and standard deviation of these pseudovalues is given in Table 6-5. Using these figures, the 90% confidence interval for the population ratio of total assessed values to total market value is

$$32.2 \pm (2.35)\frac{5.6}{\sqrt{3}} = 32.2\% \pm 7.6\%$$

where the number of degrees of freedom for determining the t value is $4 - 1 = 3$. We are pleased to see that the confidence interval includes the population ratio of 30%.

It may be noted that this confidence interval is much narrower than the one given earlier for the simple split of the original sample into two subsamples. Hence, the jackknife results in more precise estimates of population parameters than obtained from such simple splits of sample values.

The jackknife has many applications, as for example, in regression analysis in the detection of extreme observations, known as outliers. In that type of analysis, different regression equations are calculated after deleting items one at a time, as in the present example. Examination of the resulting equations assists in the discovery of outliers.

In summary, the jackknife is a technique that is easy to apply and is of use in many types of statistical analysis.

The data for the problems of this chapter are stored in the file labeled 'roa . mtw'. The following is a brief description of the data:

Column	Name	Count	Description
C1	BANKROA	193	Returns on Assets (%) in the Surveyed Banks
C2	AIRROA	43	Returns on Assets (%) in the Surveyed Firms in Air Transportation
C3	RANDOM	200	Zeros and Ones Generated from a Random Mechanism

1. Treat C1 and C2 as large samples. Obtain a 99% confidence interval for the difference between the means of AIRROA and BANKROA.

2. Past records suggest that the standard deviations for BANKROA and AIRROA are 0.6% and 4.5%, respectively. Use that information to compute separately 95% confidence intervals for the mean of BANKROA and for the mean of AIRROA. *Hint:*

   MTB > zinterval 95% 0.6 C1

   gives a 95% confidence interval for the mean of BANKROA when the standard deviation is 0.6%.

3. Suppose that the past standard deviation for AIRROA was actually 4.6%. Recompute the 95% confidence interval for the mean AIRROA. Does the larger standard deviation result in a wider confidence interval?

4. Suppose that past records are not available and that we only have 12 observations for BANKROA: 1.9, 1.7, 1.6, 1.5, 1.2, 1.2, 0.9, 0.4, 0.1, −0.3, −0.8, −1.3. Compute a 99% confidence interval for the mean of BANKROA. *Hint:* First set the data in a column, say C10,

   MTB > set C10
   DATA > 1.9 1.7 1.6 1.5 1.2 1.2 0.9 0.4 0.1 −0.3 −0.8 −1.3
   DATA > end

   and then try the command:

   MTB > tinterval 99% C10

5. Obtain a 90% confidence interval for the proportion of ones in RANDOM. Is 0.5 in the interval as it should be for a fair random mechanism?

KEY TERMS

Confidence Coefficient a figure such as 95% used to indicate the degree of confidence or credibility placed on a confidence interval estimate.

Confidence Interval Estimate an interval estimate of a population parameter with a confidence coefficient attached.

Consistent Estimator an estimator such that it approaches closer to the corresponding population parameter as sample size increases.

Efficiency the sampling variability of an estimator.

Estimate a numerical value of a sample statistic (for example, a sample mean).

Estimator a statistical measure used to estimate a population parameter e.g., the sample mean $\bar{x}$ is an estimator of the population mean, μ.

Interval Estimate a statement of two values between which we have some confidence that a population parameter lies.

Number of Degrees of Freedom roughly, the number of independent quantities. For example, in using s to estimate σ, the estimate is based on $n - 1$ independent deviations from the mean; hence, $n - 1$ degrees of freedom.

Point Estimate a single number used as an estimate of an unknown population parameter.

Statistical Inference estimation and hypothesis testing.

t-Distribution a distribution that approaches the standard normal distribution as sample size increases.

Unbiased Estimator the expected value of the estimator is equal to the corresponding population parameter.

KEY FORMULAS

Confidence Interval for a Mean (large samples)

$$\bar{x} \pm z \frac{s}{\sqrt{n}}$$

Confidence Interval for a Mean (small samples)

$$\bar{x} \pm t \frac{s}{\sqrt{n}}$$

Confidence Interval for a Proportion (large samples)

$$\bar{p} \pm z \sqrt{\frac{\bar{p}\bar{q}}{n}}$$

Confidence Interval for a Difference Between Two Means (large samples)

$$(\bar{p}_1 - \bar{p}_2) \pm z \sqrt{\frac{\bar{p}_1\bar{q}_1}{n_1} + \frac{\bar{p}_2\bar{q}_2}{n_2}}$$

Unbiased Estimator of θ

$$E(\hat{\theta}) = \theta.$$

Unbiased Estimator of a Population Variance

$$s^2 = \frac{\Sigma(x - \bar{x})^2}{n - 1}$$

Sample Size for Estimation of a Proportion

$$n = \frac{z^2(p)(1 - p)}{D^2}$$

Sample Size for Estimation of a Population Mean

$$n = \frac{z^2\sigma^2}{D^2}$$

7

HYPOTHESIS TESTING

We will now focus on hypothesis testing—the second basic subdivision of statistical inference. **Hypothesis testing** addresses the important question of how to choose among alternative propositions or courses of action, while controlling or minimizing the risks of making wrong decisions. We will now briefly and informally summarize the rationale involved in testing hypotheses and then explain the details of these testing procedures by means of examples.

7.1
THE RATIONALE OF HYPOTHESIS TESTING

To gain some insight into the reasoning involved in statistical hypothesis testing, we will consider a nonstatistical hypothesis-testing procedure with which we are all familiar. The basic process of inference involved is strikingly similar to that employed in statistical methodology.

Consider the process by which an accused individual is judged in a court of law. Under Anglo-Saxon law, the person before the bar is assumed to be innocent. The burden of proof of his guilt rests on the prosecution. Using the language of hypothesis testing, let us say that we want to test the **null hypothesis**, which we denote by H_0, that the person before the bar is innocent. The **alternative hypothesis** H_1 is that the defendant is guilty. The jury examines the evidence to determine whether the prosecution has demonstrated that this evidence is inconsistent with the basic hypothesis H_0 of innocence. If the jurors decide the evidence is inconsistent with H_0, they reject that hypothesis and accept its alternative H_1 that the defendant is guilty.

If we analyze the situation that results when the jury makes a decision, we find that four possibilities exist in terms of the basic hypothesis H_0:

1. The defendant is innocent (H_0 is true), and the jury finds that he is innocent (retains H_0); hence, the correct decision is made.
2. The defendant is innocent (H_0 is true), but the jury finds him guilty (rejects H_0); hence, an error is made.
3. The defendant is guilty (H_0 is false), and the jury finds that he is guilty (rejects H_0); hence, the correct decision is made.
4. The defendant is guilty (H_0 is false), but the jury finds him innocent (retains H_0); hence, an error is made.

In possibilities (1) and (3), the jury reaches the correct decision; in possibilities (2) and (4), it makes an error. Let us consider these errors in terms of conventional statistical terminology. In possibility (2), hypothesis H_0 is erroneously rejected. The basic hypothesis H_0 that is being tested for possible rejection is generally referred to as the **null hypothesis**. Hypothesis H_1 is designated the **alternative hypothesis**.

Type I error

> To reject the null hypothesis when in fact it is true is referred to as a Type I error.

In possibility (4), hypothesis H_0 is retained in error.

Type II error

> To retain the null hypothesis when it is false is termed a **Type II error**.

Under our legal system, the commission of a Type I error is considered far more serious than the commission of a Type II error. Thus, we feel that it is a more grievous mistake to convict an innocent person than to let a guilty person go free.

Had we made H_0 the hypothesis that the defendant is guilty, the meaning of Type I and Type II errors would have been the reverse of the first formulation; what had previously been a Type I error would become a Type II error, and a Type II would become a Type I error.

> In the statistical formulation of hypotheses, how we choose to exercise control over Type I and Type II errors serves as a basic guide in stating the hypotheses to be treated.

In this chapter, we will see how errors are controlled in hypothesis testing.

TABLE 7-1

The relationship between actions concerning a null hypothesis and the truth or falsity of the hypothesis

Action Concerning Hypothesis H_0	State of Nature	
	H_0 Is True (innocent)	H_0 Is False (guilty)
Retain H_0	Correct decision	Type II error
Reject H_0	Type I error	Correct decision

The four possible jury decisions in our example are summarized in Table 7-1. Here, the headings are stated in modern decision-theory terminology and require brief explanations. When hypothesis testing is viewed as a problem in decision making, two alternative actions can be taken: "retain H_0" or "reject H_0." The two alternatives, truth or falsity of hypothesis H_0, are viewed as "states of nature" or "states of the world" that affect the consequences, or "payoff," of the decision. The payoffs are indicated in the table in terms of the correctness of the decision or the type of error incurred. We can see from the framework of this hypothesis-testing problem that what we need is some criterion on which to base the decision to reject the null hypothesis H_0 or fail to reject H_0. Classical hypothesis testing attacks this problem by establishing **decision rules** based on data derived from simple random samples. The sample data are analogous to the evidence investigated by the jury. The decision procedure attempts to assess the risks of making incorrect decisions, and, in a sense, which we will examine, to minimize them.

Terminology

Decision rules

We emphasize at this point one matter of terminology in the hypothesis testing procedure discussed in this chapter and used in general in statistical analysis. The **null hypothesis H_0** is viewed as the *status quo* hypothesis and is the one that will be retained unless convincing evidence is produced to the contrary. In the preceding legal example, the null hypothesis H_0 is that the person before the bar is innocent.

That hypothesis will be retained unless very strong evidence is produced to reject it. The **alternative hypothesis H_1**, also referred to as the **research hypothesis**, is accepted or adopted when the null hypothesis H_0 is rejected. Although we may occasionally refer to the acceptance of the alternative hypothesis, we will avoid the use of the terminology that a null hypothesis H_0 is "accepted." Since the null hypothesis H_0 is the status quo hypothesis, it is merely retained tentatively until convincing evidence appears to reject it. The alternative or research hypothesis H_1 is the one that the investigator attempts to demonstrate by rejecting the null hypothesis H_0 on the basis of the data presented.

In summary, we will generally use the terminology in hypothesis-testing procedures that we reject the null hypothesis H_0 or that we fail to reject it. Sometimes we will state that the null hypothesis H_0 is retained, but we repeat that we will not say that a null hypothesis is accepted.

The Hypothesis-testing Procedure

Hypothesis-testing procedures can be used to solve two basic types of decision problems. In the first type of problem, we want to know whether a population parameter has changed from or differs from a particular value. Here, we are interested in detecting whether the population parameter is *either* larger than or smaller than a particular value. For example, suppose that the mean family income in a certain city was determined from a census to be $36,500 for a particular year, and two years later we want to discover whether the mean income has *changed*. If it is not feasible to take another census, we may draw a simple random sample of families and try to reach a conclusion based on this sample. As we did with the judicial decisions, we can set up two competing hypotheses and choose between them.

- The null hypothesis H_0 would simply be an assertion that the mean family income was unchanged from the $36,500 figure; in statistical language, we write this hypothesis $H_0: \mu = \$36,500$, and μ is the mean family income in the city.

- The alternative hypothesis is that the mean family income *has* changed or, in statistical terminology, $H_1: \mu \neq \$36,500$.

In this example, we would observe the mean family income in a simple random sample of, say, 1,000 families. If the value of the sample mean $\bar{x}$ differs from the population mean $\mu = \$36,500$ by more than we would be willing to attribute to chance sampling error, we will reject the null hypothesis H_0. On the other hand, if the difference between the sample mean and the population mean assumed under H_0 is small enough to be attributed to chance sampling error, we will retain H_0. How do we know for what values of the sample statistic to reject H_0 and for what values to retain H_0? The answer to this question is the essence of hypothesis testing.

> The hypothesis-testing procedure is simply a decision rule that specifies whether the null hypothesis H_0 should be rejected or retained for every possible value of a statistic observable in a simple random sample of size n.

The set of possible values of the sample statistic is referred to as the **sample space**. Therefore, the test procedure divides the sample space into mutually exclusive parts, one constituting the **rejection region**, the region within which H_0 will be rejected, and another region within which H_0 is not rejected. The rejection region will consist of two parts in a so-called **two-tailed test**, as in the present example.

The nature of the division of the sample space for the example we have been discussing is illustrated in Figure 7-1. From the sampling theory developed in Chapter 5, we know that, given a population with a mean of $36,500 and a known standard deviation, there would be sampling variation among the means

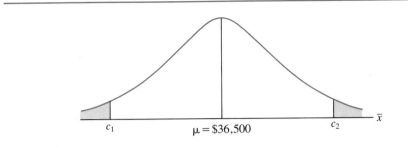

FIGURE 7-1
Two-tailed test: sampling distribution of the mean, with the rejection region for a null hypothesis

of samples of the same size drawn from that population. According to the central limit theorem, the sampling distribution of the mean for large sample sizes may be assumed to be normal, regardless of the shape of the population distribution. To decide whether family mean income has changed from $36,500, we determine two values, denoted by c_1 and c_2 in Figure 7-1, that set limits on the amount of sampling variation we feel is consistent with the null hypothesis. The decision rule in this case would be: (1) If the mean income of the sample of 1,000 families lies below c_1 or above c_2, we will reject the null hypothesis and conclude that the mean family income of the city has changed from $36,500; (2) if the sample mean lies between c_1 and c_2, we cannot reject H_0 and we will not be able to conclude that the city's mean family income has changed.

> A test in which we want to determine whether a population parameter has *changed*—regardless of the direction of change—is referred to as a **two-tailed test**, because the null hypothesis can be rejected by observing a statistic that falls in either of the two tails of the appropriate sampling distribution.

Two-tailed test

In the second type of hypothesis test, we wish to find out whether a sample comes from a population that has a parameter *less* than or *more* than a hypothesized value.

> Decision problems in which attention is focused on the direction of change give rise to **one-tailed tests**.

One-tailed test

The following example illustrates such a test. Suppose that in the past, under carefully specified driving conditions, the gas mileage of a compact car has averaged 30 miles per gallon (mpg) or less. The manufacturer of this car has redesigned the engine and wishes to test its claim that the average gas mileage is now greater than 30 miles per gallon. The company's engineers decide to

FIGURE 7-2

One-tailed test: sampling distribution of the mean, with acceptance and rejection regions for a null hypothesis

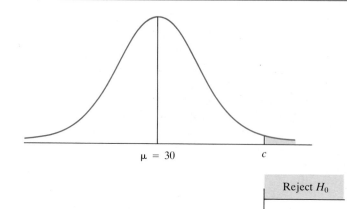

$\mu = 30$ c

Reject H_0

test the claim using a simple random sample of 50 new cars. The null hypothesis H_0 to be tested is "the true mean is equal to or less than 30 miles per gallon"; the alternative hypothesis H_1 is "the true mean is greater than 30 miles per gallon." Mathematically, these two hypotheses may be expressed

$$H_0: \mu \leqslant 30 \text{ mpg}$$

$$H_1: \mu > 30 \text{ mpg}$$

where μ denotes the mean number of miles per gallon.

From long experience, the automobile manufacturer has found that the population standard deviation is 5 miles per gallon. In this problem, we will assume that it is *known* from experience that the population standard deviation $\sigma = 5$ miles per gallon.

In the random sample of 50 automobiles included in the test, the mean number of miles per gallon observed is 32 (that is, $\bar{x} = 32$ mpg). Should we conclude that the automobile manufacturer's claim is valid—that the average gas mileage of this car is now greater than 30 mpg? Or, should we conclude that $\bar{x} = 32$ does not differ significantly from 30 mpg and that the difference may be attributable to chance sampling error?

In this case, the decision rule would be to reject H_0 if the mean of the sample $\bar{x}$ is more than some appropriate number c, and not to reject H_0 if $\bar{x}$ is equal to or less than c. The decision rule is diagrammed in Figure 7-2. Again, the number c represents the limit on the amount of sampling variation that we feel is consistent with the null hypothesis.

Concept of the Null Hypothesis

A null hypothesis is a statement about a population parameter, as in our first example ($H_0: \mu = \$36{,}500$) and in our second example ($H_0: \mu \leqslant 30$ mpg). Note that the equality sign is included in the null hypotheses, which is standard practice. By having the null hypothesis assert that the population parameter is equal to some specific value, we are able to decide for which values of the observed sample statistic we will reject or not reject that hypothesis. For example, when we hypothesized that $\mu = \$36,500$ in the first example and included in the second hypothesis that $\mu = 30$ mpg, we were able to establish sampling distributions of the sample mean $\bar{x}$ and to decide which values of $\bar{x}$ would cause us to reject the null hypothesis and which values of $\bar{x}$ would cause us not to reject it. Furthermore, we can specify how much risk of making Type I errors we are willing to tolerate. For example, if we hypothesize that $\mu = 30$ mpg, we can compute the probability of making a Type I error (that is, the probability of erroneously rejecting that hypothesis). This concept will be applied later in the chapter. These points explain why null hypotheses are set up in the form stated earlier, rather than in the form H_0: $\mu \neq \$36,500$ or H_0: $\mu > 30$ miles per gallon.

We will now turn to the application of some of these ideas. First, we will consider **one-sample tests**, which are tests of hypotheses based on data contained in a single sample. Tests involving means and proportions will be studied in that order.

One-sample tests

7.2
ONE-SAMPLE TESTS (LARGE SAMPLES)

In the one-sample tests discussed in this section, we will assume three things.

1. The sample size is large $(n > 30)$.

Assumptions

2. The sample size is determined before the test is conducted.

3. The population standard deviation is known.

The assumption of a large sample size $(n > 30)$ is made so that we can assume that the sampling distributions used in the testing procedure are normal. This is a useful simplification compared to the small-sample case, as we will see at the end of this chapter.

To construct a more detailed illustration of how the hypothesis-testing procedure can be used, we will return to the example of testing an automobile manufacturer's claim that after redesigning the engine of a compact car, the car's gasoline mileage is now in excess of 30 mpg.

We proceed to convert the testing of the automobile manufacturer's claim to a hypothesis-testing framework. This case is an example of a one-tailed test, or a **one-sided alternative**.

There are six steps involved in a test of statistical hypothesis.

1. Determine the null and alternative hypotheses.

Six steps

2. Select a level of significance for the test.

3. Choose a test statistic.

4. Select a sample size.

5. Determine the decision rule.

6. Reach a conclusion based on the sample drawn. This involves the rejection or retention of the null hypothesis.

We will now conduct the test, taking these steps in order. For ease of discussion, steps 3 and 4 and steps 5 and 6 have been combined.

State the null and alternative hypotheses (1)   We begin by stating the null and alternative hypotheses in statistical terms. As we have seen earlier, if we let μ represent the average number of miles per gallon of a compact car with the redesigned engine, the null and alternative hypotheses can be stated

$$H_0: \mu \leqslant 30 \text{ mpg}$$

$$H_1: \mu > 30 \text{ mpg}$$

Specify the level of significance (2)   Our decision will be based on the data observed in the simple random sample of compact cars produced by the automobile manufacturer. The question we want to answer is "Are the sample data so inconsistent with the null hypothesis that we must reject that hypothesis?" In designing the test, we must specify the risk that we are willing to run of rejecting the null hypothesis when it is true. In other words, we must specify the probability of committing a Type I error—or, as it is commonly designated, the **level of significance** of the test. Levels of significance such as 0.05 or 0.01 are conventionally used. Of course, such figures are rather arbitrary, but low levels are ordinarily used so that the probability of committing a Type I error will be quite low. In this problem, we will assume that we do not want the risk of erroneously rejecting the null hypothesis to be greater than 0.05. The level of significance is denoted by the Greek letter α. Hence, in this problem, $\alpha = 0.05$. In a one-tailed test, α represents the *maximum* probability of a Type I error. In a two-tailed test in which the null hypothesis consists of only one value of a population parameter, α represents *the* probability of a Type I error.

Table 7-2 summarizes the alternative hypotheses tested in our gasoline mileage problem and the possible actions related to these hypotheses. After a

---

TABLE 7-2

Gasoline mileage problem: the relationship between possible actions and hypotheses related to the number of miles per gallon delivered by a certain compact car

Action Concerning Hypothesis H_0	State of Nature	
	$H_0: \mu \leqslant 30 \text{ mpg}$	$H_1: \mu > 30 \text{ mpg}$
Retain H_0	No error	Type II error (Retain $H_0 \mid H_0$ False)
Reject H_0	Type I error (Reject $H_0 \mid H_0$ True)	No error

brief discussion of the meanings of Type I and Type II errors in this problem, we will identify the remaining steps in the hypothesis-testing procedure and carry them out quantitatively.

From Table 7-2, we can observe that a Type I error in this problem (the incorrect rejection of the null hypothesis) takes the form of concluding that the compact car delivers more than 30 mpg on the average when the true average is 30 mpg or less. In this problem, a Type II error takes the form of concluding that the compact car delivers 30 mpg or less on the average when actually the true average is greater than 30 mpg.

Since the hypotheses in this problem are related to the *mean* number of miles per gallon that the compact car yields, the **test statistic** we want to observe is the mean number of miles per gallon delivered by the sample of cars. In this problem, we are assuming that a *simple random sample of 50 cars* is used in the test. Of course, as the sample size increases, the precision of the test increases (the expected amount of sampling error decreases). The choice of sample size is essentially an economic decision, because as the sample size increases, the cost involved in performing the test increases.

Choose the test statistic and sample size (3, 4)

We will now turn to the question of how to establish a decision rule on which to base our rejection of the null hypothesis. As indicated earlier, a hypothesis-testing rule is simply a procedure that specifies the action to be taken for each possible sample outcome. Thus, we are interested in partitioning the sample space into a region in which we will reject the null hypothesis and a region in which we will not reject it. In every hypothesis-testing problem, the partitioning of the appropriate sample space is accomplished by assuming that the null hypothesis is true and considering the appropriate sampling distribution. This follows from the fact that specifying the probability of making a Type I error determines how the sample space will be partitioned.

Determine the decision rule and conclusion (5, 6)

In this particular problem, the question concerns the *mean* number of miles per gallon, so that the sampling distribution of the mean is the appropriate distribution. Because our sample size is large ($n > 30$), we use the central limit theorem and assume that the sampling distribution of means is normal. The normal distribution of samples of size 50 from a population in which $\mu = 30$ mpg (the upper limit of the hypothesis the manufacturer must reject to demonstrate the claim that $\mu > 30$) and $\sigma = 5$ mpg is shown in Figure 7-3. (Recall that we are assuming we *know* from past experience that the population standard deviation σ is equal to 5 mpg.) As indicated in the graph, the shaded region represents 5% of the area under the normal curve. Referring to Table A-5 in Appendix A, we see that 5% of the area in a normal distribution lies to the right of $z = 1.65$ (45% of the area lies between $z = 0$ and $z = 1.65$). Therefore, in this problem, the point above which we would reject the null hypothesis H_0 and conclude that the automobile manufacturer's claim has been verified is a sample mean with a value greater than $\mu + 1.65\sigma_{\bar{x}}$. The standard error of the mean $\sigma_{\bar{x}}$ is

$$\sigma_{\bar{x}} = \frac{\sigma}{\sqrt{n}} = \frac{5}{\sqrt{50}} = 0.71 \text{ mpg}$$

FIGURE 7-3

Sampling distribution of the mean, showing regions of acceptance and rejection of H_0 (population parameters $\mu = 30$ mpg and $\sigma = 5$ mpg; sample size $n = 50$)

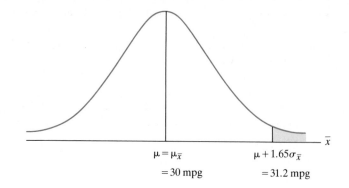

$\mu = \mu_{\bar{x}}$
$= 30$ mpg

$\mu + 1.65\sigma_{\bar{x}}$
$= 31.2$ mpg

Reject H_0

Thus, the critical value above which we would reject H_0 is

$$\mu + 1.65\sigma_{\bar{x}} = 30 + 1.65(0.71) = 31.2 \text{ mpg}$$

We can now see that this type of hypothesis-testing situation is referred to as a "one-tailed test" or a "one-sided alternative" because rejection of the null hypothesis takes place in only *one tail* of the sampling distribution.

Summary   In summary, on testing a simple random sample of 50 automobiles and observing $\bar{x}$, the sample mean number of miles per gallon, the automobile manufacturer should proceed according to the following decision rule:

DECISION RULE

1.   If $\bar{x} > 31.2$ mpg, reject H_0 (claim is valid).
2.   If $\bar{x} \leqslant 31.2$ mpg, do not reject H_0 (claim is invalid).[1]

We can now answer our original question. Should we conclude that the automobile manufacturer's claim is valid—that the average gasoline mileage of the compact car is now greater than 30 mpg? Since the sample yields a mean of $\bar{x} = 32$ mpg, which exceeds the critical value of 31.2 mpg, the null hypothesis H_0 should be rejected, and we can conclude that the manufacturer's claim is valid. Another way of stating this conclusion is to say that "the sample mean

---

[1] There is some ambiguity about whether the equal sign should appear in the reject portion or the do not reject portion of the decision rule. From the theoretical point of view, this positioning is inconsequential. The normal curve is a continuous probability distribution. Thus, the probability of observing exactly $\bar{x} = 31.2$ mpg is 0.

of 32 is significantly greater than 30"; that is, it is unlikely that the difference between those two figures can be attributed merely to chance sampling error. Hence, again we conclude that the manufacturer's claim is valid.

It is instructive to examine an alternative method of stating the decision rule. Instead of working with the original units (in this case, miles per gallon), we could calculate the z value in a standard normal distribution that corresponds to $\bar{x} = 31.2$ mpg. If the sample z value lies to the right of the critical value of 1.65, H_0 will be rejected; if the sample z value lies to the left of 1.65, H_0 will not be rejected. Thus, the decision rule can be rephrased as follows:

Alternative method

DECISION RULE

1. If $z > 1.65$, reject H_0 (claim is valid).
2. If $z \leqslant 1.65$, do not reject H_0 (claim is invalid).

The $\bar{x}$ value of 32 mpg corresponds to

$$z = \frac{32 - 30}{0.71} = \frac{2.0}{0.71} = 2.82$$

Therefore, because a mean of 32 falls 2.82 standard error units above the mean of the sampling distribution and the dividing line is at 1.65 units, the null hypothesis H_0 is rejected. This situation is graphed in Figure 7-4.

It is instructive when considering the rationale of the test to observe the implications of the computed z value. In this case, for example, an $\bar{x}$ value of 32 mpg corresponds to a z value of 2.82. We can observe from Table A-5 in Appendix A that about 0.0024 of the area in a normal distribution lies to the right of a z value of 2.82. Thus, interpreting the z value of 2.82 in terms of the

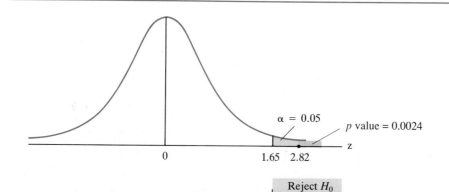

FIGURE 7-4
Standard normal curve for the gasoline mileage problem

gasoline mileage problem, if a sample of 50 cars is drawn at random from a population of automobiles that delivers an average gasoline mileage of 30 mpg with a standard deviation of 5 mpg, the probability of observing a sample mean of 32 mpg or greater is 0.0024. Because this sample result is so unlikely given the hypothesis that $\mu = 30$ mpg, we reject that hypothesis.

In modern applications of hypothesis testing, probabilities such as 0.0024 are termed *p values*.

> The *p* **values** represent the probability that if H_0 is true, we will observe a statistic ($\bar{x}$, in this case) that deviates by chance from the parameter being tested ($\mu = 30$ mpg, in this case) by a greater degree than is observed.

If a significance level of $\alpha = 0.05$ is used, as in the present problem, a *p* value of less than 0.05 would represent a significant result—that is, a rejection of the null hypothesis. The use of *p* values reflects the desire to report the results obtained and not to rely too heavily on arbitrarily significance levels such as 0.05 and 0.01, so that users of the results can draw their own conclusions. The *p* value gives us the probability of observing a value of the test statistic, that is, the *z* value, that contradicts the null hypothesis H_0 at least as much as the one computed from the sample data. We can see that as the test statistic falls farther into the rejection region, the strength of the evidence for rejecting the null hypothesis H_0 becomes stronger, and the *p* value gets smaller. The *p* value

Attained significance level is sometimes referred to as the attained significance level. Thus, in the present problem the *p* value of 0.0018, which is less than 0.05 and 0.01, indicates that the test statistic of $z = 2.82$ is significant at those significance levels and indeed has an attained significance level of 0.0018. So small a *p* value indicates that there is very strong evidence for rejecting the null hypothesis H_0, because if the null hypothesis is true, the probability of the observed sample evidence is indeed very small.

For values of $\mu < 30$ mpg in our problem, the probability of a Type I error is less than 0.05. We can see this from Figure 7-3. If $\mu < 30$ mpg (that is, if the sampling distribution shifts to the left), then less than 5% of the area will lie in the rejection region above 31.2 mpg. As the value of μ becomes smaller, the probability of committing a Type I error becomes lower. This makes sense in terms of the gasoline mileage problem. If the true number of miles per gallon delivered by the cars with redesigned engines is lower, then the probability that the null hypothesis H_0: $\mu \leqslant 30$ mpg will be erroneously rejected is lower. For $\mu = 30$ mpg, the probability of committing a Type I error is 0.05. Now the meaning of $\alpha = 0.05$, the significance level in this problem, becomes clear: it is the maximum probability of committing a Type I error. This sort of interpretation is typical in one-tailed tests.

Until this point, we have considered only situations in which the retention and the rejection of the null hypothesis result in just two possible actions. Furthermore, we have concentrated on the determination of decision rules stemming from control of Type I errors without reference to the corresponding implications for Type II errors. We will deal with these matters subsequently, and it is advisable not to clutter the present discussion with too many details. However, the following four points summarize the hypothesis-testing procedure discussed thus far.

Summary of the procedure

1. A null hypothesis and its alternative are drawn up. The null hypothesis is framed in such a way that we can compute the probability of committing a Type I error.

2. A level of significance α is determined. This controls the risk of committing a Type I error.

3. A decision rule is established by partitioning the relevant sample space into regions of rejection and retention of the null hypothesis. This partitioning is accomplished by considering the relevant sampling distribution. The nature of the null hypothesis and the choice of α determine the partition.

4. The decision rule is applied to a sample of size n. The null hypothesis is rejected or retained. Rejection of the null hypothesis implies acceptance of the alternative hypothesis.

A number of points can be made concerning the statistical theory involved in the preceding problem. First, the normal curve is used as the appropriate sampling distribution of the mean. If the population distribution of mean gasoline mileage is normal, the normal curve is the theoretically correct sampling distribution. Note that no statement at all is made in the gasoline mileage problem about the population distribution. The normal curve is used for the sampling distribution of $\bar{x}$, based on the specification of the central limit theorem that no matter what the shape of the population, the sampling distribution of $\bar{x}$ will be approximately normal for a sample as large as $n = 50$.

Remarks

Second, no finite population correction factor is used in the calculation of the standard error of the mean, despite the fact that the sample of 50 automobiles has been drawn without replacement from a finite population. However, the population size may be assumed to be very large relative to the sample size. Therefore, the finite population correction factor can be assumed to be approximately equal to 1 in this case.

Third, the population standard deviation σ is assumed to be known. If the population standard deviation is unknown and the sample size is large (say $n > 30$), then the sample standard deviation s may be substituted for σ. Hence, instead of calculating $\sigma_{\bar{x}} = \sigma/\sqrt{n}$, an estimated standard error of the mean is computed $s/\sqrt{n}$. In all other respects, the decision procedure remains the same. (In section 7.4, we will discuss how to deal with the situation in which the sample size is small and the population standard deviation is unknown.)

Fourth, the nature of the z value computed in the problem is worth noting. In Chapter 5, the concept of a standard score was discussed in the context of a normally distributed *population*. In that case, $z = (x - \mu)/\sigma$ represents a deviation of the value of an individual item from the mean of the population, expressed as a multiple of the population standard deviation. In our hypothesis-testing problem, the z values are of the form $z = (\bar{x} - \mu)/\sigma_{\bar{x}}$. Such a z value represents a deviation of a sample mean from the mean of the sampling distribution of $\bar{x}$, stated in multiples of the standard deviation of that distribution $\sigma_{\bar{x}}$. As we noted previously, the mean of the sampling distribution of $\bar{x}$, denoted by $\mu_{\bar{x}}$, is equal to the population mean μ. Thus, we use μ and $\mu_{\bar{x}}$ interchangeably. As a generalization, in hypothesis-testing problems, z values take the form

$$z = \frac{\text{Statistic} - \text{Parameter}}{\text{Standard error}}$$

For example, in the hypothesis-testing problem just discussed, the $\bar{x}$ value is the sample statistic, the population mean μ is the parameter, and the standard error of the mean $\sigma_{\bar{x}}$ is the appropriate standard error.

Fifth, we note that the size of the sample, $n = 50$, has been predetermined in our illustration. Thus, the sample is large and predetermined, and the construction of the decision rule is based on the control of only one type of incorrect decision (Type I errors). The next section dealing with the power curve discusses the measurement of Type II errors for such a test.

> Finally, it is important to realize that we cannot *prove* that a null hypothesis is false or that a null hypothesis is true. All that we can do is discredit a null hypothesis or fail to discredit it on the basis of sample data.

Actually, a single sample statistic such as $\bar{x}$ is consistent with an infinite number of hypotheses concerning μ.[2] From the standpoint of decision making and subsequent behavior, if sample data do not discredit a null hypothesis, we will act as though that hypothesis is true.

The hypothesis-testing procedure outlined thus far has concentrated on the control of Type I errors. The question of how well this test controls Type II errors naturally arises. When the null hypothesis is false, how frequently does the decision rule lead us to accept it erroneously? This question is answered

The power curve by means of the **power curve**, also called the **power function**, which can be computed from the information given in the problem and the decision rule. The Greek letter β is used to denote the probability of committing a Type II error; thus, β represents the probability of retaining the null hypothesis when it is false. In the gasoline mileage problem, the null hypothesis H_0 is false for each value of μ satisfying the alternative hypothesis $H_1: \mu > 30$ mpg. Therefore, for each particular value of μ greater than 30 mpg, we can determine a β value. Actually, by convention, the power curve gives the complementary probability to β (that is, $1-\beta$) for each value of the alternative hypothesis. Thus, it in-

---

[2] You have probably had the disconcerting experience of watching a number of experts in disagreement after they have observed ostensibly the same basic set of data. Perhaps you share the experience of finding it easier to reject and retain hypotheses when no data are available at all.

dicates the probability of rejecting the null hypothesis for each value for which the null hypothesis is false, which, of course, represents the probability of selecting the correct course of action in each case. $1-\beta$ is referred to as the "power of the test" for each particular value of the alternative hypothesis. For completeness, in a power curve, the probabilities of rejection are also shown for each value for which the null hypothesis is true.

Summary

A power curve is a function that gives the probabilities of rejecting the null hypothesis H_0 for all possible values of the parameter tested. Therefore, it measures the ability of the decision rule to discriminate between true and false hypotheses.

The power curve is useful in assessing the risks of making both Type I and Type II errors when a decision rule is employed.

The power curve for the gasoline mileage problem (shown in Figure 7-5) has the typical S shape of a power curve in a one-tailed test, with the rejection region in the right tail. For a one-tailed test with the rejection region in the left tail, the curve would be reverse S-shaped, dropping from the upper left corner to the lower right corner on the graph. Rejection probabilities for the null hypothesis are shown on the vertical axis, and possible values of the population parameter μ appear on the horizontal axis. Specifically, the figures

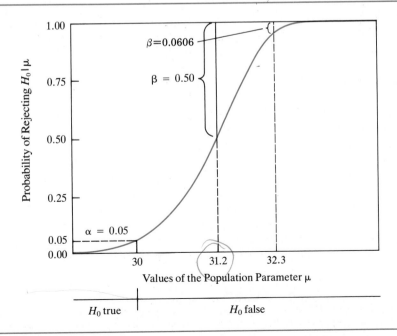

FIGURE 7-5

Power curve for the gasoline mileage problem

plotted on the vertical axis are conditional probabilities of the form P(rejection of $H_0|\mu) = P(\bar{x} > 31.2 \text{ mpg}|\mu)$.

The nature of the power curve in Figure 7-5 can be determined by considering a couple of the plotted values. The value of $\alpha = 0.05$ is shown for $\mu = 30$ mpg, indicating the significance level of the test. We can see that this is the maximum probability of erroneously rejecting the null hypothesis H_0, because the ordinates of the curve drop off to the left as μ decreases in the region where H_0 is true. The heights of the ordinates of the power curve to the left of $\mu = 30$ mpg represent the probabilities of making Type I errors. The heights of the ordinates to the right of $\mu = 30$ mpg represent the values of $1-\beta$, or the probabilities of rejecting the null hypothesis when it is false. Therefore, the complementary distances from points on the curve to 1.0 are values of β, or probabilities of making Type II errors. One such value is displayed in the graph for $\mu = 31.2$ mpg. Recall that our decision rule requires the rejection of H_0 if the sample mean $\bar{x}$ is greater than 31.2 mpg. If the population mean is 31.2 mpg, the probability of observing $\bar{x}$ values less than 31.2 mpg and therefore of rejecting H_0 is obviously 0.50. This situation is graphed in Figure 7-6.

Some computation is required to obtain the β value for $\mu = 32.3$ mpg. To compute this figure, we must refer to the sampling distribution of $\bar{x}$, given that $\mu = 32.3$, and calculate the probability that a sample mean would lie in the acceptance region $\bar{x} \leqslant 31.2$ mpg. The z value for 32.3 mpg is

$$z = \frac{31.2 - 32.3}{0.71} = -1.55$$

Thus, if $\mu = 32.3$ mpg, the critical $\bar{x}$ value of 31.2 lies 1.55 standard errors of the mean below 32.3 mpg. Referring to Table A-5 of Appendix A, we find a figure of 0.4394, which when subtracted from 0.5000 (the area to the left of $\mu = 32.3$ mpg) gives 0.0606 for β, the probability of erroneously retaining H_0.

For values of μ slightly more than 30 mpg, the probability of a Type II error is very high. In fact, we can see in Figure 7-5 that these probabilities exceed 0.50 for μ values between 30 mpg and 31.2 mpg. This simply indicates

FIGURE 7-6

Graphs illustrating Type II error probabilities for $\mu = 31.2$ mpg and $\mu = 32.3$ mpg

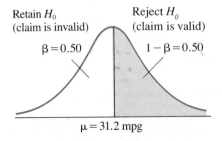

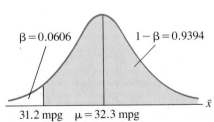

that the power of the test is low when the value of μ satisfies the alternative hypothesis ($H_1: \mu > 30$ mpg) but is close to values of μ satisfying the null hypothesis ($H_0: \mu \leqslant 30$ mpg). For a fixed sample of size n, β can be decreased only by increasing α and vice versa. If α is fixed, then as sample size is increased, β is reduced for all values of μ in the region where H_0 is false. In this type of one-tailed test, the ideal power curve would be $\int$ shaped, with the vertical line occurring at $\mu = 30$ mpg. Thus, the probability of rejecting H_0 would always be equal to 0.0 when H_0 is true and to 1.0 when H_0 is false. However, this ideal curve is clearly unattainable when sample data are used to test hypotheses, because sampling error will always be present.

The trade-off relationship between Type I and Type II errors for a sample of fixed size is such that the level of significance should be decided *by considering the relative seriousness of the two types of errors* before conducting hypothesis tests.

Deciding the level of significance

How does the decision maker use power curves in setting up an appropriate hypothesis-testing procedure? In the preceding discussion, the decision rule was determined for a fixed sample size n; α was specified, and the critical value of $\bar{x} = 31.2$ mpg was computed. Suppose that on examination of the resulting power curve, the decision maker feels that the β values are too high. For instance, in the present example, we determined that if an automobile actually delivers 31.2 miles per gallon, then $\beta = 0.50$. To reduce this risk, the decision maker (the automobile manufacturer) could decrease the critical value of $\bar{x}$, thereby raising the level of significance, which had been previously set at $\alpha = 0.05$. This would shift the entire power curve to the left. Thus, we see that if the sample size n is unchanged, the only way to reduce β levels is to increase α. On the other hand, if the decision maker is unwilling to increase the significance level α, then the only way to reduce β values is to increase the sample size n. Of course, larger sample sizes involve increased costs.

Using the power curve

To illustrate that larger sample sizes provide more discriminating tests, let us consider the effect of increasing the sample size in the gasoline mileage example from $n = 50$ to $n = 500$, while retaining the critical value at 31.2 mpg. The corresponding curves for the tests provided by these two sample sizes appear in Figure 7-7. Note from the figure that the larger sample size provides a more discriminating test because its power curve implies lower probabilities both of retaining false hypotheses and of rejecting true hypotheses. Thus, the power curve for the larger sample size lies much closer to the ideal power curve.

It is usual practice in industrial quality control to use **operating characteristic curves** (succinctly referred to as **O-C curves**), rather than power curves, to evaluate the discriminating power of a test. The O-C curve is simply the complement of the power curve. The probability of retaining, rather than the probability of rejecting the null hypothesis is plotted on the vertical axis of the O-C curve.

Operating characteristic curves

FIGURE 7-7

Comparison of power curves
for two sample sizes ($n = 50$
and $n = 500$) for the gasoline
mileage problem

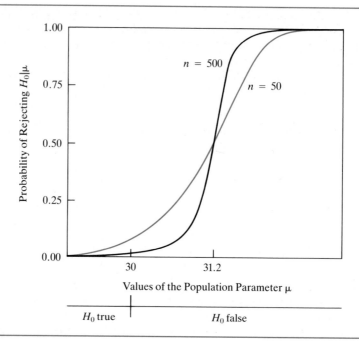

Test Involving a Proportion: Two-tailed Test

The preceding discussion dealt with a test of a hypothesis about a mean. We now turn to hypothesis testing for a proportion.

EXAMPLE 7-1

Let us consider the case of an advertising agency that developed a general theme for the commercials on a certain TV show based on the assumption that 50% of the show's viewers were over 30 years of age. The agency was interested in determining whether the percentage had changed in either an upward or downward direction. If we use the symbol p to denote the proportion of all viewers of the show, we can state the null and alternative hypotheses as follows:

$$H_0: p = 0.50 \text{ viewers over 30 years of age}$$

$$H_1: p \neq 0.50 \text{ viewers over 30 years of age}$$

Let us assume that the agency wished to run a 5% risk of erroneously rejecting the null hypothesis of "no change," or, $H_0: p = 0.50$. That is, the agency decided to test the null hypothesis at the 5% significance level ($\alpha = 0.05$).

SOLUTION

In order to test the hypothesis, the agency conducted a survey of a simple random sample of 400 viewers of the TV show. Of the 400 viewers, 210 were over 30 years of age and 190 were 30 years of age or less. What conclusion should the agency reach?

In this problem, as contrasted with the previous one, the null hypothesis concerns a single value, that is, a hypothetical population parameter of $p = 0.50$. The alternative hypothesis includes all other possible values of p. We can understand the reason for setting up the hypotheses this way by reflecting on how the test will be conducted. The hypothesized parameter under the null hypothesis is $p = 0.50$. We have observed in a sample a certain proportion, denoted $\bar{p}$, who were over 30 years of age. The testing procedure involves a comparison of $\bar{p}$ with the hypothesized value of p to determine whether a significant difference exists between them. If $\bar{p}$ does not differ significantly from p, and we do not reject the null hypothesis that $p = 0.50$, what we really mean is that the sample is consistent with a hypothesis that half the viewers of the TV show are over 30. On the other hand, if $\bar{p}$ is greater than 0.50 and a significant difference between $\bar{p}$ and p is observed, we will conclude that more than half the viewers are over 30. If the observed $\bar{p}$ is less than 0.50 and a significant difference from $p = 0.50$ is observed, we will conclude that fewer than half the viewers are over 30.

> It is important to note that in hypothesis-testing procedures, the two hypotheses and the significance level of the test must be selected before the data are examined.

We can easily see the difficulty with a procedure that would permit the investigator to select α after examination of the sample data. It would always be possible to fail to reject a null hypothesis simply by choosing a sufficiently small significance level, thereby setting up a large enough region of retention of H_0. Thus, the first step in our problem is setting up the competing hypotheses, with the null hypothesis stated in such a way that the probability of a Type I error can be calculated. We have accomplished this by a single-valued null hypothesis, $H_0: p = 0.50$. Our next step is to set the significance level, which we have taken as $\alpha = 0.05$.

We proceed with the test. The simple random sample of size 400 is drawn, the statistic $\bar{p}$ is observed, and we can now establish the appropriate decision rule. Since the sample size is large, the theory developed in Chapter 5 allows us to use the normal curve as an appropriate approximation for the sampling distribution of the percentage $\bar{p}$. As in the preceding problem, for illustrative purposes, we will establish the decision rule in two different forms, first in terms of $\bar{p}$ values, then in terms of the corresponding z values in a standard normal distribution. Under the assumption that the null hypothesis is true, the sampling distribution of $\bar{p}$ has a mean of p and a standard deviation $\sigma_{\bar{p}} = \sqrt{pq/n}$. Again, we ignore the finite population correction, because the population is large relative to the sample size. The sampling distribution of $\bar{p}$ for the present problem, in which $p = 0.50$, is shown in Figure 7-8. Also shown is the horizontal axis of the corresponding standard normal distribution, scaled in z values.

Since the null hypothesis will be rejected by an observation of a $\bar{p}$ value that lies significantly below or significantly above $p = 0.50$, we clearly must use a two-tailed test. The critical regions (rejection regions) are displayed in Figure 7-8. The arithmetic involved in establishing regions of acceptance and rejection of H_0 is as follows. The standard error of $\bar{p}$ is

$$\sigma_{\bar{p}} = \sqrt{\frac{(0.50)(0.50)}{400}} = 0.025$$

FIGURE 7-8

Sampling distribution of a proportion, with $p = 0.50$ and $n = 400$. This is a two-tailed test with $\alpha = 0.05$

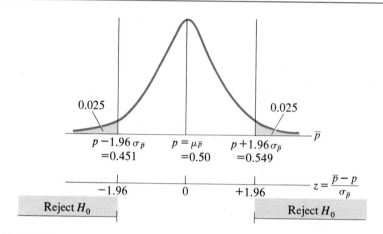

Referring to Table A-5 of Appendix A, we find that 2.5% of the area in a normal distribution lies to the right of $z = +1.96$, and therefore 2.5% also lies to the left of $z = -1.96$. Thus, we establish a significance level of 5% by marking off a non-rejection range for H_0 of $p \pm 1.96\sigma_{\bar{p}}$. The calculation is

$$p + 1.96\sigma_{\bar{p}} = 0.50 + (1.96)(0.025) = 0.50 + 0.049 = 0.549$$

$$p - 1.96\sigma_{\bar{p}} = 0.50 - (1.96)(0.025) = 0.50 - 0.049 = 0.451$$

We can now state the decision rule. The agency draws a simple random sample of 400 viewers of the TV show, observes $\bar{p}$ (the proportion of persons in the sample who are over 30 years of age), and then applies the following decision rule.

DECISION RULE

1. If $\bar{p} < 0.451$ or $\bar{p} > 0.549$, reject H_0
2. If $0.451 \leqslant \bar{p} \leqslant 0.549$, do not reject H_0

Again, we restate the decision rule in terms of z values.

DECISION RULE

1. If $z < -1.96$ or $z > +1.96$, reject H_0
2. If $-1.96 \leqslant z \leqslant +1.96$, do not reject H_0

Let us apply this decision rule to the present problem. The observed sample $\bar{p}$ was

$$\bar{p} = \frac{210}{400} = 0.525$$

and

$$z = \frac{\bar{p} - p}{\sigma_{\bar{p}}} = \frac{0.525 - 0.500}{0.025} = +1.0$$

Therefore, the null hypothesis H_0 is not rejected. This leads us to a rather vague conclusion. If $\bar{p}$ fell in the rejection region in the right tail of the sampling distribution in Figure 7-8 (that is, if $\bar{p}$ were greater than 0.549), we would conclude that more than half the viewers of the TV show are over 30 years of age. If $\bar{p}$ had fallen in the left tail of the rejection region, we would conclude that less than half are over 30. However, $\bar{p}$ lies in the do not reject region, so we *cannot* conclude that more than half of the viewers are over 30 or that less than half are over 30. The sample evidence is *consistent with the hypothesis of a 50-50 split*. Thus, not rejecting the null hypothesis means that on the basis of the available evidence, we simply are not in a position to conclude that more than half of the viewers are over 30 years of age or that less than half of them are. In some instances, the best course is to reserve judgement.

An alternative way of stating the sample evidence is in terms of the p value. In this case, the p value that corresponds to $z = 1.0$ is from the normal curve Table A-5, $P(z > |1.0|) = 1 - 2(0.3413) = 0.3174$. In terms of the z scale shown in Figure 7-8, we have computed the probability that the test statistic z lies to the left of -1.0 or to the right of $+1.0$. As in the preceding example that involved a one-tailed test, we have computed in this two-tailed test the probability that if H_0 is true, we would have observed a test statistic that disagreed with the null hypothesis at least as much as is observed in this sample. Since the p value of 0.3174 is greater than the selected significance level of 0.05, we cannot reject the null hypothesis H_0: $p = 0.50$.

Further Remarks

Power curves can be computed for two-tailed tests in an analogous way to that for one-tailed tests. However, such calculations will not be illustrated here. The power curve for the two-tailed test in the TV viewer problem is shown in Figure 7-9. The curve has the characteristic U shape of a power function for a two-tailed test. The possible values of the parameter p are shown on the horizontal axis. The height of the ordinate at $p = 0.50$ is 0.05, which is the value of α. Since the only value of p for which the null hypothesis is true is 0.50, the ordinates at all other p values denote probabilities of rejecting the null hypothesis when it is false. The complements of these ordinates are equal to the values of β, the probabilities of making Type II errors.

The hypothesized proportion according to the null hypothesis in this problem was 0.50, because the agency was interested in whether or not more than 50% of the viewers were over 30 years of age. On the other hand, if we wanted to test the assertion that the population proportion was 0.55, 0.60, or some other number, we would have used these figures as the respective hypothesized parameters. Also in this problem, the null hypothesis was single-valued ($p = 0.50$), whereas the alternative hypothesis was many-valued ($p \neq 0.50$). This resulted in a two-tailed test.

FIGURE 7-9

Power curve for the TV viewer
problem

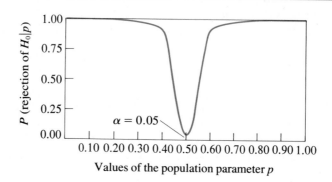

For example, suppose an assertion had been made that less than 50% of the viewers of the TV show were over 30.

We place this assertion in the alternative hypothesis, i.e., $H_1: p < 0.50$. In hypothesis testing, we ordinarily place the burden of proof on the person(s) who has made the assertion to demonstrate it convincingly. This is the same point that we made in section 7.1, when we referred to the alternative hypothesis as the *research hypothesis*. We stated there that the research or alternative hypothesis H_1 would be accepted by producing convincing evidence to reject the null hypothesis H_0. Hence, in this problem, we will conclude the assertion is correct only if the sample proportion $\bar{p}$ is significantly less than 0.50. Assume a test at the 5% significance level. The null and alternative hypotheses in this instance would be

$$H_0: p \geqslant 0.50$$

$$H_1: p < 0.50$$

This would involve a one-tailed test with a rejection region lying in the left tail and containing 5% of the area under the normal curve. The critical region would be in the left tail, since only a significant difference for a $\bar{p}$ value lying *below* 0.50 could result in the rejection of the stated null hypothesis. The decision rule in terms of z values follows.

DECISION RULE

1. If $z < -1.65$, reject H_0
2. If $z \geqslant -1.65$, do not reject H_0

On the other hand, if the assertion had been that more than 50% of the viewers were over 30 and if we performed a test at the 5% significance level, the rejection region would be the 5% area in the right tail. The corresponding hypotheses and decision rule would be

$$H_0: p \leqslant 0.50$$

$$H_1: p > 0.50$$

DECISION RULE

1. If $z \geqslant +1.65$, reject H_0
2. If $z < +1.65$, do not reject H_0

Standard normal distributions with the decision rules for these one-tailed tests are depicted in Figure 7-10. When tests are conducted for means or other

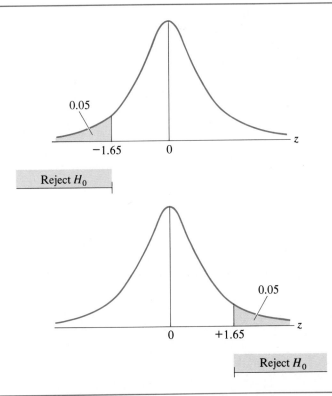

FIGURE 7-10

Standard normal distribution with decision rules for one-tailed tests in terms of z values; $\alpha = 0.05$

statistical measures, they may also be either one- or two-tailed, depending on the context of the problem. An illustration of a one-tailed test for a proportion is given in Example 7-2.

EXAMPLE 7-2

An airline is considering the addition of a midnight budget fare for one of its routes. Its analysis has shown that the midnight fare would be profitable if more than 35% of its current customers would use the route. The airline selected 400 of its passengers at random, and of these, 160 indicated that they would use the new route if it were established. What is the probability of drawing a random sample of size 400 with 160 or more potential customers from a statistical universe that in reality contains exactly 35% potential customers? The airline concluded that more than 35% of its customers were potential passengers on the new route. Should it have so concluded, in your opinion? Use $\alpha = 0.05$.

In answering the above question, you had to locate a sample statistic (proportion) on a random sampling distribution. Draw a rough sketch of this distribution showing the following:
a.   the value and location of the hypothesized parameter.
b.   the value and location of the statistic.
c.   the vertical and horizontal scale descriptions.
d.   the portion of the distribution corresponding to the computed probability.

SOLUTION

$$P\left(\bar{p} \geqslant \frac{160}{400}\right) = P\left(z \geqslant \frac{0.40 - 0.35}{\sqrt{(0.35)(0.65)/400}}\right) = 0.0179$$

Note that 0.0179 is the p value or attained significance level for the sample $\bar{p}$ figure of 0.40.

$$H_0: p \leqslant 0.35$$

$$H_1: p > 0.35$$

$$\alpha \doteq 0.05$$

Critical value $= 0.35 + (1.65)\sqrt{(0.35)(0.65)/400} = 0.389$

DECISION RULE
1.   Reject H_0 if $\bar{p} > 0.389$
2.   Do not reject H_0 otherwise

Reject H_0, and add the new route.
Yes, the airline concluded correctly.

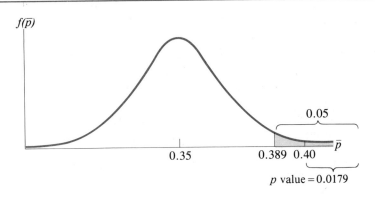

FIGURE 7-11

Sampling distribution of a proportion for the airline potential passengers example

Note: In Figure 7-11, the tail of the curve has been magnified to show the desired items more clearly.

EXERCISES 7.2

1. Distinguish between a parameter and a statistic.

2. Support or criticize the following statement: "You will make very few *TRUE, VERY FEW TYPE I* errors in hypothesis testing if you set the significance level (α) very low, *ERRORS, BUT MANY* say 0.001. Therefore, you should always do so." *TYPE II ERRORS*

3. Distinguish briefly between
   a. One-tailed test and two-tailed test
   b. Type I and Type II errors

4. What is meant by "the power of a test"? How is the power related to the Type II error?

5. A city council polled a random sample of 300 taxpayers on the issue of constructing a new high school. Its results showed 180 people were against the proposal while 120 favored the idea. Let p be the proportion of all taxpayers in the city opposed to the proposal, and consider the following hypotheses:

$$H_0: p \leqslant 0.50$$

$$H_1: p > 0.50$$

The following test is proposed: Reject H_0 if the sample proportion $\bar{p}$ is greater than 0.55.
   a. What is the greatest probability of a Type I error with this test? That is, what is the p value or attained significance level for a random sample of 180 people with a sample proportion of $\bar{p} = 0.55$?

b. Design a test such that α is at most 0.01, and test the hypothesis with the given sample data.

c. If you want to minimize the probability of a Type II error, which of the two tests would you use? Why?

6. A state highway commission is debating whether to widen a bridge on one of its intrastate highways. To determine whether the traffic load warrants such a large expenditure of state funds, the commission decided to count the motor vehicles crossing the bridge during 15-minute periods. Let us estimate the standard deviation of this variable as $\sigma = 40$ vehicles per 15 minutes. Suppose the commission feels that it should widen the bridge if the population mean number of motor vehicles exceeds 250 per 15-minute period. The commission takes a random sample of 200 15-minute intervals, and it is willing to run a risk of no more than 1% of erroneously widening the bridge.

a. What sample mean values will lead to the commission's widening the bridge?

b. What would be the p value or attained significance level if a sample mean of 255 motor vehicles were observed in a random sample of 200 15-minute intervals?

7. A year ago, a business firm instituted a morale improvement plan in an attempt to reduce absenteeism among its employees. Prior to the implementation of this plan, an average of 12 employees was absent each day with a standard deviation of two. A random sample of 80 days over the past year yielded an absenteeism rate of 11.5 employees per day.

a. Is there reason to believe that the morale improvement plan has decreased the absenteeism rate? Use a 0.05 significance level and you need not apply a finite population correction.

b. What is the p value or attained significance level for the absentee rate of 11.5 employees per day in the 80-day sample?

8. The annual mortality rate for a particular age group in the United States is 12 deaths per 1,000 persons. A simple random sample of individuals in this age group in 50 cities was taken and the result obtained was $\bar{x} = 15$ deaths per 1,000 persons. The population standard deviation is known to be two deaths per 1,000.

a. Would you conclude that the average mortality rate in these cities is above the national rate? Use a 0.01 level of significance.

b. Calculate the p value or attained significance level for $\bar{x} = 15$ deaths per 1,000 persons.

9. A fast-food restaurant advertises that it puts a quarter pound of beef in its hamburgers. A consumer protection agency has received complaints that the restaurant's burgers actually contain less than the advertised amount. The agency selects and weighs a simple random sample of 400 burgers. From past experience, it is known that the population standard deviation of hamburger weights is 0.9 ounce. Set up a test such that if the restaurant, on the average, actually places the advertised amount in

the hamburgers, it would be accused unjustly only once in 20 times. That is, let $\alpha = 0.05$.

10. In a certain year, the arithmetic mean interest rate on loans to large retailers (that is, those with assets of $5,000,000 or more) was 8.0% and the standard deviation was 0.3 percentage points. Two years later, a simple random sample of 200 loans to large retailers yielded an arithmetic mean interest rate of 8.03%.

   a. Would you conclude that the average level of interest rates for large retailers has changed? Assume that you are willing to run a 5% risk of erroneously concluding there has been a change.

   b. Using the same decision-making rule as in part (a), find the probability of making a Type II error if the average interest rate for all large retailers was 8.02%.

   c. Interpret your answer to part (b).

11. In the past, the gas mileage of a compact car has been 30 miles per gallon with a standard deviation of four miles per gallon. Engineers redesigned the engine of this car and tested the gas mileage of a simple random sample of 100 new cars.

   a. On the average, how many miles per gallon must this sample of 100 cars get for the automobile manufacturer to conclude that the new engine yields a higher gas mileage than the old engine? The manufacturer is willing to run a risk of no more than 0.02 of drawing such a conclusion in error.

   b. Suppose the universe mean for the new engine is 32 miles per gallon, and you employ the critical value(s) established in part (a).
      (1) What is the probability of reaching an incorrect conclusion?
      (2) If you reach an incorrect conclusion in this situation, is it a Type I or Type II error? Why?

12. The education department in a midwestern university selected 300 students at random from a large group of students who participated in an experimental teaching program. In this program, the students received their instruction from computers in place of professors. The university felt that at least 30% of all students should consider the program a success in order for the university to continue it. Therefore, the university decided that if 75 or fewer students felt the program was a success, it would discontinue the program. State in words the nature of the Type I error involved here. How large is the risk of such an error?

13. All students entering Old Ivy University are required to take an examination in mathematics. In the past, the mean score has been 70 with a standard deviation of 10 points. This year, the administration will draw a random sample of 100 freshmen to test the hypothesis that the mean score has not changed. The following decision rule was used for the test:

1. If $\bar{x} < 68.04$ or $\bar{x} > 71.96$, reject H_0
2. If $68.04 < \bar{x} < 71.96$, retain H_0

   a. Suppose that this year $\mu = 71$. What is the probability of retaining H_0?

   b. Calculate and explain the meaning of the test when $\mu = 71$.

   c. What level of significance was used for the test?

14. The major oil companies report that last year 6% of all credit charges for gasoline, car repairs, and parts were never collected and had to be written off as bad debts. Recently, the oil companies installed a central computer credit check system. For any credit purchase over $15, a gas station must call the computer center, and after a computer search of the customer's payment record, the purchase is either given a credit acceptance number or rejected. Any sales over $15 not given a credit acceptance number will not be honored by the oil company. To test the computer's effectiveness, the systems manager selected 1,000 approved credit charges at random and found that 38 charges were uncollected and written off as bad debts. Is the system effective? Use a 5% significance level.

15. A manufacturing firm extended its workers' coffee breaks by 10 minutes in the hope that daily productivity would subsequently increase. In the past, each worker produced an average of 240 machine parts with a standard deviation of 25. After the change, management took a random sample of 30 days and found that the average productivity per worker was 245 machine parts. Assuming $\alpha = 0.05$, would you conclude that average daily productivity has increased?

16. The J.T. McClay Company, a manufacturer of quality stereo receivers, recently held a managers meeting to review its marketing program. During the review, management decided to confirm a previous study indicating that 35% of its customers were university students. If the percentage is now different from 35%, the company will require a new marketing plan. The company wished to run a 2% risk of erroneously rejecting the hypothesis that 35% of its customers were university students. The company polled a simple random sample of 600 potential customers. Of the 600, 175 were university students.

   a. Should the J.T. McClay Company alter its marketing campaign on the basis of this sample? Roughly sketch the sampling distribution and label the horizontal axis properly. Indicate the acceptance region for the test and the location of the test statistic on which your conclusion was based.

   b. Suppose the company desired to test the null hypothesis that at least 35% of its customers are university students. On the basis of the same sample data, should the company reject the hypothesis at the 2% significance level? Justify your answer statistically.

17. The Constitution of the United States requires a two-thirds majority of both the House and the Senate to override a presidential veto. In a given situation, the necessary majority was secured in the Senate, but the action of the House was uncertain. Prior to the actual vote in the House, news-

paper reporters took a straw vote of 85 representatives, and 52 of them indicated their intention to override the veto. If you wished to be wrong no more than 5% of the time, would you conclude (assuming that the sample was taken at random) that the veto will not be overruled? Justify your conclusion statistically. Use the finite population correction, assuming that there are 435 representatives in the House.

7.3
TWO-SAMPLE TESTS (LARGE SAMPLES)

The discussion thus far has involved testing of hypotheses using data from a single random sample. Another important class of problems involves the question of whether statistics observed in two simple random samples differ significantly. Recalling that all statistical hypotheses are statements concerning population parameters, we see that this question implies a corresponding question about the underlying parameters in the populations from which the samples were drawn. For example, if the statistics observed in the two samples are arithmetic means (say, $\bar{x}_1$ and $\bar{x}_2$), the question is whether we are willing to attribute the difference between these two sample means to chance sampling errors or whether we will conclude that the populations from which the samples were drawn have **equal means**. We shall illustrate these tests, first for differences between means and then for differences between proportions. In both cases, we assume (as we did in section 7.2) that we are using large samples.

Considering unequal means

Test for Difference between Means: Two-tailed Test

A consulting firm conducting research for a client was asked to test whether the wage levels of unskilled workers in a certain industry were the same in two different geographical areas, referred to as Area A and Area B. The firm took simple random simples of the unskilled workers in the two areas and obtained the following sample data for weekly wages:

Area	Mean	Standard Deviation	Size of Sample
A	$\bar{x}_1 = \$300.01$	$s_1 = \$4.00$	$n_1 = 100$
B	$\bar{x}_2 = 295.21$	$s_2 = \$4.50$	$n_2 = 200$

If the client wished to run a risk of 0.02 of incorrectly rejecting the hypothesis that the population means in these two areas were the same, what conclusion should be reached? Note that the samples need not be the same size; different sample sizes have been assumed in this problem in order to keep the example completely general.

Let us refer to the means and standard deviations of *all* unskilled workers in this industry in Areas A and B, respectively, as μ_1 and μ_2, and σ_1 and σ_2.

These are the population parameters corresponding to the sample statistics $\bar{x}_1$, $\bar{x}_2$, s_1, and s_2. The hypotheses to be tested are

$$H_0: \mu_1 - \mu_2 = 0$$

$$H_1: \mu_1 - \mu_2 \neq 0$$

That is, the null hypothesis asserts that the population parameters μ_1 and μ_2 are equal. We form the statistic $\bar{x}_1 - \bar{x}_2$, the difference between the sample means. If $\bar{x}_1 - \bar{x}_2$ differs significantly from zero, the hypothesized value for $\mu_1 - \mu_2$, we will reject the null hypothesis and conclude that the population parameters μ_1 and μ_2 are indeed different.

Since the risk of a Type I error has been set, we turn now to determining the decision rule based on the appropriate random sampling distribution. Let us examine some of the important characteristics of this distribution. The two random samples are independent; that is, the probabilities of selection of the elements in one sample are not affected by the selection of the other sample. Hence, $\bar{x}_1$ and $\bar{x}_2$ are independent random variables. It has been shown that the mean and standard deviation of the sampling distribution of $\bar{x}_1 - \bar{x}_2$ are, respectively,

(7.1)
$$\mu_{\bar{x}_1 - \bar{x}_2} = 0$$

and

(7.2)
$$\sigma_{\bar{x}_1 - \bar{x}_2} = \sqrt{\frac{\sigma_1^2}{n_1} + \frac{\sigma_2^2}{n_2}}$$

For large, independent samples, the sampling distribution of $\bar{x}_1 - \bar{x}_2$ is approximately normal by the Central Limit Theorem.

Summary

> If $\bar{x}_1$ and $\bar{x}_2$ are the means of two large independent samples from populations with means μ_1 and μ_2 and standard deviations σ_1 and σ_2, and if we hypothesize that μ_1 and μ_2 are equal, then the sampling distribution of $\bar{x}_1 - \bar{x}_2$ may be approximated by a normal curve with mean $\mu_{\bar{x}_1 - \bar{x}_2} = 0$ and standard deviation $\sigma_{\bar{x}_1 - \bar{x}_2} = \sqrt{\sigma_1^2/n_1 + \sigma_2^2/n_2}$.

It is helpful to think of this sampling distribution as the frequency distribution that would be obtained by grouping the $\bar{x}_1 - \bar{x}_2$ values observed in repeated pairs of samples drawn independently from two populations with the same means.

The standard deviation $\sigma_{\bar{x}_1 - \bar{x}_2}$ is referred to as the **standard error of the difference between two means**. We see from equation 7.2 that we must know the population standard deviations in order to calculate this standard error. However, for *large samples*, we can approximate σ_1 and σ_2 using the sample standard deviations s_1 and s_2. The resulting estimated (or approximate) standard error is symbolized $s_{\bar{x}_1 - \bar{x}_2}$ and may be written

(7.3)
$$s_{\bar{x}_1 - \bar{x}_2} = \sqrt{\frac{s_1^2}{n_1} + \frac{s_2^2}{n_2}}$$

We can now establish the decision rule for the problem. The test is clearly two-tailed, because the hypothesis of equal population means would be rejected if $\bar{x}_1 - \bar{x}_2$ differed significantly from zero in either the positive or the negative direction. The sampling distribution of $\bar{x}_1 - \bar{x}_2$ is shown in Figure 7-12. The horizontal scale of the distribution shows the difference between the sample means $\bar{x}_1 - \bar{x}_2$. As indicated, the mean of the distribution is equal to zero; in other words, under the null hypothesis, the expected value of $\bar{x}_1 - \bar{x}_2$ is zero. Another way of interpreting the zero is that under the null hypothesis $H_0: \mu_1 - \mu_2 = 0$, we have assumed that the mean wages of the populations of unskilled workers are the same in Area A and Area B for the industry in question. Since the significance level is 0.02, 1% of the area under the normal curve is shown in each tail. From Table A-5 of Appendix A we find that in a normal distribution 1% of the area lies to the right of $z = +2.33$ and (by symmetry) 1% to the left of $z = -2.33$. Thus, we would reject the null hypothesis if the sample difference $\bar{x}_1 - \bar{x}_2$ fell more than 2.33 standard errors from the expected value of zero. The estimated standard error of the difference between means, $s_{\bar{x}_1 - \bar{x}_2}$, is, by equation 7.3,

$$s_{\bar{x}_1 - \bar{x}_2} = \sqrt{\frac{(\$4.00)^2}{100} + \frac{(\$4.50)^2}{200}} = \$.51$$

and

$$2.33 s_{\bar{x}_1 - \bar{x}_2} = (2.33)(\$.51) = \$1.19$$

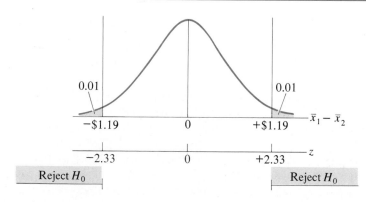

FIGURE 7-12

Sampling distribution of the difference between two means. This is a two-tailed test with $\alpha = 0.02$

Thus, the decision rule may be stated as follows:

DECISION RULE

1. If $\bar{x}_1 - \bar{x}_2 < -\$1.19$ or $\bar{x}_1 - \bar{x}_2 > \$1.19$, reject H_0
2. If $-\$1.19 \leqslant \bar{x}_1 - \bar{x}_2 \leqslant \$1.19$, do not accept H_0

In terms of z values, we have

DECISION RULE

1. If $z < -2.33$ or $z > +2.33$, reject H_0
2. If $-2.33 \leqslant z \leqslant +2.33$, do not accept H_0

---

where

$$z = \frac{\bar{x}_1 - \bar{x}_2}{s_{\bar{x}_1 - \bar{x}_2}}$$

Note that this z value is in the usual form of the ratio [(Statistic − Parameter)/ Standard error]. The difference $\bar{x}_1 - \bar{x}_2$ is the statistic. The parameter under test is $\mu_1 - \mu_2$, which by the null hypothesis is zero and thus need not be shown in the numerator of the ratio. As previously indicated, we have substituted an approximate standard error for the true standard error in the denominator.

Applying this decision rule to the problem, we have

$$\bar{x}_1 - \bar{x}_2 = \$300.01 - \$295.21 = \$4.80$$

and

$$z = \frac{\bar{x}_1 - \bar{x}_2}{s_{\bar{x}_1 - \bar{x}_2}} = \frac{\$4.80}{\$.51} = 9.4$$

Since \$4.80 far exceeds \$1.19 and correspondingly, 9.4 far exceeds 2.33, the null hypothesis is rejected. Since the observed test statistic $z = 9.4$ is larger than any of the tabulated z values in Tables A-5, the corresponding p value is 0.00000. Hence, it is extremely unlikely that these two samples were drawn from populations having the same mean. We conclude that the sample mean wages of unskilled workers in this industry *differed significantly* between Areas A and B and thus that the population means *differ* between Areas A and B. Note that it is incorrect to use the term "significant difference" when referring to the relationship between two population parameters (in this case, the population means). The student should also keep in mind that, in this as in all other hypothesis-testing situations, we are assuming random sampling. Obviously, if the samples were not randomly drawn from the two populations, the foregoing procedure and conclusion would be invalid.

Test for Difference between Proportions: Two-Tailed Test

Another important case of two-sample hypothesis testing is one in which the observed statistics are proportions. The decision procedure is conceptually the same as when the sample statistics are means; only the computational details differ. In order to illustrate the technique, let us consider the following example. Workers in the Stanley Marino Company and Rock Hayden Company, two firms in the same industry, we asked whether they preferred to receive a specified package of increased fringe benefits or a specified increase in base pay. For brevity, we will refer to the companies as the S.M. Company and the R.H. Company and the proposed increases as "increased fringe benefits" and "increased base pay." In a simple random sample of 150 workers in the S.M. Company, 75 indicated that they preferred increased base pay. In the R.H. Company, 103 out of a simple random sample of 200 preferred increased base pay. In each company, the sample was less than 5% of the total number of workers. It was desirable to have a low probability of erroneously rejecting the hypothesis of equal proportions of workers in the two companies who preferred increased base pay. Therefore, a 1% level of significance was used for the test. Can it be concluded at the 1% level of significance that these two companies differed in the proportion of workers who preferred increased base pay?

Using the subscripts 1 and 2 to refer to the S.M. Company and R.H. Company, respectively, we can organize the sample data in a table, where $\bar{p}_1$ and $\bar{q}_1$ refer to the sample proportions in the S.M. Company in favor of and opposed to increased base pay, respectively. The sample size in the S.M. Company is denoted n_1. Corresponding notation is used for the R.H. Company.

If we designate the population proportions in favor of increased pay in the two companies as p_1 and p_2, then in a manner analogous to that of the preceding problem, we set up the two hypotheses

$$H_0: p_1 - p_2 = 0$$

$$H_1: p_1 - p_2 \neq 0$$

	S.M. Company	R.H. Company
	$\bar{p}_1 = \frac{75}{150}$ $= 0.50$	$\bar{p}_2 = \frac{103}{200}$ $= 0.515$
	$\bar{q}_1 = \frac{75}{150}$ $= 0.50$	$\bar{q}_2 = \frac{97}{200}$ $= 0.485$
	$n_1 = 150$	$n_2 = 200$

The underlying theory for the test is similar to that in the two-sample test for the difference between two means. If $\bar{p}_1$ and $\bar{p}_2$ are the observed sample proportions in large simple random samples drawn from populations with parameters p_1 and p_2, then the sampling distribution of the statistic $\bar{p}_1 - \bar{p}_2$ has a mean

(7.4)
$$\mu_{\bar{p}_1 - \bar{p}_2} = p_1 - p_2$$

and a standard deviation

(7.5)
$$\sigma_{\bar{p}_1 - \bar{p}_2} = \sqrt{\sigma_{\bar{p}_1}^2 + \sigma_{\bar{p}_2}^2}$$

where $\sigma_{\bar{p}_1}^2$ and $\sigma_{\bar{p}_2}^2$ are the variances of the sampling distributions of $\bar{p}_1$ and $\bar{p}_2$. Assuming a binomial distribution, $\sigma_{\bar{p}_1}^2 = p_1 q_1 / n_1$ and $\sigma_{\bar{p}_2}^2 = p_2 q_2 / n_2$.

Although the sampling was conducted without replacement, each of the samples constituted only a small percentage of the corresponding population (less than 5%), and the binomial distribution assumption appears reasonable. Thus, equation 7.5 becomes

$$(7.6) \qquad \sigma_{\bar{p}_1 - \bar{p}_2} = \sqrt{\frac{p_1 q_1}{n_1} + \frac{p_2 q_2}{n_2}}$$

If we hypothesize that $p_1 = p_2$ (the null hypothesis in this problem) and refer to the common value of p_1 and p_2 as p, equations 7.4 and 7.6 become

$$(7.7) \qquad \mu_{\bar{p}_1 - \bar{p}_2} = p - p = 0$$

and

$$(7.8) \qquad \sigma_{\bar{p}_1 - \bar{p}_2} = \sqrt{\frac{pq}{n_1} + \frac{pq}{n_2}} = \sqrt{pq\left(\frac{1}{n_1} + \frac{1}{n_2}\right)}$$

Since the common proportion p hypothesized under the null hypothesis is unknown, we estimate it for the hypothesis test by taking a weighted mean of the observed sample percentages. If we refer to this "pooled estimator" as $\hat{p}$, we have

$$(7.9) \qquad \hat{p} = \frac{n_1 \bar{p}_1 + n_2 \bar{p}_2}{n_1 + n_2}$$

The numerator of equation 7.9 is simply the total number of "successes" in the two samples combined, and the denominator is the total number of observations in the two samples. The standard deviation in equation 7.8, $\sigma_{\bar{p}_1 - \bar{p}_2}$, is referred to as the **standard error of the difference between two proportions**. Substituting the "pooled estimator" $\hat{p}$ for p in equation 7.8, we have the following formula for the *estimated* or *approximate* standard error $s_{\bar{p}_1 - \bar{p}_2}$

$$(7.10) \qquad s_{\bar{p}_1 - \bar{p}_2} = \sqrt{\hat{p}\hat{q}\left(\frac{1}{n_1} + \frac{1}{n_2}\right)}$$

Summary   We can now summarize these results. Let $\bar{p}_1$ and $\bar{p}_2$ be proportions of successes observed in two large, independent samples from populations with parameters p_1 and p_2. If we hypothesize that $p_1 = p_2 = p$, we obtain a pooled estimator $\hat{p}$ for p, where $\hat{p} = (n_1 \bar{p}_1 + n_2 \bar{p}_2)/(n_1 + n_2)$.

Then, the sampling distribution of $\bar{p}_1 - \bar{p}_2$ may be approximated by a normal curve with mean $\mu_{\bar{p}_1 - \bar{p}_2} = 0$ and estimated standard deviation

$$s_{\bar{p}_1 - \bar{p}_2} = \sqrt{\hat{p}\hat{q}\left(\frac{1}{n_1} + \frac{1}{n_2}\right)}$$

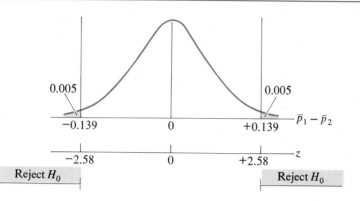

FIGURE 7-13
Sampling distribution of
the difference between two
proportions. This is a two-tailed
test with $\alpha = 0.01$

We may think of this sampling distribution as the frequency distribution of $\bar{p}_1 - \bar{p}_2$ values observed in repeated pairs of samples drawn independently from two populations having the same proportions.

Proceeding to the decision rule, we see again that the test is two-tailed, because the hypothesis of equal population proportions would be rejected for $\bar{p}_1 - \bar{p}_2$ values significantly above or below zero. The sampling distribution of $\bar{p}_1 - \bar{p}_2$ for the present problem is shown in Figure 7-13. Since the significance level is $\alpha = 0.01$, the area under the normal distribution curve shown in each tail is $\alpha/2$ or $\frac{1}{2}(1\%)$. Referring to Table A-5 of Appendix A, we find that 0.005 of the area under a normal curve lies above a z value of $+2.58$, and thus the same percentage lies below $z = -2.58$. Hence, rejection of the null hypothesis $H_0: p_1 - p_2 = 0$ occurs if the sample difference $\bar{p}_1 - \bar{p}_2$ falls more than 2.58 standard error units from zero. By equation 7.10, the standard error of the difference between proportions is

$$s_{\bar{p}_1 - \bar{p}_2} = \sqrt{\hat{p}\hat{q}\left(\frac{1}{n_1} + \frac{1}{n_2}\right)} = \sqrt{(0.51)(0.49)\left(\frac{1}{150} + \frac{1}{200}\right)}$$

$$= 0.054$$

where

$$\hat{p} = \frac{n_1\bar{p}_1 + n_2\bar{p}_2}{n_1 + n_2} = \frac{(150)(0.50) + (200)(0.515)}{150 + 200} = \frac{75 + 103}{150 + 200} = 0.51$$

Hence,

$$2.58s_{\bar{p}_1 - \bar{p}_2} = (2.58)(0.054) = 0.139$$

Therefore, the decision rule is

DECISION RULE

1. If $\bar{p}_1 - \bar{p}_2 < -0.139$ or $\bar{p}_1 - \bar{p}_2 > +0.139$, reject H_0
2. If $-0.139 \leqslant \bar{p}_1 - \bar{p}_2 \leqslant 0.139$, do not reject H_0

In terms of z values, the rule is

DECISION RULE

1. If $z < -2.58$ or $z > +2.58$, reject H_0
2. If $-2.58 \leqslant z \leqslant +2.58$, do not reject H_0

where

$$z = \frac{\bar{p}_1 - \bar{p}_2}{s_{\bar{p}_1 - \bar{p}_2}}$$

Applying this decision rule yields

$$\bar{p}_1 - \bar{p}_2 = 0.500 - 0.515 = -0.015$$

and

$$z = \frac{\bar{p}_1 - \bar{p}_2}{s_{\bar{p}_1 - \bar{p}_2}} = \frac{-0.015}{0.054} = -0.28$$

Summary   The corresponding p value is from the normal curve Table A-5 $P(z < |0.28|) = 1 - 2(0.1103) = 0.7794$. Thus, the null hypothesis cannot be rejected. In summary, the sample proportions $\bar{p}_1$ and $\bar{p}_2$ did not differ significantly, and therefore we cannot conclude that the two companies differed with respect to the proportion of workers who preferred increased base pay. This conclusion may be useful, for example, in testing a claim of a participant in labor negotiations that such a difference exists. Our reasoning is based on the finding that if the population proportions were equal, a difference between the sample proportions as large as the one observed could not at all be considered unusual.

Test for Differences between Proportions: One-Tailed Test

The two preceding examples illustrated two-tailed tests for cases in which data are available for samples from two populations. Just as in the one-sample case, the question we wish to answer may give rise to a one-tailed test. In order to illustrate this point, let us examine the following problem.

Two competing drugs are available for treating a certain physical ailment. There are no apparent side effects from administration of the first drug, whereas there are some definite side effects (nausea and mild headaches) from use of the second. A group of medical researchers has decided that it would nevertheless be willing to recommend use of the second drug in preference to the first if the proportion of cures effected by the second were higher than those by the first drug. The group felt that the potential benefits of achieving increased cures of the ailment would far outweigh the disadvantages of the possible side effects. On the other hand, if the proportion of cures effected by the second drug was equal to or less than that of the first drug, the group would recommend use of the first one. In terms of hypothesis testing, we can state the alternatives and consequent actions as

$$H_0: p_2 \leqslant p_1 \text{ (use the first drug)}$$

$$H_1: p_2 > p_1 \text{ (use the second drug)}$$

where p_1 and p_2 denote the population proportions of cures effected by the first and second drugs. Another way we may write these alternatives is

$$H_0: p_2 - p_1 \leqslant 0 \text{ (use the first drug)}$$

$$H_1: p_2 - p_1 > 0 \text{ (use the second drug)}$$

For purposes of comparison, note that the hypotheses in the preceding problem, a two-tailed testing situation, were

$$H_0: p_1 = p_2$$

$$H_1: p_1 \neq p_2$$

or in the alternative form (in terms of differences)

$$H_0: p_1 - p_2 = 0$$

$$H_1: p_1 - p_2 \neq 0$$

Clearly, the present problem involves a one-tailed test, in which we would reject the null hypothesis only if the sample difference $\bar{p}_2 - \bar{p}_1$ differed significantly from zero and was a positive number.

The medical researchers used the drugs experimentally on two random samples of persons suffering from the ailment, administering the first drug to a group of 80 patients and the second drug to a group of 90 patients. By the end of the experimental period, 52 of those treated with the first drug were classified as "cured," whereas 63 of those treated with the second drug were so classified. The sample results may be summarized in a table, with p denoting "proportion cured" and q denoting "proportion not cured."

First Drug	Second Drug
$\bar{p}_1 = \dfrac{52}{80} = 0.65$ cured	$\bar{p}_2 = \dfrac{63}{90} = 0.70$ cured
$\bar{q}_1 = \dfrac{28}{80} = 0.35$ not cured	$\bar{q}_2 = \dfrac{27}{90} = 0.30$ not cured
$n_1 = 80$	$n_2 = 90$

The pooled sample proportion cured is

$$\hat{p} = \frac{52 + 63}{80 + 90} = \frac{115}{170} = 0.676$$

and the estimated standard error of the difference between proportions is

$$s_{\bar{p}_2 - \bar{p}_1} = \sqrt{(0.676)(0.324)\left(\frac{1}{80} + \frac{1}{90}\right)} = 0.0719$$

Since the medical group wished to maintain a low probability of erroneously adopting the second drug, it selected a 1% significance level for the test. This means that 1% of the area under the normal curve lies to the right of $z = +2.33$. Therefore, the null hypothesis would be rejected if $\bar{p}_2 - \bar{p}_1$ falls at least 2.33 standard error units above zero. In terms of proportions,

$$2.33 s_{\bar{p}_2 - \bar{p}_1} = 2.33(0.0719) = 0.168$$

Hence, the decision rule is

DECISION RULE

1. If $\bar{p}_2 - \bar{p}_1 > 0.168$, reject H_0
2. If $\bar{p}_2 - \bar{p}_1 \leqslant 0.168$, do not reject H_0

In terms of z values, the rule is

DECISION RULE

1. If $z > +2.33$, reject H_0
2. If $z \leqslant +2.33$, do not reject H_0

---

where

$$z = \frac{\bar{p}_2 - \bar{p}_1}{s_{\bar{p}_2 - \bar{p}_1}}$$

In the present problem,

$$\bar{p}_2 - \bar{p}_1 = 0.70 - 0.65 = 0.05$$

so

$$z = \frac{0.70 - 0.65}{0.0719} = 0.70$$

The corresponding p value is from the normal curve Table A-5

$$p = P(z > 0.70) = 0.5000 - 0.2580 = 0.2420.$$

Thus, the null hypothesis is not rejected. On the basis of the sample data, we cannot conclude that the second drug accomplishes a greater proportion of cures than the first. The sampling distribution of $\bar{p}_2 - \bar{p}_1$ is given in Figure 7-14. Note that it is immaterial whether we state the difference between propor-

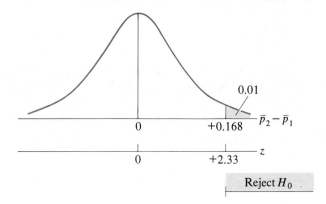

FIGURE 7-14
Sampling distribution of
the difference between two
proportions. This is a one-tailed
test with $\alpha = 0.01$

0.01

$\bar{p}_2 - \bar{p}_1$

0   +0.168

0   +2.33 z

Reject H_0

tions as $\bar{p}_1 - \bar{p}_2$ or $\bar{p}_2 - \bar{p}_1$, but we must exercise care concerning the correspondence between the way the hypothesis is stated, the sign of the difference between the sample proportions, and the tail of the sampling distribution in which rejection of the null hypothesis takes place.

EXERCISES 7.3

1. Match the correct test statistics with the following four null hypotheses:
   a. $H_0: \mu = \mu_0$ (1) $\bar{p}$
   b. $H_0: p = p_0$ (2) $\bar{x}_1 - \bar{x}_2$
   c. $H_0: \mu_1 = \mu_2$ (3) $\bar{x}_1 + \bar{x}_2$
   d. $H_0: p_1 = p_2$ (4) $\bar{p}_1 - \bar{p}_2$
                               (5) $\bar{x}$
                               (6) $\bar{p}/\bar{x}$

2. A medical research team decided to study the relationship between nutrition and height for adult males. As part of its study, the team drew two simple random samples, one from a prosperous country and the other from an underdeveloped nation. The following information on heights of adult males was obtained:

Country 1	Country 2
$\bar{x}_1 = 69.1$ inches	$\bar{x}_2 = 68.3$ inches
$s_1 = 3$ inches	$s_2 = 2$ inches
$n_1 = 2{,}000$	$n_2 = 1{,}500$

Country 1 is the prosperous country and country 2 is the underdeveloped one. Formulate an appropriate null hypothesis and then test it at the 5%

significance level. You may assume that there was no presumption concerning whether the relationship between quality of nutrition and height was positive or negative.

3. The Stillwater Brewing Company felt that two of its market areas exhibited equivalent sales patterns. The company wished to experiment with a reduced advertising budget in one of the areas, using the other as a control. In preparing the experiment it decided to test the hypothesis that sales patterns were identical prior to any advertising change. The following random sample data were obtained for daily sales in the two areas measured in appropriate units.

Area	Mean Daily Sales	Standard Deviation	Sample Size
1	2,000	150	150
2	1,960	130	200

a. Would you conclude that areas 1 and 2 have equivalent mean daily sales rate? Justify your answer using a 2% significance level. Calculate the p value, i.e., the attained significance level. In answering part (a), you had to locate a sample statistic (the arithmetic mean) on a random sampling distribution. Draw a rough sketch of this distribution, showing the following:

b. the value and location of the hypothetical parameter.

c. the value and location of the statistic.

d. the horizontal scale description.

e. the portion of the distribution corresponding to the probability $\alpha = 0.02$.

f. the p value denoting the attained significance level.

4. A computer tabulation census of the balance due on credit customers' accounts in two department stores had the following results:

Department Store	Mean Balance
Zimbels	$38
Bombergers	56

Can you conclude statistically that the average balance due on accounts at Bombergers is greater than that due at Zimbels? Would you test a hypothesis? If so, what hypothesis? If not, why not?

5. A firm was debating whether to continue its two-week training program for new salespeople. Some members of the firm felt that salespeople do not benefit from the training program and that it is a waste of time and money. The firm decided to test the program by randomly assigning the new salespeople to two groups, trained and untrained, and then measuring

their performance after a six-month period. A random sample of 60 sales-people from each group produced the following results:

Group	Mean Performance Rating (10 = highest)	Standard Deviation
1. Trained	7.5	1.5
2. Untrained	6.6	0.8

Would you conclude that the trained and untrained salespeople have equivalent performance ratings? Justify your answer using a 5% significance level.

6. A manufacturer of cookies and crackers has just developed two new types of cookies and wishes to test consumer reactions to both types. Two simple random groups of 100 persons each were chosen, and each group was asked to try one type of cookie. Each person was then asked whether he or she would purchase the cookie tasted. The following results were obtained:

Cookie Type	Number Who Would Purchase the Cookie
A	72
B	84

At a 0.01 significance level, would you conclude that the two groups differed with respect to the proportion who would purchase the type of cookie tasted?

7. Suppose that in a simple random sample of 400 patients, 325 responded favorably to medication A, and in a similar sample of 200 patients, 150 responded favorably to medication B. A random method had been used to determine which of the two medications was assigned to each patient. At the 0.05 significance level, is there reason to doubt the hypothesis that equal proportions of patients responded favorably to the two types of medication?

8. In a simple random sample of 225 foreign exchange traders, 135 expressed the opinion that a flexible exchange rate system would enhance the stability of the international monetary system. In a simple random sample of 200 professors of economics, 154 expressed the same opinion.
   a. Do you believe that there is a real difference in the attitude of the two groups with regard to the effect of flexible exchange rates on the stability of the international monetary system? Justify your answer statistically and indicate which significance level you used.
   b. Explain specifically the meaning of a Type I error in this particular problem.

9. Workers in two different industries were asked what they considered the most important labor-management problem in their industry. In industry A, 200 out of a random sample of 400 workers felt that a fair adjustment of grievances was the most important problem. In Industry B, 60 out of a random sample of 100 workers felt that this was the most important problem.

    a. Would you conclude that these two industries differed with respect to the proportion of workers who believed that a fair adjustment of grievances was the most important problem? Support your answer statistically, and give a brief statement of your reasoning. Use a 1% significance level.

    b. Make a rough sketch (labeling the horizontal scale) of the random sampling distribution on which your answer to (a) depends. Show on the sketch the approximate positions of the values pertinent to the solution.

10. Suppose that in a simple random sample of 400 people from one city, 188 preferred a particular brand of tea to all others, and in a similar sample of 500 people from another city, 210 preferred the same product. At the 0.05 significance level, is there reason to doubt the hypothesis that equal proportions of persons in the two cities prefer this brand of tea?

7.4
THE t DISTRIBUTION: SMALL SAMPLES WITH UNKNOWN POPULATION STANDARD DEVIATION(S)

The hypothesis-testing methods discussed in the preceding sections are appropriate for large samples. In this section, we concern ourselves with the case when the sample size is small. The underlying theory is exactly the same as that given in section 6.3, in which confidence interval estimation for small samples was discussed.

> In hypothesis testing, as in confidence interval estimation, the distinction between large and small sample tests becomes important when the population standard deviation is *unknown* and therefore must be *estimated* from the sample observations.

The main principles will be reviewed here. The statistic $(\bar{x} - \mu)/s_{\bar{x}}$, where $s_{\bar{x}}$ denotes an estimated standard error, is not approximately normally distributed for all sample sizes. As we have noted earlier, $s_{\bar{x}}$ is computed by the formula $s_{\bar{x}} = s/\sqrt{n}$ where s represents an estimate of the true population standard deviation. For large samples, the ratio $(\bar{x} - \mu)/s_{\bar{x}}$ is approximately normally distributed, and we may use the methods discussed in sections 7.2 and 7.3. However, since this statistic is not approximately normally distributed for small samples, the t distribution should be used instead. The use of the t distribution for testing a hypothesis concerning a population mean is demonstrated in Example 7-3. A small sample ($n \leqslant 30$) is assumed, and the population standard deviation is unknown.

EXAMPLE 7-3

One-sample Test of a Hypothesis about the Mean: Two-tailed Test
    The personnel department of a company developed an aptitude test for a certain type of semiskilled worker. The individual test scores were assumed to be normally distributed. The developers of the test asserted a tentative hypothesis that the arithmetic mean grade obtained by this type of semiskilled worker would be 100. It was agreed that this hypothesis would be subjected to a two-tailed test at the 5% level of significance. The aptitude test was given to a simple random sample of 16 semiskilled workers with the following results:

$$\bar{x} = 94$$

$$s = 5$$

$$n = 16$$

The competing hypotheses are

$$H_0: \mu = 100$$

$$H_1: \mu \neq 100$$

To carry out the test, the following quantities were calculated:

SOLUTION

$$s_{\bar{x}} = \frac{s}{\sqrt{n}} = \frac{5}{\sqrt{16}} = 1.25$$

and

$$t = \frac{\bar{x} - \mu}{s_{\bar{x}}} = \frac{94 - 100}{1.25} = -4.80$$

    The significance of this t value is judged from Table A-6 of Appendix A. The meaning of the table was discussed in section 6.4. We will explain the use of the table for this hypothesis-testing problem. Let us set up the rejection region for the hypothesis. Since the sample size is 16, the number of degrees of freedom is $v = 16 - 1 = 15$. Looking along the row of Table A-6 for 15 under the column 0.05, we find the t value, 2.131. This means that in a t distribution for $v = 15$, the probability is 5% that t is greater than 2.131 or is less than -2.131. Thus, in the present problem, at the 5% level of significance, the null hypothesis $H_0: \mu = 100$ is rejected if a t value exceeding 2.131 or less than -2.131 is observed. Since the computed t value in this problem is -4.80, we reject the null hypothesis. In other words, we are unwilling to attribute the difference between our sample mean of 94 and the hypothesized population mean of 100 merely to chance errors of sampling. The t distribution for this problem is shown in Figure 7-15.
    A few remarks can be made about this problem. Since the computed t value of -4.80 was so much less than zero, the null hypothesis would have been rejected even at the 2% or 1% level of significance (see Table A-6 of Appendix A). Had the test been

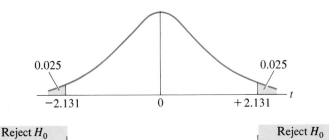

FIGURE 7-15
The *t* distribution for
$v = 15$

one-tailed at the 5% level of significance, we would have had to obtain the critical *t* value by looking under 0.10 in Table A-6, since the 0.10 figure is the combined area in both tails. Thus, for a one-tailed test at the 5% level of significance and a rejection region in the lower tail, the critical *t* value would have been -1.753.

It is interesting to compare these critical *t* values with analogous critical *z* values for the normal curve. From Table A-5 of Appendix A, we find that the critical *z* values at the 5% level of significance are -1.96 and $+1.96$ for a two-tailed test and -1.65 for a one-tailed test with a rejection region in the lower tail. As we have just seen, the corresponding figures for the critical *t* values in a test involving 15 degrees of freedom are -2.131 and $+2.131$ for a two-tailed test and -1.753 for a one-tailed test.

An underlying assumption in applying the *t* test is that the population is closely approximated by a normal distribution. Since the population standard deviation σ is unknown, the *t* distribution is the theoretically correct sampling distribution. However, if the sample size had been large, even with an unknown population standard deviation, the normal curve could have been used as an approximation to the *t* distribution. As we saw in this problem, for $v = 15$, a total of 5% of the area in the *t* distribution falls to the right of $t = +2.131$ and to the left of $t = -2.131$. The corresponding *z* values in the normal distribution are $+1.96$ and -1.96. From Table A-6, we see that the *t* value for $v = 30$ is 2.042. The closeness of this figure to $+1.96$ gives rise to the usual rule of thumb, which uses $n > 30$ as the arbitrary dividing line between large-sample and small-sample methods. We have used this convenient rule in Chapters 6 and 7. However, what constitutes a suitable approximation really depends on the context of the particular problem. Furthermore, if the population is highly skewed, a sample size as large as 100 may be required for assumption of a normal sampling distribution of $\bar{x}$ to be appropriate.

EXAMPLE 7-4

Two-sample Test for Means: Two-tailed Test

A marketing research effort has been launched to determine how different names for a particular new diet food product could possibly affect sales. The food products in the market research were identical, but they were given two different names, Slimall and Supersate.

In an initial study, 10 randomly selected potential consumers were asked to taste the product named Slimall and to rate it on a scale of 1 (very poor) to 10 (outstanding). Similarly, another 10 potential consumers were asked to taste the product named Supersate and to rate it on a scale of 1 to 10. The statistics were calculated:

Slimall	Supersate
$\bar{x}_1 = 8.6$	$\bar{x}_2 = 6.8$
$s_1 = 1.1$	$s_2 = 1.3$
$n_1 = 10$	$n_2 = 10$

Do the names produce different ratings? That is, should the difference observed between the two sample means be considered significant? Use $\alpha = 0.01$.

The alternative hypotheses are

SOLUTION

$$H_0: \mu_1 - \mu_2 = 0$$

$$H_1: \mu_1 - \mu_2 \neq 0$$

To test the null hypothesis, we have the t statistic

$$t = \frac{(\bar{x}_1 - \bar{x}_2) - 0}{s_{\bar{x}_1 - \bar{x}_2}} = \frac{\bar{x}_1 - \bar{x}_2}{s_{\bar{x}_1 - \bar{x}_2}}$$

where $s_{\bar{x}_1 - \bar{x}_2}$ is the estimated standard error of the difference between the two means.

Unlike the case with large samples, here we must assume equal population variances. An estimate of this common variance is obtained by pooling the two sample variances into a weighted average, using the numbers of degrees of freedom, $n_1 - 1$ and $n_2 - 1$, as weights. This pooled estimate of the common variance, which we denote as $\hat{s}^2$, is given by

(7.11)
$$\hat{s}^2 = \frac{(n_1 - 1)s_1^2 + (n_2 - 1)s_2^2}{n_1 + n_2 - 2}$$

The estimated standard error of the difference between two means is then

(7.12)
$$s_{\bar{x}_1 - \bar{x}_2} = \sqrt{\frac{\hat{s}^2}{n_1} + \frac{\hat{s}^2}{n_2}} = \hat{s}\sqrt{\frac{1}{n_1} + \frac{1}{n_2}}$$

A number of alternative mathematical expressions are possible for equation 7.12, but because of its similarity in appearance to previously used standard error formulas, we shall use it in this form.

We now work out the present problem. Substitution into equation 7.11 gives

$$\hat{s}^2 = \frac{(10 - 1)(1.1)^2 + (10 - 1)(1.3)^2}{10 + 10 - 2} = 1.45$$

and

$$\hat{s} = \sqrt{1.45} = 1.204$$

Thus, the estimated standard error is

$$s_{\bar{x}_1 - \bar{x}_2} = (1.204)\sqrt{\frac{1}{10} + \frac{1}{10}} = 0.538$$

and the t value is

$$t = \frac{8.6 - 6.8}{0.538} = 3.35$$

The number of degrees of freedom in this problem is $n_1 + n_2 - 2$, that is, $10 + 10 - 2 = 18$. We can explain the number of degrees of freedom in this case as follows. In the one-sample case, when the sample standard deviation is used as an estimate of the population standard deviation, there is a loss of one degree of freedom; hence, the number of degrees of freedom is $n - 1$. In the two-sample case, each of the sample variances is used in the pooled estimate of the population variance; hence, two degrees of freedom are lost, and the number of degrees of freedom is $n_1 + n_2 - 2$.

The critical t value at the 1% significance level for 18 degrees of freedom is 2.878 (see Table A-6 of Appendix A). Since the observed t value, 3.35, exceeds this critical t value, the null hypothesis is rejected, and we conclude on the basis of the sample data that the population means are indeed different. In terms of the problem, we are unwilling to attribute the difference in the Slimall and Supersate average taste ratings to chance errors of sampling.

EXERCISES 7.4

1. A simple random sample of 10 students at a university yielded a mean grade point average of 3.00 (on a 4.0 scale) and a standard deviation of 0.35.
   a. Are these data consistent with a claim that the mean grade point average at this university is 2.80? Use a two-tailed test at a 0.05 level of significance.
   b. Instead of the above sample size, assume a sample of size 100 and the same sample standard deviation. Would you reach the same conclusion as in part (a)?
   c. Comment on the results of the tests you applied in parts (a) and (b).

2. In a study of the amount spent on breakfast by a simple random sample of 14 individuals at The Morning Call restaurant, the arithmetic mean was $1.22 per person with a standard deviation of $.25. In a similar study, 10 individuals at The Lucky Bee restaurant spent an arithmetic mean of $1.48 with a standard deviation of $.10.
   a. At the 10% level of significance, would you conclude that a statistically significant difference exists in the sample averages of amounts spent at the restaurants?
   b. For this problem, state the null and alternative hypotheses and the decision rule employed.

3. The management of a motel chain is considering building a motel along a state highway. The owner of the property for the proposed motel claims that more than 1,200 cars pass this site each day. The management of the chain feels that any number less than 1,200 would be inadequate to ensure the success of the motel. They decided to take a random sample of 10 days to test the claim of the property owner. The following data resulted:

Day	Number of Cars
1	1,150
2	1,225
3	1,190
4	1,180
5	1,195
6	1,210
7	1,200
8	1,185
9	1,160
10	1,205

square & sum

Formulate an appropriate test using $\alpha = 0.01$.

4. Two random samples of 15 cars each were used in a study of two brands of gasoline. The cars in each sample were given the same number of gallons of one of the two brands, and the average miles per gallon for each car was noted as follows:

	Brand A		Brand B
Car	Average Miles per Gallon	Car	Average Miles per Gallon
1	25	1	15
2	32	2	18
3	27	3	17
4	24	4	24
5	26	5	22
6	18	6	19
7	20	7	26
8	23	8	24
9	20	9	18
10	16	10	17
11	19	11	23
12	15	12	22
13	20	13	21
14	27	14	19
15	18	15	15

At a 0.05 significance level, would you conclude that there is a significant difference in the sample means?

5. In a study of use of bar soap by a simple random sample of 15 suburban families, the consumption of such soap was found to have an arithmetic mean of 50 ounces per family per month with a standard deviation of 10 ounces. In another similar study of 11 urban families, consumption was found to average 55 ounces with a standard deviation of 12 ounces.

   a. At the 10% level of significance, would you conclude that there was a statistically significant difference in the sample averages of consumption of bar soap?

   b. For this problem, state the null and alternative hypotheses and the decision rule employed.

6. A company seeking to sell a large bond issue is trying to determine which investment banking firm should head the syndicate formed to handle the issue. The financial vice-president of the firm has instructed an assistant to take a small sample of past syndicates headed by Marriot Leech and Company, a prominent New York firm, to determine the average percentage of the issues it has sold at the initial offering price. Using a random sample of 15 issues, the assistant finds that on the average (unweighted mean), Marriot Leech has successfully sold 93% of each issue during the initial offering, with a standard deviation of 5%. Based on a random sample of 13, the assistant finds that the Third Boston Corporation has averaged 91%, with a standard deviation of 3%. Assuming a 2% significance level, should the assistant conclude that the sample average success rates of the two investment banks differ significantly? State the alternative hypotheses and the decision rule.

7. Management wants to determine whether production will be increased by instituting a bonus plan. A total of 20 plants are randomly divided into two groups of 10. A mean of 22 units per day and a standard deviation of 4 units per day are calculated for the 10 plants that are offered the bonus plan. Similarly, a mean of 18 units per day and a standard deviation of 5 units per day are calculated for the 10 plants that are not offered the bonus plan. Do these data confirm the claim that the mean daily production rate with the bonus plan exceeds the mean daily production rate without the bonus plan? (Use $\alpha = 0.01$.)

7.5
THE DESIGN OF A TEST TO CONTROL TYPE I AND TYPE II ERRORS*

Relatively simple hypothesis-testing situations have been considered in this chapter in order to convey the basic principles of classical hypothesis testing. These tests have assumed that sample size was predetermined and that only the risks of Type I errors were to be controlled formally by the decision procedure.

* This section is optional. It may be omitted without interfering with the continuity of the discussion.

Of course, we assumed that a power curve would be constructed in all cases, and therefore, we could make an examination of the risks of Type II errors for parameter values not included in the null hypothesis. However, nothing was included in the formal testing procedure to control the level of risk of a Type II error for any specific parameter value.

In this section, we consider a method of controlling the levels of both Type I and Type II errors simultaneously in the same test. In the previous tests, which controlled only Type I errors—that is, where only α was specified— one point on the power curve was determined. In a test designed to control both Type I and Type II errors, two specific points on the power curve are determined.

As an illustration, we consider a one-tailed test problem similar to the one discussed in section 7.2. This problem involves an acceptance sampling proce- dure in which $\alpha = 0.05$ for an incoming shipment whose parts have a mean heat resistance of 2250°F. We call this the "producer's risk"; that is, the producer runs a 5% risk of having a "good" shipment with $\mu = 2250°F$ rejected in error. Because H_0 is $\mu \geqslant 2250°F$, the maximum risk of rejecting a good shipment is 0.05. The probability of erroneous rejection drops below 5% for shipments with means in excess of 2250°F. Suppose now it is agreed to fix the "consumer's risk" or β at 0.03 for a shipment whose mean was 2150°F. That is, a shipment whose parts have a mean heat resistance of 2150°F does not meet specifications, and we want the probability of erroneously accepting such a shipment to be 0.03. We make the assumption that the population standard deviation is $\sigma = 300°F$. The solution to this problem involves the determination of the *sample size* re- quired to give the desired levels of control of both types of errors. After the required sample size has been calculated, the appropriate decision rule can be specified. We assume the sample size will be large enough to use normal sampling distributions for $\bar{x}$.

Figure 7-16 gives a graphic representation of the error controls specified in the preceding paragraph. In a standard solution to this problem involving control of only Type I errors, the critical value for rejection or retention of the null hypothesis $H_0: \mu \geqslant 2250°F$ is 2200°F. The critical value is now unknown

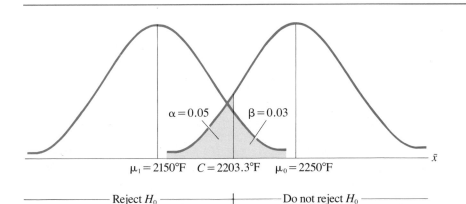

FIGURE 7-16
Acceptance sampling
problem: Control of Type I and
Type II errors

$\alpha = 0.05$ $\beta = 0.03$

$\mu_1 = 2150°F$ $C = 2203.3°F$ $\mu_0 = 2250°F$

———— Reject H_0 ————————————— Do not reject H_0 ————

and will have to be evaluated. Let us denote the new critical value as C, the mean of 2250°F under the null hypothesis as μ_0, and the mean of 2150°F under the alternative hypothesis as μ_1. The area in the left tail of the sampling distribution of $\bar{x}$ when $\mu_0 = 2250°F$ is shown as 0.05, denoting the Type I error. As is readily determined, the critical point C lies 1.65 standard error units to the left of the mean $\mu_0 = 2250°F$. Therefore, $z_0 = 1.65$. Under the alternative $\mu_1 = 2150°F$, we want the probability that an $\bar{x}$ value lies in the retention region to be 0.03. From Table A-5, we ascertain that 0.03 of the area in a normal sampling distribution lies to the right of a value 1.88 standard error units above the mean. Hence, $z_1 = 1.88$. Therefore, we can write the following relationships for the critical point C:

(7.13)
$$C = \mu_0 - z_0 \sigma_{\bar{x}} = \mu_0 - z_0 \frac{\sigma}{\sqrt{n}}$$

(7.14)
$$C = \mu_1 + z_1 \sigma_{\bar{x}} = \mu_1 + z_1 \frac{\sigma}{\sqrt{n}}$$

Substituting the numerical values for this problem, we obtain

$$C = 2250°F - 1.65 \frac{(300°F)}{\sqrt{n}}$$

$$C = 2150°F + 1.88 \frac{(300°F)}{\sqrt{n}}$$

Setting the right-hand sides of these two equations equal to one another yields the solution $n = 112$ (to the nearest integer). Therefore, a simple random sample of 112 parts from the incoming shipment would be required in order to obtain the desired levels of error control.

We can express this required sample size by solving the simultaneous equations 7.13 and 7.14. The solution is

(7.15)
$$n = \left[\frac{(z_0 + z_1)\sigma}{(\mu_0 - \mu_1)} \right]^2$$

Or course, substitution of the numerical values into this formula again yields $n = 112$. Note that both z_0 and z_1 are taken as positive, since the matter of whether C lies above or below μ_0 and μ_1 is taken care of by the signs in equations 7.13 and 7.14.

Now the decision rule can be stated. The critical value C can be obtained by substituting into either of the two simultaneous equations. Substituting into the first equation yields[3]

$$C = 2250°F - 1.65 \frac{(300°F)}{\sqrt{112}} = 2203.3°F$$

---

[3] Actually the values $C = 2203.2°F$ and 2203.4°F are obtained from the first and second equations, respectively. The discrepancy is due to rounding off n to an integral value.

where

$$s_d = \sqrt{\frac{\Sigma(d - \bar{d})^2}{n - 1}}$$

In this problem, $s_{\bar{d}} = 0.82$ pounds, as shown by the calculations in Table 7-3. Assuming that the population of differences (d values) is normally distributed, the ratio $(\bar{d} - 0)/s_{\bar{d}}$ is distributed according to the t distribution.

$$t = \frac{\bar{d} - 0}{s_{\bar{d}}} = \frac{-4}{0.82} = -4.88$$

The number of degrees of freedom is $n - 1$, where n is the number of d values. Hence, in this problem, $n - 1 = 10 - 1 = 9$. The test is one-tailed, because only a $\bar{d}$ value that is negative and significantly different from zero could result in acceptance of the alternative hypothesis, H_1: $\mu_2 - \mu_1 < 0$. Since $\alpha = 0.05$ and the test is one-tailed, we look in Table A-6 under the heading 0.10. The critical t value for $v = 9$ is 1.833, which for our purposes is interpreted as -1.833. Since the observed t value of -4.88 is less than (lies to the left of) this critical point, the null hypothesis is rejected and we accept its alternative. Therefore, on the basis of this experiment, we conclude that the special diet does result in an average weight loss over a two-week period.

Minitab output is given in Table 7-4 for displaying the paired observations on weights of the sample of 10 men, a graph of the weights, in which "A" is the weight before the diet and "B" is the weight after the diet. A "2" is shown for the second observation, in which the same weight of 172 appeared, both before and after the diet. By using the DESCRIBE command, we obtain the standard error $s_{\bar{d}}$, denoted SEMEAN in Minitab. The Minitab figure is 0.816, which has been rounded in the text presentation to 0.82. Since the mean difference of -4 pounds is also shown, we have the information needed for the t test.

This method of pairing observations is also used to reduce the effect of extraneous factors that could cause a significant difference in means, whereas the factor whose effect we are really interested in may not have resulted in such a difference if the extraneous factors had not been present.

For example, if medical experimenters wanted to test two different treatments to judge which was better, they might administer one treatment to one group of persons and the other treatment to a second group. Suppose on the basis of the usual significance test for means, it is concluded that one treatment is better than the other. Let us also assume that the group receiving the supposedly better treatment was much younger and much healthier at the beginning of the experiment than the other group and that these factors could have an effect on the reaction to the treatments. Then clearly, the relative effectiveness of the two treatments would be obscured.

On the other hand, assume that individuals were selected in pairs in which both members were about the same age and in about the same health condition. If the first treatment is given to one member of a pair and the second treatment

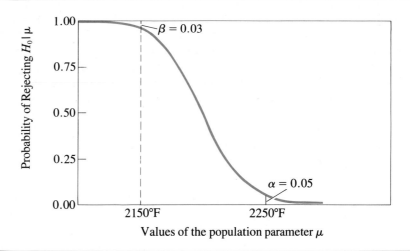

FIGURE 7-17
Power curve to control both
Type I and Type II errors

Hence, the required decision rule is

> **DECISION RULE**
>
> 1. If $\bar{x} < 2203.3°F$, reject H_0 (reject the shipment)
> 2. If $\bar{x} \geq 2203.3°F$, do not reject H_0 (accept the shipment)

The power curve for this test is displayed in Figure 7-17.

This one-tailed test for a mean illustrates the general method for controlling the levels of both Type I and Type II errors. Analogous tests can be constructed for proportions and for two-tailed tests as well.

In the type of quality control problem described here, the producer and the consumer might negotiate to determine the levels of Type I and Type II risks that they would agree to tolerate. These tolerable risks in turn would primarily depend on the costs involved to the producer and consumer of these two types of errors.

EXERCISES 7.5

1. A particular drug is supposed to have a mean period of effectiveness of at least 50 hours. Assume that the standard deviation of the period of effectiveness was known to be 10 hours. In an acceptance sampling procedure, a significance level of $\alpha = 0.01$ was adopted for an incoming shipment in which the mean period of effectiveness was 50 hours. Suppose it was agreed

to fix the "consumer's risk" or β at 0.05 for a shipment whose mean period of effectiveness was only 47 hours of effectiveness. Determine the sample size required to give the desired levels of control for both types of errors, and state the decision rule in terms of the sample mean period of effectiveness.

2. The owner of a small business has determined that in the past the mean amount of sales per sales invoice has been $30 and that the standard deviation was $9. She wishes to run a test to find out whether the mean has decreased from $30, using a Type I error probability of 0.05. In this test, she wants to maintain a Type II error probability of 0.01 of reaching an incorrect conclusion if the mean sales per sales invoice has actually dropped to $28. Determine the number of invoices to be sampled and the decision rule in terms of the sample mean.

3. Management wants to determine whether production will be increased by instituting a bonus plan. A mean production rate of 50 units per day and a standard deviation of 8 units per day had been established before the proposed bonus plan. A sample test was agreed upon in which a random sample of workers would be given a bonus plan to test the null hypothesis $H_0: \mu \leqslant 50$ versus the alternate hypothesis $H_1: \mu > 50$ using an $\alpha = 0.03$. It was desired to maintain a Type II error probability of 0.02 if the mean production rate had risen to 54 units per day under the bonus plan. What size sample should be used in this situation? Specify the decision rule to be followed in terms of the sample mean production rate.

7.6

THE t TEST FOR PAIRED OBSERVATIONS

In the two-sample tests considered so far, the two samples had to be independent. That is, the values of observations in one sample had to be *independent* of the values in the other. Situations arise in practice in which this condition does not hold. In fact, the two samples may consist of pairs of observations made on the same individual, the same object, or more generally, the same selected population elements. Clearly, the independence condition is violated in these cases.

As a concrete example, let us consider a case in which a group of 10 men was given a special diet, and it was desired to test weight loss in pounds at the end of a two-week period. The observed data are shown in Table 7-3.

As indicated in Table 7-3, X_1 denotes the weight before the diet and X_2 the weight after the diet. It would be incorrect to run a t test to determine whether there is a significant difference between the mean of the X_1 values and the mean of the X_2 values, because of the nonindependence of the two samples. An individual's weight after the test is certainly not independent of his weight before the test. Each of the d values, $d = X_2 - X_1$, represents a difference between two observations on the same individual. The assumption is made that the subtraction of one value from the other removes the effect of factors other

TABLE 7-3

Weights before and after a special diet for a simple random sample of 10 men

Man	Weight before Diet X_1	Weight after Diet X_2	Difference in Weight $d = X_2 - X_1$	$d - \bar{d}$	$(d - \bar{d})^2$
1	184	181	-3	$+1$	1
2	172	172	0	$+4$	16
3	190	185	-5	-1	1
4	187	184	-3	$+1$	1
5	210	201	-9	-5	25
6	202	201	-1	$+3$	9
7	166	160	-6	-2	4
8	173	168	-5	-1	1
9	183	180	-3	$+1$	1
10	184	179	-5	-1	1
			$\overline{-40}$		$\overline{60}$

$$\bar{d} = \frac{-40}{10} = -4 \text{ pounds}$$

$$s_d = \sqrt{\frac{\Sigma(d - \bar{d})^2}{n - 1}} = \sqrt{\frac{60}{10 - 1}} = 2.58 \text{ pounds}$$

$$s_{\bar{d}} = \frac{s_d}{\sqrt{n}} = \frac{2.58}{\sqrt{10}} = 0.82 \text{ pounds}$$

than that of the diet; that is, we assume that these other factors affect each member of any pair of X_1 and X_2 values in the same way.

We can state the hypotheses to be tested as

$$H_0: \mu_2 - \mu_1 = 0$$

$$H_1: \mu_2 - \mu_1 < 0$$

where μ_1 and μ_2 are the population mean weights before and after the diet, respectively. Let us assume that the test is to be carried out with $\alpha = 0.05$. The null hypothesis states that there is no difference between mean weight after the diet and mean weight before the diet, whereas the alternative hypothesis says that mean weight after the diet is less than mean weight before. We can visualize the situation as one in which, if the null hypothesis is true, a population of numbers represents differences in the weight after the diet and before the diet, and the mean of these numbers is zero. Hence, we wish to test the hypothesis that our simple random sample of $d = X_2 - X_1$ values comes from this universe. The procedure used in these cases is to obtain the mean of the sample differences and to test whether this average $\bar{d}$ differs significantly from zero. The estimated standard error of $\bar{d}$, denoted $s_{\bar{d}}$, is given by

$$s_{\bar{d}} = \frac{s_d}{\sqrt{n}}$$

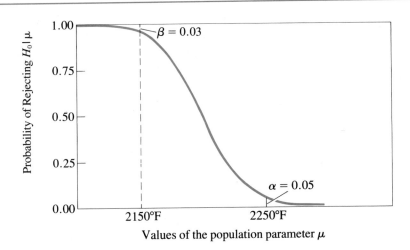

FIGURE 7-17

Power curve to control both
Type I and Type II errors

Values of the population parameter μ

Hence, the required decision rule is

DECISION RULE

1. If $\bar{x} < 2203.3°F$, reject H_0 (reject the shipment)
2. If $\bar{x} \geqslant 2203.3°F$, do not reject H_0 (accept the shipment)

The power curve for this test is displayed in Figure 7-17.

This one-tailed test for a mean illustrates the general method for controlling the levels of both Type I and Type II errors. Analogous tests can be constructed for proportions and for two-tailed tests as well.

In the type of quality control problem described here, the producer and the consumer might negotiate to determine the levels of Type I and Type II risks that they would agree to tolerate. These tolerable risks in turn would primarily depend on the costs involved to the producer and consumer of these two types of errors.

EXERCISES 7.5

1. A particular drug is supposed to have a mean period of effectiveness of at least 50 hours. Assume that the standard deviation of the period of effectiveness was known to be 10 hours. In an acceptance sampling procedure, a significance level of $\alpha = 0.01$ was adopted for an incoming shipment in which the mean period of effectiveness was 50 hours. Suppose it was agreed

to fix the "consumer's risk" or β at 0.05 for a shipment whose mean period of effectiveness was only 47 hours of effectiveness. Determine the sample size required to give the desired levels of control for both types of errors, and state the decision rule in terms of the sample mean period of effectiveness.

2. The owner of a small business has determined that in the past the mean amount of sales per sales invoice has been $30 and that the standard deviation was $9. She wishes to run a test to find out whether the mean has decreased from $30, using a Type I error probability of 0.05. In this test, she wants to maintain a Type II error probability of 0.01 of reaching an incorrect conclusion if the mean sales per sales invoice has actually dropped to $28. Determine the number of invoices to be sampled and the decision rule in terms of the sample mean.

3. Management wants to determine whether production will be increased by instituting a bonus plan. A mean production rate of 50 units per day and a standard deviation of 8 units per day had been established before the proposed bonus plan. A sample test was agreed upon in which a random sample of workers would be given a bonus plan to test the null hypothesis $H_0: \mu \leqslant 50$ versus the alternate hypothesis $H_1: \mu > 50$ using an $\alpha = 0.03$. It was desired to maintain a Type II error probability of 0.02 if the mean production rate had risen to 54 units per day under the bonus plan. What size sample should be used in this situation? Specify the decision rule to be followed in terms of the sample mean production rate.

7.6
THE t TEST FOR PAIRED OBSERVATIONS

In the two-sample tests considered so far, the two samples had to be independent. That is, the values of observations in one sample had to be *independent* of the values in the other. Situations arise in practice in which this condition does not hold. In fact, the two samples may consist of pairs of observations made on the same individual, the same object, or more generally, the same selected population elements. Clearly, the independence condition is violated in these cases.

As a concrete example, let us consider a case in which a group of 10 men was given a special diet, and it was desired to test weight loss in pounds at the end of a two-week period. The observed data are shown in Table 7-3.

As indicated in Table 7-3, X_1 denotes the weight before the diet and X_2 the weight after the diet. It would be incorrect to run a t test to determine whether there is a significant difference between the mean of the X_1 values and the mean of the X_2 values, because of the nonindependence of the two samples. An individual's weight after the test is certainly not independent of his weight before the test. Each of the d values, $d = X_2 - X_1$, represents a difference between two observations on the same individual. The assumption is made that the subtraction of one value from the other removes the effect of factors other

TABLE 7-3
Weights before and after a special diet for a simple random sample of 10 men

Man	Weight before Diet X_1	Weight after Diet X_2	Difference in Weight $d = X_2 - X_1$	$d - \bar{d}$	$(d - \bar{d})^2$
1	184	181	-3	$+1$	1
2	172	172	0	$+4$	16
3	190	185	-5	-1	1
4	187	184	-3	$+1$	1
5	210	201	-9	-5	25
6	202	201	-1	$+3$	9
7	166	160	-6	-2	4
8	173	168	-5	-1	1
9	183	180	-3	$+1$	1
10	184	179	-5	-1	1
			$\overline{-40}$		$\overline{60}$

$$\bar{d} = \frac{-40}{10} = -4 \text{ pounds}$$

$$s_d = \sqrt{\frac{\Sigma(d - \bar{d})^2}{n - 1}} = \sqrt{\frac{60}{10 - 1}} = 2.58 \text{ pounds}$$

$$s_{\bar{d}} = \frac{s_d}{\sqrt{n}} = \frac{2.58}{\sqrt{10}} = 0.82 \text{ pounds}$$

than that of the diet; that is, we assume that these other factors affect each member of any pair of X_1 and X_2 values in the same way.

We can state the hypotheses to be tested as

$$H_0: \mu_2 - \mu_1 = 0$$

$$H_1: \mu_2 - \mu_1 < 0$$

where μ_1 and μ_2 are the population mean weights before and after the diet, respectively. Let us assume that the test is to be carried out with $\alpha = 0.05$. The null hypothesis states that there is no difference between mean weight after the diet and mean weight before the diet, whereas the alternative hypothesis says that mean weight after the diet is less than mean weight before. We can visualize the situation as one in which, if the null hypothesis is true, a population of numbers represents differences in the weight after the diet and before the diet, and the mean of these numbers is zero. Hence, we wish to test the hypothesis that our simple random sample of $d = X_2 - X_1$ values comes from this universe. The procedure used in these cases is to obtain the mean of the sample differences and to test whether this average $\bar{d}$ differs significantly from zero. The estimated standard error of $\bar{d}$, denoted $s_{\bar{d}}$, is given by

$$s_{\bar{d}} = \frac{s_d}{\sqrt{n}}$$

where

$$s_d = \sqrt{\frac{\Sigma(d - \bar{d})^2}{n - 1}}$$

In this problem, $s_{\bar{d}} = 0.82$ pounds, as shown by the calculations in Table 7-3. Assuming that the population of differences (d values) is normally distributed, the ratio $(\bar{d} - 0)/s_{\bar{d}}$ is distributed according to the t distribution.

$$t = \frac{\bar{d} - 0}{s_{\bar{d}}} = \frac{-4}{0.82} = -4.88$$

The number of degrees of freedom is $n - 1$, where n is the number of d values. Hence, in this problem, $n - 1 = 10 - 1 = 9$. The test is one-tailed, because only a $\bar{d}$ value that is negative and significantly different from zero could result in acceptance of the alternative hypothesis, $H_1: \mu_2 - \mu_1 < 0$. Since $\alpha = 0.05$ and the test is one-tailed, we look in Table A-6 under the heading 0.10. The critical t value for $\nu = 9$ is 1.833, which for our purposes is interpreted as -1.833. Since the observed t value of -4.88 is less than (lies to the left of) this critical point, the null hypothesis is rejected and we accept its alternative. Therefore, on the basis of this experiment, we conclude that the special diet does result in an average weight loss over a two-week period.

Minitab output is given in Table 7-4 for displaying the paired observations on weights of the sample of 10 men, a graph of the weights, in which "A" is the weight before the diet and "B" is the weight after the diet. A "2" is shown for the second observation, in which the same weight of 172 appeared, both before and after the diet. By using the DESCRIBE command, we obtain the standard error $s_{\bar{d}}$, denoted SEMEAN in Minitab. The Minitab figure is 0.816, which has been rounded in the text presentation to 0.82. Since the mean difference of -4 pounds is also shown, we have the information needed for the t test.

This method of pairing observations is also used to reduce the effect of extraneous factors that could cause a significant difference in means, whereas the factor whose effect we are really interested in may not have resulted in such a difference if the extraneous factors had not been present.

For example, if medical experimenters wanted to test two different treatments to judge which was better, they might administer one treatment to one group of persons and the other treatment to a second group. Suppose on the basis of the usual significance test for means, it is concluded that one treatment is better than the other. Let us also assume that the group receiving the supposedly better treatment was much younger and much healthier at the beginning of the experiment than the other group and that these factors could have an effect on the reaction to the treatments. Then clearly, the relative effectiveness of the two treatments would be obscured.

On the other hand, assume that individuals were selected in pairs in which both members were about the same age and in about the same health condition. If the first treatment is given to one member of a pair and the second treatment

```
MTB > NAME C1 'MAN' C2 'WGT-1' C3 'WGT-2' C4 'B-A'
MTB > PRINT C1-C4
```

ROW	MAN	WGT-1	WGT-2	B-A
1	1	184	181	-3
2	2	172	172	0
3	3	190	185	-5
4	4	187	184	-3
5	5	210	201	-9
6	6	202	201	-1
7	7	166	160	-6
8	8	173	168	-5
9	9	183	180	-3
10	10	184	179	-5

```
MTB > MPLOT 'WGT-1' 'MAN' 'WGT-2' 'MAN'
```

```
        -
        -
        -                              A
        -                         B       A
  200 +                                   B
        -
        -                  A
        -               B     A
        -    A                B                        A       A
  180 +    B                                         B       B
        -                                       A
        -         2                             B
        -                                  A
        -                                  B
  160 +
        -
        -
   ----- + ---------- + ---------- + ---------- + ---------- + ---------- + ----
       0.0         2.0         4.0         6.0         8.0        10.0
```

A = WGT-1 VS. MAN   B = WGT-2 VS. MAN

```
MTB > DESCRIBE 'B-A'
```

	N	MEAN	MEDIAN	TRMEAN	STDEV	SEMEAN
B-A	10	-4.000	-4.000	-3.875	2.582	0.816

	MIN	MAX	Q1	Q3
B-A	-9.000	0.000	-5.250	-2.500

to the other, and then a difference measure is calculated for the effect of treatment, neither age nor health condition would affect this measurement. Ideally, we would like to select pairs that are identical in all characteristics other than the factor whose effects we are attempting to measure. Obviously, as a practical matter this is impossible, but the guiding principle is clear. Once differences are taken between members of each pair, the t test proceeds exactly as in the preceding example. Note that in the weight example, the differences were measured on the same individual, whereas in the present illustration the differences are derived from the two members of each pair.

The method of paired observations is a useful technique. Compared with the standard two-sample t test, in addition to the advantage that we do not have to assume that the two samples are independent, we also need not assume that the variances of the two samples are equal.

EXERCISES 7.6

1. A research group ran an experiment to determine if a certain cold tablet increases the user's reaction time to various stimuli. The research group felt that if the tablet was found to have an effect on the user's reaction time, it would be necessary to include a warning on the package against operating a vehicle while taking the medication. The research team selected a random sample of 12 students at a university for the experiment. Each student's reaction time to a stimulus was measured before and after taking the medication. The results of the experiment were as follows:

Student	Reaction Time without Medication (in seconds)	Reaction Time with Medication (in seconds)
1	0.75	0.84
2	0.82	0.78
3	1.04	1.15
4	0.77	0.81
5	0.92	0.95
6	1.11	1.08
7	0.69	0.82
8	0.84	0.96
9	0.91	0.95
10	0.98	0.83
11	0.83	0.91
12	0.75	0.81

Using a paired t test, would you conclude that the medication does increase the user's reaction time? Use $\alpha = 0.05$.

2. Blair Good's advertising claims that individuals who complete the six-week "Blair Good Reading Comprehension Course" will increase their under-

standing of written materials while maintaining the same reading rate. A simple random sample of 13 persons enrolled in the course was taken. The reading comprehension rate for each individual was measured both before the course and on completion of it, with the following results (1 = lowest, 9 = highest):

Individual	Reading Comprehension Rating before Course	Reading Comprehension Rating after Course
1	6.2	7.8
2	5.4	5.9
3	7.0	7.6
4	6.6	6.6
5	6.9	7.6
6	7.2	7.7
7	5.5	6.0
8	7.1	7.0
9	7.9	7.8
10	5.9	6.4
11	8.4	8.7
12	6.5	6.5
13	5.6	5.8

Using the paired t test, would you conclude that taking the "Blair Good Reading Comprehension Course" is effective in increasing reading comprehension ratings? Using a 5% significance level.

3. Eleven middle managers from the Mammoth Manufacturing Company attended a three-week session of group meetings and workshops to increase their managerial effectiveness and sensitivity. The firm's top management evaluates its middle managers using the "grid" methodology. Each manager is given a rating consisting of two numbers between 1 (worst) and 9 (best): The first number indicates the manager's concern for production, and the second reflects the manager's concern for people. The following table shows the ratings given to each manager before and after attending the session:

Manager	Rating before Session	Rating after Session
1	5,9	7,9
2	6,4	9,4
3	3,6	4,7
4	3,4	9,8
5	8,6	9,8
6	5,6	4,7
7	3,4	4,5
8	1,6	5,7
9	3,8	4,8
10	6,7	3,7
11	7,3	8,4

Assuming consistency in top management's ratings, would you conclude that the session had a positive effect on the managers' (a) concern for production? (b) concern for people? Use a 2.5% significance level.

7.7
USE OF MINITAB FOR HYPOTHESIS TESTING

The various hypothesis-testing procedures discussed in this chapter are easily carried out with the use of the Minitab statistical package. We illustrate the application of Minitab using the data bank for computer exercises in Appendix D that contains economic and demographic data for 100 families. These data are in a file named 'databank . mtw' on the computer disk that accompanies this text. The Minitab printout of selected descriptive statistics for the 100 families is given in Table 7-5.

Our first example is a one-sample test of a hypothesis for the population mean μ when the population standard deviation σ is known. In this example, a simple random sample of 35 families has been drawn and we wish to test the following two-tailed hypothesis concerning the population mean income: $H_0: \mu = \$44,000$ vs. $H_1: \mu \neq \$44,000$. The population standard deviation is assumed to be known and equal to $\sigma = \$19,000$.

TABLE 7-5
Selected descriptive statistics for the 100 families included in the data bank for computer exercises

```
MTB > INFORMATION

COLUMN NAME     COUNT
C1      FOOD      100
C2      INCOME    100
C3      SIZE      100
C4      AGE       100
C5      OWNER     100

CONSTANTS USED: NONE

MTB > DESCRIBE C1 C2 C3
```

	N	MEAN	MEDIAN	TRMEAN	STDEV	SEMEAN
FOOD	100	7.364	6.300	7.106	3.651	0.365
INCOME	100	41.72	41.00	40.62	18.98	1.90
SIZE	100	2.750	3.000	2.656	1.493	0.149

	MIN	MAX	Q1	Q3
FOOD	2.400	20.000	4.800	9.675
INCOME	12.00	112.00	28.00	52.75
SIZE	1.000	7.000	1.000	4.000

TABLE 7-6

Minitab output for a one-sample test for a mean μ when σ is known; two-tailed test

```
MTB > SAMPLE 35 OBS FROM C2, PUT INTO C12
MTB > PRINT C12

C12
    47  57  51  28  36  17  31  46  72  51  28  42  26
    47  27  59  32  75  48  12  31  19  30  25  78  28
    27  54  45  30  54  30  51  15  50

MTB > ZTEST 44 19 C12

TEST OF MU = 44.000 VS MU N.E. 44.000
THE ASSUMED SIGMA = 19.0
```

	N	MEAN	STDEV	SEMEAN	Z	P VALUE
C12	35	39.971	16.748	3.212	−1.25	0.21

One-sample Test for a Mean μ; σ is Known; Two-tailed Test

As can be seen in the Minitab printout in Table 7-6 we draw a sample of 35 observations of family income (data are in thousands of dollars). We specify a z test because σ is known where $z = \dfrac{\bar{x} - \mu}{\sigma/\sqrt{n}}$. Using Minitab's ZTEST command, we specify the population mean (44), population standard deviation (19) and the column containing the data (C12). If we wish to run a one-tailed test, we must use a subcommand as in the next illustration. The abbreviation N.E. in a two-tailed test stands for "not equal to."

The value of $\bar{x} = 39.971$ differs from 44 by -4.029. It has a z value of

$$z = \frac{\bar{x} - \mu}{\sigma/\sqrt{n}} = \frac{39.971 - 44}{19/\sqrt{35}} = -1.25$$

The p value of 0.21 indicates that we cannot reject the null hypothesis at conventional levels of significance such as 0.05 or 0.01.

If we wish to use the one-tailed test $H_0: \mu = 44$ against $H_1: \mu < 44$, in which the rejection region is in the lower tail, we use the subcommand *ALTERNATIVE − 1*. For the one-tailed test $H_0: \mu = 44$ against $H_1: \mu > 44$, in which the rejection region is in the upper tail, we use *ALTERNATIVE 1*.

One-sample Test for a Mean μ; σ is Known; One-tailed Test

Table 7-7 gives the Minitab output for the one-tailed test $H_0: \mu = 44$ against $H_1: \mu < 44$.

As indicated by the p value, the null hypothesis cannot be rejected at conventional levels such as $\alpha = 0.05$.

TABLE 7-7

Minitab output for a one-sample test for a mean μ when σ is known; one-tailed test

```
MTB > ZTEST 44 19 C12;
SUBC> ALTERNATIVE -1.

TEST OF MU = 44.000 VS MU L.T. 44.000
THE ASSUMED SIGMA = 19.0
```

	N	MEAN	STDEV	SEMEAN	Z	P VALUE
C12	35	39.971	16.748	3.212	-1.25	0.10

One-sample t tests are run in a similar manner to z tests. However, in the t tests, as we have seen, the t statistic is $t = \dfrac{\bar{x} - \mu}{s/\sqrt{n}}$. That is, the sample standard deviation s is used rather than the population standard deviation σ, which is usually unknown.

One-sample Test for a Mean μ; Two-tailed and One-tailed Test; Large Sample Comparison of ZTEST and TTEST

Table 7-8 shows the Minitab output for a two-tailed test of the hypothesis $H_0: \mu = 38$ versus $H_1: \mu \neq 38$. The sample size is large, in this case, 100 observations. Table 7-9 shows the Minitab output for the same data for a one-tailed test of the hypothesis $H_0: \mu = 38$ versus $H_1: \mu > 38$. The results are very similar for the z test and t test as indicated by the closeness of the p values in both the

TABLE 7-8

Minitab output for a one-sample test for a mean μ; two-tailed tests by the ZTEST and TTEST; sample size is 100

```
MTB > ZTEST 38 18.983 C2

TEST OF MU = 38.000 VS MU N.E. 38.000
THE ASSUMED SIGMA = 19.0
```

	N	MEAN	STDEV	SEMEAN	Z	P VALUE
INCOME	100	41.720	18.983	1.898	1.96	0.050

```
MTB > TTEST 38 C2

TEST OF MU = 38.000 VS MU N.E. 38.000
```

	N	MEAN	STDEV	SEMEAN	T	P VALUE
INCOME	100	41.720	18.983	1.898	1.96	0.053

TABLE 7-9

Minitab output for a one-sample test for a mean μ; one-tailed tests by the ZTEST and TTEST;
sample size is 100

```
MTB > ZTEST 38 18.983 C2;
SUBC> ALTERNATIVE 1.

TEST OF MU = 38.000 VS MU G.T. 38.000
THE ASSUMED SIGMA = 19.0

                 N        MEAN       STDEV      SEMEAN          Z      P VALUE
INCOME         100      41.720      18.983       1.898       1.96        0.025

MTB > TTEST 38 C2;
SUBC> ALTERNATIVE 1.

TEST OF MU = 38.000 VS MU G.T. 38.000

                 N        MEAN       STDEV      SEMEAN          T      P VALUE
INCOME         100      41.720      18.983       1.898       1.96        0.026
```

two-tailed and one-tailed tests. This is an empirical confirmation of the point
that the t distribution (used in the t test) approaches the normal distribution
(used in the z test) as sample size increases.

Two-sample Tests for Means; Two-tailed Test; Large Samples

To illustrate how Minitab carries out two sample tests for means, we continue
to use the data bank for computer exercises given in Appendix D. We would
like to construct a test for whether the population mean food expenditures of
homeowners and renters differ. Food expenditures are in column 1 and homeowner status is in column 5 (code: homeowner (0); renter (1)). If two sets of
data are both in the same column, with another column used to identify each
set, the data are said to be "stacked." Since we would like to have the food
expenditures for homeowners and renters in separate columns we "unstack"
the food expenditures data by the following Minitab command and subcommand:

MTB > unstack C1, put into C11 and C21;
SUBC > subscripts in C5.

We can get a count of the numbers of homeowners and renters in columns 11
and 21, respectively by the command:

MTB > info C11 C21

COLUMN	NAME	COUNT
C11		49
C21		51

TABLE 7-10

Minitab output for a two-sample *t*-test for means; two-tailed test

---

```
MTB > TWOSAMPLE C11 C21;
SUBC> POOLED.

TWOSAMPLE T FOR C11 VS C21
                 N           MEAN            STDEV          SE MEAN
C11             49           9.56            3.47             0.50
C21             51           5.25            2.36             0.33

95 PCT CI FOR MU C11 - MU C21: (3.13, 5.48)

TTEST MU C11 = MU C21 (VS NE): T=7.29 P=0.0000 DF=98

POOLED STDEV = 2.95
```

We now carry out a two-tailed t test for the hypothesis $H_0 : \mu_1 = \mu_2$ versus the alternative $H_1 : \mu_1 \neq \mu_2$ where μ_1 and μ_2 are the mean amounts of food expenditures for homeowners and renters, respectively. We use the subcommand "pooled" to pool the two sample variances into a weighted average as was done in section 7.4, Example 7-4. Table 7-10 gives the Minitab output for the test.

TABLE 7-11

Minitab output for one sample tests for proportions; two-tailed test and one-tailed test

---

```
MTB > DESCRIBE C5

              N        MEAN       MEDIAN      TRMEAN       STDEV       SEMEAN
OWNER        100      0.5100      1.0000      0.5111      0.5024      0.0502

             MIN        MAX          Q1          Q3
OWNER      0.0000      1.0000      0.0000      1.0000

MTB > ZTEST 0.5 0.5 C5

TEST OF MU = 0.5000 VS MU N.E. 0.5000
THE ASSUMED SIGMA = 0.500

              N        MEAN       STDEV       SEMEAN          Z       P VALUE
OWNER        100      0.5100      0.5024      0.0500        0.20        0.84

MTB > ZTEST 0.5 0.5 C5;
SUBC> ALTERNATIVE 1.

TEST OF MU = 0.5000 VS MU G.T. 0.5000
THE ASSUMED SIGMA = 0.500

              N        MEAN       STDEV       SEMEAN          Z       P VALUE
OWNER        100      0.5100      0.5024      0.0500        0.20        0.42
```

The difference between the average amounts spent on food by the samples of homeowners and renters are highly significant as indicated by p value of 0.0000. Minitab also gives a 95% confidence interval for the difference between the two population means. One-tailed tests are carried out by placing semicolon after "pooled" and typing ALTERNATIVE -1 or 1 after the subcommand prompt.

Testing for Proportions

There is no separate test procedure for proportions in Minitab. However, tests can be carried out using tests for mean, if we consider that a proportion is the mean of 0's and 1's for the pertinent classification. For example, in column 5, home ownership, there are 49 homeowners (coded 0) and 51 renters (coded 1). If we take the mean of the 100 0's and 1's, we get $51/100 = 0.51$, interpreted as 51% renters.

Table 7-11 shows a two-tailed test of the hypothesis $H_0 : p = 0.50$ versus $H_1: p \neq 0.50$, followed by a one-tailed test of the hypothesis $H_0: p = 0.50$ versus $H_1: p > 0.50$. In both cases, as indicated by the p values, we cannot reject the null hypotheses at any reasonable significance levels.

7.8
SUMMARY AND A LOOK AHEAD

In this chapter, we have considered some classical hypothesis-testing techniques. These tests represent only a few of the simplest methods. All the cases discussed thus far have involved only one or two samples, but methods are available for testing hypotheses concerning three or more samples. The cases we have dealt with also have tested only one parameter of a probability distribution. However, techniques are available for testing whether an entire frequency distribution is in conformity with a theoretical model, such as a specified probability distribution. Finally, the tests we have considered involved a final decision on the basis of the sample evidence; that is, a decision concerning acceptance or rejection of hypotheses was reached on the basis of the evidence contained in one or two samples. Some of the broader decision procedures are discussed in subsequent chapters.

Although classical hypothesis-testing techniques of the type discussed in this chapter have been widely applied in a great many fields, it would be incorrect to infer that their use is noncontroversial and that they can simply be employed in a mechanical way. At this point, it suffices to indicate that the methods discussed are admittedly incomplete and that Bayesian decision theory addresses itself to the required completion. Thus, for example, in hypothesis testing, establishing significance levels such as 0.05 to 0.01 inevitably appears to be a rather arbitrary procedure, despite the fact that the relative seriousness of Type I and Type II errors is supposed to be considered in designing a test.

Although costs of Type I and Type II errors can theoretically be considered in the classical formulation, as a matter of practice, they are rarely included explicitly in the analysis. In Bayesian decision theory, the costs of Type I and Type II errors, as well as the payoffs of correct decisions, are an explicit part of the formal analysis.

Moreover, in classical hypothesis testing, decisions are reached solely on the basis of current sample information without reference to prior knowledge concerning the hypothesis under test. On the other hand, Bayesian decision theory provides a method for combining prior knowledge with current sample information for decision-making purposes. These Bayesian decision-theory methods are discussed in subsequent chapters.

The exercises of this chapter are based on the data in the file named 'data-bank . mtw'. A description of the data is as follows:

Column	Name	Count	Description
C1	FOOD	100	Annual Food Expenditures in the Surveyed Families ($000)
C2	INCOME	100	Annual Incomes in the Surveyed Families ($000)
C3	SIZE	100	Family Sizes in the Surveyed Families
C4	AGE	100	Ages of the Highest Income Earners in the Surveyed Families (in years)
C5	OWNER	100	Homeowners (0) or Renters (1)

1. Test the hypothesis that the mean of FOOD is 7 ($000) against the hypothesis that it is not, at a 0.15 significance level. Explain why the Minitab commands

   MTB > ttest mean = 7 for C1

   and

   MTB > let k1 = stdev(C1)
   MTB > ztest mean = 7 with sigma = k1 for C1

   give us the same p value.

2. Use the computer outputs in Exercise 1 to obtain the p value for the test of the hypothesis that the mean of FOOD is 7 ($000) against the hypothesis that the mean of FOOD is greater than 7 ($000). Verify your answer with command:

   MTB > ttest mean = 7 for C1;
   SUBC > alternative 1.

3. Past records indicate that the standard deviation of INCOME is 20 ($000). Use that information to test the hypothesis that the mean of INCOME is greater than or equal to 45 ($000) against the hypothesis that the mean of INCOME is less than 45 ($000), at a 0.10 significance level.

4. Minitab commands

   MTB > boxplot C2;
   SUBC > by C5.

   draw two boxplots on the same graph. From the boxplots, is there a clear indication that the average income of families with their own homes is greater than that of families without their own homes? Formulate a hypothesis-testing problem and actually test the hypothesis, using alpha = 0.05. *Hint:* try

MTB > twot C2, groups in C5;
SUBC > alternative — 1;
SUBC > pooled.

5. Carry out a two-tailed test of the null hypothesis that the proportion of families with their own homes is 0.5, using alpha = 0.05. Note that you may treat C5 as a large sample.

KEY TERMS

Acceptance Region   values of the sample statistic for which the null hypothesis is retained.

Alternative Hypothesis (H_1)   the alternative to the null hypothesis (H_0). Rejection of the null hypothesis constitutes acceptance of the alternative hypothesis.

Hypothesis-Testing Procedure   a decision rule that specifies the values of a sample statistic (for example, $\bar{x}$) for which the null hypothesis will be retained or rejected.

Level of Significance (α)   in one-tailed tests, the maximum probability of a Type I error. In two-tailed tests, the probability of a Type I error.

Null Hypothesis (H_0)   the basic hypothesis that is tested for possible rejection.

One-Sample Tests   tests of hypotheses based on data contained in a single sample.

One-Sample Test (one-sided alternative)   a test in which the null hypothesis is rejected by observing a statistic that falls in one tail of the appropriate sampling distribution.

Pooled Estimator ($\hat{p}$)   a weighted mean of the two observed sample proportions in a two-sample test of the difference between proportions.

Pooled Estimator ($\hat{s}^2$)   a weighted mean of the two observed sample variances in a two-sample test of the difference between means (t-test for small samples).

Power Curve   a function that gives the probabilities of rejecting the null hypothesis for all possible values of the parameter tested.

Power of a Test   the probability of rejecting the null hypothesis for a particular value of the alternative hypothesis.

p Value   given that H_0 is true, this is the probability that a statistic will differ from the parameter being tested by a greater degree than is observed.

Rejection (critical) Region   values of the sample statistic for which the null hypothesis is rejected.

Sample Space   the set of possible values of the sample statistic.

t Test for Paired Observations   a test in which pairs of observations are made on the same sample elements, such as the same individual or object.

Two-Tailed Test (two-sided alternative)   a test in which the null hypothesis is rejected by observing a statistic that falls in either of the two tails of the appropriate sampling distribution.

Type I Error   Rejection of the null hypothesis when it is true.

Type II Error   Acceptance of the null hypothesis when it is false.

KEY FORMULAS

z Values for Testing Hypotheses

$z = (\bar{x} - \mu)/\sigma_{\bar{x}}$; about a mean

$z = (\bar{p} - p)/\sigma_{\bar{p}}$; about a proportion

$z = (\bar{x}_1 - \bar{x}_2)/\sigma_{\bar{x}_1 - \bar{x}_2}$; about a difference between means

$z = (\bar{p}_1 - \bar{p}_2)/\sigma_{\bar{p}_1 - \bar{p}_2}$; about a difference between proportions

z Values in Hypothesis Testing

$$z = \frac{\text{Statistic} - \text{Parameter}}{\text{Standard error}}$$

Estimated Standard Error of the Difference Between Two Proportions

$$s_{\bar{p}_1 - \bar{p}_2} = \sqrt{\hat{p}\hat{q}\left(\frac{1}{n_1} + \frac{1}{n_2}\right)}$$

where

$$\hat{p} = \frac{n_1\bar{p}_1 + n_2\bar{p}_2}{n_1 + n_2}$$

Standard Error of the Difference Between Two Means

$$\sigma_{\bar{x}_1 - \bar{x}_2} = \sqrt{\frac{\sigma_1^2}{n_1} + \frac{\sigma_2^2}{n_2}}$$

$$s_{\bar{x}_1 - \bar{x}_2} = \sqrt{\frac{s_1^2}{n_1} + \frac{s_2^2}{n_2}}; \text{estimated standard error}$$

t-Test (one sample)

$$t = \frac{\bar{x} - \mu}{s_{\bar{x}}}$$

t-Test (two samples)

$$t = \frac{\bar{x}_1 - \bar{x}_2}{s_{\bar{x}_1 - \bar{x}_2}}$$

where

$$s_{\bar{x}_1 - \bar{x}_2} = \sqrt{\frac{\hat{s}^2}{n_1} + \frac{\hat{s}^2}{n_2}}$$

t-Test for Paired Observations

$$t = \frac{\bar{d}}{s_{\bar{d}}}$$

where

$$s_{\bar{d}} = \frac{\sqrt{\dfrac{\Sigma(d - \bar{d})^2}{n - 1}}}{\sqrt{n}}$$

8

CHI-SQUARE TESTS
AND ANALYSIS OF VARIANCE

In Chapter 7, procedures were discussed for testing hypotheses with data obtained from a single simple random sample or from two such samples. For example, we considered tests of whether two population proportions or two population means were equal. Obvious generalizations of such techniques are tests for the equality of more than two proportions or more than two means. The two topics discussed in this chapter supply these generalizations. **Chi-square tests**[1] provide the basis for judging whether more than two population proportions may be considered to be equal; **analysis of variance** techniques provide ways to test whether more than two population means may be considered to be equal.

We discuss χ^2 tests first, considering the topics of (1) goodness of fit and (2) independence. **Tests of goodness of fit** provide a means for deciding whether a particular theoretical probability distribution—such as the binomial distribution—is a close enough approximation to a sample frequency distribution for the population from which the sample was drawn to be described by the theoretical distribution. **Tests of independence** constitute a method for deciding whether the hypothesis of independence between different variables is tenable. This procedure provides a test for the equality of more than two population proportions. Both types of χ^2 tests furnish a conclusion on whether a set of

Tests of goodness of fit

Tests of Independence

---

[1] The procedures are referred to as "χ^2 tests" or "chi-square tests," where the symbol χ is the Greek lower-case letter "chi" (pronounced "kye").

observed frequencies differs so greatly from a set of *theoretical* frequencies that the hypothesis under which the theoretical frequencies were derived should be rejected.

8.1
TESTS OF GOODNESS OF FIT

One of the major problems in the application of probability theory, statistics, and mathematical models in general is that the real-world phenomena to which they are applied usually depart somewhat from the assumptions embodied in the theory or models. For example, let us consider use of the binomial probability distribution in a particular problem. As indicated in section 3.4, two of the assumptions involved in the derivation of the binomial distribution are

Assumptions   1. The probability of a success p remains constant from trial to trial.

2. The trials are independent.

We consider whether these assumptions are met in the following problem.

A firm bills its accounts at a 2% discount for payment within 10 days and for the full amount due for payment after 10 days. In the past, 40% of all invoices have been paid within 10 days. In a particular week, the firm sends out 20 invoices. Is the binomial distribution appropriate for computing the probabilities that 0, 1, 2, ..., 20 firms will receive the discount for payment within 10 days?

Considering the possible use of the binomial distribution, we can let $p = 0.40$ represent the probability that a firm will receive the discount and $n = 20$ firms represent the number of trials. Does it seem reasonable to assume that p, the probability of receiving the discount, is 0.40 for each firm? Past relative frequency data for *each* firm could be brought to bear on this question. In most practical situations, we would probably find that the practices of individual firms vary widely, with some firms nearly always receiving discounts, some firms rarely receiving discounts, and most firms falling somewhere between these two extremes.

Does the assumption of independence seem tenable in this problem? That is, does it seem reasonable that whether one firm receives the discount is independent of whether another firm does? Probably not, since general monetary conditions affect many firms in a similar way. For example, when money is "tight" and it is difficult to acquire adequate amounts of working capital, the fact that one firm does not receive the discount is related to, rather than *independent* of, whether other firms have done so. Moreover, there may be traditional practices in certain industries concerning whether or not discounts are taken. Other factors that would interfere with the independence assumption may also be present.

How great a departure from the assumptions underlying a probability distribution—or, more generally, from the assumptions embodied in any theory or mathematical model—can be tolerated before we should conclude that the distribution, theory, or model is no longer applicable? This complex question cannot be readily answered by any simple universally applicable rule.

> The purpose of χ^2 "goodness of fit" tests is to provide one type of answer to this question by comparing *observed frequencies* with *theoretical*, or *expected, frequencies* derived under specified probability distributions or hypotheses.

The sequence of steps in performing goodness of fit tests is similar to previously discussed hypothesis-testing procedures.

Goodness of fit tests

1. Null and alternative hypotheses are established, and a significance level is selected for rejection of the null hypothesis.

2. A random sample of observations is drawn from a relevant statistical population or process.

3. A set of expected, or theoretical, frequencies is derived under the assumption that the null hypothesis is true. This is generally an assumption that a particular probability distribution is applicable to the statistical population under consideration.

4. The observed frequencies are compared with the expected, or theoretical, frequencies.

5. If the aggregate discrepancy between the observed and theoretical frequencies is too great to attribute to chance fluctuations at the selected significance level, the null hypothesis is rejected.

We now illustrate goodness of fit tests and discuss some of the underlying theory for an example involving a uniform probability distribution (see section 3.3).

> Suppose a consumer research firm wished to determine whether any of 5 brands of coffee was preferred by coffee drinkers in a certain metropolitan area. The firm took a simple random sample of 1,000 coffee drinkers in the area and conducted the following experiment. Each consumer was given 5 cups of coffee, one of each brand (A, B, C, D, and E), without identification of the individual brands. The cups were presented to each consumer in a random order determined by sequential selection from 5 paper slips, each containing one of the letters A, B, C, D, and E. Table 8-1 shows the numbers of coffee drinkers who stated that they liked the indicated brands best.

Coffee-tasting problem

TABLE 8-1

Number of coffee drinkers in a certain metropolitan area who most preferred the specified brand of coffee

Brand Preference	Number of Consumers
A	210
B	312
C	170
D	85
E	223
	1,000

Denoting the true proportions of preference for each brand as p_A, p_B, p_C, p_D, and p_E, we can state the null and alternative hypotheses as follows:

$$H_0\text{: } p_A = p_B = p_C = p_D = p_E = 0.20$$

$$H_1\text{: The } p \text{ values are not all equal}$$

That is, if in the population from which the sample was drawn there were no differences in preference among the 5 brands, 20% of coffee drinkers would prefer each brand. An equivalent way of stating these hypotheses is

$$H_0\text{: The probability distribution is uniform}$$

$$H_1\text{: The probability distribution is not uniform}$$

In other words, we want to know whether the sample of 1,000 coffee drinkers can be considered a random sample from a population in which the proportions who prefer each of the 5 brands are equal. Of course, this hypothesis is only one of many that could conceivably be formulated.

> One of the strengths of the goodness of fit test is that it permits a variety of different hypotheses to be raised and tested.

If the null hypothesis of no difference in preference were true, the *expected* or *theoretical* number of the 1,000 coffee drinkers in the sample who would prefer each brand would be $0.20 \times 1,000 = 200$. Hence, the expected frequency corresponding to each of the observed frequencies in Table 8-1 is 200. We can now compare the set of observed frequencies with the set of theoretical frequencies derived under the assumption that the null hypothesis is true. The test statistic we compute to make this comparison is known as chi-square, denoted χ^2. The computed value of χ^2 is

(8.1)
$$\chi^2 = \Sigma \frac{(f_o - f_t)^2}{f_t}$$

where f_o = an observed frequency
        f_t = a theoretical (or expected) frequency

TABLE 8-2

Calculation of the χ^2 statistic for the coffee-testing problem

Brand Preference	(1) Observed Frequency f_o	(2) Theoretical (expected) Frequency f_t	(3) $(f_o - f_t)$	(4) $(f_o - f_t)^2$	(5) $\dfrac{\text{Column 4}}{\text{Column 2}}$ $\dfrac{(f_o - f_t)^2}{f_t}$
A	210	200	10	100	0.5
B	312	200	112	12,544	62.7
C	170	200	− 30	900	4.5
D	85	200	− 115	13,225	66.1
E	223	200	23	529	2.6
Total	1,000	1,000			136.4

$$\chi^2 = \Sigma \frac{(f_o - f_t)^2}{f_t} = 136.4$$

As we can see from equation 8.1, if every observed frequency is exactly equal to the corresponding theoretical frequency, the computed value of χ^2 is zero. This is the smallest value χ^2 can have. The larger the discrepancies between the observed and theoretical frequencies, the larger is χ^2.

The computed value of χ^2 is a random variable that takes on different values from sample to sample. That is, χ^2 has a sampling distribution just as do the test statistics discussed in Chapter 7. We wish to answer the following question: Is the computed value of χ^2 so large that we must reject the null hypothesis? In other words, are the aggregate discrepancies between the observed frequencies f_o and theoretical frequencies f_t so large that we are unwilling to attribute them to chance, and have to reject the null hypothesis? The calculation of χ^2 for the present problem is shown in Table 8-2.

The χ^2 Distribution

Before we can answer the questions raised in the preceding paragraph, we must digress for a discussion of the appropriate sampling distribution. Then we will complete the solution to the coffee-tasting problem. It can be shown that for large sample sizes the sampling (probability) distribution of χ^2 can be closely approximated by the χ^2 distribution whose probability function is

(8.2) $$f(\chi^2) = c(\chi^2)^{(v/2) - 1} e^{-\chi^2/2}$$

where $e = 2.71828\ldots$
  v = number of degrees of freedom
  c = a constant depending only on v

The χ^2 distribution has only one parameter v, the number of degrees of freedom. This is similar to the case of the t distribution, discussed in section 6.4.

FIGURE 8-1

The χ^2 distributions for $v = 1$, 5, and 10

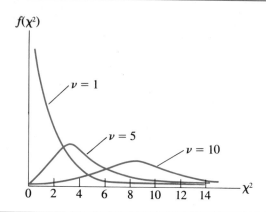

Hence, $f(\chi^2)$ is a family of distributions, one for each value of v. χ^2 is a continuous random variable greater than or equal to zero. For small values of v, the distribution is skewed to the right. As v increases, the distribution rapidly becomes symmetrical. In fact, for large values of v, the χ^2 distribution is closely approximated by the normal curve. Figure 8-1 depicts the χ^2 distributions for $v = 1$, 5, and 10.

Since the χ^2 distribution is a probability distribution, the area under the curve for each value of v equals one. Because there is a separate distribution for each value of v, it is not practical to construct a detailed table of areas. Therefore, for compactness, a χ^2 table generally shows the relationship between areas and values of χ^2 for only a few levels of significance in different χ^2 distributions.

Table A-7 of Appendix A shows χ^2 values corresponding to selected areas in the right tail of the χ^2 distribution. These tabulations are shown separately for the number of degrees of freedom listed in the left column. The χ^2 values are shown in the body of the table, and the corresponding areas are shown in the column headings.

As an illustration of the use of the χ^2 table, let us assume a random variable having a χ^2 distribution with $v = 8$. In Table A-7, we find a χ^2 value of 15.507 corresponding to an area of 0.05 in the right tail. The relationships described in this illustrative problem are shown in Figure 8-2. Hence, if the random variable has a χ^2 distribution with $v = 8$, the probability that χ^2 is greater than 15.507 is 0.05. Let us give the corresponding interpretation in a hypothesis-testing context. If the null hypothesis being tested is true, the probability of observing a χ^2 figure greater than 15.507 because of chance variation is equal to 0.05. Therefore, for example, if the null hypothesis were tested at the 0.05 level of significance and we calculated $\chi^2 = 16$, we would reject the null hypothesis, because so large a χ^2 value would occur less than 5% of the time if the null hypothesis is true. We now turn to a discussion of the rules for determining the number of degrees of freedom involved in a χ^2 test.

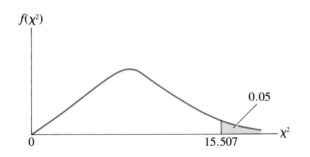

FIGURE 8-2
χ^2 distribution for eight
degrees of freedom

Number of Degrees of Freedom

We have seen that a χ^2 goodness of fit test involves a comparison of a set of observed frequencies, denoted f_o, with a set of theoretical frequencies, denoted f_t. Let k be the number of classes for which these comparisons are made. For example, in the coffee-tasting problem $k = 5$, because there are 5 classes for which we computed relative deviations of the form $(f_o - f_t)^2/f_t$. To determine the number of degrees of freedom, we must reduce k by one for each restriction imposed. In the coffee-tasting example, the number of degrees of freedom is equal to $v = k - 1 = 5 - 1 = 4$. The rationale for this computation follows.

In the calculations shown in Table 8-2, there are 5 classes for which f_o and f_t values are to be compared. Hence, we start with $k = 5$ degrees of freedom. However, we have forced the total of the theoretical frequencies, Σf_t, to be equal to the total of the observed frequencies, Σf_o; that is, 1,000. Therefore, we have reduced the number of degrees of freedom by one and there are now only 4 degrees of freedom. That is, once the total of the theoretical frequencies is fixed, only 4 of the f_t values may be freely assigned to the classes; when these 4 have been assigned, the fifth class is immediately determined, because the theoretical frequencies must total 1,000.

The number of degrees of freedom is reduced by one for each restriction imposed in the calculation of the theoretical frequencies. An additional degree of freedom is lost for each parameter value that must be estimated from the sample. For instance, if the mean μ of a Poisson distribution must be estimated from the mean $\bar{x}$ of a sample, there would be a reduction of one degree of freedom. In summary, we can state the following rules for determining v, the number of degrees of freedom in a χ^2 test in which k classes of observed and theoretical frequencies are compared:

1. If the only restriction is $\Sigma f_t = \Sigma f_o$, the number of degrees of freedom is $v = k - 1$.

2. If, in addition to the previous restriction, m parameters are replaced by sample estimates, the number of degrees of freedom is $v = k - 1 - m$.

Rules for determining v, the number of degrees of freedom

Decision Procedure

We can now return to the coffee-tasting example to perform the goodness of fit test. Let us assume we wish to test the null hypothesis at the 0.05 level of significance. Since the number of degrees of freedom is 4, we find the critical value of χ^2, which we denote as $\chi^2_{0.05}$, to be 9.488 (Table A-7 of Appendix A). This means that if the null hypothesis is true, the probability of observing a χ^2 value greater than 9.488 is 0.05. Specifically in terms of the coffee problem, this means that an aggregate discrepancy between the observed and theoretical frequencies larger than 9.488 will occur only 5% of the time if there is no difference in preference among brands. We can state the decision rule for this problem in which $\chi^2_{0.05} = 9.488$ as follows:

DECISION RULE

1. If $\chi^2 > 9.488$, reject H_0
2. If $\chi^2 \leqslant 9.488$, do not reject H_0

Since the computed χ^2 value in this problem is 136.4 and is thus much larger than the critical $\chi^2_{0.05}$ value of 9.488, we reject the null hypothesis. Therefore, we conclude that real differences exist in consumer preference among the brands of coffee involved in the experiment. In statistical terms, we cannot consider the 1,000 coffee drinkers in the experiment to be a simple random sample from a population whose members prefer each of the 5 brands in equal proportions. In terms of goodness of fit, we reject the null hypothesis that the probability distribution is uniform. Hence, we conclude that the uniform distribution is decidedly not a "good fit" to the sample data.

We now turn to other examples of χ^2 goodness of fit tests.

In Example 8-1, the hypothesis that the population probability distribution is Poisson is tested and retained. A similar procedure could be used to test whether a shift away from a previously established Poisson distribution has occurred. The number of degrees of freedom would be $k - 1$ if μ was estimated from previous results or $k - 1 - 1$ if μ was estimated from the mean $\bar{x}$ of the present sample data.

EXAMPLE 8-1

Jonathan Falk, a management scientist, was developing an inventory control system for a manufacturer of a diversified product line. He wanted to determine whether the Poisson distribution was an appropriate model for the demand for a particular product. He obtained the frequency distribution of the number of units of this product demanded per day for the past 200 business days. That distribution is shown in columns (1) and (2) of Table 8-3(a). The mean number of units per day is shown at the bottom of the table; it is obtained by dividing the total of column (3) by the total of column (2).

TABLE 8-3

(a) Number of units of a particular product demanded per day for the past 200 business days			(b) Theoretical distribution of demand assuming a Poisson distribution		
(1)	(2) Observed Number of Days f_o	(3) Column 1 × Column 2 $f_o x$	(1)	(2)	(3) Column 2 × 200 Expected Number of Days f_t
Number of Units Demanded per Day x			Number of Units Demanded per Day x	Probability $f(x)$	
0	11	0	0	0.050	10.0
1	28	28	1	0.149	29.8
2	43	86	2	0.224	44.8
3	47	141	3	0.224	44.8
4	32	128	4	0.168	33.6
5	28	140	5	0.101	20.2
6	7	42	6	0.050	10.0
7	0	0	7	0.022	4.4
8	2	16	8	0.008	1.6
9	1	9	9	0.003	0.6
10	1	10	10	0.001	0.2
Total	200	600	Total	1.000	200.0

$$\bar{x} = \frac{600}{200} = 3 \text{ units per day}$$

Using the mean of this sample of observations, $\bar{x} = 3$ units demanded per day, as an estimate of the parameter μ of the corresponding theoretical Poisson distribution, Falk calculated the Poisson probability distribution shown in the first two columns of Table 8-3(b). Multiplying the probabilities in column (2) by 200 days, he obtained the theoretical, or expected, frequencies if the demand were distributed according to the Poisson distribution. These theoretical frequencies are shown in column (3). For example, if the *probability* that zero units would be demanded is 0.050, then in 200 days the *expected number* of days in which zero units would be demanded is $0.050 \times 200 = 10.0$ days, the first entry in column (3). The analyst was now able to apply a χ^2 goodness of fit test using the actual number of days in column (2) of Table 8-3(a) as the observed frequencies f_o and the expected number of days in column (3) of Table 8-3(b) as the theoretical frequencies f_t. These two sets of frequencies are shown in columns (2) and (3) of Table 8-4, where the calculation of the χ^2 value is carried out. The hypotheses under test in this problem may be stated as follows:

H_0: The population probability distribution is Poisson with $\mu = 3$

H_1: The population probability distribution is not Poisson with $\mu = 3$

Assume that we wish to test the null hypothesis at the 0.05 level of significance.

We can see in Table 8-4 that the last 4 classes of Table 8-3 for 7, 8, 9, and 10 units of demand have been combined into one class titled "7 or more." Both f_o and f_t values have been cumulated for the 4 classes, and a single relative deviation of the form $(f_o - f_t)^2/f_t$ has been calculated for the combined class. There are now 8 classes, $k = 8$,

TABLE 8-4
Calculation of the χ^2 statistic for the demand distribution problem

(1) Number of Units Demanded per Day	(2) Observed Number of Days f_o	(3) Theoretical Number of Days f_t	(4) $f_o - f_t$	(5) $(f_o - f_t)^2$	(6) $\dfrac{(f_o - f_t)^2}{f_t}$
0	11	10.0	1.0	1.00	0.10
1	28	29.8	-1.8	3.24	0.11
2	43	44.8	-1.8	3.24	0.07
3	47	44.8	2.2	4.84	0.11
4	32	33.6	-1.6	2.56	0.08
5	28	20.2	7.8	60.84	3.01
6	7	10.0	-3.0	9.00	0.90
7 or more	4	6.8	-2.8	7.84	1.15
Total	200	200.0	0		5.53

$$\chi^2 = \Sigma \frac{(f_o - f_t)^2}{f_t} = 5.53$$

in Table 8-4 for which the χ^2 value has been computed and from which the number of degrees of freedom will be determined. The reason for this combination of classes will be explained at the completion of the problem.

Let us now compute the number of degrees of freedom in the test. As indicated in the earlier discussion, the number of degrees of freedom is given by $v = k - 1 - m$, where m is the number of parameters that have been replaced by sample estimates. Since the sample mean $\bar{x}$ was used as the estimate of the parameter μ in the Poisson distribution, $m = 1$. Hence, the number of degrees of freedom is $v = 8 - 1 - 1 = 6$. For 6 degrees of freedom, the critical value of χ^2 at the 0.05 level of significance is $\chi^2_{0.05} = 12.592$ (Table A-7 of Appendix A). Therefore, since the observed χ^2 value of 5.53 is less than 12.592, we do not reject the null hypothesis.

In other words, the aggregate discrepancy between the observed and theoretical frequencies is sufficiently small for us to *conclude that the Poisson distribution with $\mu = 3$ is a good fit*. Based on this result, the Poisson distribution can reasonably be used as a model for demand for the product.

Rule concerning size of theoretical frequencies

As indicated earlier, for large sample sizes, the probability function of the computed χ^2 values can be closely approximated by the χ^2 distribution given in equation 8.2, which is the distribution of a *continuous* random variable. However, there are only a finite number of possible combinations of f_t values, and hence only a finite number of computed χ^2 values. Thus, a computed χ^2 figure is one value of a *discrete* random variable. If the sample size is large, the approximation of the probability distribution of this discrete random variable by the continuous chi-square distribution will be a good one. This is analogous to approximating the binomial distribution, which is discrete, by the normal curve, which is continuous (see section 5.7).

When the expected frequencies (the f_t values) are small, the approximation discussed in the preceding paragraph is inadequate. A frequently used rule is that each f_t value should be equal to or greater than 5. This is why the classes for 7, 8, 9, and 10 units of demand were combined in Example 8-1 in the computation of the χ^2 value. As shown in Table 8-4, the computed f_t value for the combined class is equal to 6.8, which satisfies the commonly used rule of thumb for a minimum expected frequency.

In Example 8-2, the investigator tested a hypothesis using the χ^2 goodness of fit procedure, and she tested another related hypothesis using a standard one-tailed test for a proportion. The conclusions from these two tests were consistent. However, Example 8-2 illustrates the tentative nature of conclusions drawn from hypothesis-testing procedures. It is conceivable that a different sample could have led to the retention of a hypothesis involving $p = 0.522$ when used in the χ^2 test, but if that same sample were used in the hypothesis test of a proportion, it could have led to the retention of the hypothesis that $p = 0.50$, even at the same significance level.

EXAMPLE 8-2

Researcher Lana Mauro hypothesized that the determination of sex in human births could be considered a Bernoulli process. However, she suspected that male and female births were not equally likely. Specifically, she believed that large families tended to have more male children than female. She had data for a simple random sample of 320 families with 5 children each, which had been drawn for another purpose, so she decided to conduct a partial test of her theory using these data. The frequency distribution of male children in these 320 families is shown in Table 8-5.

Since Mauro hypothesized that human births could be treated as a Bernoulli process, she decided to fit a binomial distribution to the data. She further decided to

TABLE 8-5
Calculation of $\bar{p}$, the proportion of male children in the sample of 320 families

(1) Number of Male Children x	(2) Observed Number of Families f_o	(3) Column 1 × Column 2 $f_o x$
0	12	0
1	42	42
2	92	184
3	108	324
4	46	184
5	20	100
Total	320	834

$$\bar{x} = \tfrac{834}{320} = 2.61 \text{ male children per family}$$

$$\bar{p} = \tfrac{2.61}{5} = 0.522$$

estimate p, the probability of a male birth, by using $\bar{p}$, the proportion of male births in the sample. As indicated in Table 8-5, she calculated $\bar{x}$, the mean number of male children per family, and divided that figure by 5. These computations are shown in columns 1, 2, and 3 and at the bottom of Table 8-5. As shown, the sample proportion of male children was $\bar{p} = 0.522$ for the sample of 320 families. Therefore, the investigator stated the hypotheses as follows:

H_0: The population probability distribution is binomial with $p = 0.522$

H_1: The population probability distribution is not binomial with $p = 0.522$

Since 0.05 was the conventional level of significance ordinarily used by other researchers in her field, Mauro decided to use that level of significance in testing the hypothesis. Letting

$$p = \text{the probability of a male birth} = 0.522$$

$$q = \text{the probability of a female birth} = 0.478$$

and

$$n = \text{number of children per family} = 5$$

she computed the binomial probability distribution given in columns (1) and (2) of Table 8-6(a). Multiplying the probabilities in column (2) by 320 families, she determined the expected number of families with 0, 1, 2, 3, 4, and 5 male children each. These theoretical frequencies are shown in column (3) of Table 8-6(a). She then proceeded with the χ^2 test

TABLE 8-6

(a) Calculation of expected frequencies of number of male children: Assumed binomial distribution with $p = 0.522$ and $n = 5$

(b) Calculation of the χ^2 statistic for the number of male children problem: Fit of binomial distribution with $p = 0.522$ and $n = 5$

(1)	(2)	(3)	(1)	(2)	(3)	(4)	(5)	(6)
		Column 2 × 230						
Number of Male Children x	Probability* $f(x)$	Expected Number of Families f_t	Number of Male Children	Observed Number of Families f_o	Theoretical Number of Families f_t	$f_o - f_t$	$(f_o - f_t)^2$	$\dfrac{(f_o - f_t)^2}{f_t}$
0	0.025	8.00	0	12	8.00	4.00	16.00	2.00
1	0.136	43.52	1	42	43.52	−1.52	2.31	0.05
2	0.298	95.36	2	92	95.36	−3.36	11.29	0.12
3	0.324	103.68	3	108	103.68	4.32	18.66	0.18
4	0.178	56.96	4	46	56.96	−10.96	120.12	2.11
5	0.039	12.48	5	20	12.48	7.52	56.55	4.53
	1.000	320.00	Total	320	320.00	0		8.99

$$\chi^2 = \Sigma \frac{(f_o - f_t)^2}{f_t} = 8.99$$

* These probabilities can only be approximated by using Tables A-1 and A-2 in Appendix A because those tables give probabilities only for p values of 0.05, 0.10, 0.15, and so on.

of goodness of fit using the observed frequencies in column (2) of Table 8-5 as the f_o values and the expected frequencies in column (3) of Table 8-6(a) as f_t. Table 8-6(b) shows the calculation of the χ^2 value, 8.99.

The number of classes is $k = 6$, and $m = 1$ because one sample estimate $\bar{p}$ was used to replace a population parameter p. Therefore, the number of degrees of freedom is $v = k - 1 - m = 6 - 1 - 1 = 4$. For $v = 4$, the critical χ^2 value is $\chi^2_{0.05} = 9.488$. Since the computed χ^2 value was only 8.99, the null hypothesis that the population probability distribution is binomial with $p = 0.522$ was not rejected. Hence, Mauro concluded that the evidence represented by the observed frequency distribution of the number of male children in the sample families was consistent with the hypothesis of the operation of a Bernoulli process with $p = 0.522$ for births.

As a further test of her belief that large families have more male than female children, the investigator carried out a one-tailed test of the null hypothesis $H_0: p = 0.50$ males against the alternative hypothesis $H_1: p > 0.50$ males at the 0.05 level of signifi- cance. The sample statistic $\bar{p} = 0.522$ males for a sample of 1,600 births (5 children per family times 320 families) differed significantly from $p = 0.50$. She was pleased to observe this additional evidence in favor of her belief that male births tended to occur more frequently than female births in large families.

There is a subtle point involved in the example.

> Under classical hypothesis-testing procedures, the hypothesis should be set up before the data are gathered.

However, note that in the null hypothesis of the χ^2 test (H_0: The population probability distribution is binomial with $p = 0.522$), the hypothesized value of p was derived from the sample statistic $\bar{p}$.

> In practice, many hypotheses are tested after sample evidence is obtained, sometimes by persons who controlled the collection and tabulation of the data, and sometimes by others.

Classical statistics does not provide separate techniques for testing hypotheses before and after sample evidence has been collected. However, Bayesian deci- sion theory, discussed in Chapters 14 through 17, provides different techniques for decision making prior to obtaining sample data and for the incorporation of sample information with prior knowledge.

EXERCISES 8.1

1. In an investigation of trust departments of U.S. commercial banks, a simple random sample of portfolios containing investments in five different stocks was examined to determine how many of the stocks declined during the

past year. The results were as follows:

Number of Stocks Declining	Number of Portfolios
0	50
1	405
2	485
3	585
4	475
5	0
Total	2,000

Determine whether the binomial distribution is a good fit to these data. Assume a 0.50 probability that a particular stock declined in price during the past year. Use a 0.01 significance level.

2. Firerock, Inc., a major producer of tires, has plants in 100 countries worldwide as follows:

Number of plants	0	1	2	3	4	5
Number of countries	6	7	40	24	14	9

Use a χ^2 test to determine whether the Poisson distribution is a good fit. Use a 0.01 significance level.

3. Assume that between 1980 and 1990 the Brazilian cruzeiro declined in value against most major currencies and that its performance against four major currencies was as follows:

Currency	Number of Times Cruzeiro Declined in Value
German mark	10
Swiss franc	10
Japanese yen	7
U.S. dollar	5

Test the hypothesis that these currencies performed equally well against the Brazilian cruzeiro.

4. On a scale of 0 to 5, 50 major companies are rated as follows for their attractiveness as stock market investments:

Rating	0	1	2	3	4	5
Number of companies	10	18	13	6	2	1

Is the Poisson distribution a suitable model for the ratings?

5. The number of typist errors per page of a 2,000-page manuscript is as follows:

Number of Errors per Page	Observed Number of Pages
0	1,102
1	657
2	193
3	35
4	13
Total	2,000

Fit a Poisson distribution to these data, and use a χ^2 test to determine the goodness of fit. Use a 0.01 significance level.

6. A department store classifies its credit customers into six categories based on the size of their account balances due. The number in each classification is as follows:

$$f_o \qquad f_t \qquad f_o - f_t \qquad \frac{(f_o - f_t)^2}{f_t}$$

Classification	Number of Credit Customers	f_t	$f_o - f_t$
A	46	50	−4
B	42	50	
C	56	56	
D	64	50	
E	48	50	
F	44	50	
Total	300		

$$\frac{300}{6} = 50$$

$$\chi^2 = \Sigma$$

Test the hypothesis that these credit customers are drawn from a population that has an equal distribution among the six classes.

7. A Hartford-based life insurance company was considering a change in the premium charges on its policies. One of its actuaries, who was reformulating life expectancy tables, gathered some data from the records of 20 Boston-area hospitals. The number of deaths resulting from childbirth from 1970 to 1975, tabulated by number of hospital-years, was as follows:

Number of Deaths:	0	1	2	3	4	5	6	7	8	9
Number of Hospital-Years:	12	22	30	15	7	6	4	2	1	1

Use a χ^2 test to determine whether the Poisson distribution is a good fit.

8. A leading New York commercial bank employs a large number of MBA graduates but limits its recruiting and hiring to graduates of four top business schools, which we call A, B, C, and D. A simple random sample of 24 MBA holders drawn from the bank's employee population reveals that four have degrees from school A, nine from school B, eight from school C, and three from school D. Test the hypothesis that the bank's employees are equally distributed among the four schools.

8.2
TESTS OF INDEPENDENCE

Another important application of the χ^2 distribution is in testing for the independence of two variables on the basis of sample data. The general nature of the test is best explained with a specific example.

Automobile-ownership problem

> In an investigation of the socioeconomic characteristics of the families in a certain city, a market research firm wished to determine whether the number of telephones owned was independent of the number of automobiles owned. The firm obtained this ownership information from a simple random sample of 10,000 families who lived in the city.

The results are shown in Table 8-7. This type of table, which has one basis of classification vertically across the rows (in this case, number of telephones owned) and another basis of classification horizontally across the columns (in this case, number of automobiles owned), is known as a **contingency table**. If the table has three rows and three columns, as Table 8-7 has, it is called a **three-by-three** (often written 3 × 3) **contingency table**. In general, in an *r* × *c*

r × c contingency table

contingency table, where *r* denotes the number of rows and *c* denotes the number of columns, there are *r* × *c* **cells**. For example, in the 3 × 3 table under discussion, there are 3 × 3 = 9 cells with observed frequencies. In a 3 × 2 table, there are 3 × 2 = 6 cells, and so on.

TABLE 8-7

A simple random sample of 10,000 families classified by number of automobiles and telephones owned

Number of Telephones Owned	Number of Automobiles Owned			
	(A_1) Zero	(A_2) One	(A_3) Two	Total
(B_1) Zero	1,000	900	100	2,000
(B_2) One	1,500	2,600	500	4,600
(B_3) Two or More	500	2,500	400	3,400
Total	3,000	6,000	1,000	10,000

The χ^2 test consists of calculating expected frequencies under the hypothesis of independence and comparing the observed and expected frequencies.

The competing hypotheses under test in this problem may be stated as follows:

H_0: The number of automobiles owned is independent of the number of telephones owned

H_1: The number of automobiles owned is not independent of the number of telephones owned

Calculation of Theoretical (Expected) Frequencies

Since we are interested in determining whether the hypothesis of independence is tenable, we calculate the theoretical, or expected, frequencies by assuming that the null hypothesis is true. We observe from the marginal totals in the last column of Table 8-7 that $\frac{2,000}{10,000}$, or 20%, of the families do not own telephones. If the null hypothesis H_0 is true—that is, if ownership of automobiles is independent of ownership of telephones—then 20% of the 3,000 families owning no automobiles, 20% of the 6,000 families owning one automobile, and 20% of the 1,000 families owning 2 automobiles would be expected to have no telephones.

Thus, the expected number of "no-car" families who do not own telephones is

$$\frac{2,000}{10,000} \times 3,000 = 600$$

This *expected* frequency corresponds to 1,000, the *observed* number of "no-car" families who do not own telephones.

Similarly, the expected number of "one-car" families who do not own telephones is

$$\frac{2,000}{10,000} \times 6,000 = 1,200$$

This figure corresponds to the 900 shown in the first row.

In general, the theoretical or expected frequency for a cell in the ith row and jth column is calculated as follows:

(8.3)
$$(f_t)_{ij} = \frac{(\Sigma \text{ row } i)(\Sigma \text{ column } j)}{\text{Grand total}}$$

where $(f_t)_{ij}$ = the theoretical (expected) frequency for a cell in the ith row and j the column

Σ row i = the total of the frequencies in the ith row

Σ column j = the total of the frequencies in the jth column

grand total = the total of all of the frequencies in the table

TABLE 8-8

Expected frequencies for the problem on the relationship between telephone and automobile ownership

| Number of Telephones Owned | Number of Automobiles Owned | | | |
	Zero	One	Two	Total
Zero	600	1,200	200	2,000
One	1,380	2,760	460	4,600
Two or More	1,020	2,040	340	3,400
Total	3,000	6,000	1,000	10,000

For example, the theoretical frequency in the first row and first column of Table 8-8 (whose rationale of calculation was just explained) is computed by equation 8.3 as

$$(f_t)_{11} = \frac{(2,000)(3,000)}{10,000} = 600$$

In order to keep the notation uncluttered, we will drop the subscripts denoting rows and columns for f_t values in the subsequent discussion.

The expected frequencies for the present problem are shown in Table 8-8. Because of the method of calculating the expected frequencies, the totals in the margins of the table are the same as the totals in the margins of the table of observed frequencies (Table 8-7). Note that the method of computing the expected frequencies under the null hypothesis of independence is simply an application of the multiplication rule for independent events given in equation 2.10. For example, in Table 8-7, the "no-car" and "no telephone" categories have been denoted A_1 and B_1, respectively. Under independence, $P(A_1$ and $B_1) = P(A_1)P(B_1)$. The marginal probabilities $P(A_1)$ and $P(B_1)$ are given by

$$P(A_1) = \frac{3,000}{10,000} = 0.30$$

$$P(B_1) = \frac{2,000}{10,000} = 0.20$$

$$P(A_1 \text{ and } B_1) = P(A_1)P(B_1) = (0.30)(0.20) = 0.06$$

Multiplying this joint probability by the total frequency (10,000), we obtain the expected frequency previously derived for the upper left cell.

$$0.06 \times 10,000 = 600$$

The χ^2 Test

How great a departure from the theoretical frequencies under the assumption of independence can be tolerated before we reject the hypothesis of independence? The purpose of the χ^2 test is to provide an answer to this question by comparing observed frequencies with the theoretical, or expected, frequencies derived under the hypothesis of independence. The test statistic used to make this comparison is the chi-square statistic, $\chi^2 = \Sigma (f_o - f_t)^2/f_t$, as defined in equation 8.1, where f_o is an observed frequency and f_t is a theoretical frequency.

Number of Degrees of Freedom

The number of degrees of freedom in the contingency table must be determined in order to apply the χ^2 test. The number of degrees of freedom in a 3 × 3 contingency table is 4, calculated as follows. In determining the expected frequencies, we used the marginal row and column totals. With 3 rows, only 2 row totals are "free," since the row totals must sum to Σf_o, which is 10,000 in the present illustration. Correspondingly, with 3 columns, 2 column totals are "free." This gives us the freedom to specify 4 cell totals, where the "free" columns and "free" rows intersect.

> In general, in a contingency table containing r rows and c columns, there are $(r - 1)(c - 1)$ degrees of freedom.

Thus, in a 2 × 2 table, $v = (2 - 1)(2 - 1) = 1$; in a 3 × 2 table, $v = (3 - 1)(2 - 1) = 2$; in a 3 × 3 table, $v = (3 - 1)(3 - 1) = 4$.

We now return to our example to perform the χ^2 test of independence. Again denoting the observed frequencies as f_o and the expected frequencies as f_t, we have shown the calculation of the χ^2 statistic in Table 8-9. No cell designations are indicated, but of course every f_o value is compared with the corresponding f_t figure. As shown at the bottom of the table, the computed value

TABLE 8-9
Calculation of the χ^2 statistic for the telephone and automobile ownership problem

Observed Number of Families f_o	Expected Number of Families f_t	$f_o - f_t$	$(f_o - f_t)^2$	$\dfrac{(f_o - f_t)^2}{f_t}$
1,000	600	400	160,000	266.7
1,500	1,380	120	14,400	10.4
500	1,020	−520	270,400	265.1
900	1,200	−300	90,000	75.0
2,600	2,760	−160	25,600	9.3
2,500	2,040	460	211,600	103.7
100	200	−100	10,000	50.0
500	460	40	1,600	3.5
400	340	60	3,600	10.6
Total 10,000	10,000	0		$\chi^2 = 794.3$

of χ^2 is equal to 794.3. The number of degrees of freedom is $(r - 1)(c - 1)$ or $(3 - 1)(3 - 1) = 4$. In Table A-7 of Appendix A, we find a critical value at the 0.01 level of significance of $\chi^2_{0.01} = 13.277$. This means that if the null hypothesis is true, the probability of observing a χ^2 value greater than 13.277 is 0.01. Specifically in terms of the problem, this means that if ownership of telephones was independent of ownership of automobiles, an aggregate discrepancy between the observed and theoretical frequencies larger than a χ^2 value of 13.277 would occur only 1% of the time. We can state the decision rule for this problem as follows:

DECISION RULE

1. If $\chi^2 > 13.277$, reject H_0
2. If $\chi^2 \leqslant 13.277$, accept H_0

Since the computed χ^2 value of 794.3 so greatly exceeds this critical value, the null hypothesis of independence between telephone and automobile ownership is rejected.

Calculation of Chi-Square Using Minitab

The χ^2 statistic can be very easily computed on a personal computer using the Minitab software. Table 8-10 shows the Minitab calculation for the automobile and telephone ownership example we have just discussed. Note that the data are typed in by rows as shown originally in Table 8-7. As shown in Table 8-10, the χ^2 statistics and the number of degrees of freedom are printed out in Minitab, and then we must do the usual table lookup to test for significance.

Further Comments

We have seen how the χ^2 test for independence in contingency tables is a means of determining whether a relationship exists between two bases of classification, or in other words, whether a relationship exists between two variables. Although this type of tabulation provides a basis for testing whether there is a dependence between the two classificatory variables, it does not yield a method for estimating the values of one variable from known values or assumed values of the other. In the next chapter, which deals with regression and correlation analysis, methods for providing such estimates are discussed. For example, **regression analysis** provides a method for estimating or predicting the number of telephones owned by a family with a specific number of automobiles. Regression analysis, in particular, provides a powerful tool for stating in explicit mathe-

---

TABLE 8-10
Minitab printout for the problem on the relationship between telephone and automobile ownership

---

```
KEY: TABLE 8-10

MTB > READ C1-C3
DATA> 1000  900 100
DATA> 1500 2600 500
DATA>  500 2500 400
DATA> END
      3 ROWS READ

MTB > CHISQUARE C1-C3

EXPECTED COUNTS ARE PRINTED BELOW OBSERVED COUNTS

              C1       C2       C3 TOTAL
    1       1000      900      100  2000
           600.00  1200.00  200.00

    2       1500     2600      500  4600
          1380.00  2760.00  460.00

    3        500     2500      400  3400
          1020.00  2040.00  340.00

TOTAL       3000     6000     1000 10000

CHISQ =266.667 +  75.000 +  50.000 +
         10.435 +   9.275 +   3.478 +
        265.098 +103.725 +  10.588 = 794.267
DF = 4
```

---

matical form the nature of the relationship that exists between two or more variables.

However, we may obtain at least some indication of the nature of the relationship between the two variables in a contingency table. Equivalently to the null hypothesis of independence rejected in our example, we have rejected the null hypothesis H_0: $p_1 = p_2 = p_3$, where p_1, p_2, and p_3 denote the population proportions of zero-, one-, and two-car families who do not have telephones. Reference to Table 8-8 makes it obvious why the null hypothesis was rejected. Of the 3,000 familes who did not own automobiles, $\frac{1,000}{3,000} = 0.33$ did not own a telephone. Let $\bar{p}_1 = 0.33$. The corresponding proportions of one- and two-car families who did not own telephones were $\bar{p}_2 = \frac{900}{6,000} = 0.15$ and $\bar{p}_3 = \frac{100}{1,000} = 0.10$. Hence, we have concluded that it is highly unlikely that these three statistics represent samples drawn from populations that have the same proportions ($p_1 = p_2 = p_3$). Clearly, the proportion of no-telephone families declines as automobile ownership increases. The data suggest a strong relationship between the ownership of telephones and automobiles for the families studied.

A powerful generalization develops from the preceding discussion. It can be shown that a χ^2 test applied to a 2×2 contingency table is algebraically identical to the two-sample test for difference between proportions by the methods of section 7.3 using equation 7.10 to calculate the estimated standard error of the difference. This means that the test of the hypothesis of independence carried out in a χ^2 test for a 2×2 contingency table is identical to the testing of the following hypotheses:

$$H_0: p_1 = p_2$$

$$H_1: p_1 \neq p_2$$

As we have seen, in our illustrative problem involving a 3×3 contingency table, we tested the null hypothesis

$$H_0: p_1 = p_2 = p_3$$

against the alternative that the p values were not all equal. The analogous test can be applied in general to c categories, where $c \geqslant 2$.

Additional Comments

Since the sampling distribution of the χ^2 statistic, $\chi^2 = \Sigma(f_o - f_t)^2/f_t$, is only an approximation to the theoretical distribution defined in equation 8.2, the sample size must be large to yield a good approximation. As in the goodness of fit tests, in contingency tables, cells with expected frequencies of less than five should be combined.

Furthermore, in 2×2 tables (that is, when there is one degree of freedom), an adjustment known as **Yates' correction for continuity** may be used. We introduce this correction because the theoretical χ^2 distribution is continuous, whereas the tabulated values in Table A-7 of Appendix A are based on the distribution of the discrete χ^2 statistic of equation 8.1. We apply the correction by computing the following χ^2 statistic:

(8.4)
$$\chi^2 = \Sigma \frac{(|f_o - f_t| - \frac{1}{2})^2}{f_t}$$

In this correction, $\frac{1}{2}$ is subtracted from the absolute value of the difference between f_o and f_t before squaring. The effect is to reduce the calculated value of χ^2 compared with the corresponding calculation by equation 8.1 without the correction.[2] In an example such as the one just discussed, where the expected

[2] For a more complete discussion of Yates' correction, see F. Yates, "Contingency Tables Involving Small Numbers and the χ^2 Test." Suppl. *J. Royal Stat. Soc.*, 1, 1934, 217–235, and Snedecor, George W. and William G. Cochran, *Statistical Methods*, 7th ed., Iowa State University Press, 1980.

frequencies are large, the effect of this correction is clearly unimportant, but it may be of greater significance for smaller samples.

We have seen that in both χ^2 goodness of fit tests and tests of independence, the null hypothesis is rejected when large enough values of χ^2 are observed. Some investigators have raised the question whether the null hypothesis should also be rejected when the computed value of χ^2 is too low, that is, too close to zero. This is a situation in which the observed frequencies f_o appear to *agree too well* with the theoretical frequencies f_t. The recommended course of action is to examine the data closely to see whether errors have been made in recording them. Perhaps the data rather than the null hypothesis should be rejected. One researcher's experience is relevant to this point. He was analyzing some data on oral temperatures and found that a disturbingly large number of the recorded temperatures were equal to the "normal" figure of 98.6°F. He suspected that these data were "too good to be true." Upon investigation, he found that the temperatures were recorded by relatively untrained nurses' aides. Several of them had misread temperatures by recording the number to which the arrow on the thermometer pointed, namely, 98.6°! Clearly, this was a case in which the data, rather than an investigator's null hypothesis, should be rejected.

EXERCISES 8.2

1. A subscription service stated that reader preferences for different national magazines were independent of geographical location. A survey was taken in which 300 persons randomly chosen from each of three areas were asked to choose their favorite from among three different magazines. The following results were obtained:

Region	Magazine X	Magazine Y	Magazine Z	Total
New England	75	50	175	300
Middle Atlantic	120	85	95	300
South	105	110	85	300
Total	300	245	355	900

Do you agree with the subscription service? Use a 0.05 significance level.

2. A firm is testing to determine if a buyer's sex is independent of his or her preference for ice creams in each of the following situations, where the sample results have been summarized. Find the degrees of freedom and the critical value of the test statistic. Then indicate whether the null hypothesis of independence should be accepted or rejected.
   a. Vanilla, strawberry, and chocolate are considered ($\chi^2 = 7.50$ and $\alpha = 0.05$).

b. Pistachio, banana, nutty and grandeur are considered ($\chi^2 = 10.50$ and $\alpha = 0.01$).

3. An operations researcher wants to determine how Miser Lane Corporation's cash flows are related to its operating leverage. From the historical records the analyst has prepared the following table:

| Operating Leverage | Cash Flows (millions of dollars) | | | |
	0–24	25–49	50–74	Total
1	6	12	2	20
2	14	28	4	46
3	10	20	4	34
Total	30	60	10	100

Do these data support the hypothesis that the cash flows are independent of the company's operating leverage? Use a 0.01 significance level.

4. Mony Television, Inc. wished to determine whether a relationship exists between the nature of its radio advertisements and audience recall. The company selected two small communities, similar in many respects (demography, types of occupations, climate, and so forth), and broadcast two different types of advertisement, one in each locality. One advertisement was humorous, the other serious. After three months of running the advertisements, random samples of these stations' regular listeners were drawn from each community. Each listener was asked whether he or she remembered the advertisement presented in his or her locality. The following results were recorded:

| Audience Recall | Nature of Advertisement | | |
	Humorous	Serious	Total
Number who recalled	75	25	100
Number who did not recall	23	75	98
Total	98	100	198

Test the hypothesis that the impact of the advertisements is independent of the nature of the advertisement. Use $\alpha = 0.01$.

5. A research organization obtained data concerning a sample of 200 stock investments made by each of four highly regarded mutual funds between 1985 and 1990. The number of stocks that suffered losses were 6, 8, 9, and 12. Test whether the proportions of stocks suffering losses are the same for these mutual funds. (Let $\alpha = 0.05$).

6. A security analyst is attempting to determine whether the market prices of Superior Technology Corporation common stock are influenced by the

company's declared dividends per share. The analyst examines 100 quarters of the company's history and tabulates the following data:

Declared Dividend per share	Market Price per Share			Total
	$0–$9	$10–$19	$20–$29	
$0–$0.15	6	12	2	20
$0.16–$0.31	14	28	4	46
$0.32–$0.47	10	20	4	34
Total	30	60	10	100

Do these data support the hypothesis that share price is independent of declared dividend at the 0.01 significance level?

7. Securities Management Corporation, a holding company of four common stock mutual funds, gathers samples of 200 stock investments from each of its four affiliates. For each grouping of 200 investments, the corporation determines the number of stocks that have suffered losses in 1985. The numbers are 6, 8, 9, and 12. Test whether or not the affiliates have the same proportions of stocks that have suffered losses. (Use $\alpha = 0.05$.)

8.3
ANALYSIS OF VARIANCE:
TESTS FOR EQUALITY OF SEVERAL MEANS

In section 8.2, we saw that

> the χ^2 test provides a generalization of the two-sample test for proportions

and enables us to test for the significance of the difference among c ($c > 2$) sample *proportions*. Conceptually, this represents a test of whether the c samples can be treated as having been drawn from the same population or, in other words, from populations having the same proportions.

In this section, we consider an ingenious technique known as the analysis of variance.

> The analysis of variance is a generalization of the two-sample test for means

and enables us to test for the significance of the difference among c ($c > 2$) sample *means*. As with the χ^2 test, this technique represents a test of whether the c samples can be treated as having been drawn from the same population or, more precisely, from populations having the same means.

Analysis of variance

The **analysis of variance** technique uses sample information to determine whether three or more treatments yield different results. The term *"treatment"* (conventionally used in the analysis of variance) derives from work in the field of agricultural experimentation—in which the treatments may be different types of fertilizer applied to plots of farm land, different types of feeding methods for animals, and so forth. We will use the conventional terminology. In the example that follows, the treatments are different teaching methods.

A central point is that the analysis of variance—literally a technique that analyzes or tests variances—provides us with a test for the significance of the difference among *means*. The rationale by which a test of variances is, in fact, a test for means will be explained shortly.

Teaching methods problem

As an example, suppose we wish to test whether 3 methods of teaching a mathematics course differ in effectiveness.

Method 1. The lecturer neither works out nor assigns problems.

Method 2. The lecturer works out and assigns problems.

Method 3. The lecturer works out and assigns problems. Students are also required to construct and solve their own problems.

The same professor teaches 3 different sections of students, using one of the 3 methods in each class. All of the students are sophomores at the same university and are randomly assigned to the 3 sections. There are only 12 students in the experiment, 4 in each of the 3 different sections.

In practice, a substantially larger number of observations would be required to furnish convincing results; however, limiting the group to 12 permits

---

TABLE 8-11

(a) Final examination grades of 12 students taught by 3 different methods

(b) Notation corresponding to the data listed in part (a)

Student	Teaching Method 1	2	3		i	X_{i1}	X_{i2}	X_{i3}
1	16	19	24		1	X_{11}	X_{12}	X_{13}
2	21	20	21		2	X_{21}	X_{22}	X_{23}
3	18	21	22		3	X_{31}	X_{32}	X_{33}
4	13	20	25		4	X_{41}	X_{42}	X_{43}
Total	68	80	92		Total	$\sum_i X_{i1}$	$\sum_i X_{i2}$	$\sum_i X_{i3}$
Mean	17	20	23		Mean	$\bar{X}_1$	$\bar{X}_2$	$\bar{X}_3$

$$\text{Grand mean} = \frac{17 + 20 + 23}{3} = 20 \qquad \bar{\bar{X}} = \frac{\bar{X}_1 + \bar{X}_2 + \bar{X}_3}{3}$$

us to examine the principles of the analysis without cumbersome computational detail.

It was agreed that student grades on a final examination covering the work of the entire course would be used as the measure of effectiveness. The final examination was graded using 25 as the maximum score and zero as the minimum score. The final examination grades of the 12 students in the 3 sections are given in Table 8-11. As shown in the table, the mean grades for students taught by methods (1), (2), and (3) were 17, 20, and 23, respectively, and the overall average of the 12 students, referred to as the "grand mean," was 20. (Note that the grand mean of 20 is the same figure that would be obtained by adding up all 12 grades and dividing by 12.)

At this point, we introduce some useful notation. In Table 8-11(a), there are 4 rows and 3 columns. As in the discussion of χ^2 tests for contingency tables, r represents the number of rows and c represents the number of columns. There is a total of $r \times c$ observations in the table, in this case $4 \times 3 = 12$. Let X_{ij} be the score of the ith student taught by the jth method (treatment), where $i = 1$, 2, 3, 4 and $j = 1, 2, 3$. (Thus, for example, X_{12} denotes the score of student 1 taught by method 2 and is equal to 19; $X_{23} = 21$, and so on.) In this problem, the 3 different methods of instruction are indicated in the columns of the table, and interest centers on the differences among the scores in the 3 columns. This is typical of the so-called **one-factor** (or **one-way**) **analysis of variance**, in which an attempt is made to assess the effect of only one factor (in this case, instructional method) on the observations. In the present problem, there are 3 columns. Hence, we denote the *values* in the columns as X_{i1}, X_{i2}, and X_{i3}, and the *totals* of these columns are denoted as

Notation

$$\sum_i X_{i1}, \quad \sum_i X_{i2}, \quad \text{and} \quad \sum_i X_{i3}.$$

The subscript i under the summation signs indicates that the total of each of the columns is obtained by summing the entries over the row. Adopting a simplified notation, we will refer to the means of the 3 columns as $\bar{X}_1$, $\bar{X}_2$, and $\bar{X}_3$, or in general, $\bar{X}_j$. Finally, we denote the *grand mean* as $\bar{\bar{X}}$ (pronounced "X double-bar"), where $\bar{\bar{X}}$ is the mean of all $r \times c$ observations. Since each column in our example contains the same number of observations, $\bar{\bar{X}}$ can be obtained by taking the mean of the 3 sample means $\bar{X}_1$, $\bar{X}_2$, and $\bar{X}_3$. This notation is summarized in Table 8-11(b) so that you can compare the notation with the corresponding entries in Table 8-11(a).

The Hypothesis to Be Tested

As indicated earlier, we want to test whether the 3 methods of teaching a mathematics course differ in effectiveness. We have calculated the following mean final examination scores of students taught by the 3 methods, $\bar{X}_1 = 17$, $\bar{X}_2 = 20$, and $\bar{X}_3 = 23$. The statistical question is: Can the three samples represented by these 3 means be considered as having been drawn from populations having the same mean? Denoting the population means corresponding to $\bar{X}_1$, $\bar{X}_2$, and $\bar{X}_3$ as μ_1, μ_2, and μ_3, respectively, we can state the null hypothesis as

$$H_0: \mu_1 = \mu_2 = \mu_3$$

This hypothesis is to be tested against the alternative hypothesis

$$H_1: \text{The means } \mu_1, \mu_2, \text{ and } \mu_3 \text{ are not all equal.}$$

What we wish to determine is whether the differences among the sample means $\bar{X}_1$, $\bar{X}_2$, and $\bar{X}_3$ are too great to be attributed to the chance errors of drawing samples from populations having the same means. If we decide that the sample means differ significantly, our substantive conclusion is that the teaching methods differ in effectiveness.

Assumptions   The underlying assumptions in the analysis of variance procedure and their interpretation in the teaching methods problem are as follows:

1. *Normality.*   The samples are drawn from normally distributed populations. This means that the population of examinations graded for each teaching method is assumed to be normally distributed.

2. *Independence.*   The samples are independent. This means that the grades obtained under any teaching method do not affect the grades obtained under any other teaching method.

3. *Homoscedasticity.*   The populations have equal variances. This property is referred to as *homoscedasticity.* This means that the three population variances of examination grades are equal, that is, $\sigma_1^2 = \sigma_2^2 = \sigma_3^2$.

Decomposition of Total Variation

Before discussing the procedures involved in the analysis of variance, we consider the general rationale underlying the test.

> If the null hypothesis that the 3 population means (μ_1, μ_2, and μ_3) are equal is true, then both the variation among the sample means ($\bar{X}_1$, $\bar{X}_2$, and $\bar{X}_3$) and the variation within the 3 groups reflect chance errors of the sampling process.

● The first of these types of variation is conventionally referred to as between-treatment variation (the word "between" rather than "among" is used even when there are more than two groups present). **Between-treatment variation** is variation of the sample means $\bar{X}_1$, $\bar{X}_2$, and $\bar{X}_3$ around the grand mean $\bar{\bar{X}}$. This variation is sometimes referred to as variation between the c means, between-group variation, and between-column variation.

- The second type of variation is referred to as within-treatment variation (also known as within-group variation and within-column variation). **Within-treatment variation** is variation of the individual observations within each column from their respective means $\bar{X}_1$, $\bar{X}_2$, and $\bar{X}_3$.

Under the null hypothesis that the population means are equal, the between-treatment variation and the within-treatment variation would be expected not to differ significantly from one another after adjustment for degrees of freedom, since they both reflect the same type of chance sampling errors. If the null hypothesis is false and the population column means are indeed different, then the between-treatment variation should significantly exceed the within-treatment variation. This follows from the fact that the between-treatment variation would now be produced by the inherent differences among the treatment means as well as by chance sampling error. On the other hand, the within-treatment variation would still reflect chance sampling errors only.

> A comparison of between-treatment variation and within-treatment variation yields information concerning differences among the treatment means.

This is the central insight provided by the analysis of variance technique.

Terms

The term **variation** is used in statistics in a specific way to refer to a sum of squared deviations and is often referred to simply as a **sum of squares**. When a measure of variation is divided by an appropriate number of degrees of freedom (as we have seen earlier in this text), it is referred to as a *variance*, and in the analysis of variance, a variance is referred to as a **mean square**. For example, the variation of a set of sample observations, denoted X, around their mean $\bar{X}$ is $\Sigma(X - \bar{X})^2$. Dividing this sum of squares by the number of degrees of freedom $n - 1$ (where n is the number of observations), we obtain $\Sigma(X - \bar{X})^2 / (n - 1)$, the sample variance, which is an unbiased estimator of the variance of an infinite population as indicated in section 1.15. This sample variance can also be referred to as a mean square.

We now proceed with the analysis of variance by calculating the between-treatment variation and within-treatment variation for our problem.

Between-Treatment Variation

As indicated earlier, the between-treatment variation, or **between-treatment sum of squares**, measures the variation among the sample column means. It is calculated as follows:

(8.5)
$$\sum_{j} r(\bar{X}_j - \bar{\bar{X}})^2$$

Between-treatment sum of squares

TABLE 8-12
Calculation of the between-treatment sum of squares for the teaching methods problem

$$(\bar{X}_1 - \bar{\bar{X}})^2 = (17 - 20)^2$$
$$= 9$$
$$(\bar{X}_2 - \bar{\bar{X}})^2 = (20 - 20)^2$$
$$= 0$$
$$(\bar{X}_3 - \bar{\bar{X}})^2 = (23 - 20)^2$$
$$= 9$$
$$\sum_j r(\bar{X}_j - \bar{\bar{X}})^2 = 4(9) + 4(0) + 4(9)$$
$$= 72$$

---

where r = number of rows (sample size involved in the calculation of each column mean)[3]

  $\bar{X}_j$ = the mean of the jth column (treatment)

  $\bar{\bar{X}}$ = the grand mean

  $\sum_j$ = summation taken over all columns

As indicated in equation 8.5, the between-treatment sum of squares is calculated by the following steps:

1. Compute the deviation of each treatment mean from the grand mean.
2. Square the deviations obtained in step 1.
3. Weight each deviation by the sample size involved in calculating the respective mean. In our example, all sample sizes are the same and are equal to the number of rows, $r = 4$.
4. Sum over all columns the products obtained in step 3.

The calculation of the between-treatment sum of squares for the example involving 3 different teaching methods is given in Table 8-12. As indicated in the table, the between-treatment variation is 72.

Within-treatment Variation

The **within-treatment sum of squares** is a summary measure of the random errors of the individual observations around their column (treatment) means. The

---

[3] In this example, equal sample sizes (equal numbers of rows) are assumed. Subsequently, we generalize this approach to allow for different sample sizes.

TABLE 8-13
Calculation of the within-treatment sum of squares for the teaching methods problem

i	$(X_{i1} - \bar{X}_1)$	$(X_{i1} - \bar{X}_1)^2$	$(X_{i2} - \bar{X}_2)$	$(X_{i2} - \bar{X}_2)^2$	$(X_{i3} - \bar{X}_3)$	$(X_{i3} - \bar{X}_3)^2$
1	$(16 - 17) = -1$	1	$(19 - 20) = -1$	1	$(24 - 23) = 1$	1
2	$(21 - 17) = 4$	16	$(20 - 20) = 0$	0	$(21 - 23) = -2$	4
3	$(18 - 17) = 1$	1	$(21 - 20) = 1$	1	$(22 - 23) = -1$	1
4	$(13 - 17) = -4$	16	$(20 - 20) = 0$	0	$(25 - 23) = 2$	4
		$\overline{34}$		$\overline{2}$		$\overline{10}$

$$\sum_j \sum_i (X_{ij} - \bar{X}_j)^2 = 34 + 2 + 10 = 46$$

formula for its computation is

(8.6)
$$\sum_j \sum_i (X_{ij} - \bar{X}_j)^2$$
Within-treatment sum of squares

where X_{ij} = the value of the observation in the ith row and jth column

$\bar{X}_j$ = the mean of the jth column

$\sum_j \sum_i$ = means that the squared deviations are first summed over all sample observations within a given column, then summed over all columns

As indicated in equation 8.6, the within-treatment sum of squares is calculated as follows:

1. Calculate the deviation of each observation from its column mean.
2. Square the deviations obtained in step 1.
3. Add the squared deviations within each column.
4. Sum over all columns the figures obtained in step 3.

The computation of the within-treatment variation for the teaching methods problem is given in Table 8-13.

Total Variation

The between-treatment variation and within-treatment variation represent the two components of the total variation in the overall set of experimental data. The total variation, or **total sum of squares**, is calculated by adding the squared deviations of all of the individual observations from the grand mean $\bar{\bar{X}}$. Hence, the formula for the total sum of squares is

(8.7)
$$\sum_j \sum_i (X_{ij} - \bar{\bar{X}})^2$$
Total sum of squares

The total sum of squares is computed by the following steps:

1. Calculate the deviation of each observation from the grand mean.
2. Square the deviations obtained in step 1.
3. Add the square deviations over all rows and columns.

The total sum of squares, or total variation of the 12 observations in the teaching methods problem, is $(16 - 20)^2 + (21 - 20)^2 + \cdots + (25 - 20)^2 = 118$. Referring to the results obtained in Tables 8-12 and 8-13, we see that the total sum of squares, 118, is equal to the sum of the between-treatment sum of squares, 72, and the within-treatment sum of squares, 46. In general, the following relationship holds:

(8.8)
$$\frac{\text{Total}}{\text{variation}} = \frac{\text{Between-treatment}}{\text{variation}} + \frac{\text{Within-treatment}}{\text{variation}}$$

Although, as we have indicated earlier, the test of the null hypothesis in a one-factor analysis of variance involves only the between-treatment variation and the within-treatment variation, it is useful to calculate the total variation as well. This computation is helpful as a check procedure and is instructive in indicating the relationship between total variation and its components.

Shortcut Computational Formulas

The formulas we have given for calculating the between-treatment sum of squares (8.5), the within-treatment sum of squares (8.6), and the total sum of squares (8.7) clearly reveal the rationale of the analysis of variance procedure. However, the following shortcut computation formulas are often used to calculate these sums of squares.

Between-treatment sum of squares   (8.9)
$$\frac{\sum_j T_j^2}{r} - C$$

instead of 8.6

Within-treatment sum of squares   (8.10)
$$\sum_j \sum_i X_{ij}^2 - \sum_j \frac{T_j^2}{r}$$

Total sum of squares   (8.11)
$$\sum_j \sum_i X_{ij}^2 - C$$

instead of 8.7

where C, the so-called **correction term**, is given by

Correction term   (8.12)
$$C = \frac{T^2}{rc}$$

T_j is the total of the r observations in the jth column, and T is the grand total of all rc observations, that is,

(8.13)
$$T = \sum_j \sum_i X_{ij}$$

Grand total of all rc observations

All other terms are as previously defined.

These formulas are especially useful when the column means and grand mean are not integers. The shortcut formulas not only save time and computational labor, but also are more accurate because of avoidance of rounding problems, which usually occur with the use of equations 8.5, 8.6, and 8.7. The shortcut computations for the teaching methods example are as follows:

$$C = \frac{(240)^2}{(3)(4)} = 4{,}800$$

$$\text{Between-treatment sum of squares} = \frac{(68)^2 + (80)^2 + (92)^2}{4} - 4{,}800$$

$$= 72$$

$$\text{Within-treatment sum of squares} = (16)^2 + (21)^2 + \cdots + (25)^2 - \frac{(68)^2 + (80)^2 + (92)^2}{4}$$

$$= 46$$

$$\text{Total sum of squares} = (16)^2 + (21)^2 + \cdots + (25)^2 - 4{,}800$$

$$= 118$$

It is recommended that the shortcut formulas be used, particularly when carrying out computations by hand.

Number of Degrees of Freedom

Although the preceding discussion was in terms of *variation* or *sums of squares* rather than *variance*,

> the actual test of the null hypothesis in the analysis of variance involves a comparison of the *between-treatment variance* with the *within-treatment variance*.

or, in equivalent terminology, a comparison of the *between-treatment mean square* with the *within-treatment mean square*. Hence, the next step in our procedure is to determine the number of degrees of freedom associated with each of the measures of variation. As stated earlier in this section, if a measure of variation—that is, a sum of squares—is divided by the appropriate number of degrees of freedom, the resulting measure is a variance—that is, a mean square.

Between-treatment
sum of squares

The number of degrees of freedom associated with the between-treatment sum of squares is $c - 1$. We can see the reason for this by applying the same general principles indicated earlier for determining number of degrees of freedom in t tests and χ^2 tests. Since there are c columns, or c group means, there are c sums of squares involved in measuring the variation of these column means around the grand mean. Because the sample grand mean is only an estimate of the unknown population mean, we lose one degree of freedom. Hence, there are $c - 1$ degrees of freedom present. The number of degrees of freedom in our example, which has 3 different teaching methods (that is, 3 treatments), is $c - 1 = 3 - 1 = 2$.

Within-treatment variation

The number of degrees of freedom associated with the within-treatment variation is $rc - c = c(r - 1)$. This may be reasoned as follows. There are a total of rc observations. In determining the within-treatment variation, the squared deviations within each treatment were taken around the treatment (column) mean. There are c treatment means, each of which is an estimate of the true unknown population treatment mean. Hence, there is a loss of c degrees of freedom, and c must be subtracted from rc, the total number of observations.

Alternatively, there are r squared deviations in each treatment taken around the treatment mean and a total sum of squares for the treatment. We can assign $r - 1$ of the sums of squares arbitrarily, and the last becomes fixed in order for the sum to equal the column sum. Since there are c treatments, we have $c(r - 1)$ degrees of freedom. In the present problem, the number of degrees

Within-treatment
sum of squares

of freedom associated with the within-treatment sum of squares is $c(r - 1) = 3(4 - 1) = 9$.

Within-treatment variance

A simpler way to denote the number of degrees of freedom associated with the within-treatment variance is $n - c$, where n is the total number of observations. We observe that $n = rc$. For example, in the present illustration, $n = (4)(3) = 12$. Thus, we see that the number of degrees of freedom associated with the within-treatment variance may be written

$$c(r - 1) = cr - c = n - c$$

Total variation

The number of degrees of freedom associated with the total variation is $rc - 1 = n - 1$. The are rc squared deviations taken from the sample grand mean $\bar{\bar{X}}$. Since $\bar{\bar{X}}$ is an estimate of the true unknown population mean, there is a loss of one degree of freedom. Alternatively, in the determination of the total sum of squares, there are rc squared deviations. We may assign $rc - 1$ squared deviations arbitrarily, but the last one is constrained in order for the sum to be equal to the total sum of squares. In the example, the number of degrees of freedom associated with the total variation is $rc - 1 = (4)(3) - 1 = 11$, or $n - 1 = 12 - 1 = 11$.

Just as the between-treatment and the within-treatment variations sum to the total variation, the numbers of degrees of freedom associated with the between-treatment and within-treatment variations add to the number associated with the total variation. In symbols,

(8.14)

$$rc - 1 = (c - 1) + (rc - c)$$

and

$$n - 1 = (c - 1) + (n - c)$$

In the teaching methods problems, the numerical values corresponding to equation 8.14 are $11 = 2 + 9$.

The Analysis of Variance Table

An analysis of variance table for the teaching methods problem is given in Table 8-14. The table uses the standard form to summarize the results of an analysis of variance. In columns (1), (2), and (3) are listed the possible sources of variation, the sum of squares for each of these sources, and the number of degrees of freedom associated with each of the sums of squares. The term **error** has been included in parentheses after "within treatments" to indicate that such variation is attributed to chance sampling error. "Error" is a frequently used term in computer printouts of analyses of variance. We again note that both sums of squares and numbers of degrees of freedom are additive; that is, these figures for between-treatment and within-treatment sources of variation add to the corresponding figure for total variation. Dividing the sums of squares in column (2) by the numbers of degrees of freedom in column (3) yields the between-treatment and within-treatment variances shown in column (4). As indicated earlier, **mean square** is another name for a sum of squares divided by the appropriate number of degrees of freedom, and it is conventional to use this term in an analysis of variance table. Thus, in our problem, the between-treatment mean square is equal to $\frac{72}{2} = 36$. The within-treatment mean square is equal to $\frac{46}{9} = 5.11$. The test of the null hypothesis that the population treatment means are equal is carried out by a comparison of the between-treatment mean square with the within-treatment mean square.

TABLE 8-14
Analysis of variance table for the teaching methods problem

	(1) Source of Variation	(2) Sum of Squares	(3) Degrees of Freedom	(4) Mean Square
Between treatments		72	2	36
Within treatments (error)		46	9	5.11
Total		118	11	

$$F(2, 9) = \frac{36}{5.11} = 7.05$$

$$F_{0.05}(2, 9) = 4.26$$

Since $7.05 > 4.26$, reject H_0

TABLE 8-15

General format of a one-factor analysis of variance table

(1) Source of Variation	(2) Sum of Squares	(3) Degrees of Freedom	(4) Mean Square
Between treatments	SSA	$v_1 = c - 1$	$SSA/(c - 1)$
Within treatments (error)	SSE	$v_2 = n - c$	$SSE/(n - c)$
Total	SST	$n - 1$	

$$F(v_1, v_2) = \frac{SSA/(c - 1)}{SSE/(n - c)}$$

Table 8-15 gives the general format of a one-factor analysis of variance table. The sums of squares are denoted as follows:

$$SSA = \text{between-treatment sum of squares}$$

$$SSE = \text{within-treatment sum of squares}$$

$$SST = \text{total sum of squares}$$

The rationale for the SSA symbol for the treatment sum of squares is as follows: The SS denotes "sum of squares," and the A indicates the first treatment. Hence, if there were additional treatments, the notation for their sums of squares would be SSB, SSC, and so forth. In the symbol SSE for the within-treatment sum of squares, the E denotes error while the SS again stands for sum of squares. The SSE symbol, referred to as the "error sum of squares," is a widely used notation. Other notation is as previously defined or as given in the next subsection.

The F Test and F Distribution

The comparison of the between-treatment mean square (variance) with the within-treatment mean square (variance) is made by computing their ratio, referred to as F. Hence, F is given by

(8.15) $$F = \frac{\text{Between-treatment variance}}{\text{Within-treatment variance}} = \frac{SSA/(c - 1)}{SSE/(n - c)}$$

F ratio   In the F ratio, the between-treatment variance is always placed in the numerator and the within-treatment variance in the denominator. Under the null hypothesis that the population treatment means are equal, the F ratio would tend to equal one. If the population treatment means do indeed differ, then the between-treatment mean square will tend to exceed the within-treatment mean square, and the F ratio will be greater than one. In terms of our problem concerning different teaching methods, if F is large, we will reject the null hypothesis that the population mean examination scores are all equal; that is, we will reject

$H_0: \mu_1 = \mu_2 = \mu_3$. If F is close to one, we will accept the null hypothesis. We can determine how large the test statistic F must be in order to reject the null hypothesis by referring to the probability distribution of the F random variable. This distribution is complex, and its mathematical expression is given here for reference only. Fortunately, critical values of the F ratio have been tabulated for frequently used significance levels analogous to the case of the χ^2 distribution. The probability density function of F is

F test

(8.16)
$$f(F) = kF^{(v_1/2)-1}\left(1 + \frac{v_1 F}{v_2}\right)^{-(v_1+v_2)/2}$$

where $v_1 =$ the number of degrees of freedom of the numerator of F
$v_2 =$ the number of degrees of freedom of the denominator of F
$k =$ a constant depending only on v_1 and v_2

Unbiased estimators $\hat{\sigma}_1^2$ and $\hat{\sigma}_2^2$ of the population variances are constructed from the sample, and

$$F = \frac{\hat{\sigma}_1^2}{\hat{\sigma}_2^2}$$

The F distribution is similar to the distributions of t and χ^2 in that it is actually a family of distributions. Each pair of values of v_1 and v_2 specifies a different distribution. F is a continuous random variable that ranges from zero to infinity. Since the variances in both the numerator and denominator of the F ratio are squared quantities, F cannot take on negative values. The F distribution has a single mode, and although the specific distribution depends on the values of v_1 and v_2, its shape is generally asymmetrical and skewed to the right. The distribution tends towards symmetry as v_1 and v_2 increase. We will use the notation $F(v_1, v_2)$ to denote the F ratio defined in equation 8.15, where the numerator and denominator are between-treatment mean squares and within-treatment mean squares with v_1 and v_2 degrees of freedom, respectively. Table A-8 of Appendix A presents the critical values of the F distribution for two selected significance levels, $\alpha = 0.05$ and $\alpha = 0.01$. In this table, v_1 values are listed across the columns and v_2 values are listed down the rows. There are two entries in the table corresponding to every pair of v_1 and v_2 values. The upper figure (in lightface type) is an F value that corresponds to an area of 0.05 in the right tail of the F distribution with v_1 and v_2 degrees of freedom. That is, it is an F value that would be exceeded only 5 times in 100 if the null hypothesis under consideration were true. The lower figure (in boldface type) is an F value corresponding to a 0.01 area in the right tail.

F distribution

F table

We will illustrate the use of the F table in terms of the teaching methods problem.

F table used for teaching methods problem

Assuming that we wish to test the null hypothesis $H_0: \mu_1 = \mu_2 = \mu_3$ at the 0.05 level of significance, we find in Table A-8 of Appendix A that for $v_1 = 2$ and

FIGURE 8-3

The *F* distribution for the teaching methods problem indicating the critical *F* value at the 5% level of significance

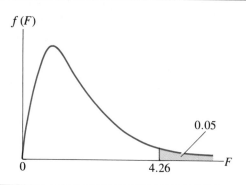

$v_2 = 9$ degrees of freedom, an *F* value of 4.26 would be exceeded 5% of the time if the null hypothesis were true. As indicated at the bottom of Table 8-14, we denote this critical value as $F_{0.05}(2, 9) = 4.26$. This relationship is depicted in Figure 8-3. Again referring to Table 8-14, since the computed value of the *F* ratio (the between-treatment mean square over the within-treatment mean square) is 7.05, and therefore greater than the critical value of 4.26, we reject the null hypothesis. Hence, we conclude that the treatment means (that is, the sample means of final examination scores in classes taught by the 3 teaching methods) differ significantly. The inference about the corresponding population means is that they are not all the same. Referring to Table 8-11(a), we see that average grades under method (3) exceed those under method (2), which are higher than those under method (1). Note that we would not be able to reject the null hypothesis at $\alpha = 0.01$ because $F_{0.01}(2, 9) = 8.02$. Since $7.05 < 8.02$, we would retain the null hypothesis, $H_0: \mu_1 = \mu_2 = \mu_3$ at the 1% significance level.

Conclusion on teaching methods

Based on these data, our inference is that the 3 teaching methods are not equally effective, and there is evidence that method (3) is the most effective and method (1) is the least effective.

Caveat

The foregoing example was used to illustrate the rationale involved in the analysis of variance, the statistical technique employed and the nature of the conclusions that can be drawn. However, we repeat the warning that the sample sizes in this illustration are too small for safe conclusions to be drawn about differences in effectiveness of the 3 teaching methods. After all, only 4 observations were made under each teaching method. Although the risk of a Type I error was controlled at 0.05 in this problem, the risk of Type II errors may be intolerably high. Suffice it to say that larger sample sizes are generally required. Also, the example was in terms of a treatment (teaching methods) applied to groups of equal size. In the next subsection, we discuss the use of a computer program for carrying out an analysis of variance.

Analysis of Variance and the Computer

The computer output for a Minitab analysis of variance program is shown in Table 8-16. The output pertains to the teaching methods problem discussed in this section. In Minitab, a one factor or one-way analysis of variance can be entered in different ways. The method shown here is with the use of the ONEWAY command in which the data are unstacked. That is, all of the data are typed in

---

TABLE 8-16

Computer output from a Minitab analysis of variance for the teaching methods problem

---

```
                              MTB > PRINT C1 C2

                              ROW  GRADE  TM

                               1    16    1
                               2    21    1
                               3    18    1
                               4    13    1
                               5    19    2
                               6    20    2
                               7    21    2
                               8    20    2
                               9    24    3
                              10    21    3
                              11    22    3
                              12    25    3

MTB > ONEWAY C1 C2

ANALYSIS OF VARIANCE ON GRADE
SOURCE        DF          SS         MS        F         P
TM             2       72.00      36.00     7.04     0.014
ERROR          9       46.00       5.11
TOTAL         11      118.00

                              INDIVIDUAL 95 PCT CI'S FOR MEAN
                              BASED ON POOLED STDEV

LEVEL    N               STDEV     ------ + ----- + ------ + ---------------
   1     4      MEAN      3.367     (----- * -----)
   2     4     17.000     0.816            (----- * ----- )
   3     4     20.000     1.826                   (----- * ----- )
                23.000              ------ + ----- + ------ + ---------------
POOLED STDEV = 2.261                     17.5      21.0     24.5
MTB >
```

one column. Note that all of the grades given originally in Table 8-11(a) are typed in one column. A second column is used to indicate the different "levels" of each observation. In this case, the levels are the different teaching methods, 1, 2, and 3 as in Table 8-11(a). The analysis of variance table shows the same numerical results as were shown in Table 8-14, except that the Minitab output includes a p value. In this case, the p value of 0.014 confirms our earlier conclusion that the null hypothesis is rejected at the five percent significance level but not at the one percent level.

Below the analysis of variance table, the Minitab output give the sample sizes and means for each teaching method. Also displayed are 95% confidence limits for each population mean. As indicated the confidence intervals are based on a pooled standard deviation, which is the square root of the mean square error, in this case, $2.261 = \sqrt{5.11}$.

One-Factor Analysis of Variance: Unequal Sample Sizes

In the teaching methods example, the samples were all the same size; the same number of students (4) were taught by each method. Although it is often simplest in the collection of data to work with samples of the same size, this is not always feasible. The analysis of variance computational procedure given earlier is easily adjusted for differing sample sizes. The general format for a one-factor analysis of variance for unequal sample sizes is the same as for equal sample sizes, as given in Table 8-15. However, the shortcut computational formulas differ slightly.

> Let us now consider the teaching methods problem as an example of unequal sample sizes. However, we shall assume that there were only nine students in the experiment: two in the first section, three in the second section, and four in the third section.

The final examination grades of these students are given in Table 8-17(a), and the analysis of variance is carried out in Table 8-17(b).

Columns (1), (2), and (3) of Table 8-17(b) list the possible sources of variation, the sums of squares for each of these sources, and the number of degrees of freedom associated with the sums of squares. Note that both sums of squares and degrees of freedom are additive; that is, these figures for between-treatment and within-treatment sources of variation add to the corresponding total variation figure.

The sums of squares have been computed by the shortcut formulas given in equations 8.17, 8.18, and 8.19, which are similar to the shortcut formulas of equations 8.9, 8.10, and 8.11 for equal-sized samples.

Between-treatment   (8.17)
sum of squares

$$\sum_j \frac{T_j^2}{r_j} - C = \frac{(37)^2}{2} + \frac{(60)^2}{3} + \frac{(92)^2}{4} - \frac{(189)^2}{9}$$

$$= 31.5$$

TABLE 8-17

(a) Final examination grades of 9 students taught by 3 different methods

Teacher	Teaching Method 1	2	3
1	16	19	24
2	21	20	21
3	—	21	22
4	—	—	25
Total	37	60	92
Mean	18.5	20.0	23.0

(b) Analysis of variance table for the teaching methods problem with unequal sample sizes

(1) Source of Variation	(2) Sum of Squares	(3) Degrees of Freedom	(4) Mean Square
Between treatments	31.5	2	15.75
Within treatments (error)	24.5	6	4.08
Total	56.0	8	

$$F(2, 6) = \frac{15.75}{4.08} = 3.86$$

$$F_{0.05}(2, 6) = 5.14$$

Since $3.86 < 5.14$, reject H_0

(8.18)
$$\sum_j \sum_i X_{ij}^2 - \sum_j \frac{T_j^2}{r_j} = (16)^2 + (21)^2 + (19)^2 + \cdots + (25)^2$$

$$-\frac{(37)^2}{2} - \frac{(60)^2}{3} - \frac{(92)^2}{4}$$

$$= 24.5$$

Within-treatment sum of squares

(8.19)
$$\sum_j \sum_i X_{ij}^2 - C = SSA + SSE = 31.5 + 24.5$$

$$= 56$$

Total sum of squares

where C, the correction term, is given by

(8.20)
$$C = \frac{T^2}{\sum_j r_j} = \frac{(189)^2}{9}$$

where T_j and r_j are the total and number, respectively, of observations in the jth column, and T is the grand total of all observations as given in equation 8.13.

The number of degrees of freedom associated with the between-treatment sum of squares is $c - 1 = 3 - 1 = 2$. The number of degrees of freedom associated with the within-treatment variation is $n - c = 9 - 3 = 6$. As shown in Table 8-15(b), the analysis of variance computations are carried out in similar fashion to the equal-sized sample case. Since the F test results in the acceptance of the null hypothesis, we cannot conclude that there is any difference in the effectiveness of the three teaching methods as measured by final examination grades.

The one-factor teaching methods illustration used in this chapter has another characteristic: This type of example is conventionally designated as a

completely randomized experimental design and it involves the random selection of independent samples from c (the number of columns) normal populations. This implies that the students taught by each of the three methods were randomly selected from the sophomore class.

Two-factor Analysis of Variance

Let us look more critically at the possible interpretations of our findings. We assumed that the same teacher taught three different sections of the basic statistics course using three specified methods—1, 2, and 3. Final examination scores seemed to indicate that method 3 was the most effective and method 1 the least effective teaching method.

New data for the teaching-methods problem

> Suppose the method 3 class is given early in the morning, when the teacher (and students, too?) is fresh, wide awake, and enthusiastic. On the other hand, suppose the classes taught by methods 2 and 1 are in the middle of the day and the late afternoon, respectively. Let us assume that by late afternoon the instructor is tired, sleepy, and rather unenthusiastic. In this case, the differences in teaching effectiveness may not be attributable solely to the different teaching methods but rather to some unknown mixture of the difference in teaching methods and the aforementioned time-of-day factors.

Time of day may be thought of as an extraneous factor that influences the variable of interest, teaching methods. A **two-factor** or **two-way analysis of variance** isolates the influence of this extraneous factor so the effects of the main variable may be more accurately judged.

Randomized block design

There are two basic types of the two-factor analysis of variance: the *randomized block design* and the *completely randomized design*. The **randomized block design** is appropriate for the situation just described; we are interested in the influence of a certain factor (for example, teaching methods), but we also wish to isolate the effects of a second extraneous factor, such as time of day.

> The term "block" derives from experimental design work in agriculture, in which parcels of land are referred to as blocks. In a randomized block design, treatments are randomly assigned to units within each block. In testing the yield of different fertilizers, for example, this design ensures that the best fertilizer is applied to all types of soil, not just the best soil.

In the teaching methods illustration, the times of day could be treated as the blocks, and the teaching methods (treatments) would be randomly assigned within the blocks. Thus, methods 1, 2, and 3 would each be equally represented in the early morning, middle of the day, and late afternoon. The null hypothesis is the same as in the one-factor analysis of variance, namely $H_0: \mu_1 = \mu_2 = \mu_3$. However, the design removes the variation in time of day from the comparison of the three teaching methods. No inference is attempted about the effect of time of day, since time of day is viewed as an extraneous factor. The extraneous blocking factor usually represents time, location, or experimental material. Just as the one-factor analysis of variance represents a generalization of the t test for means of two independent samples, the randomized block design represents a generalization of the t test for paired observations discussed in section 7.5.

The second type of two-factor analysis of variance is used when inferences about both factors are desired.

> Suppose that 4 different instructors were teaching the statistics course and that inferences were desired about differences in effectiveness among the 3 methods of teaching *and* among the 4 different teachers.

Of course, the sample sizes in this experiment and in the randomized block design would have to be much larger than the 12 students used earlier. This two-factor study would use a completely randomized design, rather than the randomized block design. In the **completely randomized design**, the sample units (students, in this case; sections of students in a more realistic example) would be randomly assigned to each factor combination. For example, teacher 1 using method 1, teacher 1 using method 2, and so on, would represent factor combinations. This experimental design then permits the testing of two null hypotheses:

Completely randomized design

1. H_0: No difference in population mean final examination scores among the different teaching methods
2. H_0: No difference in population mean final examination scores among the different teachers

Table 8-18 shows final examination scores for the three teaching methods by four different teachers.

TABLE 8-18
Final examination grades of 12 students for 3 different teaching methods by 4 different teachers

Teacher	Teaching Method 1	2	3	Row Sum T_i
1	16	19	24	59
2	21	20	21	62
3	18	21	22	61
4	13	20	25	58
Column sum (T_j)	68	80	92	240

TABLE 8-19
General format of a two-factor analysis of variance table

(1) Source of Variation	(2) Sum of Squares	(3) Degrees of Freedom	(4) Mean Square
Teaching methods	SSA	$c - 1$	$SSA/(c - 1)$
Teachers	SSB	$r - 1$	$SSB/(r - 1)$
Error	SSE	$(r - 1)(c - 1)$	$SSE/(r - 1)(c - 1)$
Total	SST	$rc - 1$	

To test the two hypotheses, we set up a general format of a two-factor analysis of variance table as shown in Table 8-19. As before,

$$SSA = \sum_j \frac{T_j^2}{r} - \frac{T^2}{rc}$$

$$= \frac{(68)^2}{4} + \frac{(80)^2}{4} + \frac{(92)^2}{4} - \frac{(240)^2}{(4)(3)}$$

$$= 72.00$$

$$SST = \sum_j \sum_i X_{ij}^2 - \frac{T^2}{rc}$$

$$= (16)^2 + (21)^2 + (18)^2 + \cdots + (25)^2 - \frac{(240)^2}{(4)(3)}$$

$$= 118.00$$

And analogous to SSA, the sum of squares within columns is

$$SSB = \sum_i \frac{T_i^2}{c} - \frac{T^2}{rc}$$

$$= \frac{(59)^2}{3} + \frac{(62)^2}{3} + \frac{(61)^2}{3} + \frac{(58)^2}{3} - \frac{(240)^2}{(4)(3)}$$

$$= 3.33$$

Since $SST = SSA + SSB + SSE$, the sum of squares of the error term, SSE, can be found as a residual by subtraction. $SSE = 118.00 - 72.00 - 3.33 = 42.67$.

An analysis of variance table for the two-factor teaching methods and teacher problem is given in Table 8-20, and the two hypotheses are tested as shown.

TABLE 8-20
Analysis of variance table for the two-factor teaching problem

(1) Source of Variation	(2) Sum of Squares	(3) Degrees of Freedom	(4) Mean Squares
Teaching methods	72.00	$3 - 1 = 2$	36.00
Teachers	3.33	$4 - 1 = 3$	1.11
Error	42.67	$(3 - 1)(4 - 1) = 6$	7.11
Total	118.00	11	

$$F(2, 6) = \frac{36.00}{7.11} = 5.06$$

Hypothesis test for 3 teaching methods

$$F_{0.05}(2, 6) = 5.14$$

Since $5.06 < 5.14$, retain H_0.

$$F(3, 6) = \frac{1.11}{7.11} = 0.16$$

Hypothesis test for 4 teachers

$$F_{0.05}(3, 6) = 4.76$$

Since $0.16 < 4.76$, retain H_0.

> Since neither of the two null hypotheses was rejected, we cannot conclude from these data on examination grades that there were real differences in effectiveness among the 3 teaching methods or among the 4 teachers.

Summary

Two-factor analyses of variance have some distinct advantages. Note from the example that we were able to test two separate null hypotheses from the same set of experimental data. We did not need to run two one-factor experiments to get information about two factors. Furthermore, certain types of questions can be answered by two-factor designs but cannot be treated in one-factor analyses. For example, the interaction, or joint effects, of the two factors may be examined as well as their separate effects.

Advantages of two-factor analyses

In the next subsection, we discuss the method by which these interactions or joint effects of different factors can be measured.

Two-Factor Analyses with Interaction

In order to explain the meaning of interaction in terms of a simple example, we consider a case in which management has analyzed productivity, measured as mean output per day, for two different departments by number of years of experience of employees. As we can see from the data given in Table 8-21 the productivity of Department B is greater than that of Department A, but the

TABLE 8-21

Mean output per day of two departments of a company by number of years of experience of employees

	Years of Experience		
	Less than 4 Years	4 through 6 Years	More than 6 Years
Department A	85	88	110
Department B	87	95	130
Difference (B − A)	2	7	20

difference depends on the number of years of experience. For employees with less than four years of experience the difference is 2 units; for those with four through six years of experience the difference is 7 units; for those with more than six years of experience the difference is 20 units.

If these differences in productivity were the same for the three levels of experience, there would be no interaction between the two factors "department" and "years of experience." Because the differences in the departmental population means of productivity depend on years of experience, there is said to be interaction between the "department" and "years of experience" factors. In general, if the differences between population means of one factor vary by levels of the other factor, the two factors interact.

Figure 8-4 shows a plot of the department population mean productivities by years of experience of employees. If the differences in productivity between departments had been constant for the three levels of experience, two parallel straight lines would have been obtained. Since the lines for the two departments are not parallel, interaction is indicated.

We turn now to a discussion of an example of two-factor analysis of variance with a test for interaction as carried out in Minitab. We consider the case of a food products company that wanted to determine what effect, if any, three different food plant treatments and three water temperatures have on the weights of an experimental type of large pumpkin. Therefore, there were $9 = 3 \times 3$ combinations of levels of food plant treatments and water temperatures. A sample was used in which 3 pumpkins were assigned randomly to each combination. Thus, there were 27 pumpkins whose weights could be measured to determine the effects of the two factors. The questions to be answered were as follows:

- Are the mean weights of pumpkins for the three food treatments equal?
- Are the mean weights of pumpkins for the three water temperatures equal?
- Is there a significant interaction between food treatments and water temperature?

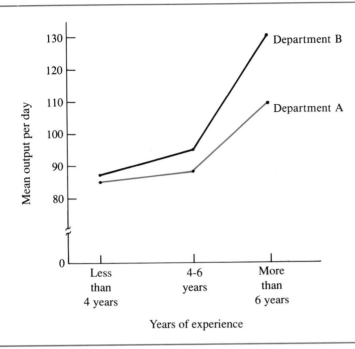

FIGURE 8-4

Departmental mean outputs per day by years of experience of employees

The basic data are given in Table 8-22. The weights appear in column 1, water temperatures in column 2 (code: cool = 1, warm = 2, hot = 3), and type of food treatment in column 3 (code: 1, 2, 3). The data are summarized in tabular form in Table 8-23 with the command TABLE C2 C3; and the subcommand DATA in C1.

Table 8-24 gives means for the nine cells of a table formed from the three levels of water temperature (rows) and three levels of food treatments (columns). Note that each cell contains the same number of observations, three in this example. The experimental design must be balanced in this sense. If there is only one observation per cell, then the sum of squares for interaction cannot be computed. We can get some suggestions concerning what the analysis of variance might reveal by observing this table of cell means. Noting the average effect of each factor, we see that average weights increase with water temperature. That is, as we move from 1 (cool) to 3 (hot), average weights increase steadily from 18.778 to 25.444. Average weights also increase as the level of food treatment moves from 1 to 3, although the level 2 figure 23.000 is almost the same as the level 3 figure 23.333. Hence, we would not be surprised if type of food treatment turns out to be not as significant as water temperature.

Turning to the effect of interaction we note that for food treatment 1, average weights increase as water temperature increases. However, for food treatment 2 and food treatment 3 average weights increase as water temperature level is increased from cool (1) to warm (2), but then decrease as the temperature

TABLE 8-22
Data for pumpkin weights classified by water temperature and type of food treatment

```
MTB > INFORMATION

COLUMN NAME    COUNT
C1     WEIGHT    27
C2     TEMP      27
C3     FOOD      27

CONSTANTS USED:NONE

MTB > PRINT C1 C2 C3

ROW WEIGHT TEMP FOOD
```

ROW	WEIGHT	TEMP	FOOD
1	11	1	1
2	15	1	1
3	12	1	1
4	21	1	2
5	22	1	2
6	17	1	2
7	26	1	3
8	22	1	3
9	23	1	3
10	12	2	1
11	14	2	1
12	13	2	1
13	28	2	2
14	29	2	2
15	18	2	2
16	29	2	3
17	27	2	3
18	26	2	3
19	40	3	1
20	28	3	1
21	32	3	1
22	24	3	2
23	26	3	2
24	22	3	2
25	19	3	3
26	18	3	3
27	20	3	3

TABLE 8-23
Data for pumpkin weights in table form

```
MTB > TABLE C2 C3;
SUBC> DATA IN C1.

ROWS: TEMP   COLUMNS: FOOD
                 1           2           3

     1        11.000      21.000      26.000
              15.000      22.000      22.000
              12.000      17.000      23.000

     2        12.000      28.000      29.000
              14.000      29.000      27.000
              13.000      18.000      26.000

     3        40.000      24.000      19.000
              28.000      26.000      18.000
              32.000      22.000      20.000

   CELL CONTENTS --
                         WEIGHT: DATA
```

TABLE 8-24
Table of cell means for the pumpkin weights data

```
MTB > TABLE C2 C3;
SUBC> MEAN C1.

ROWS: TEMP   COLUMNS: FOOD
              1           2           3          ALL

   1        12.667      20.000      23.667      18.778
   2        13.000      25.000      27.333      21.778
   3        33.333      24.000      19.000      25.444
 ALL        19.667      23.000      23.333      22.000

CELL CONTENTS --
                   WEIGHT: MEAN
```

TABLE 8-25
Notation for a two-factor analysis of variance with interaction

Source of Variation	Sum of Squares	Degrees of Freedom
Rows	$cn\Sigma_i(\bar{x}_{i..} - \bar{x})^2$	$r - 1$
Columns	$rn\Sigma_j(\bar{x}_{.j.} - \bar{x})^2$	$c - 1$
Interaction	$n\Sigma_i\Sigma_j(\bar{x}_{ij.} - \bar{x}_{i..} - \bar{x}_{.j.} + \bar{x})^2$	$(r - 1)(c - 1)$
Error	$\Sigma_i\Sigma_j\Sigma_k(x_{ijk} - \bar{x}_{ij.})^2$	$rc(n - 1)$
Total	$\Sigma_i\Sigma_j\Sigma_k(x_{ijk} - \bar{x})^2$	$rcn - 1$

level is increased to hot (3). Since the effect of water temperature appears to depend on type of food treatment, we suspect that the analysis of variance may indicate that this interaction will be significant.

We carry out the analysis of variance with the use of the command TWOWAY. The total variation or total sum of squares can now be conceived as being partitioned as follows:

Total variation = Variation due to rows (water temperature)

+ variation due to columns (food treatment)

+ variation due to interaction

+ variation due to error

To indicate the nature of the analysis of variance table, we use the following notation:

r is the number of levels of the first factor (number of rows)

c is the number of levels of the second factor (number of columns)

n is the number of observations in each cell

x_{ijk} is the kth observation in cell (i, j) (i.e., in the cell in row i and column j)

$\bar{x}_{ij.}$ is the mean of the n observations in cell (i, j)

$\bar{x}_{i..}$ is the mean of the cn observations in row i

$\bar{x}_{.j.}$ is the mean of the rn observations in column j

$\bar{x}$ is the mean of the rcn observations

With this notation, the two-factor analysis of variance (with interaction) table depicted in Table 8-25. The analysis of variance using Minitab is given in Table 8-26.

The F tests for the effects of the two factors, usually referred to as tests of main effects follow from our earlier discussions. However, the test for interaction should be carried out before the tests for the effects of the two factors. If a significant interaction is present, care must be taken concerning the interpretation of the tests for the main effects. We should always construct a graph

TABLE 8-26
Minitab analysis of variance table for the pumpkin weights data

```
MTB > TWOWAY DATA IN C1, SUBSCRIPTS IN C2 AND C3

ANALYSIS OF VARIANCE WEIGHT
```

SOURCE	DF	SS	MS
TEMP	2	200.7	100.3
FOOD	2	74.0	37.0
INTERACTION	4	786.7	196.7
ERROR	18	196.7	10.9
TOTAL	26	1258.0	

of the effects of the factors such as the profile plot in Figure 8-4. In that case a comparison of the department effect may be meaningful because Department B average productivity is always greater than that of Department A even though the difference depends on years of experience of employees. Similarly the effects of the experience factor would be of interest even though the difference between levels depends upon departments.

Let us now examine the results of the pumpkin weights analysis. As shown in Table 8-26, the interaction sum of squares is 786.7 with 4 degrees of freedom yielding a mean square for interaction of $786.7/4 = 196.7$. We carry out an F test for the significance of the interaction by computing

$$F = \frac{MS \text{ INTERACTION}}{MS \text{ ERROR}} = \frac{196.7}{10.9} = 18.05$$

Referring to Appendix Table A-8, using 4 and 18 degrees of freedom we find critical values at $\alpha = 0.05$ and $\alpha = 0.01$ of 2.93 and 4.58. Since the observed F value of 18.05 substantially exceeds these critical values, a significant interaction effect is present. As we have observed informally, the effect of one factor depends on the level of the other factor. In such a situation, we must use judgment as to whether the row and column means are of interest.

For completeness, we carry out the F tests for the two factors. For the water temperature effect, we have from Table 8-26

$$F = \frac{MS \text{ TEMP}}{MS \text{ ERROR}} = \frac{100.3}{10.9} = 9.20$$

Using 2 and 18 degrees of freedom, we find critical values of F in Appendix Table A-8 to be at $\alpha = 0.05$ and $\alpha = 0.01$ to be 3.55 and 6.01, respectively. Since the observed $F = 9.20$ exceeds these values, this results in a rejection of the null hypothesis of no main effect for water temperatures. More specifically, the test results in a rejection of the null hypothesis that the means for water temperatures (averaged over the categories of food treatments) are equal. For the food treatment effect, we have

$$F = \frac{\text{MS FOOD}}{\text{MS ERROR}} = \frac{37.0}{10.9} = 3.39$$

Since the numbers of degrees of freedom are again 2 and 18, we have critical F values at $\alpha = 0.05$ and $\alpha = 0.01$ of 3.55 and 6.01. Hence, the food treatment effect is not significant at either the 5% or 1% levels of significance. Note

---

TABLE 8-27

Tabular arrangement to obtain a profile plot for the pumpkin weights data

---

```
MTB > TABLE C2 C3;
SUBC> MEAN C1.

ROWS: TEMP    COLUMNS: FOOD

                  1              2              3            ALL

     1        12.667         20.000         23.667         18.778
     2        13.000         25.000         27.333         21.778
     3        33.333         24.000         19.000         25.444
   ALL        19.667         23.000         23.333         22.000

CELL CONTENTS --
                        WEIGHT: MEAN


MTB > TABLE C2 C3;
SUBC> DATA IN C10.

ROWS: TEMP    COLUMNS: FOOD

                1            2            3

  1        12.667       20.000       23.667
           12.667       20.000       23.667
           12.667       20.000       23.667

  2        13.000       25.000       27.333
           13.000       25.000       27.333
           13.000       25.000       27.333

  3        33.333       24.000       19.000
           33.333       24.000       19.000
           33.333       24.000       19.000

CELL CONTENTS --
                   C10: DATA
```

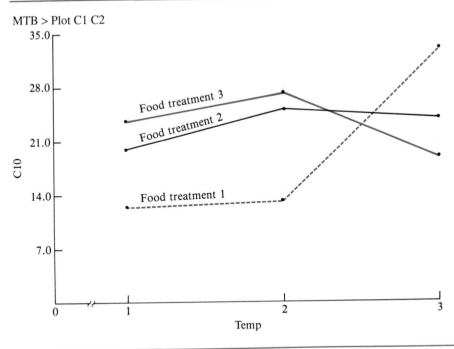

FIGURE 8-5
Profile plot for the pumpkin
weights problem

that the general nature of these tests confirms the suggestions given in our informal analysis of the table of cell means, Table 8-24.

A profile plot of the cell means should always be examined to observe the nature of the interaction effect. There are a number of ways of plotting the data in Minitab. Table 8-27 and Figure 8-5 indicate one method for obtaining the plot. In Table 8-27, the cell averages are replicated three times to indicate that three observations are present in each average. The profile plot in Figure 8-5 depicts one line for each food treatment level. Water temperature levels are on the horizontal axis and mean weights on the vertical axis. We observe visually the interaction effect referred to earlier in an informal examination of the cell means data, that is, the differences in effects of one factor depend on the levels of the other factor.

Only a brief introduction to the analysis of variance has been given in this chapter. More elaborate designs than those considered in this book are available; they attempt to control and test for the effects of more factors, both qualitative and quantitative.

Further Remarks

As we saw in the one-factor teaching methods example, observed differences in teaching effectiveness may not have been attributable solely to the different teaching methods, but rather to some unknown mixture of teaching methods

Importance of careful experimental design

and factors associated with time of day, and perhaps other factors as well. Hence, unless careful thought is given to the experimental design from which data are to be collected, erroneous inferences may be drawn. This applies equally to the hypothesis-testing methods considered earlier, for example in Chapter 7, because we might have had only two teaching methods to compare rather than three. Thus, we must guard against mechanical or rote application of statistical techniques such as hypothesis-testing methods. In this book, we consider the general principles involved in some of the simpler, basic procedures. More refined and sophisticated techniques may very well be required in particular instances.

Alternative interpretations

One of the points we have attempted to convey in the preceding discussion is that statistical results are virtually always consistent with more than one interpretation. The researcher must avoid naively leaping to conclusions and must give careful consideration to alternative interpretations and explanations. We conclude this chapter with two anonymous humorous stories that are relevant to the point that alternative interpretations and explanations of experimental results are often possible.

An investigator wished to determine the differential effects involved in drinking various types of mixed drinks. Therefore, he had subjects drink substantial quantities of scotch and water, bourbon and water, and rye and water. All of the subjects became intoxicated. The investigator concluded that since water was the one factor common to all of these drinks, the imbibing of water makes people drunk.

The heroine of our second story is a grammar school teacher, who wished to explain the harmful effects of drinking liquor to her class of 8-year-olds. She placed two glass jars of worms on her desk. Into the first jar, she poured some water. The worms continued to move about, and did not appear to have been adversely affected at all by the contact with the water. Then she poured a bottle of whiskey into the second jar. The worms became still and appeared to have been mortally stricken.

The teacher then called on a student and asked, "Johnny, what is the lesson to be learned from this experiment?" Johnny, looking thoughtful, replied, "I guess it proves that it is good to drink whiskey, because it will kill any worms you may have in your body."

EXERCISES 8.3

Note: In Exercises 8.3 you may ignore interaction effects.

1. A consumer research organization was interested in the influence of type of water on the effectiveness of a detergent. Test batches of washings were run in four randomly chosen machines having a particular type of water— soft, moderate, and hard. All batches had equal numbers of oil-stained

rags, and after each washing the number of rags still stained was determined. The following results were obtained:

Sample Observation	Number of Rags with Stains		
	Soft	Moderate	Hard
1	0	4	10
2	1	8	5
3	2	3	7
4	1	9	6

Using an $\alpha = 0.01$ level of significance, would you conclude that the type of water influences the effectiveness of the detergent?

2. The credit department supervisor of Racy's Department Store is evaluating collection of delinquent accounts. The following figures show the number of accounts each collector successfully removed from delinquency during each week of a five-week period. The accounts to be investigated were equally divided among the collectors each week. Should the supervisor conclude that there is no difference among these collectors in the average number of successful investigations per week? Use a 0.05 significance level.

Week	Collectors		
	A	B	C
1	25	45	45
2	35	45	40
3	30	25	35
4	25	30	35
5	20	30	45

3. An economist wishes to compare the profit levels of firms which she has classified as "competitors," "monopolistic competitors," and "monopolists." She has randomly chosen 10 firms in a large city and has recorded their profit levels (profit-to-investment ratios) below:

Competitors	Monopolistic Competitors	Monopolists
0%	5%	9%
2	7	11
5	9	13
		25

Should the economist conclude that no difference exists among the profit levels of these three classes of firms? Use a 0.01 significance level.

4. The president of Finplan Corporation has asked his financial analyst to compare the expected return and risk levels of nine investments in three industries. The analyst proceeds by comparing the ratios of the expected return to the corresponding standard deviation for these investments. These ratios were then tabulated as shown below:

Industry A	Industry B	Industry C
0.5	0.3	0.7
0.6	0.4	0.8
	0.6	1.1
		1.2

Should the financial analyst conclude that there is no difference in the average ratios of expected return to standard deviation of the three industries? Use a 0.05 significance level.

5. China Air Company wants to determine the effect of advertising expenditures and sales discounts on sales. The sales manager has collected the following information regarding sales at different levels of advertising and sales discount.

Advertising Expenditure	Sales Discount Rates		
	0–4.9%	5–9.9%	10–14.9%
$0–$50,000	$2,000,000	$2,300,000	$3,400,000
$50,000–$100,000	2,100,000	2,500,000	3,400,000
$100,000–$150,000	2,200,000	2,900,000	3,600,000

Test the following hypotheses: (1) Sales at the three levels of advertising expenditure are identical. (2) Sales for the three sales discount rates are identical. Use a 0.01 significance level.

6. The Song Song government wants to determine the effects of national income and relative local prices (to those overseas) on the national trade surplus (or deficit). The government tabulated the following historical data:

National Income	Relative Price Ratio			
	1.0	1.1	1.2	1.3
$ 90 billion	($0.9 billion)	$1.0 billion	$0.8 billion	$1.0 billion
100 billion	(2.1 billion)	(0.9 billion)	0.6 billion	2.5 billion

Test the following hypotheses: (1) National trade surpluses (or deficits) at the three levels of national income are the same. (2) National trade surpluses (or deficits) for the four levels of relative prices are the same. Use 0.05 significance levels.

7. The treasurer of Specific Motors Company wished to compare the performances of 3 pension funds. To do so, she selected 30 observations, 10 corresponding to each pension fund. Assuming identical contributions and the following results (with a base index of 100), would you conclude that there was no real difference in performance among the three funds?

Pension fund I	71	60	83	70	90	62	73	74	65	88
Pension fund II	123	140	111	129	127	136	130	119	121	149
Pension fund III	133	118	141	132	127	138	133	129	122	144

8. Labor and management disagree over productivity of a machine shop with regard to labor time and machine time on a job. They have collected the following data regarding output times in minutes.

| Machine | Worker | | | |
	1	2	3	4
A	12	13	14	11
B	16	15	13	12
C	14	12	13	13
D	15	13	15	12

Perform a two-way analysis of variance at the 0.01 significance level.

9. The treasurer of a multinational corporation wishes to compare the debt-to-asset ratios of the corporation's three subsidiaries in Australia, Germany, and Taiwan over a 10-year period at 5-year intervals. He collected the following data:

| Country | Debt-to-Asset Ratios (percent) | | |
	1980	1985	1990
Australia	50	40	45
Germany	48	39	45
Taiwan	52	44	48

a. Using a two-way analysis of variance, test whether the debt-to-asset ratios differ among the three countries. Use $\alpha = 0.05$.

b. Test whether the debt-to-asset ratio differs over time. Use $\alpha = 0.05$.

10. An investment analyst wishes to compare the performance of three money market funds. To do so, he selects 30 observations, 10 corresponding to each fund. An index of performance is developed with a reference base of

100. The following results are obtained:

Fund I	71	60	83	70	90	62	73	74	65	88
Fund II	123	140	111	129	127	136	130	119	121	149
Fund III	133	118	141	132	127	138	133	129	122	144

Can you conclude that no real differences in performance exist among the three funds if $\alpha = 0.05$?

11. Three personnel officers rate potential employees on a scale from 1 to 10. The following data are collected on the basis of the scores given to 11 potential employees by the three officers:

$$\bar{x}_1 = 8.0 \qquad s_1^2 = 3 \qquad n_1 = 11$$

$$\bar{x}_2 = 7.0 \qquad s_2^2 = 4 \qquad n_2 = 11$$

$$\bar{x}_3 = 6.0 \qquad s_3^2 = 5 \qquad n_3 = 11$$

Can you conclude that average ratings differ among the three personnel officers if $\alpha = 0.05$?

1. A survey of 1,000 people under 40 years of age has been carried out and the result is as follows:

Number of People	Age Range
116	Under 5 Years of Age
138	Between 5–9 Years of Age
125	Between 10–14 Years of Age
123	Between 15–19 Years of Age
132	Between 20–24 Years of Age
117	Between 25–29 Years of Age
121	Between 30–34 Years of Age
~~127~~ 128	Between 35–39 Years of Age

Calculate the chi-square statistic for the testing problem:

H_0: The proportions of the population in the above age groups are equal

H_1: The proportions of the population in the above age groups are not equal

Test H_0 against H_1 at alpha $= 0.05$.

2. Minitab command

MTB > random 200 C1;

SUBC > binomial with n = 3 and p = 0.5.

supposedly generates a sample of 200 observations from the binomial distribution with $n = 3$ and $p = 0.5$. Try to produce such a sample and test the hypothesis that your sample came from the binomial distribution with $n = 3$ and $p = 0.5$. Assume alpha $= 0.10$.

3. Given the following data:

Numbers of workers classified by city and method of going to workplace

City	Method of Going to Workplace			
	Carpooling	PublicTrans	Walking	Home
City A	994	33	476	226
City B	1556	86	964	308
City C	1614	61	887	279
City D	840	28	1041	415

Use Minitab to carry out a chi-square test of independence between locations of the cities and the ways of going to work, at alpha $= 0.01$. If you multiply all the data by 100, do you expect to get the same chi-square statistic? *Hint:* First store the data in C1 — C4, and then try

MTB > chisquare C1 — C4

4. Minitab file 'salary . mtw' stores the following information:

Column	Count	Description
C1	30	Rate of Increase (%) in Annual Salary of Surveyed Workers in Manufacturing Industry
C2	23	Rate of Increases (%) in Annual Salary of Surveyed Workers in Energy Industry
C3	28	Rate of Increase (%) in Annual Salary of Surveyed Workers in Government
C4	25	Rate of Increase (%) in Annual Salary of Surveyed Workers in Banking Industry

Note that the following command

MTB > dotplot C1 C2 C3 C4;
SUBC > same.

produces dotplots of C1 to C4 on the same graph. Guess the tenability of the hypothesis that the mean rates of increase in annual salary in the 4 industries are equal. Now carry out a test of the hypothesis at alpha = 0.05, with the help of the Minitab command:

MTB > aovoneway C1 — C4

5. Consider the data in the file labeled 'dmotor . mtw':

Column	Name	Description
C1	STATES	Selected Counties in California, coded by 1; Those in Illinois, Coded by 2; Alaska, by 3
C2	DMOTOR	Number of Deaths from Motor Vehicle Accidents per 10,000 Population in One Year

Test the hypothesis that the means of DMOTOR in the 3 states are equal at alpha = 0.01. *Hint:* try

MTB > oneway for data in C2, subscripts in C1

*6. In Minitab file 'lifetime . mtw', you will find

Column	Name	Description
C1	DEGREE	The Highest Degree Obtained: Bachelor Degree (1), Master Degree (2), or Ph.D Degree (3)
C2	SEX	Male (0), or Female (1)
C3	LIFE	Length of Life (in *years*)

Note that the Minitab command

MTB > table C1 C2;
SUBC > data in C3.

* Optional exercise.

can help you better view the data. Also try the command

MTB > table C1 C2;
SUBC > mean of C3.

to see if there is an interaction between SEX and DEGREE. Construct a two-way analysis of variance table for data in C3 with factors in C1 and in C2, via command

MTB > two-way analysis, obs in C3, factors in C1 and C2

and formally test the null hypothesis that no interaction exists between SEX and DEGREE, at alpha = 0.05.

KEY TERMS

Analysis of Variance   a test to determine on the basis of sample data whether more than two population means may be considered to be equal.

Between-Treatment Variation   variation of the sample means ($\bar{X}_1$, $\bar{X}_2$, etc.) around the grand mean ($\bar{\bar{X}}$).

Chi-Square Tests   tests of goodness of fit and tests of independence.

Completely Randomized Design   a two-factor analysis of variance in which the sample units, eg., sections of students are randomly assigned to each combination of the two factors, eg., teaching methods and teachers.

F-Ratio   the ratio of the between-treatment variance to the within-treatment variance.

Goodness of Fit Test   a test in which observed sample frequencies are compared to theoretical frequencies to determine whether a particular probability distribution, such as the binomial distribution, constitutes a "good fit."

Grand Mean ($\bar{\bar{X}}$)   the mean of all of the observations.

Homoscedasticity   equal variances.

Mean Square   a variance.

One-Factor (one-way) Analysis of Variance   a test in which the effect of a single factor is assessed.

Randomized Block Design   a two-factor analysis of variance in which the effect of one factor, considered extraneous, is isolated in order that the effects of the main factor may be more accurately measured.

Test of Independence   a test for the independence of two variables on the basis of sample data.

Total Variation (total sum of squares)   the sum of the squared deviations of all of the individual observations from the grand mean $\bar{\bar{X}}$.

Two-Factor (two-way) Analysis of Variance   a test in which the effects of two factors are measurable.

Variance   a measure of variation divided by an appropriate number of degrees of freedom.

Variation   a sum of squared deviations.

Within-Treatment Variation   variation of the individual observations within each treatment from their respective means $\bar{X}_1$, $\bar{X}_2$, etc.

KEY FORMULAS

Chi-Square

$$\chi^2 = \sum \frac{(f_o - f_t)^2}{f_t}$$

Within-Treatment Variation

$$\text{SSE} = \sum_j \sum_i (X_{ij} - \bar{X}_j)^2$$

Between-Treatment Variation

$$\text{SSA} = \sum_j r(\bar{X}_j - \bar{\bar{X}})^2$$

Total Sum of Squares

$$\text{SST} = \sum_j \sum_i (X_{ij} - \bar{\bar{X}})^2$$

F Test

$$F = \frac{\text{Between-treatment variance}}{\text{Within-treatment variance}} = \frac{\text{SSA}/(c-1)}{\text{SSE}/(n-c)}$$

Between-Treatment Variance (mean square)

$$\frac{\text{SSA}}{(c-1)} = \sum_j \frac{r\,(\bar{X}_j - \bar{\bar{X}})^2}{(c-1)}$$

Within-Treatment Variance (mean square)

$$\frac{\text{SSE}}{(n-c)} = \sum_j \sum_i \frac{(X_{ij} - \bar{X}_j)^2}{(n-c)}$$

REVIEW EXERCISES FOR
CHAPTERS 6 THROUGH 8

1. In section 1 of an MBA course, a module was added with cases that deal with ethical issues. This module was not given to section 2 of the course. Both sections were given the same case to analyze on the final examination. Of the 60 students in section 1, 33 students mentioned ethical issues in analyzing the case. In contrast, only 27 of the 60 students in section 2 mentioned ethical issues.

   a. Do the data support the claim that more than 50% of students would be made sensitive to ethical issues if given the module? Use $\alpha = 0.05$.

   b. How well would the test in part (a) perform if, in fact, 60% of students are made sensitive to ethical issues by the module? How would you improve the performance of the test?

   c. Do the data suggest that students who are given the module with cases on ethical issues are more sensitive to those issues than students who are not given the module? Use $\alpha = 0.02$.

2. Universal Steel Company is in financial trouble. The company is currently negotiating with the steelworker's union for pay increases in a new contract. The president of the company suggests a profit-sharing arrangement with no pay increase. The officials of the union want to determine quickly the sentiment of the union members in regard to this offer. To this end, the officials telephone 100 union members at random and find that 60 of them are favorable to the offer.

   a. Can the union be fairly certain (i.e., use $\alpha = 0.01$) that the proportion of all union members that are favorable exceeds 0.5?

   b. Assume that the proportion of all union members that support the claim is 0.64. What is the probability that the union will erroneously conclude that the proportion does not exceed 0.5 if the union uses the decision rule from part (a)?

3. A company has been accused of practicing sex discrimination in its salary scales. In particular, for a certain entry-level position the following data have been collected over a period of 1 year.

	Men	Women
Number of employees observed	60	34
Mean starting salary	$15,500	$14,600
Standard deviation of starting salary	$650	$700

The company does not dispute the accuracy of the data. However, it contends that the difference in average salaries is simply due to chance and does not reflect a systematic bias. If the population mean salaries for men and women were the same, what would be the probability of observing a difference in sample means as large or larger than the difference observed here? What assumptions did you have to make?

4. Four hundred purchasers of a new product were asked whether they liked the product. The company conducting the survey felt that at least 20% of all purchasers must like the product in order for the firm to continue marketing it. The firm wished to run a low risk of incorrectly discontinuing the marketing of the product. Therefore, the firm decided that if fewer than 68 people responded favorably, it would stop marketing the product.
    a. State briefly in words the nature of the Type I error involved here. Calculate the level of significance (α) associated with the company's decision procedure.
    b. If the true (but unknown) percentage of purchasers of this new product who liked it was 15%, what is the probability that the firm's decision procedure would erroneously lead to continuance of marketing of the product?

5. The Educational Testing Association (ETA) claims the GMAT scores follow a normal distribution with mean $\mu = 500$ and standard deviation $\sigma = 80$.
    a. If ETA is correct, what proportion of scores is above 600?
    b. One hundred students are selected at random. Of these, 20 scored below 400, 60 scored between 400 and 600, and 20 scored above 600. Do these results support ETA's statement about the distribution of GMTA scores? Show your work. Use $\alpha = 0.05$.

6. Dietrich Investments is a large firm which specializes in developing portfolios for nonprofit institutions. Each portfolio is assigned to one of its analysts. In 1990 fifteen analysts were hired. Five of these attended a special program, but otherwise only had a high school diploma. Five other analysts had a college degree but no graduate training, and the

remaining five had MBA degrees. One portfolio is randomly selected from each analyst. The fifteen percentage returns are combined into three groups as follows:

High school $\bar{x}_1 = 6, s_1^2 = 0.36$

College $\bar{x}_2 = 7, s_2^2 = 0.64$

MBA $\bar{x}_3 = 8, s_3^2 = 0.25$

a. Dietrich Investments wants to be confident that the mean percentage return is greater than five. Is it safe to hire high school graduates? Show your work. Use $\alpha = 0.05$.

b. Can it be shown that there is a difference in performance between high school graduates and MBAs? Show your work. Use $\alpha = 0.05$.

c. Are there differences in performance among the three groups? Show your work. Use $\alpha = 0.05$.

7. A simple random sample of 22 active charge accounts at a large store yielded the following dollar amounts (to the nearest dollar) of purchases of clothing over the past six months:

34 12 68 24 56 19 83   8 27 44 97
22 42 16 74   0 30 18 33 59 119 24

($\bar{x} = 41.32$ and $s = 30.84$)

a. Historically, the mean clothing purchase has been $5.00 per month. Assuming that the true standard deviation $\sigma = \$30.84$, what is the probability that in a simple random sample of 22 active charge accounts you will get an $\bar{x}$ of at least $41.32? What assumption(s) did you make in the above calculation? Do the data support the assumption(s)?

b. Give a 95% confidence interval for the population mean of six-month clothing purchases using the above sample data.

c. The store manager suspected that the average has increased for the six-month period in question above the mean of $5.00 per month. Do the data conclusively support this belief? Use $\alpha = 0.01$.

8. A consumer lobby wishes to have an accurate estimate of voter sentiment concerning a forthcoming referendum. The lobby has data from a simple random sample of 290 voters in which 150 replied "yes," 100 replied "no," and 40 said "no opinion." Assume the lobby ignores the "no opinion" responses.

a. Construct a 90% confidence interval for the population proportion of "yes" voters.

b. The above confidence interval was regarded as not sufficiently precise. The lobby requires an estimate accurate to within $\pm 1\%$ with 98% confidence. Sampling costs are $2.50 per voter. If the lobby assumes that there will be zero "no opinions," what will be the cost of the survey?

9. In an earlier year the mean interest rate charged on its loans by Loans Unlimited, Inc., was 0.075 and the standard deviation was 0.02. A random sample of 100 loans made this year yielded an arithmetic mean interest rate of 0.078.

   a. If the manager of the loan company believed that there was no possibility that the rate had decreased, would you conclude that there has been a change in the average interest rate? Use a 0.05 level of significance.

   b. What is the power of the test when the true rate is 0.0775?

10. A company is considering two computer systems. Thirty computer jobs are chosen at random and are run on both systems. Twenty of the computer jobs run faster on System A and the remaining ten jobs run faster on System B.

    a. Do the data support the claim that more than 50% of computer jobs run faster on System A? Use $\alpha = 0.05$.

    b. If 60% of jobs run faster on System A, what is the probability that the test in part (a) will lead to an incorrect conclusion?

11. Four advertisements for television of a soft drink are tested by inviting 25 individuals into a studio, showing the four advertisements to the 25 individuals and asking each individual to rate each of the advertisements. The first two advertisements are similar in that they focus on thirst quenching. The last two advertisements are similar in that they compare the soft drink to the main competitor's product. The following data are collected:

$$\text{Advertisement 1:} \quad \bar{x}_1 = 62; \; s_1^2 = 466.0$$

$$\text{Advertisement 2:} \quad \bar{x}_2 = 56; \; s_2^2 = 420.0$$

$$\text{Advertisement 3:} \quad \bar{x}_3 = 54; \; s_3^2 = 422.0$$

$$\text{Advertisement 4:} \quad \bar{x}_4 = 68; \; s_4^2 = 508.0$$

    a. An advertisement is acceptable if it is fairly certain that its mean rating exceeds fifty. In particular, is the claim that the mean rating for the first advertisement exceeds fifty supported by the above data? Use $\alpha = 0.05$.

    b. It is also of interest to see whether the first two advertisements differ significantly and if the last two advertisements differ significantly. In particular, do the above data support the claim that the mean ratings of the third and fourth advertisements are different? Use $\alpha = 0.05$.

    c. Do the data support the claim that the means of the four advertisements are not all the same? Use $\alpha = 0.05$.

12. A financial analyst is doing research on the performance of stocks over the last two years in three industries: I (oil companies), II (high technology companies), and III (large industrial companies). The following

data on annual percentage yields are collected for ten stocks in each of the three industries:

											$\bar{x}$	s^2
I	30	20	10	20	30	5	25	20	15	35	21	87.78
II	5	10	20	40	20	10	10	15	10	20	16	98.89
III	-5	10	15	-10	0	20	10	5	-5	10	5	94.44

a. Do the above data indicate that the mean percentage yields of the three industries differ? Use $\alpha = 0.05$.

b. Instead of comparing the mean percentage yields, the analyst classified the stocks into two groups:

> A: stocks that earned less than 15%
> B: stocks that earned at least 15%

Would you conclude that there is dependence between percentage yield and industry classifications? Perform an appropriate test using $\alpha = 0.05$.

13. A cigarette manufacturer wishes to estimate the average nicotine content of the company's cigarettes. A random sample of 144 cigarettes is chosen and the nicotine of each is determined. A sample arithmetic mean of 3.15 milligrams per cigarette is obtained. Based on past experience, it appears that 0.5 milligrams is a reasonable estimate of the population standard deviation.

a. Determine a 95% confidence interval for the population mean nicotine level.

b. It is decided that the interval estimate determined in part (a) is not sufficiently precise. An estimate is required which is within 0.06 milligrams of the true population mean value. Using the same confidence level, determine how many cigarettes must be sampled to satisfy this requirement.

14. In a survey of the coffee-drinking habits of the American people, 100 adults were randomly sampled in each region. The following means and standard deviations were obtained from these two samples:

Pounds of Coffee Used per Adult per Year

New England Region	North Central Region
$\bar{x}_1 = 19.0$	$\bar{x}_2 = 24.5$
$s_1 = 5.0$	$s_2 = 12.0$

a. Test the hypothesis that there is no difference in coffee consumption between the two regions at the 0.01 level of significance.

Decide whether the following statements are true or false. Give a concise explanation for each decision. These 3 questions are based on the test performed in part (a).

b. The probability of a Type II error is the same when $\mu_1 - \mu_2 = 25$ lbs. $- 21$ lbs. as when $\mu_1 - \mu_2 = 18$ lbs. $- 22$ lbs.

c. The size of the critical (i.e., rejection) region increases when we decrease the level of significance, holding everything else constant.

d. Since $\alpha + \beta = 1$, increasing α brings about decreases in the probabilities of making Type II errors.

15. A computer manufacturer produces large numbers of microprocessor chips with great care, but only a certain proportion p meets acceptable standards for use in the company's computers. Let us make the rather strong assumption that the production of each chip is a Bernoulli trial, with probability p of making a good chip, and probability $1 - p$ of making an unacceptable one. The n trials under consideration will be treated as independent events.

a. Suppose that $n = 200$ chips were produced, and suppose $x = 70$ met the acceptance standard. What is your estimated value of p?

b. The proportion of acceptable chips in the sample is approximately normally distributed when n is large. What are your estimates of the mean and variance of that distribution based on the sample data in part (a)?

c. Find the confidence interval that will include the unknown parameter p 98% of the time for many repeated random samples of $n = 200$ chips from the population. Use the large-sample normal distribution specified in part (b) and the estimate of its variance that you calculated in part (b).

d. At a later date, the computer manufacturer wishes to make another estimate of p. He is convinced that p is equal to 42% or less. How large a simple random sample would be required to estimate p with 90% confidence that the estimate would be within 3 percentage points of the true population figure?

16. In a recent investigation at a major U.S. airport, researchers studied the arrivals of aircraft scheduled to land between 7:00 P.M. and 7:15 P.M. Over a 5-day span with ideal weather conditions, there were 64 landings with an average delay time of 10 minutes. The sample variance was found to be 20.25 (minutes)^2.

a. Construct a 97% confidence interval estimate of the average delay time for aircraft under the stated weather conditions.

b. Assume that the standard deviation of the delay time for landings under the stated weather conditions is in fact 6 minutes. How large a sample size is required to estimate average delay time to within 3 minutes with probability 0.99?

c. The research team reported a confidence interval estimate of (9.279, 10.721) after analyzing the data. What confidence level did the research team use?

17. *Financial Week* magazine recently conducted a survey of its readers in selected metropolitan areas. One of the questions asked for annual income.

A rough summary of the data for 3 of the areas includes the following number of respondents.

Annual Income in 3 Metropolitan Areas

	Area A	Area B	Area C	Total
Less than $30,000	15	50	5	70
$30,000–$60,000	50	20	20	90
More than $60,000	5	5	35	45
Total	70	75	60	205

a. *Financial Week's* director of advertising believes that the readers in area A in the stated income ranges, from lowest to highest, are in the ratio 2:3:2. Do the data support this hypothesis? Use $\alpha = 0.05$.

b. Are the 3 metropolitan areas alike in their income profiles for the readers of *Financial Week*? Test using $\alpha = 0.01$.

18. A candy wholesaler has found from extensive experience that his monthly sales to drugstores of a certain size has a mean of $200 and a standard deviation σ of $42. Recently the wholesaler acquired $n = 9$ additional drugstores of the same sort when a competitor went out of business. During the first complete month of business the wholesaler's average sales to the new stores was $\bar{x} = \$188$.

a. Test the hypothesis at the 5% level of significance that the new stores came from a population with mean $200 as opposed to an alternative of a lower mean. You may assume that monthly sales is a normal random variable, and that the 9 stores constitute a simple random sample. Your answer should include a statement of the hypotheses and a clear indication of the decision rule used.

b. Now suppose that the population standard deviation $\sigma = \$42$ was not known, but the wholesaler calculated the unbiased sample variance to be $s^2 = 350$. Test the same hypotheses on mean sales under these assumptions at the 5% level.

c. Assume that the *true population mean* of monthly sales for the new stores was $162.90. With the decision rule you derived in part (a), what is the probability that you would erroneously accept the hypothesis that the population mean was $200?

19. You are approached by an entrepreneur who offers you a limited partnership in a real estate development he is organizing. From past experience you are willing to assume that the risk of the venture can be measured by $\sigma = 5\%$ and that the rate of return of the venture follows a normal distribution. You investigate 25 previous projects of this developer in order to estimate the mean percentage return on investment, μ. The average return from these 25 projects x was 12%.

    a. Find a 95% confidence interval for μ.

    b. If you want to be 95% confident that the maximum error of your estimate will not exceed 0.5%, what sample size should you take?

    c. This venture will earn less than what level of return only 10% of the time if, in fact, the true mean $\mu = 12\%$?

20. It has been claimed that about 20% of all offers to prospective executives include short-run stock options (options good for 3 years or less) and another 20% include long-run options (good for more than 3 years). It is also claimed that these options have no effect on the likelihood of an executive accepting the offer. A headhunter checked her recent files and found a total of 200 offers with data broken down as shown in the table.

Type of Stock Option

	No Options	Short-run Options	Long-run Options	Total
Accepted	61	25	34	120
Rejected	49	15	16	80
Total	110	40	50	200

    a. Are the data from the 200 offers significantly different from the 60% no options, 20% short-run, 20% long-run claim? Justify your answer using $\alpha = 0.01$.

    b. Would you conclude that the type of option has an effect on acceptance or rejection of offers? Justify your answer using $\alpha = 0.05$.

21. Ventures Unlimited is a firm that buys ideas for new products from inventors, designs the products, and then markets them. One inventor sends in his idea for a new computer terminal that is faster than any terminal on the market but is expensive to manufacture. Ventures Unlimited feels that if it were likely that more than 25% of large firms are willing to consider buying the product, it would be worthwhile to design and manufacture a prototype.

    a. A survey is sent out to a simple random sample of 100 of these firms and 35 firms show an interest in buying the terminal. Is this a statistically significant result indicating that the company should proceed with the manufacturing of the prototype assuming $\alpha = 0.05$? Show your work, stating the alternative hypotheses and decision rule that you have used.

    b. The prototype is manufactured and the sales in the test market are measured for 36 weeks. The arithmetic mean for weekly sales was 111 terminals. We assume that the standard deviation for weekly sales is $\sigma = 21$ terminals. Ventures Unlimited has decided that if it is conclusive that the mean for weekly sales exceeds 105 terminals, then manufacturing should commence. Can this claim be made using $\alpha = 0.01$? Show your work.

c. If, in fact, the true mean for weekly sales is 110 terminals, what is the probability that Ventures Unlimited would erroneously decide not to manufacture the product using the test constructed in part (b)?

22. The Burger Chief fast food chain has a number of restaurants located in the Philadelphia area. The chain claims that it dispenses hamburgers of virtually identical content, quality, and size at all locations. A total of 7 hamburgers purchased at the chain's Main Line location contained beef weighing an average of 3.7 ounces per hamburger, with a sample variance of 0.081. At the North Philadelphia restaurant, however, a sample of 11 hamburgers yielded corresponding values equal to 3.4 and 0.072.

    a. Set up and test a hypothesis versus a suitable alternative to check the chain's claim. Use $\alpha = 0.01$.

    b. Test the null hypothesis that the average weight of the beef in the hamburgers at the North Philadelphia location is at least that of the hamburgers at the Main Line location, versus the alternative that it is not. Use $\alpha = 0.05$.

    c. A confidence interval estimate of (3.362, 4.038) ounces was reported for the average weight of beef in the hamburgers at the Main Line location based on the data in part (a). What confidence level was used? Show your work.

23. A production process is supposed to make steel bars with an average weight of 18.0 pounds. To ascertain whether or not the production process is turning out steel bars with this mean weight, management has adopted the following procedure: Periodically select 36 bars from the production process and weigh them. If the sample mean is between 17.25 and 18.75 pounds, conclude that the process is performing satisfactorily. Otherwise conclude that the process mean has changed. (Assume that the process standard deviation is 1.8 pounds.)

    a. What is the probability of making a Type I error? What constitutes a Type I error in this case?

    b. Management decides to evaluate this testing procedure by plotting the power curve. Without doing any computations, sketch the shape of this power curve, label the axes, and indicate on the curve what the powers are at the critical points 17.25 and 18.75.

    c. True or false? Give a concise explanation for each.

       (i) The hypotheses for this test are $H_0: \bar{x} = 18$ lb. vs. $H_1: \bar{x} \neq 18$ lb.

       (ii) Both Type I and Type II errors are constants in this testing procedure.

      (iii) Suppose a new decision rule is stated: If the sample mean is between 17.4 and 18.6 pounds, conclude that the process is performing satisfactorily. Otherwise conclude that the mean has changed. The sample size remains at 36. The probability of a Type II error increases under the new decision rule.

      (iv) The null hypothesis is usually set forth to be rejected, since proving it is difficult or impossible in many cases.

(v)   Under the null hypothesis, and assuming that the weights of the bars are normally distributed, we can state that "for practical purposes" *all* steel bars produced are between 12.6 and 23.4 pounds.

MINITAB REVIEW EXERCISES FOR CHAPTERS 6 THROUGH 8

Problems 1 through 3 are based on the following data: A gubernatorial election in a certain state was held between a Republican and a Democrat. Before the voting took place, a news organization conducted a survey of 500 potential voters. The data were entered into a file labeled 'vote . mtw'.

Column	Name	Description
C1	VOTE	Favor Democrate (1); Favor Republican (−1); Undecided (0)

1. Carry out a test of the hypothesis that the proportions of residents in the state who favored the Democrat, favored the Republican or were undecided are equal, at a 0.15 significance level.

2. Construct a 99% confidence interval for the proportion of the residents in the state who favored the Democrat. At alpha = 0.01, conduct a two-sided test of the hypothesis that the proportion of the residents in the state who favored the Democrat was 33%. *Hint:* To transform C1 into a column of ones and zeroes, with one indicating 'for the Democrat' and zero indicating 'not for the Democrat', type

    MTB > code (−1) to 0 in C1, put into C2

3. The residents in the state were mainly interested in the person who would win the election. Formulate a hypothesis-testing problem and test your hypothesis at a 0.05 significance level. State your conclusion.

Problems 4 and 5 are based on the following data: Different management methods were adopted in two similar firms for a 3-year period. The monthly profits (in $000) of the two firms have been recorded and stored in a file named 'profit . mtw'.

Column	Name	Description
C1	Method 1	Monthly Profits ($000): Using Method 1
C2	Method 2	Monthly Profits ($000): Using Method 2

4. Construct a 97% confidence interval for the mean monthly profit when using Method 1. At a 0.03 significance level, carry out a two-sided test of the hypothesis that the mean monthly profit using Method 1 is $200,000.

5. Make dotplots of C1 and C2 on the same graph. Construct a 98% confidence interval for the difference between the population means. Formulate a hypothesis-testing problem to judge which method is more effective. Test your hypothesis at a 0.02 significance level and state your conclusion.

6. A first-class sprinter has run 10 practice heats of 100 meters. The times (in seconds) were: 10.3, 10.5, 9.9, 10.1, 10.2, 10.6, 9.8, 10.2, 10.4, and 10.0.
   a. Construct a 95% confidence interval for mean time to run 100 meters.
   b. Past records indicate that the population standard deviation is 0.2 seconds. Construct a 95% confidence interval for the population mean and carry out a two-sided test of the hypothesis that the population mean is 10 seconds at a 0.05 significance level.

7. Retrieve the data in 'databank . mtw'. Is there a significant difference in per capita income among families of one to four people? Delete unnecessary data by typing:

   MTB > copy C2 C3 into C2 C3;
   SUBC > use C3 = 1:4.

   Formulate a hypothesis-testing problem and test the hypothesis at a 0.20 significance level. Compute a 95% confidence interval for each of the mean per capita incomes among families of sizes one through four.

8. To study potential sex discrimination in promotions among middle managers in a large company, let us consider the following data:

   Number of Employees Classified by Sex and Promotion

	Male	Female
Promoted	45	31
Not Promoted	68	57

   Under the assumption of independence, find the expected numbers in all cells. Carry out a test of independence at alpha = 0.10.

Problems 9 through 11 are based on the following data: A very large company tests the overall abilities of MBA's of two famous business schools. Twelve MBA's are randomly chosen from each business school and assigned posts in the east, west, and the middle of the United States. The overall ability of a graduate is evaluated by academic experts and industry experts and is measured by a number from 0 to 100, with higher numbers representing greater ability. The data are:

Overall Evaluation

	East	Middle	West
School 1	85	82	89
	88	75	91
	92	86	95
	83	79	87
School 2	76	82	93
	81	71	83
	85	69	84
	80	75	79

Note that the data have been entered into a computer file named 'ability . mtw'.

Column	Name	Description
C1	SCHOOL	School 1, Coded by 1; School 2, Coded by 2
C2	REGION	East (1), Middle (2), and West (3)
C3	ABILITY	Overall Evaluation: 0–100

9.   Construct a table of cell means by typing

MTB > table C1 and C2;
SUBC > mean of C3.

From the table, can you see a clear interaction between SCHOOL and REGION? Formally carry out a test of interaction at a 0.20 significance level. State your conclusion.

10.   Carry out separate tests for the significance of the factors SCHOOL and REGION at the 0.01 significance level. Is there a consistent superiority of one business school over the other? Use Minitab to calculate the grand mean, REGION effects and SCHOOL effects.

11.   If you had analyzed the same data as a one-factor ANOVA, ignoring the REGION effects, would you have obtained a different conclusion about the superiority of one business school over the other?

9

REGRESSION ANALYSIS AND CORRELATION ANALYSIS

Prediction is required in virtually every aspect of the management of enterprises. Indeed, business planning and decision making are inseparable from prediction. Forecasting of sales, earnings, costs, production, personnel requirements, inventories, purchases, and capital requirements is the foundation of company planning and control. We could give many other illustrations of the need for prediction in the management of private and public organizations.

In this chapter, we concern ourselves with a broad class of techniques for prediction, namely, *regression* and *correlation analysis*. This type of analysis is undoubtedly one of the most widely used statistical methods. **Regression analysis** provides the basis for predicting the values of a variable from the values of one or more other variables; **correlation analysis** enables us to assess the strength of the relationships (correlations) among the variables.

Equations are used in mathematics to express the relationships among variables. In fields such as geometry or trigonometry, these mathematical equations, or functions, express the **deterministic (exact) relationships** among the variables of interest.

Deterministic relationships

453

- The equation $A = s^2$ describes the relationship between s (the length of the side of a square) and A (the area of the square).

- The equation $A = ab/2$ expresses the relationship between b (the length of any side of a triangle), a (the altitude or perpendicular distance to that side from the angle opposite it, and A (the area of the triangle).

By substituting numerical values for the variables on the right-hand sides of these equations, we can *determine* the *exact values* of the quantities on the left-hand sides.

Statistical relationships   In the social sciences and in fields such as business and government administration, exact relationships are not generally observed among variables, but rather **statistical relationships** prevail. That is, certain average relationships may be observed among variables, but these average relationships do not provide a basis for perfect predictions.

- If we know how much money a corporation spends on television advertising, we cannot make an exact prediction of the amount of sales this promotional expenditure will generate.

- If we know a family's net income, we cannot make an exact forecast of the amount of money that family saves, but we can measure statistically how family savings vary, on the average, with differences in income. On the other hand, we can measure statistically how sales vary, on the average, with differences in television advertising.

We can also determine to what extent actual figures vary from these average relationships. On the basis of these relationships, we may be able to estimate the values of the variables of interest closely enough for decision-making purposes. The techniques of regression and correlation analysis are important statistical tools in this measurement and estimation process.

Techniques of Analysis

The term **regression analysis** refers to the methods by which estimates are made of the values of a variable from a knowledge of the values of one or more other variables, and to the measurement of the errors involved in this estimation process. The term **correlation analysis** refers to methods for measuring the degree of association among these variables.

Two-variable linear regression and correlation analysis   We begin by discussing **two-variable linear regression and correlation analysis**. The term **linear** means that an equation of a straight line of the form $Y = A + BX$, where A and B are fixed numbers, is used to describe the average relationship between the two variables and to carry out the estimation process. The factor whose values we wish to estimate is referred to as the **dependent variable** and is denoted by the symbol Y. The factor from which these estimates are made is called the **independent variable** and is denoted by X.

In the formulation of a regression analysis, the investigator should use prior knowledge and the results of past research to select independent variables that are potentially helpful in predicting the values of the dependent variable. Hence, the investigator should use any available information concerning the direction of cause and effect in the selection of relevant variables. As indicated earlier, correlation analysis can be used to measure the strength or degree of correlation among the variables of interest.

Let us consider some illustrative cases of variables that it is reasonable to assume are related to one another, that is, correlated. If suitable data were available, we might attempt to construct an equation that would permit us to estimate the values of one variable from the values of the other. The first named factor in each pair is the variable to be estimated, that is, the dependent variable, and the second one is independent. Consumption expenditures might be estimated from a knowledge of income; investment in telephone equipment from expenditures on new construction; personal net savings from disposable income; commercial bank interest rates from Federal Reserve Bank discount rates; and success in college from Scholastic Aptitude Test scores.

EXAMPLES OF INDEPENDENT + DEPENDENT VARIABLE USE.

Of course, additional definitions are required to attach meaning to the estimation problems listed above. Thus, in the illustration of consumption expenditures and income, we would have to specify whose expenditures and whose income are involved. If we wanted to estimate family consumption expenditures from family income, the family would be the "unit of association." The estimating equation would be constructed from data representing observations of these two variables for individual families. We would have to define the variables more specifically. For example, we might be interested in estimates of annual family consumption expenditures from annual family net income, where again these terms would require precise definitions.

In each these examples, we can specify other independent variables that might be included to obtain good estimates of the dependent variable. Hence, in estimating a family's consumption expenditures, we might wish to use knowledge of the size of the family in addition to information on the family's income. This would be an illustration of **multiple regression analysis**, where two independent variables (family income and family size) are used to obtain estimates of a dependent variable. In this chapter, we shall consider two-variable problems involving a dependent factor and only one independent factor. Chapter 10 covers multiple regression problems involving more than one independent variable.

Multiple regression analysis

The Simple Two-variable Linear Regression Model

The use of a variable to predict the values of another variable may be viewed as a problem of statistical inference. The population consists of all relevant pairs of observations of the dependent and independent variables. Generally, estimates or predictions must be made from only a sample of that population. For example, suppose we are interested in estimating a family's annual food

FIGURE 9-1

The linear regression model
and three conditional probability
distributions of annual family
food expenditures for selected
annual family net incomes

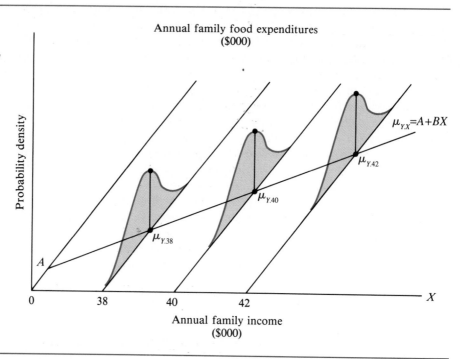

where the meaning of $\mu_{Y.X}$ is explained in the next paragraph.

expenditures (the dependent variable) from a knowledge of the family's annual
net income (the independent variable). Furthermore, let us assume that we are
interested in these estimates for families in a metropolitan area in 1990. Figure
9-1 shows what the population relationship between these two variables might
look like for families in the given income range if we assume that this relationship is linear (that is, in the form of a straight line). The fact that the line runs
from the lower left to the upper right side of the graph indicates that as family
annual net income increases, family food expenditures also increase. We denote this **population regression line** or **true regression line** as

Population regression line   (9.1)
$$\mu_{Y.X} = A + BX$$

where the meaning of $\mu_{Y.X}$ is explained in the next paragraph.

In order to understand the assumptions involved in the linear regression
model, let us consider families with certain specified incomes (say, $38,000
$40,000, and $42,000), as indicated in Figure 9-1. If we now focus on families
at a given income level (say, $40,000), the probability distribution of the Y vari-
able, food expenditures, is a **conditional probability distribution** of Y given
$X = \$40,000$. Such a conditional probability distribution may be symbolized
in the usual way as $f(Y|X = \$40,000)$ or, in general, $f(Y|X)$. This distribution
has a mean, which may be denoted $\mu_{Y.X}$, and a standard deviation, which may

be denoted $\sigma_{Y.X}$. In the linear regression model, the assumed relationship be-tween $\mu_{Y.X}$ and X can be graphed as a straight line. That is, the means of the conditional probability distributions (i.e., $\mu_{Y.38}$, $\mu_{Y.40}$, and $\mu_{Y.42}$) are assumed to lie along a straight line. In this model of a linear regression function, A and B are population parameters that must be *estimated* from sample data.

The conditional probability distributions describe the variability of the Y values. Thus, for families with \$40,000 annual net income in the given metro-politan area, the conditional probability distribution indicates the variability of family annual food expenditures around the conditional mean food expendi-tures, $\mu_{Y.X}$ (in this case $\mu_{Y.40}$). The symbol $\mu_{Y.X}$ represents the mean of the Y variable for a given X value. The line that connects the conditional means in the X, Y plane is the true population regression line $\mu_Y = A + BX$. The con-ditional distributions are shown as rising above the X, Y plane into the third dimension. Each conditional distribution is a normal curve centered on its own mean $\mu_{Y.X}$.

In addition to the assumption of a linear relationship, the following assumptions are involved in the use of linear regression model:

Assumptions of the linear regression model

1. The Y values are independent of one another.
2. The conditional probability distributions of Y given X are normal.
3. The conditional standard deviations $\sigma_{Y.X}$ are equal for all values of X.

A few comments about these assumptions are in order. The first assump-tion means that there is independence between successive observations. This means, for example, that a low value for Y on the first observation does not imply that the second Y value will also be low. In certain situations, such an assumption may not be particularly valid. We see this clearly in time-series data that move in cycles around a fitted trend line. Observations in the expan-sion phase of a business cycle will tend to lie above the trend line and to have relatively high values; the opposite is true for observations in a contraction or recession phase. Hence, successive observations in this case tend to be related rather than independent.

The second assumption means that for each value of X, we are assuming that the Y values are normally distributed around $\mu_{Y.X}$. As we will see, this assumption is useful for making probability statements about estimates of the dependent variable Y.

The third assumption implies that there is the same amount of variability around the regression line at each value of the independent variable, X. As in section 8.3, where analogous assumptions were discussed for the analysis of variance, this characteristic is referred to as **homoscedasticity**. Note that in re-gression analysis, according to the second assumption, only Y is considered a random variable. X is considered fixed. Hence, if we attempt to predict a Y value (for example, a family's food expenditures) from a knowledge of X (for example, a family's income), the predicted Y value is subject to error. X is

Homoscedasticity

assumed to be known without error. On the other hand, in correlation analysis, both X and Y are treated as normally distributed random variables.

Clearly these three assumptions are never perfectly met in the real world. In many situations, however, these asssumptions are approximately true and the model is useful. In this chapter, we develop the standard linear regression model under these assumptions.

Formulation of the problem

Before turning to the methodology of regression and correlation analysis beginning in the next section, it is important to emphasize that the *formulation of the problem* is of critical importance. In the formulation stage, the analyst must

- identify the dependent and independent variables

    specify the relationships among these variables (for example, linear or non-linear)

- decide what the relevant population is (for example, low-income families in a particular metropolitan area)

- supply the data on the variables for a sample or for the entire relevant population.

9.2
SCATTER DIAGRAMS

Plotting data on a graph is useful in studying the relationship between two variables. A graph permits visual examination of the extent to which the variables are related and aids in choosing the appropriate type of model for estimation. The chart used for this purpose is known as a *scatter diagram*, which is a graph on which each plotted point represents an observed pair of values of the dependent and independent variables. We will illustrate this by plotting a scatter diagram for the data given in Table 9-1, which are observations for the first 20 families in the computer data bank given in Appendix D. We will assume that the 20 families comprise a simple random sample of families in a metropolitan area in 1990. The figures represent observations of annual family expenditures on food, which we shall treat as the dependent variable, Y (the factor to be estimated), and annual net income, X, which is the independent variable (the factor from which the estimates are to be made). Although only 20 families is a relatively small sample from which to draw conclusions that would apply to all families in an entire metropolitan area, we will use such a small sample for convenience of exposition.

Figure 9-2 presents the data of Table 9-1 plotted as a scatter diagram by Minitab. The 2 in the diagram represents two points that have identical values (families 1 and 13; 5.2, 28). The food expenditures and income variables were entered in columns 1 and 2, respectively. Note that in response to the Minitab

TABLE 9-1

Annual food expenditures and annual net income of a sample of 20 families in a metropolitan area in 1990

Family	Annual Food Expenditures ($000) Y	Annual Income ($000) X
1	5.2	28
2	5.1	26
3	5.6	32
4	4.6	24
5	11.3	54
6	8.1	59
7	7.8	44
8	5.8	30
9	5.1	40
10	18.0	82
11	4.9	42
12	11.8	58
13	5.2	28
14	4.8	20
15	7.9	42
16	6.4	47
17	20.0	112
18	13.7	85
19	5.1	31
20	2.9	26

command "plot C1 C2", the first named variable is plotted along the vertical axis.

Thus, on the Y axis are plotted the figures on food expenditures and on the X axis annual net income. This follows the standard convention of plotting the dependent variable along the Y axis and the independent variable along the X axis. The pair of observations for each family determines one point on the scatter diagram. For example, for family 1, a point is plotted corresponding to $X = 28$ along the horizontal axis and $Y = 5.2$ along the vertical axis; for family 2, a point is plotted corresponding to $X = 26$ and $Y = 5.1$. An examination of the scatter diagram gives some useful indications of the nature and strength of the relationship between the two variables. For example, depending on whether the Y values tend to increase or to decrease as the values of X increase, there is a *direct* or *inverse* relationship, respectively, between the two variables. The configuration in Figure 9-2 indicates a general tendency for the points to run from the lower left to the upper right side of the graph. Hence, as noted in section 9.1, as income increases, food expenditures tend to increase. This is an example of a **direct relationship** between the two variables. On the

Direct relationship

FIGURE 9-2

A scatter diagram of annual food expenditures and annual net income of a sample of 20 families in a metropolitan area in 1990

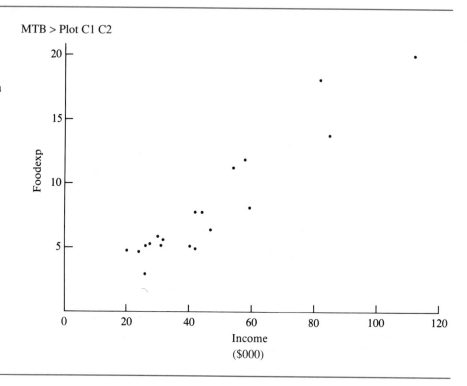

MTB > Plot C1 C2

Inverse relationship

other hand, if the scatter of points runs from the upper left to the lower right (that is, if the Y variable tends to decrease as X increases), there is an **inverse relationship** between the variables. Also, an examination of the scatter diagram gives an indication of whether a straight line appears to be an adequate description of the average relationship between the two variables. If a straight line is

Linear relationship

used to describe the average relationship between Y and X, a **linear relationship** is present. However, if the points on the scatter diagram appear to fall along a curved line rather than a straight line, a **curvilinear relationship** exists. Figure

Curvilinear relationship

9-3 presents illustrative combinations of the foregoing types of relationships. Parts (a), (b), (c), and (d) of Figure 9-3 show, respectively, direct linear, inverse linear, direct curvilinear, and inverse curvilinear relationships. The points tend to follow a straight line sloping upward in (a), a straight line sloping downward in (b), a curved line sloping upward in (c), and a curved line sloping downward in (d). Of course, the relationships are not always so obvious. In (e), the points appear to follow a horizontal straight line. Such a case depicts "no correlation" between the X and Y variables, or no evident relationship, since the horizontal line implies no change in Y, on the average, as X increases. In (f), the points follow a straight line sloping upward as in (a), but there is a much wider scatter of points around the line than in (a).

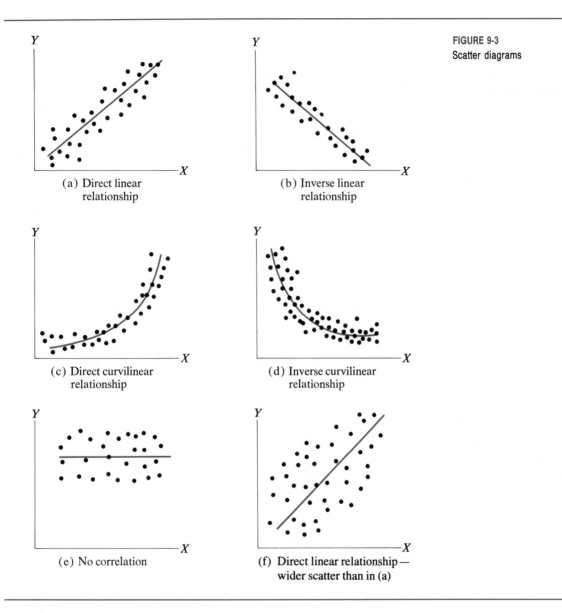

FIGURE 9-3
Scatter diagrams

(a) Direct linear
relationship

(b) Inverse linear
relationship

(c) Direct curvilinear
relationship

(d) Inverse curvilinear
relationship

(e) No correlation

(f) Direct linear relationship—
wider scatter than in (a)

In our present problems, we assume that our prior expectation was for a direct linear relationship between the two variables and that a visual examination of Figure 9-2 confirmed this expectation. In the next section, we discuss the procedures used in regression and correlation analysis. We begin by considering what we hope to accomplish through the use of such techniques.

9.3
PURPOSES OF REGRESSION AND CORRELATION ANALYSIS

What does a regression and correlation analysis attempt to accomplish in studying the relationship between two variables, such as expenditures on food and net annual income of families? We will concentrate on these basic goals, which emphasize the relationships contained in the particular sample under study. Later, we will consider other objectives involving statistical inference, that is, inferences concerning the population from which the sample was drawn.

The first two objectives and the statistical procedures involved in their accomplishment fall under the heading of *regression analysis*, whereas the third objective and related procedures are classified as *correlation analysis*. These objectives are stated below, and the statistical measures used to achieve the objectives are named. However, the mathematical definitions of these measures are postponed until the discussion of their use in the problem involving family expenditures on food and family income.

Regression analysis

> The first purpose of regression analysis is to provide estimates of values of the dependent variable from values of the independent variable.

The device used to accomplish this estimation procedure is the *sample regression line*, which is a line fitted to the data by a method we shall describe in the next section. The **sample regression line** describes the average relationship between the X and Y variables in the sample data. The equation of this line, known as the **sample regression equation**, provides estimates of the mean value of Y for each value of X. Hence, in the illustration given in section 9.1 we could obtain estimates from the sample regression equation of the mean family food expenditures for each level of family income.

> A second goal of regression analysis is to obtain measures of the error involved in using the regression line as a basis of estimation.

For this purpose, the *standard error of estimate* and related measures are calculated. The **standard error of estimate** measures the scatter, or spread, of the observed values of Y around the corresponding values estimated from the fitted regression line. **Measures of forecast error** take into account this scatter as well as the probable difference between the regression line fitted to sample data and the true but unknown population regression line.

Correlation analysis

> The third objective, which we have classified as correlation analysis, is to obtain a measure of the degree of association or correlation between the two variables.

The **coefficient of correlation** and the **coefficient of determination**, calculated for this purpose, measure the strength of the relationship between the two variables.

9.4
ESTIMATION USING THE REGRESSION LINE

As indicated in the preceding section, to accomplish the first objective of a regression analysis, we must obtain the mathematical equation of a line that describes the average relationship between the dependent and independent variables. We can then use this line to estimate values of the dependent variable. Since the present discussion is limited to *linear* regression analysis, we are referring to a straight line. Ideally, we would like to obtain the equation of the straight line that best fits the data. Let us defer for the moment what we mean by "best fits" and review the concept of the equation of a straight line.

> The equation of a straight line is $Y = a + bX$, where a is the so-called "Y intercept," or the computed value of Y when $X = 0$, and b is the slope of the line, or the amount by which the computed value of Y changes with each unit change in X.

Let us review by means of a simple illustration the relationship between the equation $Y = a + bX$ and the straight line that represents the graph of the equation. Suppose the equation is

(9.2)
$$Y = 2 + 3X$$

Thus, $a = 2$ and $b = 3$. If we substitute a value of X into this equation, we can obtain the corresponding computed value of Y. Each pair of X and Y values represents a single point. Although only two points are required to determine a straight line, several pairs of X and Y values for the line $Y = 2 + 3X$ are shown next to the graph corresponding to the line shown in Figure 9-4. On this graph, since the a value in the equation of the line is two, the line intersects the Y axis at a height of two units. Also, since the b value, or slope of the line, is three, we note that the Y values increase by three units each time X increases by one unit. This is shown graphically in Figure 9-4 as a rise of three units in the line when X increases by one unit.

The terms "regression line" and "regression equation" for the estimating line and equation stem from the pioneer work in regression and correlation analysis of the British biologist Sir Francis Galton in the nineteenth century. The lines that he fitted to scatter diagrams of data on heights of fathers and sons in this early work came to be known as "regression lines" and the equations of these lines as "regression equations," because Galton found that the heights of the sons "regressed" toward an average height. Unfortunately,

FIGURE 9-4
Graph of the line
$\hat{Y} = 2 + 3X$, based on the
following calculations

X	Y
0	2
1	5
2	8
3	11
4	14

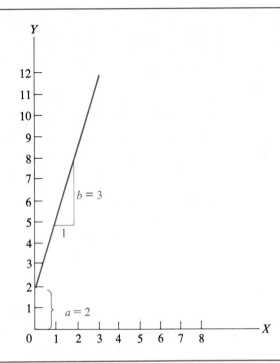

the terminology has persisted. Thus, these terms for the estimating line and estimating equation are used in the wide variety of fields in which regression analysis is applied, despite the fact that the original implication of a regression toward an average is not necessarily present for the phenomena under investigation.

We now turn to the question of obtaining a best-fitting line to the data plotted on a scatter diagram in a two-variable linear regression problem. The fitting procedure discussed is the method of least squares, undoubtedly the most widely applied curve-fitting technique in statistics.

The Method of Least Squares

In discussing the linear regression model in section 9.1, we denoted the population or true regression line as $\mu_{Y \cdot X} = A + BX$. Correspondingly, the sample regression line, which is the best-fitting line to the sample data, is denoted as

Sample regression line   (9.3)

$$\hat{Y} = a + bX$$

where a and b represent estimates of A and B in the population regression line.

Goodness of fit   In order to establish a best-fitting line to a set of data on a scatter diagram, we must have criteria concerning what constitutes **goodness of fit**. A number of criteria that might at first seem reasonable turn out to be unsuitable.

For example, we might entertain the idea of fitting a straight line to the data in such a way that half of the points fall above the line and half below. However, such a line may represent a quite poor fit to the data if, say, the points that fall above the line lie very close to it whereas the points below deviate considerably from it.

Let us now consider the most generally applied curve-fitting technique in regression analysis, namely, the **method of least squares**. This method imposes the requirement that the *sum of the squares* of the deviations of the observed values of the dependent variable from the corresponding computed values on the regression line must be a minimum. Thus, if a straight line is fitted to a set of data by the method of least squares, it is a "best fit" in the sense that the sum of the squared deviation, $\Sigma(Y - \hat{Y})^2$, is less than it would be for any other possible straight line. Another useful characteristic of the least squares straight line is that it passes through the point of means $(\bar{X}, \bar{Y})$, and therefore makes the total of the positive and negative deviations equal to zero. In summary, the least squares straight line possesses the following mathematical properties:

Method of least squares

USED TO FIT CURVE
IN A SCATTER DIAGRAM

(9.4)
$$\Sigma(Y - \hat{Y})^2 \text{ is a minimum}$$

(9.5)
$$\Sigma(Y - \hat{Y}) = 0$$

Figure 9-5 presents graphically the nature of the least squares property. A scatter plot is shown around a least squares sample regression line, denoted $\hat{Y} = a + bX$. Let us consider the first point, whose coordinates are $X = 2$ and $Y = 10$. The vertical distance of the point from the X axis is its Y value, in this case, 10. The $\hat{Y}$ value for the same point is the vertical distance from the X axis to the regression line. In this case, $\hat{Y} = 7$. The vertical distance of the point from the corresponding value on the regression line is the deviation, or residual, $Y - \hat{Y}$. In this illustration, $Y - \hat{Y} = 10 - 7 = 3$. Since the Y value is above the regression line, the deviation is positive; for Y values below the

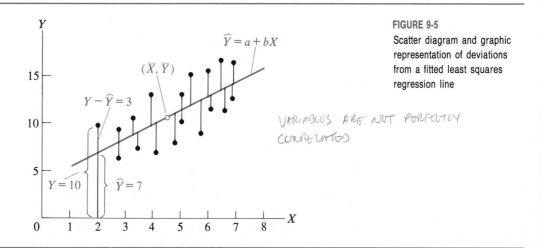

FIGURE 9-5
Scatter diagram and graphic representation of deviations from a fitted least squares regression line

VARIABLES ARE NOT PERFECTLY CORRELATED

line, the deviations are negative. The square of the deviation for the first point is $3^2 = 9$. As mentioned above, the line fitted by this method has the property that the total of all squared deviations is less than the corresponding totals for any other straight line that could have been fitted to the same data.

The regression line is depicted as passing through the point established by the means of the X and Y variables, $(\bar{X}, \hat{Y})$, which in this case is (4.5, 11). It can be shown that a least squares straight line must include this point of means and that this property makes the algebraic sum of the deviations above and below the line equal to zero. That is,

---

the sum of the deviations of the points lying above the regression line, which are positive, and the deviations of the points lying below the regression line, which are negative, is zero.

It can be shown that the a and b values derived by the method of least squares are unbiased, efficient, and consistent estimators of the corresponding parameters A and B in the regression line. Considering the "best fit" properties mentioned earlier and these other desirable properties of this estimation method, it is perhaps not surprising that the method of least squares is the standard method of curve-fitting in regression analysis.

Computational Procedure

By using calculus methods to apply the condition that the sum of the squared deviations from a straight line must be a minimum, two equations known as the **normal equations** are derived.[1] These equations can be solved for the values

---

[1] The derivation of the equations is as follows. We denote the sum of squared deviations that must be minimized as some function of the unknown quantities a and b. Thus, let

$$F(a, b) = \Sigma(Y - \hat{Y})^2$$

Substituting $a + bX$ for $\hat{Y}$ into the above equation gives

$$F(a, b) = \Sigma(Y - a - bX^2)$$

We impose the condition of a minimum value for $F(a, b)$ by obtaining its partial derivatives with respect to a and b and setting them equal to zero. Thus,

$$\frac{\partial F(a, b)}{\partial a} = -2\Sigma(Y - a - bX) = 0$$

$$\frac{\partial F(a, b)}{\partial b} = -2\Sigma(Y - a - bX)(X) = 0$$

Solving these equations yields

$$\Sigma Y = na + b\Sigma X$$

$$\Sigma XY = a\Sigma X + b\Sigma X^2$$

from which equations 9.6 and 9.7 can be derived. A check reveals that the second derivatives of $F(a, b)$ are positive, so a minimum has been found.

of a and b in the regression equation $\hat{Y} = a + bX$. The values of a and b can then be determined from the following general solutions for a and b:

(9.6)
$$a = \bar{Y} - b\bar{X}$$

Normal equations

(9.7)
$$b = \frac{\Sigma XY - n\bar{X}\bar{Y}}{\Sigma X^2 - n\bar{X}^2}$$

FORMULA FOR REGRESSION LINE

where $\bar{X}$ and $\hat{Y}$ are the arithmetic means of the X and Y variables.

Although most regression analyses are currently carried out with software packages, such as Minitab, for purposes of explaining the underlying theory and methodology, we will carry out illustrative computational procedures. At various stages of the explanation, we will also indicate the commands and output for obtaining the corresponding results in Minitab.

Let us return to the problem involving the sample of 20 families and assume that we have decided to fit a *straight line* to the data. From the original observations, we can determine the various quantities (n, ΣY, ΣX, ΣXY, and ΣX^2) required in equations 9.6 and 9.7, where n is the number of pairs of X and Y values (in this case 20). For our illustration, the computation of the required totals is shown in Table 9-2. Although ΣY^2 is not needed for the calculation of

TABLE 9-2
Computations for a regression and correlation analysis for the data shown in Table 9-1

Family	Y	X	XY	X^2	Y^2
1	5.2	28	145.6	784	27.04
2	5.1	26	132.6	676	26.01
3	5.6	32	179.2	1,024	31.36
4	4.6	24	110.4	576	21.16
5	11.3	54	610.2	2,916	127.69
6	8.1	59	477.9	3,418	65.61
7	7.8	44	343.2	1,936	60.84
8	5.8	30	174.0	900	33.64
9	5.1	40	204.0	1,600	26.01
10	18.0	82	1,476.0	6,724	324.00
11	4.9	42	205.8	1,764	24.01
12	11.8	58	684.4	3,364	139.24
13	5.2	28	145.6	784	27.04
14	4.8	20	96.0	400	23.04
15	7.9	42	331.8	1,764	62.41
16	6.4	47	300.8	2,209	40.96
17	20.0	112	2,240.0	12,544	400.00
18	13.7	85	1,164.5	7,225	187.69
19	5.1	31	158.1	961	26.01
20	2.9	26	75.4	676	8.41
	159.3	910	9,255.5	52,308	1,682.17

a and *b*, its computation is also shown. This figure is useful for calculating the standard error of estimate, to be discussed shortly.

From Table 9-2, we compute the means of X and Y as

$$\bar{X} = \frac{\Sigma X}{n} = \frac{910}{20} = 45.50 \text{ (thousands of dollars)}$$

$$\bar{Y} = \frac{\Sigma Y}{n} = \frac{159.3}{20} = 7.965 \text{ (hundreds of dollars)}$$

Substituting the additional quantities $n = 10$, $\Sigma XY = 1,159$, and $\Sigma X^2 = 406$ from Table 9-2 into equations 9.6 and 9.7, we obtain the following values for *a* and *b*:

$$b = \frac{9,255.5 - (20)(45.50)(7.965)}{52,308 - (20)(45.50)^2} = 0.18411$$

$$a = 7.965 - (0.18411)(45.50) = -0.4120$$

Hence, the least square regression line is

(9.8) $\hat{Y} = -0.412 + 0.184X$

If a family drawn from the same population had an annual net income of $30,000 in 1990, its estimated annual food expenditures from equation 9.8 would be

$$\hat{Y} = -0.412 + 0.184(30) = 5.111 \text{ (thousands of dollars)}$$

Because the best-fitting line to the sample data also estimates the population regression line $\mu_{Y.X} = A + BX$, where $\mu_{Y.X}$ is a conditional mean, we can state that the $\hat{Y}$ value of $5,111 is the best estimate of *mean* annual family food expenditures for an $X = 30(\$30,000)$ annual family income. Thus, the sample regression line, or prediction line as it is often called, may be used to predict a future value of Y or to estimate a mean value of Y for a given value of X.

Terms   Incidentally, when the context is clear, the terms **regression line** and **regression equation** are often used for **sample regression line** and **sample regression equation**. That practice is used in this chapter and in Chapter 10.

By plotting the point thus determined ($X = 30$, $Y = 5.111$) and one other point, or by plotting any two points derived from the regression equation, we can graph the regression line. The line is shown in Figure 9-6, along with the original data. Hence, in this case $a = -0.412$ means that the estimated annual food expenditures for a family whose income is zero dollars in 1990 is -0.412 (thousands of dollars) or $-\$412$. Since no families in the original sample had incomes less than $20,000, it would be extremely hazardous to make predictions for families with incomes less than the $20,000 figure and, of course, the $-\$412$ figure for estimated food expenditures is ridiculous. Prediction outside the range of the original observations is discussed in a later section of this chapter.

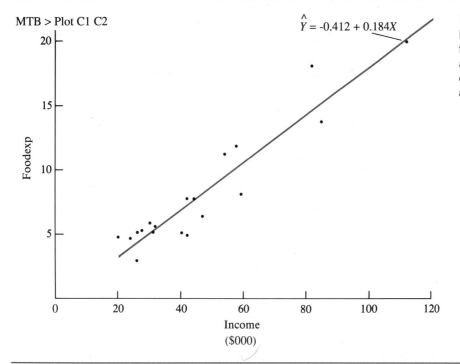

MTB > Plot C1 C2

$\hat{Y} = -0.412 + 0.184X$

FIGURE 9-6
Least squares regression line
for food expenditures and
annual net income of a sample
of 20 families in a metropolitan
area in 1990

    The b value in the regression equation is often referred to as the **sample regression coefficient** or **sample slope coefficient**. The figure of $b = 0.184$ indicates first of all that the slope of the regression line is positive. Thus, as income increases, estimated food expenditures increase. Taking into account the units in which the X and Y variables are stated, $b = 0.184$ means that for two families whose annual net incomes differ by $1,000, the *estimated difference* in their annual food expenditures is $184. This is an interpretation in terms of the *regression line*. If we think of the figure $b = 0.184$ in terms of the *sample studied*, we can say that for two families whose annual net incomes differ by $1,000, their annual expenditures differ, *on the average*, by $184.

*Sample regression
coefficient*

EXERCISES 9.4

1. A scatter diagram and regression line appear in the figure below showing the relationship between earnings during 1989 and price per share at the end of 1989 for selected common stocks. On the basis of this chart, estimate the values of a and b in the equation of the regression line $\hat{Y} = a + bX$. Use specific numbers.

Graph for exercise 1

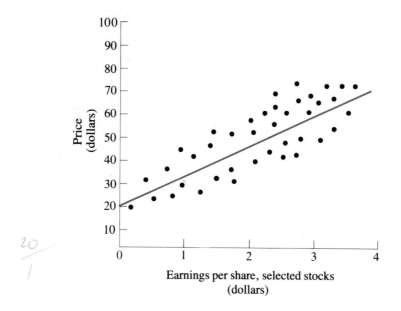

Earnings per share, selected stocks
(dollars)

2. A personnel director of a large manufacturing firm wishes to examine the relationship between employees' aptitude test scores and merit ratings after two years of service. The director drew a simple random sample of 10 employees and obtained the following results:

Employee	Score on Aptitude Test (0–100)	Merit Rating (1–10)
A	75	7
B	64	7
C	92	9
D	80	8
E	76	4
F	58	6
G	96	9
H	89	7
I	98	8
J	79	5

a. Derive the linear regression equation relating merit rating (Y) to aptitude text score (X). Predict the merit rating for each employee's test score listed above.
b. What is the meaning of the regression coefficient b in this case?
c. Estimate the merit rating given to an employee who has an aptitude test score of 90.

3. List the three main objectives of regression and correlation analysis. What statistical measures are used to achieve these objectives?

4. A machine that can be operated at different speeds produces a varying number of defective articles each hour. With the machine operating at different speeds, a simple random sample of 45 hour-long observations was selected.

X = speed of machine in revolutions per second (rps)

Y = number of defective articles produced by the machine during observation

$n = 45$

$\Sigma X = 637$

$\Sigma Y = 234$

$\Sigma XY = 4,470$

$\Sigma X^2 = 10,124$

a. Estimate a least squares regression equation, obtaining estimates for a and b.

b. The machine was operated at 12 rps during one hour. Estimate the number of defectives for this period using the above equation.

5. Martin Breit, the financial vice-president of Lemming Lubricating Liquids, Ltd., wished to estimate the interest rate that his firm would have to pay on a bond issue to be floated soon. Because of past experience, he believes that the size of the issue has a significant effect on the rate and that an estimated relationship between these two variables would be of value in prediction. The following data on bond issues of Lemming, Ltd. have been collected:

Issue Size ($millions)	Rate (%)	Issue Size ($millions)	Rate (%)
$5	5.5	$25	7.5
5	6.5	30	7.0
10	6.0	30	8.0
15	6.5	30	8.5
15	6.0	40	9.0
20	7.0	45	9.5
20	7.5	60	10.5

a. Determine the linear regression equation by the method of least squares, using issue size as the independent variable.

b. If the size of the uncoming bond issue will be $35 million, what is the estimated interested rate?

c. Describe the main assumptions of the linear regression model.

6. A strategic planner of Specified Instruments, Inc., a medium-sized conglomerate, believes that the profitability of any product is largely determined
by the product's market share. To test this belief, the planner selects five
major products manufactured by the company and computes each product's market share and return on average investment. The resulting data
are given below:

Product	Market Share	Return on Average Investment
1	4%	9.1%
2	36%	23.4%
3	25%	17.6%
4	15%	14.1%
5	47%	30.0%

a. Use the method of least squares to determine the linear regression
equation if market share is the independent variable. Calculate $\hat{Y}$ for
each market-share level given.
b. What is the meaning of the regression coefficient b in this case?
c. Estimate the average return on investment of a product with a market
share of 30%.

9.5
CONFIDENCE INTERVALS AND PREDICTION INTERVALS IN REGRESSION ANALYSIS

Standard Error of Estimate

Now that we have seen that the regression equation is used for estimation, we
can turn to the second objective in section 9.3—obtaining measures of the error
involved in using the regression line for estimation. If there is a great deal of
scatter of the observed Y values around the fitted line, estimates of Y values
based on computed values from the regression line will not be close to the observed Y values. On the other hand, if every point falls on the regression lines,
insofar as the **sample observations** are concerned, perfect estimates of the Y
values can be made from the fitted regression line. Just as the standard deviation was used as a measure of the scatter of a set of observations about the
mean, an analogous measure of dispersion of observed Y values around the
regression line is desirable. The measure of dispersion, referred to as the **standard error of estimate** (or the **standard deviation of the residuals**), is obtained

by solving the equation

(9.9)
$$s_{Y.X} = \sqrt{\frac{\Sigma(Y - \hat{Y})^2}{n - 2}}$$

where, as before, n is the sample size.

Although $s_{Y.X}$ is called the *standard error* of estimate, it is not a standard error in the sense that the term was used in Chapters 6 and 7, that is, as a measure of sampling error. The standard error of estimate simply measures the scatter of the observed values of Y around the corresponding computed $\hat{Y}$ values on the regression line. The sum of the squared deviations is divided by $n - 2$ because this divisor makes $s_{Y.X}^2$ an unbiased estimator of the conditional variance around the population regression line,[2] denoted as $\sigma_{Y.X}^2$. The $n - 2$ represents the number of degrees of freedom around the fitted regression line. In general, the denominator is $n - k$ where k is the number of constants in the regression equation. In the case of a straight line, the denominator is $n - 2$ because two degrees of freedom are lost when a and b are used as estimates of the corresponding constants in the population regression line.

It is useful to consider the nature of the notation for the standard error of estimate $s_{Y.X}$. In the discussion of dispersion in section 1.15, we used the symbol s to denote the standard deviation of a sample of observations. The use of the letter s in $s_{Y.X}$ is analogous, since, as explained in the preceding paragraph, $s_{Y.X}$ is also a measure of dispersion computed from a sample. However, since both the variables Y and X are present in a two-variable regression and correlation analysis, subscript notation is required to distinguish among the various possible dispersion measures. Hence, the notation for the standard error of estimate, where Y and X are, respectively, the dependent and independent variables, is $s_{Y.X}$. The letter to the left of the period in the subscript is the dependent variable, and the letter to the right denotes the independent variable. Subscripts are also required to distinguish standard deviations around the means of the two variables. Thus, s_Y denotes the standard deviation of the Y values of a sample around the mean $\bar{Y}$, and s_X denotes the standard deviation of the X values around their mean $\bar{X}$.

Notation $s_{Y.X}$

In a problem containing large numbers of observations, the computation of the standard error of estimate using equation 9.9 clearly involves a great deal of arithmetic. Calculation of Y for each X value in the sample is required, and then the arithmetic implied by the formula must be carried out. A useful shortcut formula involving only quantities already computed is given by

(9.10)
$$s_{Y.X} = \sqrt{\frac{\Sigma Y^2 - a\Sigma Y - b\Sigma XY}{n - 2}}$$

---

[2] In using $s_{Y.X}$, to estimate $\sigma_{Y.X}$, the deviation of each Y value is taken around its own estimated conditional mean $\hat{Y}$. We can combine the squares of these deviations, which are obtained from different conditional probability distributions, because of the assumption of the linear regression model that the conditional standard deviations $\sigma_{Y.X}$ are all equal.

All quantities required by equation 9.10 were calculated for our illustrative problem in Table 9-2, or were computed in obtaining the constants of the regression line. Hence, the standard error of estimate for these data is

$$s_{Y \cdot X} = \sqrt{\frac{1682.17 - (-0.412)(159.3) - (0.18411)(9255.5)}{20 - 2}}$$

$$= 1.559 \text{ (thousands of dollars)}$$

Since the standard error of estimate $s_{Y \cdot X}$ is an estimate of $\sigma_{Y \cdot X}$ (the standard deviation around the true but unknown population regression line), $s_{Y \cdot X}$ may be used and interpreted as a standard deviation. If every sample point falls on the regression line—that is, if there is no scatter around the line—then $s_{Y \cdot X} = 0$. This indicates that the regression line is a perfect fit to the sample data.

> The larger the value of $s_{Y \cdot X}$, the greater is the scatter around the regression line.

As indicated earlier, in the case of a perfect linear relationship between X and Y for a *sample* of data, given a value of X, we could estimate or predict the corresponding *sample* Y value perfectly. However, to make predictions about values in the population not included in our sample, we would have to take account of the fact that the sample regression line we have fit to the data plotted on a scatter diagram may differ from the unknown regression line because of chance errors of sampling. That is, because of chance sampling errors, the a and b values in the sample regression line, $\hat{Y} = a + bX$, may differ from the A and B values in the true population regression line, $\mu_{Y \cdot X} = A + BX$. This means that the height and slope of the sample regression line may differ from the height and slope of the population regression line. Therefore, in making predictions about items in the population that are not included in the sample, we cannot simply use $s_{Y \cdot X}$ in establishing confidence intervals. We must use appropriate standard errors that take the aforementioned chance errors into account. We now turn to the discussion of such confidence intervals.

Types of Interval Estimates

Two types of estimates, or predictions, of values of the dependent variable are ordinarily made in regression analysis. Interval estimates are generally made for both types of estimates, with the *width of the intervals* indicating the precision of the estimation procedure.

Confidence intervals   Interval estimates for a *conditional mean* are usually referred to as **confidence intervals**. For example, in our food expenditures illustration, we may be interested in estimating *average* food expenditures for families with annual net incomes of $30,000; that is, we may want an estimate of the mean of the conditional probability distribution of food expenditures for families with annual net incomes of $30,000.

Estimates that involve predicting an *individual value* of the dependent variable Y are referred to as **prediction intervals**. Sometimes we wish to predict a *single value* of the dependent variable Y, rather than a conditional average value as in the first type of estimate. Thus, for example, we may wish to predict food expenditures for a *particular* family whose income is $30,000. Such single values cannot be predicted with as much precision as for conditional means.

Prediction intervals

We consider these two types of estimates in the following subsections.

Confidence Interval Estimate of a Conditional Mean

In Chapter 6, we discussed methods for estimating a population parameter from a statistic observed in a simple random sample. Thus, for example, we can establish a *confidence interval* for the population mean food expenditures for the sample of 20 families for whom data were shown in Table 9-1. Assuming that the dependent variable Y is normally distributed, and using equation 6.15, we can calculate the confidence limits for population mean food expenditures as

Confidence limits of population mean

$$\bar{Y} \pm ts_{\bar{Y}} = \bar{Y} \pm t\,\frac{s_Y}{\sqrt{n}}$$

where $\bar{Y}$ is the sample mean for the desired confidence level, t is determined for $n - 1$ degrees of freedom for the desired confidence level, $s_{\bar{Y}}$ is the estimated standard error of the mean, s_Y is the sample standard deviation of the Y data, and n is the sample size (in this case, 20). The symbol Y is used for the variable of interest here rather than the X used in equation 6.15. Carrying out the arithmetic in this illustration for a 95% confidence interval, we have:

$$\bar{Y} = 7.965 \text{ (thousands of dollars)}$$

$$s_Y = \sqrt{\frac{\Sigma(Y - \bar{Y})^2}{n - 1}} = \sqrt{\frac{413.35}{19}} = 4.664 \text{ (thousands of dollars)}$$

We have $t = 2.093$ for $v = 20 - 1 = 19$ degrees of freedom (obtained from Table A-6 of Appendix A for a 95% confidence coefficient, that is, 5% area in both tails, or 2.5% in each tail), and

$$s_{\bar{Y}} = \frac{s_Y}{\sqrt{n}} = \frac{4.664}{\sqrt{20}} = 1.043$$

Therefore, $\bar{Y} \pm 2.093s_{\bar{Y}} = 7.965 \pm 2.093(1.043)$ (thousands of dollars).

Hence, the 95% confidence limits are 5.782 and 10.148 (thousands of dollars). That is, population mean food expenditures would be included in 95% of the intervals so constructed.

Interval estimates can be determined from a regression equation in an essentially similar way. Let us consider the construction of an interval estimate for a conditional mean $\mu_{Y.X}$. Returning to our illustration, suppose we wanted to estimate the *mean* or expected food expenditures for a given family with an annual net income of $30,000. As we have seen, another way of stating this is that we wish to estimate the mean of the conditional probability distribution of food expenditures for $X = 30$ (thousands of dollars). An unbiased estimate of this conditional mean, denoted $\mu_{Y.30}$, is given by $\hat{Y}$ from our sample regression line for $X = 30$ (thousands of dollars). We previously calculated $\hat{Y}$ in section 9.4 as

$$\hat{Y} = -0.412 + 0.184(30) = 5.111 \text{ (thousands of dollars)}$$

This figure of $\hat{Y} = 5.111$ (thousands of dollars) is the point on the sample regression line that corresponds to $X = 30$ (thousands of dollars), and it is our best estimate of the conditional mean of Y, the vertical coordinate of the point on the population regression line when $X = 30$. Of course we do not ordinarily know the position of the population regression line. This situation is depicted in Figure 9-7 for an assumed population regression line. The confidence interval estimate for the conditional mean (also called the confidence interval for the regression line) is given by

Confidence interval estimate for the conditional mean

(9.11)
$$\hat{Y} \pm ts_{\hat{Y}}$$

where $s_{\bar{Y}}$ is the estimated standard error of the conditional mean and t is the t multiple determined for $n - 2$ degrees of freedom for the desired confidence level. The number of degrees of freedom is $n - 2$ because two degrees of freedom are lost by using a and b from the sample regression equation, $\hat{Y} = a + bX$, to estimate A and B in the population regression line, $\mu_{Y.X} = A + BX$.

The estimated standard error of the conditional mean is given by

Estimated standard error of the conditional mean

(9.12)
$$s_{\hat{Y}} = s_{Y.X} \sqrt{\frac{1}{n} + \frac{(X_0 - \bar{X})^2}{\Sigma(X - \bar{X})^2}}$$

This expression for $s_{\hat{Y}}$ can be used to obtain a confidence interval estimate for the mean or expected value of annual food expenditures for a given value of X, say X_0.

Continuing with our illustration, for a family with an annual net income of $30,000, we compute the estimated standard error of the condition mean. A more convenient form of equation 9.12 for computation purposes is

A more convenient form of equation 9.12

(9.13)
$$s_{\hat{Y}} = s_{Y.X} \sqrt{\frac{1}{n} + \frac{(X_0 - \bar{X})^2}{\Sigma X^2 - \frac{(\Sigma X)^2}{n}}}$$

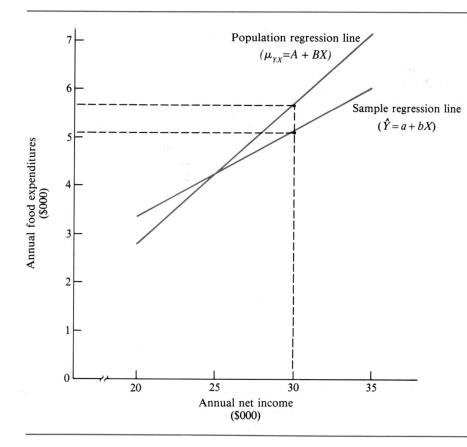

FIGURE 9-7

Estimated and population mean food expenditures when income equals $30,000

Substituting into equation 9.13 for $X_0 = 30$ (thousands of dollars), we get

$$s_{\hat{Y}} = (1.559)\sqrt{\frac{1}{20} + \frac{(30 - 45.50)^2}{52,308 - \frac{(910)^2}{20}}}$$

$$s_{\hat{Y}} = 0.418 \text{ (thousands of dollars)}$$

Hence, a 95% confidence interval for the conditional mean obtained by substitution into equation 9.11 is

$$5.111 \pm 2.101(0.418) \quad \text{or} \quad (4.233, 5.989) \text{ in thousands of dollars}$$

where the t value of 2.101 for a 95% confidence coefficient was obtained from Table A-6 of Appendix A for $n - 2 = 20 - 2 = 18$ degrees of freedom. Therefore, the desired 95% confidence limits are about $4,233 to $5,989 for the mean food expenditure for a family with an annual net income of $30,000.

FIGURE 9-8

95% confidence interval for the
conditional mean

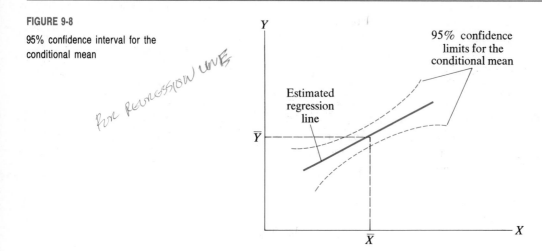

Note from equation 9.12, the formula for $s_{\hat{Y}}$, that the farther an X value lies from the mean, the larger is $(X_0 - \bar{X})^2$ and the wider is the confidence interval at that value of X. This concept is intuitively plausible because the greater the distance from the mean, the greater would be the effect of an error in the estimated slope of the regression line. This situation is depicted in Figure 9-8.

Prediction Interval for an Individual Value of Y

Just as in the case of estimating a conditional mean for a given X value, we obtain an unbiased estimate in predicting an *individual Y value* for a given X from the point on the estimated regression line corresponding to X. That is, the best point prediction for the individual Y value is obtained from the regression $\hat{Y} = a + bX$ by substituting the given X value.

However, for a given X value, we cannot predict an individual Y value with as much precision as we can estimate a conditional mean. This follows from the fact that the sampling error in predicting an individual value of Y is greater than the sampling error involved in estimating a conditional mean value of Y.

It can be shown that the standard deviation of the error $(Y - \hat{Y})$ of predicting an individual value of Y when $X = X_0$ is

Standard error of forecast   **(9.14)**
(of an individual value of Y)

$$s_{\text{IND}} = s_{Y \cdot X} \sqrt{1 + \frac{1}{n} + \frac{(X_0 - \bar{X})^2}{\Sigma(X - \bar{X})^2}}$$

As in the case of the standard error of the conditional mean, a more convenient form of equation 9.14 for purposes of calculation is

(9.15)
$$s_{IND} = s_{Y.X} \sqrt{1 + \frac{1}{n} + \frac{(X_0 - \bar{X})^2}{\sum X^2 - \frac{(\sum X)^2}{n}}}$$

A more convenient form of equation 9.14

The prediction of an individual value of Y is usually referred to as an **individual forecast**, and s_{IND} is called the **standard error of forecast**.[3] The standard error of forecast can be used to set up a prediction interval for an individual Y value in a manner analogous to setting up confidence intervals for the conditional mean.

Returning to the food expenditures example, assume that we want to predict food expenditures for a particular family with an annual net income of $30,000. The prediction interval for an individual Y value is given by

(9.16)
$$\hat{Y} \pm t s_{IND}$$

Prediction interval of an individual Y value

where $\hat{Y}$ is the individual forecast as determined from the sample regression line, s_{IND} is the standard error of forecast, and t is the t multiple determined for $n - 2$ degrees of freedom for the desired confidence level.

Substituting into formula 9.15 for $X_0 = 30$ (thousands of dollars), we obtain for the standard error of forecast

$$s_{IND} = 1.559 \sqrt{1 + \frac{1}{20} + \frac{(30 - 45.50)^2}{52,308 - \frac{(910)^2}{20}}}$$

$$s_{IND} = 1.614 \text{ (thousands of dollars)}$$

Therefore, by formula 9.16, a 95% prediction interval for the individual forecast of the Y value for a family with a $30,000 annual net income is

$$5.111 \pm 2.101(1.614)$$

where the t value of 2.101 for a 95% confidence coefficient was obtained earlier from Table A-6 for $n - 2 = 20 - 2 = 18$ degrees of freedom.

Hence, the prediction interval is from 1.720 to 8.502 (thousands of dollars), or approximately $1,720 to $8,502 of food expenditures. We note that, as expected, these limits are wider than those previously computed for the conditional mean. Also, as was true for confidence limits for the conditional mean, the farther the X value is from $\bar{X}$, the wider is the prediction interval. This characteristic of prediction intervals for individual forecasts is depicted in Figure 9.9.

---

[3] Actually, s_{IND} should be referred to as the "estimated standard error of forecast" just as $s_{\hat{Y}}$ is the "estimated standard error of the conditional mean." In both of these cases, "estimated" is appropriate because $s_{Y.X}$ is an estimate of the population $\sigma_{Y.X}$. However, it is conventional to refer to the "standard error of forecast" without the use of the word "estimated."

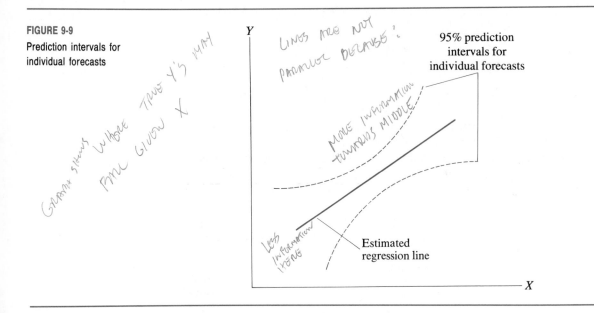

FIGURE 9-9

Prediction intervals for individual forecasts

(Handwritten annotations on figure:)

GRAPH SHOWS WHERE TRUE Y's MAY FALL GIVEN X

LINES ARE NOT PARALLEL BECAUSE:

MORE INFORMATION TOWARDS MIDDLE

LESS INFORMATION HERE

95% prediction intervals for individual forecasts

Estimated regression line

Width of Confidence and Prediction Intervals

We can see from equations 9.12 and 9.14 that a number of factors affect the sizes of the estimated standard error of the conditional mean and the standard error of forecast and, hence, the width of confidence and prediction intervals. The larger the sample size n, the smaller are these standard errors, and therefore the narrower are the widths of the intervals. This agrees with our intuitive notion that larger sample sizes provide greater precision of estimation.

Second, as noted earlier, the greater the deviation of X from $\bar{X}$, the greater is the standard error and the wider are the intervals for the given X value. This means that confidence and prediction intervals may be quite wide for very small or very large X values compared with the analogous intervals for average-sized X values.

Third, the larger the estimated standard error of estimate $s_{Y \cdot X}$, the larger are the confidence and prediction intervals. This agrees with the intuitive idea that the more variable (less uniform) the data, the less precise would be predictions from these data.

Finally, note that the more variability there is in the sample of X values (in our illustrative problem, the more variability there is in family incomes), the larger is $\Sigma(X - \bar{X})^2$, and hence the smaller are the standard errors and the narrower are the confidence and prediction intervals. This is consistent with the idea that, for example, if we observe a relationship for families with large variation in incomes, we would expect to make a better estimate of food expenditures than if the variation in incomes was small.

We have discussed estimation procedures for confidence intervals for conditional means and for prediction intervals for individual forecasts. Setting up confidence intervals for conditional means is appropriate whenever we are interested in estimating the *average value* of Y for a given X. In this chapter, we illustrated the estimation of *average food expenditures* for families with $30,000 of income. The same procedure would be used if in other regression analyses we were interested, for example, in confidence intervals for *average* or *expected sales* for an expenditure of $1 million on advertising, for *average number of accidents* for a company with a certain number of workers exposed to some hazard, or *average weight* for black males in the United States who are 70 inches tall.

On the other hand, prediction intervals for individual forecasts are appropriate when we are interested in predicting an *individual Y value* for a given X. Conceptually, we wish to predict the Y value for a new observation drawn from the same population as the sample from which we have constructed the regression line. Hence, in this chapter, we illustrated the prediction interval estimation of *actual food expenditures* of a *particular* family with a $30,000 income. Analogously, we would use this procedure if we were interested in prediction intervals for the *specific amount of sales* for the *next company* that spends $1 million on advertising, for the *specific number* of *accidents* for a *particular company* with 150 employees exposed to some hazard, or the *actual weight* of a *particular* black male in the United States who is 70 inches tall.

In terms of statistical theory, when we estimate a conditional mean, we estimate the value of a *parameter*; when we predict an individual value of Y, we predict the value of a *random variable*. Stated differently, in the former case we are trying to estimate a point on the population regression line; in the latter case we are trying to predict an individual value of Y.

The usefulness of confidence intervals or prediction intervals depends on the purposes for which they are used. For example, for long-range planning purposes, relatively wide limits may be appropriate and useful. On the other hand, for short-term operational decision making, narrower and therefore more precise intervals may be required. In a two-variable regression analysis, the standard error of estimate $s_{Y.X}$ may be so large as to yield confidence and prediction intervals that are too wide for the investigator's purposes. In such a case, the investigator may introduce additional independent variables to obtain greater precision of estimation and prediction. We discuss the use of two or more independent variables in Chapter 10, which deals with *multiple* regression and correlation analysis.

EXERCISES 9.5

1. A business school is conducting a study of the relationship between a student's score on the GMAT examination and his or her starting salary

on completion of the MBA program. The school has collected data on 100 graduates with the following results:

$$X = \text{GMAT test score}$$

$$Y = \text{starting salary (thousands of dollars)}$$

$$n = 100$$

$$\Sigma X = 55{,}000$$

$$\Sigma X^2 = 38{,}124{,}400$$

$$\Sigma Y = 2{,}100$$

$$\Sigma Y^2 = 50{,}500$$

$$\Sigma XY = 1{,}364{,}600$$

a. Estimate the linear regression equation, using the method of least squares, with GMAT test score as the independent variable.
b. Calculate the standard error of estimate $s_{Y \cdot X}$.
c. Estimate the starting salary for a graduate with a GMAT score of 655. Do the same for a graduate with a score of 500.
d. Obtain 95% confidence interval estimates for the conditional mean starting salary for graduates with GMAT score of 655 and 500.

2. An accounting standards board investigating the treatment of research and development expenses by the nation's major electronic firms was interested in the relationship between a firm's research and development expenditures and its earnings. The board compiled the following data on 15 firms for the year 1990.

Firm	Earnings (1990) ($millions)	Research and Development Expenditures for Previous Year (1989) ($millions)
A	$221	$15.0
B	83	8.5
C	147	12.0
D	69	6.5
E	41	4.5
F	26	2.0
G	35	0.5
H	40	1.5
I	125	14.0
J	97	9.0
K	53	7.5
L	12	0.5
M	34	2.5
N	48	3.0
O	64	6.0

a. Estimate the linear regression equation, using the method of least squares, with research and development expenditures as the independent variable.

b. What are the estimated 1990 earnings for a firm that spent $10 million on research and development in 1989?

c. Calculate the standard error of estimate, $s_{Y.X}$. What does this measure?

d. Obtain a 95% confidence interval for the conditional mean 1990 earnings for the firm in part (b).

3. A congressional committee studying federal tax reform is taking a close look at the taxes that large corporations pay on reported income. A random sample of 12 firms yielded the following statistics:

Firm	Reported Income in 1990 ($millions)	Income Taxes Paid in 1990 ($millions)
Excelsior	$300	$120
Diamond International	250	110
General Nuts and Bolts	425	200
Scolding, Inc.	210	100
American Abacus	170	75
Stronghold Safes, Inc.	125	65
Dooley Brothers	100	40
Marathon Motors	280	125
Slick Oil Company	375	175
United Pickle	115	50
Agony Airlines	80	40
Stalwart Steel, Inc.	210	100

a. Estimate the linear regression equation, using the method of least squares, with reported 1990 income as the independent variable.

b. Calculate the standard error of estimate, $s_{Y.X}$.

c. Interpret the estimated regression coefficient, b, in terms of this problem. Calculate a weighted average tax rate. How close are the two estimates?

d. Obtain 95% confidence intervals for conditional mean income taxes paid for the following levels of reported income: $100 million, $220 million, and $340 million. Calculate the width of these intervals. What do you observe?

e. Follow the instructions in part (d) for estimating prediction intervals for income taxes paid for each income level. Calculate and comment upon the interval widths. t_{SY} vs. t_{SIND} E= width

4. Texecon Products, Inc. a major producer of household products, has introduced 14 new products during the past two years. The marketing research unit needs an estimate of the relationship between first-year sales and an appropriate independent variable; this relationship will be used in future planning in marketing and advertising. The researchers have constructed

TABLE 9-3
Texecon Products, Inc., marketing research data

Product	First-Year Sales ($millions)	Customer Awareness (%)
E-Z-Kleen	82	50
Sud-Z-Est	46	45
Alumofoil	17	15
Backscratcher	21	15
Pest Killer	112	70
Liquid Lush	105	75
Bubbly Bath	65	60
Wipe Away	55	40
Whirlwind	80	60
Magic Mop	43	25
Cobweb Cure	79	50
Oven Eater	24	20
Dirt-B-Gone	30	30
Dustbowl	11	5

a variable called "customer awareness," measured by the proportion of consumers who had heard of the product by the third month after its introduction. The data are shown in Table 9.3.

a. Find the least squares regression equation, with customer awareness as the independent variable.

b. Calculate the standard error of estimate, $s_{Y.X}$. What assumption is necessary in using $s_{Y.X}$ to estimate $\sigma_{Y.X}$, the conditional standard deviation around the population regression line?

c. Calculate a 95% confidence interval estimate for conditional mean sales for a customer awareness level of 35%.

d. Obtain a 95% prediction interval for new product sales for the same customer awareness level as in part (c).

e. Which interval was wider: that obtained in part (c) or that obtained in part (d)? Why?

9.6
CORRELATION ANALYSIS: MEASURES OF ASSOCIATION

In the preceding two sections, regression analysis was discussed, with emphasis on estimation and measures of error in the estimation process. We now turn to correlation analysis, in which the basic objective is to obtain a measure of the degree of association between two variables. In this analysis, interest centers on the strength of the relationship between the variables, that is, on how well the variables are correlated. The assumptions of the two-variable correlation model are as follows:

1. Both X and Y are random variables.
2. Both X and Y are normally distributed. The two distributions need not be independent.
3. The standard deviations of the Ys are assumed to be equal for all values of X, and the standard deviations of the Xs are assumed to be equal for all values of Y.

Note that in the correlation model, both X and Y are assumed to be random variables. On the other hand, in the regression model, only Y is a random variable, and the Y observations are treated as a random sample from the conditional distribution of Y for a given X.

The Coefficient of Determination

A measure of the amount of correlation between Y and X can be explained in terms of the relative variation of the Y values around the regression line and the corresponding variation around the mean of the Y variable. The term **variation**, as used in statistics, conventionally refers to a sum of squared deviations.

The variation of Y values around the regression line is measured by

(9.17)
$$\Sigma(Y - \hat{Y})^2$$

Variation of Y values around regression line

The variation of Y values around the mean of the Y variable is measured by

(9.18)
$$\Sigma(Y - \bar{Y})^2$$

Variation of Y values around the mean

Equation 9.17 is the sum of the squared vertical deviations of the Y values from the regression line. Equation 9.18 is the sum of the squared vertical deviations of the Y values from the horizontal line $Y = \bar{Y}$. The relationship between the variations around the regression line and the mean can be summarized in a single measure to indicate the degree of association between X and Y. The measure used for this purpose is the **sample coefficient of determination**, defined as follows:

(9.19)
$$r^2 = 1 - \frac{\Sigma(Y - \hat{Y})^2}{\Sigma(Y - \bar{Y})^2}$$

Sample coefficient of determination

As we shall see from the subsequent discussion, r^2 may be interpreted as the proportion of variation in the dependent variable Y that has been accounted for, or "explained," by the relationship between Y and X expressed in the regression line. Hence, it is a measure of the degree of association or correlation between Y and X.

To present the rationale of this measure of strength of the relationship between Y and X, we will consider two extreme cases, zero linear correlation and perfect direct linear correlation. The term *linear* indicates that a straight line has been fitted to the X and Y values, and the term *direct* indicates that the line has a positive slope.

Two sets of data are labeled (a) and (b) and are shown in the scatter diagrams in Figure 9-10. The data in (a) and (b) illustrate the cases of zero linear correlation and perfect direct linear correlation, respectively. In the discussion that follows, we will assume that the observations shown in (a) and (b) represent simple random samples from their respective universes. Therefore, we employ notation appropriate to samples. If we assumed that the observations represent population data, the notation would change correspondingly. The calculations given below the scatter diagrams in Figure 9-10 will be explained in terms of the data displayed in the charts.

Case (a) represents a situation in which $\bar{Y}$, the mean of the Y values, coincides with a least squares regression line fitted to these data. Even without doing the arithmetic, we can see why this is so. The slope of the regression line is zero, because the same Y values are observed for $X = 1, 2, 3,$ and 4. Thus, the regression line would coincide with the mean of the Y values, balancing deviations above and below the regression line. Another way of observing this

FIGURE 9-10

Scatter diagrams representing zero linear correlation and perfect direct linear correlation based on the following sets of data for observations A through H

	(a)		(b)	
	X	Y	X	Y
A	1	4	1	2
B	1	6	2	4
C	2	4	3	6
D	2	6	4	8
E	3	4	5	10
F	3	6	6	12
G	4	4	7	14
H	4	6	8	16
	$\bar{Y} = 5$		$\bar{Y} = 9$	

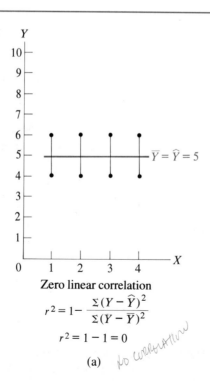

Zero linear correlation

$$r^2 = 1 - \frac{\Sigma(Y - \hat{Y})^2}{\Sigma(Y - \bar{Y})^2}$$

$$r^2 = 1 - 1 = 0$$

(a)   *No correlation*

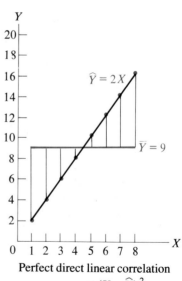

Perfect direct linear correlation

$$r^2 = 1 - \frac{\Sigma(Y - \hat{Y})^2}{\Sigma(Y - \bar{Y})^2}$$

$$r^2 = 1 - 0 = 1$$

(b)   *Perfect correlation*

relationship is in terms of the first of the two equations used to solve for a and b. In equation 9.6, $a = \bar{Y} - b\bar{X}$, and since $b = 0$, then $a = \bar{Y}$. Hence, the regression line has a Y intercept equal to $\bar{Y}$. Since it is also a horizontal line, the regression line coincides with $\bar{Y}$. From the point of view of estimation of the Y variable, the regression line represents no improvement over the mean of the Y values. This can be shown by a comparison of $\Sigma(Y - \hat{Y})^2$, the variation around the regression line, and $\Sigma(Y - \bar{Y})^2$, the variation around the mean of the Y values. In this case, the two variations are equal. These variations may be interpreted graphically as the sum of the squares of the vertical distances between the points on the scatter diagram and $\bar{Y}$. shown in Figure 9-10(a).

Now, let us consider case (b). In this situation, the regression line is a perfect fit to the data. The regression equation is a simple one, which can be determined by inspection. The Y intercept is zero, since the line passes through the origin (0, 0). The slope is two, because for every unit increase in X, Y increases by two units. Hence, the regression equation is $\hat{Y} = 2X$, and all the data points lie on the regression line. As far as the data in the sample are concerned, perfect predictions are provided by this regression line. Given a value of X, the corresponding value of Y can be correctly estimated from the regression equation, indicating a perfect linear relationship between the two variables. Again, a comparison can be made of $\Sigma(Y = \hat{Y})^2$ and $\Sigma(Y - \bar{Y})^2$. Since all points lie on the regression line, the variation around the line $\Sigma(Y - \hat{Y})^2$ is zero. On the other hand, the variation around the mean, $\Sigma(Y - \hat{Y})^2$, is some positive number, in this case, 168.

As indicated in Figure 9-10(a), when there is no linear correlation between X and Y, the sample coefficient of determination, r^2, is zero. This follows from the fact that since $\Sigma(Y - \hat{Y})^2$ and $\Sigma(Y - \bar{Y})^2$ are equal, the ratio $\Sigma(Y - \hat{Y})^2/\Sigma(Y - \bar{Y})^2$ equals one. Hence, from equation 9.19, $r^2 = 0$, because the computation of the coefficient of determination requires subtraction of this ratio from one.

On the other hand, as indicated in Figure-10(b), when there is perfect linear correlation between X and Y, the sample coefficient of determination, r^2, is one. In this case, the variation around the regression line is zero, while the variation around the mean is some positive number. Thus, the ratio $\Sigma(Y - \hat{Y})^2/\Sigma(Y - \bar{Y})^2$ is zero. Hence, from equation 9.19, $r^2 = 1$ when the value of this ratio is subtracted from one.

In realistic problems, r^2 falls somewhere between the two limits zero and one. An r^2 value close to zero suggests not much linear correlation between X and Y; an r^2 value close to one connotes a strong linear relationship between X and Y.

Population Coefficient of Determination

The measure r^2, called the **sample coefficent of determination**, pertains only to the sample of n observations studied. The regression line computed from the sample may be viewed as an estimate of the true population regression line, which may be denoted

The true population (9.20)
regression line

$$\mu_{Y.X} = A + BX$$

The corresponding population coefficient of determination is defined as

Population coefficient (9.21)
of determination

$$\rho^2 = 1 - \frac{\sigma^2_{Y.X}}{\sigma^2_Y}$$

The use of the symbol ρ^2 (rho squared) adheres to the usual convention of employing a Greek letter for a population parameter corresponding to the same letter in our alphabet that denotes a sample statistic. In the definition of ρ^2, $\sigma^2_{Y.X}$ is the variance around the population regression line $\mu_{Y.X} = A + BX$, and σ^2_Y is the variance around the population mean of the Ys, denoted μ_Y. Both the sample and population coefficients of determination are equal to one minus a ratio of the variability around the regression line to the variability around the mean of the Y values.

A slightly different form of the *sample* coefficient of determination, which is directly parallel to equation 9.21, is

(9.22)
$$r_c^2 = 1 - \frac{s^2_{Y.X}}{s^2_Y} = 1 - \frac{\Sigma(Y - \hat{Y})^2/(n-2)}{\Sigma(Y - \bar{Y})^2/(n-1)}$$

The quantity r_c^2 is the **corrected** (or adjusted) **sample coefficient of determination**. This terminology is used because $s^2_{Y.X}$ and s^2_Y are estimators of $\sigma^2_{Y.X}$ and σ^2_Y that make the appropriate corrections or adjustments for degrees of freedom.[4] Since $s^2_{Y.X}$ and s^2_Y are unbiased estimators of $\sigma^2_{Y.X}$ and σ^2_Y, the adjusted sample coefficient of determination r_c^2, rather than the unadjusted coefficient r^2, is ordinarily used in estimating the population coefficient of determination, ρ^2. However, in the discussion that follows we use only the unadjusted value r^2, because of the complication associated with the divisors $n - 2$ and $n - 1$ in the adjusted measure.

Interpretation of the Coefficient of Determination

Let us consider in more detail the specific interpretations that may be made of coefficients of determination. For convenience, only the sample coefficient r^2 will be discussed, but the corresponding meanings for ρ^2 are obvious.

An important interpretation of r^2 may be made in terms of variation in the dependent variable Y, which has been explained by the regression line. We conceive of the problem of estimation in terms of "explaining" or accounting for the variation in the dependent variable Y. Figure 9-11, on which a single point is shown, gives a graphic interpretation of the situation. In this context, if $\bar{Y}$, the mean of the Y values, were used to estimate the value of Y, the total

---

[4] The relationship between r_c^2 and r^2 is given by $r_c^2 = 1 - (1 - r^2)\left(\frac{n-1}{n-2}\right)$. For large sample sizes $\left(\frac{n-1}{n-2}\right)$ is close to 1 and r_c^2 and r^2 are approximately equal.

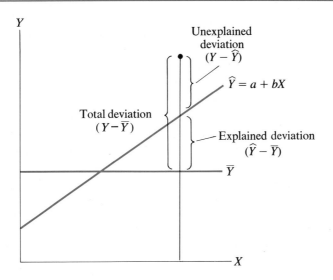

FIGURE 9-11
Graphic representation of total, explained, and unexplained variation

deviation would be $Y - Y$. We can think of total deviation as being composed of the following elements:

Total deviation = Explained deviation + Unexplained deviation
$$(Y - \bar{Y}) \quad = \quad (\hat{Y} - \bar{Y}) \quad + \quad (Y - \hat{Y})$$

Total deviation

If the regression line were used to estimate the value of Y, we would now have a closer estimate. As shown in the figure, there is still an "unexplained deviation" of $(Y - \hat{Y})$, but we have explained $(\hat{Y} - \bar{Y})$ out of the total deviation by assuming that the relationship between X and Y is given by the regression line.

In an analogous manner, we can partition the total variation of the dependent variable (or total sum of squares), $\Sigma(Y - \bar{Y})^2$, as follows:

Total variation = Explained variation + Unexplained variation
$$\Sigma(Y - \bar{Y})^2 \quad = \quad \Sigma(\hat{Y} - \bar{Y})^2 \quad + \quad \Sigma(Y - \hat{Y})^2$$

Total variation

The ratio $\Sigma(Y - \hat{Y})^2/\Sigma(Y - \bar{Y})^2$ is the proportion of total variation that remains unexplained by the regression equation; correspondingly, $1 - [\Sigma(Y - \hat{Y})^2/\Sigma(Y - \bar{Y})^2]$ represents the *proportion of total variation in Y that has been explained* by the regression equation. These ideas may be summarized as follows:

(9.23)
$$r^2 = 1 - \frac{\Sigma(Y - \hat{Y})^2}{\Sigma(Y - \bar{Y})^2} = 1 - \frac{\text{Unexplained variation}}{\text{Total variation}}$$

$$r^2 = \frac{\text{Explained variation}}{\text{Total variation}}$$

A simple numerical example helps to clarify these relationships. Let $\Sigma(Y - \bar{Y})^2 = 10$ and $\Sigma(Y - \hat{Y})^2 = 4$. Thus, $r^2 = 1 - \frac{4}{10} = \frac{6}{10} = 60\%$. In this problem, 10 units of total variation in Y have to be accounted for. After we fit the regression line, the residual variation or unexplained variation amounts to four units. Hence, 60% of the total variation in the dependent variable is explained by the relationship between Y and X expressed in the regression line.

Calculation of the Sample Coefficient of Determination

The computation of r^2 from the definitional formula, equation 9.19, becomes quite tedious, particularly with a large sample. Just as in the case of the standard error of estimate, shorter methods of calculation are ordinarily used. These shortcut formulas are particularly helpful when computations are carried out by hand or on a calculator; even when computers are used, they represent more efficient methods of computation. Such a formula, which involves only quantities already calculated, is

(9.24)
$$r^2 = \frac{a\Sigma Y + b\Sigma XY - n\bar{Y}^2}{\Sigma Y^2 - n\bar{Y}^2}$$

Substituting into equation 9.24, we obtain
$$r^2 = \frac{(-0.412)(159.3) + (0.184)(9255.5) - 20(7.965)^2}{(1,682.17) - 20(7.965)^2} = 0.892$$

Thus, for our sample of 20 families, about 89% of the variation in annual food expenditures was explained by the regression equation, which related such expenditures to annual net income.

The Coefficient of Correlation

A widely used measure of the degree of association between two variables is the coefficient of correlation, which is simply the square root of the coefficient of determination. Thus, the population and sample coefficients of correlation are

Population coefficient of correlation (9.25)
$$\rho = \pm\sqrt{\rho^2}$$

and

Sample coefficient of correlation (9.26)
$$r = \pm\sqrt{r^2}$$

Again, for convenience, our discussion will relate only to the sample value.

The algebraic sign attached to r is the same as that of the regression coefficient, b. Thus, if the slope of the regression line b is positive, then r is also positive; if b is negative, r is negative.[5] Hence r ranges in value from -1 to

---

[5] An interesting relationship between r and b is given by $b = r(\sigma_Y/\sigma_X)$. Since σ_Y and σ_X are positive numbers, b has the same sign as r. We also note that when the regression line is horizontal, $b = 0$ and $r = 0$.

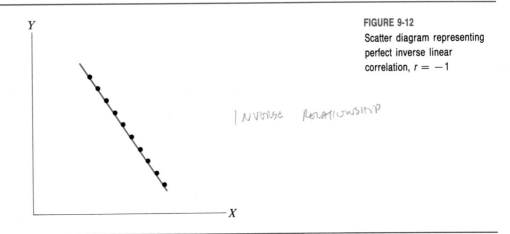

INVERSE RELATIONSHIP

FIGURE 9-12
Scatter diagram representing
perfect inverse linear
correlation, $r = -1$

$+1$. A figure of $r = -1$ indicates a perfect inverse linear relationship, $r = +1$ indicates a perfect direct linear relationship, and $r = 0$ indicates no linear relationship.

Scatter diagrams for the cases of $r = +1$ ($r^2 = 1$) and $r = 0$ ($r^2 = 0$) were given in Figure 9-10. A corresponding scatter diagram for $r = -1$, the case of perfect inverse linear correlation, is shown in Figure 9-12. As indicated in the graph, the slope of the regression line is negative and every point falls on the line. Thus, for example, if the slope of the regression line, b, were equal to -2, with each increase of one unit in X, Y would decrease by two units. Since all points fall on the regression line in the case of perfect inverse correlation, $\Sigma(Y - \hat{Y})^2 = 0$. Therefore, substituting into equation 9.19 to compute the sample coefficient of determination, we have $r^2 = 1 - 0 = 1$. Taking the square root, we obtain $r = \pm \sqrt{1} = \pm 1$. However, since the b value is negative (that is, X and Y are inversely correlated), we assign the negative sign to r, and $r = -1$.

In our problem,

$$r = \sqrt{0.892} = 0.944$$

The sign is positive because b was positive, indicating a direct relationship between food expenditures and net income.

Despite the rather common use of the coefficient of correlation, it is preferable for interpretation purposes to use the coefficient of determination. As we have seen, r^2 can be interpreted as a proportion or a percentage figure. When the square root of a percentage is taken, the specific meaning becomes obscure. Furthermore, since r^2 is a decimal value (unless it is equal to zero or one), its square root, or r, is a larger number. Thus, the use of r values to indicate the degree of correlation between two variables tends to give the impression of a stronger relationship than is actually present. For example, an r value of $+0.7$ or -0.7 seems to represent a reasonably high degree of association. However, since $r^2 = 0.49$, less than half of the total variance in Y has been explained by the regression equation.

Observe that the values of r and r^2 do not depend on the units in which X and Y are stated nor on which of these variables is selected as the dependent or independent variable. Whether a value of r or r^2 is considered high depends somewhat on the specific field of application. With some types of data, r values in excess of about 0.80 are relatively unusual. On the other hand, particularly in the case of time-series data, r values in excess of 0.90 are quite common. In the following section, we consider the matter of determining whether the observed degree of correlation in a sample is sufficiently large to justify a conclusion that correlation between X and Y actually exists in the population.

9.7
INFERENCE ABOUT POPULATION PARAMETERS IN REGRESSION AND CORRELATION

In the procedures discussed to this point, computation and interpretation of *sample* measures have been emphasized. However, from our study of statistical inference, we know that sample statistcis ordinarily differ from corresponding population parameters because of chance errors of sampling. Therefore, it is useful to have a protective procedure against the possible error of concluding from a sample that an association exists between two variables, while actually no such relationship exists in the population from which the sample was drawn. Hypothesis-testing techniques, such as those discussed in Chapter 7, can be employed for this purpose.

Inference about the Population Correlation Coefficient, ρ

Let us assume a situation in which we take a simple random sample of n units from a population and make paired observations of X and Y for each unit. The sample correlation coefficient r, as defined in equation 9.26, is calculated. The procedure tests the hypothesis that the population correlation coefficient, ρ, is zero in the universe from which the sample was drawn. In keeping with the language used in Chapter 7, we wish to test the null hypothesis that $\rho = 0$ versus the alternative that $\rho \neq 0$. Symbolically, we may write

$$H_0: \rho = 0$$

$$H_1: \rho \neq 0$$

If the computed r values in successive samples of the same size from the population were distributed normally around $\rho = 0$, we would only have to know the standard error of r, σ_r, to perform the usual test involving the normal distribution. Although r values are not normally distributed, a similar procedure is provided by the following statistic:

(9.27)
$$t = \frac{r - \rho}{s_r} = \frac{r}{\sqrt{(1 - r^2)/(n - 2)}}$$

which has a t distribution for $n - 2$ degrees of freedom. The estimated standard error of r is $s_r = \sqrt{(1 - r^2)/(n - 2)}$, and we have $\rho = 0$ by the null hypothesis. Note that despite the previous explanation that r^2 is easier to interpret than r, the hypothesis-testing procedure is in terms of r rather than r^2. The reason is that under the null hypothesis, H_0: $\rho = 0$, the sampling distribution of r leads to the t statistic, a well-known distribution that is relatively easy to work with. On the other hand, under the same hypothesis of no correlation in the universe, r^2 values, which range from zero to one, would not even be symmetrically distributed, and the sampling distribution would be more difficult to deal with. Suppose we wish to test the hypothesis that $\rho = 0$ at the 5% level of significance for our problem involving 20 families. Since $r = 0.944$ and $n = 20$, substitution into equation 9.27 yields

$$t = \frac{0.944}{\sqrt{\dfrac{1 - 0.892}{20 - 2}}} = 12.2$$

Referring to Table A-6, we find a critical t value of 2.101 at the 5% level of significance for eighteen degrees of freedom. Therefore, the decision rule is

---

DECISION RULE

1. If $-2.101 \leqslant t \leqslant 2.101$, retain H_0
2. If $t < -2.101$ or $t > 2.101$, reject H_0

Since our computed t value is 12.2, far in excess of the critical value, we conclude that the sample r value differs significantly from zero. We reject the hypothesis that $\rho = 0$ and conclude that a positive relationship exists between annual food expenditures and annual net income in the population from which our sample was drawn. Since the critical t value is 2.878 at the 1% level of significance (the smallest level shown in Table A-6 of Appendix A), it is extremely unlikely that an r value as high as 0.94 would have been observed in a sample of 20 items drawn from a population in which X and Y were uncorrelated.

A few comments may be made concerning this hypothesis-testing procedure. First of all, this technique is valid only for a hypothesized universe value of $\rho = 0$. Other procedures must be used for assumed universe correlation coefficients other than zero.[6]   *Comments*

Second, only Type I errors are controlled by this testing procedure. That is, when the significance level is set at, say, 5%, the test provides a 5% risk of

---

[6] Fisher's z transformation may be used when ρ is hypothesized to be nonzero. In this procedure, a change of variable is made from the sample r to a statistic z, defined as $z = \frac{1}{2} \log_e[(1 + r)/(1 - r)]$. This statistic is approximately normally distributed with mean $\mu_z = \frac{1}{2} \log_e[(1 + \rho)/(1 - \rho)]$ and standard deviation $\sigma_z = 1/\sqrt{n - 3}$.

incorrectly rejecting the null hypothesis of no correlation. No attempt is made to fix the risks of Type II errors (that is, the risk of accepting H_0: $\rho = 0$ when $\rho \neq 0$) at specific levels.

Third, even though the sample r value is significant according to this test, in some instances the amount of correlation may not be considered substantively important. For example, in a large sample, a low r value may be found to differ significantly from zero. However, since relatively little correlation has been found between the two variables, we may be unwilling to use the relationship observed between X and Y for decision-making purposes. Furthermore, prediction intervals based on the use of the applicable standard errors of estimate may be too wide to be of practical use.

Fourth, the distributions of t values computed by equation 9.27, approach the normal distribution as sample size increases. Hence, for large sample sizes, the t value is approximately equal to z in the standard normal distribution, and critical values applicable to the normal distribution may be used instead. For example, in the preceding illustration, in which the critical t value was 2.101 for 18 degrees of freedom at the 5% level of significance, the corresponding critical z value would be 1.96 at the same significance level. For large sample sizes, these values would be much closer.

Fifth, as noted earlier, in correlation analysis we assume that both X and Y are normally distributed *random variables*. Hence, in order to use a hypothesis-testing procedure about the value of ρ, such as the one illustrated above, X and Y should both be random variables. On the other hand, in the regression model, the independent variable X is not a random variable. In a particular analysis, if the X values are indeed fixed or predetermined, it is not proper to use the sample correlation coefficient r to test hypotheses about ρ. However, in that case, as we have seen earlier, r or r^2 may be used to measure the effectiveness of the regression equation in explaining variation in the dependent variable Y.

Inference about the Population Regression Coefficient, B

In many cases, a great deal of interest is centered on the value of b, the slope of the regression line computed from a sample. Statistical inference procedures involving either hypothesis testing or confidence interval estimation are often useful for answering questions concerning the size of the population regression coefficient B in the population regression equation $\mu_{Y \cdot X} = A + BX$.

In order to illustrate the hypothesis-testing procedure for a regression coefficient, let us return to the data in our problem, in which $b = 0.184$. We interpreted this figure to mean that the estimated difference in annual food expenditures for two families whose annual net income in 1990 differed by $1,000 was $184. Suppose that on the basis of similar studies in the same metropolitan area, it had been concluded that in previous years, a valid assumption for the true population regression coefficient was $B = 0.2$. Can we conclude that the population regression coefficient has changed?

To answer this question, we use a familiar hypothesis-testing procedure. We establish the following null and alternative hypotheses:

$$H_0: B = 0.2$$

$$H_1: B \neq 0.2$$

Assume that we were willing to run a 5% risk of erroneously rejecting the null hypothesis that $B = 0.2$. The procedure involves a two-tailed t test in which the estimated standard error of the regression coefficient, denoted s_b, is given by

(9.28)
$$s_b = \frac{s_{Y.X}}{\sqrt{\Sigma(X - \bar{X})^2}}$$

Estimated standard error of the regression coefficient

Hence, s_b, the estimated standard deviation of the sampling distribution of b values, is a function of the scatter of points around the regression line and the dispersion of the X values around their mean. The t statistic, computed in the usual way, is given by

(9.29)
$$t = \frac{b - B}{s_b}$$

t distribution

We calculate s_b according to equation 9.28 as follows:

$$t = \frac{0.184 - 0.2}{0.0149} = -1.07$$

Substituting this value for s_b into equation 9.29 gives

$$s_b = \frac{1.559}{\sqrt{10,903}} = 0.0149$$

Since the same level of significance and the same number of degrees of freedom $(n - 2)$ are involved as in the preceding test for the significance of r, the decision rule is identical. With critical t values of ± 2.101 at the 5% level of significance, we cannot reject the null hypothesis that $B = 0.2$. Hence, we cannot conclude that the regression coefficient has changed from $B = 0.2$ for families in the given metropolitan area. That is, we retain the hypothesis that if a family had an annual net income $1,000 higher than that of a second family, then on the average, the first family's food expenditures would be $200 higher.

The corresponding confidence interval procedure involves setting up the interval

(9.30)
$$b \pm ts_b$$

In this problem, the 95% confidence interval for B is $0.184 \pm (2.101)(0.0149) = 0.184 \pm 0.031$. Therefore, we can assert that the population B figure is included in the interval 0.153 to 0.215 with an associated confidence coefficient of 95%.

Note that we used the t distribution in both the hypothesis-testing and confidence interval procedures just discussed. As in previous examples, normal curve procedures can be used for large sample sizes. Hence, for two-tailed hypothesis testing at the 5% level of significance and for 95% confidence interval estimation, the 2.101 t value given in the preceding examples would be replaced by a normal curve z value of 1.96.

A frequently used hypothesis test concerning the parameter B is for the null hypothesis, H_0: $B = 0$. This test determines whether the slope of the sample regression line differs significantly from a hypothesized value of zero. A slope of zero for the population regression coefficient B implies that there is no linear relationship between the variables X and Y and that the population regression line is horizontal. In other words, a B value of zero in the linear model implies that all of the conditional probability distributions have the same mean. Hence, regardless of the value of X, all conditional probability distributions of Y are identical, so the same value of Y would be predicted for all values of X.

If B is assumed to be zero, equation 9.29 reduces to

(9.31)
$$t = \frac{b}{s_b}$$

Substitution into equation 9.31 in the present example yields

$$t = \frac{0.184}{0.0149} = 12.3$$

Of course, the same number of degrees of freedom, $n - 2 = 20 - 2 = 18$, is involved in this test as in the preceding test that $B = 2$, and at the 5% level of significance, the critical t values are again ± 2.101. Hence, with a t value of 12.3, we reject the hypothesis that $B = 0$. We conclude that the slope of the population regression line is not zero. Based on our simple random sample, our best estimate of the slope of the population regression line is 0.184. Note that the null hypotheses H_0: $\rho = 0$ and H_0: $B = 0$ are equivalent assumptions. In packaged computer programs for regression and correlation analysis, the printouts generally do not show the t test for H_0: $\rho = 0$. However, the printouts for simple two-variable and multiple regression analysis either display the t values for all regression coefficients (b values) or display all regression coefficients and their corresponding standard errors, so that t values may easily be computed.

Caution   A final caution is appropriate. The tests discussed here pertain to the simple two-variable linear regression model. Although we might conclude on the basis of such tests that there is no linear relationship between X and Y, some other statistically significant relationship (such as curvilinear or logarithmic) may exist: moreover, if additional variables are present but unnoticed, they may obscure the relationship (linear or otherwise) between X and Y.

9.8
USE OF MINITAB FOR TWO VARIABLE REGRESSION ANALYSIS

Computers are a powerful tool for carrying out regression analyses, particularly multiple regression analysis, in which two or more independent variables are present. Even for two variable analysis, hand calculations can become quite tedious and time consuming for large samples. Figures 9-13 and 9-14 present

```
MTB > INFO

COLUMN                     NAME                    COUNT
C1                         FOOD                      20
C2                         INCOME                    20
C3                         SIZE                      20

CONSTANTS USED: NONE

MTB > PRINT C1-C3

ROW            FOOD               INCOME             SIZE

  1             5.2                 28                 3
  2             5.1                 26                 3
  3             5.6                 32                 2
  4             4.6                 24                 1
  5            11.3                 54                 4
  6             8.1                 59                 2
  7             7.8                 44                 3
  8             5.8                 30                 2
  9             5.1                 40                 1
 10            18.0                 82                 6
 11             4.9                 42                 3
 12            11.8                 58                 4
 13             5.2                 28                 1
 14             4.8                 20                 5
 15             7.9                 42                 3
 16             6.4                 47                 1
 17            20.0                112                 6
 18            13.7                 85                 5
 19             5.1                 31                 2
 20             2.9                 26                 2

MTB > BRIEF 3
MTB > REGRESS C1 1 C2;
SUBC> PREDICT 30.

THE REGRESSION EQUATION IS
FOOD = - 0.412 + 0.184 INCOME

PREDICTOR        COEF        STDEV       T-RATIO         P
CONSTANT       -0.4120      0.7638       -0.54        0.596
INCOME          0.18411     0.01493      12.33        0.000

S = 1.559        R-SQ = 89.4%         R-SQ (ADJ) = 88.8

ANALYSIS OF VARIANCE

SOURCE         DF         SS          MS          F         P
REGRESSION      1       369.57      369.57      151.97    0.000
ERROR          18        43.77        2.43
TOTAL          19       413.35
```

FIGURE 9-14

Minitab computer output for the regression analysis of family food expenditures and family net income (a continuation of Figure 9-13)

OBS.	INCOME	FOOD	FIT	STDEV.FIT	RESIDUAL	ST.RESID
1	28	5.200	4.743	0.436	0.457	0.31
2	26	5.100	4.375	0.454	0.725	0.49
3	32	5.600	5.480	0.403	0.120	0.08
4	24	4.600	4.007	0.474	0.593	0.40
5	54	11.300	9.530	0.371	1.770	1.17
6	59	8.100	10.450	0.403	−2.350	−1.56
7	44	7.800	7.689	0.349	0.111	0.07
8	30	5.800	5.111	0.419	0.689	0.46
9	40	5.100	6.952	0.358	−1.852	−1.22
10	82	18.00	14.685	0.647	3.315	2.34R
11	42	4.900	7.321	0.353	−2.421	−1.59
12	58	11.800	10.266	0.396	1.534	1.02
13	28	5.200	4.743	0.436	0.457	0.31
14	20	4.800	3.270	0.516	1.530	1.04
15	42	7.900	7.321	0.353	0.579	0.38
16	47	6.400	8.241	0.349	−1.841	−1.21
17	112	20.000	20.208	1.053	−0.208	−0.18X
18	85	13.700	15.237	0.685	−1.537	−1.10
19	31	5.100	5.295	0.410	−0.195	−0.13
20	26	2.900	4.375	0.454	−1.475	−0.99

R DENOTES AN OBS. WITH A LARGE ST.RESID.
X DENOTES AN OBS. WHOSE X VALUE GIVES IT LARGE INFLUENCE.

FIT	STDEV.FIT	95% C.I.	95% P.I.
5.111	0.419	(4.232, 5.991)	(1.718, 8.504)

the Minitab commands and computer output for the illustrative problem discussed in this chapter dealing with the relationship between family food expenditures and family income for a sample of 20 families in a metropolitan area in 1990.

After retrieving the data for family food expenditures (FOOD), family net income (INCOME), and family size (SIZE) for the first 20 families in the computer data bank listed in Appendix D, we use the commands INFO and PRINT C_1-C_3 to examine the data (Figure 9-13). Although only family food expenditures and family net income were used in the two variable analysis, the data for family size were also retrieved for use in the multiple regression analysis given in Chapter 10.

The next command BRIEF 3 asks for the full output of a regression analysis. The BRIEF command specifies the amount of detail of output obtained from all REGRESS commands that follow. BRIEF 1 provides the regression line, table of coefficients, s (the standard error of estimate), R^2 and R^2 adjusted (denoted

r^2 and r_c^2 in this chapter for two variable analysis), and the first part of an analysis of variance table. BRIEF 2 presents additional information representing an intermediate amount of output, and BRIEF 3 provides a full output. The command REGRESS C1 1 C2 asks for a regression analysis of the dependent variable (FOOD) in column 1 on one independent variable (INCOME) in column 2. We typed the semicolon appearing after the REGRESS command as a request for a subcommand prompt. We responded here with PREDICT 30. (don't forget the period at the end of the subcommand). This is a request for predictions for an observation in which net income equals 30 thousand dollars.

The regression equation printed out by Minitäb is the same as the one calculated in section 9.4. The minor differences shown in the computer printout for other calculations and those given earlier in this chapter are due to rounding. Below the regression equations is more detailed information about the regression equation. We state the headings first and items under the headings in the following explanation:

1.   Predictor—the names of the independent variables; Constant is the Y intercept or a value.

2.   Coef—the coefficients of the regression equation, that is, the estimated a and b values.

3.   Stdev.—the standard errors of the coefficients of the regression equation (s_a and s_b).

4. t-ratio—the t statistics for the coefficients of the regression equation. For example,

$$t = \frac{b}{s_b} = \frac{0.18411}{0.01493} = 12.33$$

as calculated in equation 9.31 in section 9.7.

5. p—the p values for the t-ratios. For example, for the slope coefficient, b, the p value of 0.000 indicates that the estimated slope coefficient rejects the null hypothesis $H_0: B = 0$ at conventional significance levels such as 0.05, 0.01, and even 0.001. The p value of 0.596 for the constant or a value retains the null hypothesis $H_0: A = 0$ at conventional significance levels. This implies that the hypothesis that the population regression line $\hat{Y} = A + BX$ passes through the origin (0, 0) is tenable. Usually, interest inheres primarily in the slope coefficients and not in the a value, especially when the Y intercept figure represents a value outside the range of observed data. For example, as noted earlier, the Y intercept in this problem (-0.412) is the estimated amount of food expenditures for a family whose income is $0, which is below the observed income figures.

The next line of output shows

1. s—the standard error of estimate, symbolized in section 9.5 as $s_{Y.X}$

2.   R-sq—R^2, the coefficient of determination, symbolized in section 9.6 as r^2

3. *R*-sq (adj)—the corrected or adjusted coefficient of determination, symbolized in section 9.6 as r_c^2

We postpone the discussion of the Analysis of Variance table until Chapter 10 to forestall unnecessary repetition.

We turn now to Figure 9-14 in which is listed a table of observed and predicted values, residuals, and standardized residuals for each observation in the data set, in this example, for each family.

The first three columns list the observation number, the independent variable INCOME, and the dependent variable FOOD. The remaining columns contain the following information:

1. Fit—these are the fitted values of the dependent variable, that is, the $\hat{Y}$ values for each observation. For example, the first listed value is $4.743 = -0.4120 + 0.18411 (28)$

2. Stdev.Fit—these are the standard deviations of the fitted values that we referred to in section 9.5 as the estimated standard error of the conditional mean, $s_{\hat{Y}}$. As indicated in section 9.5,

$$s_{\hat{Y}} = s_{Y.X} \sqrt{\frac{1}{n} + \frac{(X_0 - \bar{X})^2}{\Sigma(X - \bar{X})^2}}$$

Thus, these values can be used to compute confidence intervals for the population mean of all *Y* values corresponding to the listed values of *X*.

3. Residual—there are the differences between the actual values and fitted values of the dependent variable, that is, the $Y - \hat{Y}$ values. For example, the first residual is $0.457 = 5.200 - 4.743$.

4. St.Resid—these are referred to as "standardized residuals." One's first guess as to what these figures are might be the residuals divided by the standard error of estimate, $s_{Y.X}$. However, Minitab calculates a unique standard deviation associated with each residual.[7] Standardized residuals indicate the number of standard deviations that each residual is away from the mean of zero. The purpose of this calculation is to identify extreme observations, or outliers more sharply than using the method of dividing each residual by $s_{Y.X}$. Many analysts use a rule of thumb that residuals which lie more than two standard deviations away from zero represent outlier observations that are atypical and deserving of investigation. Minitab prints an *R* next to standardized residuals that are greater than $|2|$. In the present example,

---

[7] Standardized residuals are calculated by (residual)/(standard deviation of residual), where,

$$\text{Standard deviation of residual} = \sqrt{\text{Mean square error} - \left(\text{Standard deviation of fit}\right)^2}$$

an R is shown for family number 10 which has a standardized residual of 2.34. Minitab also prints an X for each observation whose X value gives it substantial influence. That is, if this particular case were removed, the coefficients of the fitted line would be substantially changed from those for the line fitted with the inclusion of this observation. In the present example, family number 17, which has an income figure of 112 is designated with an X.

The last line of Figure 9-14 gives the information that was asked for by the subcommand Predict 30. Note that 30 is the income figure for family number 8 and that some output from the subcommand can be checked on the row corresponding to that family. The output for the PREDICT 30 subcommand is as follows:

1. Fit—This is the fitted Y value for the X value of 30. That is, $5.111 = -0.4120 + 0.18411 (30)$

2. Stdev.Fit—as explained earlier in the table, this is the standard error of the conditional mean, $s_{\hat{Y}}$.

3. 95% C.I.—this is the 95% confidence interval for the population mean of all Y values corresponding to $X = 30$. That is $\hat{Y} \pm ts_Y = 5.111 \pm 2.101(0.419) = (4.232, 5.991)$

4. 95% P.I.—this is the 95% prediction interval for an individual value of Y. That is, $\hat{Y} \pm ts_{IND} = 5.111 \pm 2.101(1.614) = (1.718, 8.504)$. The calculation for s_{IND} was given in section 9.5.

9.9
CAVEATS AND LIMITATIONS

Regression analysis and correlation analysis are useful and widely applied techniques. However, it is important to understand the limitations of these methods and to interpret the results with care.

Cause and Effect Relationships

In correlation analysis, the value of the coefficient of determination r^2 is calculated. This statistic measures the degree of association between two variables. Neither this quantity nor any other statistical technique that measures or expresses the relationship among variables can prove that one variable is the *cause* and one or more other variables are the *effects*. Indeed, through the centuries philosophical speculation and debate have considered the meaning of cause and effect and whether such a relationship can ever be demonstrated by experimental methods. In any event, a measure such as r^2 does not prove the existence of a cause and effect relationship between two variables X and Y.

When a high value of r^2 is obtained, X may be producing variations in Y, or third and fourth variables W and Z may be producing variations in both X and Y. Numerous examples, frequently humorous in nature, have been given to demonstrate the pitfalls in attempting to draw conclusions about cause and effect in such cases. For example, if the average salaries of ministers are associated with the average price of a bottle of Scotch whiskey over time, a high degree of correlation between these two variables will probably be observed. Doubtless, we would be reluctant to conclude that the fluctuations in ministers' salaries cause the variations in the price of a bottle of Scotch, or vice versa. In this case, a third variable, which we may conveniently designate as the general level of economic activity, produces variations in both of the variables. From the economic standpoint, salaries of ministers represent the price paid for a particular type of labor; the cost of a bottle of Scotch is also a price. When the general level of economic activity is high, both of these prices tend to be high. When the general level of economic activity is low, as in periods of recession or depression, both of these prices tend to be lower than during more prosperous times. Thus, the high degree of correlation between the two variables of interest is produced by a third variable (and possibly others); certainly, neither variable is *causing* the variations in the other.

Furthermore, it is important to keep in mind the problem of sampling error. As we have seen, in a particular sample a high degree of correlation, either direct or inverse, may be observed, when in fact no correlation (or very little correlation) exists between the two variables in the population.

Finally, in applying critical judgment to the evaluation of observed relationships, we must be on guard against "nonsense correlations" in which no meaningful unit of association is present. For example, suppose we record in a column labeled X the distance from the ground of the skirt hemlines of the first 100 women who pass a particular street corner. In a column labeled Y, we record 100 observations of the heights of the Himalaya mountains along a certain latitude at 5-mile intervals. It is possible that a high r^2 value might be obtained for these data. Clearly, the result is nonsensical, because there is no meaningful unit or entity through which these data are related. In the illustrative example in this chapter, expenditures and income were observed for the same family. Hence, the family may be referred to as the *unit of association*. A **unit of association** might be a time period or some other entity, but it must provide a reasonable link between the variables studied.

Extrapolation beyond the Range of Observed Data

In regression analysis, an estimating equation is established on the basis of a particular set of observations. A great deal of care must be exercised in predicting values of the dependent variable based on values of the independent variable outside the range of the observed data. Such predictions are referred to as **extrapolations**. For example, in the problem considered in this chapter, a regression line was computed for a sample of families whose annual net incomes ranged from $20,000 to $112,000. It would be extremely unwise to make a prediction of food expenditures for a family with an annual net income of $200,000 using the computed regression line. To do so would imply that the straight-line relationship could be projected up to a value of $200,000 for the independent variable. Clearly, in the absence of other information, we simply do not know whether the same functional form of the estimating equation is valid outside the range of the observed data. In fact, in certain cases, unreasonable or even impossible values may result from such extrapolations. For example, suppose a regression line with a negative slope had been computed relating the percentage of defective articles produced (Y) with the number of weeks of on-the-job training received (X) by a group of workers. An extrapolation for a large enough number of weeks of training would produce a negative value for the percentage of articles produced, which is an impossible result. Clearly in this case, although the computed estimating equation may be a good description of the relation between X and Y within the range of the observed data, an equation with different parameters or even a completely different functional form is required outside this range. Without a specific investigation or a pertinent theory, we simply do not know what the appropriate estimating device is outside the range of observed data.

However, sometimes the exigencies of a situation require an estimate, even though it is impractical or impossible to obtain additional data. Extrapolations and alternative methods of prediction have to be used, but the limitations and risks involved must be kept constantly in mind.

Other Regression Models

So far, we have considered only one form of the regression model, namely, a straight-line equation relating the dependent variable Y to the independent variable X. Sometimes, theoretical considerations indicate that this is the required model. On the other hand, a linear model is often used either because the theoretical form of the relationship is unknown and a linear equation appears to be adequate or because the theoretical form is known but rather complex and a linear equation may provide a sufficiently good approximation. In all cases, the determination of the most appropriate regression model should result from a combination of theoretical reasoning, practical considerations, and careful scrutiny of the available data.

Often, the straight-line model $\hat{Y} = a + bX$ is not an adequate description of the relationship between the two variables. In some situations, models involving transformations of one or both of the variables may provide better fits to the data. For example, if the dependent variable Y is transformed to a new variable, log Y, a regression equation of the form

(9.32) $$\log Y = a + bX$$

may yield a better fit. Insofar as arithmetic is concerned, log Y is substituted for Y everywhere that Y appeared previously in the formulas. However, care must be used in the interpretation. The antilogarithm of log $\hat{Y}$ must be taken to provide an estimate of the dependent variable Y for a given value of X. Note that different assumptions are involved in this model than in the model $\hat{Y} = a + bX$. We now assume that log Y rather than Y is a normally distributed random variable.

> Furthermore, the logarithmic model implies that there is constant *percentage* change in $\hat{Y}$ per unit change in X, whereas the $\hat{Y} = a + bX$ model implies a constant *amount* of change in $\hat{Y}$ per unit change in X.

Possible transformations include the use of square roots, reciprocals, and logarithms of one or both of the variables. As an example of one such useful transformation, if a straight-line equation is fitted to the logarithms of both variables, the model takes the form

(9.33)
$$\log \hat{Y} = a + b \log X$$

The regression coefficient b in this model has an interesting interpretation, if as in the illustrative example used in this chapter, Y is a consumption variable and X is income. For such variables, the regression coefficient b in the model $\hat{Y} = a + bX$ can be interpreted as a marginal propensity-to-consume coefficient; that is, it estimates the dollar change in consumption per dollar change in income. Analogously, in the model $\log \hat{Y} = a + b \log X$, the coefficient b can be interpreted as an income elasticity of consumption coefficient; that is, it estimates the *percentage change* in consumption per 1% *change* in income. Of course, fitting an equation of the form $\log \hat{Y} = a + b \log X$ in the illustration under discussion implies that the income elasticity of consumption is constant over the range of income observed. Similarly, a model of the form $\hat{Y} = a + bX$ implies that the marginal propensity to consume is constant over the range of observed income. In fact, according to Keynesian economic theory, the marginal propensity to consume (for total consumption expenditures) decreases with increasing income. Fitting such regression models clearly cannot be merely a mechanistic procedure, but must involve a combination of knowledge of the field of application, good judgment, and experimentation.

In some applications, a curvilinear regression function may be more appropriate than a linear one. Polynomial functions are particularly convenient to fit by the method of least squares. The straight-line regression equation $\hat{Y} = a + bX$ is a polynomial of the first degree, since X is raised to the first power. A second-degree polynomial would involve a regression function of the form

(9.34)
$$\hat{Y} = a + bX + cX^2$$

in which the highest power to which X is raised is 2. This is the equation of a second-degree parabola, which is characterized by *one change in direction* in

$\hat{Y}$ as X increases, whereas in the case of a straight line, no changes in direction can take place. A third-degree polynomial permits *two changes in direction*, and so on. In the straight-line function, the amount of change in $\hat{Y}$ is constant per unit change in X. In the second-degree parabola, the amounts of change in $\hat{Y}$ may decrease or increase per unit change in X, depending on the shape of the function. Figure 9-15 shows two scatter diagrams for situations in which a second-degree regression function of the form of equation 9.34 may provide a good fit. The probable shape of the regression function has been indicated. Analogous situations could be portrayed for cases of inverse relationships between X and Y.

In the case of the straight-line regression function, the application of the method of least squares leads to 2 normal equations that must be solved for a and b. Analogously, to obtain the values of a, b, and c in the second-degree polynomial function, the following 3 simultaneous equations must be solved:

(9.35)
$$\Sigma Y = na + b\Sigma X + c\Sigma X^2$$
$$\Sigma XY = a\Sigma X + b\Sigma X^2 + c\Sigma X^3$$
$$\Sigma X^2 Y = a\Sigma X^2 + b\Sigma X^3 + c\Sigma X^4$$

Various computer programs have been developed to solve such normal equation systems, provide for transformations of the variables, and calculate all the regression and correlation measures we have discussed (as well as others). In applied problems in which large quantities of data are present and considerable experimentation with the form of the regression model is required, or in which more complex models than those so far discussed are appropriate, the use of computers may be the only feasible method of implementation.

The discussion in this chapter has been limited to two-variable regression and correlation analysis. In many problems, the inclusion of more than one independent variable in a regression model may be required to provide useful estimates of the dependent variable. Suppose, for example, that in the illustration involving family food expenditures and family income, poor predictions were made based on the single independent variable "income." Other factors

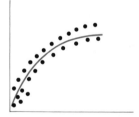 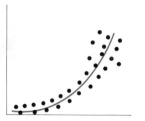

FIGURE 9-15

Scatter diagrams for 2 situations in which a second-degree polynomial regression function might be appropriate

such as family size, age of the head of the family, and number of employed persons in the family might be considered as additional independent variables to aid in the estimation of food expenditures. When two or more independent variables are utilized, the problem is referred to as a *multiple regression and correlation analysis*. A description of the technique is included in the next chapter.

EXERCISES 9.9

Note: For simplicity in these exercises, substitute r_c for r in the t test of the null hypothesis H_0: $\rho = 0$. That is, use the t statistic

$$t = \frac{r_c}{\sqrt{(1 - r_c^2)/(n - 2)}}$$

1. A regression of company sales on operating expenditures for a random sample of 48 firms yielded the following results:

$$Y = \text{company sales (millions of dollars)}$$

$$X = \text{company operating expenditures (millions of dollars)}$$

$$\hat{Y} = 1.2 + 0.416X$$

$$\Sigma(Y - \hat{Y})^2 = 2{,}025$$

$$\Sigma(Y - \bar{Y})^2 = 17{,}500$$

$$\Sigma(X - \bar{X})^2 = 12{,}800$$

   a. Calculate and intepret the standard error of estimate, $s_{Y.X}$.
   b. Calculate the unadjusted and adjusted coefficients of determination. What does the coefficient of determination measure?
   c. Describe and carry out a test of the significance of the estimated relationship between the two variables.

2. Assume that the following least squares regression equation was determined from a random sample of 62 new products:

$$\hat{Y} = 1.0 + 4.0X$$

$$Y = \text{demand measured by first-year sales (millions of dollars)}$$

$$X = \text{awareness measured by the proportion of consumers}$$
$$\text{who had heard of the product by the third month after}$$
$$\text{its introduction; } 0 \leqslant X \leqslant 1$$

$$s_{Y.X} = 0.2$$

$$s_Y = 1.00$$

$$\Sigma(X - \bar{X})^2 = 0.816$$

a. Calculate the adjusted coefficient of determination and the adjusted coefficient of correlation. Why is it preferable for interpretation purposes to use r_c^2 rather than r_c?

b. Is the estimated regression coefficient b significantly different from zero? Use a 1% significance level.

3. One of the models used in security analysis to evaluate the expected return for a particular stock is $E(R_i) = R_F + \beta_i[E(R_m) - R_F]$, where R_i is the return on the stock, R_F is the risk-free rate (return on a riskless asset), R_m is the return on a market portfolio, and β_i is the stock's beta coefficient (a measure of the market risk of the stock). For regression purposes, this equation is expressed as

$$R_i = (1 - \beta_i)R_F + \beta_i R_m \quad \text{or} \quad \hat{R}_i = a + bR_m$$

where

$$a = (1 - \beta_i)R_F$$

$$b = \beta_i$$

The following data have been collected on Technometrics Consolidated, a multinational firm:

Year	R_i	R_m
1971	0.16	0.20
1972	0.12	0.11
1973	0.16	0.14
1974	0.11	0.09
1975	0.06	0.08
1976	0.14	0.12
1977	0.18	0.16
1978	0.09	0.10
1979	0.15	0.16
1980	0.17	0.21
1981	0.19	0.23
1982	0.25	0.22
1983	0.10	0.07
1984	0.05	0.10
1985	0.07	0.18
1986	0.15	0.18
1987	0.20	0.26
1988	0.09	0.09
1989	0.03	0.06
1990	0.23	0.22

a. Estimate a and b in the equation $R_i = a + bR_m$ by the method of least squares. Assuming that the above model is correct, derive an estimate of R_F (this assumes that R_F has remained constant over time).

b. Calculate the adjusted and unadjusted coefficients of determination.

c.   Theoretically, a market portfolio has a beta coefficient of one. Is Tech-nometrics Consolidated's beta coefficient significantly different from one? Use a 1% level of significance.

d.   Obtain an interval estimate of the conditional mean return on Techno-metrics Consolidated stock if the market return (the average annual return on common stocks during the past 40 years) is 0.091. Use a 95% level of confidence.

4.   An enterprising statistics student took a random sample of 62 students en-rolled in an accounting course and compiled the following:

$$\hat{Y} = 6.4 + 7.6X$$

$$X = \text{number of hours spent studying for a test}$$

$$Y = \text{score on test (from } 0 - 100)$$

X ranges from 0–12 hours

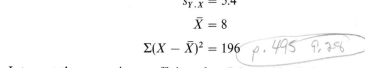

$$n = 62 \qquad S_b = \frac{5.4}{\sqrt{196}}$$

$$s_Y = 8.2$$

$$s_{Y.X} = 5.4$$

$$\bar{X} = 8$$

$$\Sigma(X - \bar{X})^2 = 196 \quad p. 495 \quad 9.28$$

*b is acceleratior how
when changed.
does that affect Y.*

a.   Interpret the regression coefficient, $b = 7.6$, specifically in terms of this problem. Is it significantly different from zero at the 1% level?

b.   Calculate the adjusted coefficient of determination.

c.   Would an accounting student who studied nine hours for the test and received a 42 be considered a "poor performer" or just an average per-former? Why?

5.   A regression of company profits on company asset size for a random sample of 42 firms yielded the following results:

$$Y = \text{company profits (in millions of dollars)}$$

$$X = \text{company asset size (in millions of dollars)}$$

$$Y_c = 2.0 + 0.15X$$

$$\Sigma(Y - \hat{Y})^2 = 2000$$

$$\Sigma(Y - \bar{Y})^2 = 17{,}000$$

$$\Sigma(X - \bar{X})^2 = 500{,}000$$

a.   Calculate and interpret the standard error of estimate, $s_{Y.X}$.

b.   Calculate the unadjusted and adjusted coefficients of determination. What does the coefficient of determination measure?

c.   Describe and carry out a test of the significance of the estimated rela-tionship between the two variables.

6. As the manager of a local bank, you are given the following information, which has been collected from a simple random sample of checking account customers:

$$\hat{Y} = 6 - 0.01X$$

where $X =$ balance in a customer's checking account
   $Y =$ number of bad checks written per month by the customer
   $-\$10 < X < \$500$ $n = 200$ $s_Y^2 = 25$ $s_{Y.X}^2 = 8$
   a. Calculate r_c^2, and interpret your answer.
   b. Explain the meaning of the negative regression coefficient $b = -0.01$.
   c. Do you think that there is any correlation (for all checking accounts) between the balance and the number of bad checks written per month? Justify your answer statistically.

7. A regression analysis is run in an attempt to predict demand from price. The following summary information is collected:
   $\bar{X} = \$2$ $\bar{Y} = 3$ (million units) $\Sigma X^2 = 102$
   $\Sigma Y^2 = 229$ $\Sigma XY = 147.5$ $\Sigma(X - \bar{X})^2 = 2$ $n = 25$
   a. Find a, b, and $s_{Y.X}$
   b. Calculate a 95% confidence interval for B.
   c. Test $H_0: B = 0$ at the $\alpha = 0.05$ significance level.

MINITAB EXERCISES

Problems 1 through 5 refer to the following data: From Minitab file 'gnp . mtw', you can retrieve:

Column	Name	Description
C1	Year	From 1981 to 1987
C2	Quarter	Spring (1), Summer (2), Fall (3), Winter (4)
C3	GNP	Gross National Product ($Bn)
C4	Consump	Personal Consumption Expenditures ($Bn)

(Source: Slater-Hall *Information Products & Business*, 1988)

1. It is often observed that there is a linear relationship between GNP and consumption at a macroeconomic level. Use the Minitab command

   MTB > plot C4 against C3

   to obtain a scatterplot of C4 versus C3. Visually, is the claimed linear relationship clear? Type:

   MTB > regress C4 on 1 predictor C3

   Based on the computer output, test the significance of the regression coefficient at alpha = 0.01. On average, how much additional was spent on consumption for each additional dollar of GNP?

2. Find the sample coefficient of determination, and interpret your answer. Since it is apparent that GNP and consumption cannot be negatively correlated, carry out a one-sided test for the significance of the correlation coefficient between GNP and consumption at a 0.10 significance level.

3. Minitab commands

   MTB > brief 3
   MTB > name C10 'st.resid' C11 'fits' C12 'residual'
   MTB > regress C4 1 C3, store st.resid in C10, fits in C11;
   SUBC > residuals into C12.

   will produce a very detailed output of the regression. To visualize the goodness of fit, plot C11 and C4 on the same graph against C3, and plot the difference of C4 and C11, that is, C12 against C3.

4. Note that the standardized residuals in C10, which are calculated by the residuals divided by the standard deviations that are uniquely associated with each residual, are different from the standardized residuals that are calculated by the residuals divided by s, the common standard deviation of residuals, i.e., the standard error of estimate. Enter the latter standardized residuals into C20. To compare, print C10 and C20. Identify any residuals that are more than two common standard deviations away from zero. Note that these residuals signal outlier observations that need further investigation.

5. At alpha = 0.05, conduct a two-tailed test of the hypothesis that on average, 70 cents were spent in consumption for each additional dollar of GNP. Find the p-value for the test and interpret it.

Problems 6 through 9 are based on the following data: A survey of 60 graduates of an undergraduate school of business was conducted to study how the starting salary of a graduate was related to the graduate's overall rating in his class. The results of the survey can be obtained from the file named 'rating . mtw'.

Column	Name	Description
C1	SALARY	Starting Salary ($000): Annual
C2	RATING	Overall Rating in a Class of 100 Students

6. Plot C1 versus C2. Fit a straight line to the starting salaries and overall ratings by the method of least squares. At alpha = 0.01, test the significance of the overall regression. Then, at a 0.05 significance level, test the significance of the constant term in the regression, and separately conduct a reasonable one-sided test for the significance of the regression coefficient.

7. Find the residuals whose distances away from zero are at least two standard errors of estimate. Delete all such 'outlier' observations and run the regression analysis again. Now do you get any residuals that signal outlier observations? Continue this process until no more observations can be deleted. *Hint:* To delete observations from columns, use the COPY command and some subcommands.

8. Note that the starting salary of the graduate who ranked number 1 was not in the survey. Predict the starting salary for this graduate. Find a 95% prediction interval for the graduate's starting salary. *Hint:* Try

   MTB > regress C1 1 C2;
   SUBC > predict 1.

9. Note that the starting salaries of the graduates who ranked number two and number three were not in the survey. Find a 95% confidence interval for the population mean of all the second-ranking graduates. Calculate a 99% confidence interval for the population mean of all the third-ranking graduates.

KEY TERMS

Coefficients of Correlation (r) and Determination (r^2)   measures of the strength of the relationship between the dependent variable (Y) and the independent variable (X).

Confidence Interval   interval estimate for a conditional mean.

Correlation Analysis   measurement of the strength of relationships among variables.

Dependent Variable   the variable (Y) whose values are to be estimated.

Direct (inverse) Relationship   the dependent variable (Y) tends to increase (decrease) as the independent variable (X) increases.

Homoscedasticity   the assumption of equal variances around the regression line at each value of the independent variable.

Independent Variable   the variable (X) from which estimates of the dependent variable (Y) are made.

Method of Least Squares   a curve fitting technique that minimizes the sum of the squared deviations, $\Sigma(Y - \hat{Y})^2$.

Prediction Interval   interval estimate for predicting an individual value of Y.

Regression Analysis   the estimation of values of a variable from the values of one or more other variables and the measurement of the errors involved in the estimation process.

Regression Equation and Line   the equation and line that describe the relationship between the dependent variable (Y) and the independent variable (X).

Scatter Diagram   a graph on which each plotted point represents an observed pair of values of the dependent and independent variables.

Slope (regression) Coefficient (b)   the estimated amount of change in $\hat{Y}$ per unit change in X.

Standard Error of Estimate or Standard Deviation of the Residuals ($s_{Y.X}$)   a measure of the scatter of the Y values around the $\hat{Y}$ values.

Standard Error of the Regression Coefficient (s_b)   the standard deviation of the sampling distribution of b values.

Extrapolation   prediction of Y based on an X value outside the range of observed data.

KEY FORMULAS

Confidence Interval Estimate for the Conditional Mean
$$\hat{Y} \pm ts_{\hat{Y}}$$

Confidence Interval for the Population Regression Coefficient
$$b \pm ts_b$$

Explained Variation
$$\Sigma(\hat{Y} - \bar{Y})^2$$

Population Regression Line
$$\mu_{Y.X} = A + BX$$

Prediction Interval for an Individual Value of Y

$$\hat{Y} \pm t s_{\text{IND}}$$

Sample Coefficient of Determination

$$r^2 = 1 - \frac{\Sigma(Y - \hat{Y})^2}{\Sigma(Y - \bar{Y})^2}$$

Sample Coefficient of Determination Adjusted for Degrees of Freedom

$$r_c^2 = 1 - \frac{\Sigma(Y - Y)^2/(n - 2)}{\Sigma(Y - \bar{Y})^2/(n - 1)}$$

Sample Regression Line

$$\hat{Y} = a + bX$$

Standard Error of Estimate

$$s_{Y \cdot X} = \sqrt{\frac{\Sigma(Y - \hat{Y})^2}{n - 2}}$$

Standard Error of Forecast (of an individual value of Y)

$$s_{\text{IND}} = s_{Y \cdot X} \sqrt{1 + \frac{1}{n} + \frac{(X_0 - \bar{X})^2}{\Sigma(X - \bar{X})^2}}$$

Standard Error of the Conditional Mean

$$s_{\hat{Y}} = s_{Y \cdot X} \sqrt{\frac{1}{n} + \frac{(X_0 - \bar{X})^2}{\Sigma(X - \bar{X})^2}}$$

Standard Error of the Regression Coefficient

$$s_b = \frac{s_{Y \cdot X}}{\sqrt{\Sigma(X - \bar{X})^2}}$$

***t*-Test for $H_0: B = 0$**

$$t = \frac{b}{s_b}$$

Total Variation

$$\Sigma(Y - \bar{Y})^2$$

Unexplained Variation

$$\Sigma(Y - \hat{Y})^2$$

10

MULTIPLE REGRESSION AND CORRELATION ANALYSIS

Multiple regression analysis represents a logical extension of two-variable regression analysis. Instead of a single independent variable, two or more independent variables are used to estimate the values of a dependent variable. However, the fundamental concepts in the analysis remain the same.

10.1
PURPOSES

Just as in the analysis involving the dependent and only one independent variable, the following three general purposes apply to multiple regression and correlation analysis:

1. To derive an equation that provides estimates of the dependent variable from values of two or more independent variables.
2. To obtain measures of the error involved in using this regression equation as a basis for estimation.
3. To obtain a measure of the proportion of variance in the dependent variable accounted for, or "explained by," the independent variables.

The first purpose is accomplished by deriving an appropriate regression equation by the method of least squares. The second purpose is achieved through the calculation of a standard error of estimate and related measures. The third purpose is accomplished by computing the multiple coefficient of determination, which is analogous to the coefficient of determination in the two-variable case and, as indicated in (3) above, measures the proportion of variation in the dependent variable explained by the independent variables.

As an example, let us return to our Chapter 9 problem of estimating family food expenditures from family net income for a sample of 20 families, both variables being stated on an annual basis. As indicated at the end of that chapter, the use of additional variables to income might improve prediction of the dependent variable. We select family size as a second independent variable. Estimates of food expenditures may now be made from the following linear multiple regression equation:

(10.1)
$$\hat{Y} = a + b_1 X_1 + b_2 X_2$$

TABLE 10-1
Annual food expenditures, annual net income, and family size for a sample of 20 families in a metropolitan area in 1990

Family	Annual Food Expenditures ($000) Y	Annual Net Income ($000) X_3	Family Size (number in family) X_2
1	5.2	28	3
2	5.1	26	3
3	5.6	32	2
4	4.6	24	1
5	11.3	54	4
6	8.1	59	2
7	7.8	44	3
8	5.8	30	2
9	5.1	40	1
10	18.0	82	6
11	4.9	42	3
12	11.8	58	4
13	5.2	28	1
14	4.8	20	5
15	7.9	42	3
16	6.4	47	1
17	20.0	112	6
18	13.7	85	5
19	5.1	31	2
20	2.9	26	2

where $\hat{Y}$ = family food expenditures (estimated)
   X_1 = family net income
   X_2 = family size

and a, b_1, and b_2 are numerical constants that must be determined from the data in a manner analogous to that of the two-variable case. For simplicity, we have assumed a linear regression function.

   We carry out a multiple regression and correlation analysis by fitting the linear regression equation 10.1 to data for the indicated variables. The basic data for family food expenditures, family income, and family size are shown in Table 10-1. The data for the first two of these variables are the same as those given in Table 9-1 for the two-variable problem solved in Chapter 9. The data on family size represent the total number of persons in each of the families in the sample.

10.2
THE MULTIPLE REGRESSION EQUATION

We begin the analysis by using the method of least squares to obtain the best-fitting three-variable linear regression equation of the form given in equation 10.1. In the two-variable regression problem, the method of least squares was used to obtain the best-fitting straight line. Analogously in the present problem, the method of least squares is used to obtain the best-fitting plane. In a three-variable regression problem, the points can be plotted in three dimensions, along the X_1, X_2, and Y axes (analogous to the case of a two-variable problem, in which the points are plotted in two dimensions along an X and Y axis). The best-fitting plane would pass through the points as shown in Figure 10-1, with some falling above and some below the plane in such a way that $\Sigma(Y - \hat{Y})^2$ is a minimum. Whereas in our previous illustration (involving two variables), two normal equations resulted from the minimization procedure, now three normal equations must be solved to determine the values of a, b_1, and b_2:[1]

(10.2)

$$\Sigma Y = na + b_1 \Sigma X_1 + b_2 \Sigma X_2$$

$$\Sigma X_1 Y = a\Sigma X_1 + b_1 \Sigma X_1^2 + b_2 \Sigma X_1 X_2$$

$$\Sigma X_2 Y = a\Sigma X_2 + b_1 \Sigma X_1 X_2 + b_2 \Sigma X_2^2$$

Normal equations

   Although the calculations to obtain the values of a, b_1, and b_2 can be simplified somewhat, solving the normal equations is clearly a laborious mathematical procedure. In the present example, three equations in three unknowns

---

[1] In a manner similar to that of the two-variable case, a function of the form

$$F(a, b_1, b_2) = \Sigma(Y - \hat{Y})^2 = \Sigma(Y - a - b_1 X_1 - b_2 X_2)^2$$

is set up. This function is minimized by the standard calculus method of taking its partial derivatives with respect to a, b_1, and b_2 and equating these derivatives to zero. This procedure results in the three normal equations (10.2).

FIGURE 10-1

Graphs of a multiple regression plane for data on the variables Y, X_1, and X_2

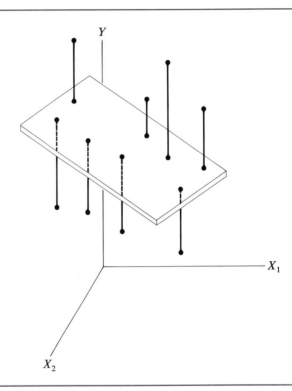

have to be solved. If we needed to determine four constants, four equations would have to be solved. Fortunately, since the advent of computers and the associated availability of numerous packaged programs for multiple regression and correlation analysis, virtually all such computations are done by computer. Hence, the computer solves the normal equations and carries out the other associated calculations. Since the use of computers is usually the only feasible method for carrying out multiple regression and correlation analyses, we will focus on the concepts involved in such analyses and on the interpretation of computer output rather than the calculations by which we obtain the output.

Returning to our problem, we show in Table 10-2 the Minitab commands for performing a multiple regression analysis. As in the corresponding two-variable problem in Chapter 9, we use the command BRIEF 3 to obtain a detailed level of output. The command REGRESS C1 2 C2 C3 means that we wish to obtain a regression analysis of the dependent variable in column 1 on two independent variables that are located in columns 2 and 3. We triggered the subcommand prompt by typing a semicolon after the REGRESS command. We then typed the subcommand PREDICT 30 2. in order to illustrate a prediction of food expenditures for a family whose income is $30,000 and whose family size is 2.

TABLE 10-2

Selected computer output for the regression equation for annual food expenditures (FOOD), annual net income (INCOME), and family size (SIZE)

```
MTB > REGRESS C1 2 C2 C3;
SUBC> PREDICT 30 2.

THE REGRESSION EQUATION IS
FOOD = - 1.12 + 0.148 INCOME + 0.793 SIZE
```

PREDICTOR	COEF	STDEV	T-RATIO	P
CONSTANT	-1.1183	0.6549	-1.71	0.106
INCOME	0.14821	0.01638	9.05	0.000
SIZE	0.7931	0.2444	3.24	0.005

That is, we wish a prediction $\hat{Y}$ for $X_1 = 30$ and $X_2 = 2$. As we have seen in Chapter 9, Minitab also provides 95% confidence intervals and prediction intervals for such predictions.

In the discussion of the family food expenditures problem, we focus on only a few selected items of computer output that pertain primarily to the accomplishment of the three general purposes of multiple regression and correlation analysis, and related matters of statistical inference. Then we discuss a number of problems that arise in regression and correlation analysis. We defer until section 10.13 a more detailed discussion of an iterative approach by which a multiple regression model can be built. In that example, as would be done in most applications of regression analysis, we analyze each run of computer output to see what problems are present, what our level of satisfaction is with the results obtained, and we decide on the steps that can be taken in the next run to improve the model under construction.

In computer output, many digits—sometimes 10 or more—are shown for the constants in the multiple regression equation and for other measures. Usually more figures are shown than are justified by the usual rules of rounding according to numbers of significant digits. This is a standard procedure in multiple regression and correlation analysis, because many of the measures computed in such analyses are particularly sensitive to rounding. For simplicity of exposition here, we show only a few decimal places for the constants in the regression equation.

Computer output

In Table 10-2, the regression equation is given as

$$\hat{Y} = -1.12 + 0.148 \text{ INCOME} + 0.793 \text{ SIZE}$$

Let us illustrate the use of the regression equation for estimation. Suppose we want to estimate food expenditures for a family from the same population as the sample studied. The family's income is $30,000, and there are 2 persons in the family. Substituting $X_1 = 30$ and $X_2 = 2$ yields the following estimated expenditures on food:

$$\hat{Y} = -1.12 + 0.148(30) + 0.793(2)$$
$$= 4.91 \text{ (thousands of dollars)}$$
$$= \$4,910$$

In two-variable analysis, we discussed the interpretation of the constants a and b in the regression equation. Let us consider the analogous interpretation of the constants a, b_1, and b_2 in the multiple regression equation. The constant a is again the Y intercept. However, now it is interpreted as the value of $\hat{Y}$ when X_1 and X_2 are both zero. The b values are referred to in multiple regression analysis as **net regression coefficients**. The b_1 coefficient measures the change in $\hat{Y}$ per unit change in X_1 when X_2 is held fixed, and b_2 measures the change in $\hat{Y}$ per unit change in X_2 when X_1 is held fixed.[2]

Net regression coefficients

Hence, in the present problem, the b_1 value of 0.148 indicates that if a family has an income \$1,000 greater than another's (a unit change in X_1) and *the families are the same size* (X_2 is held constant), then the estimated food expenditures of the higher-income family exceed those of the other by 0.148 thousands of dollars, or by about \$148. Similarly, the b_2 value of 0.793 means that if a family has one person more than another (a unit change in X_2) and *the families have the same income* (X_1 is held constant), then the estimated food expenditures of the larger family are greater than those of the smaller family by about \$793.

Two properties of these net regression coefficients are worth noting. The b_1 value of 0.148 thousands of dollars implies that an increment of one unit in X_1, or a \$1,000 increment in income, occasions an increase of \$148 in $\hat{Y}$ estimated food expenditures, regardless of the size of the family (for families of the sizes studied). Hence, an increase of \$1,000 in income adds \$148 to estimated food expenditures, regardless of whether there are two or six people in the family. An analogous interpretation holds for b_2. These interpretations are embodied in the assumption of linearity and therefore follow from the fact that we used a *linear* multiple regression equation in this example.

A second property of regression coefficients is apparent from a comparison of the b value of 0.184—in the simple regression equation 9.8, $\hat{Y} = -0.412 + 0.184X$, previously obtained when family income X was the only independent variable—with the b_1 value of 0.148, the net regression coefficient of income in the multiple regression equation $\hat{Y} = -1.12 + 0.148X_1 + 0.793X_2$, when the family size variable is included in the regression equation. The coefficient $b = 0.184$ in the simple two-variable regression equation makes no explicit allowance for family size. The net regression coefficient $b_1 = 0.148$, on the other hand, "nets out" the effect of family size. A net regression coefficient may in

---

[2] In the language of calculus, b_1 and b_2 are the partial derivatives of $\hat{Y}$ with respect to X_1 and X_2, respectively; that is,

$$\frac{\partial \hat{Y}}{\partial X_1} = b_1 \quad \text{and} \quad \frac{\partial \hat{Y}}{\partial X_2} = b_2$$

TABLE 10-3
Correlation coefficients for each pair of the three variables: expenditures on food (Y), net income (X_1), and family size (X_2)

```
    MTB > CORRELATION C1 C2 C3

                    FOOD          INCOME
    INCOME          0.946
    SIZE            0.787          0.676
```

general be greater or less than the corresponding regression coefficient in a two-variable analysis.

In this problem, the families with larger incomes were also the larger families. The positive correlation between income and family size is indicated by the correlation coefficient $r = 0.676$, shown in Table 10-3. The foregoing pattern exemplifies an important characteristic of regression coefficients, regardless of the number of independent variables included in the study.

> A regression coefficient for any specific independent variable (for example, income) measures not only the effect (of *income*) on the dependent variable, but also the effect attributable to *any other independent variables* that happen to be correlated with the independent variable but that have not been explicitly included in the analysis.

This is true for both two-variable and multiple regression analyses.

When independent variables are highly correlated, rather odd results may be obtained in a multiple regression analysis. For instance, a regression coefficient that is positive (or negative) in sign in a two-variable regression equation may change to a negative (or positive) sign for the same independent variable in a multiple regression equation containing other independent variables that are highly correlated with the one in question. For example, in this problem, the dependent variable, food expenditures (Y), is positively correlated with family size (X_2), as indicated by the correlation coefficient of $+0.787$ (Table 10-3). Hence, the regression coefficient for family size would also be positive in sign. The net regression coefficient for family size (b_2) in the three-variable regression equation is $+0.793$, but could very well have turned out to be negative in sign.

In the discussion of statistical inference in multiple regression, we shall see that the net regression coefficients for highly correlated independent variables tend to be unreliable. This is an important concept, because when independent variables are highly correlated, it is extremely difficult to separate the individual influences of each variable. Consider an extreme case. Suppose a two-variable regression and correlation analysis is carried out between a dependent variable, denoted Y, and an independent variable, denoted X_1. Furthermore, assume that we introduce another independent variable X_2, which has perfect positive correlation with X_1—that is, the correlation coefficient between X_1 and X_2 is $+1$. We now conduct a three-variable regression and correlation analysis. Clearly X_2 cannot account for or explain any additional variance in the dependent variable Y after X_1 has been taken into account. The same argument could be made if X_1 were introduced after X_2. As indicated in the ensuing discussion of statistical inference in multiple regression, the net regression coefficients, b_1 and b_2, in cases of high correlation between X_1 and X_2, will tend not to differ significantly from zero. Yet, if separate two-variable analyses had been run between Y and X_1 and Y and X_2, the individual regression coefficients might have differed significantly from zero. There is a great deal of concern in fields such as econometrics and applied statistics with this problem of correlation among independent variables, often referred to as **multi-**

Collinearity   **collinearity** or simply **collinearity**. One of the simplest solutions to the problem of two highly correlated independent variables is merely to discard one of them, but sometimes more sophisticated procedures are required.

The illustration in this section used only two independent variables. The general form of the linear multiple regression function for $k - 1$ independent variables $X_1, X_2, \ldots, X_{k-1}$ is

(10.3)
$$\hat{Y} = a + b_1 X_1 + b_2 X_2 + \cdots + b_{k-1} X_{k-1}$$

For convenience, the regression equation has been written with $k - 1$ independent variables. There are then k constants $a, b_1, b_2, \ldots, b_{k-1}$ to be determined in the regression equation. This simplifies somewhat the notation in later formulas, compared with the situation when k independent variables are included in the regression equation.

A linear function that is fitted to data for two variables is referred to as a **straight line**; a linear function for three variables is a **plane**, and a linear function for four or more variables is a **hyperplane**.

Although we cannnot visualize a hyperplane, its linear characteristics are analogous to those of the linear functions of two or three variables. With the use of electronic computers, it is possible to test and include large numbers of independent variables in a multiple regression analysis. However, good judgment and knowledge of the logical relationships involved must always be the main guides to deciding which variables to include in the construction of a regression equation.

10.3
STANDARD ERROR OF ESTIMATE

As in simple two-variable regression analysis, a measure of dispersion or scatter around the regression plane or hyperplane can be used as an indicator of the error of estimation. Probability assumptions similar in principle to those of the simple regression model must be introduced. The following are the usual assumptions made in a linear multiple regression analysis, illustrated for the case of two independent variables:

1. The Y values are assumed to be independent of one another.

2. The conditional distributions of Y given X_1 and X_2 are assumed to be normal.

3. These conditional distributions for each independent variable are assumed to have equal standard deviations.

Assumptions

The variance around the regression hyperplane is

(10.4)
$$S^2_{Y.12\ldots(k-1)} = \frac{\Sigma(Y - \hat{Y})^2}{n - k}$$

Variance around the regression hyperplane

where n is the number of observations and k is the number of constants in the regression equation. The divisor $n - k$ represents the number of degrees of freedom, and its use provides an unbiased estimator of the population variance. The subscript notation to S^2 lists the dependent variable to the left of the period and the $k - 1$ independent variables to the right. The subscripts $1, 2, \ldots, k - 1$ denote the variables $X_1, X_2, \ldots, X_{k-1}$, respectively. Hence, in our example involving the three variables Y, X_1, and X_2, the variance around the regression plane $\hat{Y} = -1.12 + 0.148X_1 + 0.793X_2$ is given by

Subscript notation

(10.5)
$$S^2_{Y.12} = \frac{\Sigma(Y - \hat{Y})^2}{n - 3}$$

Variance around the regression plane

The standard error of estimate, designated in Minitab by the letter s, which is the square root of this variance, is

(10.6)
$$S_{Y.12} = \sqrt{\frac{\Sigma(Y - \hat{Y})^2}{n - 3}}$$

Standard error of estimate

In Table 10-4, the following result is obtained for the standard error of estimate:

$$S_{Y.12} = 1.261 \text{ (thousands of dollars)}$$

As in two-variable analysis, the standard error of estimate is the standard deviation of the observed values of Y around the fitted regression equation.

TABLE 10-4
The standard error of estimate (s), and the coefficients of determination (R-sq and R-sq [adj]) for the regression equation for annual food expenditures (FOOD), annual net income (INCOME), and family size (SIZE)

$s = 1.261$	R-sq $= 93.5\%$	R-sq (adj) $= 92.7\%$

10.4
COEFFICIENT OF MULTIPLE DETERMINATION

In two-variable correlation analysis, the degree of association between the two variables was defined in equation 9.23 as

(10.7) $$r^2 = 1 - \frac{\Sigma(Y - \hat{Y})^2}{\Sigma(Y - \bar{Y})^2} = 1 - \frac{\text{Unexplained variation}}{\text{Total variation}}$$

In this form, r^2 measures the proportion of *variation* in the dependent variable explained by the regression equation relating Y to X.

An analogous measure, the *coefficient of multiple determination*, denoted R^2, quantifies the degree of association when more than two variables are present. Therefore, for the case of one dependent and two independent variables, the coefficient of multiple determination is defined as

(10.8) $$R^2 = 1 - \frac{\Sigma(Y - \hat{Y})^2}{\Sigma(Y - \bar{Y})^2} = 1 - \frac{\text{Unexplained variation}}{\text{Total variation}}$$

where R^2 measures the proportion of *variation* in the dependent variable Y explained by the regression equation relating Y to X_1 and X_2.

In the problem involving the regression equation relating annual food expenditures (FOOD) to annual net income (INCOME) and family size (SIZE), Table 10-4 gives the R^2 value, denoted R-sq in Minitab, as 93.5%.

The corresponding coefficient of determination, adjusted for degrees of freedom, in two-variable correlation analysis was defined an equation 9.22 as

$$r_c^2 = 1 - \frac{s_{Y.X}^2}{s_Y^2}$$

where

$$s_{Y.X}^2 = \frac{\Sigma(Y - \hat{Y})^2}{n - 2} \quad \text{and} \quad s_Y^2 = \frac{\Sigma(Y - \bar{Y})^2}{n - 1}$$

In this form r_c^2 measures the proportion of *variance* in the dependent variable explained by the regression equation relating Y to X.

The analogous measure, denoted R_c^2 correspondingly measure the proportion of *variance* in the dependent variable accounted for when more than two variables are present. For the case of one dependent and two independent variables, the coefficient of multiple determination, corrected for degrees of freedom, is defined as

(10.9) $$R_c^2 = 1 - \frac{S_{Y.12}^2}{s_Y^2}$$ Coefficient of multiple determination

where

(10.10) $$S_{Y.12}^2 = \frac{\Sigma(Y - \bar{Y})^2}{n - 3}$$ Unexplained variance of Y

and

(10.11)
$$s_Y^2 = \frac{\Sigma(Y - \bar{Y})}{n - 1}$$
Total variance of Y

We see from these definitions that $S_{Y.12}^2$ is the variance of Y values around the regression plane and s_Y^2 is the variance of the Y values around their mean. Just as $\Sigma(Y - \hat{Y})^2$ in equation 10.8 was earlier referred to as "unexplained variation" and $\Sigma(Y - \bar{Y})^2$ in equation 10.10 as the "total variation" of the Y variable, $S_{Y.12}^2$ is referred to as the "unexplained variance" and s_Y^2 as the "total variance." Hence, similar to the interpretation of r_c^2, we may interpret R_c^2 as the proportion of variance in the dependent variable explained by the regression equation relating Y to X_1 and X_2. Alternatively, it measures the proportion of variance in the Y variable accounted for by all the independent variables combined.

Table 10-4 gives the value for R_c^2, (denoted R-sq (adj) in Minitab), as 92.7%.

We illustrate the calculation of R_c^2 for the example with which we have been working. We defer the analogous illustration for the calculation of R^2 until the discussion of analysis of variance in section 10.6.

In Table 10-4, $S_{Y.12}$ (denoted s) is given as 1.261. Hence,

$$S_{Y.12}^2 = (1.261)^2 = 1.590121$$

The standard deviation s_Y, calculated with the Minitab command DESCRIBE is 4.66. The total variance, or variance around $\bar{Y}$, is

$$(4.66)^2 = 21.715600$$

Therefore, substituting into equation 10.9, the coefficient of multiple determination, corrected for degrees of freedom, is

$$R_c^2 = 1 - \frac{1.590121}{21.715600} = 92.7\%$$

Thus, we have found that 92.7% of the variance in food expenditures has been explained by the linear regression equation relating that variable to family income and family size. The figure obtained Table 10-3 for the two-variable correlation coefficient, unadjusted for degress of freedom, for food expenditures and family income was $r = 0.946$ or $r^2 = (0.946)^2 = 0.895$. The corresponding figure adjusted for degrees of freedom is $r_c^2 = 0.889$.[3] Comparing $R_c^2 = 0.927$ with the corresponding two-variable r_c^2 value of 0.889, we find that the value of R_c^2 is only 0.038 higher than the figure for r_c^2. This means that the addition of the second independent variable family size, has explained only a small amount of the variance in food expenditures, Y, beyond that which was already

---

[3] The r_c^2 figure is calculated as follows:

$$r_c^2 = 1 - (1 - r^2)\left(\frac{n - 1}{n - 2}\right) = 1 - (1 - 0.895)\left(\frac{20 - 1}{20 - 2}\right) = 0.889$$

accounted for by family income alone. As we noted earlier, one reason for this is the high correlation between the independent variables. Once family income has been taken into account, since family size moves together with that variable, family size does little to explain residual variation in food expenditures.

Two-Variable Correlation Coefficients

From the preceding discussion of the difficulties encountered in multiple correlation analysis when independent variables are intercorrelated, it is evident that it is good practice to compute coefficients of correlation or determination between each pair of independent variables that the analyst plans to enter into the regression equation. It is standard procedure in most multiple regression and correlation analysis computer programs to present a table of correlation coefficients for every pair of variables, including the dependent as well as all independent variables. In the printout of computer programs, the correlation coefficients are often presented in the form of a triangular table, as shown in Table 10-3.

10.5
INFERENCES ABOUT POPULATION NET REGRESSION COEFFICIENTS

In the preceding discussion of correlation and regression analysis, the various equations and measures were all stated in terms of sample values, rather than in terms of the corresponding population equations and characteristics. If the assumptions given at the beginning of the discussion of the standard error of estimate are met, then appropriate inferences and probability statements can be made concerning population parameters. In multiple regression analysis, a great deal of interest is centered on the reliability of the observed slope (net regression) coefficients. Just as in the two-variable case referred to in section 9.7, in which statistical inference about the population regression coefficient B was discussed, analogous hypothesis-testing and estimation techniques are available for slope coefficients, where three or more variables are involved.

In order to obtain data required for the discussion in the next paragraph, we show in Table 10-5 statistics provided for the present example by the Minitab command DESCRIBE. We require the values of $\Sigma(X_1 - \bar{X}_1)^2$ and $\Sigma(X_2 - \bar{X}_2)^2$, the variations around the means of INCOME and SIZE, for the subsequent discussion. We obtain these from the standard deviations shown in Table 10-5 for INCOME (23.96) and SIZE (1.605). Since the standard deviation equals

$$\sqrt{\frac{\Sigma(X - \bar{X})^2}{n - 1}}$$

it follows that $\Sigma(X_1 - \bar{X}_1)^2 = 10{,}907.55$ and $\Sigma(X_2 - \bar{X}_2)^2 = 48.944475$.

TABLE 10-5
Statistics obtained by the Minitab command DESCRIBE for the regression of annual food expenditures (FOOD) on annual net income (INCOME) and family size (SIZE)

```
MTB > DESCRIBE C1-C3

                 N         MEAN       MEDIAN       TRMEAN        STDEV       SEMEAN
FOOD            20         7.97         5.70         7.58         4.66         1.04
INCOME          20        45.50        41.00        43.22        23.96         5.36
SIZE            20        2.950        3.000        2.889        1.605        0.359

               MIN          MAX           Q1           Q3
FOOD          2.90        20.00         5.10        10.50
INCOME       20.00       112.00        28.00        57.00
SIZE         1.000        6.000        2.000        4.000
```

In the two-variable problem, the slope coefficient b in the equation $\hat{Y} = a + bX$ is an estimate of the population parameter B in the population relationship $\mu_{Y.X} = A + BX$. Correspondingly, the slope coefficients in a three-variable problem, b_1 and b_2 in the equation $\hat{Y} = a + b_1X_1 + b_2X_2$, are estimates of the parameters B_1 and B_2 in a population relationship denoted $\mu_{Y.12} = A + B_1X_1 + B_2X_2$. The standard errors of the coefficients, which represent the estimated standard deviations of the sampling distributions of b_1 and b_2 values, are given by

(10.12)
$$s_{b_1} = \frac{S_{Y.12}}{\sqrt{\Sigma(X_1 - \bar{X}_1)^2(1 - r_{12}^2)}}$$

and

(10.13)
$$s_{b_2} = \frac{S_{Y.12}}{\sqrt{\Sigma(X_2 - \bar{X}_2)^2(1 - r_{12}^2)}}$$

where all terms in equations 10.12 and 10.13 have the definitions stated above.
Substituting the required numerical values, we find

$$s_{b_1} = \frac{1.261}{\sqrt{10,907.55[1 - (0.676)^2]}}$$
$$= 0.01638$$

and

$$s_{b_2} = \frac{1.261}{\sqrt{48.944475[1 - (0.676)^2]}}$$
$$= 0.2446$$

These are the figures given in Table 10-2 for the standard deviations of the slope coefficients of INCOME and SIZE.

We can test hypotheses concerning B_1 and B_2 by computing t statistics in the usual way:

(10.14)
$$t_1 = \frac{b_1 - B_1}{s_{b_1}}$$

and

$$t_2 = \frac{b_2 - B_2}{s_{b_2}}$$

These t statistics approach normality as the sample size and number of degrees of freedom become large.

Hence, to test the hypotheses that the net regression coefficients are equal to zero, that is, that family income and family size have no effect on food expenditures, or

$$H_0: B_1 = 0 \qquad H_0: B_2 = 0$$
$$\text{and}$$
$$H_1: B_1 \neq 0 \qquad H_1: B_2 \neq 0$$

we calculate

(10.15)
$$t_1 = \frac{b_1 - 0}{s_{b_1}} = \frac{b_1}{s_{b_1}} \quad \text{and} \quad t_2 = \frac{b_2 - 0}{s_{b_2}} = \frac{b_2}{s_{b_2}}$$

In the illustrative problem, we find

$$t_1 = \frac{0.14821}{0.01638} = 9.05 \quad \text{and} \quad t_2 = \frac{0.7931}{0.2444} = 3.24$$

As indicated in Table 10-2, the p values corresponding to the t-ratios of 9.05 and 3.24 are respectively 0.000 and 0.005. We conclude that b_1 and b_2 differ significantly from zero at both the 5% and 1% levels of significance. Therefore, we reject the null hypotheses that $B_1 = 0$ and $B_2 = 0$.

Summary   In summary, we conclude that family net income X_1 and family size X_2 have statistically significant influences on family food expenditures Y. However, as noted earlier, the correlation coefficient between family net income and family size is 0.676 (see Table 10-3), and there is a consequent difficulty of measuring the separate effects of these two intercorrelated variables.

An important point concerning the interpretation of the results of a multiple regression analysis follows from the above discussion. If the basic purpose of computing a regression equation is to make predictions of values of the dependent variable, then the reliability of the individual net regression coefficients is not of great consequence. On the other hand, if the purpose of the analysis is to measure accurately the separate effects of each of the independent variables on the dependent variable, then the reliability of the individual net regression coefficients is clearly important.

10.6
THE ANALYSIS OF VARIANCE

The **analysis of variance** in multiple regression analysis appraises the overall significance of the regression equation. It tests the null hypothesis that all of the true population regression (slope) coefficients equal zero. Hence, the null and alternative hypotheses are

$$H_0\text{: All of the } B_i \text{ values equal zero}$$

$$H_1\text{: Not all of the } B_i \text{ values equal zero}$$

As usual in an analysis of variance, the null hypothesis is retained or rejected on the basis of an F test. In a simple two-variable regression analysis, the F test gives exactly the same result as the t test for the null hypothesis H_0: $B = 0$.

In the interpretation of the coefficient of determination in section 9.6, we gave the following components of the total variation of the dependent variable:

(10.16)
$$\text{Total variation} = \text{Explained variation} + \text{Unexplained variation}$$
$$\Sigma(Y - \bar{Y})^2 \quad = \quad \Sigma(\hat{Y} - \bar{Y})^2 \quad + \quad \Sigma(Y - \hat{Y})^2$$

On the right-hand side of equation 10.16, the $\Sigma(\hat{Y} - \bar{Y})^2$ term represents the variation in the dependent variable *explained* by the regression equation; the $\Sigma(Y - \hat{Y})^2$ term represents the variation in the dependent variable *not explained* by the regression equation. The *explained* variation is often referred to as the **regression sum of squares**, the *unexplained* variation is referred to as the **error sum of squares** or the **residual sum of squares**.

The analysis of variance is carried out as indicated by the standard format shown in Table 10-6.

TABLE 10-6
General format of the analysis of variance in regression analysis

(1) Source of Variation	(2) Sum of Squares	(3) Degrees of Freedom	(4) Mean Square
Regression	$\Sigma(\hat{Y} - \bar{Y})^2$	$v_1 = k - 1$	$\Sigma(\hat{Y} - \bar{Y})^2/(k - 1)$
Error	$\Sigma(Y - \hat{Y})^2$	$v_2 = n - k$	$\Sigma(Y - \hat{Y})^2/(n - k)$
Total	$\Sigma(Y - \bar{Y})^2$	$n - 1$	

$$F(v_1, v_2) = \frac{\Sigma(\hat{Y} - \bar{Y})^2/(k - 1)}{\Sigma(Y - \hat{Y})^2/(n - k)}$$

- The first column shows the sources of variation. Note that we have used the terminology usually given in computer printouts, with **error** denoting the unexplained variation.
- The second column gives the sums of squares or variations.
- The third column shows the numbers of degrees of freedom that correspond to the sums of squares, that is, $k - 1$ for the regression sum of squares, where k is the number of constants in the regression equation, and $n - k$ for error.
- Column (4) gives the mean squares or variances, in this case, the regression (explained) variance and the error (unexplained) variance.
- The F value shown at the bottom of the table is the ratio of the regression variance to the error variance, or in other words, the ratio of the explained variance to the unexplained variance. If the null hypothesis is true that all of the B_i values equal zero, then this ratio is distributed according to the F distribution given in equation 8.16 with $v_1 = k - 1$ and $v_2 = n - k$ degrees of freedom.

The decision rule for a critical F value denoted F_α is:

DECISION RULE

1. If $F(v_1, v_2) > F_\alpha$, reject the null hypothesis that all the $B_i = 0$.
2. If $F(v_1, v_2) \leq F_\alpha$, do not reject the null hypothesis.

Table 10-7 shows a computer printout of the analysis of variance for the regression problem of predicting family food expenditures from family income and family size. Note that the number of degrees of freedom is $k - 1 = 3 - 1 = 2$

TABLE 10-7
Computer output for analysis of variance for the regression equation for annual food expenditures (FOOD), annual net income (INCOME), and family size (SIZE)

ANALYSIS OF VARIANCE

SOURCE	DF	SS	MS	F	P
REGRESSION	2	386.31	193.16	121.47	0.000
ERROR	17	27.03	1.59		
TOTAL	19	413.35			

---

for the regression sum of squares because there are $k = 3$ constants to be estimated in the regression equation $\hat{Y} = a + b_1 X_1 + b_2 X_2$ relating food expenditures (Y) to income (X_1) and family size (X_2). There are $n - k = 20 - 3 = 17$ degrees of freedom for the error sum of squares because there are $n = 20$ observations in the problem and $k = 3$ constants in the regression equation. As indicated in Table 10-6, the F ratio is calculated by dividing the regression mean square (variance) by the error mean square (variance). In Table 10-7, we have

$$F = \frac{386.31/2}{27.03/17} = \frac{193.16}{1.59} = 121.47$$

We turn to Table A-8 of Appendix A, where degrees of freedom for the numerator are read across the top of the table, and degrees of freedom for the denominator are read down the side. For 2 and 17 degrees of freedom, respectively, we find critical F values of 3.59 at the 5% level of significance and 6.11 at the 1% level of significance. Since the computed F value shown for the computer output in Table 10-7, is 121.47, we reject the null hypothesis at both the 5% and 1% levels. These relationships are shown in Figure 10-2. The p value of 0.000 in Table 10-7 indicates that the null hypothesis would be rejected even at a 0.001 level of significance. Since we have rejected the null hypothesis that all of the slope coefficients are equal to zero, we have rejected a null hypothesis of no relationship between the dependent variable and all of the independent variables considered collectively.

It is instructive to consider the relationship of the analysis of variance to other measures given earlier in computer output.

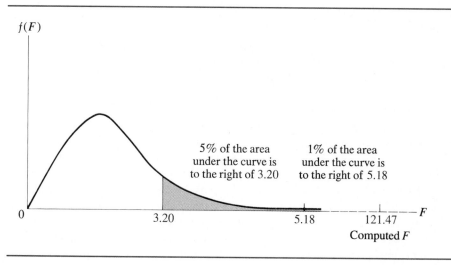

FIGURE 10-2

The critical and computed values of F for the food expenditures problem

$f(F)$

5% of the area under the curve is to the right of 3.20

1% of the area under the curve is to the right of 5.18

0   3.20   5.18   121.47 F

Computed F

The unexplained variance, or error variance, as it is often called, is the mean square error shown in Table 10-7, and in the present example is

$$(10.17) \quad S_{Y.12}^2 = \frac{\text{Error sum of squares}}{\text{Degrees of freedom}} = \frac{\Sigma(Y - \hat{Y})^2}{n - k} = \frac{27.03}{17} = 1.59$$

The standard error of estimate, denoted s in Minitab, is the square root of the unexplained variance and is given by

$$(10.18) \quad S_{Y.12} = \sqrt{\frac{\Sigma(Y - \hat{Y})^2}{n - k}} = \sqrt{\frac{27.03}{17}} = \sqrt{1.59} = 1.261$$

The coefficient of multiple determination, R^2, is related to the analysis of variance, in the following ways, as illustrated in Table 10-7 for the present problem:

$$(10.19) \quad R^2 = 1 - \frac{\text{Unexplained Variation}}{\text{Total Variation}} = 1 - \frac{\Sigma(Y - \hat{Y})^2}{\Sigma(Y - \bar{Y})^2}$$

$$= 1 - \frac{27.03}{413.35} = 93.5\%$$

or

$$(10.20) \quad R^2 = \frac{\text{Explained Variation}}{\text{Total Variation}} = \frac{\Sigma(\hat{Y} - \bar{Y})^2}{\Sigma(Y - \bar{Y})^2} = \frac{386.31}{413.35} = 93.5\%$$

Adjusting for degrees of freedom, we have

$$(10.21) \quad R_c^2 = 1 - \frac{\text{Unexplained Variance}}{\text{Total Variance}} = 1 - \frac{\Sigma(Y - \hat{Y})^2/(n - k)}{\Sigma(Y - \bar{Y})^2/(n - 1)}$$

$$= 1 - \frac{27.03/17}{413.35/19} = 92.7\%$$

and

$$(10.22) \quad R_c^2 = 1 - (1 - R^2)\left(\frac{n - 1}{n - k}\right) = 1 - (1 - 0.935)\left(\frac{19}{17}\right) = 92.7\%$$

The coefficient of multiple determination R^2 may be described as a measure of the effect on the dependent variable of all the independent variables combined. More specifically, R^2 measures the percentage of variance in the dependent variable that has been accounted for by all the independent variables combined. The square root of the coefficient of multiple determination $R = \sqrt{R^2}$ is referred to as the coefficient of multiple correlation. It is always positive. Since some of the individual independent variables may be positively correlated with the dependent variable and others negatively correlated, there would be no meaning in distinguishing between a positive and negative value for R. As in the case of r^2 and r in two-variable analysis, the R^2 figure is easier to interpret than R because R^2 is a percentage whereas R is not. We now consider some techniques and problems associated with multiple regression analysis.

Thus far in our discussion of regression analysis, the variables involved have been *quantitative*. However, variables of interest may be *qualitative* rather than quantitative. For example, in the family food expenditures example, suppose we wished to distinguish between urban and nonurban families, between home-owning and renting families, or between cases in which the head of the family is native born versus foreign born. Each of these examples represents a situation in which a classificatory or qualitative variable has two categories. If we wished to include one of these qualitative variables in the multiple regression analysis of family food expenditures versus family income and family size, we could set up a qualitative variable as a third independent variable as follows:

Variable X_3	Variable X_3	Variable X_3
$X_3 = 0$ if nonurban $X_3 = 1$ if urban	$X_3 = 0$ if renter $X_3 = 1$ if homeowner	$X_3 = 0$ if foreign born $X_3 = 1$ if native born

Such *qualitative variables*, when included in regression analyses, are referred to as **dummy variables**. We illustrate the use of a dummy variable in the following example.

Assume that an airline wanted to relate its sales revenue (Y) to the incomes of its passengers (X_1) on the Boston to Miami route. Further assume that 6 months of the year are considered "off-peak months," while 6 months are considered "peak months." To account for this qualitative distinction, an analyst at the airline set up the following dummy variable: $X_2 = 0$ for off-peak months and $X_2 = 1$ for peak months. The data for the year analyzed are shown in Table 10-8.

Figure 10-3 is a scatter diagram for the data on airline sales revenue and passenger income displayed in Table 10-8. Two distinct patterns are evident in the graph, one for the off-peak periods concentrated toward the lower left and one for the peak periods oriented toward the upper right. Therefore, it is generally evident that sales revenue varies directly with passenger income, but another factor also affects sales revenue on the Boston–Miami route, namely, whether the month of travel is a peak or off-peak period. To account for these relationships, we now set up the following linear regression equation:

(10.23)
$$\hat{Y} = a + b_1 X_1 + b_2 X_2$$

where $X_2 = 0$ for off-peak months
          $= 1$ for peak months

Using ordinary least squares methods, we can solve for the constants a, b_1, and b_2 and analyze the implications of such a regression equation. If we

TABLE 10-8

Data for regression analysis of airline sales revenue (Y), mean annual income of passengers (X_1), and a dummy variable (X_2) for off-peak (0) and peak months (1)

Month	Sales Revenue ($000)	Mean Annual Income of Passengers ($000)	Peak or Off-peak Month	Dummy Variable X_2
Jan.	4,480	45	Peak	1
Feb.	4,620	37	Peak	1
Mar.	4,550	49	Peak	1
Apr.	4,700	52	Peak	1
May	3,200	32	Off-peak	0
Jun.	3,360	35	Off-peak	0
Jul.	3,450	30	Off-peak	0
Aug.	3,400	28.5	Off-peak	0
Sep.	3,850	26	Off-peak	0
Oct.	3,940	34.5	Off-peak	0
Nov.	4,010	41.5	Peak	1
Dec.	3,700	33.5	Peak	1

substitute $X_2 = 0$ into equation 10.23, we obtain

$$(10.24) \qquad \hat{Y} = a + b_1 X_1$$

which we interpret as an equation that yields estimates of sales revenue for *off-peak* months. If we substitute $X_2 = 1$ into equation 10.23, we obtain

$$\hat{Y} = a + b_1 X_1 + b_2$$

or

$$(10.25) \qquad \hat{Y} = (a + b_2) + b_1 X_1$$

which we interpret as an equation that yields estimates of sales revenue for *peak* months.

The two regression equations 10.24 and 10.25 have the same slopes (b_1) but have different Y intercepts. Since the intercepts for the lines for off-peak periods and peak periods are, respectively, a and $a + b_2$, then b_2 can be interpreted as the extra amount of sales revenue, on the average, attributable to the peak months.

Using the method of least squares to estimate the values of the constants in equation 10.23, we obtain

$$(10.26) \qquad \hat{Y} = 26.54 + 0.2804X_1 + 4.84X_2$$

or, equivalently

$$(10.27) \qquad \hat{Y} = 26.54 + 0.2804X_1 \text{ for off-peak months}$$

$$(10.28) \qquad \hat{Y} = 31.38 + 0.2804X_1 \text{ for peak months}$$

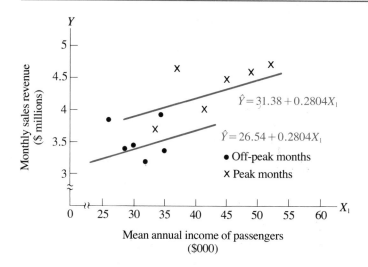

FIGURE 10-3

Scatter diagram of monthly sales revenue and mean annual income of passengers of an airline

The regression lines corresponding to equations 10.27 and 10.28 are shown on the scatter diagram in Figure 10.3. In summary, by computing the one regression equation $\hat{Y} = a + b_1X_1 + b_2X_2$, in which X_2 was a **dummy variable**, we actually obtained two regression lines, one for off-peak periods and another for peak periods.

Dummy variable

Note that in using the dummy variable X_2, we assume that it is reasonable to consider the slopes of the regression lines for off-peak and peak periods to be the same. If the slopes were markedly different, the dummy variable technique should not have been used and separate equations should have been computed for the off-peak and peak periods.

Let us now consider the implication of ignoring X_2, the dummy variable that accounted for whether a month was off-peak or peak. Referring again to the scatter diagram in Figure 10-3, we see that a single regression line fitted to the 12 points would have a much steeper slope than that of the two parallel lines. The slope of this regression line between sales revenue (Y) and passenger income (X_1) would be too great; that is, it would have an upward bias. Clearly, the effect that we would attribute to income alone should also be partially attributed to the factor of peak periods.

In the preceding example, we accounted for two qualitative categories, off-peak months and peak months. It is not appropriate to encode a dummy variable as 0, 1, 2, 3, and so forth, with more than two categories of a qualitative factor. By the use of such a code, we would imply no difference in the effect on the dependent variable of a change from zero to one, one to two, and two to three of the qualitative variable. This may be a totally unreasonable assumption. Hence, we may use a technique that establishes several dummy variables for one qualitative variable. We illustrate this technique in a simple example.

Suppose we wished to establish a linear regression model for the following variables relating to the production of a particular product in a company:

$$Y = \text{average unit cost of production in dollars}$$

$$X_1 = \text{daily number of production workers}$$

$$X_2 = \text{daily total number of hours of work} \\ \text{performed by the production workers}$$

Assume that data on Y, X_1, and X_2 are available for three plants in the company.

To establish a coding for the three plants, we set up two dummy variables, that is, one fewer dummy variable than the number of categories in the qualitative variable. Labeling the two dummy variables X_3 and X_4, we denote the plants by the following code:

X_3	X_4	
1	0	when X_1 and X_2 are from plant 1
0	1	when X_1 and X_2 are from plant 2
0	0	when X_1 and X_2 are from plant 3

The linear equation would then take the following form:

(10.29)
$$\hat{Y} = a + b_1X_1 + b_2X_2 + b_3X_3 + b_4X_4$$

Computation   The estimated average unit cost of production at plant 3, computed by substituting $X_3 = 0$ and $X_4 = 0$ in equation 10.29, would be equal to:

$$\hat{Y} = a + b_1X_1 + b_2X_2 + b_3(0) + b_4(0)$$

or

(10.30)
$$\hat{Y} = a + b_1X_1 + b_2X_2$$

The estimated average unit cost of production at plant 1 is

$$\hat{Y} = a + b_1X_1 + b_2X_2 + b_3(1) + b_4(0)$$

or

(10.31)
$$\hat{Y} = (a + b_3) + b_1X_1 + b_2X_2$$

Similarly, the estimated unit cost of production at plant 2 is

$$\hat{Y} = a + b_1X_1 + b_2X_2 + b_3(0) + b_4(1)$$

or

(10.32)
$$\hat{Y} = (a + b_4) + b_1X_1 + b_2X_2$$

Summary

In summary, using this coding technique, we establish one multiple regression equation of the form in equation 10.29. However, that single equation yields three regression equations, one for each plant.

Interpretation

Let us consider the interpretation of the slope coefficients for the dummy variables, that is, b_3 and b_4. Note that plant 3 was the category for which $X_3 = 0$ and $X_4 = 0$. Let us designate plant 3 as the "base plant." Comparing the Y intercept for plant 1 $(a + b_3)$ in equation 10.31 with that of the base plant, (a) in equation 10.30, we can state that, on the average, the unit cost of production at plant 1 is b_3 greater than at the base plant.

Analogously, referring to equation 10.32 for plant 2, and comparing the Y intercept of $(a + b_4)$ with that of the base plant, (a) in equation 10.30, we can state that, on the average, the unit cost of production at plant 2 is b_4 greater that at the base plant.

Let us assume, for example, that in this example of estimating average unit cost of production the multiple regression equation given in equation 10.33 had been derived.

$$(10.33) \qquad \hat{Y} = 3.123 - 0.0051X_1 - 0.00036X_2 + 0.470X_3 + 0.237X_4$$

The respective regression equations for plants 1, 2, and 3 are

$$\hat{Y} = 3.123 - 0.0051X_1 - 0.00036X_2 + 0.470(1) + 0.237(0) \qquad \textit{Plant 1}$$

or $\qquad \hat{Y} = 3.593 - 0.0051X_1 - 0.00036X_2$

$$\hat{Y} = 3.123 - 0.0051X_1 - 0.00036X_2 + 0.470(0) + 0.237(1) \qquad \textit{Plant 2}$$

or $\qquad \hat{Y} = 3.360 - 0.0051X_1 - 0.00036X_2$

$$\hat{Y} = 3.123 - 0.0051X_1 - 0.00036X_2 + 0.470(0) + 0.237(0) \qquad \textit{Plant 3}$$

or $\qquad \hat{Y} = 3.123 - 0.0051X_1 - 0.00036X_2$

Hence, estimating the average unit cost of production for $X_1 = 100$ workers and $X_2 = 4,100$ total hours of work at each of the three plants, we have

$$\hat{Y} = 3.593 - 0.0051(100) - 0.00036(4,100) = \$1.607 \qquad \textit{Plant 1}$$

$$\hat{Y} = 3.360 - 0.0051(100) - 0.00036(4,100) = \$1.374 \qquad \textit{Plant 2}$$

$$\hat{Y} = 3.123 - 0.0051(100) - 0.00036(4,100) = \$1.137 \qquad \textit{Plant 3}$$

Thus, we observe that the average unit cost at plant 3, the base plant, is $1.137; the figure for plant 1 is $1.607, or $0.47 higher $(b_3 = 0.47)$ than for the base plant; and the figure for plant 2 is $1.374, or $0.237 higher $(b_4 = 0.237)$ than for the base plant. Of course, if either plant 1 or plant 2 had been selected as the base rather than plant 3, the same cost estimates would have resulted and the same relationships would have been observed among the cost figures for the three plants.

10.8
MULTICOLLINEARITY

Multicollinearity—or simply, **collinearity**—describes a problem that arises in multiple regression analysis when independent variables are highly correlated. In such situations, it is not possible to separate the individual effects of the independent variables on the dependent variable. To view the problem intuitively, we consider the extreme situation in which we correlate a dependent variable Y with an independent variable X_1 and then add another independent variable X_2 that is perfectly linearly correlated with X_1. Since X_1 and X_2 move together, it is impossible to disentangle the separate effects of these variables on Y. Furthermore, from a practical standpoint, once a regression equation has been established between Y and X_1, nothing is gained by adding X_2 to the equation. With X_1 already in the equation, X_2 does not account for any of the unexplained variance in the dependent variable.

When multicollinearity exists among the independent variables in a multiple regression equation, the net regression (slope) coefficients tend to be unreliable. Usually one or more of the independent variables will have a net regression coefficient that is not significantly different from zero.

In the example in this chapter in which we related family food expenditures (Y) to family income (X_1) and family (X_2), we found that the slope coefficients for income (b_1) and family size (b_2) differed significantly from zero. The 0.676 correlation coefficient between income and family size represents a moderately high degree of association between these two variables. In such a situation it would have been possible for one or both slope coefficients *not* to have differed significantly from zero, even though in separate two-variable regression equations—$\hat{Y} = a + b_1 X_1$ and $\hat{Y} = a' + b_2 X_2$—the coefficients b_1 and b_2 were significantly different from zero. It can be seen from equations 10.12 and 10.13 that the larger the correlation coefficient between the two independent variables X_1 and X_2, the larger is the standard error of the slope coefficient. Therefore, the smaller will be the t-ratio for the slope coefficient and the more likely it will be that the observed slope coefficient will not differ significantly from zero.

Remedy 1: delete one or more variables   Although a number of steps can be taken to solve the problem of multicollinearity, the simplest is to delete one or more of the correlated variables. Judgment and significance tests for the net regression coefficients can determine which variables to drop. In general, if one of two highly correlated independent variables is dropped, the R^2 value will not change much. For example, when the food expenditures figure was regressed on income and family size, the R_c^2 value was 0.935. The r_c^2 value for food expenditures regressed on income alone was 0.895. Clearly, if the correlation between income and family size had been even higher, little predictive ability is lost by dropping the family size variable.

A rule of thumb is often used to determine whether to delete a collinear variable: Drop the variable if R_c^2 increases upon its deletion.

It can be shown that this rule is equivalent to dropping the variable if the t statistic for its net slope coefficient is less than one.

Another possible remedy for multicollinearity is to change the form of one or more of the independent variables. For example, if national income in current dollars is one of the independent variables and one or more other independent variables are highly correlated with it, then dividing income by population to yield a per capita national income variable or dividing income by a price index to yield national income in *real dollars* rather than *current dollars* may result in less correlated independent variables. Econometricians sometimes use more sophisticated measures. For example, when dealing with time-series data in estimating demand from income, prices, and other data, they may estimate one of the parameters in the regression equation independently from a cross-sectional study of family budget data. This combined used of time-series and cross-sectional data helps solve the multicollinearity problem, but gives rise to other methodological problems.

Remedy 2: alter form of one or more variables

In summary, no single simplistic solution eliminates the multicollinearity problem. If we attempt to devise an explanatory model, and we use secondary data rather than the results of an original experiment, there may be no practical way to disentangle the separate effects of intercorrelated independent variables. On the other hand, if we construct a regression model for forecasting purposes, multicollinearity is of less concern, provided we can expect the correlations among the independent variables to persist in the future.

Summary

10.9
AUTOCORRELATION

Autocorrelation or **serial correlation** is a problem of regression analysis that is usually present when time-series data are used. Specifically, it arises because of

> the violation of the regression analysis assumption that successive observations of the dependent variable Y are independent.

Autocorrelation

As noted in section 9.1, this assumption means that a low (or high) value for a particular observation does not imply that the next Y value observed will also be low (or high). We may state the assumption in another way: A positive (or negative) value of a residual $(Y - \hat{Y})$ for a particular observation does not imply that the next observation will also have a positive (or negative) residual. The validity of this assumption for *cross-sectional data* is virtually assured because of random sampling. For example, in a regression analysis of the heights (X) and weights (Y) of a random sample of women, the fact that a particular woman's weight yields an observation that is above the regression line or curve should have no bearing on whether the next woman's weight is also a point above the regression line or curve. On the other hand, in the case of economic data, which are often cyclical in nature, if the observation for a particular period lies above the regression line or curve—that is, it has a positive residual—it is likely that the next one also will. In fact, the observations will tend to be sequences or runs of positive residuals for points above the values predicted from the regression equation followed by runs of negative residuals. Stated differently, the residuals tend to be correlated among themselves (hence, the term, autocorrelation) rather than independent.

Durbin-Watson statistic   A widely used method of detecting the presence of autocorrelation involves the Durbin-Watson statistic. Before examining that statistic, let us consider a minor change in the notation that we have been using for the regression equation. We illustrate this change in terms of the two-variable linear regression equation that we have denoted $\hat{Y} = a + bX$. The following alternative symbology states the equation for the ith observation of the dependent variable:

(10.34) $$Y_i = a + bX_i + u_i$$

As indicated in Figure 10-4, in this notation $a + bX_i$ is the height of the regression line for the ith observation of the independent variable, and u_i is the residual or deviation from the regression line. To see the meaning of u_i more clearly, note that if the regression equation is expressed as $\hat{Y}_i = a + bX_i$, then equation 10-34 becomes $Y_i = \hat{Y}_i + u_i$ and u_i is equal to $Y_i - \hat{Y}_i$, or the deviation of the observation Y_i from the predicted value $\hat{Y}_i$.

Using the aforementioned notation, the Durbin-Watson statistic, denoted d, is defined as

(10.35) $$d = \frac{\sum\limits_{i=2}^{n} (u_i - u_{i-1})^2}{\sum\limits_{i=1}^{n} u_i^2}$$

where u_i = a residual from the regression equation in time period i
      u_{i-1} = a residual from the regression equation in time period $i - 1$,
             that is, the period before i

FIGURE 10-4
Alternative notation for the
linear regression equation for
the *i*th observation

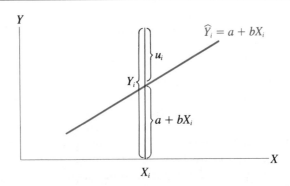

> The Durbin-Watson statistic ordinarily tests the null hypothesis that *no positive autocorrelation* is present.

This is equivalent to the null hypothesis that the residuals are random.

J. Durbin and G. S. Watson have tabulated lower and upper critical values of the d statistic, d_L and d_U, such that

1. If $d < d_L$, reject the null hypothesis of no positive autocorrelation and conclude that there is positive autocorrelation. *Critical values of d_L and d_U*

2. If $d > d_U$, accept the hypothesis of no positive autocorrelation and conclude that there is no positive autocorrelation.

3. If $d_L \leqslant d \leqslant d_U$, the test is inconclusive.

The critical values of d_L and d_U are given in Table A-11 of Appendix A for $\alpha = 0.05$ and $\alpha = 0.01$. These tabulated lower (d_L) and upper (d_U) bounds are given for n, the number of observations, and k, the number of independent variables.

For example, suppose we have a time series with $n = 25$ obervations and $k = 3$ independent variables and we want to test the null hypothesis of no positive autocorrelation at the 5% significance level. In Appendix Table A-11, we find $d_L = 1.12$ and $d_U = 1.66$. Therefore, if the observed d value is less than 1.12, we reject the null hypothesis of no positive autocorrelation and conclude that the residuals exhibit positive autocorrelation. If the observed d value is greater than 1.66, we accept the null hypothesis of randomness, or no positive autocorrelation.

To see clearly why positive autocorrelation leads to low observed values of the Durbin-Watson statistic and why randomness leads to higher values, let us consider the two illustrative situations in Table 10-9. In Example A, the

TABLE 10-9
Examples of a low and high value for the Durbin-Watson statistic (d)

Time Period	Example A: Low Value of d				Example B: High Value of d			
	u_i	$u_i - u_{i-1}$	$(u_i - u_{i-1})^2$	u_i^2	u_i	$u_i - u_{i-1}$	$(u_i - u_{i-1})^2$	u_i^2
1	-2			4	-2			4
2	-1	1	1	1	2	4	16	4
3	0	1	1	0	0	-2	4	0
4	1	1	1	1	1	1	1	1
5	2	1	1	4	-1	-2	4	1
			$\overline{4}$	$\overline{10}$			$\overline{25}$	$\overline{10}$

$$ d = \frac{\sum\limits_{i=2}^{5} (u_i - u_{i-1})^2}{\sum\limits_{i=1}^{5} u_i^2} = \frac{4}{10} = 0.4 \qquad d = \frac{\sum\limits_{i=2}^{5} (u_i - u_{i-1})^2}{\sum\limits_{i=1}^{5} u_i^2} = \frac{25}{10} = 2.5 $$

residuals (u_i) exhibit a trend. This might occur when the observations of Y are on the portion of a cycle rising above the regression line. The computed d value is 0.4. On the other hand, in Example B, the same residuals are shown as in Example A, except now they are randomly distributed. The calculated value of d in Example B is 2.5, more than 6 times the corresponding value given for Example A. Incidentally, note that there is *perfect direct autocorrelation* between u_i and u_{i-1} in Example A. That is, the correlation coefficient r equals one for the two following series:

Time Period	u_i	u_{i-1}
2	−1	−2
3	0	−1
4	1	0
5	2	1

The major difficulty in working with autocorrelated time-series data is that probability statements involving the use of standard errors assume that the $Y - \hat{Y}$ deviations are randomly distributed around the regression equation, and these statements can be seriously in error if the distribution is not random. For example, using the notation for a two-variable regression, the standard error of estimate $s_{Y.X}$ and the standard error of the regression coefficient s_b are both biased downwards. Therefore, if the residuals are serially correlated, the standard errors are deceptively small compared with what they should be if the residuals are randomly distributed. Therefore, invalid probability statements will be made about the regression equation and its slope coefficients, and F tests and t tests will not be strictly valid.

Remedy: change the variable   What techniques can remedy the problem of a considerable degree of autocorrelation? Probably the most widely used method is to work in terms of *changes* in the dependent and independent variables—referred to as **first differences**—rather than in terms of the original data themselves. For example, we might use the variable "year-to-year change in population" rather than population itself. The year-to-year changes are referred to as the *first differences in population.*

Other remedies   Other remedies include transforming the variables, (for example, using logarithms), adding another variable, and using various modified versions of the first difference transformation. These latter techniques, which are not dealt with here, are referred to as **autoregressive schemes** and are treated in most texts in econometrics. Of course, combinations of these techniques are also used.

Our discussion has dealt with the problem of **positive autocorrelation**, a situation in which the successive terms in a time series are *directly* related. **Negative autocorrelation**, in which the successive terms are *inversely* related, is not a practical problem for economic data. In such a series, high and low values alternate in a systematic fashion in successive time periods. This phenomenon is not frequently encountered in economic activity.

10.10
ANALYSIS OF RESIDUALS

It is useful to study the residuals in a regression analysis for a variety of reasons. We can analyze these deviations of actual values from predicted values to determine the magnitudes of prediction errors, to check on the departures from regression model assumptions, and to make diagnoses concerning the steps to take in continuing the analysis.

Virtually all computer programs print out actual Y values, predicted $\hat{Y}$ values, and the residuals $Y - \hat{Y}$. Furthermore, most computer programs contain various options for graphing the residuals versus the order of observations and independent variables. We now consider a number of situations concerning residuals.

Residuals May Not Be Independent

Departures from the assumption of independence can be observed in a plot of the residuals versus the order of the original observations. Plotting this type of graph is always advisable, particularly for time-series data, in which the order is chronological time. In Figure 10-5, the residuals $Y - \hat{Y}$ are plotted against the order of observations for a time series. Usually, in computer printouts, the points are not joined by lines or curves, but they are so depicted here in order to clarify the pattern. The figure shows a run of positive residuals followed by a run of negative residuals and the beginning of another run of positive residuals. Such a configuration signals a problem of positive autocorrelation (direct serial

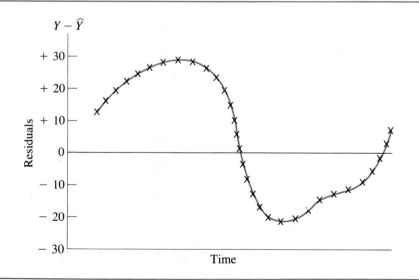

FIGURE 10-5

Positive autocorrelation in the residuals

correlation). In the section on autocorrelation, we considered some remedies for this problem.

A Functional Transformation May Be Required

Sometimes after a linear regression equation has been fitted to the data, the graph of the residuals indicates that some type of functional transformation may be appropriate. For example, Figure 10-6 depicts a curvilinear pattern when the residuals are plotted against the values of an independent variable X_1. In such a situation, if the fitted regression equation was of the form $\hat{Y} = a + b_1 X_1 + \cdots$, on the next run, we might fit a second-degree parabolic function of the form $\hat{Y} = a + b_1 X_1 + b_2 X_1^2 + \cdots$. Most computer programs for regression analysis accomplish this by defining a new variable X_1^2 and including it as an additional independent variable. Another possibility for transforming a pattern such as that in Figure 10-6 is to use logarithms. That is, on the next run, we might fit a function of the form $\log \hat{Y} = a + b_1 X_1 + \cdots$.

Outlier Observations Are Present

Sometimes, a regression equation may be a good fit to all but one or a few extreme observations, often referred to as *outliers*. For example, the plots in Figures 10-7(a) and (b) depict a sin'le outlier observation in each graph.

As in many other problems of regression analysis, there is no single simplistic solution to the problem of extreme observations. We might be tempted merely to delete the outlier observations. However, it would clearly be an unscientific procedure to delete or ignore observations that in some way did not agree with a particular hypothesis or model. On the other hand, if we can

One solution: delete outlier

FIGURE 10-6

A curvilinear pattern in the residuals

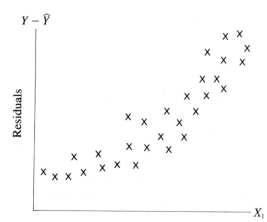

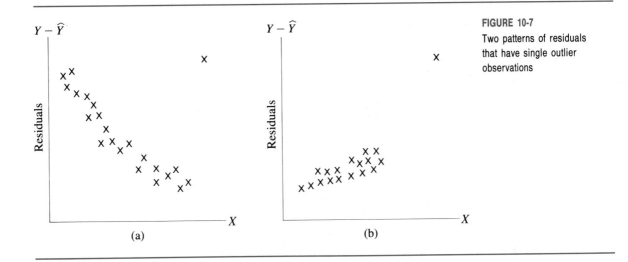

demonstrate logically that the extreme observations were derived from a different conceptual universe than the other observations, then we may be justified in excluding such outliers. For example, if all of the points except the extreme observation pertained to women and the outlier was an observation for a man, or if all but one of the points pertained to towns with populations of 10,000 and under and the extreme observation related to New York City, then a logical justification might exist for excluding the outliers.

A compromise position that is sometimes used is to run the analysis both with and without the extreme observations. Then, use of the analysis is a matter of *caveat emptor*, "let the buyer beware." That is, the analysis contains the results for both the inclusion and exclusion of outliers, and the investigator is free to draw his or her own conclusions.

Compromise solution

Independent Variables May Have Been Omitted

Patterns in residuals or very large residuals may indicate that one or more independent variables have been omitted from the regression equation. To solve this problem on the next run, we should add one or more variables that might account for this variation. For example, suppose after running the regression analysis in Chapter 9 for family food expenditures as a function of family income, we discovered that most of the observations with negative deviations were from families headed by women. Correspondingly, many observations displaying positive deviations were from families headed by men. It would then make sense to introduce a dummy variable that might take on the value of zero for families headed by women and the value of one for families headed by men. Similarly, the introduction of quantitative rather than dummy variables might assist in accounting for unexplained variation.

One solution

Another solution

FIGURE 10-8
Heteroscedasticity in residuals

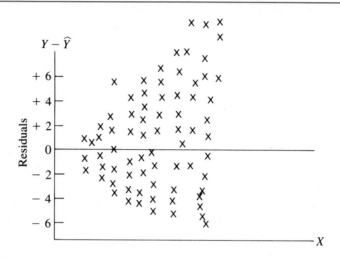

Other clues   Of course, a systematic pattern in residuals is only one of many possible indications that a variable has been omitted. Other clues would be a high standard error of estimate, a low R^2 value, and low t values for slope coefficients, which, on the basis of our knowledge of the situation, should have significant effects on the dependent variable.

The Error Terms May Not Have Constant Variance

Sometimes departures from linear regression model assumptions are clearly revealed by plots of the residuals. For example, let us assume that the equation $\hat{Y} = a + bX$ has been fitted to a set of data. The pattern of residuals displayed in Figure 10-8 indicates that the assumption of *homoscedasticity*—that is, constant variance of the residuals for all values of X—has been violated. The variance of residuals increases as the values of the independent variable increase. The corrections for **heteroscedasticity** (that is, departure from homoscedasticity) depend on whether the error variances for each set of independent variable observations are known from prior information. If not, then to estimate error variances, we need several observations of the dependent variable for each set of observations of the independent variables. One appropriate corrective technique, known as *weighted least squares*, is a special case of a general econometric technique called *generalized least squares*. These remedies for heteroscedasticity are outside the scope of our discussion, but may be found in most texts on econometrics.[4]

[4] For example, see R. S. Pindyck and D. L. Rubenfeld, *Econometric Models and Econometric Forecasts* (New York: McGraw-Hill, 1981), pages 142–46.

10.11
OTHER MEASURES IN MULTIPLE
REGRESSION ANALYSIS

A number of other measures are sometimes calculated in a multiple regression and correlation analysis. Only brief reference will be made to them here.

It is possible to calculate measures that indicate the separate effect of each of the independent variables on the dependent variable, if the influence of all the other independent variables has been accounted for. For this purpose, it is conventional to compute **coefficients of partial correlation**. For example, the partial correlation coefficient for family income in our illustrative problem, designated $r_{Y1.2}$, would show the partial correlation between Y and X_1 after the effect of X_2 on Y had been removed. The square of this coefficient $r_{Y1.2}^2$ measures the reduction in variance brought about by introducing X_1 after X_2 has already been accounted for.

Coefficients of partial correlation

Sometimes, it is difficult to compare the differences in net regression coefficients because the independent variables are stated in different units. For example, in the illustrative example, b_1 indicates the average difference in food expenditures Y *per unit difference in family income* X_1, whereas b_2 indicates the average difference in food expenditures Y *per unit difference in family size*. (In both cases, the other independent variable is held constant.) Unit differences in X_1 and X_2 are in different units—$1,000 and one person, respectively. To improve comparability, we can state the regression equation in a different form, giving each of the variables in units of its own standard deviation. The transformed net regression coefficients are called **beta coefficients**. For example, in terms of beta coefficients, the linear regression equation for three variables would be

Beta coefficients

$$(10.36) \qquad \frac{\hat{Y}}{s_Y} = \alpha + \beta_1 \frac{X_1}{s_{X_1}} + \beta_2 \frac{X_2}{s_{X_2}}$$

Thus, the beta coefficients are equal to

$$(10.37) \qquad \beta_1 = b_1 \frac{s_{X_1}}{s_Y} \quad \text{and} \quad \beta_2 = b_2 \frac{s_{X_2}}{s_Y}$$

As an illustration of the meaning of the beta coefficients, β_1 measures the number of standard deviations that $\hat{Y}$ changes with each change of one standard deviation in X_1.[5]

---

[5] The reader is alerted that the battle of notation must be continually fought. In Chapter 7, β denoted the probability of a Type II error. The specific meaning of this symbol must be determined from the context of the particular discussion.

Selected General Considerations

Analysis, knowledge, and judgment

A great deal of care must be exercised in the use of multiple regression and correlation techniques. In the development of the model, theoretical analysis, knowledge of the field of application, and logical judgment should aid in the selection of variables to be used in the study. Frequently, in business and economic applications, some of the relevant variables may not be easily quantifiable. Sometimes variables are not readily available and must be constructed from different sets of data.

Validity of assumptions

In the discussion of multiple regression and correlation analysis, we have confined ourselves to a linear model. We know that the underlying assumptions of this model should be checked for their validity. We should examine the graphs of the dependent variable against each of the independent variables at the outset of the analysis and check the plots of the $Y - \hat{Y}$ deviations against each of the independent variables after fitting the regression equation. Sometimes transformations—such as taking logarithms, reciprocals, or square roots of original observations—may provide better adherence to original assumptions and better fits of regression equations to the data. Of course, a linear regression equation simply represents a convenient approximation to the unknown "true" relationship. When linear relationships provide inadequate fits, curvilinear regression equations may be required.

Parsimony

The quest for a good fit of the regression equation to the data leads to adding more and more independent variables. However, cost considerations, difficulties of providing data in the implementation and monitoring of the model, and the search for a reasonably simple model ("parsimony") point toward the use of as few independent variables as possible. Since no mechanistic statistical procedure can resolve this dilemma and many other problems of multiple regression and correlation analysis, subjective judgment inevitably plays a large role. Statistical theory alone cannot determine which variables should be included in a regression analysis. Prior knowledge of the field of application is important in the initial selection of independent variables and in choices of variables to include or exclude based on the statistical analysis.

Dangers of extrapolation

We must guard against the dangers of extrapolation. There are subtle difficulties in multiple analysis compared with two-variable analysis. Even within the range of the data, certain combinations of values of the independent variables may not have been observed. Therefore, statistically valid estimates of the dependent variable cannot be made for these combinations of values.

10.12
THE USE OF COMPUTERS IN MULTIPLE REGRESSION ANALYSIS

The use of high-speed electronic computers has greatly simplified the testing and analysis of statistical relationships among variables. In the past, the cost and tedious labor involved in multiple regression analyses involving more than two or three independent variables severely restricted the analyst's ability to test and experiment. Using modern computer programs, the analyst now has a much wider range of choice in selecting variables, in options for performing transformations, in adding and deleting variables at various stages of the analysis, and in testing curvilinear as well as linear relationships.

Stepwise regression analysis is a versatile form of multiple regression analysis for which computers are particularly useful and for which a number of computer programs exist. In this type of analysis, at the first stage, the computer determines which of the independent variables is most highly correlated with the dependent variable. The computer printout then displays all the usual statistical measures for the two-variable relationship. At the next stage, the program selects the independent variable that accomplishes the greatest reduction in the unexplained variance remaining after the two-variable analysis. The computer printout then displays all the usual statistical measures for the three-variable relationship. The program continues in this stepwise fashion, at each stage entering the "best" independent variable in terms of ability to reduce the remaining unexplained variance. Analysis of variance tables and lists of residuals $(Y - \hat{Y})$ are provided at each stage. Obviously, without the use of computers and "canned" programs, the time needed to perform such an analysis by hand would be prohibitive.

Stepwise regression analysis

One final comment is in order. The advent of electronic computers has opened up a greater choice than was formerly available in selection of variables, the inclusion of larger numbers of variables, more options in performing transformations of variables, and more testing and experimentation with different types of statistical relationships. However, because of these increased possibilities, it becomes even more important that care and good judgment be exercised to avoid misuse of methods and misinterpretation of findings.

Caution

10.13
A COMPUTER APPLICATION

As an example of multiple regression analysis that deals with a number of the general issues and problems involved in the construction of an appropriate multiple regression model, we present a computer application that was carried out using the Minitab software package (Release 6.1.1). Minitab commands are shown wherever appropriate. We state the problem as follows: The president of a firm producing specialized electronics products is interested in predicting company sales. He has collected data on company sales and other variables by quarters of the year for 1985.1–1991.1, where this notation means from the first quarter of 1985 through the first quarter of 1991. This is a frequently used type of notation in which, for example, the quarters of 1990 would be denoted 1990.1, 1990.2, 1990.3, and 1990.4. Hence, in the present problem, there are 25 quarterly observation periods. Often, when dealing with monthly or quarterly time series data such as in this problem, analysts may adjust the original data for seasonal variations in order to deal more clearly with longer-run underlying factors. This adjustment usually takes the form of dividing an original series by an appropriate set of seasonal indexes. This topic is dealt with in Chapter 11, and we will simply assume that in the present problem, appropriate preliminary adjustments for seasonality have been made. We start by assuming that data have been collected for the following variables that have been entered in the indicated columns in Minitab:

Column Number	Variable ($millions)	Variable Name
1	Y = Company sales	SALES
2	X_1 = Company research expenditures	RES.EXP
3	X_2 = Industry sales	INDSALES

where industry sales is defined in terms of sales of companies that produce products that are competitive with those of the subject company. The company encountered a supply shortage during the last two quarters of 1987, and the president wishes to include this factor in the analysis. A printout of the data appears in Table 10-10. (As a general matter, it is a good policy after typing the input data to print out and check the data before proceeding further.)

Before running the regression analysis, it is useful to examine basic descriptive statistics and the correlations between each pair of variables. These data appear in Table 10-11. We note that on the basis of the correlation coefficients, company sales are more correlated with research expenditures (0.920) than with industry sales (0.465). Also, since the correlation coefficient between research expenditures and industry sales is low (0.299), little multicollinearity exists between the independent variables.

The first computer run involves fitting a linear regression equation, which in the standard notation that we have been using is expressed as $\hat{Y} = a + b_1X_1 + b_2X_2$. Selected output for the first run is displayed in Table 10-12. After the command for the regression of the dependent variable in column 1 on the two independent variables in columns 2 and 3, for purposes of illustration, we asked for a prediction for the 24th observation, that is, for 1990.4 using the observed values of $X_1 = 96.1$ and $X_2 = 500.8$ to obtain the predicted value $\hat{Y}$.

Interpreting the output in Table 10-12, we see the regression equation is

SALES = 25.7 + 0.688 RES.EXP + 0.0598 INDSALES

The p values for the constant and slope coefficients (0.013, 0.000, and 0.011) indicate that these values are significant at the 5% level, and that the constant and the slope coefficient for industry sales are just about borderline at the 1% level.

The standard error of estimate (s) is 11.47. In the notation that we have used in this chapter, $S_{Y.12} = 11.47$ millions of dollars. The next figure, uncorrected for degrees of freedom, is $R^2 = 88.7\%$. The R^2 value indicates a strong relationship in which about 89% of the variation in company sales has been accounted for by the regression equation relating that variable to company

TABLE 10-10
Data for 25 periods for company sales (Y), company research expenditures (X_1), and industry sales (X_2) in millions of dollars

```
MTB > INFO

COLUMN               NAME          COUNT
     C1              SALES           25
     C2             RES.EXP          25
     C3             INDSALES         25

CONSTANTS USED: NONE

MTB > PRINT C1 C2 C3

ROW          SALES          RES.EXP          INDSALES
  1          38.2             2.1             201.3
  2          45.7             2.3             334.6
  3          47.2             2.5             358.6
  4          40.3             2.9             306.1
  5          41.8             3.0             290.8
  6          62.1             3.7             448.3
  7          61.4             3.8             444.3
  8          56.1             3.7             392.0
  9          52.1             3.6             467.3
 10          73.4             4.1             591.1
 11          73.1             4.4             561.6
 12          67.3            10.2             487.1
 13          65.0            21.7             373.2
 14          85.3            41.8             728.3
 15          80.1            61.3             573.5
 16          78.5            65.4             480.2
 17          91.4            73.8             326.6
 18         107.8            81.2             552.9
 19         105.1            87.3             507.2
 20         103.8            90.1             389.7
 21         112.9            91.2             403.2
 22         121.3            92.5             481.6
 23         130.3            95.2             491.7
 24         139.2            96.1             500.8
 25         151.3            98.4             512.5
```

research expenditures and industry sales. Adjusted for degrees of freedom, R_c^2 is 87.6%, just slightly lower than the unadjusted R^2 value.

The analysis of variance table is in the standard form. In this problem, the p value of 0.000 indicates that the F ratio is significant at levels far below the conventional 0.05 and 0.01 critical values. In view of the relatively high R^2

TABLE 10-11
Descriptive statistics and correlation coefficients for company sales (Y), company research expenditures (X_1), and industry sales (X_2)

```
MTB > DESCRIBE C1 C2 C3

                  N         MEAN      MEDIAN    TRMEAN     STDEV     SEMEAN
SALES            25        81.23      73.40      80.05     32.61       6.52
RES.EXP          25        41.69      21.70      40.95     40.65       8.13
INDSALES         25        448.2      467.3      446.7     113.8       22.8

                MIN          MAX         Q1         Q3
SALES         38.20       151.30      54.10     106.45
RES.EXP        2.10        98.40       3.65      88.70
INDSALES      201.3        728.3      365.9      509.9

MTB > CORRELATION C1 C2 C3

              SALES       RES.EXP
RES.EXP       0.920
INDSALES      0.465        0.299
```

and R_c^2 values, we are not surprised that the *F* test is significant indicating that the overall regression relationship is significant.

   To perform a diagnostic check for autocorrelation in the residuals ($Y - \hat{Y}$), we can observe the Durbin-Watson statistic. We obtain this statistic in Minitab by typing a semicolon after the REGRESS prompt and then responding with DW to the subcommand prompt as follows:

MTB > regress C1 2 C2 C3;
SUBC > dw.

Of course, if you know in advance which subcommands you would like to use such as PREDICT and DW, they can all be included at the outset after the original REGRESS command, to forestall repeated output to the REGRESS command.

   In the present problem, we obtain the following output for the requested measure of autocorrelation:

$$\text{Durbin-Watson statistic} = 0.45$$

Referring to Appendix Table A-11, we find $d_L = 1.21$ and $d_U = 1.55$ for $n = 25$ observations and $k = 2$ independent variables. Since the observed Durbin-Watson statistic of 0.45 is less than $d_L = 1.21$, we reject the hypothesis of no positive autocorrelation and conclude that the residuals exhibit positive serial correlation. Referring to Table 10-13, we see clear evidence of positive autocorrelation—runs of negative and positive residuals.

   We may also note the following additional items of output shown in Table 10-13. Two periods, 15 and 25 corresponding to 1988.3 and 1991.1 were indi-

---

TABLE 10-12

Computer printout for the regression of company sales (Y) on company research expenditures (X_1) and industry sales (X_2)

```
MTB > REGRESS C1 2 C2 C3;
SUBC> PREDICT 96.1 500.8.

THE REGRESSION EQUATION IS
SALES = 25.7 + 0.688 RES.EXP + 0.0598 INDSALES
```

PREDICTOR	COEF	STDEV	T-RATIO	P
CONSTANT	25.732	9.516	2.70	0.013
RES.EXP	0.68822	0.06038	11.40	0.000
INDSALES	0.05980	0.02158	2.77	0.011

```
S = 11.47    R-SQ = 88.7%    R-SQ (ADJ) = 87.6%
```

ANALYSIS OF VARIANCE

SOURCE	DF	SS	MS	F	P
REGRESSION	2	22631	11315	85.95	0.000
ERROR	22	2896	132		
TOTAL	24	25527			

---

cated as having large standardized residuals (-2.02 and 2.53, respectively). Also shown are the predicted $\hat{Y}$ value equal to 121.82 and the 95% confidence interval and 95% prediction interval requested at the beginning of the example for $X_1 = 96.1$ and $X_2 = 500.8$ in 1990.4.

In summary, in this problem, company sales were related to a company variable (research expenditures) and an industry variable (industry sales). A linear regression equation exhibited plausible slope coefficients, that is, company sales were directly related to both research expenditures and industry sales, and the slope coefficients were significant. The two independent variables accounted for about 89% of the variation in company sales and were not particularly correlated. The analysis of variance confirmed that the overall relationship between company sales and the independent variables was significant. However, a significant degree of positive serial correlation existed among the residuals.

In an attempt to use the observed autocorrelation to improve the prediction model, we add a new variable in a second computer run. This new variable is the value of the dependent variable lagged one quarter of the year. The rationale of this procedure is that if company sales were positively autocorrelated, then the sales in any quarter might represent a good predictor of the sales in the succeeding quarter. Hence, the next computer run involves the equation

(10.38) $$\hat{Y}_T = a + b_1 X_{1T} + b_2 X_{2T} + b_3 Y_{T-1}$$

TABLE 10-13
Computer printout of actual values of the dependent variable Y (SALES), predicted values Y (FIT), residuals $Y - \hat{Y}$ (RESIDUAL), and related output

OBS.	RES.EXP	SALES	FIT	STDEV.FIT	RESIDUAL	ST.RESID
1	2.1	38.20	39.22	5.63	-1.02	-0.10
2	2.3	45.70	47.33	3.67	-1.63	-0.15
3	2.5	47.20	48.90	3.44	-1.70	-0.16
4	2.9	40.30	46.03	3.98	-5.73	-0.53
5	3.0	41.80	45.19	4.18	-3.39	-0.32
6	3.7	62.10	55.09	3.25	7.01	0.64
7	3.8	61.40	54.92	3.22	6.48	0.59
8	3.7	56.10	51.72	3.21	4.38	0.40
9	3.6	52.10	56.16	3.36	-4.06	-0.37
10	4.1	73.40	63.90	4.91	9.50	0.92
11	4.4	73.10	62.35	4.43	10.75	1.02
12	10.2	67.30	61.88	3.25	5.42	0.49
13	21.7	65.00	62.99	2.86	2.01	0.18
14	41.8	85.30	98.05	6.46	-12.75	-1.35
15	61.3	80.10	102.22	3.47	-22.12	-2.02R
16	65.4	78.50	99.46	2.68	-20.96	-1.88
17	73.8	91.40	96.05	4.35	-4.65	-0.44
18	81.2	107.80	114.68	3.58	-6.88	-0.63
19	87.3	105.10	116.15	3.52	-11.05	-1.01
20	90.1	103.80	111.05	4.20	-7.25	-0.68
21	91.2	112.90	112.61	4.11	0.29	0.03
22	92.5	121.30	118.19	3.72	3.11	0.29
23	95.2	130.30	120.66	3.84	9.64	0.89
24	96.1	139.20	121.82	3.89	17.38	1.61
25	98.4	151.30	124.10	4.01	27.20	2.53R

R DENOTES AN OBS. WITH A LARGE ST.RESID.

FIT	STDEV.FIT	95% C.I.	95% P.I.
121.82	3.89	(113.75, 129.88)	(96.69, 146.95)

where $\hat{Y}_T$ = the predicted value of Y for quarter T
Y_{T-1} = the value of Y in the preceding quarter
X_{1T} and X_{2T} = values of X_1 and X_2, respectively in period T

Introduction of a lagged relationship   It is interesting to note how the Minitab program accomplishes the lagged relationship. As shown in Table 10-14(a), we use the command LAG 1 C1 PUT INTO C4. This means that we lag the SALES variable in column 1 one period and then place the resulting series in column 4. We have named the variable in column 4 LAGGED. Note that in Table 10-14(a) there is no figure in row 1 for the LAGGED variable. That is, 38.2 in the first row is the company sales figure for 1985.1 and there is no LAGGED figure for company sales for the preceding quarter, 1984.4. In order to run a regression analysis of the four variables in Table 10-14a, we must delete the data in row 1 because of the missing item in the LAGGED column.

As indicated in Table 10-14(b), this is accomplished by the command DELETE ROWS 1 of C1-C4. The resulting data bank now contains 24 observations and four variables, namely the variables shown in equation 10.38. The first observation for the original three variables pertains to the second quarter of 1985, whereas the first observation for lagged company sales (Y_{T-1}) pertains to the first quarter of 1985.

TABLE 10-14(a)

Computer printout of the method of lagging the dependent variable

```
MTB > LAG 1 C1 PUT INTO C4
MTB > NAME C4 'LAGGED'
MTB > PRINT C1-C4
```

ROW	SALES	RES.EXP	INDSALES	LAGGED
1	38.2	2.1	201.3	*
2	45.7	2.3	334.6	38.2
3	47.2	2.5	358.6	45.7
4	40.3	2.9	306.1	47.2
5	41.8	3.0	290.8	40.3
6	62.1	3.7	448.3	41.8
7	61.4	3.8	444.3	62.1
8	56.1	3.7	392.0	61.4
9	52.1	3.6	467.3	56.1
10	73.4	4.1	591.1	52.1
11	73.1	4.4	561.6	73.4
12	67.3	10.2	487.1	73.1
13	65.0	21.7	373.2	67.3
14	85.3	41.8	728.3	65.0
15	80.1	61.3	573.5	85.3
16	78.5	65.4	480.2	80.1
17	91.4	73.8	326.6	78.5
18	107.8	81.2	552.9	91.4
19	105.1	87.3	507.2	107.8
20	103.8	90.1	389.7	105.1
21	112.9	91.2	403.2	103.8
22	121.3	92.5	481.6	112.9
23	130.3	95.2	491.7	121.3
24	139.2	96.1	500.8	130.3
25	151.3	98.4	512.5	139.2

TABLE 10-14(b)
Computer printout of the method of lagging the dependent variable: deletion of the first row of data

```
MTB > DELETE ROWS 1 OF C1-C4
MTB > PRINT C1-C4
```

ROW	SALES	RES.EXP	INDSALES	LAGGED
1	45.7	2.3	334.6	38.2
2	47.2	2.5	358.6	45.7
3	40.3	2.9	306.1	47.2
4	41.8	3.0	290.8	40.3
5	62.1	3.7	448.3	41.8
6	61.4	3.8	444.3	62.1
7	56.1	3.7	392.0	61.4
8	52.1	3.6	467.3	56.1
9	73.4	4.1	591.1	52.1
10	73.1	4.4	561.6	73.4
11	67.3	10.2	487.1	73.1
12	65.0	21.7	373.2	67.3
13	85.3	41.8	728.3	65.0
14	80.1	61.3	573.5	85.3
15	78.5	65.4	480.2	80.1
16	91.4	73.8	326.6	78.5
17	107.8	81.2	552.9	91.4
18	105.1	87.3	507.2	107.8
19	103.8	90.1	389.7	105.1
20	112.9	91.2	403.2	103.8
21	121.3	92.5	481.6	112.9
22	130.3	95.2	491.7	121.3
23	139.2	96.1	500.8	130.3
24	151.3	98.4	512.5	139.2

```
MTB > DESCRIBE C1-C4
```

	N	MEAN	MEDIAN	TRMEAN	STDEV	SEMEAN
SALES	24	83.02	75.95	81.86	32.03	6.54
RES.EXP	24	43.34	31.75	42.70	40.66	8.30
INDSALES	24	458.5	473.7	453.8	103.6	21.2
LAGGED	24	78.31	73.25	77.36	29.79	6.08

	MIN	MAX	Q1	Q3
SALES	40.30	151.30	57.42	107.12
RES.EXP	2.30	98.40	3.70	89.40
INDSALES	290.8	728.3	377.3	511.2
LAGGED	38.20	139.20	53.10	104.78

The computer printout of the regression analysis after introduction of lagged company sales as variable 4 is shown in Table 10-15. Examining the correlation matrix, we observe that company sales and lagged company sales are highly correlated ($r = 0.960$). Also, we note a high degree of collinearity between the independent variables research expenditures and lagged company sales ($r = 0.916$). In the printout for the regression equation, the signs of the slope coefficients are all positive and therefore plausible. The slope coefficients are all significant at the 5% level, but only lagged company sales is significant at the 0.01 significance level. The p values for the slope coefficients of both research expenditures and industry sales are higher (less significant) than they

TABLE 10-15
Computer printout of company sales (Y_T) on research expenditures (X_{1T}), industry sales (X_{2T}), and lagged company sales (Y_{T-1})

```
MTB > REGRESS C1 3 C2-C4, STORE ST.RESIDUALS IN C5, FITS IN C6

THE REGRESSION EQUATION IS
SALES > - 0.31 + 0.219 RES.EXP + 0.0386 INDSALES + 0.717 LAGGED
```

PREDICTOR	COEF	STDEV	T-RATIO	P
CONSTANT	-0.306	9.321	-0.03	0.974
RES.EXP	0.2185	0.1039	2.10	0.048
INDSALES	0.03857	0.01720	2.24	0.036
LAGGED	0.7173	0.1450	4.95	0.000

```
S = 8.067   R-SQ = 94.5%   R-SQ(ADJ) = 93.7%
```

ANALYSIS OF VARIANCE

SOURCE	DF	SS	MS	F	P
REGRESSION	3	22296.8	7432.3	114.22	0.000
ERROR	20	1301.4	65.1		
TOTAL	23	23598.2			

SOURCE	DF	SEQ SS
RES.EXP	1	19897.6
INDSALES	1	805.8
LAGGED	1	1593.4

```
MTB > CORRELATION C1 C2 C3 C4
```

	SALES	RES.EXP	INDSALES
RES.EXP	0.918		
INDSALES	0.398	0.238	
LAGGED	0.960	0.916	0.310

FIGURE 10-9(a)

Plot of standardized residuals against research expenditures

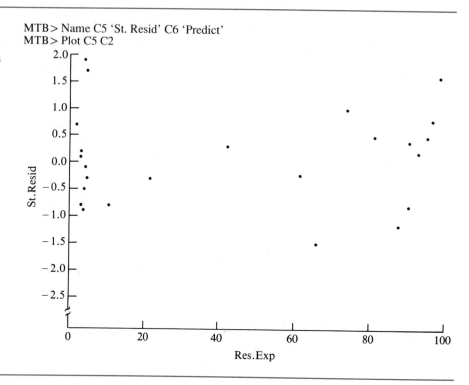

MTB > Name C5 'St. Resid' C6 'Predict'
MTB > Plot C5 C2

were in the previous regression equation without the LAGGED variable. Indeed because of the high degree of multicollinearity that is present, it would not have been surprising if one or more of the three independent variables turned out to be nonsignificant even at the 5% level.

The standard error of estimate is now less than in the previous run (8.067 versus 11.47) and the R_c^2 value is higher (93.7% versus 87.6%). Thus, the overall goodness of fit is somewhat improved by the addition of the lagged sales variable. Sometimes, because of high collinearity among the independent variables, there may be very little improvement in goodness of fit as measured by R_c^2. Note that because we introduced a lagged value of the dependent variable as an independent variable, the Durbin-Watson test is no longer valid.[6]

The commands STORE ST.RESIDUALS IN C5, FITS IN C6 in Table 10-15 are included for the purpose of doing diagnostic checking of residuals. Hence, we have plotted in Figures 10-9(a), 10-9(b), and 10-9(c) the standardized residuals

---

[6] In J. Durbin and G. S. Watson, "Testing for Serial Correlation in Least Squares Regression," *Biometrica*, vol. 38, 1951, pp. 159–78, the authors state, "It should be emphasized that the tests described . . . apply only to regression models in which the independent variables can be regarded as 'fixed variables.' They do not, therefore, apply to autoregressive schemes and similar models in which lagged values of the dependent variable occur as independent variables."

MTB> Plot C5 C3

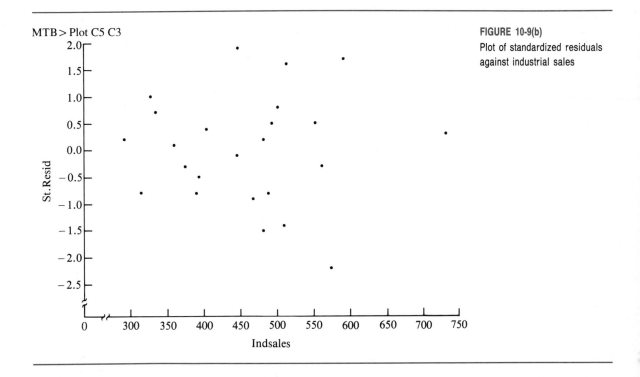

FIGURE 10-9(b)

Plot of standardized residuals against industrial sales

MTB> Plot C5 C4

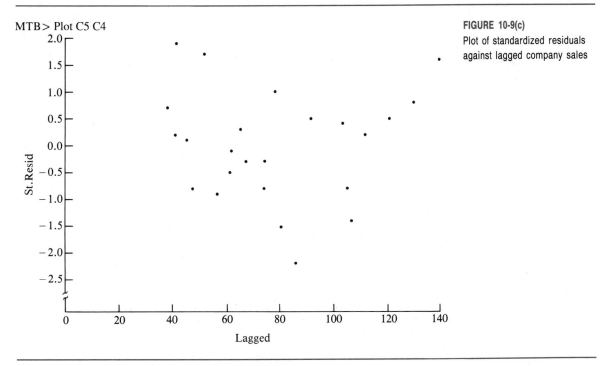

FIGURE 10-9(c)

Plot of standardized residuals against lagged company sales

FIGURE 10-10

Plot of standardized residuals against predicted ($\hat{Y}$) values of company sales

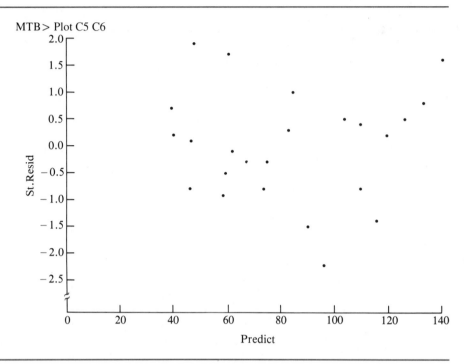

against each of the independent variables. The numbers 2 that appear on Figure 10-9(a) indicate two points plotted in the same place. Examination of these three plots indicates that the standardized residuals are fairly randomly distributed against each of the independent variables. That is, there is no systematic increase or decrease of the residuals nor is there any clear indication of a nonlinear relationship as we move from lower to higher values of the independent variables.

Continuing the diagnostic checking, we plotted in Figure 10-10 the standardized residuals (C5) against the $\hat{Y}$ or fit values (C6), now named PREDICT. Visually, there does not seem to be any clear evidence of nonrandomness in the standardized residuals as we move from lower to higher Y values. Sometimes it is useful to connect the points on a plot such as this one to obtain a clearer impression of the pattern. Table 10-16 gives the figures for the residuals and standardized residuals after fitting regression equation 10.38. Only the standardized residual for the 14th observation for 1988.2 is now indicated as being large. The last five residuals beginning with 1990.1 are positive and the last seven are generally increasing. This pattern raises a question as to whether the relationship of company sales with one or more of the independent variables may be nonlinear with a generally accelerating pattern. For example, a pattern

TABLE 10-16
Selected computer output for the regression analysis of company sales (SALES) on company research expenditures (RES.EXP), industry sales (INDSALES), and lagged company sales (LAGGED)

OBS.	RES.EXP	SALES	FIT	STDEV.FIT	RESIDUAL	ST.RESID
1	2.3	45.70	40.50	3.21	5.20	0.70
2	2.5	47.20	46.85	2.71	0.35	0.05
3	2.9	40.30	45.99	3.17	-5.69	-0.77
4	3.0	41.80	40.47	3.50	1.33	0.18
5	3.7	62.10	47.78	2.77	14.32	1.89
6	3.8	61.40	62.21	2.75	-0.81	-0.11
7	3.7	56.10	59.67	2.90	-3.57	-0.47
8	3.6	52.10	58.75	2.45	-6.65	-0.86
9	4.1	73.40	60.76	3.54	12.64	1.74
10	4.4	73.10	74.97	4.04	-1.87	-0.27
11	10.2	67.30	73.15	3.23	-5.85	-0.79
12	21.7	65.00	67.11	2.37	-2.11	-0.27
13	41.8	85.30	83.55	5.61	1.75	0.30
14	61.3	80.10	96.40	2.76	-16.30	-2.15R
15	65.4	78.50	89.97	2.69	-11.47	-1.51
16	73.8	91.40	84.73	4.00	6.67	0.95
17	81.2	107.80	104.33	3.30	3.47	0.47
18	87.3	105.10	115.66	2.48	-10.56	-1.38
19	90.1	103.80	109.81	3.03	-6.01	-0.80
20	91.2	112.90	109.63	3.00	3.27	0.44
21	92.5	121.30	119.47	2.63	1.83	0.24
22	95.2	130.30	126.47	2.95	3.83	0.51
23	96.1	139.20	133.48	3.62	5.72	0.79
24	98.4	151.30	140.82	4.41	10.48	1.55

R DENOTES AN OBS. WITH A LARGE ST.RESID.

of company sales increasing by increasing amounts as research expenditures grow would be a possible scenario.

To examine the possibility of a nonlinear relationship, we plotted in Figure 10-11 company sales (SALES) against research expenditures (RES.EXP)). This scatter diagram does indicate a concave upward slope, that is, an accelerating pattern of increasing company sales as research expenditures increase. Since the pattern appears curvilinear, we introduce a squared term for research expenditures by adding that squared value to the equation as another variable. As can be seen in Table 10-17, we carry this out in Minitab by the command LET C7 = C2**2, that is, we square the research expenditure values in column 2 and place the squared values in column 7. We name the variable in column

Introduction of a nonlinear relationship

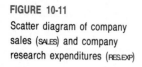

FIGURE 10-11

Scatter diagram of company sales (SALES) and company research expenditures (RES.EXP)

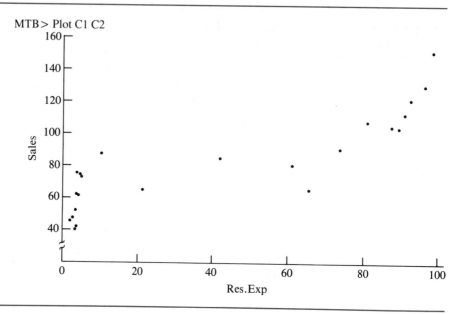

7 'RES.SQ'. Using the DESCRIBE and CORRELATION commands, we observe in Table 10-17 that our data bank still consists of 24 observations (rows), and that SALES is very highly correlated with the new RES.SQ variable ($r = 0.938$). However, we note the very high degree of collinearity between RES.EXP and RES.SQ ($r = 0.983$) and between LAGGED and RES.SQ (0.937). This is an example of a frequently encountered problem in multiple regression analysis, namely that an action taken to solve a particular problem in the analysis may introduce another problem or may make an existing problem even worse.

After the term for company research expenditures squared is included, as shown in Table 10-18, the following equation is obtained:

SALES = 7.74 − 0.390 RES.EXP + 0.0668 INDSALES + 0.412 LAGGED + 0.00845 RES.SQ

Comparing some of the associated output with results from the previous run we observe that the standard error of estimate is lower (6.693 versus 8.0676) and the R_c^2 is higher (95.6% versus 93.7%). However, we note that the slope coefficient for company research expenditures has become negative (−0.3899) and is not significant at the 5% level ($p = 0.079$). The other slope coefficients are all significant at the 5% level and have plausible positive signs. Undoubtedly the negative sign for RES.EXP has come about because of the aforementioned multicollinearity problem.

Because of the results reported in the preceding paragraph, we now consider dropping one of the collinear variables. It seems reasonable to delete the company research expenditures variable (RES.EXP), which would leave intact the curvilinear relationship implied by the company research expenditures squared

TABLE 10-17
Minitab commands for squaring the company research expenditures variable (X_i) and related computer output

```
MTB > LET C7 = C2**2
MTB > NAME C7 'RES.SQ'
MTB > DESCRIBE C1 C2 C3 C4 C7
```

	N	MEAN	MEDIAN	TRMEAN	STDEV	SEMEAN
SALES	24	83.02	75.95	81.86	32.03	6.54
RES.EXP	24	43.34	31.75	42.70	40.66	8.30
INDSALES	24	458.5	473.7	453.8	103.6	21.2
LAGGED	24	78.31	73.25	77.36	29.79	6.08
RES.SQ	24	3463	1109	3337	3893	795

	MIN	MAX	Q1	Q3
SALES	40.30	151.30	57.42	107.12
RES.EXP	2.30	98.40	3.70	89.40
INDSALES	290.8	728.3	377.3	511.2
LAGGED	38.20	139.20	53.10	104.78
RES.SQ	5	9683	14	7994

```
MTB > CORRELATION C1 C2 C3 C4 C7
```

	SALES	RES.EXP	INDSALES	LAGGED
RES.EXP	0.918			
INDSALES	0.398	0.238		
LAGGED	0.960	0.916	0.310	
RES.SQ	0.938	0.983	0.176	0.937

TABLE 10-18

Selected computer output for the regression of company sales (SALES) on company research expenditures (RES.EXP), industry sales (INDSALES), lagged company sales (LAGGED), and company research expenditures squared (RES.SQ)

```
MTB > REGRESS C1 4 C2 C3 C4 C7

THE REGRESSION EQUATION IS
SALES = 7.74 - 0.390 RES.EXP + 0.0668 INDSALES + 0.412 LAGGED + 0.00845
RES.SQ
```

PREDICTOR	COEF	STDEV	T-RATIO	P
CONSTANT	7.737	8.140	0.95	0.354
RES.EXP	-0.3899	0.2104	-1.85	0.079
INDSALES	0.06676	0.01681	3.97	0.001
LAGGED	0.4124	0.1540	2.68	0.015
RES.SQ	0.008455	0.002667	3.17	0.005

```
S = 6.693    R-SQ = 96.4%    R-SQ (ADJ) = 95.6%
```

ANALYSIS OF VARIANCE

SOURCE	DF	SS	MS	F	P
REGRESSION	4	22747.0	5686.8	126.94	0.000
ERROR	19	851.2	44.8		
TOTAL	23	23598.2			

SOURCE	DF	SEQ SS
RES.EXP	1	19897.6
INDSALES	1	805.8
LAGGED	1	1593.4
RES.SQ	1	450.2

variable (RES.SQ). Table 10-19 gives the computer output for the following resultant regression equation and related findings:

SALES = 6.53 + 0.0531 INDSALES + 0.492 LAGGED + 0.00395 RES.SQ

All slope coefficients are positive, therefore plausible, and are significant at the 1% level. As compared to the previous run that included the RES.EXP variable, the standard error of estimate is somewhat higher (7.089 versus 6.693) and the R_c^2 is somewhat lower (95.1% versus 95.6%). However, on balance considering these results alone, it seems fair to say that this version of the multiple regression equation is preferable to the immediately preceding one. Examination of plots of residuals confirm this conclusion.

Use of a dummy variable   We turn now to the consideration of the supply shortage that occurred in the last two quarters of 1987, that is, in 1987.3 and 1987.4. We take this into account by adding a dummy variable that is equal to zero for all quarters except for 1987.3 and 1987.4 when it is equal to one. Since the first observation

TABLE 10-19
Selected computer output for the regression of company sales (SALES) on industry sales (INDSALES), lagged company sales (LAGGED), and company research expenditures squared (RES.SQ)

```
MTB > REGRESS C1 3 C3 C4 C7, STORE ST, RESID IN C8, PREDICTS IN C9

THE REGRESSION EQUATION IS
SALES = 6.53 + 0.0531 INDSALES + 0.492 LAGGED + 0.00395 RES.SQ
```

PREDICTOR	COEF	STDEV	T-RATIO	P
CONSTANT	6.529	8.593	0.76	0.456
INDSALES	0.05305	0.01599	3.32	0.003
LAGGED	0.4916	0.1567	3.14	0.005
RES.SQ	0.003947	0.001158	3.41	0.003

```
S = 7.089   R-SQ = 95.7%   R-SQ (ADJ) = 95.1%
```

ANALYSIS OF VARIANCE

SOURCE	DF	SS	MS	F	P
REGRESSION	3	22593.2	7531.1	149.87	0.000
ERROR	20	1005.0	50.2		
TOTAL	23	23598.2			

SOURCE	DF	SEQ SS
INDSALES	1	3735.0
LAGGED	1	18274.1
RES.SQ	1	584.1

was for 1985.1, this means that observations 11 and 12 are encoded one and all other are equal to zero. The Minitab commands for the setting up of the dummy variable and some descriptive information are given in Table 10-20. The dummy variable was entered in column 10.

The correlations of the dummy variable with the other independent variables yield little information. Turning to Table 10-21 and the output pertaining to the regression equation, we see that all slope coefficients other than the dummy variable are significant at the 5% level. The negative correlation between company sales and the dummy variable ($r = -0.162$) agrees with our expectation. That is, since there was a supply shortage in 1987.3 and 1987.4 (DUMMY = 1), we expect that company sales would decrease during those periods and an inverse correlation would be observed between SALES and DUMMY. On the other hand, we see that when the dummy variable is included in the regression equation, its slope coefficient is positive ($b = 1.83$). However, this slope coefficient is insignificant at any conventional significant level ($p = 0.780$). It would seem reasonable to drop the dummy variable and to revert to the previous regression equation. In summary, our experiment in constructing a regression model has resulted in the following multiple regression equation and

---

TABLE 10-20
Coding used for the dummy variable that pertains to a supply shortage in the third and fourth quarters of 1987 and some related descriptive statistics

---

```
MTB > NAME C10 'DUMMY'

MTB > SET C10
DATA> 10(0) 2(1) 12(0)
DATA> END

MTB > DESCRIBE C1 C3 C4 C7 C10
```

	N	MEAN	MEDIAN	TRMEAN	STDEV	SEMEAN
SALES	24	83.02	75.95	81.86	32.03	6.54
INDSALES	24	458.5	473.7	453.8	103.6	21.2
LAGGED	24	78.31	73.25	77.36	29.79	6.08
RES.SQ	24	3463	1109	3337	3893	795
DUMMY	24	0.0833	0.0000	0.0455	0.2823	0.0576

	MIN	MAX	Q1	Q3
SALES	40.30	151.30	57.42	107.12
INDSALES	290.8	728.3	377.3	511.2
LAGGED	38.20	139.20	53.10	104.78
RES.SQ	5	9683	14	7994
DUMMY	0.0000	1.0000	0.0000	0.0000

```
MTB > CORR C1 C3 C4 C7 C10
```

	SALES	INDSALES	LAGGED	RES.SQ
INDSALES	0.398			
LAGGED	0.960	0.310		
RES.SQ	0.938	0.176	0.937	
DUMMY	-0.162	-0.084	-0.084	-0.251

---

selected associated measures. The standard errors of the slope coefficients are shown in parentheses below the corresponding coefficients.

$$\text{SALES} = 6.53 + 0.0531 \text{ INDSALES} + 0.492 \text{ LAGGED} + 0.00395 \text{ RES.SQ}$$
$$\phantom{\text{SALES} = 6.53 + }(0.016) \phantom{\text{ INDSALES} + } (0.157) \phantom{\text{ LAGGED} + } (0.00116)$$

$$s = 7.089 \qquad R^2 = 95.7\% \qquad R_c^2 = 95.1\%$$

Caveats

In this example of the development of a regression model for the prediction of company sales, the last computer run included 24 observations and 4 variables. In general, as more variables are added to the equation, the goodness of fit of

TABLE 10-21
Selected computer output for the regression of company sales (SALES) on industry sales (INDSALES), lagged company sales (LAGGED), company research expenditures squared (RES.SQ), and a dummy variable for a supply shortage (DUMMY)

```
MTB > REGRESS C1 4 C3 C4 C7 C10, STORE ST, RESID IN C11, FITS IN C12

THE REGRESSION EQUATION IS
SALES = 7.11 + 0.0544 INDSALES + 0.464 LAGGED + 0.00417 RES.SQ + 1.83 DUMMY
```

PREDICTOR	COEF	STDEV	T-RATIO	P
CONSTANT	7.106	9.031	0.79	0.441
INDSALES	0.05444	0.01709	3.19	0.005
LAGGED	0.4644	0.1871	2.48	0.023
RES.SQ	0.004170	0.001422	2.93	0.009
DUMMY	1.829	6.464	0.28	0.780

```
S = 7.258    R-SQ = 95.8%    R-SQ (ADJ) = 94.9%
```

ANALYSIS OF VARIANCE

SOURCE	DF	SS	MS	F	P
REGRESSION	4	22597.4	5649.4	107.26	0.000
ERROR	19	1000.8	52.7		
TOTAL	23	23598.2			

SOURCE	DF	SEQ SS
INDSALES	1	3735.0
LAGGED	1	18274.1
RES.SQ	1	584.1
DUMMY	1	4.2

the model will tend to improve, as measured by R^2 and the standard error of estimate. After all, if we fit a straight line of the form $\hat{Y} = a + bX$ to two observations, we will have zero degrees of freedom ($n - k = 2 - 2 = 0$), and we will have a perfect fit. If we fit a plane of the form $\hat{Y} = a + b_1X_1 + b_2X_2$ to three observations, we will again have zero degrees of freedom ($n - k = 3 - 3 = 0$) and a perfect fit. Analogously, if we have 24 points and we fit a function with 24 variables, we will force a perfect fit. Such a model would clearly be non-sensical and would probably be a poor model for prediction purposes. In fact, with the same number of both observations and variables, we could include independent variables that are logically unrelated to the dependent variable, and we would still have a perfectly fitting model.

> We must remember that our purpose in deriving a prediction model is not to obtain an equation that *at any cost* is the best fitting one to the past data, but rather to obtain a model that will predict well in the future.

Hence, we should strive for simplicity in the construction of the model. A regression equation with a small number of effective independent variables is preferable to another with a large number of less effective variables. In the development of the company sales regression model, we sequentially added and deleted variables in order to demonstrate a number of points. Prior knowledge and reasoning concerning how independent variables affect the dependent variable should be major determinants for the inclusion of variables in the development of regression models.

EXERCISES 10.13

Note: A number of different forms of computer output for multiple regression analyses are illustrated in these exercises.

1. A traffic commission conducted a regression analysis using the following data:

$$Y = \text{number of cars owned}$$

$$X_1 = \text{total income of family (thousands of dollars)}$$

$$X_2 = \text{size of family}$$

The commission selected a random sample of 123 families. The computer output yielded the following data:

Variable	Mean	Regression Coefficient	Standard Error of Regression Coefficient
X_1	12.550	$0.088(b_1)$	0.006
X_2	3.975	$0.065(b_2)$	0.055

Intercept (a): 0.570

Analysis of Variance Table

Source	Sum of Squares	Degrees of Freedom
Regression	1216.652	2
Error	419.384	120
Total	1636.036	

a. Construct and carry out a test of the significance of the net regression coefficients. Use a 1% significance level.

b. State exactly the conclusions to be drawn from your results in part a.

   c. Describe two properties of net regression coefficients in a multiple regression analysis.

   d. Calculate and interpret the coefficient of multiple determination adjusted for degrees of freedom.

   e. Estimate the number of cars owned by a family of six with an income of $12,000. Assume that this situation has been observed.

   f. Test the overall significance of the regression.

2. In a random sample of 23 firms in the same industry, the following quantities were obtained for each:

X_1 = research and development expenditures (millions of dollars)

X_2 = advertising expenditures on television (millions of dollars)

X_3 = all other advertising expenditures (millions of dollars)

Y = annual sales (millions of dollars)

Regression analysis yielded the following results:

$$\hat{Y} = -2.3 + 5.8X_1 + 4.2X_2 + 7.4X_3$$

$$(1.20) \quad (1.31) \quad (1.56)$$

The quantities in parentheses are the standard errors of the net regression coefficients. The standard error of estimate $S_{Y.123}$ was 12.4. The standard deviation of the dependent variable s_Y was 25.

   a. Interpret the net regression coefficient b_1.

   b. Test at the 1% level of significance whether each of the net regression coefficients is significantly different from zero.

   c. What is the expected effect when highly correlated independent variables are included in a multiple regression equation?

   d. Calculate the coefficient of multiple determination.

   e. Estimate the average annual sales for a firm that has research and development expenditures of $6 million, television advertising expenditures of $10 million, and all other advertising expenditures of $7 million.

   f. Assume the following standard deviations:

$$s_{X_1} = 15 \qquad s_{X_2} = 10 \qquad s_{X_3} = 5$$

Calculate the beta coefficients, β_1, β_2, and β_3. Interpret β_1.

3. The American Society of Monetarists is studying the relationship between the unemployment rate in percent (Y), the per capita money supply in dollars (X_1), and the logarithm of federal expenditures (X_2). This last variable was used as an indicator of the yearly percentage change in federal expenditures. The following information is an attempt to establish a least squares linear relationship using annual data for a 21-year period. The estimated relation was

$$\hat{Y} = a + b_1X_1 + b_2X_2$$

The following statistics were computed:

$$n = 21 \qquad\qquad a = 17$$
$$\Sigma(Y - \hat{Y})^2 = 28 \qquad\qquad b_1 = -0.003$$
$$\Sigma(Y - \bar{Y})^2 = 120 \qquad\qquad b_2 = -4$$
$$r_{12} \doteq 0.85 \qquad \Sigma(X_1 - \bar{X}_1)^2 = 800{,}000$$
$$r_{Y1} = -0.5 \qquad \Sigma(X_2 - \bar{X}_2)^2 = 100$$
$$r_{Y2} = -0.3$$
$$\Sigma(\hat{Y} - \bar{Y})^2 = 92$$

a. Calculate and interpret the standard error of estimate. What assumptions are necessary in using this measure? How many degrees of freedom are there?

b. Calculate the coefficient of multiple determination adjusted for degrees of freedom.

c. Calculate the standard errors of the net regression coefficients. Carry out appropriate tests of the statistical significance of the net regression coefficients. For each case, state the null and alternative hypotheses.

d. Estimate the percentage unemployed if the per capita money supply is $200 and the logarithm of federal expenditures is 2.00. Assume that these X_1 and X_2 values have been observed.

e. Carry out an appropriate test of the overall significance of the regression using a 1% level of significance.

4. The research department of the Inexperienced Investment Company provides the firm's account executives with information on options for use in managing clients' investment portfolios. An analysis of a valuation formula for a random sample of 43 options yielded the following results:

$$Y = \text{value of option (dollars)}$$
$$X_1 = \text{volatility of underlying stock}$$
$$X_2 = \text{time remaining before expiration of the option (months)}$$
$$\hat{Y} = a + b_1 X_1 + b_2 X_2$$

The statistics computed were

$$a = 3.0 \qquad \Sigma(Y - \hat{Y})^2 = 236$$
$$b_1 = 0.3 \qquad \Sigma(Y - \bar{Y})^2 = 741$$
$$b_2 = 0.8 \qquad \Sigma(\hat{Y} - \bar{Y})^2 = 622$$

a. Interpret the net regression coefficient b_2 in this multiple regression.

b. Under what circumstances is the reliability of the individual net regression coefficients of particular importance?

c. Calculate the standard error of estimate.

d. Calculate the coefficient of multiple determination adjusted for degrees of freedom.

e. Estimate the value of an option whose underlying stock has a price volatility of 20 and whose expiration date is six months away. Assume that such an option is included in the sample.

f. Using a 1% level of significance, perform an appropriate test of the overall significance of the regression.

5. A random sample of 16 graduates of a prestigious eastern business school was analyzed in an attempt to establish a relationship between starting salary (Y) and Graduate Management Admissions Test score (X_1). Y is expressed in thousands of dollars. The analysis resulted in the following estimated equation:

$$\hat{Y} = 4.2 + 0.031X_1$$

The statistics computed were

$$s_{Y.X} = 3$$
$$s_Y = 8$$
$$s_{b_1} = 0.008$$

In a second analysis, a second independent variable—number of years of work experience (X_2)—was added in an attempt to improve the explanatory power of the regression. The results were

$$\hat{Y} = 4.2 + 0.030X_1 + 1.05X_2$$

The statistics computed in the second analysis were

$$S_{Y.12} = 2.6$$
$$s_Y = 8$$
$$s_{b_1} = 0.009$$
$$s_{b_2} = 0.75$$

a. Compute the coefficient of determination for the first regression.

b. Test the significance of b_1 in the first regression. How many degrees of freedom are there?

c. Compute the coefficient of multiple determination adjusted for degrees of freedom in the second regression. Has the inclusion of the second independent variable explained very much of the variance in Y beyond that already accounted for by X_1 alone?

d. Interpret the coefficient b_1 in the second regression.

e. Test the significance of the regression coefficients in the second regression. Interpret the results carefully. What could account for these results?

6. The president of channel WTV divides advertising (in minutes per hour) into two classes; advertising of sponsors (X_1) and advertising of future network programs (X_2). As a consultant to the president, you collect data from 33 periods and run a regression with the number of viewers in tens

of thousands (Y) as the dependent variable and X_1 and X_2 as the independent variables. The results are summarized below:

Variable	Regression Coefficient	Standard Error of the Regression Coefficient
X_1	-0.25120	0.03411
X_2	-0.03015	0.00712

Intercept (a): 6.81255

Analysis of Variance Table

Source	Sum of Squares	Degrees of Freedom
Regression	3.5512	2
Error	2.8712	30
Total	6.4224	

a. Are the regression coefficients significantly different from zero? Test at the 1% level of significance.
b. Calculate the coefficient of multiple determination and the coefficient of multiple correlation. Which is easier to interpret?
c. Test the overall significance of the regression. Express the F statistic in mathematical terms. Describe in words the two quantities that are compared in the calculation of this statistic.

7. In a study of U.S. domestic oil companies, a regression analysis was carried out for 1990 on the following variables:

$$Y = \text{sales level}$$

$$X_1 = \text{asset size}$$

$$X_2 = \text{net income}$$

$$X_3 = \text{stockholders' equity}$$

a. What sources of multicollinearity do you see? The following results were found when regression analyses were carried out:

(1) $\hat{Y} = -3.29 + 0.8905X_1 + 7.9591X_2 - 0.4236X_3$

  t values: (2.47)   (1.07)   (0.40)

(2) $\hat{Y} = 1.433 + 0.8149X_1 - 5.561X_2$

  t values: (2.72)   (1.92)

Note: The numbers in parentheses under the slope coefficients are t values, not standard errors.

$$r_{13}^2 = 0.880 \qquad r_{23}^2 = 0.944$$

	(1)	(2)
F value	15.41	23.81
s_Y^2	4,621.00	4,621.00
Durbin-Watson	1.07	1.10

b.   What procedure was used to deal with multicollinearity?
c.   Was the procedure referred to in part (b) effective? Explain.

CHAPTER 10
REVIEW EXERCISES

Note: Access to a computer and a standard computer program for multiple regression analysis are needed to carry out these review problems.

1.   A multiple regression model was constructed to assist in analyzing freshman-year grade point averages (0 to 4) for a class at a certain midwestern university. A simple random sample of 20 students was drawn from the class that had just completed the freshman year. Freshman-year grade point averages (GPA) were designated as the dependent variable, and the independent variables were verbal SAT scores (V-SAT), mathematics SAT scores (M-SAT), rank in high school class (RANK), and high school class size (C-SIZE). The following observations were obtained from the sample:

STUDENT	GPA	V-SAT	M-SAT	RANK	C-SIZE
1	2.37	41	46	29	280
2	3.79	55	64	2	63
3	2.42	41	43	46	434
4	3.40	62	60	2	10
5	2.11	45	57	1	5
6	2.25	50	51	123	284
7	3.42	48	62	9	275
8	3.26	47	52	2	145
9	1.77	43	45	9	198
10	1.84	48	60	5	10
11	3.32	47	60	32	320
12	2.91	48	56	12	130
13	2.55	39	49	25	228
14	1.46	45	53	21	302
15	2.91	45	42	4	130
16	1.20	42	40	34	130
17	1.04	42	58	83	377
18	2.60	45	44	7	213
19	2.21	43	41	2	10
20	3.40	56	69	6	186

a. Using a computer, fit a multiple linear regression function to the data given above and obtain associated computer output.

b. Interpret the slope coefficients (b values) specifically in terms of the variables and units in this problem.

c. Test the significance of the slope coefficients using $\alpha = 0.05$.

d. Do the signs of the coefficients make sense? Explain your answer for each coefficient in one short sentence.

e. Test the overall significance of this regression model using $\alpha = 0.05$. State your conclusion.

f. Interpret the coefficient of multiple determination specifically in terms of this problem.

g. Suggest a reason why the slope coefficient for the mathematics SAT score variable is insignificant at the 0.05 level.

2. Revise the regression model used in Review Exercise 1 by deleting the mathematics SAT score variable and by creating a new variable, the ratio of rank to class size. That is, fit a multiple linear regression function to the following variables: GPA (dependent) and V-SAT and RANK/C-SIZE (independent).

a. Obtain associated computer output.

b. Using the computer output obtained in part (a), answer questions (b) through (f) of Exercise 1.

c. What can you conclude in the comparison of the models used in Exercises 1 and 2?

3. A multiple regression analysis was carried out to determine which team statistics (batting average, earned run average, and so on) for professional baseball teams were the most significant in determining win-loss percentages (proportion of games won). The following model was fitted using the indicated variables:

$$\hat{Y} = a + b_1 X_1 + b_2 X_2 + b_3 X_3 + b_4 X_4 + b_5 X_5 + b_6 X_6$$

$$Y = \text{win-loss percentage (W-L)}$$

$$X_1 = \text{batting average (BA)}$$

$$X_2 = \text{earned run average (ERA)}$$

$$X_3 = \text{total stolen bases (SB)}$$

$$X_4 = \text{total runs (R)}$$

$$X_5 = \text{total hits (H)}$$

$$X_6 = \text{home runs (HR)}$$

All variables represent team statistics for an entire season. The data follow:

TEAM	W-L	BA	ERA	SB	R	H	HR
1:	0.556	0.258	3.33	152	708	1404	133
2:	0.547	0.257	3.41	213	684	1390	115
3:	0.488	0.264	4.05	110	664	1461	72
4:	0.469	0.254	3.42	80	633	1404	121
5:	0.426	0.249	3.58	97	600	1351	79
6:	0.407	0.245	3.87	100	607	1332	86
7:	0.586	0.264	3.12	137	727	1435	149
8:	0.571	0.256	3.81	137	710	1378	136
9:	0.549	0.248	3.3	87	613	1331	117
10:	0.519	0.252	3.28	152	591	1349	75
11:	0.457	0.258	3.63	178	605	1408	70
12:	0.426	0.244	4.08	90	600	1312	123
13:	0.613	0.267	3.18	98	735	1489	125
14:	0.607	0.267	3.54	74	796	1463	172
15:	0.574	0.276	3.67	95	804	1530	173
16:	0.559	0.258	3.57	75	659	1397	154
17:	0.531	0.271	3.64	90	714	1520	129
18:	0.434	0.261	3.97	64	639	1400	106
19:	0.366	0.25	4.55	28	590	1358	98
20:	0.568	0.268	3.44	216	743	1469	98
21:	0.537	0.259	3.65	86	691	1417	108
22:	0.537	0.253	3.36	196	692	1353	132
23:	0.451	0.267	3.69	99	666	1472	82
24:	0.441	0.264	4.22	83	634	1423	106
25:	0.426	0.245	3.62	144	532	1304	100
26:	0.35	0.248	4.7	123	614	1327	97

a. Using a computer, fit the multiple linear regression model to the given data and obtain associated computer output.

b. Test the significance of the slope coefficients using $\alpha = 0.01$.

c. Test the overall significance of this regression model using $\alpha = 0.01$.

d. What percentage of the variation in win-loss percentages is accounted for by the regression relationship?

e. Revise the regression model that you have fitted, now using only earned run average, total stolen bases, and total runs in a linear regression equation.

f. What can you conclude in the comparison of the models fitted above?

MINITAB EXERCISES

Problems 1 through 3 are based on the data in the file named 'invest . mtw'. From the file, you can retrieve:

Column	Name	Description
C1	Y	Net Exports. Quarterly 1978–1985
C2	X	Gross National Product. Quarterly 1978–1985
C3	R	Interest Rate: US Treasury Constant Maturities, 30 years (% per year). Quarterly 1978–1985
C4	I	Gross Private Domestic Investment. Quarterly 1978–1985

(Source: *Citibase Series Data*, 1988)

Note: Economists generally denote Net Exports by X and GNP by Y. However, for consistency, we reverse this notation because Net Exports is treated here as a dependent variable and GNP as an independent variable.

1. One of the trade theories claims that $Y = a - bX - cR$, where $b > 0$ and $c > 0$. Obtain a regression equation by using

   MTB > regress C1 on 2 predictors C2 and C3

   and check whether the signs of the regression coefficients coincide with the theory. Test $H_0: c = 0$ against $H_1: c$ is not 0, at alpha $= 0.05$.

2. Consider Investment Theory (I):

   $$I = a - bR + cX \qquad \text{where} \quad b > 0 \quad \text{and} \quad c > 0$$

   Estimate all the coefficients of the above model with the data available. Test the significance of the slope for GNP, at alpha $= 0.01$.

3. Consider Investment Theory (II):

   $$I = a + b(X(-1) - X(-2))$$

   where $b > 0$

   $X(-i)$ denotes X lagged by i periods ($i = 1, 2$).

   Use Minitab to obtain a regression equation and test the significance of b, at alpha $= 0.10$. Which theory gets more support from the data, Investment Theory (I) or Investment Theory (II)? *Hint:* you may create a column to represent $X(-1) - X(-2)$ via

   MTB > let C10 = lag(C2) — lag(lag(C2))

4. File 'dummy . mtw' contains the following information:

Column	Name	Description
C1	IP	Industrial Production: Total Index (1977 $= 100$). Quarterly 1948.1–1985.2
C2	PXI	Expenditures for New Plant and Equipment: Total All Industries (Bil$). Quarterly 1948.1–1985.2

(Source: *Citibase Series Data*, 1988)

Obtain a scatterplot of C1 versus C2 and judge whether a simple straight line would fit the data well. Define a column, called DUMMY, by

```
MTB > set C3
DATA > 57(1) 93(0)
DATA > end
MTB > name C3 'DUMMY'
```

This dummy variable separates the series into two parts: 1948.1–1962.2 and 1962.3–1985.2. Then regress C1 on C2 and C3. Test the significance of the DUMMY variable and explain its role in the regression model.

Problems 5 through 8 are based on the data in the file named 'dow . mtw'. A description of the data is as follows:

Column	Name	Description
C1	U	Unemployment Rate. Monthly 1981.01–1987.12
C2	CPI	Consumer Price Index. 1981.01–1987.12
C3	E	Total Exports. 1981.01–1987.12
C4	I	Total Imports. 1981.01–1987.12
C5	DOW	Dow Jones Averages: Industrial (30 Stocks) Index. 1981.01–1987.12

(Source: *Slater-Hall Information & Business*, 1988)

5. Find the correlation of each pair of the independent variables U, CPI, E, and I, with the help of the command:

   ```
   MTB > correlation C1–C4
   ```

   and comment on the collinearity among the independent variables.

6. Regress C5 on the 4 predictors C1 to C4. Delete the predictor that yields the largest p value, and regress C5 on the remaining predictors to see whether the p values change.

7. Select a pair of predictors for DOW by means of stepwise regression analysis. *Hint:*

   ```
   MTB > stepwise regress C5 on 4 predictors C1–C4
   ```

8. Regress C5 on C1 through C4, only using the data before September, 1987. Compare the signs of regression coefficients with those in Problem 60 and comment on the influence of the last 4 observations. *Hint:* to delete the data after August, 1987, use

   ```
   MTB > copy C1–C5 to C1–C5;
   SUBS > use 1:80.
   ```

KEY TERMS

Note: Terms that have been included at the end of Chapter 9 for two-variable regression analysis have not been repeated here for multiple regression analysis.

Analysis of Variance   provides a test of the null hypothesis of no relationship between the dependent variable and all of the independent variables considered collectively.

Autocorrelation (serial correlation)   dependence among the successive observations of the dependent variable Y.

Dummy Variable   a qualitative variable included in regression analysis.

Durbin-Watson Test   a test for the presence of autocorrelation.

Multicollinearity (collinearity)   correlation among independent variables.

Net Regression Coefficient   slope coefficient (b value) in multiple regression analysis.

Outlier   an extreme observation in a regression analysis.

Stepwise Multiple Regression Analysis   a type of multiple regression analysis in which variables are entered (or deleted) in a one-at-a-time stepwise fashion.

KEY FORMULAS

Coefficient of Multiple Determination

$$R^2 = 1 - \frac{\Sigma(Y - \hat{Y})^2}{\Sigma(Y - \bar{Y})^2}$$

Coefficient of Multiple Determination Adjusted for Degrees of Freedom

$$R_c^2 = 1 - (1 - R^2)\left(\frac{n - 1}{n - k}\right)$$

Durbin-Watson statistic

$$d = \frac{\sum_{i=2}^{n} (u_i - u_{i-1})^2}{\sum_{i=1}^{n} u_i^2}$$

F-Test for the Analysis of Variance in Regression Analysis

$$F(v_1, v_2) \frac{\Sigma(\hat{Y} - \bar{Y})^2/(k - 1)}{\Sigma(Y - \hat{Y})^2/(n - k)}$$

Linear Multiple Regression Equation ($k - 1$ Independent Variables)

$$\hat{Y} = a + b_1 X_1 + b_2 X_2 + \cdots + b_{k-1} X_{k-1}$$

Standard Error of Estimate ($k - 1$ Independent Variables)

$$S_{Y.12\ldots(k-1)} = \sqrt{\frac{\Sigma(Y - \hat{Y})^2}{n - k}}$$

11

TIME SERIES

Decisions in private and public sector enterprises depend on perceptions of future outcomes that will affect the benefits and costs of possible alternative courses of action. Not only must managers forecast these future outcomes, but they must also plan and think through the nature of the activities that will permit them to accomplish their objectives. Clearly, managerial planning and decision making are inseparable from forecasting.

11.1
METHODS OF FORECASTING

Methods of forecasting vary considerably. For example, in forecasting customer demand, the prevailing methods include informal "seat of the pants" estimating, executive panels and composite opinions, consensus of sales force opinions, combined user responses, statistical techniques, and various other methods.

Because of such factors as the increased complexity of business operations, the need for greater accuracy and timeliness, the dependence of outcomes on so many different variables, and the demonstrated utility of the techniques, management is increasingly turning to formal models—such as those provided by statistical methods—for assistance in the difficult task of peering into the future. A widely applied and extremely useful set of procedures is time-series analysis.

Statistical method of forecasting

> A **time series** is a set of statistical observations arranged in chronological order.

Examples include a weekly series of end-of-week stock prices, a monthly series of amounts of steel production, and an annual series of national income. Such time series are essentially historical series, whose values at any given time result from the interplay of large numbers of diverse economic, political, social, and other factors.

A first step in the prediction of any series involves an examination of past observations.

> Time-series analysis deals with the methods of analyzing past data and then projecting the data to obtain estimates of future values.

The traditional, or "classical," methods of time-series analysis, also referred to as "time-series decomposition methods," are descriptive in nature and do not provide for probability statements concerning future events. However, these time-series models, although admittedly only approximate and not highly refined, have proven their worth when cautiously and sensibly applied. It is important to realize that these methods cannot simply be used mechanistically but must at all times be supplemented by sound subjective judgment.

Although this introduction has referred to the use of time-series analysis for the purposes of forecasting, planning, and control, these procedures also are often used for the simple purpose of historical description and analysis. For example, they may be usefully employed in an analysis in which interest centers on the differences in the nature of variations in different time series. The classical time-series model is described in the next section.

11.2
THE CLASSICAL TIME-SERIES MODEL

If we wished to construct an ideally satisfying mathematical model of an economic time series, we might seek to define and measure the many factors determining the variations in the time series and then proceed to state the mathematical relationships between these and the particular series in question.

Determinants
of change

However, the determinants of change in an economic time series are multitudinous, including such factors as changes in population, consumer tastes, technology, investment or capital-goods formation, weather, customs, and numerous other economic and noneconomic variables. The enormity and impracticability of the task of measuring all these factors and then relating them mathematically to an economic time series militates against the use of this direct approach to time-series analysis. Hence, it is not surprising that a more indirect and practical approach has come into use. **Classical time-series analysis** is essentially a descriptive method that attempts to break down an economic time series into distinct components representing the effects of groups of explanatory factors such as those given earlier. These component variations are

1. Trend
2. Cyclical fluctuations
3. Seasonal variations
4. Irregular movements

Components of classical time series

Trend refers to a smooth upward or downward movement of a time series over a long period of time. Such movements are thought of as requiring a minimum of *about 15 or 20 years* to describe, and as being attributable to factors such as population change, technological progress, and large-scale shifts in consumer tastes.

Trend

Cyclical fluctuations, or business cycle movements, are recurrent upward and downward movements around trend levels that have a duration of anywhere from *about 2–15 years*. There is no single simple explanation of business cycle activity, and there are different types of cycles of varying length and size. Not surprisingly, no satisfactory mathematical model has been constructed for either describing or forecasting these cycles, and perhaps none will ever be.

Cyclical fluctuations

Seasonal variations are cycles that complete themselves within the period of *a calendar year* and then continue to repeat this basic pattern. The major factors producing these annually repetitive patterns of seasonal variations are weather and customs, the latter term broadly interpreted to include observance of various holidays such as Easter and Christmas. Series of monthly and quarterly data are ordinarily used to examine these seasonal variations. Hence, regardless of trend or cyclical levels, we can observe in the United States that each year more ice cream is sold during the summer months than during the winter, whereas more fuel oil for home heating is consumed in the winter than during the summer months. Both of these cases illustrate the effect of weather or climatic factors in determining seasonal patterns. Department store sales generally reveal a minor peak during the month in which Easter occurs and a larger peak in December, when Christmas occurs, reflecting the shopping customs of consumers. The techniques of measurement of seasonal variations that we will discuss are particularly well suited to the measurement of relatively stable patterns of seasonal variations, but they can be adapted to cases of changing seasonal movements as well.

Seasonal variations

Irregular movements are fluctuations in time series that are short in duration, erratic in nature, and follow no regularly recurrent or other discernible pattern. These movements are sometimes referred to as **residual variations**, since, by definition, they represent what is left over in an economic time series after trend, cyclical, and seasonal elements have been accounted for. Irregular fluctuations result from sporadic, unsystematic occurrences such as erratic shifts in purchasing habits, accidents, strikes, and the like. Whereas in the classical

Irregular movements

FIGURE 11-1

The components of a
time series

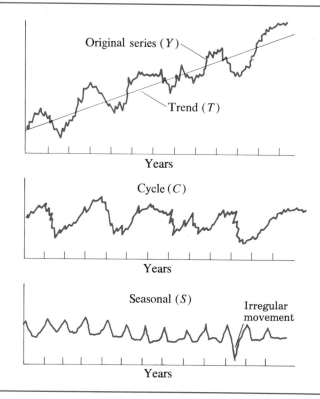

time-series model, the elements of trend, cyclical fluctuations, and seasonal variations are viewed as resulting from systematic influences leading to either gradual growth, decline, or recurrent movements, irregular movements are considered to be so erratic that it would be fruitless to attempt to describe them in terms of a formal model.

Figure 11-1 presents the typical patterns of the trend, cyclical, and seasonal components of a time series. Time is plotted on the horizontal axes of the graphs, and the values of the particular series (which might be retail sales, ice cream production, or airline revenues) are plotted on the vertical axes.

11.3
DESCRIPTION OF TREND

In the preceding section, we pointed out that the classical model involves the separate statistical treatment of the component elements of a time series. We begin our discussion by indicating how the description of the underlying trend is accomplished.

Before the trend of a particular time series can be determined, it is generally necessary to subject the data to some preliminary treatment. The amount of such adjustment depends to some extent on the period for which the data are stated. For example, if the time series is in monthly form, it may be necessary to revise the monthly data to account for the differing number of days per month. This may be accomplished by dividing the monthly figures by the number of days in the respective months, or by the number of working days per month, to state the data for each month on a per day basis. *Preliminary treatment of data*

Even when the original data are in annual form, which is often the case when primary interest is centered on the long-term trend of the series, the data may require a considerable amount of preliminary treatment before a meaningful analysis can be carried out. Adjustments for changes in population size are often made by dividing the original series by population figures to state the series in per capita form. Frequently, comparisons of trends in per capita figures are far more meaningful than corresponding comparisons in unadjusted figures.

It is particularly important to scrutinize a time series and adjust it for differences in definitions of statistical units, the consistency and coverage of the reported data, and similar items. It is important to realize that we cannot simply proceed in a mechanical fashion to analyze a time series. Careful and critical preliminary treatment of such data is required to ensure meaningful results.

Matching a Measuring Technique to the Purpose of the Analysis for Fitting Trend Lines

The trend in a time series can be measured by the free-hand drawing of a line or curve that seems to fit the data, by fitting appropriate mathematical functions, or by the use of "moving average" methods. Moving averages are discussed later in this chapter in connection with seasonal variations.

A **free-hand curve** may be fitted to a time series by visual inspection. When this type of characterization of a trend is employed, the investigator is usually interested in a quick description of the underlying growth or decline in a series, without any careful further analysis. In some instances, this rapid graphic method may suffice. However, it clearly has certain disad6antages. Different investigators would surely obtain different results for the same time series. Indeed, even the same analyst would probably not sketch exactly the same trend line in two different attempts on the same series. This excessive amount of s5bjectivity in choice of a trend line is especially problematic if further quantitative analysis is planned. In the ensuing discussion, we will concentrate on mathematically fitted trend lines. *Quick graphic description*

Even in the **mathematical measurement** of trend, the *purpose* of the analysis is of considerable importance in the selection of the appropriate trend line. Several different types of purposes can be specified.

Historical description   ● Trend lines may be fitted for the purpose of historical description. If so, any line that fits well will suffice. The line need not have logical implications for forecasting purposes, nor should it be evaluated primarily by characteristics that might be desirable for other purposes.

Future projection   ● If prediction or projection into the future is the purpose, particularly if long-term projection is desired, the selected line should have logical implications when it is extended into the future. The analyst must always carefully weigh the implications of the models being projected into the future as regards their reasonableness for the phenomena being described and predicted. For example, constant amounts of growth per unit time are implied if we project a *straight-line* trend into the future. This may not be a reasonable long-term projection for many series.

Study of nontrend elements   ● Trend lines are fitted to economic data to describe and eliminate trend movements from the series in order to study nontrend elements. Thus, if the analyst's primary interest is to study cyclical fluctuations, freeing the original data of trend makes it possible to examine cyclical movements without the presence of the trend factor. For this purpose, any type of trend line that does a reasonably good job of bisecting the individual business cycles in the data would be appropriate.

Types of Trend Movements

There have been considerable variations in the trend movements of different economic and business series. Over long periods of time, some companies and industries have experienced periods of growth and then have gone into steep declines when more modern competitive processes and products have emerged in other companies and industries. Real GNP in the United States has exhibited a relatively constant rate of growth of about 3% per year since the early part of this century. Since real GNP is a measure of overall economic activity, it represents a type of average of all series for production of goods and services. Although many series have shown more or less similar trends to that of real GNP, sharp divergences have also occurred. One example of these differential movements is that the service sector of the economy has been growing relative to the agricultural sector. While employment in the service industries is increasing, the number of persons employed in agriculture is decreasing.

For a large number of American industries, numerous studies have revealed a trend that may be characterized as increasing at a decreasing percentage rate. Indeed, some investigators have adapted growth curves originally used to describe biological growth to depict the past change of many industrial series. *Growth curves*   These **growth curves** (of which the *Gompertz* and *logistic* are the best known) are S-shaped for increasing series plotted on graph paper with an arithmetic vertical scale, and are concave downward on a semilogarithmic chart. By convention, we graph time series on either arithmetic or semilogarithmic paper by plotting the variable of interest on the vertical axis and time on the horizontal

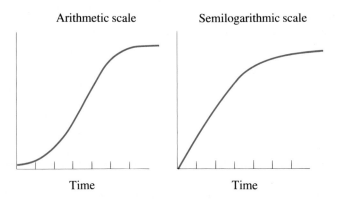

Arithmetic scale   Semilogarithmic scale

Time   Time

FIGURE 11-2

A growth curve plotted
on arithmetic and on
semilogarithmic paper

axis.[1] The general shape of such a growth curve, plotted on arithmetic graph paper and on semilogarithmic graph paper, is depicted in Figure 11-2. The so-called "law of growth" has been used to describe this type of change over time in an industry. On the arithmetic chart, in the early stages of the industry, the growth is slow at first and then becomes rapid, with the series increasing by increasing amounts. Then the industry moves through a point beyond which it increases by decreasing amounts, and finally it tapers off into a period of "maturity." Throughout all stages, as seen on the semilogarithmic chart, although the industry is growing, the increases are at a decreasing *percentage rate*. Various reasons for this type of industrial growth were propounded by the investigators who discerned analogous changes in biological and industrial time-series data. Since the equations of these growth curves are exponential in character, it is difficult to fit some of these curves by the method of least squares described later in this chapter.[2]

Law of growth

We indicated earlier that a great variety of types of trend movements exist in economic time series, and many of these trends cannot be adequately described or projected by means of growth curves. The growth curves have a number of desirable characteristics. For example, they have finite lower and upper limits, which are determined by the data to which the curves are fitted. However, no one family of curves is apt to be generally satisfactory for trend fitting purposes. In fact, the growth curves have been found to be quite inadequate for industrial growth prediction. The most commonly used polynomial-type trend lines, which are fitted by the method of least squares, are discussed in this chapter.

[1] The familiar arithmetic graph has an arithmetically ruled vertical scale on which equal vertical distances represent equal *amounts* of change. Semilogarithmic graphs have an arithmetic horizontal scale, but a logarithmically ruled vertical scale on which equal vertical distances represent equal *percentage rates* of change. For example, on an arithmetic graph, a straight line inclined upward depicts a series that is increasing with constant amounts of change. On a semilogarithmic graph, a straight line inclined upward depicts a series that is increasing at a constant percentage rate.

[2] A special technique known as "the method of selected points" is sometimes used for this purpose. Also, logarithmic transformations are often used.

> The purpose of fitting, the goodness of fit obtained, knowledge of the growth and decay processes involved, and trial-and-error experimentation are all essential ingredients in the selection of the appropriate trend line.

11.4
FITTING TREND LINES BY THE METHOD OF LEAST SQUARES

For situations in which it is desirable to have a mathematical equation to describe the trend of a time series, the most widely used method is to fit some form of polynomial function to the data. In this section, we illustrate the general method by means of simple examples, fitting a straight line and a second-degree parabola to time-series data by the method of least squares. We use the Minitab regression analysis program for the fitting process, treating time as the independent variable and the factor whose trend we wish to describe as the dependent variable.

The Method of Least Squares

Major difficulty

The method of least squares, when used to fit trend lines to time-series data, is employed mainly because it is a simple, practical method that provides the best fits according to a reasonable criterion. However, we should recognize that the method of least squares does not have the same type of theoretical under-pinning when applied to fitting trend lines as when used in regression and correlation analysis, described in Chapters 9 and 10. The major difficulty is that the usual probabilistic assumptions made in regression and correlation analysis are simply not met in the case of time-series data. For example, the illustrative problem in Chapter 9 involving the relationship between food expenditures and income for a sample of families included two possible theoretical models. In the first, both food expenditures and income were random variables; in the second, income was a controlled variable, that is, families of prespecified incomes were selected, and food expenditures was a random variable. In each model, the dependent variable was a random variable, and the model assumed conditional probability distributions of this random variable around the computed values of the dependent variable, which fell along the regression line. These computed Y values were the means of the conditional probability distributions. A number of assumptions are implicit in this type of model: Deviations from the regression line are considered to be random errors describable by a probability distribution. The successive observations of the dependent variable are assumed to be independent. For example, Family B's expenditures were assumed to be independent of Family A's, and so on.

> In fitting trend lines to time-series data, the probabilistic assumptions of the method of least squares are not met.

Fitting trend lines to time-series data

If a trend line is fitted, for example, to an annual time series of department store sales, time is treated as the independent variable X and department store sales as the dependent variable Y. It is not reasonable to think of the deviation of actual sales in a given year from the computed trend value as a random error. Indeed, if the original data are annual, then deviations from trend would represent the operation of cyclical and irregular factors. (Seasonal factors would not be present in annual data, because by definition they complete themselves within a year.) Finally, the assumption of independence is not met in the case of time-series data. A department store's sales in a given year surely are not independent of what they were in the preceding year. In summary, we return to the point made at the outset of this discussion:

> The method of least squares when used to fit trend lines is employed primarily because of its practicality, simplicity, and good fit characteristics rather than because of its justification from a theoretical viewpoint.

Fitting an Arithmetic Straight-Line Trend

As an example we will fit a straight line by the method of least squares to an annual series on local advertising sales of United States television stations from 1980 to 1994. Although we wrote the equations of a straight line in the discussion of regression analysis in Chapter 9 as $\hat{Y} = a + bX$, in time-series analysis we will use the equation

(11.1)
$$\hat{Y}_t = a + bt$$

The notation $\hat{Y}_t$ means computed trend value for time period "t." When a regression equation is used as a trend line, usually the independent variable "time" is transformed to a simple variable with fewer digits. For example, in our first trend illustration for annual data, instead of using the years 1980, 1981, ... as the values of the time variable, we simply use the digits 1, 2, ... Table 11-1 shows selected output of a Minitab regression analysis for the aforementioned fit of a straight line trend to an annual series on local advertising sales of United States television stations, 1980–1994. Using the notation of equation 11.1, we can write the trend equation LOCSALES = 1435.4 + 560.17 TIME as

$$\hat{Y}_t = 1435.4 + 560.17t$$

where $\hat{Y}_t$ is the trend figure for local advertising sales in millions of dollars

$$t = 1 \text{ in } 1980$$

t is in one-year intervals

TABLE 11-1
Straight-line trend fitted to local advertising sales of United States television stations, 1980–1994

```
MTB > BRIEF 3
MTB > REGRESS C2 1 C1

THE REGRESSION EQUATION IS
LOCSALES = 1435 + 560 TIME
```

PREDICTOR	COEF	STDEV	T-RATIO	P
CONSTANT	1435.4	159.0	9.03	0.000
TIME	560.17	17.49	32.03	0.000

S = 292.6   R-SQ = 98.7%   R-SQ (ADJ) = 98.7%

(1) OBS.	(2) YEAR	(3) TIME	(4) LOCSALES	(5) FIT	(6) %TREND
1	1980	1.0	2484.0	1995.6	124.5
2	1981	2.0	2767.0	2555.8	108.3
3	1982	3.0	3088.0	3115.9	99.1
4	1983	4.0	3611.0	3676.1	98.3
5	1984	5.0	4216.0	4236.3	99.5
6	1985	6.0	4665.0	4796.5	97.3
7	1986	7.0	5275.0	5356.6	98.5
8	1987	8.0	5616.0	5916.8	94.9
9	1988	9.0	6165.0	6477.0	95.2
10	1989	10.0	6720.0	7037.1	95.5
11	1990	11.0	7400.0	7597.3	97.4
12	1991	12.0	7975.0	8157.5	97.8
13	1992	13.0	8800.0	8717.7	100.9
14	1993	14.0	9520.0	9277.8	102.6
15	1994	15.0	10450.0	9838.0	106.2

Sources: FCC television financial data (1980); TELEVISION/RADIO AGE "TV Business Barometer" (1981–1987); Dick Gideon Enterprises (1988–1994)

The identifying notes below the trend equation, such as $t = 1$ in 1980, are useful in communicating information about the trend line. The figures for R^2 and R_c^2, both of which are 98.7% indicate that a straight line is a very good fit to this TV advertising sales series. For convenience, the actual years have been included in the table next to the time variable used in the regression analysis. The FIT column displays the computed trend figures $(\hat{Y}_t)$ for the specified time periods.

The constants in the trend equations are interpreted in a similar way to those in the straight line discussed in regression analysis: a is the computed trend figure for the period when $t = 0$, in this case, 1979: b is the slope of the trend line, or the amount of change in $\hat{Y}_t$ per unit change in t (per year in the present example). The trend figures (FIT) are determined by substituting the appropriate

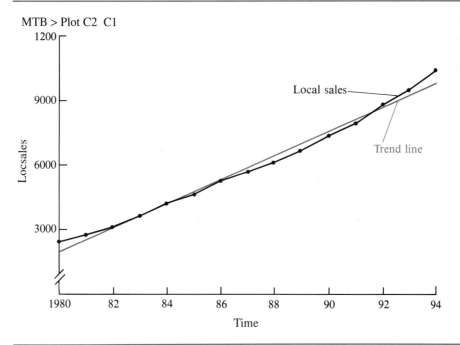

MTB > Plot C2 C1

FIGURE 11-3
Straight-line trend fitted to local advertising sales of United States television stations, 1980–1994

values of t into the trend equation. For example, the trend figure for 1991 is

$$\hat{Y}_{t,1991} = 1435.4 + 560.17(12) = 8157.4 \text{ millions of dollars}$$

The trend line is graphed in Figure 11-3. For convenience, the points plotted by Minitab have been connected by lines to show the configuration of the annual data.

Projection of the Trend Line

Projections of the computed trend line can be obtained by substituting the appropriate values of t into the trend equation. For example, the projected trend figure for 1995 for local advertising sales of U.S. television stations would be computed by substituting $t = 16$ in the previously determined trend equation. Hence,

$$\hat{Y}_{t,1995} = 1435.4 + 560.17(16) = \$10,398.12 \text{ (million)}$$

A rougher estimate of this trend figure would be obtained by extending the straight line graphically in Figure 11-3 to the year 1995. Remember that these projections are estimates of only the trend level in 1995 and not of the actual figure for local advertising sales of U.S. television stations in that year. If a prediction of the latter figure were desired, estimates of the nontrend factors would have to be combined with the trend estimate. This means that a prediction of cyclical fluctuations would have to be made and incorporated with the trend figure. Accurate forecasts of this type are difficult to make over extended time periods. However, insofar as managerial applications of trend analysis are concerned, for long-range planning purposes often all that is desired is a projection of the trend level of the economic variable of interest. For example, a good estimate of the trend of demand would be adequate for a business planning a plant expansion to anticipate demand many years into the future. Accompanying predictions of business cycle standings many years into the future would not be required; nor, for that matter, would they be realistically feasible.

Purpose

Cyclical Fluctuations

As previously indicated, when a time series consists of annual data, it contains trend, cyclical, and irregular elements. The seasonal variations are absent, since they occur within a year. Hence, deviations of the actual annual data from a computed trend line are attributable only to cyclical and irregular factors. Since the cyclical element is the dominant factor, a study of these deviations from trend essentially represents an examination of business cycle fluctuations. The deviations from trend are most easily observed by dividing the original data by the corresponding trend figures for the same period. By convention, the result of dividing an original figure by a trend value is multiplied by 100 to express the figure as a **percentage of trend**. Hence, if the original figure is exactly equal to the trend figure, the percentage of trend is 100; if the original figure exceeds the trend value, the percentage of trend is above 100; and if the original figure is less than the trend value, the percentage of trend is below 100.

Percentage of trend

The formula for percentage of trend figures is

$$(11.2) \qquad \text{Percentage of trend} = \frac{Y}{\hat{Y}_t} \cdot 100$$

where Y = annual time-series data
$\hat{Y}_t$ = trend values (FIT in Minitab)

When converted to percentage of trend, the data contain only cyclical and irregular movements, since the division by trend eliminates that factor. A so-called multiplicative model for the analysis provides the rationale of this procedure. That is, the original annual figures are viewed as representing the combined effects of trend, cyclical, and irregular factors. In symbols, let T, C, and I represent trend, cyclical, and irregular factors, respectively, and let Y and $\hat{Y}_t$ mean the same as in equation 11.2. Dividing the original time series by the corresponding trend values yields

$$(11.3) \qquad \frac{Y}{\hat{Y}_t} = \frac{T \times C \times I}{T} = C \times I$$

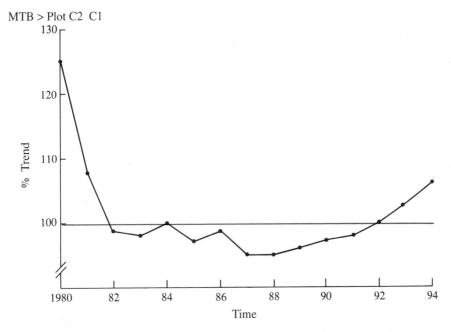

MTB > Plot C2 C1

The percentages of trend for the series on network compensation to television stations are given in column (6) of Table 11-1 and are plotted in Figure 11-4. In the figure, the upward trend movement is no longer present. Instead, the percentage of trend series fluctuates about the line labeled 100, which is the **trend level**. These percentages of trend are sometimes referred to as cyclical **relatives**; that is, the original data are stated relative to the trend figure. (Of course, strictly speaking, $Y/\hat{Y}_t$ is the cyclical relative, and the multiplication by 100 converts the relative to a percentage figure.) Another way of depicting cyclical fluctuations is in terms of relative cyclical residuals, which are percentage deviations from trend and are computed by the formula

Relative cyclical residuals

(11.4)
$$\text{Relative cyclical residual} = \frac{Y - \hat{Y}_t}{\hat{Y}_t} \cdot 100$$

For example, if we refer to the local advertising sales data in Table 11-1 for 1980, the actual figure is 2484.0, the computed trend value is 1995.6, and the percentage of trend is 124.5. The relative cyclical residual in this case is $+24.5\%$ indicating that actual network compensation is 24.5% above the trend figure because of cyclical and irregular factors. These residuals are positive or negative depending on whether the actual time-series figures fall above or below the computed trend values. The graph of relative cyclical residuals is visually identical to that of the percentage of trend values except that relative cyclical residuals are shown as fluctuations around a zero base line rather than around a base line of 100%.

Familiar charts   The familiar charts of business cycle fluctuations that often appear in publications such as the financial pages of newspapers and business periodicals are usually graphs of either percentages of trend or relative cyclical residuals. These charts may be studied for timing of peaks and troughs of cyclical activity, for amplitude of fluctuations, for duration of periods of expansion and contraction, and for other items of interest to the business cycle analyst.

Fitting a Second-Degree Trend Line

The preceding discussion on the fitting of a straight line pertains to the case in which the trend of a time series can be characterized as increasing or decreasing by constant amounts per time period. Actually very few economic time series exhibit this type of constant change over a long period of time (say, over a period of several business cycles). Therefore, it generally is necessary to fit other types of lines or curves to the given time series. Polynomial functions are particularly convenient to fit by the method of least squares. Frequently, a second-degree parabola provides a good description of the trend of a time series. In this type of curve, the amounts of change in the trend figures $\hat{Y}_t$ may increase or decrease per time period.

Second-degree
parabola

> A second-degree parabola may provide a good fit to a series whose trend is increasing by increasing amounts, increasing by decreasing amounts, and so on.

The procedure of fitting a parabola by the method of least squares involves the same general principles as fitting a straight line. We illustrate the method of fitting a second-degree parabola to a time series in Example 11-1.

EXAMPLE 11-1

A time series on national/regional spot sales of United States television stations, 1961–1990 appears in Table 11-2. This time series is also graphed in Figure 11-5. The trend of these data may be described as increasing by increasing amounts and can be represented as a second-degree parabola. The general form of a second-degree parabola is $\hat{Y}_t = a + bt + ct^2$.[3]

We have computed the trend equation for the national/regional spot sales (natreg) using Minitab by placing the sales data in column 1, time in column 2, and time squared

---

[3] The derivative of the second-degree parabola trend equation is

$$\frac{d\hat{Y}_t}{dt} = b + 2ct$$

Hence, the slope of the curve differs at each period t. When $t = 0$, $\frac{d\hat{Y}_t}{dt} = b$. Therefore, the slope at the time origin is b. The second derivative is $\frac{d^2\hat{Y}_t}{dt^2} = 2c$. Thus, the acceleration, or rate of change in the slope, is $2c$ per time period.

TABLE 11-2
Second-degree parabola fitted to national/regional spot sales of United States television stations, 1961–1990

```
MTB > BRIEF 3
MTB > REGRESS C1 2 C2,C3

THE REGRESSION EQUATION IS
NATREG = 1210 - 187 TIME + 14.2 TIMESQ
```

PREDICTOR	COEF	STDEV	T-RATIO	P
CONSTANT	1210.5	155.6	7.78	0.000
TIME	-186.80	23.14	-8.07	0.000
TIMESQ	14.1766	0.7244	19.57	0.000

```
S = 265.4   R-SQ = 98.9%   R-SQ (ADJ) = 98.8%
```

OBS.	YEAR	TIME	NATREG	FIT
1	1961	1.0	480.0	1037.8
2	1962	2.0	554.0	893.6
3	1963	3.0	616.0	777.6
4	1964	4.0	711.0	690.1
5	1965	5.0	786.0	630.9
6	1966	6.0	872.0	600.0
7	1967	7.0	872.0	597.5
8	1968	8.0	998.0	623.3
9	1969	9.0	1108.0	677.5
10	1970	10.0	1092.0	760.1
11	1971	11.0	1013.0	871.0
12	1972	12.0	1167.0	1010.2
13	1973	13.0	1221.0	1177.9
14	1974	14.0	1329.0	1373.8
15	1975	15.0	1441.0	1598.1
16	1976	16.0	1920.0	1850.8
17	1977	17.0	1960.0	2131.8
18	1978	18.0	2326.0	2441.2
19	1979	19.0	2564.0	2778.9
20	1980	20.0	2920.0	3145.0
21	1981	21.0	3302.0	3539.5
22	1982	22.0	3846.0	3962.3
23	1983	23.0	4211.0	4413.4
24	1984	24.0	4716.0	4892.9
25	1985	25.0	5221.0	5400.8
26	1986	26.0	5769.0	5937.0
27	1987	27.0	6329.0	6501.5
28	1988	28.0	7221.0	7094.4
29	1989	29.0	8044.0	7715.7
30	1990	30.0	8881.0	8365.3

Sources: FCC television financial data (1961–1980) "TELEVISION/RADIO AGE" TV Business Barometer (1981–1983) Dick Gideon Enterprises (1984–1990)

FIGURE 11-5

Second-degree parabola fitted to national/regional spot sales of United States television stations, 1961–1990

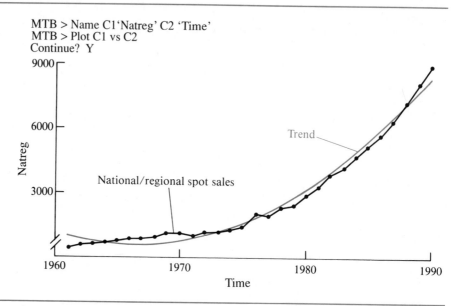

```
MTB > Name C1'Natreg' C2 'Time'
MTB > Plot C1 vs C2
Continue? Y
```

in column 3. Then, as shown in Table 11-2, we regressed the sales data in column 1 on the time and time squared figures in columns 2 and 3 respectively. In Table 11-2 are shown the Minitab commands for the regression, the regression equation, and associated output. Also listed are the years, the time variable (1, 2, ..., 30), the national/regional sales variable and the computed trend figures (Fit).

Dangers of mechanistic projections   Although the second-degree parabola appears from Figure 11-5 to provide a reasonably good fit to the data in this example, the dangers of a mechanistic projection of a trend line are clearly illustrated. The projected trend figures imply annual increases by increasing amounts. Therefore, we should not entertain the notion of extending the trend line into the future for forecasts unless an analysis of the underlying factors determining the trend of this series revealed reasons for a continuation of this type of trend.

Fitting Logarithmic Trend Lines

As discussed earlier, the equations of trend lines embody assumptions concerning the type of change that takes place over time. The arithmetic straight line assumes a trend that increases or decreases by constant amounts, whereas the second-degree parabola assumes that the change in these amounts of change is constant per unit time. It is often useful to describe the trend of an economic time series in terms of the percentage rates of change that are taking place, for example, by the use of logarithmic trend lines.

> If a time series increases at exactly a constant percentage rate, a straight line fitted to the logarithms of the data constitutes a perfect fit.

Constant percentage rate of increase or decrease

For example, suppose a time series has successive values of 10, 100, 1,000, and 10,000. This series may be characterized as increasing at a constant percentage rate of 900%. The logarithms of these values are log 10 = 1, log 100 = 2, log 1,000 = 3, and log 10,000 = 4. Hence, the logarithms are increasing by a constant amount, namely one unit, and a straight line could be drawn through the numbers 1, 2, 3, and 4. Some economic time series in the United States, although not changing exactly at a constant rate, have exhibited trends of approximately constant percentage increases over substantial periods of time. The equation of the logarithmic straight line that would describe the trend of such series is

(11.5)
$$\log \hat{Y}_t = a + bt$$

The method of fitting this line is the same as for the arithmetic straight line, except that logarithms of the data series have to be taken and then a straight line is fitted to these log Y values.

After the trend line has been calculated, trend figures are determined by substituting values of t into the trend equation, computing log $\hat{Y}_t$, and taking the antilogarithm to obtain $\hat{Y}_t$. Although we will not present another example of the fitting process, since there really are no new principles involved, we will illustrate the calculation of a trend figure for the type of trend line under discussion.

Suppose the logarithmic trend line for a particular series had been determined as

$$\log \hat{Y}_t = 2.3657 + 0.0170t$$

Then the logarithm of the trend figure for the year in which $t = 2$ would be given by substituting this value of t into the trend equation to obtain

$$\log \hat{Y}_t = 2.3657 + 0.0170(2)$$
$$\log \hat{Y}_t = 2.3997$$

Taking the antilog of this value yields the trend figure

$$\hat{Y}_t = \text{antilog } 2.3997 = 251.0$$

The rate of change implied by this trend line can be obtained by calculating antilog $b - 1$. For example, in the above illustration the antilog of the slope coefficient b is

$$\text{antilog } 0.0170 = 1.040$$

This figure is the ratio of each trend figure to the preceding one. Subtracting 1.00 from this figure yields 1.04 − 1.00 = 0.04. Hence, the trend figures increase by 4% per time period. If the series had been a declining one and the result of the above calculation was −0.04, this would mean that the trend figures decrease by 4% per time period.

Selecting constant-rate segments	A time series sometimes can be broken down into segments during which the rate of change has been approximately constant—even though the complete series does not exhibit a trend with constant rates of change throughout.

It is often useful in such cases to make comparisons of the rates of change similarly determined from different economic time series of interest.

Logarithmic second-degree parabolas   Logarithmic second-degree parabolas can also be fitted to time series in which the trend is increasing at an increasing percentage rate, increasing at a decreasing percentage rate, and so on. Ordinarily polynomials of third or higher degree are not fitted to time series in either arithmetic or logarithmic form because such curves permit too many changes in direction and tend to follow the cyclical fluctuations in the data as well as the trend. Therefore, these curves often do not have the required trend line characteristic of depicting the smooth, continuous movement underlying the cyclical swings in a time series.

Finding an appropriate trend line   In the attempt to find an appropriate trend line, the analyst should always plot the time series on both arithmetic and semilogarithmic graph paper. These two types of graphs may help in determining whether an arithmetic or logarithmic line would provide a better description of the trend.

EXERCISES 11.4

1. a. Discuss the nature and causes of the component variations of an economic time series.
    b. What are the typical ways of expressing the cyclical component of a time series?

2. The following table gives the total output of services for the United States for the period 1975–1988 in billions of 1982 constant dollars.

Year	GNP-Services (billions of 1982 dollars)	Year	GNP-Services (billions of 1982 dollars)
1975	$1286.4	1982	$1547.5
1976	1324.4	1983	1585.5
1977	1368.7	1984	1625.2
1978	1426.9	1985	1684.3
1979	1478.6	1986	1738.1
1980	1511.1	1987	1801.1
1981	1533.4	1988	1855.4

   a. Fit a straight line by the method of least squares to this annual series.
   b. Compute the percentage of trend for each year. What is accomplished in the conversion of the annual series into a percentage of trend series?
   c. Graph the trend line and the actual data against time.

3. Follow the instructions given in parts (a) and (b) of exercise 2 for the following economic series. The series shown is more inclusive than that in exercise 2, and the data are expressed in current dollars rather than constant dollars.

Year	GNP-Services (current dollars)	Year	GNP-Services (current dollars)
1975	$ 725.2	1982	$1547.5
1976	803.5	1983	1682.5
1977	895.9	1984	1813.9
1978	1003.0	1985	1968.3
1979	1121.9	1986	2118.4
1980	1265.0	1987	2295.7
1981	1415.4	1988	2478.4

Graph this trend line and the actual data against time. Which trend line appears to fit the data more closely, the trend line for the constant-dollar series or that for the current-dollar series?

4. Obtain annual trend figures for the 1990–1995 period by projecting a straight-line trend for U.S. federal government purchases of goods and services in 1982 dollars using the data in the following table:

Year	Purchases ($ billions)	Year	Purchases ($ billions)
1980	246.9	1985	326.0
1981	259.6	1986	333.4
1982	272.7	1987	339.0
1983	275.1	1988	326.1
1984	290.8		

5. The following table presents the total population of the United States taken during census years:

Year	Total Population (billions)
1930	123
1940	132
1950	152
1960	181
1970	205
1980	222

a. What was the average increase per decade of U.S. population between 1930 and 1980? (Do not use a trend line.)
b. Fit a linear trend by the method of least squares to the original data.
c. Is the answer to part (a) consistent with the slope obtained for the original data in part (b)? Why or why not?

d. In 1975, total U.S. population was 214 billion. Compute and interpret the absolute cyclical residual and the relative cyclical residual for that year, using the straight line fitted to the original data in part (b).

6. The following trend equation resulted from the fitting of a least squares second-degree parabola to labor force data from an eastern county:

$$\hat{Y}_t = 202.53 + 6.85t - 0.52t^2$$

where $t = 1$ in 1975
   t is in 1 year intervals
   Y is the size of the labor force (thousands)

a. Assume that this trend line is "a good fit," What generalizations can you make concerning the way in which the labor force of this county has grown in absolute amounts? Concerning the percentage rate at which it has grown?

b. In part (a), you were instructed to assume that the trend line was a good fit. However the actual size of the labor force in 1990 was 205,000, striking evidence that the equation is not a good fit. Do you agree? Discuss.

7. Assume that you are analyzing the loan portfolio of a company and that you wish to determine the cyclical fluctuations of short-term versus long-term interest rates. From the company files, you obtain the following annual series on (1) interest rates for prime commercial paper (4–6 months) and (2) interest rates on new issues of high-grade corporate bonds (Moody Aaa Index).

Year	Prime Commercial Paper (4–6 months)	Corporate Bonds (Moody Aaa)
1	2.97	4.35
2	3.26	4.33
3	3.55	4.26
4	3.97	4.40
5	4.38	4.49
6	5.55	5.13
7	5.10	5.51
8	5.90	6.18
9	7.83	7.03
10	7.72	8.04
11	5.11	7.39
12	4.69	7.21
13	8.15	7.44
14	9.87	8.57
15	6.33	8.83
16	5.35	8.43
17	5.60	8.02
18	7.99	8.73
19	10.91	9.63

a. Fit a straight line by the method of least squares to each of these series.
b. Calculate relative cyclical residuals. What can you conclude about the cyclical behavior of short-term and long-term interest rates?

8. A Public Policy Fellow was assigned to Washington, D.C., for her summer internship. She was given an assignment to establish the trend for state and local government expenditures for 1975–1988 from the following data:

Year	Expenditures ($ billions)	Year	Expenditures ($ billions)
1975	235.2	1982	414.3
1976	254.9	1983	440.2
1977	273.2	1984	475.9
1978	301.3	1985	516.7
1979	327.7	1986	561.9
1980	363.2	1987	602.8
1981	391.4	1988	646.6

a. Fit a straight line to this series by the method of least squares.
b. Fit a second-degree parabola to this series by the method of least squares.
c. Test which trend line seems to lie closer to the actual data by computing the percentages of trend for each series.

11.5
MEASUREMENT OF SEASONAL VARIATIONS

For long-range planning and decision making, in terms of time-series components, executives of a business or government enterprise concentrate primarily on forecasts of trend movements. For intermediate planning—say, from about two to five years—business cycle fluctuations are of critical importance, too. For short-range planning, and for purposes of operational decisions and control, seasonal variations must also be taken into account.

Seasonal movements, as indicated in section 11.2, are periodic patterns of variation in a time series. Strictly speaking, the terms **seasonal movements** and **seasonal variations** can be applied to any regularly repetitive movements that occur in a time series when the interval of time for completion of a cycle is one year or less. Hence, this classification includes movements such as *daily* cycles in utilization of electrical energy and *weekly* cycles in the use of public transportation vehicles. However, these terms generally refer to the annual repetitive patterns of economic activity associated with climate and custom. As noted earlier, these movements are generally examined by using series of *monthly* or *quarterly* data.

Daily

Weekly

Monthly

Quarterly

Purpose of Analyzing Seasonal Variations

Focus of interest

Just as was true in the study of trend movements, seasonal variations may be studied because *interest is primarily centered on these movements*, or they may be measured merely *in order to eliminate them*, so that business cycle fluctuations can be more clearly revealed. As an illustration of the first purpose, a company might analyze the seasonal variations in sales of a product it produces in order to iron out variations in production, scheduling, and personnel requirements. Another reason a company's interest may be primarily focused on seasonal variations is to budget a predicted annual sales figure by monthly or quarterly periods based on seasonal patterns observed in the past. Short range forecasts of this sort are often made by time series decomposition methods that project a moving average that describes the trend and cyclical movements in the series and then multiplies these figures by a seasonal index. The Census X-11 method is probably the best known of such methods. This U.S. Bureau of the Census method provides short range forecasts including seasonal adjustments.

Elimination to reveal other factors

On the other hand, as an illustration of the second purpose, an economist may wish to eliminate the usual month-to-month variations in series such as personal income, unemployment rates, and housing starts in order to study the underlying business cycle fluctuations present in these data.

Rationale of the Ratio to Moving Average Method

Seasonal variations can be measured by a number of techniques, but only the most widely used one, the so-called **ratio to moving average method**, will be discussed here. It is most frequently applied to monthly data, but we will illustrate its use for a series of quarterly figures, thus reducing substantially the required number of computations.

Obtaining seasonal indices

In understanding the rationale of the measurement of seasonal fluctuations, it is helpful to begin with the final product, the seasonal indices. When the raw data are for quarterly periods and a stable or regular seasonal pattern is present, the object of the calculations is to obtain four **seasonal indices**, each one indicating the seasonal importance of a quarter of the year. The arithmetic mean of these four indices is 100.0. Hence, a seasonal index of 105 for the first quarter means that the first quarter averages 5% higher than the average for the year as a whole. If the original data had been monthly, there would be 12 seasonal indices, which average 100.0, and each index would indicate the seasonal importance of a particular month. These indices are descriptive of the recurrent seasonal pattern in the original series.

As an example of how these seasonal indices might be used, we can refer to budgeting a predicted annual sales figure by quarterly periods. Suppose that $40 million of sales of particular products was budgeted for the next year, or an average of $10 million per quarter. If the quarterly seasonal indices based on an observed stable seasonal pattern were 97.0, 110.0, 85.0, and 108.0, then the amount of sales budgeted for each quarter would be

First quarter: $0.97 \times \$10$ million $= \$ \; 9.7$ million

Second quarter: $1.10 \times \;\; 10$ million $= \;\; 11.0$ million

Third quarter: $0.85 \times \;\; 10$ million $= \;\;\; 8.5$ million

Fourth quarter: $1.08 \times \;\; 10$ million $= \;\; 10.8$ million

A problem

The essential problem in the measurement of seasonal variations is elimi-nating from the original data the nonseasonal elements in order to isolate the stable seasonal component. In trend analysis, when annual data were used and we wanted to arrive at cyclical fluctuations, a similar problem existed. It was solved by obtaining measures of trend and using them as base line or reference figures. Deviations from trend were then measures of cyclical (and irregular) movements. Analogously, when we have monthly or quarterly original data, which consist of all of the components of trend, cyclical, seasonal, and irregular movements, ideally we would like to obtain a series of base line figures that contain all the nonseasonal elements. Then deviations from the base line would represent the pattern of seasonal variations. Unsurprisingly, this ideal method of measurement is not feasible. However, the practical method is to obtain a series of moving averages that roughly include the trend and cyclical compo-nents. Dividing the original data by these moving average figures eliminates the trend and cyclical elements and yields a series of figures that contain sea-sonal and irregular movements. These data are then averaged by months or by quarters to eliminate the irregular disturbances and to isolate the seasonal factor. This method of describing a pattern of stable seasonal movements is explained below.

The solution

Ratio to Moving Average Method

In order to derive a set of seasonal indices from a series characterized by a stable seasonal pattern, about five to nine years of monthly or quarterly data are required. A stable seasonal pattern means that the peaks and troughs gen-erally occur in the same months or quarters of each year.

The ratio to moving average method of computing seasonal indices for quarterly data may be summarized as consisting of the following steps:

Computing seasonal indices for quarterly data

1. Derive a four-quarter moving average that contains the trend and cyclical components present in the original quarterly series.

> A **four-quarter moving average** is simply an annual average of the original quarterly data successively advanced one quarter at a time.

For example, the first moving average figure contains the first four quarters. Then the first quarter is dropped, and the second through fifth quarterly figures

Deriving a four-quarter moving average

are averaged. The computation proceeds this way until the last moving average is calculated, containing the last four quarters of the original series. In the actual calculation, an adjustment is made to center the moving average figures so their timing corresponds to that of the original data.

Characteristics of moving averages   The reason these moving averages include the trend and cyclical components may perhaps be most easily understood by considering what these averages do *not* contain. Since they are annual averages, they do not contain seasonal movements (such fluctuations, by definition, average out over a one-year period). Furthermore, the irregular movements that raise the figures for certain months or quarters and lower them in others tend to cancel out when averaged over the year. Thus, only the trend and cyclical elements tend to be present in the moving averages.

2. Divide the original data for each quarter by the corresponding moving average figure.

These "ratio to moving average" numbers contain only the seasonal and irregular movements, since the trend and cyclical components are eliminated in the division by the moving average.

3. Arrange the ratio to moving average figures by quarters, that is, all the first quarters in one group, all the second quarters in another, and so forth.

Average these ratio to moving average figures for each quarter to eliminate the irregular movements, and thus to isolate the stable seasonal component. One type of average used for this purpose is referred to as a **modified mean**—an arithmetic mean of the ratio to moving average figures taken after dropping the highest and lowest extreme values.

4. Make an adjustment to force the four modified means to total 400, and thus average out to 100.0. The resulting four figures, one for each quarter of the year, constitute the seasonal indices for the series in question.

In symbols, this procedure may be summarized as follows. Let Y be the original quarterly observations; MA the moving average figures; and T, C, S, and I the trend, cyclical, seasonal, and irregular components, respectively. Dividing the original data by the moving average values gives

(11.6)
$$\frac{Y}{MA} = \frac{T \times C \times S \times I}{T \times C} = S \times I$$

Averaging these ratio to moving average figures (Y/MA) eliminates the irregular movements that tend to make the Y/MA values too high in certain years and too low in others. Hence, if the eliminations of the nonseasonal elements were perfect, the final seasonal indices would reflect only the seasonal variations. Of course, since the entire method is a rather rough and approximate procedure, the nonseasonal elements are generally not completely eliminated. The moving average usually contains the trend and *most* of the cyclical fluctuations. Therefore, the cyclical component may not be completely absent in the Y/MA values. Moreover, the modified means do not ordinarily remove all of the erratic disturbances attributed to the irregular component. Nevertheless, in the case of series with a stable seasonal pattern, the computed seasonal indices generally isolate the underlying seasonal pattern quite well.

Table 11-3 gives a quarterly series of national/regional spot TV time sales in the United States from 1984 to 1988. Examination of this series reveals that national/regional spot TV time sales tend to be highest during the second quarter and lowest during the first quarter. The calculation of quarterly seasonal indices will be illustrated in terms of this series.

TABLE 11-3

National/regional spot TV time sales in the United States by quarters, 1984–1988: computations for seasonal indices and deseasonalizing of original data

(C1)		(C2)	(C3)	(C4)	(C5)	(C6)	(C7)	(C8)
TIME		Spot * TVS	FQMT	EQMT	MOVAVE	ODPMA	SEASIND	DESEAS
1984	I	942.0					82.02	1148.53
1984	II	1355.4					114.48	1183.99
1984	III	1168.8	4714.5	9485.5	1185.69	98.576	99.84	1170.70
1984	IV	1248.3	4771.0	9656.8	1207.10	103.413	103.67	1204.14
1985	I	998.5	4885.8	9899.9	1237.49	80.688	82.02	1217.42
1985	II	1470.2	5014.1	10091.1	1261.39	116.554	114.48	1284.27
1985	III	1297.1	5077.0	10269.0	1283.63	101.050	99.84	1299.21
1985	IV	1311.2	5192.0	10479.0	1309.87	100.101	103.67	1264.81
1986	I	1113.5	5287.0	10687.8	1335.97	83.347	82.02	1357.63
1986	II	1565.2	5400.8	10974.4	1371.80	114.098	114.48	1367.26
1986	III	1410.9	5573.6	11186.1	1398.26	100.904	99.84	1413.20
1986	IV	1484.0	5612.5	11313.3	1414.16	104.938	103.67	1431.50
1987	I	1152.4	5700.8	11433.3	1429.16	80.635	82.02	1405.06
1987	II	1653.5	5732.5	11516.8	1439.60	114.858	114.48	1444.39
1987	III	1442.6	5784.3	11683.3	1460.41	98.780	99.84	1444.95
1987	IV	1535.8	5899.0	11823.2	1477.90	103.918	103.67	1481.47
1988	I	1267.1	5924.2	11932.8	1491.60	84.949	82.02	1544.91
1988	II	1678.7	6008.6	12106.7	1513.34	110.927	114.48	1466.41
1988	III	1527.0	6098.1				99.84	1529.49
1988	IV	1625.3					103.67	1567.80

* The full headings (and formulas) for Table 11-3:

TIME:	Year/Quarter
Spot * TVS:	National/Regional Spot TV Time Sales ($ millions)
FQMT:	Four-Quarter Moving Total
EQMT:	Two-of-a-Four-Quarter Moving Total
MOVAVE:	Moving Average (col. 4 × 1/8)
ODPMA:	Original Data as Percentage of Moving Average (100 × col. 2 ÷ col. 5)
SEASIND:	Seasonal Index
DESEAS:	Deseasonalized National/Regional Spot TV Time Sales (100 × col. 2 ÷ col. 7)

A problem

The national/regional TV times sales figures have been listed in column (2) of Table 11-3 from the first quarter of 1984 through the fourth quarter of 1988. Our first task is the calculation of the four-quarter moving average. As indicated above, this moving average is calculated by averaging four quarters at a time, continually moving the average up by a quarter. However, because of a problem of centering of dates, a slightly different type of average, called a **two-of-a-four-quarter moving average**, is calculated. This problem is as follows. An average of four quarterly figures would be centered halfway between the dating of the second and third figures and would thus not correspond to the date of either of those figures. For example, the average of the four quarters of 1984, the first figures shown in column (2) of Table 11-3, would be centered midway between the second and third quarter dates, or at the center of the year, July 1, 1984. The original quarterly figures are centered at the middle of their respective time periods, for simplicity, say, February 15, May 15, August 15, and November 15. Hence, the dates of a simple four-quarter moving average would not correspond to those of the original data. This problem is easily solved by averaging the moving averages two at a time. For example, as we have seen, the first moving average obtainable from Table 11-3 is centered at July 1, 1984. The second moving average, which contains the last three quarters of 1984 and the first quarter of 1985, is centered at October 1, 1984. Averaging these two figures yields a figure centered at August 15, the same as the dating of the third quarter.

The solution

Calculating

The easiest way to calculate this properly centered moving average is given in columns (3) through (5) of Table 11-3. Column (3) gives a four-quarter moving total. The first figure, 4714.5 is the total of the first four quarterly figures—942.0, 1355.4, 1168.8, 1248.3. This figure is listed opposite the third quarter, 1984, although actually it is centered at July 1. The next four-quarter moving total is obtained by dropping the figure for the first quarter of 1984 and including the figure for the first quarter of 1985. Hence, 4771.0 is the total of 1355.4, 1168.8, 1248.3, and 998.5. The total of 4714.5, and 4771.0, or 9485.5, is the first entry in column (4). This represents the total for the eight months that would be present in the averaging of the first two simple four-quarter moving averages. Dividing this total by eight yields the first two-of-a-four-quarter moving average figure of 1185.69, properly centered at the middle of the third quarter, 1984.

Using graphs

The moving averages given in column (5) of Table 11-3 are shown in Figure 11-6 along with the original data. It is useful to examine graphs in the calculation of seasonal indices, because we can observe visually what is accomplished in each major step of the procedure. We have noted earlier that the original data, if stated in monthly or quarterly form, contain all of the components of trend, cyclical, seasonal, and irregular movements. Although the period in this example is somewhat short for trend to be revealed, we can observe that national/regional spot TV time sales tend to rise throughout the entire period. The repetitive annual rhythm of the seasonal movements is clearly discernible. Irregular movements are also present. The moving average, which runs smoothly through the original data, can be observed to follow the trend

```
MTB >Set C1
DATA > 1:20
DATA > End

MTB >Mplot C2 C1 C5 C1
```

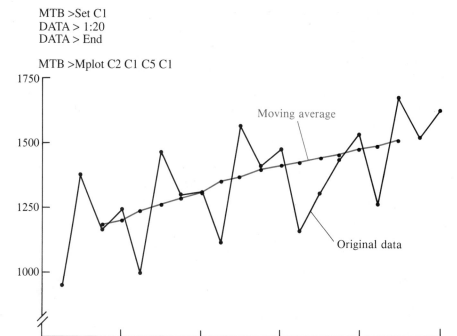

FIGURE 11-6

National/regional spot TV time sales, 1984–1988

movements rather closely, and if cyclical fluctuations were clearly indicated, we would be able to see how the moving average describes them as well. Another way to view this point is to note that the seasonal variations and (to a large degree) the irregular movements are absent from the smooth line that traces the path of the moving average. Note also that there are no moving average figures corresponding to the first two and the last two quarters of original data. Correspondingly, if the original data were in monthly form and a 12-month moving average were computed, there would be no moving averages to correspond to the first six months of data or to the last six months of data.

Percentage of moving average

The ratio to moving average figures—that is, original data in column (2) divided by the moving averages in column (5) are given in column (6) of Table 11-3. As in customary, these figures have been multiplied by 100 to express them in percentage form. They are often referred to as **percentage of moving average** values and may be represented symbolically as $(Y/MA) \times 100$. These values are graphed in Figure 11-7. The graph was plotted in Minitab. Lines were drawn between the points and the 100-base line was also drawn in manually. We can see in the graph that the trend and cyclical movements are no longer present in these figures. The 100-base line represents the level of the moving average, or the trend-cyclical base. The fluctuations above and below

FIGURE 11-7

Percentage of moving average
for national/regional spot TV
time sales, 1984–1988

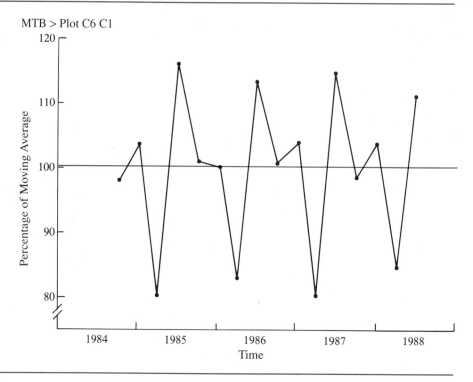

MTB > Plot C6 C1

this base line clearly reveal the repetitive seasonal movement of the spot TV time sales. As noted earlier, the irregular component is also present in these figures.

The next step in the procedure is to remove the effect of irregular movements from the $(Y/MA) \times 100$ values. This is accomplished by averaging the percentages of moving average figures for the same quarter. That is, the first-quarter $(Y/MA) \times 100$ values are averaged, the second-quarter values are averaged, and so on. The average customarily used in this procedure is a modified mean, which is simply the arithmetic mean of the percentages of moving average figures for each quarter over the different years, after eliminating the lowest and highest figures. It is desirable to make these deletions particularly when the highest and lowest figures tend to be atypical because of erratic or irregular factors such as strikes, work stoppages, or other unusual occurrences. We have deleted the lowest and highest figure in the present example to illustrate the standard method, although the highest and lowest figures are not particularly atypical in this example.

The percentages of moving average figures for each quarter are listed in Table 11-4. The highest and lowest figures have been deleted by a line drawn through them, and the modified means of the remaining values are shown for each quarter. These means are 82.02, 114.48, 99.84, 103.67, respectively, for the

TABLE 11-4

National/regional spot TV time sales, 1984–1988: Calculation of quarterly seasonal indices from percentage of moving average figures

```
MTB > NAME C10 'YEAR' C11 'QUARTER'

MTB > SET C10
DATA> 4(1984) 4(1985) 4(1986) 4(1987) 4(1988)
DATA> END

MTB > SET C11
DATA> 5(1:4)
DATA> END

MTB > TABLE C10 C11;
SUBC> DATA IN C6.

ROWS: YEAR  COLUMNS: QUARTER
```

PERCENTAGE OF MOVING AVERAGES BY QUARTER

YEAR	1	2	3	4
1984	--	--	98.58	103.41
1985	80.69	116.55	101.05	100.10
1986	83.35	114.10	100.90	104.94
1987	80.63	114.86	98.78	103.92
1988	84.95	110.93	--	--
MODIFIED MEANS	82.02	114.48	99.84	103.67

```
CELL CONTENTS --
        ODPMA:DATA

TOTAL OF MODIFIED MEANS = 400.01

ADJUSTMENT FACTOR = 400/400.01 = 0.999975
```

SEASONAL INDICES	I	II	III	IV
	82.018	114.477	99.837	103.667

first through fourth quarters. The total of these modified means is 400.1. Since it is desirable that the four indices total 400, in order that they average 100%, each of them is multiplied by an adjustment factor of $(\frac{400}{400.1})$. This adjustment forces a total of 400 by lowering each of the unadjusted figures by the same percentage. The final quarterly seasonal indices are shown on the bottom line of Table 11-4. In the case of monthly seasonal indices, a similar adjustment is made in order for the 12 monthly indices to total 1,200; thus, the average monthly index equals 100%.

As indicated earlier, if interest centers on the pattern of seasonal variations itself, the four quarterly indices repesent the final product of the analysis. On the other hand, sometimes the purpose of measuring seasonal variations is to eliminate them from the original data in order to examine, for example, the cyclical movements. To **deseasonalize** the original data, or adjust these figures for seasonal movements, we simply divide them by the appropriate seasonal indices. This adjustment is shown in Table 11-3 for the spot TV time sales data by the division of the original figures in column (2) by the seasonal indices in column (7). The result is multiplied by 100, since the seasonal index is stated as a percentage rather than as a relative.

Deseasonalized figures

Let us illustrate the meaning of a deseasonalized figure using the first line of figures in Table 11-3. The national/regional spot TV time sales in the first quarter of 1984 was $942.0 million. Dividing this figure by the seasonal index for the first quarter, 82.02, and multiplying by 100 yields 1148.53 million. This figure is the national/region spot TV time sales for the first quarter of 1984 adjusted for seasonal variations.

> The figure, adjusted for seasonal variations, represents the level that such sales would have attained if the depressing effect of seasonality in the first quarter of the year had not been present.

All time-series components other than seasonal variations are present in these deseasonalized figures. This idea can be expressed symbolically as follows in terms of the multiplicative model of time-series analysis:

(11.7)
$$\frac{Y}{SI} = \frac{T \times C \times S \times I}{S} = T \times C \times I$$

The figures for national/regional spot TV time sales adjusted for seasonal movements are graphed in Figure 11-8. We can see the underlying trend and irregular movements in these data, and if cyclical fluctuations had been clearly present in the original data, we would see them as well. Note that compared with the plot of the original data in Figure 11-6, most of the repetitive seasonal movements are no longer present in the deseasonalized figures. However, ordinarily the adjustment for seasonality is not perfect, as is the case here. To the extent that seasonal indices do not completely portray the effect of seasonality, division of original data by seasonal indices will not entirely remove these influences.

Seasonal indices are often used for the purpose just discussed. Economic time series adjusted for seasonal variations are often charted in the *Federal Reserve Bulletin,* the *Survey of Current Business,* and other publications. Quarterly gross national product figures are often given as "seasonally adjusted at annual rates." These are simply deseasonalized quarterly figures multiplied by four to state the result in annual terms.

MTB > Plot C8 C1

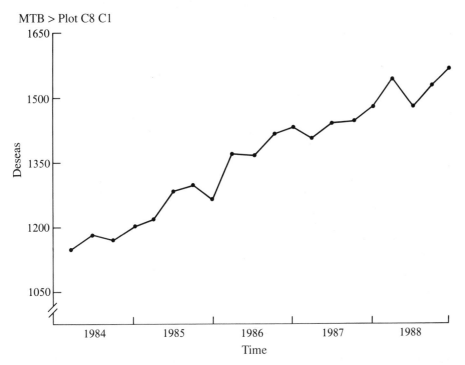

FIGURE 11-8
Deseasonalized figures for
national/regional spot TV time
sales in the United States by
quarters, 1984–1988

EXERCISES 11.5

1. Using only trend analysis, Mid-South Airlines has estimated that it will carry 325,000 passengers in December 1991.
    a. How many passengers should the company anticipate if the December seasonal index is 132? Will the number of passengers be closer to 325,000 or 429,000 (325,000 × 1.32)? Discuss.
    b. If the actual number of passengers in December 1991 is 415,000, calculate the seasonally adjusted number of passengers as a percentage of trend.
    c. What is the meaning of the trend value?
2. The sales manager of the Dolly Adams Ice Cream Plant uses trend projection to estimate that annual sales of ice cream for 1991 will be 15% higher than 1990 levels. He expects that no sharp cyclical fluctuation will occur during the year, that the effect of trend within the year will be negligible,

and that the past pattern of seasonal variation will continue. The 1990 quarterly sales volume of ice cream and quarterly seasonal indices are given below:

	Spring	Summer	Fall	Winter
Ice cream sales (thousands of gallons)	25,500	31,875	25,725	20,160
Seasonal index	90	125	105	80

   a. Is the deseasonalized sales volume for the 1990 quarters consistent with the 1990 trend projection of 25,000,000 gallons per quarter? Show calculations.

   b. Prepare a forecast of quarterly ice cream sales for 1991 from the above figures.

3. The vice-president of marketing of the Bonnet Dairy Plant was upset over the sales figures for the company's milk product for the third quarter of 1990. Despite heavy advertising and promotion of fresh milk and the absence of any sharp cyclical downturn, sales figures for July to September 1990 fell continuously. As the company's market analyst, you were asked to explain this dismal performance. The sales volume of fresh milk and its monthly seasonal indices are given below:

Month	Fresh Milk (thousands of gallons)	Seasonal Index
July 1990	19,680	131.2
August 1990	17,250	109.5
September 1990	16,120	98.7

   a. Was the vice-president correct in her observation that sales for the third quarter declined? Explain your answer.

   b. Explain how you would determine the effect of business cycle fluctuations on sales of fresh milk. Specify any information you would need in addition to that given above.

4. The trend equation for newsprint production of the MCB Company is as follows:

$$Y_t = 641.04 + 0.96t$$

where $t = 1$ in January 1985

             t is in monthly intervals

             Y = monthly production (millions of tons)

Actual production figures for selected months in 1990 are given below:

Month	Production (millions of tons)	Seasonal Index
March	570	86
April	708	101
May	804	110
June	1176	165

a. In which of these months is newsprint production the highest (1) if seasonal variation is not considered? (2) If seasonal variation is considered?

b. In which of these months is production most typical of the monthly average?

c. For March 1990 isolate the effect of each component of a time series (trend, cyclical, seasonal, and irregular).

5. The following data pertain to the number of automobiles sold by the Atoyot Corporation:

Year	Quarter	Number of Automobiles Sold
1985	I	362
	II	386
	III	437
	IV	427
1986	I	405
	II	433
	III	470
	IV	443
1987	I	402
	II	420
	III	456
	IV	440
1988	I	416
	II	440
	III	508
	IV	492
1989	I	455
	II	480
	III	525
	IV	498
1990	I	436
	II	472
	III	536
	IV	529

a. Using the ratio to moving average method, determine constant seasonal indices for each of the four quarters.
b. Do you think constant seasonal indices should be employed in this problem? Why or why not?
c. Assuming that constant seasonal indices are appropriate, adjust the quarterly sales figures between 1985 and 1990 for seasonal variations.
d. Assume that the trend in sales for Atoyot Corporation can be described by the following equation:

$$Y_t = 385 + 20t$$

where $t = 1$ in the second quarter of 1985

t is in one-year intervals

Y = the number of automobiles sold

Do you see any evidence of cyclical fluctuations in the data between 1985 and 1990? What is the basis for your answer?

6. Data Processing, Inc., a computer firm specializing in software engineering, has determined seasonal indices and a trend line for its monthly sales. The seasonal index for December is 110. The trend line for monthly sales is $Y_t = 1.85 + 0.16t + 0.03t^2$ with t in one-year intervals and $t = 1$ in December 15, 1988; Y_t is in millions of dollars. Actual sales in December 1990 were $2.5 million.
a. What is the relative cyclical residual for December 1990 adjusted for seasonal variation?
b. What does the relative cyclical residual mean?
c. Forecast company sales for December 1991, adjusting for seasonal variations.

7. The dean of admissions of a small eastern college has compiled the following quarterly enrollment figures for 1986–1990:

		Enrollment (hundreds)		
Year	Fall	Winter	Spring	Summer
1986	270	253	243	134
1987	285	258	256	126
1988	286	256	259	123
1989	291	265	256	142
1990	289	271	263	165

Using the ratio to moving average method, determine constant seasonal indices for each of the four quarters. Adjust the quarterly enrollment figures for seasonal variation.

8. The following sentences refer to the ratio to moving average method of measuring seasonal variation when applied to monthly gasoline sales in United States from 1978 to 1990. Determine which phrase will complete the sentence most appropriately.
a. A 12-month moving total is computed because
    (1) trend is thus eliminated.

      (2) this will give column totals equal to 1,200.

      (3) seasonal variation cancels out over a period of 12 months.

b. A two-item total is then taken of the 12-month totals in order to

      (1) obtain moving average figures for the first and last six months.

      (2) center the moving average figures properly.

      (3) eliminate the rest of the irregular movements.

c. This two-item total of a 12-month moving total is divided by

      (1) 14

      (2) 2

      (3) 12

      (4) 24

d. This moving average contains

      (1) all of the trend, most of the cyclical, all of the seasonal, and some irregular variation.

      (2) all of the trend, most of the cyclical, and possibly some irregular variation.

      (3) most of the trend, most of the cyclical, and all of the irregular variation.

e. The original data are then divided by the moving figures. These specific seasonal relatives (Y/MA values) contain

      (1) all of the seasonal, possibly some of the cyclical, and almost all of the irregular variation.

      (2) seasonal variation only.

      (3) all of the trend, most of the cyclical, and none of the irregular variation.

f. Modified means of the specific seasonal relatives (Y/MA) are calculated in order to

      (1) eliminate the nonseasonal elements from the specific seasonal relatives (Y/MA values).

      (2) get rid of seasonal elements.

      (3) eliminate the trend in the specific seasonal relatives (Y/MA).

      (4) compensate for a changing seasonal pattern.

g. To adjust the original data for seasonal variation, we

      (1) multiply the original data by the seasonal index.

      (2) divide the seasonal index by the original data.

      (3) divide the original data by the seasonal index.

11.6
METHODS OF FORECASTING

Classical methods used in analyzing the separate components of an economic time series involve an implicit assumption that the various components act independently of one another.

For example, no specific procedures were established for taking into account cyclical influences on seasonal variations or long-run changes in the structure of business cycles. Special procedures can be established to gauge some of these interactions, but basically the model used in classical time-series analysis assumes independent sources of variation in economic time series and measures these sources separately. This *decomposition* or *separation* process, although often useful for descriptive or analytical purposes, is nevertheless artificial. Therefore, it is not surprising that for a complex problem such as economic forecasting, making mechanistic extrapolations based on classical time-series analysis alone will not suffice. However, time-series analysis frequently is a helpful starting point and an extremely useful supplement to other analytical and judgmental methods of forecasting.

Short-term forecasting   In short-term forecasting, a combined trend-seasonal projection often provides a convenient first step. For example, as a first approximation in a company's forecast of next year's sales by months, a trend projection for annual sales might be obtained. Then this figure might be allocated among months based on an appropriate set of seasonal indices. Of course, the underlying assumption in this procedure is the persistence of the historical pattern of trend and seasonal variations of this company's sales into the next year.

More complete forecasting   A more complete forecast might involve superimposing a cyclical prediction as well. For example, the first step may again involve a projection of trend to obtain an annual sales figure. Then an adjustment of this estimate may be made based on judgment of recent cyclical growth rates. Suppose that the past few years represented the expansion phase of a business cycle and the cyclical growth rate for the economy during the next year was predicted at about 4% by a group of economists. Assume further that the company in question had found in the past that these forecasts were quite accurate and applicable to the company's own cyclical growth rate—over and above its own forecast of trend levels. Then the company might increase its trend forecast by this 4% figure to obtain a trend-cyclical prediction. Again, if predictions by months are required, a monthly average could be obtained from the trend-cyclical forecast and seasonal indices could be applied to yield the monthly allocation. Ordinarily, no attempt would be made to predict the irregular movements. An alternative short-term forecasting technique that is similar to the one just described involves a projection of a 12-month moving average to obtain trend-cyclical forecasts, either monthly or as annual averages. Then seasonal indices are applied to obtain the monthly predictions.

Cyclical Forecasting and Business Indicators

Cyclical movements are more difficult to forecast than trend and seasonal elements. Cyclical fluctuations in a specific time series are strongly influenced by the general business cycle movements characteristic of large sectors of the overall economy. However, since there is considerable variability in the timing and amplitude with which many individual economic series trace out their cyclical swings, there is no simple mechanical method of projecting these movements.

Relatively "native" methods such as the extension of the same percentage rate of increase or decrease in, say, sales for last year or during the past few years are often used. These may be quite accurate, particularly if the period for which the forecast is made occurs during the same phase of the business cycle as the periods from which the projections are made. However, the most difficult and most important items to forecast are the cyclical turning points at which reversals in direction occur. Obviously, managerial planning and implementation that presuppose a continuation of a cyclical expansion phase can give rise to serious problems if an unpredicted cyclical downturn occurs during the planning period.

Many statistical series produced by government and private sources have been extensively used as business indicators. Some of these series represent activity in specific areas of the economy such as employment in nonagricultural establishments or average hours worked per week in manufacturing. Others are broad measures of aggregate activity pertaining to the economy as a whole, as for example, gross national product and personal income. We noted earlier that economic series, while exhibiting a certain amount of resemblance in business cycle fluctuations, nevertheless display differences in timing and amplitude. The National Bureau of Economic Research and the U.S. Department of Commerce have studied these differences carefully and have specified a number of time series as statistical indicators of cyclical revivals and recessions.

Business indicators

These time series have been classified into three groups.

- The first group consists of the so-called **leading series**. These series have usually reached their cyclical turning points prior to the analogous turns in general economic activity. The group includes series such as the layoff rate in manufacturing, the average workweek of production workers in manufacturing, the index of net business formation, the Standard & Poor's index of prices of 500 common stocks, and the index of homebuilding permits.

Time series as statistical indicators

- The second group comprises **coinciding series** whose cyclical turns have roughly coincided with those of the general business cycle. Included are such series as the unemployment rate, the industrial production index, gross national product, and dollar sales of retail stores.

- Finally, the third group consists of the **lagging series**, those whose arrivals at cyclical peaks and troughs usually lag behind those of the general business cycle. This group includes series such as labor cost per unit of output, consumer installment debt, and bank interest on short-term business loans.

Note that rational explanations stemming from economic theory can be given for placing the various series into the respective groups, in addition to the empirical observations themselves. These statistical indicators are adjusted for seasonal movements. They are published monthly in *Business Conditions Digest* by the U.S. Department of Commerce. Another publication that carries the statistical indicators, as well as other time series with accompanying analyses, is *Economic Indicators*, published by the Council of Economic Advisors.

Probably the most widespread application of these cyclical indicators is as an aid in the prediction of the timing of *cyclical turning points*. If, for example, most of the leading indicators move in an opposite direction from the prevailing phase of the cyclical activity, this is taken to be a possible harbinger of a **cyclical turning point**. A subsequent similar movement by a majority of the roughly coincident indices would be considered a confirmation that a cyclical turn was in progress. These cyclical indicators, like all other statistical tools, have their limitations and must be used carefully. They are not completely consistent in their timing, and leading indicators sometimes give incorrect signals of forthcoming turning points because of erratic fluctuations in individual series. Furthermore, it is not possible to predict with any high degree of assurance the length of time between a signal given by the leading series group of an impending cyclical turning point and the turning point itself. There has been considerable variation in this lead time during past cycles of business activity.

Cyclical turning point

Other Methods of Forecasting

Most individuals and companies engaged in forecasting do not depend on any single method, but rather utilize a variety of different approaches.

> We can place greater reliance on the consensus of a number of forecasts arrived at by relatively independent methods than on the results of any single technique.

Other methods of prediction range from informal judgmental techniques to highly sophisticated mathematical models. At the informal end of this scale, for example, sales forecasts are sometimes derived from the combined outlooks of the sales force of a company, from panels of executive opinion, or from a composite of both of these. At the other end of the scale are the formal mathematical models such as regression equations or complex econometric models. There is widespread use of various types of regression equations, by which firms attempt to predict the movements of their own company's or industry's activity on the basis of relationships to other economic and demographic factors. Often, for example, a company's sales are predicted on the basis of relationships with other series whose movements precede those of the sales series to be forecasted.

In recent years a wide variety of forecasting methods have been introduced, ranging from the relatively simple to the highly sophisticated. The availability of computers has made possible more extensive processing of data necessary in the derivation of forecasts. Although detailed discussion of forecasting techniques is outside the scope of this book, brief comments on several of these methods are given here.

Perhaps the most detailed refinements of the classical time-series decomposition method can be found in **Census Method II**, developed by the U.S. Bureau of the Census. The elaborations of the classical decomposition method

Census II decomposition method

include better descriptions of trend-cyclical movements, permitting the analyst to make more accurate estimates of the combined effect of these components. The procedure provides for the derivation of changing seasonal indices as well as for stable or constant indices. The Census method removes extremely large or small values (outliers) and smooths out irregular movements more effectively than classical decomposition. The Census method also includes tests for determining how well the decomposition has been carried out.

The **Box-Jenkins methods** represent a three-stage set of procedures for modeling a time series. The first stage, known as *identification*, attempts to determine whether the time series can be described through a combination of moving average and autocorrelation terms. In the second stage, *estimation*, the time-series data are used to estimate the parameters of a tentative model. The third stage, *diagnostic testing*, consists of tests that examine the deviations from fitted models to determine the adequacy of the models. The forecasting model results from the modifications accomplished in stage three.

Box-Jenkins methods

All time-series forecasting methods involve the processing of past data of the series to be predicted. If we project a 12-month moving average to obtain a forecast of trend-cyclical levels, past observations have been given equal weight. That is, the monthly figures are given equal weight in calculating the 12-month moving average figures. In the exponential smoothing method that is discussed at the end of this section, the most recent observations are weighted most heavily. **Generalized adaptive filtering** attempts to establish the optimal set of weights to be applied to past observations. No fixed weighting scheme is assumed. The techniques differ from those of Box-Jenkins in the methods of optimizing the parameters in the models and in the procedures used in the selection of models.

Generalized adaptive filtering

Among the most formal and mathematically sophisticated methods of forecasting in current use are econometric models. An **econometric model** is a set of two or more simultaneous mathematical equations that describe the interrelationships among the variables in the system. Some of the more complex models for prediction of movements in overall economic activity include dozens of individual equations. Special methods of solution for the parameters of these equation systems have been developed, since in many instances ordinary least squares techniques are not appropriate. Econometric models have in the past been used primarily at the first three levels in the hierarchy of forecasts, which includes the national economy, geographical regions, industries, companies, product lines, and products. However, in recent years econometric models have been used for prediction at the company level as well. Many companies, for example, have developed multiple regression equations for sales and other variables that are functions of industry and economy variables predicted by econometric models.

Econometric models

Management uses forecasting as an important ingredient of its planning, operational, and control functions. Invariably, judgment is applied to the results of various forecasting methods rather than a single method. Often, formal prediction techniques make a useful contribution by narrowing considerably the area within which intuitive judgment is applied.

Exponential smoothing   As stated earlier, all methods of forecasting involve processing past information in order to extract insights for prediction. Some methods, such as the classical time-series analysis discussed in this chapter, require extensive manipulation of past data. Other techniques, such as regression models or econometric models, require not only the use of considerable amounts of past data but also the fitting of formal mathematical models to these data. However, in some situations, we need a fast, simple, and rather mechanical way of forecasting for large numbers of items, without the computational burden of working with large quantities of data for each item for which forecasts are required. For example, in connection with inventory control, predictions of demand for thousands of items may be required for the determination of ordering points and quantities. **Exponential smoothing** is a forecasting method that meets the aforementioned requirements and is particularly appropriate for application on high-speed electronic computers. The technique simply weights together the current (most recent) actual and the current forecast figures to obtain the new forecast value. The basic formula for exponential smoothing is

(11.8)
$$F_t = wA_{t-1} + (1 - w)F_{t-1}$$

where F_t is the forecast for time period t
         A_{t-1} is the actual figure for time period $t - 1$
         F_{t-1} is the forecast for time period $t - 1$
         w is a constant whose value is between 0 and 1

The time period $t - 1$ is the *current period*, and the prediction F_t is the *new forecast* for the next period. Therefore, equation 11.8 may be written in words as

(11.9)   New forecast $= w$(Current actual) $+ (1 - w)$(Current forecast)

Note that the new forecast F_t is made in the current period $t - 1$; the current forecast F_{t-1} was made in the preceding period $t - 2$. The new forecast can be considered a weighted average of the current actual figure A_{t-1} and the forecast for the current period F_{t-1}. Another way of writing 11.8 is

(11.10)
$$F_t = A_{t-1} + (1 - w)(F_{t-1} - A_{t-1})$$

In this form, the new forecast may be viewed as the current actual figure A_{t-1} plus some fraction of the most recent forecast error $(F_{t-1} - A_{t-1})$, that is, the difference between the current forecast and current actual figures.

An important advantage   An important advantage of this method of a forecasting is that the only quantity that must be stored until the next period is the new forecast F_t. This modest storage requirement is an extremely useful feature for computer-based inventory management systems where, as mentioned earlier, there may be thousands of items for which forecasts are required. Despite the fact that exponential smoothing requires the storage of only one piece of data, that is, the new forecast, a bit of algebra demonstrates that each new forecast takes into account all past actual figures.

For example, we may write

$$F_2 = wA_1 + (1 - w)F_1$$

$$F_3 = wA_2 + (1 - w)F_2$$

$$F_4 = wA_3 + (1 - w)F_3$$

Substituting the first equation value for F_2 into the second equation and then the resulting value of F_3 into the third equation yields

(11.11) $$F_4 = wA_3 + (1 - w)wA_2 + (1 - w)^2 wA_1 + (1 - w)^3 F_1$$

Note that the last term of equation 11.11 is the only one that is not expressed in terms of actual figures (A_i). However, F_1 can be written to summarize the results of the actual figures A_0, A_{-1}, A_{-2}, and so on. Therefore, we can think of the forecast for any period as a weighted sum of all past actual figures, with the heaviest weight assigned to the most recent actual figure, the next highest to the second most recent one, and so on. We now see why the method is referred to as *exponential smoothing*, since the weighting factor $(1 - w)$ has successively higher exponents the farther the data recede into the past. From a management viewpoint, the fact that the most recent experience gets the greatest weight, with earlier periods getting respectively less weight, makes a great deal of sense.

Selecting the value of the weighting factor

One choice that the analyst must make in exponential smoothing is the value of the weighting factor, w. We can see from equations 11.8 and 11.11, if w is assigned a value close to zero, the most recent actual figure for, say, demand or sales will receive relatively little weight. Thus, forecasts made with a small w factor tend to have little variation from period to period. Hence, low values for w would tend to minimize the reactions of forecasts to relatively brief or erratic fluctuations in demand. This points up the basic trade-off involved in the choice of w.

If w is too low, the forecasts respond too slowly to basic long-lived changes in the demand pattern. On the other hand, if w is too high, the forecasts tend to react too sharply to random shifts is demand.

The values of w used in practice are relatively low, usually 0.3 or less, with a value of 0.1 often proving to be quite effective in a variety of forecasting situations. When policy decisions that are expected to change the demand pattern are made, appropriate changes can be made in w values.

As an example of the exponential smoothing procedure, let us examine Table 11.5. Columns (1) and (2) give the data for the weekly number of units of demand for a certain product. This series is characterized by a constant demand of 100 units per week except for a sudden sharp jump to 150 units in week 5. Columns (3) and (4) give the exponential smoothing forecasts for weight factors of $w = 0.1$ and $w = 0.8$, respectively. Note that the first actual demand figure A_1 is assumed to be the first forecast figure F_2. This is often the way an

TABLE 11-5
Exponential smoothing forecasts for weekly demand using weight factors of $w = 0.1$ and $w = 0.8$

(1) Week	(2) Actual Demand	(3) Next Forecast $w = 0.1$	(4) Next Forecast $w = 0.8$
1	$100 = A_1$	$100 = F_2$	$100 = F_2$
2	$100 = A_2$	$100 = F_3$	$100 = F_3$
3	$100 = A_3$	$100 = F_4$	$100 = F_4$
4	$100 = A_4$	$100 = F_5$	$100 = F_5$
5	$150 = A_5$	$105 = F_6$	$140 = F_6$
6	$100 = A_6$	$104.5 = F_7$	$108 = F_7$
7	$100 = A_7$	$104.05 = F_8$	$101.6 = F_8$
8	$100 = A_8$	$103.645 = F_9$	$100.32 = F_9$
9	$100 = A_9$	$103.2805 = F_{10}$	$100.064 = F_{10}$
10	$100 = A_{10}$	$102.95245 = F_{11}$	$100.0128 = F_{11}$

exponential smoothing set of forecasts is begun. Note too that the forecasts on each line of the table are the forecasts made in the given week for the following week.

As an illustration of the calculations, for example, the value of F_6 in columns (3), which is 105, is derived as follows:

$$F = wA_5 + (1 - w)F_5$$
$$= (0.1)(150) + (0.9)(100)$$
$$= 105$$

The effects of low and high values for the weighting factor w can be seen in Table 11-5. With the low value of $w = 0.1$, the forecasts show a sluggish response to the sudden spurt in demand to 150 units in week 5, with F_6 rising only to a value of 105. Furthermore, the effect of this spurt in actual demand lasts longer in the forecast series with $w = 0.1$ than with $w = 0.8$. With the higher value of $w = 0.8$, the forecast for F_6 jumps to 140 from the previous level of 100, and the effect dies out more rapidly than in the case of the lower weighting factor, as seen in columns (3) and (4). All the decimal places in the smoothing calculations have been shown in Table 11-5 for checking purposes. In practice, if demand can occur only in whole numbers of units, then only such integral values would be used as forecasts.

Summary   The method described above is the simplest and most basic form of exponential smoothing. It is particularly appropriate for economic time series that are relatively stable and do not have pronounced trend or seasonal movements. More complex forms of exponential smoothing are used, sometimes combined with spectral analysis (a method for determining which of various component cycles are most influential in a time series), to handle more complicated series.

Exponential smoothing techniques are often useful in situations in which a large number of economic time series must be forecast frequently.

It is a relatively simple method characterized by economy of storage and computational time.

1. A California-based manufacturer of heavy earth-moving equipment has recorded the following sales (in numbers of units) over the past 10 months:

   10   10   10   15   10   10   10   10   10   10

   Use the exponential smoothing procedure to prepare next-month forecasts, using weighting factors of $w = 0.1$ and $w = 0.6$. Assume that the first actual sales figure is the first forecast.

2. Referring to exercise 1, assume that the monthly amounts of heavy earth-moving equipment sold were 10, 10, 10, 15, 15, 15, 15, 15, 15, and 15. Again use the exponential smoothing procedure to prepare next-month forecasts using weighting factors of $w = 0.1$ and $w = 0.6$. As in exercise 1, assume that the first actual figure is the first forecast.

3. Discuss the effects of using low versus high values of the weighting factor for the sales series given in exercises 1 and 2.

4. Using the exponential smoothing procedure with a weighting factor of 0.5, derive a forecast for period 20 for the following annual series for interest rates on new issues of high-grade corporate bonds (Moody Aaa).

Year	Corporate Bonds Interest Rate	Year	Corporate Bonds Interest Rate
1	4.35%	11	7.39%
2	4.33	12	7.21
3	4.26	13	7.44
4	4.40	14	8.57
5	4.49	15	8.83
6	5.13	16	8.43
7	5.51	17	8.02
8	6.18	18	8.73
9	7.03	19	9.63
10	8.04		

Does this procedure seem to be appropriate for forecasting this series? Use the first actual interest rate figure as the first forecast.

5. Exponential smoothing is used in security analysis as a mechanical means of forecasting company earnings. The accuracy of analysts' forecasts can then be compared with the accuracy of these mechanical forecasts in order to evaluate the analysts' performances. Earnings per share for a diversified electronics company for 1974–1989 are given below:

Year	Actual Earnings per Share	Year	Actual Earnings per Share
1974	$3.50	1982	$4.79
1975	3.56	1983	5.25
1976	3.545	1984	4.74
1977	3.69	1985	4.83
1978	3.445	1986	4.50
1979	3.92	1987	5.86
1980	4.705	1988	6.25
1981	5.38	1989	8.48

Use the exponential smoothing procedure to prepare next-year forecasts using a weighting factor of 0.4. Forecast earnings per share for 1990 and compare the forecasted value with the actual value of $7.26 per share. Use the first actual figure to begin the set of forecasts.

MINITAB EXERCISES

Problems 1 through 4 are based on the following data: In a university bookstore, the sales of books for 27 months have been observed and the data (in thousands of copies) have been stored in the Minitab file labeled 'sales . mtw'.

Column	Name	Count	Description
C1	SALES	27	Sales of Books (in thousands of copies)

1. Make a time-series plot of SALES via

    MTB > tsplot C1

    and answer: (1) does the time-series have a trend? and (2) does it show a cyclical movement?

2. Create a column for time by using

    MTB > name C2 'TIME'
    MTB > set C2
    DATA > 1:27
    DATA > end

    Regress SALES over TIME and store the residuals in a column, say C10, with the help of the following commands

    MTB > regress C1 1 C2;
    SUBC > residuals in C10.
    MTB > name C10 'RESIDUAL'

    Obtain a scatterplot of RESIDUAL versus TIME. Does RESIDUAL show cyclical movements?

3. (*optional*) Calculate the autocorrelations of RESIDUAL in exercise 2 by

    MTB > acf C10

    and regress C1 on C2 and C3, where C3 is defined by

    MTB > lag 3 C1, put into C3

    The reason for the last command is that the third order autocorrelation coefficient is the largest one. Show that TIME is not significant in the regression, at alpha = 0.05. Regress C1 on C3 only, and enter the residuals into C11. Does C11 show a cyclical movement?

4. (*optional*) Find the autocorrelations of C11 and try

    MTB > arima 3 0 0 C11

    to show that C11 is autocorrelated, at alpha = 0.05.

There are different attitudes about the residual autocorrelation problem. Problems 5 through 7 are designed to reflect some of those attitudes. The relevant time-series are saved in the file named 'butter . mtw':

Column	Name	Description
C1	BUTTER	Butter Production (Mn Lbs)
C2	PRICE	Producer Price Index: Butter. (1982 = 100)

5. (*optional*) The first attitude is called the Laissez-Faire Policy: it simply ignores the effect of autocorrelation. Use Minitab to regress C1 on C2 and obtain the regression coefficient, the *R*-square, the Durbin-Watson statistic, and store the residuals in column C10. Make a time-series plot of the residuals and comment on the severity of the autocorrelation problem. Use the Durbin-Watson statistic to test the significance of positive autocorrelation, at alpha = 0.05. *Hint:* Minitab commands

   MTB > regress C1 1 C2;
   SUBC > residuals in C10;
   SUBC > dw.
   MTB > name C10 'RESIDUAL'

   will print out, among others, the Durbin-Watson statistic.

6. (*optional*) The second attitude is called the Generalized Difference Policy: it changes the original model

   BUTTER(t) = a + b ∗ PRICE(t) + e(t)

   to a more complex one

   BUTTER(t + 1) − p ∗ BUTTER(t)
   = a + b ∗ (PRICE(t + 1) − p ∗ PRICE(t)) + e(t + 1), where

   p is set to be $(2 - d)/2$ and d stands for the Durbin-Watson statistic. Use Minitab to obtain a regression equation for the new model, and the corresponding *R*-square. *Hint:* a column to represent the variable BUTTER($t + 1$) − p BUTTER(t) can be obtained from

   MTB > let C3 = C1 − p ∗ lag(C1)

7. (*optional*) The third attitude is called the First Difference Policy: it differs from the one in exercise 6 only in that p is always set equal to 1. Use Minitab to obtain the corresponding regression equation and *R*-square. Save the residuals in column C10 and analyze the residuals from their autocorrelations obtained from the Minitab command:

   MTB > acf C10

KEY TERMS

Box-Jenkins Method   a three-stage set of procedures (identification, estimation, and diagnostic testing) for modeling and forecasting a time series.

Classical (decomposition) Time Series Model   a descriptive method that breaks down an economic time series into components representing the effects of trend, cyclical, seasonal, and irregular movements.

Coinciding Series   a time series whose cyclical turns roughly coincide with those of the general business cycle.

Cyclical Fluctuations   recurrent upward and downward movements around trend levels with a duration of two to fifteen years.

Deseasonalized Figures   original data adjusted for seasonal movements.

Econometric Model   a set of two or more simultaneous mathematical equations that describe the interrelationships among variables in a system.

 Exponential Smoothing   a forecasting method that weights together the current actual figures and current forecast figures to obtain new forecast values.

 Irregular (residual) Movements   short, erratic fluctuations in time series that have no discernible pattern.

Lagging Series   a time series that reaches its cyclical turning points after those of the general business cycle.

Leading Series   a time series that reaches its cyclical turning points before the analogous turns in general business activity.

Ratio to Moving Average Method   a method for measuring seasonal variations by means of monthly or quarterly indexes.

Relative Cyclical Residual   the percentage deviation from trend because of cyclical and irregular factors.

 Seasonal Movements   repetitive cycles that complete themselves within a one-year period.

 Time Series   a set of statistical observations arranged in chronological order.

 Trend   the smooth upward or downward movement of a time series over a long period of time.

KEY FORMULAS

Arithmetic Straight-Line Trend
$$\hat{Y}_t = a + bt$$

Deseasonalized Figure
$$\frac{Y}{SI} \cdot 100$$

Exponential Smoothing
$$F_t = wA_{t-1} + (1 - w)F_{t-1}$$

Logarithmic Straight-Line Trend
$$\log \hat{Y}_t = a + bt$$

Percentage of Trend (cyclical relative)
$$\frac{Y}{\hat{Y}_t} \cdot 100$$

Relative Cyclical Residual
$$\frac{Y - \hat{Y}_t}{\hat{Y}_t} \cdot 100$$

Second-Degree Parabola Trend Line
$$\hat{Y}_t = a + bt + ct^2$$

12

INDEX NUMBERS

In our daily lives, we often make judgments that involve summarizing how an economic variable changes with *time* or *place*. As an example of variation over time, a family's income may have increased by 40% over a five-year period. Suppose this was a period of inflation (generally rising prices). Has the family's "real income" increased? That is, can the family purchase more goods and services with its income than it could five years ago? If the general price level of items has increased more than the 40% figure, the family clearly cannot purchase as much. On the other hand, if prices have increased less than 40%, the family's real income has increased.

As an example of variation with changes in place, let us consider a company that wishes to transfer an executive from St. Louis to New York City. What should be the executive's minimum salary increase to allow for the higher cost of living in New York?

12.1
THE NEED FOR AND USE OF INDEX NUMBERS

Both cases mentioned above require measurements of general price levels. The prices of the numerous items purchased by the family have doubtless increased at different rates; a few may even have decreased. Similarly, although some prices in New York City are much higher than in St. Louis, some may be lower. How can we summarize in a single composite figure the average differences between the two time periods or the two cities? Index numbers answer questions of this type.

> An **index number** is a summary measure that states a relative comparison between groups of related items. In its simplest form, an index number is nothing more than a **price relative**—a percentage figure that expresses the relationship between two numbers with one of the numbers used as the base.

Calculating price relatives   For example, in a time series of prices of a particular commodity, the prices may be expressed as percentages by dividing each figure by the price in the base period. In the calculation of economic indices, it is conventional to state these price relatives as numbers lying above and below 100, where the base period is 100%. For example, suppose the price of one pound of a certain brand of coffee was $2.00 in 1978, $3.22 in 1984, and $6.04 in 1990. Then the price relatives of the three figures, with 1978 as the base (written, 1978 = 100), are

Year	Price Relatives (1978 = 100)
1978	$\left(\dfrac{\$2.00}{\$2.00}\right)(100) = 100$
1984	$\left(\dfrac{\$3.22}{\$2.00}\right)(100) = 161$
1990	$\left(\dfrac{\$6.04}{\$2.00}\right)(100) = 302$

As indicated, the price relative for any given year is obtained by dividing the price for that year by the base period figure. The resulting figure is multiplied by 100 to express the price relative in percentage form.

Interpreting price relatives   The price relatives may be interpreted as follows. In 1984, it would have cost 161% of the price in 1978 to purchase a pound of this brand of coffee; that is, the price increased 61% from 1978 to 1984. Similarly, in 1990, the price was 302% of the 1978 price, so the price had risen 202% from 1978 to 1990.

Composite index numbers   Of course, we are usually interested in price changes for more than one item. For example, in the cases of the family and the transferred executive cited earlier, we were interested in the prices of all commodities and services included in the cost of living; this means combining the price relatives for many different items into a single summary figure for each period. Similarly, we may wish to compute a food price index, a clothing price index, or an index of medical costs. Such summary figures constitute **composite index number** series.

Since our discussion will pertain solely to composite indices, we will ordinarily use the term "index number" to mean "composite index number."

Series of index numbers are extremely useful in the study and analysis of economic activity. Every economy, regardless of the political and social structure within which it operates, is engaged in the production, distribution, and consumption of goods and services. Convenient methods of aggregation, averaging, and approximation are required to summarize the myriad individual activities and transactions. Index numbers have proved to be useful tools in this connection. Thus, we find indices of industrial production, agricultural production, stock market prices, wholesale prices, consumer prices, prices of exports and imports, incomes of various types, and so on, in common use. Economic indices can be conveniently classified as indices of price, quantity, or value. The present discussion concentrates primarily on price indices, because most problems of construction, interpretation, and use of indices may be illustrated in terms of such measures. First, we deal with methods of index number construction, using the illustrative data of a simple example. Then we consider some general problems of index number construction.

Uses for index numbers

12.2
AGGREGATIVE PRICE INDICES

In order to illustrate the construction and interpretation of price indices, we consider the artificial problem of constructing a price index for a list of four food commodities. A realistic counterpart of this problem is the Consumer Price Index (CPI) produced by the Bureau of Labor Statistics (BLS) of the U.S. Department of Labor. This index is used in many ways and provides the basis for many economic decisions. For example, fluctuations in the wages of millions of workers and the Social Security benefits of millions of people are partially based on changes that occur in the index figures. The series is also closely watched by monetary authorities as an indicator of inflationary price movements.

Unweighted Aggregates Index

In our simple illustration, we use a base period of 1980, and we are interested in the change in these prices from 1980 to 1990 for a typical family of four that purchased these products at retail prices in a certain city. The universe and other basic elements of the problem should be carefully defined. (For example, the price of a dozen eggs might be the price of a dozen Grade A, large white eggs.) However, in this problem we purposely leave these matters indefinite and concentrate on the construction and interpretation of the various indices. Table 12-1 shows the basic data of the problem and the calculation of the unweighted aggregates index. As indicated at the bottom of Table 12-1, the prices

TABLE 12-1

Calculation of the unweighted aggregates index for food prices in 1980 and 1990

| | Unit Price | |
Food Commodity	1980 P_{80}	1990 P_{90}
Coffee (pound)	$2.82	$3.53
Bread (loaf)	0.89	1.39
Eggs (dozen)	0.90	1.50
Hamburger (pound)	1.90	2.64
	$6.51	$9.06

Unweighted Aggregates Index for 1990 on 1980 Base

$$\frac{\Sigma P_{90}}{\Sigma P_{80}} \cdot 100 = \frac{\$9.06}{\$6.51} \times 100 = 139.2$$

per unit are summed (or aggregated) for each year. Then one year (in this case, 1980) is selected as a base. The price index for any given year is obtained by dividing the sum of prices for that year by the sum for the base period. The resulting figure is multiplied by 100 to express the index in percentage form. (Hence, the index takes the value 100 in the base period.) If the symbol P_0 is used to denote the price in a base period and P_n the price in a nonbase period, the general formula for the unweighted aggregates index may be expressed as follows:

Unweighted aggregates   (12.1)
price index

$$\frac{\Sigma P_n}{\Sigma P_0} \cdot 100$$

Let us interpret the index figure of 139.2 for 1990. In 1980, it would have cost $6.51 to purchase one pound of coffee, one loaf of bread, one dozen eggs, and one pound of hamburger. The corresponding cost in 1990 was $9.06. Expressing $9.06 as a percentage of $6.51, we find that in 1990 it would have cost 139.2% of the cost in 1980 to purchase one unit each of the specified commodities. In terms of percentage change, it would have cost 39.2% *more* in 1990 than in 1980 to purchase this "market basket" of goods.

The interpretation of the unweighted aggregates index is very straightforward. However, this type of index suffers from a serious limitation.

Limitation of unweighted
aggregates index

> The unweighted aggregates index is unduly influenced by the price variations of high-priced commodities.

The price totals in 1980 and 1990, respectively, were $6.51 and $9.06, an increase of $2.55. If we added to the list of commodities one that declined from

$6.00 to $3.00 per unit from 1980 to 1990, the totals for 1980 and 1990 would then be $12.51 and $12.06. The price index figure for 1990 would be 96.4, indicating a 3.6% decline in prices. Although the prices of four commodities increased and only one decreased, the overall index shows a decline, because of the dominance of the price change in one high-priced commodity. Furthermore, this high-priced commodity may be relatively unimportant in the consumption pattern of the group to which the index pertains. We conclude that this so-called "unweighted index" has an inherent haphazard weighting scheme.

Another deficiency of this type of index is the arbitrary nature of the units for which the prices are stated. For example, if the price of eggs were stated per half-dozen rather than per dozen, the calculated price index figure would change. However, even if the prices of each commodity were stated in the same unit (say, per pound), the problem of the inherent haphazard weighting scheme would still remain; the index would be dominated by the commodities that happened to have high prices per pound. These may be the very commodities that are purchased least, because they are expensive. Clearly, explicit weights are needed to convert a simple aggregative index into an economically meaningful measure. We now turn to weighted aggregative price indices.

Weighted Aggregates Indices

To attribute the appropriate importance to each of the items included in an aggregative index, some reasonable weighting plan must be used. The weights to be used depend on the purposes of the index calculation, that is, on the economic question the index attempts to answer. In the case of a consumer food price index such as the one we have been discussing, reasonable weights would be the amounts of the individual food commodities purchased by the consumer units to whom the index pertains. These would constitute **quantity weights**, since they represent quantities of commodities purchased. The specific types of quantities use in an aggregative index would depend, of course, on the economic nature of the index computed. For example, an aggregative index of export prices would use quantities of commodities and services exported, whereas an index of import prices would use quantities imported.

Table 12-2 shows the prices of the same food commodities given in Table 12-1, but quantities consumed during the base period, 1980, are also shown. The figures in column Q_{80} (the symbol Q denotes quantity) represent average quantities consumed per week in 1980 by the consumer units to which the index pertains. Hence, they indicate an average consumption of one pound of coffee, three loaves of bread, and so on. The figures in the column labeled $P_{80}Q_{80}$ indicate the dollar expenditures for the quantities purchased in 1980. Correspondingly, the numbers in the column headed $P_{90}Q_{80}$ specify what it would cost to purchase these amounts of food in 1990. Therefore, the sums, $\Sigma P_{80}Q_{80} = 9.19 and $\Sigma P_{90}Q_{80} = 13.34, indicate what it would have cost to purchase the specified quantities of food commodities in 1980 and 1990. The index number

---

TABLE 12-2

Calculation of the weighted aggregates index for food prices, using base period quantities consumed as weights (Laspeyres method)

Food Commodity	1980 P_{80}	1990 P_{90}	Quantity 1980 Q_{80}	$P_{80}Q_{80}$	$P_{90}Q_{80}$
Coffee (pound)	$2.82	3.53	1	$2.82	$3.53
Bread (loaf)	0.89	1.39	3	2.67	4.17
Eggs (dozen)	0.90	1.50	2	1.80	3.00
Hamburger (pound)	1.90	2.64	1	1.90	2.64
				$9.19	$13.34

Weighted aggregates index, with base period weights, for 1990 on 1980 base

$$\frac{\Sigma P_{90}Q_{80}}{\Sigma P_{80}Q_{80}} \cdot 100 = \frac{\$13.34}{\$9.19} \times 100 = 145.2$$

---

for 1990 on a 1980 base is given by expressing the figure for $\Sigma P_{90}Q_{80}$ as a percentage of the $\Sigma P_{80}Q_{80}$ figure, yielding in this case a figure of 145.2, shown at the bottom of Table 12-2. Of course, the index number for the base period, 1980, would be 100.0.

This type of index measures the change in the total cost of a fixed market basket of goods. For example, the 145.2 figure indicates that in 1990 it would have cost 145.2% of what it cost in 1980 to purchase the weekly market basket of commodities representing a typical consumption pattern in 1980. Roughly speaking, this indicates an average price rise of 45.2% for this food market basket from 1980 to 1990. We can see that the corresponding index figure of 139.2% for the simple, or unweighted, index is quite close to the 145.2% figure for the weighted index. The reason for this is that in our example we have assumed that the prices of all four commodities moved in the same direction with the percentage changes all falling between about 25% and 67%. If there had been more dispersion in price movements (for example, if some prices increased while others decreased, as is often the case), the weighted index would have tended to differ more from the unweighted one.

The weighted aggregative index using base period weights is known as the **Laspeyres index**. The general formula for this type of index may be expressed as follows:

Weighted aggregates price index, base period weights (Laspeyres method)

(12.2)

$$\frac{\Sigma P_n Q_0}{\Sigma P_0 Q_0} \cdot 100$$

where P_0 = price in a base period
   P_n = price in a nonbase period
   Q_0 = quantity in a base period

The Laspeyres index clearly illustrates the basic dilemma posed by the use of any weighting system. Since an aggregative price index attempts to measure price changes and contains data on both prices and quantities, it appears logical to hold the quantity factor constant in order to isolate change attributable to price movements. If both prices and quantities were permitted to vary, their changes would be entangled and it would be impossible to ascertain what part of the movement was due to price changes alone. However, by keeping quantities fixed as of the base period, the Laspeyres index assumes a frozen consumption pattern. As time goes on, this assumption becomes more unrealistic and untenable. The consumption pattern of the current period would seem to represent a more realistic set of weights from the economic viewpoint.

However, let us consider the implications of an aggregative index using current period (nonbase period) weights, which is known as the **Paasche index**. The general formula for a Paasche index is

(12.3)
$$\frac{\Sigma P_n Q_n}{\Sigma P_0 Q_n} \cdot 100$$

Weighted aggregates price index, current period weights (Paasche method)

If such an index is prepared on an annual basis, the weights would change each year, since they would consist of current year quantity figures. The 1988 Paasche index would be computed by the formula $\Sigma P_{88} Q_{88} / \Sigma P_{87} Q_{88}$, the 1989 index would be $\Sigma P_{89} Q_{89} / \Sigma P_{87} Q_{89}$, and so on. The interpretation of any one of the resulting figures (as price change from the base period, assuming the consumption pattern of the current period) is clear. However, the use of changing current period weights destroys the possibility of obtaining unequivocal measures of year-to-year price change. For example, if we compare the Paasche formulas for the 1988 and 1989 indices (above), we can see that both prices and quantities have changed. Therefore, we can make no clear statement about price movements from 1988 to 1989. The use of current year weights makes year-to-year comparisons of price changes impossible.

Another practical disadvantage of using current period weights is the necessity of obtaining a new set of weights in each period. Let us consider the U.S. Bureau of Labor Statistics CPI to illustrate the difficulty of obtaining such weights. In order to obtain an appropriate set of weights for this index, the BLS conducts a massive sample survey of the expenditure patterns of families in a large number of cities. Such surveys have been carried out in 1917–1919, 1934–1936, 1950–1951, 1960–1961, 1972–1973, and 1982–1984. They are large-scale, expensive undertakings, and it would be infeasible to conduct such surveys once a year or more frequently. Because of these disadvantages, the current period weighted aggregative method is not used in any well-known price index number series.

Because of these considerations and other factors, the most generally satisfactory type of price index is probably the weighted relative of aggregates index using a fixed set of weights. The term "fixed set of weights" rather than "base period weights" is used here, because the weights may pertain to a period different from the one that represents the base for measuring price changes. For example, one of the base periods for the CPI was 1957–1959, whereas the corresponding weights were derived from a 1960–1961 survey of consumer expenditures. The base period for the current CPI is $1982–1984 = 100$. The BLS revises its weighting system about every 10 years and also changes the reference base period for the measurement of price changes with about the same frequency. This procedure constitutes a workable solution to the need for both constant weights (in order to isolate price change) and up-to-date weights (in order to have a recent realistic description of consumption patterns).

The weighted [relative of] aggregates index using a fixed set of weights is referred to as the **fixed-weight aggregative index** and is defined by the formula

Weighted aggregates price   (12.4)
index with fixed weights

$$\frac{\Sigma P_n Q_f}{\Sigma P_0 Q_f}$$

where Q_f denotes a fixed set of quantity weights. The Laspeyres method may be viewed as a special case of this index, in which the period to which the weights refer is the same as the base period for prices. In order to clarify discussion of the two different periods, the term **weight base** is used for the period to which the quantity weights pertain, whereas the term **reference base** is used to designate the period from which the price changes are measured. Of course, a distinct advantage of a fixed-weight aggregative index is that the reference base period for measuring price changes may be changed without a corresponding change in the weight base. This is sometimes a useful and practical procedure, particularly in the case of some U.S. government indices that utilize data from censuses or large-scale sample surveys for changes in weights.

EXERCISES 12.2

1. Krystin Foster, the marketing manager of Sound Dynamics, a manufacturer of blank cassette tapes, wanted to assess the competitiveness of the company's product in the market. She compiled the following table of average prices and quantities of blank cassettes sold in 1983 and 1990:

| | 1983 | | 1990 | |
Length of Tape	Price	Quantity	Price	Quantity
30-minute	$2.10	32	$3.35	40
60-minute	2.50	150	3.75	160
90-minute	3.00	95	4.15	120
120-minute	3.20	12	4.50	30

a. Compute an appropriate weighted aggregates index for 1990 on a 1983 base.

b. A competitor's 1990 index for the same lengths of tapes (1983 base year) was 130. Would you conclude that Sound Dynamics has a higher pricing structure in 1990 than its competitor? Explain.

2. A security analyst wishes to construct an index to reflect the changes in the prices of five stocks in a certain portfolio. The following figures represent the average price and number of shares purchased for each stock for 1988–1990:

	1988		1989		1990	
Stock	Average Price	Shares Purchased	Average Price	Shares Purchased	Average Price	Shares Purchased
A	35	110	45	210	40	215
B	55	210	50	160	55	115
C	65	160	85	260	75	115
D	50	310	55	260	60	215
E	25	200	30	410	40	310

a. Construct an unweighted aggregates index using 1988 as a base year. What are the disadvantages of this type of index?

b. Construct a Laspeyres index using 1988 as a base year. What are the disadvantages of this type of index?

c. Construct a Paasche index using 1988 as a reference base year. What are the disadvantages of this type of index?

12.3
AVERAGE OF RELATIVES INDICES

A second basic method of price index construction is the *average of relatives* procedure. In an **average of relatives** index, the first step involves the calculation of a price relative for each commodity by dividing its price in a nonbase period by the price in a base period. Then an average of these price relatives is calculated. As in the case of aggregative indices, an average of relatives index may be either unweighted or weighted. We consider first the unweighted indices, using the same data on prices as in the preceding section.

Unweighted Arithmetic Mean of Relatives Index

The price data previously shown in Tables 12-1 and 12-2 are repeated in Table 12.3. The first step in the calculation of any average of relatives price index is the calculation of price relatives, which express the price of each commodity as a percentage of the price in the base period. These price relatives for 1990

TABLE 12-3
Calculation of the unweighted arithmetic mean of relatives index of food prices
for 1990 on a 1980 base

	(1)	(2)	(3)	(4)
		Unit Price		Price Relative
		1980	1990	$\dfrac{P_{90}}{P_{80}} \cdot 100$
	Food Commodity	P_{80}	P_{90}	
	Coffee (pound)	$2.82	$3.53	125.2
	Bread (loaf)	0.89	1.39	156.2
	Eggs (dozen)	0.90	1.50	166.7
	Hamburger (pound)	1.90	2.64	138.9
				587.0

Unweighted arithmetic mean of relatives index for 1990, on a 1980 base

$$\frac{\Sigma\left(\dfrac{P_{90}}{P_{80}} \cdot 100\right)}{4} = \frac{587.0}{4} = 146.8$$

on a 1980 base, denoted $(P_{90}/P_{80}) \times 100$, are shown in column (4) of Table
12-3. Theoretically, once the price relatives are obtained, any average (including
the arithmetic mean, median, and mode) could be used as a measure of their
central tendency. The arithmetic mean has been most frequently used, doubtless
because of its simplicity and familiarity. The calculation of the unweighted
arithmetic mean of relatives for 1990 on a 1980 base appears at the bottom of
Table 12-3. The formula for this unweighted arithmetic mean of relatives is
equation 1.1, with the price relatives as the items to be averaged.

Unweighted **(12.5)**
arithmetic mean of
relatives index

$$\frac{\Sigma\left(\dfrac{P_{n}}{P_{0}} \cdot 100\right)}{n}$$

where $\dfrac{P_{n}}{P_{0}} \cdot 100$ = the price relative for a commodity or service

n = the number of commodities and services

As in the case of the unweighted aggregative index, this "unweighted"
index has an inherent weighting pattern. It is useful to consider the implications of this inherent weighting system. In the unweighted arithmetic mean of
relatives, percentage increases are balanced against equal percentage decreases.
For example, if we consider two commodities, one whose price increased by
50% and one whose price declined by 50% from 1980 to 1990, the respective

price relatives for 1990 on a 1980 base would be 150 and 50. The unweighted arithmetic mean of these two figures is 100, indicating that, on the average, prices have remained unchanged. Thus, the unweighted arithmetic mean attaches the same weight to equal percentage changes in opposite directions. However, this method does not provide for explicit weighting in terms of the importance of the commodities whose prices have changed. Since it is widely recognized that explicit weighting is required to permit the individual items in an index to exert their proper influence, virtually none of the important government or private organization price indices are "unweighted." We now consider weighted average of relatives indices.

Weighted Arithmetic Mean of Relatives Indices

Although several averages can theoretically be used for calculating weighted averages of relatives, only the weighted arithmetic mean is ordinarily employed. The general formula for a weighted arithmetic mean of price relatives is

(12.6)
$$\frac{\Sigma\left(\frac{P_n}{P_0} \cdot 100\right)w}{\Sigma w}$$

Weighted arithmetic mean of relatives, general form

where w = the weight applied to the price relatives

Customarily, the weights used in this type of index are values, such as values consumed, produced, purchased, or sold. For example, in the food price index used as our illustrative problem, the weights are values consumed, that is, dollar expenditures on the individual food commodities by the typical family to whom the index pertains. It seems reasonable that the importance attached to the price change for each commodity be determined by the amounts spent on these commodities.

> In index number construction, value = price × quantity.

For example, if a commodity has a price of $0.10 and the quantity consumed is three units, then the *value* of the commodity consumed is $0.10 × 3 = $0.30. Since prices and quantities can pertain to either a base period or a current period, the following systems of weights are all possibilities: $P_0 Q_0$, $P_0 Q_n$, $P_n Q_0$, and $P_n Q_n$. The weights $P_0 Q_0$ and $P_n Q_n$ are, respectively, base period values and current period values; the other two are mixtures of base and current period prices and quantities. Interestingly, the weighting systems $P_0 Q_0$ and $P_0 Q_n$, when used in the weighted arithmetic mean of relatives, result in indices that are algebraically identical to the Laspeyres and Paasche aggregative indices, respectively. This point illustrated in equation 12.7, where base period weights $P_0 Q_0$ are used.

TABLE 12-4

Calculation of the weighted arithmetic mean of relatives index of food prices for 1990 on a 1980 base, using base period weights

(1)	(2)	(3)	(4)	(5)	(6)	(7)
			Price			Weighted Price Relatives
	Prices		Relatives	Quantity		Col. 4 × Col. 6
Food	1980	1990	$\frac{P_{90}}{P_{80}} \cdot 100$	1980		$\left(\frac{P_{90}}{P_{80}} \cdot 100\right)(P_{80}Q_{80})$
Commodity	P_{80}	P_{90}		Q_{80}	$P_{80}Q_{80}$	
Coffee (pound)	$2.82	3.53	125.2	1	$2.82	$353.064
Bread (loaf)	0.89	1.39	156.2	3	2.67	417.054
Eggs (dozen)	0.90	1.50	166.7	2	1.90	300.060
Hamburger (pound)	1.90	2.64	138.9	1	$9.19	263.910
						$1334.088

Weighted arithmetic mean of relatives for 1990, on 1980 base, using base period value weights

$$\frac{\Sigma\left(\frac{P_{90}}{P_{80}} \cdot 100\right)(P_{80}Q_{80})}{\Sigma P_{80}Q_{80}} = \frac{\$1,334,088}{\$9.19} = 145.2$$

Weighted arithmetic mean of relatives, with base period weights

(12.7)

$$\frac{\Sigma\left(\frac{P_n}{P_0}\right)P_0Q_0}{\Sigma P_0Q_0} \cdot 100 = \frac{\Sigma P_nQ_0}{\Sigma P_0Q_0} \cdot 100$$

As is clear from equation 12.7, the P_0's in the numerator cancel, yielding the Laspeyres index. The calculation of the weighted arithmetic mean of relatives using 1980 base period value weights is given in Table 12-4 for the data of our illustrative problem. The numerical value of the index is, of course, exactly the same as that obtained previously for the weighted aggregative index with base period quantity weights (Laspeyres method) in Table 12-2.

Since the two indices in equation 12.7 are algebraically identical, it would seem immaterial which is used, but there are instances when it is more feasible to compute one than the other. For example, it is more convenient to use the weighted average of relatives than the Laspeyres index

● when value weights are easier to obtain than quantity weights
● when the basic price data are more easily obtainable in the form of relatives than absolute values
● when an overall index is broken down into a number of component indices and we wish to compare the individual components in the form of relatives.

As an illustration of the first situation, it is usually easier for manufacturing firms to furnish value of production weights in the form of "value added by manufacturing" (sales minus cost of raw materials) than to provide detailed data on quantities produced.

As indicated earlier, the Paasche index and the weighted arithmetic mean of relatives with a $P_0 Q_n$ weighting system are algebraically identical. The reasons given for the wider use of the Laspeyres than the Paasche index apply to a similarly wider usage of weighted means of relatives with $P_0 Q_0$ than with $P_0 Q_n$ weights. The other two possible value weighting systems, $P_n Q_0$ and $P_n Q_n$, create interpretational difficulties and therefore are not utilized in any important indices.

EXERCISES 12.3

1. An automobile manufacturer produces three types of cars. The average unit selling prices and quantities sold in 1986 and 1990 follow:

| | 1986 | | 1990 | |
Model	Price	Quantity (thousands)	Price	Quantity (thousands)
A	$14,000	200	$16,000	250
B	14,800	400	17,500	380
C	15,500	200	19,000	150

   a. Calculate the index of car prices for 1986 on a base year of 1990. Use the weighted arithmetic mean of relatives index with base-period weights.
   b. Explain what the value of your index means in terms that can be understood by a layperson.

2. The average weekly stock prices of Overseas Trading, Inc., a U.S.-based multinational company, can be estimated from stock-price movements of its Asian and European subsidiaries for the previous week. The average weekly prices and the number of shares traded for the Asian and European subsidiaries are given for the weeks in October 1990:

| | Average Weekly Prices | | Shares Traded (thousands) | |
Week Starting	Asian	European	Asian	European
October 4	$35.00	$32.00	1,020	2,000
October 11	36.50	33.00	1,200	2,500
October 18	36.00	31.00	1,500	2,800
October 25	36.25	32.50	1,300	2,200

a. Compute the stock-price index for Overseas Trading, Inc., using the weighted arithmetic mean of relatives index with October 4 = 100, using base-period weights.

b. Compute a price index by the weighted aggregate method with October 4 = 100, using base-period weights.

3. A grain trader in the futures market constructed a price index of barley and wheat from the following data:

Commodity	Unit Price (dollars per bushel)		Quantities for Future Delivery (thousands of bushels)	
	1988	1989	1988	1989
Wheat	$2.40	$2.64	40	50
Barley	3.30	3.00	60	50

A simple unweighted arithmetic mean of the two price relatives for 1989 on a 1988 base indicates that prices in 1989 were, on the average, 0.5% higher than in 1988. A simple unweighted arithmetic mean of the two price relatives for 1988 on a 1989 base indicates that prices in 1988 were, on the average, 0.5% higher than in 1989.

a. How do you explain these paradoxical results?

b. Compute what you consider to be the most generally satisfactory price index for 1989, using 1988 as a base year. Use any of the given data that seem appropriate to you.

c. Explain your answer in part (b).

12.4
GENERAL PROBLEMS OF INDEX NUMBER CONSTRUCTION

In a brief treatment, it is not feasible to discuss all the problems of index number construction. However, many of the important matters are included in the following two categories: (1) selection of items to be included and (2) choice of a base period.

Selection of items to be included

> In the construction of price indices, as in other problems involving statistical methods, the definition of the problem and the statistical universe to be investigated are of paramount importance.

Most of the widely used price index number series are produced by government agencies or sizable private organizations and are used in many ways. Hence, it is not feasible to state a simple purpose for each price index that will clearly define the problem and the statistical universe. However, every index attempts to answer meaningful questions, and these general purposes of an index determine the specific items to be included. For example, the CPI attempts to answer a question concerning the average movement of certain prices over time. The specific nature of this question about price movements determines which items are included in the index. Similarly, many limitations of the use of the index stem from what the index does and does not attempt to measure.

Let us pursue the illustration of the CPI. Essentially, this index attempts to measure how much it would cost to purchase (at retail) a particular combination of goods and services compared with what it would have cost in a base period. More specifically, the combination of goods and services consists of items that represent a typical market basket of purchases by all urban consumers. Prior to 1978, the index covered only city wage earners and city clerical workers and their families, but in 1978 the index was expanded to include all urban consumers. The BLS continues to monitor the narrower index as well. By means of periodic consumer surveys, the BLS determines the goods and services purchased by the specified consumers and how they spread their spending among these items. The more broadly based CPI for All Urban Consumers (CPI-U) takes into account the buying patterns of professional and salaried workers, part-time workers, the self-employed, the unemployed, and retired people, in addition to wage earners and clerical workers. The most recent expenditures weights for both the CPI and the more broadly based index CPI-U were derived from the 1982–1984 Consumer Expenditures Survey.

The general question the index purports to answer determines the items to be included.

Summary

Obviously, indices constructed for other purposes—such as indices of export prices or agricultural prices—would be based on different lists of items.

However, even when the general purpose of an index is clearly defined, many problems remain concerning the choice of items to be included. In the case of the CPI, the BLS has determined that urban consumers purchase thousands of items.

However, the BLS includes only about 400 of these goods and services, having found that these few hundred accurately reflect the average change in the cost of the entire market basket. The choice of the commodities to be included in a price index is ordinarily not determined by usual sampling procedures. Each good or service cannot be considered a random sampling unit that is as representative as any other unit. Rather, an attempt is made to include practically all of the most important items and, by pricing these, to obtain a representative portrayal of the movement of the entire population of prices. If subgroup indices are required (for example, indices of food, housing, or medical care), as well as an overall consumers' price index, more items must be included than if only the overall index were desired. After the decisions have been made concerning the commodities to be included, sophisticated sampling procedures are often utilized to determine which specific prices will be included.

Choice of a base period

A second problem in the construction of a price index is the choice of a base period, that is, a period whose level of prices represents the base from which changes in prices are measured. As indicated earlier, the level of prices in the base period is taken as 100%. Price levels in nonbase periods are stated as percentages of the base period level.

Conventional calendar time with "normal" price levels

> The base period may be a conventional calendar time interval such as a month, a year, or even a period of years. Ideally, a period with "normal" price levels should be used.

Of course, it is virtually impossible to devise a meaningful definition of "normal" in almost any area of economic experience, but the time period selected should not be at or near the peaks or troughs of price fluctuations. Although there is nothing *mathematically* incorrect about using a base period with unusually low or high price levels, the use of such time intervals tends to produce distorted concepts, since comparisons are made with atypical periods.

The use of a period of years as a base produces an averaging effect on year-to-year variations. Any particular year may have unusual influences present, but if, say, a three- to five-year base period is used, these variations will tend to even out. Most of the U.S. government indices have used such time intervals (for example, 1935–1939, 1947–1949, 1957–1959, and 1982–1984) as base periods.

Another point to consider in choosing a base time interval is suggested by the three periods just mentioned:

Base period should not be too distant

> The base period should not be too distant from the present.

The farther away we move from the base period, the less we know about economic conditions prevailing at that time. Consequently, comparisons with these remote periods lose significance and become rather tenuous in meaning. This is why producers of index number series, such as U.S. government agencies, shift their base periods every decade or so, in order to make comparisons with a base time interval in the recent past. Furthermore, it is desirable to shift the base from time to time because a period previously thought of as normal or average may no longer be so considered after a long lapse of time.

Other considerations

Other considerations may also be involved in choosing a base period for an index. If a number of important existing indices have a certain base period, then newly constructed indices may use the same time period for ease of comparisons. Moreover, as new commodities are developed and indices are revised to include them, the base period may be shifted to a time interval that reflects the newer economic environment.

12.5
QUANTITY INDICES

The discussion in the preceding sections referred to price indices. Another important group of summary measures of economic change is represented by **quantity indices**, which measure changes in physical *quantities* such as the volume of industrial production, physical volume of imports and exports, quantities of goods and services consumed, and volume of stock transactions.

> Virtually all currently used *quantity indices* measure the change in the *value* of a set of goods from the base period to the current period attributed only to changes in *quantities*, prices being held constant.

This corresponds to what is measured in weighted price indices as the change in the *value* of a set of good attributed only to changes in *prices*, quantities being held constant. The same types of procedures used to calculate price indices are also employed to obtain quantity indices. Except for the case of the unweighted aggregates index, which would not be meaningful for a quantity index, corresponding quantity indices may be obtained by interchanging P's and Q's in the formulas given earlier in this chapter.

An unweighted average of relatives quantity index can be determined by establishing quantity relatives $(Q_n/Q_o) \times 100$ and calculating the arithmetic mean of these figures. An unweighted aggregative quantity index would not be meaningful, because it does not make sense to add up quantities stated in different units.

As was true for price indices, weighted quantity indices are preferable to unweighted ones. A weighted aggregative index of the Laspeyres type is given by the following formula:

(12.8)
$$\frac{\Sigma Q_n P_0}{\Sigma Q_0 P_0} \cdot 100$$

Weighted relative of aggregates quantity index, base period weights (Laspeyres method)

Just as the corresponding Laspeyres price index measures the change in price levels from a base period assuming a fixed set of quantities produced or consumed in the base period, and so on, this quantity index measures the change in quantities produced or consumed, and so on, assuming a fixed set of prices that existed in the base period. As with price indices, a quantity index computed by the weighted average of relatives method, using the base period value weights given in equation 12.9, is algebraically identical to this Laspeyres index.

(12.9)
$$\frac{\Sigma \left(\dfrac{Q_n}{Q_0} \cdot 100 \right) Q_0 P_0}{\Sigma Q_0 P_0}$$

Weighted arithmetic mean of relatives quantity index, base period weights

Let us interpret these two equivalent weighted indices by considering the Laspeyres version, given in equation 12.9. We continue the assumption that the raw data refer to quantities of food items consumed (during one week) and prices paid by a typical family in an urban area. The numerator of the index shows the value of the specified food items consumed in the nonbase year at base year prices. The denominator refers to the value of the food items consumed in the base year. Suppose a figure of 125 resulted from such an index. Since prices were kept constant, the increase would be solely attributed to an average increase of 25% in the quantity of these items consumed.

FRB Index of Industrial Production

Probably the most widely used and best-known quantity index in the United States is the Federal Reserve Board (FRB) Index of Industrial Production. This index measures changes in the physical volume of output of manufacturing, mining, and utilities. In addition to the overall index of industrial production, component indices are published by industry groupings such as Manufactures and Minerals, and by subcomponents such as Durable Manufactures and Nondurable Manufactures. Separate indices are reported for the output of consumer goods, output of equipment for business and government use, and output of materials. Based on the groupings used by the Standard Industrial Classification of the U.S. Bureau of the Budget, indices are also prepared for major industrial groups and subgroups. The indices are issued monthly, with 1977 as both reference base and weight base. The Index of Industrial Production is closely watched by business executives, economists, and financial analysts as a major indicator of the physical output of the economy.

The method of construction is the weighted arithmetic mean of relatives, using the base periods mentioned above. Numerous problems have had to be resolved concerning both the quantity relatives and value weights. Since many industries cannot easily provide physical output data for the quantity relatives, related data that tend to move more or less parallel to output (such as shipments and employee-hours worked) are sometimes used instead. The weights used are value-added data, which at the individual company level represent the sales of the firm minus all purchases of materials and services from other business firms.

Double counting

> The index uses value-added rather than value of final product weights to avoid the problem of **double counting**.

For example, if the value of the final product were used for a steel company that sells its steel to an automobile company, and the value of the final product of the automobile company were also used, the steel that went into making the automobile would be counted twice. Hence, the weights used follow the value-added approach, in which the values of so-called "intermediate products" produced at all stages prior to the final product are excluded. From the viewpoint of the economist, a firm's value added is conceptually equivalent to the total of its factor of production payments—wages, interest, rent, and profits.

12.6
DEFLATION OF VALUE SERIES BY PRICES INDICES

One of the most useful applications of price indices is in adjusting series of dollar figures for changes in levels of prices. The result of this adjustment procedure, known as **deflation**, is a restatement of the original dollar figures in terms of so-called "constant dollars." The rationale of the procedure can be illustrated in terms of the simple example given in Table 12-5. Column (2) shows average (arithmetic mean) weekly wage figures for factory workers in a large city in 1980 and 1990. Such unadjusted dollar figures are said to be stated in "current dollars." Column (3) shows a CPI for the given city, with 1980 as reference base period. For simplicity of interpretation, let us assume that the CPI was computed by the Laspeyres method. As we note from column (2), average weekly wages of the given workers have increased from $210.00 in 1980 to $407.60 in 1990, a gain of 94.1%. But can these workers purchase 94.1% more goods and services with this increased income? If prices of the goods and services had remained unchanged between 1980 and 1990, then all other things being equal, the answer would be yes. But as we can see in column (3), prices rose 51% over this period. To determine what average weekly wages are in terms of 1980 constant dollars (dollars with 1980 purchasing power), we carry out the division $407.60/1.51 = $269.93. That is, we divide the 1990 weekly wage figure in current dollars by the 1990 CPI stated as a decimal figure (using a base of 1.00 rather than 100) to obtain the figure $269.93 for average weekly wages in 1980 constant dollars. As indicated in the heading of column (4), the result of this adjustment for price change is referred to as **real wages** (in this case, "real average weekly wages"). The term *real* implies that a portion of the dollar increase in wages is absorbed by the increase in prices, so the adjustment attempts to isolate the "real change" in the volume of goods and services the weekly wages can purchase at base year prices. In summary, the dollar value figures in column (2) are divided by the price index figures in column (3) (stated on a base of 1.00) to obtain the real value figures in column (4). The same

Real wages

TABLE 12-5
Calculation of average weekly wages in 1980 constant dollars for factory workers in a large city, 1980 and 1990

(1) Year	(2) Average Weekly Wages	(3) Consumer Price Index (1980 = 100)	(4) "Real" Average Weekly Wages (1980 constant dollars)
1980	$210.00	100	$210.00
1990	$407.60	151	269.93

procedure would have been followed if there had been a series of figures in columns (2) and (3) (perhaps for several consecutive years) rather than just the current and base period figures.

> This division of a dollar figure by a price index is referred to as a **deflation of the current dollar value series**, whether a decrease or an increase occurs in going from figures in current dollars to constant dollars.

Value aggregates

The rationale of the deflation procedure stems from the basic relationship value = price × quantity. The weekly wages in current dollars are value figures: they may be viewed as **value aggregates**, that is, as sums of prices of labor times quantities of such labor. By dividing such a figure by a price index, we attempt to isolate the change attributable to quantity or physical volume. Hence, we may think of the real average weekly wage figures as reflecting the changes in quantities of goods over which the wage figures have command.

This deflation procedure is widely used in business and economics. One interesting application is in measuring economic well-being and growth. For example, in comparing growth rates among countries, one of the most important indicators used is per capita growth in real GNP. Dividing GNP by population to yield per capita figures may be viewed as an adjustment for differences in population size. The division of the figures by a relevant price index to obtain real GNP is an adjustment for change in price levels. Per capita real GNP is an extremely useful measure of physical volume of production.

Of course, there are numerous limitations to the use of the deflation procedure. For example, in the weekly wages illustration, the market basket of commodities and services implicit in the consumer price index may not refer specifically to the factory workers to whom the weekly wages pertain. Even if the index had been constructed for this specific group of factory workers, it is still only an average, subject to all the interpretational problems of any measure of central tendency. Furthermore, inferences from such data must be made with care. For instance, an increase in real average weekly wages from one period to another does not imply an increase in economic welfare for the factory worker group in question; in the later period, there may be a less equitable distribution of this income, taxes may be higher (leading to lower disposable income), and so on. Nevertheless, despite such limitations and caveats, the deflation procedure is a useful, practical, and widely utilized tool of business and economic analysis.

We have seen how a price index may be used to remove from a value aggregate the change attributable to price movements. We may also view the deflation procedure as a method of adjusting a value figure for changes in the purchasing power of money. In this connection, we should note that

Purchasing power index

> a purchasing power index is conceptually the reciprocal of a price index.

For example, suppose you have $40 to purchase shoes in a certain year when a pair of shoes cost $20. The $40 enables you to purchase two pairs of shoes. But if in a later year the price of shoes rises to $40, you can then purchase only one pair of shoes. Let us imagine a price index composed solely of the price of this pair of shoes. If the earlier year is the base period, the base period price index is 100 and the later period figure is 200. On the other hand, if the price of these shoes has doubled, the purchasing power of the dollar relative to shoes has halved. Hence, a purchasing power index that was 100 in the earlier base year should stand at 50 in the later period. If the indices are stated using 1.00 rather than 100 in the base period, the reciprocal relationship can be expressed as $2 \times \frac{1}{2} = 1$. That is, the doubling in the price index and the halving in the purchasing power index are reciprocals. This relationship between price and purchasing power indices is implied in such popular statements as that a dollar today is worth only 50 cents in terms of the dollar in some earlier period.

EXERCISES 12.6

1. The Farmers Cooperative Association boasted that under its intensive farming program the growth in total feed grain production from 1988–1990 was at least twice its growth from 1988–1989. The prices and quantities produced for each type of feed grain appear in the following table for 1988–1990:

Type of Feed Grain	Total Production (bushels)			Price (per bushel)		
	1988	1989	1990	1988	1989	1990
A	40,930	42,160	45,500	$3.58	$3.80	$3.70
B	30,090	31,600	30,020	2.32	2.25	2.40
C	47,950	50,827	54,000	4.47	4.61	4.06

Compute an index of feed grain production by the weighted aggregates method, using 1988 = 100 and base year weights. Is the association's claim valid?

2. The MFX Corporation would like you to assess its market share for three recent years. Provide reasonable estimates of the company's market share using the following data on industry sales, industry price indices, and the company's sales revenue. Note that the company's sales figures are given in 1985 dollars.

Year	Industry Sales (thousands of dollars)	Wholesale Price Index (1985 = 100)	Company Sales in Constant 1985 Dollars (thousands of dollars)
1989	$20,550	186.4	$2,205
1990	32,600	195.6	3,666
1991	43,000	218.9	4,910

3. To provide incentives for international assignments, the International Marketing Corporation adjusts the current salaries of its U.S. employees assigned to foreign countries using the indices of foreign country living costs published by the State Department. If a U.S. employee earns an annual salary of $100,000 in Washington, D.C., determine his or her equivalent salary if assigned to any of the following countries: Japan, Philippines, Brazil, Canada, and Switzerland. The cost of living indices for these countries appears in the following table with a base index of 100 for Washington, D.C.:

Country	City	Cost of Living Index
United States	Washington, D.C.	100
Japan	Tokyo	179
Philippines	Manila	87
Canada	Ottawa	100
Switzerland	Bern	187

4. The U.S. gross national product was $3,721.7 billion in 1986 and $3,847.0 billion in 1987. Using the following price index published in the *Economic Report of the President, 1988*, as a price deflator, calculate the percentage increase in "real" GNP from 1986 to 1987.

Year	Price Index
1982	100.0
1986	113.9
1987	117.7

5. A new graduate is considering two job offers. One job requires that she relocate to city A; the other, to city B. She provides you with the following information:

City	Salary Offer	City Cost-of-Living Index (U.S. average = 100)
A	$54,000	116
B	49,000	90

a. Based solely on a salary criterion, which job offer is preferable?
b. Comment on the procedure you followed to answer part (a).

12.7
SOME CONSIDERATIONS IN THE USE
OF INDEX NUMBERS

Numerous problems arises in connection with the use of index numbers for analysis and decision purposes. A few of these are discussed below.

For a variety of reasons, it is often necessary to change the reference base of an index number series from one period to another without returning to the original raw data and recomputing the entire series. This change of reference base period is usually referred to as **shifting the base**.	*Shifting the base*

For example, we may want to compare several index number series computed on different base periods. Particularly if the several series are to be shown on the same graph, we may want them to have the same base period. In other situations, the shifting of a base period may simply reflect the desire to state the series in terms of a more recent period. The simple procedure for accomplishing the shift is illustrated in the following example, in which a food price index for a certain city with a reference base of 1980 is shifted to a new base period of 1986.

In Table 12-6, the original price index is shown in the first column stated on a base period of 1983. The shift to a 1989 base period is accomplished by dividing each figure in the original series by the index number for the desired new base period stated in decimal form. Hence, in this illustration, each index number on the old 1983 base is divided by 1.237, the 1989 figure stated as a decimal. Thus, the index number of 1982 shifted to the new base of 1989 is 78.7 $(\frac{97.3}{1.237})$; for 1983, the new figure is 80.8 $(\frac{100.0}{1.237})$.

TABLE 12-6

	Food Price Index (1983 = 100)	Food Price Index (1989 = 100)
1982	97.3	78.7
1983	100.0	80.8
1984	103.7	83.8
1985	106.9	86.4
1986	109.4	88.4
1987	113.7	91.9
1988	118.3	95.6
1989	123.7	100.0
1990	129.8	104.9

Note that the relationships among the new index figures after the base is shifted are the same as in the old series. For example, the index number for 1983 exceeds that of 1982 by the same percentage in both series. That is, $(\frac{100.0}{97.3}) = (\frac{80.8}{78.7}) = 1.027$, and so on throughout the two series. However, a subtle problem arises with the weighting scheme. Let us suppose that the old series was computed using a Laspeyres type index. Hence, both the reference base and weight base periods are 1983. The procedure of dividing the series by the 1989 index number changes the reference period, but the weights still pertain to 1983. That is, the raw data originally collected for quantity weights pertained to 1983. The mere procedure of dividing the index numbers in the old series by one of its members does nothing to change these weights. Indeed, obtaining new weights for 1989 would involve a new data collection process. In summary, the new series has been shifted to a reference base period of 1989 for measuring price changes, but the weights are fixed at 1983. This point may be demonstrated algebraically as follows. Consider the Laspeyres price index for 1984 computed on the reference base of 1980. The formula for computing this figure may be written as $\Sigma P_{84}Q_{83}/\Sigma P_{83}Q_{83}$. Correspondingly, the formula for the 1989 price index on the 1983 base is $\Sigma P_{89}Q_{83}/\Sigma P_{83}Q_{83}$. (The multiplication by 100 has been dropped in these formulas to simplify the discussion.) To obtain the new price index figure for 1984 on a 1989 base, we divide the old 1984 figure by the old 1989 figure.

$$\frac{\Sigma P_{84}Q_{83}}{\Sigma P_{83}Q_{83}} \div \frac{\Sigma P_{89}Q_{83}}{\Sigma P_{83}Q_{83}} = \frac{\Sigma P_{84}Q_{83}}{\Sigma P_{89}Q_{83}}$$

Since the $\Sigma P_{83}Q_{83}$ values cancel, we see that the resulting index figure for 1984 is stated on a reference base of 1989, but the weights still pertain to 1983.

> Despite the fact that weights are not changed by the procedure discussed here, this method of shifting reference bases is widely employed. It often represents the only practical way of shifting a base, and analysts ordinarily do not view as a matter of serious concern the fact that the weighting system remains unchanged.

Splicing   Sometimes an index number series is available for a period of time and then undergoes substantial revision, including a shift in the reference base period. In these cases, if we desire a continuous series going back through the period of the older series, the old and revised series must be *spliced* together. Splicing involves a similar arithmetic procedure to that for shifting a base. For example, suppose a price index number series was revised by inclusion of certain new products, exclusion of some old products, and change in definition of some other products. Table 12-7 shows such an old series on a reference base of 1984 and the revised series on a base of 1986. There must be an overlapping period of the old and revised series to provide for splicing them, and the period of overlap in this example is 1986. The splicing of the two series to obtain a

TABLE 12-7

	Old Price Index (1984 = 100)	Revised Price Index (1986 = 100)	Spliced Price Index (1986 = 100)
1983	96.9		88.9
1984	100.0		91.7
1985	105.2		96.5
1986	109.0	100.0	100.0
1987		104.3	104.3
1988		106.8	106.8
1989		108.1	108.1
1990		110.1	110.1

continuous series on the new base of 1986 is accomplished by dividing each figure in the old series by the old index figure for 1986 stated in decimal form, that is, by 1.09. This restates the old series on the new base of 1986. The resulting spliced series is shown in the last column of Table 12-7. Had it been desired to state the continuous series on the old reference base of 1984, each figure in the revised index would be multiplied by 1.09.

> Although the arithmetic procedure involved in splicing is simple, the interpretation of the resulting continuous series may be extremely difficult, particularly if long time periods are involved.

For example, it is difficult to specify precisely what is measured if a price index in the later period contains prices of frozen foods, clothing made from synthetic fibers, television, and similar recently developed products, whereas the spliced indices for the earlier period (before these products were on the market) do not contain these products.

> Despite such conceptual difficulties, splicing frequently represents the only practical method of comparing similar phenomena measured by indices over different time periods.

Quality changes

In the construction of an index such as the CPI, the basic data on prices are collected by trained investigators who price goods for which detailed specifications have been made. The same items are always priced in the same stores. However, as a result of technological and other improvements, a corresponding improvement in the quality of many commodities often occurs over time. It is difficult, and in many cases impossible, to make suitable adjustments in a price index for **quality changes**.

> The artificial but practical procedure adopted by the Bureau of Labor Statistics considers a product's quality improved only if changes that increase the cost of producing the product have occurred.

For example, an automobile tire is not considered improved if it delivers increased mileage at the same cost of production. Because of such actual improvements in product quality, many analysts feel that over reasonably long periods of time, indices such as the CPI that have shown steady rises in price levels overstate the actual price increases of a fixed market basket of goods.

Uses of Indices Index number series are widely used in connection with decision making and analysis in business and government. One of the best known applications of a price index is the use of the CPI as an escalator in collective bargaining contracts. In this connection, millions of workers are covered by contracts that specify periodic changes in wage rates depending on the amount by which the CPI moves up or down. The Bureau of Labor Statistics Producers Price Index is similarly used for escalation clauses in contracts between business firms.

Much use is made of index numbers by individual companies as well as at the levels of entire industries and the overall economy. In certain industries, it is standard practice to key changes in selling prices to changes in indices of prices of raw materials and wage earnings. Assessments of past trends and current status and projection of future economic activity are made on the basis of appropriate indices. Economists follow many of the various indices in order to appraise the performance of the economy and to analyze its structure and behavior.

KEY TERMS

Aggregates Price Index an index number obtained by dividing a sum of prices (unweighted or weighted) by a corresponding base period figure.

Average of Relatives Price Index an index number obtained by averaging price relatives (unweighted or weighted).

Deflation the restatement of "current dollar" figures in terms of "constant dollars."

Index Number a summary measure that presents a relative comparison between groups of related items

Purchasing Power Index the reciprocal of a price index.

Price Relative the figure obtained by dividing a price by a corresponding base price figure

Quantity Index an index that measures the change in the value of a set of goods attributed only to changes in quantities, with prices being held constant.

Splicing the combination of an old index and a revised index to obtain a continuous series on a new base.

KEY FORMULAS

Laspeyres Index (weighted aggregates price index, base period weights)

$$\frac{\Sigma P_n Q_0}{\Sigma P_0 Q_0} \cdot 100$$

Paasche Index (weighted aggregates price index, current period weights)

$$\frac{\Sigma P_n Q_n}{\Sigma P_0 Q_n} \cdot 100$$

Unweighted Aggregates Price Index

$$\frac{\Sigma P_n}{\Sigma P_0} \cdot 100$$

Unweighted Arithmetic Mean of Price Relatives Index

$$\frac{\Sigma \left(\dfrac{P_n}{P_0} \cdot 100 \right)}{n}$$

Weighted Arithmetic Mean of Price Relatives

$$\frac{\Sigma \left(\dfrac{P_n}{P_0} \cdot 100 \right) w}{\Sigma w}$$

Weighted Relative of Aggregates Quantity Index, Base Period Weights

$$\frac{\Sigma Q_n P_0}{\Sigma Q_0 P_0} \cdot 100$$

13

NONPARAMETRIC STATISTICS

Most of the methods discussed thus far have involved assumptions about the distributions of the populations sampled. For example, when certain hypothesis-testing techniques are used, it assumed that the observations are drawn from normally distributed populations. A number of useful techniques that do not make these restrictive assumptions have been developed. Such procedures are referred to as **nonparametric** or **distribution-free** tests. Many writers prefer the latter term, because it emphasizes the fact that the techniques are free of assumptions concerning the underlying population distribution. However, the two terms are generally used synonymously.

13.1
CHARACTERISTICS

In addition to making less restrictive assumptions than the corresponding so-called "parametric" methods, nonparametric procedures are generally easy to carry out and understand. Furthermore, as is implied by the lack of underlying assumptions, they are applicable under a wide range of conditions. Many non-parametric tests are in terms of the *ranks* or *order* rather than the numerical values of the observaions. Sometimes, even ordering is not required.

     However, when distribution-free procedures are applied where parametric techniques are possible, the nonparametric methods have one disadvantage. When these nonparametric procedures use ordering or ranking rather than the actual numerical values of the observations, they are ignoring a certain amount

Advantages

A "disadvantage"

of information. As a result, nonparametric tests are somewhat less efficient than the corresponding standard tests. This means that in testing at a given level of significance, say $\alpha = 0.05$, the probability of a Type II error, β, would be greater for the nonparametric than for the parametric test.

> Advocates of nonparametric tests argue that, despite the lessened efficiency of nonparametric tests, the analyst can have more confidence in these tests than in the standard ones that often require restrictive and somewhat unrealistic assumptions.

In this chapter, we consider a few simple and widely applied nonparametric techniques.

13.2
THE SIGN TEST

We saw in Chapter 7 that—in business and public administration and in social science research—the solution to many problems centers on a comparison between two different samples. In some of the hypothesis-testing techniques previously discussed, restrictive assumptions about the populations sampled were necessary. For example, the t test for the difference between two sample means assumes that the populations are normally distributed and have equal variances. Sometimes, one or both of these assumptions may be unwarranted. Furthermore, situations often exist in which quantitative measurements are impossible. In such cases, it may be possible to assign ranks or scores to the observations in each sample. In these situations, the **sign test** can be used. The name of the test indicates that the signs of observed differences (that is, positive or negative signs) are used rather than quantitative magnitudes.

We will illustrate the sign test in terms of data obtained from a panel of 60 beer-drinking consumers. Let us assume a blindfold test in which the tasters were asked to rate two brands of beer (Wudbeiser and Diller) on a scale from 1 to 5, with 1 representing the best taste (excellent) and 5 representing the worst taste (poor) and the other scores denoting the appropriate intermediates. Table 13-1 shows a partial listing of the scores assigned by the panel members in this taste test.

Column (4) shows the signs of the difference between the scores assigned by each participant in columns (2) and (3). As indicated, a plus sign means a higher numerical score was assigned to Wudbeiser than to Diller beer, a minus sign means Diller beer was rated higher than Wudbeiser, and a zero denotes a tie score. Let us assume that the following results were obtained:

+ Scores	35
− Scores	15
0 Scores	10
Total	60

TABLE 13-1
Ranking scores assigned to taste of 2 brands of beer by a panel

(1) Panel Member	(2) Score for Wudbeiser	(3) Score for Diller	(4) Sign of Difference
A	3	2	+
B	4	1	+
C	2	4	−
D	3	3	0
E	1	2	−
⋮	⋮	⋮	⋮

Note: Best score = 1; worst score = 5. Hence, a plus sign means Diller is preferred; a minus sign means Wudbeiser is preferred.

Method

By means of the sign test, we can test the null hypothesis of no difference in rankings of the two brands of beer. More specifically, we can test the hypothesis that plus and minus signs are equally likely for the differences in rankings. If this null hypothesis were true, we would expect about equal numbers of plus and minus signs. We would reject the null hypothesis if too many of one type of sign occurred. If we use p to denote the probability of obtaining a plus sign, we can indicate the hypotheses as

$$H_0: p = 0.50$$

$$H_1: p \neq 0.50$$

Since tied cases are excluded in the sign test, the data used for the test consist of 35 pluses and 15 minuses. The problem is conceptually the same as one in which a coin has been tossed 50 times, yielding 35 heads and 15 tails, and we wish to test the hypothesis that the coin is fair. The binomial distribution is the theoretically correct one. However, we can use the large-sample method of section 7.2 consisting of the normal curve approximation to the binomial distribution. In terms of proportions, the mean and standard deviation of the sampling distribution are

$$\mu_{\bar{p}} = p = 0.50$$

$$\sigma_{\bar{p}} = \sqrt{\frac{pq}{n}} = \sqrt{\frac{(0.50)(0.50)}{50}} = 0.071$$

Assuming that the test is performed at the 5% level of significance ($\alpha = 0.05$), we would reject the null hypothesis if $z < -1.96$ or $z > 1.96$.

Since, in this problem, the observed proportion of plus signs is $\bar{p} = \frac{35}{50} = 0.70$, then

$$z = \frac{\bar{p} - p}{\sigma_{\bar{p}}} = \frac{0.70 - 0.50}{0.071} = 2.82$$

FIGURE 13-1
Sample distribution of a proportion for beer-tasting problem: $p = 0.50$, $n = 50$. This is a two-tailed test with $\alpha = 0.05$

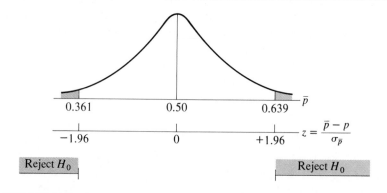

Hence, we reject the null hypothesis that plus and minus signs are equally likely. Since the plus signs exceeded minus signs in the observed data, our interpretation of the experimental data is that Diller beer is preferred to Wudbeiser, according to the rank scores given by the consumer panel.

Note that the arithmetic could have been carried out in terms of critical limits of $\bar{p}$, rather than for z values. The critical limits for $\bar{p}$ are

$$p + 1.96\sigma_{\bar{p}} = 0.50 + (1.96)(0.071) = 0.639$$

$$p - 1.96\sigma_{\bar{p}} = 0.50 - (1.96)(0.071) = 0.361$$

Since the observed $\bar{p}$ of 0.70 exceeds 0.639, we reach the same conclusion and reject H_0. The testing procedure is shown in the usual way in Figure 13-1.

Two points may be made concerning the techniques used in this illustration of the sign test.

- Although a two-tailed test was appropriate for this problem, the sign test can also be used in one-tailed test situations.
- A normal curve approximation to the binomial distribution was used. For small samples, binomial probability calculations and tables should be used.

General Comments

As we have seen, the sign test is simple to apply. In the example, it was not feasible to obtain quantitative data for beer tasting, so rank scores were used.

> Because of its simplicity, the sign test is sometimes used instead of a standard test even when quantitative data are available.

For example, suppose we refer to the beer taste rankings for any particular individual as observations from a "matched pair." We might also have matched-pair observations when the data are, say, weights before and after a diet, grades on a preliminary scholastic aptitude test and on the regular aptitude test, or other pairs of quantitative values. In applying the sign test, only the signs of the differences in the matched-pair observations would be used, rather than the actual magnitudes of the differences. Of course, as noted earlier, some loss in the efficiency of the test would result.

> In addition to being simple to apply, the sign test is applicable in a wide variety of situations.

As in the example given above, the two samples do not even have to be independent. Indeed, when matched-pair observations are used, the elements in the first sample are usually matched as closely as possible with the corresponding elements in the second sample. Furthermore, the sign test may be used in cases of qualitative classifications in which it may be difficult to use a ranking scheme such as that used in the beer-tasting problem. For example, after a treatment has been applied, an experimenter may classify subjects as improved ($+$), worse ($-$), or the same (0) and then use the sign test.

EXERCISES 13.2

1. The management of a large manufacturing firm decided to test a new incentive program designed to increase the productivity of the firm's factory workers. The workers reported the following results for the new program compared with the previous incentive program:

Effectiveness of Incentive Programs	Numbers of Workers
New program better	123
Previous program better	62
No difference	15
Total	200

Use the method developed in this section to test the null hypothesis of no difference in effectiveness of the two incentive programs. Use a one-tailed test at the 0.01 significance level to determine whether the new program is more effective than the old program.

2. Workers on an assembly line at the Colorful Crayon Company are required to put 100 crayons in each box. In a sample of 400 boxes packaged by these workers, 260 had more than 100 crayons and 122 had less. Replacing sample values in excess of specification by a plus sign and those less than specification by a minus sign, test the null hypothesis that p, the proportion of plus signs, equals 0.50. Use a two-tailed test at the 0.05 significance level.

3. A market research team wanted to compare the effectiveness of two television commercials for a new brand of shampoo. A randomly selected sample of 35 consumers assigned ratings from zero to 10 for the effectiveness of each commercial, with 10 being the highest possible score. The ratings follow. For example, the first consumer assigned a seven to the first commercial and a six to the second.

    First commercial: 7, 10, 9, 6, 4, 7, 8, 5, 6, 9, 8, 7, 9, 9, 7, 5, 10, 6,
    3, 7, 6, 6, 7, 8, 8, 7, 6, 9, 6, 7, 2, 8, 8, 9, 7

    Second commercial: 6, 10, 8, 4, 5, 6, 7, 3, 9, 10, 7, 5, 10, 8, 6, 4, 9, 7,
    6, 6, 5, 4, 8, 6, 8, 8, 4, 10, 5, 6, 3, 7, 9, 6, 7

    Wishing to place the burden of proof on the first commercial, the marketing research team decided to use the following one-tailed test:

    $$H_0: p \leqslant 0.50$$

    $$H_1: p > 0.50$$

    where p is the probability of obtaining a plus sign when subtracting the score on the second commercial from the score on the first. Carry out the appropriate sign test using a 0.05 significance level.

4. A research agency was interested in determining whether American economists felt that the influence of Federal Reserve stabilization policies had increased over the past decade. In a simple random sample of 500 economists, 252 felt that its influence had increased, 160 felt that its influence had decreased, and 88 felt that there had been no change. Use a plus sign to represent a perceived increase, a minus sign for a perceived decrease, and a zero for "no change," and apply a one-tailed test at the 0.02 significance level. Use the null hypothesis $H_0: p \leqslant 0.50$, where p is the probability of a perceived increase in the influence of Federal Reserve stabilization policies.

5. After a month of trial use of supermarket housebrand A and the leading national brand, a panel of 200 detergent users have responded as follows:

Prefer the national brand	80
Prefer household A	45
Find no difference	75

Construct a two-tailed test at the 0.01 level of significance to test the null hypothesis that there is no difference in consumer preference for the two brands.

6. The admissions office is examining the verbal SAT scores of the entering class. Of the 250 students in the survey, 135 score over 600, 103 score under 600, and the remainder score exactly 600. Replace the sample values in excess of 600 with plus signs and the sample values less than 600 with minus signs. Then construct a two-tailed test at the 0.05 level of significance to test the null hypothesis that the proportion of pluses $p = 0.50$.

7. Residents of an area served by two supermarkets have been asked to evaluate each market on the basis of certain criteria (quality of products, prices, layout, etc.) and to score each criterion on a scale of 0 to 5 (0 = poor; 5 = excellent). The 30 respondents have produced the following scores for "quality of products":

Supermarket A: 4, 5, 3, 1, 3, 4, 2, 2, 3, 4, 5, 3, 3, 1, 5, 4, 3, 2, 3, 1, 0, 3, 5, 2, 1, 3, 5, 4, 3, 2

Supermarket B: 2, 3, 3, 3, 4, 5, 4, 2, 5, 4, 4, 5, 1, 4, 3, 4, 2, 1, 4, 4, 4, 3, 1, 5, 3, 0, 4, 5, 4, 5

Match the scores, assigning a plus sign when the rating for supermarket A is higher and a minus sign when the rating for supermarket B is higher. Construct a two-tailed test at the 0.05 level of significance to test the null hypothesis that the respondents exhibit no preference between supermarkets A and B.

13.3
THE WILCOXON MATCHED-PAIRS SIGNED RANK TEST

In section 13.2, we discussed the sign test, which is useful in testing for significant difference between paired observations. As we have seen, the sign test is carried out in terms of the signs of the differences between matched pairs of observations, without regard to the magnitudes of these differences. We turn now to another nonparametric test for significant differences between paired observations, the **Wilcoxon matched-pairs signed rank test**, which *does* take account of the magnitudes of the differences.

The Wilcoxon matched-pairs signed rank test is preferable to the sign test when the differences between paired observations can be *quantitatively* measured rather than merely assigned rankings.

Both the sign test and the Wilcoxon matched-pairs signed rank test may be considered substitutes for the analogous parametric t test for paired observations (see section 7.6). The nonparametric tests have the advantage of making no assumption about the population distribution, while the parametric t test for paired observations requires the assumption that the underlying population of differences is normally distributed.

We illustrate the Wilcoxon matched-pairs signed rank test by an example of the numerical grades on the midterm and final examinations in a certain course for 10 students selected at random from the course rolls. These data are given in Table 13-2. The Wilcoxon test is carried out in terms of the same paired differences as the parametric test, that is, $d = X_2 - X_1$, where the d values represent differences between two observations on the same individual or object. However, in the Wilcoxon test, the **absolute values** of the differences are obtained, pooled, and ranked from 1 to n, with the smallest difference being assigned the rank 1. These ranks are then given the sign ($+$ or $-$) of the corresponding value of d. If rankings are tied, the mean rank value is assigned to the tied items. (For example, in Table 13-2, because the sixth and seventh ranked items are tied, a rank of $\frac{6+7}{2} = 6.5$ is assigned to each item. The analogous procedure is used if more than two items are tied.) If the difference between the paired observations for an item is zero, as in the case of student 6 in Table 13-2, that item is dropped, and the number of differences is correspondingly reduced. Since there is one such item in the present example, the effective sample size is $n = 10 - 1 = 9$.

TABLE 13-2

Calculations for the Wilcoxon matched-pairs signed rank test of significance of differences in grades on 2 examinations for a simple random sample of 10 students

Student	Grade on Midterm Examination X_1	Grade on Final Examination X_2	Difference $d = X_2 - X_1$	Rank of $\lvert d \rvert$	Signed Rank Rank $(+)$	Signed Rank Rank $(-)$
1	75	72	-3	3		3
2	87	94	$+7$	6.5	6.5	
3	72	92	$+20$	9	9	
4	65	67	$+2$	2	2	
5	93	86	-7	6.5		6.5
6	85	85	0			
7	59	58	-1	1		1
8	73	79	$+6$	5	5	
9	64	69	$+5$	4	4	
10	71	82	$+11$	8	8	
Total (Σ)					34.5	10.5

As indicated in the last two colums of Table 13-2, the sums of the ranks are obtained separately for the positive and negative differences. These sums, denoted Σ rank $(+)$ and Σ rank $(-)$, form the basis for the null hypothesis H_0: Σ rank $(+) = \Sigma$ rank $(-)$. The null hypothesis is often referred to as one of "identical population distributions." More specifically, the hypothesis states that the population positive and negative differences are symmetrically distributed about a mean of zero. The smaller of the two ranked sums, conventionally known as **Wilcoxon's T statistic**, is the test statistic. Hence, in Table 13-2, the test statistic is $T = \Sigma$ rank $(-) = 10.5$.

Wilcoxon's T statistic

It can be shown that when n is large (at least 25), T is approximately normally distributed with the following mean and standard deviation:

(13.1)
$$\mu_T = \frac{n(n+1)}{4}$$

(13.2)
$$\sigma_T = \sqrt{\frac{n(n+1)(2n+1)}{24}}$$

Therefore, we can compute

$$z = \frac{T - \mu_T}{\sigma_T}$$

and carry out the test in the usual way.

The critical values of T are shown in Table A-9 of Appendix A, where, in keeping with the aforementioned convention, T is the smaller of the positive or negative rank sums. The table indicates that with $n = 9$ pairs, the null hypothesis of identical population distributions would be rejected at the 5% significance level using a two-tailed test at $T \leqslant 5$ ($T_{0.05} = 5$). Note that Table A-9 presents the *maximum* values that T can have and still be considered significant at the significance levels. In this example, since the calculated value of T (10.5) exceeds 5, the evidence on the midterm and final examinations does not permit us to reject the null hypothesis of identical population distributions. Hence, we conclude that the performances of this sample of students on the midterm and final examinations did not differ significantly.

EXERCISES 13.3

1. A large number of high school seniors take college Scholastic Aptitude Tests (SAT) more than once in an effort to improve their scores. The following data represent the scores of a simple random sample of twelve students in a certain high school on the SAT mathematics test.

Student	Score on First Test	Score on Second Test
A	550	560
B	490	530
C	610	580
D	570	650
E	670	660
F	480	500
G	720	680
H	630	570
I	500	570
J	660	710
K	540	520
L	570	570

Use the Wilcoxon matched-pairs signed rank test to determine whether the scores on the second test are significantly higher than the scores on the first test. Use $\alpha = 0.05$ in a one-tailed test.

2. The following data represent the monthly numbers of sales by ten scales representatives for one-month periods before and after a special company promotional campaign.

Sales Representative	Sales before Campaign	Sales after Campaign
A	22	25
B	10	14
C	15	17
D	21	31
E	28	33
F	27	27
G	33	32
H	18	19
I	17	28
J	16	22
K	15	22

Use the Wilcoxon matched-pairs signed rank test to determine whether the promotional campaign was accompanied by an increase in number of sales. Use $\alpha = 0.01$ in a one-tailed test.

3. A company made an improvement on a new product after experiencing what it felt were a large number of customer returns of the product. The following data present the numbers of returns for six-month periods before and after the product improvement, by cities in which sales took place.

City	Returns before Improvement	Returns after Improvement
Atlanta	110	42
Boston	122	70
Chicago	467	301
Detroit	206	325
Los Angeles	340	283
Miami	76	38
New Orleans	134	75
New York	643	397
Philadelphia	389	227
San Francisco	291	183

a. Would you conclude that the product improvement resulted in a decrease in the numbers of product returns? Use a Wilcoxon matched-pairs signed rank test in a one-tailed test at the 5% significance level.

b. Assuming that the normal distribution assumptions of the parametric t test for paired observations (discussed in section 7.6) are satisfied, test the hypothesis of no difference between the average numbers of returns before and after the product improvement using this t test. Again, use a one-tailed test at the 5% significance level.

4. In a test involving 55 paired observations of portfolio performance before and after adoption of a new investment strategy, the lower of the two ranked sums was $T = \Sigma$ rank $(+) = 510$. Use the normal approximation to the Wilcoxon matched-pair signed rank test to determine whether the null hypothesis of identical population distributions should be rejected. Use $\alpha = 0.05$ in a two-tailed test.

13.4
MANN-WHITNEY U TEST (RANK SUM TEST)

Another useful nonparametric technique involving a comparison of data from two samples is the **Mann-Whitney U test.**

> This procedure, often referred to as the **rank sum test**, is used to test whether two independent samples have been drawn from populations having the same *mean*.

Hence, the rank sum test may be viewed as a substitute for the parametric t test or the corresponding large-sample normal curve test for the difference between two means. The rank sum test uses more information than does the sign test in that it explicitly takes into account the rankings of measurements in each sample.

TABLE 13-3

Aptitude test grades obtained by graduates of 2 universities

H University	50	51	53	56	57	63	64	65	71	73	74	78	89	90	95
W University	70	76	77	80	81	82	83	86	87	88	92	93	96	98	99

As an illustration of the use of the rank sum test, consider the data shown in Table 13-3. These data represent the grades obtained by management training program applicants on an aptitude test given by a large corporation. The samples consist of graduates of two different universities, referred to as H and W.

Method

The first step in the rank sum test is to merge the two samples, arraying the individual scores in rank order as shown in Table 13-4. The test is then carried out in terms of the sum of the ranks of the observations in either of the two samples. The following symbolism is used:

n_1 = number of observations in sample number one

n_2 = number of observations in sample number two

R_1 = sum of the ranks of the items in sample number one

R_2 = sum of the ranks of the items in sample number two

Treating the data for H University as sample number one, we find that R_1 is the sum of ranks 1, 2, 3, 4, 5, 6, 7, 8, 10, 11, 12, 15, 23, 24, and 27, which is 158. Correspondingly, $R_2 = 307$.

If the null hypothesis that the two samples were drawn from the same population were true, we would expect the totals of the ranks (or equivalently, the mean ranks) of the two samples to be about the same. In order to carry out the test, a new statistic, U, is calculated. This test statistic, which depends only on the number of items in the samples and the total of the ranks in one of the samples, is defined as follows:

(13.3)
$$U = n_1 n_2 + \frac{n_1(n_1 + 1)}{2} - R_1$$

The statistic U provides a measurement of the difference between the ranked observations of the two samples and yields evidence about the difference between the two population distributions.

> Very large or very small U values constitute evidence of the separation of the ordered observations of the two samples.

TABLE 13-4
Array of aptitude test grades obtained by graduates of 2 universities

Rank	Grade	University	Rank	Grade	University	Rank	Grade	University
1	50	H	11	73	H	21	87	W
2	51	H	12	74	H	22	88	W
3	53	H	13	76	W	23	89	H
4	56	H	14	77	W	24	90	H
5	57	H	15	78	H	25	92	W
6	63	H	16	80	W	26	93	W
7	64	H	17	81	W	27	95	H
8	65	H	18	82	W	28	96	W
9	70	W	19	83	W	29	98	W
10	71	H	20	86	W	30	99	W

Under the null hypothesis stated above, it can be shown that the sampling distribution of U has a mean equal to

(13.4)
$$\mu_U = \frac{n_1 n_2}{2}$$

and a standard deviation of

(13.5)
$$\sigma_U = \sqrt{\frac{n_1 n_2 (n_1 + n_2 + 1)}{12}}$$

Furthermore, it can be shown the sampling distribution approaches normality rapidly and may be considered approximately normal when both n_1 and n_2 are in excess of about 10 items.

For the data in the present problem, substituting into equations 13.3, 13.4, and 13.5, we have

$$U = (15)(15) + \frac{(15)(15 + 1)}{2} - 158 = 187$$

$$\mu_U = \frac{(15)(15)}{2} = 112.5$$

$$\sigma_U = \sqrt{\frac{(15)(15)(15 + 15 + 1)}{12}} = 24.1$$

Proceeding in the usual manner, we calculate the standardized normal variate

$$z = \frac{U - \mu_U}{\sigma_U} = \frac{187 - 112.5}{24.1} = 3.09$$

Thus, if the test was originally set up as a two-tailed test at, say, the 1% significance level with a critical absolute value for z of 2.58, we would reject the null hypothesis that the samples were drawn from the same populations. On the other hand, if our alternative hypothesis predicted the direction of the difference, we would be dealing with a one-tailed test. For example, suppose our alternative hypothesis had stated that applicants from W University had a higher average aptitude test score ranking than did applicants from H University. The calculated z value would be the same as previously ($z = 3.09$), but the critical absolute value for z for a one-tailed test at the 1% significance level is 2.33. Since $3.09 > 2.33$, again we would reject the null hypothesis that the samples were drawn form the same population. Referring to the original sample data in Tables 13-3 and 13-4, we observe that W University has the higher average ranking ($R_2/n_2 = \frac{307}{15} = 20.5$ against $R_1/n_1 = \frac{158}{15} = 10.5$.). We now accept the alternative hypothesis of the one-tailed test and conclude that the population of applicants from W University has a higher average aptitude test score than does the corresponding population from H University.

General Comments

The above test was carried out in terms of the sum of the ranks for sample number one. That is, the U statistic was defined in terms of R_1. It could similarly have been defined in terms of R_2 as

(13.6)
$$U = n_1 n_2 + \frac{n_2(n_2 + 1)}{2} - R_2$$

The subsequent test would have yielded the same z value as the one previously calculated, except that the sign would change. Of course, the conclusion would be exactly the same.

There were no ties in rankings in the example given. However, if such ties occur, the average rank value is assigned to the tied items. A correction is available for the calculation of σ_U when ties occur, but the effect is generally negligible for large samples.

As mentioned earlier,

> the rank sum test may be viewed as a substitute for the t test for the difference between two means. The rank sum test may be particularly useful in this connection, because it does not require the restrictive assumptions of the t test.

For example, the t test assumption of population normality may not be valid in the preceding example.

EXERCISES 13.4

1. The job performance scores received by two groups of employees after their first year of work at a large corporation are shown below. The job performance scores, assigned by each employee's supervisor, range from $0 - 10$,

with 10 being the highest score. Employee group one had no prior work experience upon entering the employ of this company, while the second had significant work experience.

Group 1: 7.8, 8.2, 7.3, 6.8, 5.9, 9.3, 7.6, 9.5, 8.7, 9.2, 7.9

Group 2: 6.5, 7.4, 9.0, 8.1, 6.9, 7.5, 7.1, 9.4, 7.7, 8.5, 5.5

Use the rank sum procedure to test the hypothesis that the two samples were drawn from populations with the same average job performance. Use $\alpha = 0.01$.

2. Assume you select a simple random sample of 12 companies from each of two industries. The firms' research and development expenditures (hundreds of thousands of dollars) for last year are recorded below.

    Industry A: 12.5, 14, 5, 7.8, 11.2, 22, 10.7, 18.9, 17, 12, 13.9, 19.2

    Industry B: 6.5, 8.5, 9.6, 8, 15.3, 19.1, 9.5, 4.1, 7.5, 10, 6.8, 7.7

Use the rank sum procedure to test the null hypothesis that no difference exists between the average levels of research and development expenditures in the two industries. Use $\alpha = 0.05$.

3. A women's sportswear company drew a random sample of 14 of its stores from each of two sales regions and studied their sales figures. The sportswear company had advertised heavily in region A, while in region B, it had not advertised at all. The sales figures were merged and ranked to obtain the following results:

    Region A: 1, 3, 4, 5, 7, 9, 10, 11, 12, 14, 15, 16, 19, 20

    Region B: 2, 6, 8, 13, 17, 18, 21, 22, 23, 24, 25, 26, 27, 28

Use the rank sum test at the 0.01 significance level to determine whether there is a significant difference in the average sales level in the two samples.

4. An efficiency expert was examining the amounts of time needed by an assembly line to produce a television set by two different methods. The times for both methods were recorded under essentially similar situations. When the times were merged and ranked, the following results were observed:

    Method A: 1, 2, 4, 6, 7, 8, 10, 13, 14, 15, 18, 20, 25

    Method B: 3, 5, 9, 11, 12, 16, 17, 19, 21, 22, 23, 24, 26, 27, 28

Use the rank sum procedure to test the null hypothesis of no difference between the true average times for the two different methods. Use $\alpha = 0.05$.

5. The following data represent the profits ($ thousands) generated by 28 foreign exchange traders during the past two months. The first group traded only European currencies, while the second group dealt with only Asian and South American currencies.

Group 1:   10.4, 9.7, 9.6, 9.3, 8.9, 8.7, 8.2, 7.7, 7.5, 6.9, 6.2, 5.8, 5.5, 5.1

Group 2:   9.8, 9.5, 8.8, 8.6, 8.4, 8.3, 7.9, 7.8, 7.6, 7.2, 7.1, 6.8, 5.4, 5.3

Use the rank sum procedure to test the hypothesis that the two samples were drawn from populations generating the same average profit. Use $\alpha = 0.01$.

6. A machinist was interested in comparing the times needed to complete a certain project using two different methods. The times were recorded under essentially similar situations for both methods. When the times were merged and ranked, the following results were observed:

Method 1:   1, 2, 3, 6, 10, 11, 13, 16, 17, 18, 21, 23, 27

Method 2:   4, 5, 7, 8, 9, 12, 14, 15, 19, 20, 22, 24, 25, 26, 28

Use the rank sum procedure to test the null hypothesis that there is no difference between the true average times for the two different methods. Use $\alpha = 0.05$.

13.5
ONE-SAMPLE TESTS OF RUNS

We have seen in Chapters 6, 7, and 8 that estimation procedures and parametric tests of hypotheses are predicated on the assumption that the observed data have been obtained from random samples. Indeed, in many instances, evidence of nonrandomness can represent an important phenomenon. As an example, in the frontier days of the Wild West, rather serious consequences were predictable if a card player questioned the randomness of the hands of cards dealt by another player. In many less exotic contexts as well, the randomness of selection of sampled items is of considerable import.

Let us consider a rather oversimplified situation. Suppose that in a certain city, the rolls of persons eligible for jury duty consisted of about 50% people 40 or older and 50% people under 40. Further, let us assume that the following sequence represents the order in which the first 48 persons were drawn from the rolls (B = below 40, F = 40 or older).

BBBBBBBBBBBB FFFFFFFFFFFF BBBBBBBBBBBB FFFFFFFFFFFF

On an intuitive basis, would you question the randomness of selection of these persons? Undoubtedly your answer is in the affirmative, but why? Note that there are 24 B's and 24 F's. Hence, the observed proportion of below-40's (and the proportion of 40-or-older's) is 50%, which does not differ from the known population proportion. Your suspicions concerning nonrandomness doubtless stem from the order of the items listed, rather than from their frequency of occurrence. Similarly, we would find a perfectly alternating sequence, BFBFBFBF ..., suspect with respect to randomness of order of occurrence.

The **theory of runs** has been developed to test samples of data for randomness, with emphasis on the *order* in which these events occur.

Theory of runs

A **run** is defined as a sequence of identical occurrences (symbols) that are followed and preceded by different occurrences (symbols) or by none at all.

Hence, in the listing of 48 symbols, there are four runs, the first run consisting of the first 12 B's, the second run consisting of 12 F's, and so on. Our intuitive feeling is that this represents too few runs. Analogously, in the perfectly alternating series, BFBFBFBF ..., we would feel that there are too many runs to have occurred on the basis of chance alone.

Method

We illustrate the analytical procedure for the test of runs in terms of a somewhat less extreme illustration than the jury selection example. Let us assume that the following 42 symbols represent the successive occurrences of births of males (M) and females (F) in a certain hospital.

MM F M FFF MM FF M F MMM FF M FFF MM FF MM FF
MMM FF MM FF MMM

Using the symbol r to denote the number of runs, we have $r = 21$. The runs (which, of course, may be of differing lengths) have been indicated by separation of sequences. As we have noted, if there are too few or too many runs, we have reason to doubt that their occurrences are random. The runs test is based on the idea that if there are n_1 symbols of one type and n_2 symbols of a second type, and r denotes the total number of runs, the sampling distribution of r has a mean of

(13.7)
$$\mu_r = \frac{2n_1 n_2}{n_1 + n_2} + 1$$

and a standard deviation of

(13.8)
$$\sigma_r = \sqrt{\frac{2n_1 n_2 (2n_1 n_2 - n_1 - n_2)}{(n_1 + n_2)^2 (n_1 + n_2 - 1)}}$$

If either n_1 or n_2 is larger than 20, the sampling distribution of r is closely approximated by the normal distribution. Hence, we can compute

$$z = \frac{r - \mu_r}{\sigma_r}$$

and proceed with the test in the usual manner.

In the present problem, where there are $n_1 = 22$ Ms, $n_2 = 20$ Fs, and $r = 21$, we have

$$\mu_r = \frac{(2)(22)(20)}{22 + 20} + 1 = 21.95$$

and

$$\sigma_r = \sqrt{\frac{(2)(22)(20)[(2)(22)(20) - 22 - 20]}{(22 + 20)^2(22 + 20 - 1)}} = 3.19$$

Therefore,

$$z = \frac{21 - 21.95}{3.19} = -0.30$$

Testing at a significance level of, say, 5%, where a critical absolute value of 1.96 for z is required for rejection of the null hypothesis, we find that the randomness hypothesis cannot be rejected. In other words, the number of runs is neither small enough nor large enough for us to conclude that the sequence of male and female births is nonrandom.

General Comments

The runs test has many applications, including cases in which the sequential data are *numerical* in form rather than *symbolical* representations of attributes such as the letters used in the preceding illustrations.

- runs tests could be applied to sequences of random numbers, such as those in Table 4-1 on page 213. Such tests might be applied to sequences of random numbers generated on computers.
- One form of the test might be in terms of runs of numbers above the median and runs of numbers below the median. For the digits 0, 1, 2, 3, 4, 5, 6, 7, 8, and 9, the median is 4.5. Hence, runs could be determined for digits that fall above and below the median.
- The test could also be applied in terms of runs of odd-numbered digits and runs of even-numbered digits.
- Another alternative is to group the numbers into pairs of digits, so that the possible occurrences are 00, 01, ..., 99. Here the median is 49.5, and runs tests similar to those suggested for the one-digit case could be applied.

Clearly, with a bit of imagination, an analyst can devise many useful and easily applied versions of the runs test.

EXERCISES 13.5

1. In the table of random numbers (Table 4-1 on page 213), consider the 50 digits in the first line. Label the even digits **a** and the odd digits **b**. Carry out a runs test at both the 0.05 and 0.01 levels of significance.

2. Consider the same 50 digits as in exercise 1. Now label the digits **a** or **b** depending on whether they are above or below the theoretical mean, 4.5. Carry out a runs test at both the 0.05 and 0.01 levels of significance.

3. The following figures represent the unemployment figures (in percentages) in a certain small municipality for a 20-week period:

> 6.2, 6.3, 6.2, 6.2, 6.6, 6.3, 6.0, 6.3, 6.3, 6.1,
> 6.2, 6.4, 6.6, 6.8, 6.9, 6.6, 6.5, 6.7, 6.4, 6.6

Determine the median of this set of 20 figures. Label the numbers above and below the median **a** and **b**, respectively, and perform a runs test at the 0.01 level of significance. This test of runs above and below the median is particularly useful for determining the existence of trend patterns in data. If there is a trend, **a**'s will tend to appear in the early part of the series and **b**'s in the later part, or vice versa.

4. The following figures represent the weekly overtime hours worked by an employee over a 26-week period:

> 5, 10, 0, 15, 6, 0, 7, 8, 0, 0, 18, 7, 3, 0, 12, 4, 5, 8, 0, 3, 2, 5, 6, 0, 0, 10

Find the mean overtime hours worked. Label the numbers above and below the mean A and B, respectively. Perform a runs test on the series of As and Bs at the 0.05 significance level.

5. Boxes of candy shipped from the packaging department of a chocolate factory are weighed to determine whether they were above or below the specified weight for the box. The sequences of observed weights above (A) and below (B) are

> AAAABAABAABBAAAABBBB
> AABBBBBBAABBBAAABBAA

Carry out the runs test at the 0.01 significance level.

6. Toss a coin 40 times and record heads and tails as H and T, respectively. Test these data for randomness at the 0.05 level of significance.

13.6
KRUSKAL-WALLIS TEST

In section 8.3, we noted that the one-factor analysis of variance represents an extension of the two-sample test for means and provides a test for whether several independent samples can be considered to have been drawn from populations having the same mean. Analogously, the Kruskal-Wallis one-factor analysis of variance by ranks is a nonparametric test that represents a generalization of the two-sample Mann-Whitney U rank sum test. The parametric analysis of variance discussed in section 8.3 assumes that the populations are normally distributed; otherwise, the F-test procedure is invalid. The Kruskal-Wallis test makes no assumptions about the population distribution.

The Kruskal-Wallis test is based on a test statistic calculated from ranks established by pooling the observations from c independent simple random samples, where $c > 2$. The null hypothesis is that the populations are identically distributed or alternatively, that the samples were drawn from c identical populations. We illustrate the procedure for the test by the following example.

Simple random samples of corporate treasurers in a certain industry were drawn from firms classified into three size categories (large, medium, and small). These executives, after being assured of the confidentiality of their replies, were asked to rate the overall quality of the Federal Reserve Board's performance in discount rate policy during the past six-month period on a scale from 0 to 100, with 0 denoting the lowest quality rating and 100 denoting the highest. The scores, classified by size of firm, and the rankings of the pooled sample scores are shown in Table 13-5. The result was the following pooled ranking, with the lowest score that was actually given represented by rank 1 and the highest by rank $n = 20$ (where n denotes the total number of pooled sample observations):

Score:	50	60	61	62	65	68	70	72	73	75	77	78	80	82	84	85	87	90	93	95
Rank:	1	2	3	4	5	6	7	8	9	10	11	12	13	14	15	16	17	18	19	20

The test statistic involves a comparison of the variation of the ranks of the sample groups. The Kruskal-Wallis test statistic is

$$(13.9) \qquad K = \frac{12}{n(n + 1)}\left(\Sigma \frac{R_j^2}{n_j}\right) - 3(n + 1)$$

where n_j = the number of observations in the jth sample
$n = n_1 + n_2 + \cdots + n_c$
    = the total number of observations in the c samples
R_j = the sum of the ranks for the jth sample

TABLE 13-5

Calculations for the Kruskal-Wallis one-factor analysis of variance: scores and ranks classified by size of firm

Large Firms (group 1)		Medium-sized Firms (group 2)		Small Firms (group 3)	
Score	Rank	Score	Rank	Score	Rank
78	12	68	6	82	14
95	20	77	11	65	5
85	16	84	15	50	1
87	17	61	3	93	19
75	10	62	4	70	7
90	18	72	8	60	2
80	13			73	9
$n_1 = 7$		$n_2 = 6$		$n_3 = 7$	
$R_1 = 106$		$R_2 = 47$		$R_3 = 57$	

The sample sizes and rank sums are shown in Table 13-5 for each sample group. Substituting into equation 13.9, we compute the K statistic in the present example.

$$K = \frac{12}{20(20 + 1)}\left(\frac{106^2}{7} + \frac{47^2}{6} + \frac{57^2}{7}\right) - 3(20 + 1) = 6.641$$

It can be shown that the sampling distribution of K is approximately the same as the χ^2 distribution with $v = c - 1$ degrees of freedom (where c is the number of sample groups). In this example, where there are three sample groups, the number of degrees of freedom is $v = c - 1 = 3 - 1 = 2$. Testing the null hypothesis at the 5% level of significance ($\alpha = 0.05$) and using Table A-7 of Appendix A, we find the critical value of χ^2 to be $\chi^2_{0.05} = 5.991$. Hence, our decision rule is

If $K > 5.991$, reject the null hypothesis

If $K \leqslant 5.991$, accept the null hypothesis

Since $K = 6.641$ is greater than the critical value of 5.991, we reject the null hypothesis of identically distributed populations. Therefore, we conclude that there are significant differences by size of firm in the scores assigned by these three samples of corporate treasurers. Looking back at the scores given in Table 13-5, we find that treasurers of large firms tended to assign higher scores than did their counterparts in medium-sized or small firms.

Further Remarks

As in other nonparametric tests, when there are ties, observations are assigned the mean of the tied ranks. In the case of ties, a corrected K value K_c should be computed as follows:

(13.10)

$$K_c = \frac{K}{1 - \left[\dfrac{\Sigma(t_j^3 - t_j)}{(n^3 - n)}\right]}$$

where t_j is the number of tied scores in the jth sample.

Furthermore, for the χ^2 distribution to be applicable, the sample sizes (that is, the n_j values) should all be greater than five.

EXERCISES 13.6

1. As insurance company wished to determine whether there was a difference in the amount of whole life insurance held by members of different professions. The following table gives the values of the policies held by simple random samples of members of four different professions. Would you conclude that there was no difference in the values of the policies held by members of the different professions? Use the Kruskal-Wallis analysis of variance by ranks test at a 1% significance level.

	Value of Policies (thousands of dollars)		
Physicians	Engineers	Dentists	Lawyers
18	54	74	125
105	111	190	87
57	68	134	118
145	174	115	56
67	205	49	122
28	34	206	96
198			

2. The credit department supervisor of Racy's Department Store is evaluating the performances of three collectors, whose responsibilities include the collection of delinquent accounts. The following figures show the number of accounts each collector successfully removed from delinquency during each

week of a six-week period. The accounts to be investigated were equally divided among the collectors each week.

Week	Collectors A	B	C
1	25	45	46
2	35	48	40
3	30	26	36
4	31	38	39
5	20	32	42
6	50	49	52

a. Using the Kruskal-Wallis analysis of variance by ranks, test the hypothesis of no difference in effectiveness among the three collectors in terms of the number of accounts removed from delinquency. Use $\alpha = 0.01$.

b. Assuming that the normal distribution assumptions of the analysis of variance are satisfied, use the parametric analysis of variance discussed in section 8.3 to test the hypothesis of no difference in average number of accounts removed from delinquency by the three collectors. Use $\alpha = 0.01$.

3. A manufacturer purchased large batches of a product from four subcontractors. There were the same number of articles in each batch. The table gives the numbers of defective articles in each batch. Would you conclude that there was no difference among subcontractors in numbers of defectives per batch? Use the Kruskal-Wallis analysis of variance by ranks test at a 1% significance level.

Number of Defectives per Batch

Subcontractor A	Subcontractor B	Subcontractor C	Subcontractor D
12	30	15	18
6	28	17	27
10	7	20	13
0	25	3	22
2	24	19	16
4	29	21	23
	31	8	
	14		

4. In a test, the marketing research department of a firm sent 3 differently designed advertisements to equal-sized simple random samples of potential

customers in 6 cities. The numbers of units of sales that resulted from business reply cards attached to these advertisements are as follows:

City	Design A	Design B	Design C
	Numbers of Units Sold		
1	38	64	55
2	59	75	82
3	30	36	80
4	52	77	66
5	61	69	73
6	43	67	47

a. Using the Kruskal-Wallis analysis of variance by ranks, test the hypothesis of no difference in effectiveness among the 3 advertisement designs in terms of numbers of sales that resulted. Use $\alpha = 0.01$.
b. Assuming that the normal distribution assumptions of analysis of variance are satisfied, use the parametric analysis of variance discussed in section 8.3 to test the hypothesis of no difference in average numbers of sales resulting from the 3 advertisements. Use $\alpha = 0.01$.

13.7
RANK CORRELATION

Nonparametric procedures can be useful in correlation analysis when the basic data are not available in the form of numerical magnitudes but rankings can be assigned. If two variables of interest can be ranked in separate ordered series, a **rank correlation coefficient** can be computed; this is a measure of the degree of correlation that exists between the two sets of ranks. We illustrate the method in terms of a simple random sample of individuals for whom rankings have been established for two variables concerning ability in two different sports activities.

Method

For illustrative purposes, we consider two extreme cases, the first case representing perfect *direct* correlation between two series, the second case representing perfect *inverse* correlation. Table 13-6 displays data on the rankings of a simple random sample of 10 individuals according to playing abilities in baseball and tennis. Clearly, this represents a case in which it would be extremely difficult, if not impossible, to obtain precise quantitative measures of these abilities, but in which rankings may be feasible. In rank correlation analysis, the rankings may be assigned in order from high to low (with one representing the highest

TABLE 13-6
Rank correlation of baseball-playing ability with tennis-playing ability (perfect correlation case)

Individual	Rank in Baseball Ability X	Rank in Tennis Ability Y	Difference in Ranks $d = X - Y$	$d^2 = (X - Y)^2$
A	1	1	0	0
B	2	2	0	0
C	3	3	0	0
D	4	4	0	0
E	5	5	0	0
F	6	6	0	0
G	7	7	0	0
H	8	8	0	0
I	9	9	0	0
J	10	10	0	0
Total				$\overline{0}$

$$r_r = 1 - \frac{6\Sigma d^2}{n(n^2 - 1)} = 1 - \frac{6(0)}{10(10^2 - 1)} = 1$$

rating, two the next highest, and so on) or from low to high (with one representing the lowest rank, two the next lowest, and so on). The computed rank correlation coefficient will be the same, regardless of the rank ordering used. Let us assume in this case that one represents the highest or best rank, two the second highest, and so on.

The rank correlation coefficient (also referred to as the Spearman rank correlation coefficient) can be derived mathematically from one of the formulas for r, the sample correlation coefficient discussed in Chapter 9, where ranks are used for the observations of X and Y. We will use the symbol r_r to denote the rank correlation coefficient, computed by the following formula:

(13.11)
$$r_r = 1 - \frac{6\Sigma d^2}{n(n^2 - 1)}$$

Computing the rank correlation coefficient

where d = difference between the ranks of the paired observations
n = number of paired observations

The calculations of the rank correlation coefficients for the two extreme cases mentioned earlier are shown in Tables 13-6 and 13-7. In Table 13-6, there is perfect direct correlation in the rankings. That is, the individual who ranks highest in baseball playing ability is also best in tennis, and so on. On the other hand, in Table 13-7, there is perfect inverse correlation. That is, the individual who ranks highest in baseball playing ability is worst in tennis, and so on. Note from the calculations shown in the two tables that in the case of perfect direct

TABLE 13-7
Rank correlation of baseball-playing ability with tennis-playing ability
(perfect inverse correlation case)

Individual	Rank in Baseball Ability X	Rank in Tennis Ability Y	Difference in Ranks $d = X - Y$	$d^2 = (X - Y)^2$
A	1	10	−9	81
B	2	9	−7	49
C	3	8	−5	25
D	4	7	−3	9
E	5	6	−1	1
F	6	5	1	1
G	7	4	3	9
H	8	3	5	25
I	9	2	7	49
J	10	1	9	81
Total				330

$$r_r = 1 - \frac{6\Sigma d^2}{n(n^2 - 1)} = 1 - \frac{6(330)}{10(10^2 - 1)} = -1$$

correlation between the ranks, $r_r = 1$; in perfect inverse correlation, $r_r = -1$. This is not surprising, because the rank correlation coefficient is derived mathematically from the sample correlation coefficient r. Hence, the range of possible values of these coefficients is the same. An r_r value of zero would analogously indicate no correlation between the rankings. Tied ranks are handled in the calculations by averaging in the usual way.

The significance of the rank correlation may be tested in the same way as for the sample correlation coefficient r. That is, we compute the statistic

(13.12)
$$t = \frac{r_r}{\sqrt{(1 - r_r^2)/(n - 2)}}$$

which has a t distribution for $n - 2$ degrees of freedom. For example, suppose in a situation such as the one above, r_r had been computed to be 0.90. Then substitution into (13.12) would yield

$$t = \frac{0.90}{\sqrt{(1 - 0.81)/(10 - 2)}} = 5.84$$

Let us assume that we are using a two-tailed test of the null hypothesis of zero correlation in the ranked data of the population. Then referring to Table A-6 of Appendix A, we find critical t values of 2.306 and 3.355 at the 5% and 1% levels of significance, respectively. We would reject the hypothesis of no rank correlation at both levels and conclude that a positive linear relationship exists between the rankings in baseball-playing ability and tennis-playing ability.

1. The following table gives the rankings of a simple random sample of companies with regard to total sales and after-tax profits:

Company	Sales Rank	After-tax Profits Rank
A	7	5
B	3	4
C	6	7
D	2	1
E	4	3
F	5	6
G	12	10
H	11	12
I	8	9
J	10	11
K	9	8
L	1	2

Calculate the rank correlation for this set of data.

2. The following table shows the rankings of a simple random sample of students with regard to scores on the mathematics and verbal sections of the Scholastic Aptitude Test (SAT):

Student	Mathematics Rank	Verbal Rank
A	3	4
B	1	5
C	8	3
D	10	9
E	2	8
F	6	2
G	9	10
H	4	6
I	5	1
J	7	7

Calculate the rank correlation coefficient for this set of data.

3. The following data represent the ranking of ten companies, based on net assets and revenues:

Company	Rank Based on Net Assets	Rank Based on Revenues
A	2	3
B	3	1
C	1	4
D	6	9
E	8	10
F	4	2
G	9	7
H	7	5
I	10	8
J	5	6

Calculate the rank correlation coefficient.

4. The following data represent the rankings of 12 college football teams in pre-season and regular-season competition:

Team	Pre-season Ranking	Regular Ranking
A	4	7
B	6	4
C	12	9
D	2	3
E	7	8
F	1	5
G	9	11
H	11	12
I	3	1
J	8	6
K	5	2
L	10	10

Calculate the rank correlation coefficient for this set of data, and test its significance at $\alpha = 0.05$.

Problems 1 through 4 use the information saved in the file named 'lab . mtw'.

Column	Name	Description
C1	LIFEINSU	Net Income Per Dollar Corporate Sales in Life Insurance Industry in 1987
C2	AIRTRANS	Net Income Per Dollar Corporate Sales in Air Transportation Industry in 1987
C3	BROKERS	Net Income Per Dollar Corporate Sales in Security Brokers & Dealers Industry in 1987

(Source: *Slater-Hall Information & Business*, 1988)

1. Use the *t*-distribution to carry out a two-tailed test of H_0: mean of LIFEINSU is zero, at a 0.10 significance level. To check for normality of C1, type

   MTB > nscores of C1, put into C11
   MTB > plot C1 versus C11

   If sample C1 comes from a normal distribution, then the above plot is approximately a straight line. Does C1 appear to come from a normal distribution? If not, comment on the validity of using the *t*-test.

2. Use the sign test procedures to carry out a two-tailed test of H_0: the proportion of LIFEINSU values above (or below) a median of zero is 50%, at a 0.10 significance level. Obtain a 99% confidence interval for the mean of LIFEINSU. *Hint*:

   MTB > stest 0 C1
   MTB > sinterval 99% C1

3. (*optional*) Assume that the population distribution of C2 is symmetric. Use the Wilcoxon signed rank procedures to conduct a two-tailed test of H_0: median of AIRTRANS is 0.05, at alpha = 0.01. Obtain a 95% confidence interval for the median of AIRTRANS. *Hint*:

   MTB > wtest 0.05 C2
   MTB > winterval 95% C2

4. Assume that the populations from which C2 and C3 come have roughly the same shape so that we can apply the Mann-Whitney *U* test. Conduct a two-sided test of H_0: the populations have the same median, at a 0.10 significance level. Find a 98% confidence interval for the difference of the population medians. *Hint*:

   MTB > Mann-Whitney 98% for C2 and C3

5. Each of three MBA graduates of the same program was given 8 hours several times to sell the same product under similar conditions. The results are tabulated below:

Number of Units Sold		
Graduate 1	Graduate 2	Graduate 3
27	31	25
35	34	29
30	28	32
33	36	32
26	29	28
37	39	38
28	31	33
	32	29
		30

Use Kruskal-Wallis procedures to determine whether the three populations may be thought of as identical. Use alpha = 0.10. Also try the one-way analysis of variance for the above data, and compare the results with those obtained from the Kruskal-Wallis procedures. *Hint*: First put the data into C1 to C3, and then

MTB > stack C1-C3 C10;
SUBC > subscripts into C11.
MTB > Kruskal-Wallis for data in C10, indices in C11
MTB > aovoneway for samples C1 C2 C3

6. Two Master chess players have had a series of games. The results of one player were: W W L D D W L L L W D D W W L, where W stands for win, D for draw, and L for lose. Use runs procedures to test for the randomness of the order, at alpha = 0.05. *Hint*: Code W as 2, D as 1, and L as 0. Store the outcomes in C1, and then type

MTB > runs C1

7. Randomly generate 20 zeros and ones by:

MTB > random 20 C1;
SUBC > integers 0 and 1.

Conduct a runs test for C1 at a 0.10 significance level. Create a new column via

MTB > stack C1 C1 C1 C1 C1 C2

and carry out a runs test for C2 at the same significance level. Comment on the conclusions of the two tests.

Kruskal-Wallis Test a nonparametric analysis of variance test in terms of ranked observations.

Mann-Whitney U (rank sum) Test a test that sums the ranks of observations of two samples that have been merged.

Nonparametric (distribution-free) Tests tests that make less restrictive and fewer assumptions than parametric tests.

Rank Correlation Coefficient a correlation coefficient for ranked (rather than numerical magnitude) observations.

Runs Test a test in terms of sequences of identical occurrences that are followed and preceded by different occurrences or by none at all.

Sign Test a test that uses the sums of the positive or negative differences in ranks of observations in two samples.

Wilcoxon Matched-Pairs Signed Rank Test a test that uses the absolute values of the differences in the magnitudes of matched-pairs of observations in two samples.

Wilcoxon T Statistic the smaller of the two ranked sums in the Wilcoxon matched-pairs signed rank test.

Kruskal-Wallis Test Statistic

$$K = \frac{12}{n(n+1)} \left(\Sigma \frac{R_j^2}{n_j} \right) - 3(n+1)$$

where R_j = the sum of the ranks for the jth sample

Mean of the Distribution of the Proportion ($\bar{p}$) of Plus Signs in the Sign Test

$$\mu_{\bar{p}} = p$$

Mean of the Mann-Whitney U Distribution

$$\mu_U = \frac{n_1 n_2}{2}$$

Mean of the Total Number of Runs (r)

$$\mu_r = \frac{2n_1 n_2}{n_1 + n_2} + 1$$

Mean of the Wilcoxon T Distribution

$$\mu_T = \frac{n(n+1)}{4}$$

Rank Correlation Coefficient

$$r_r = 1 - \frac{6\Sigma d^2}{n(n^2 - 1)}$$

Standard Deviation of the Distribution of the Proportion of Plus Signs in the Sign Test

$$\sigma_{\bar{p}} = \sqrt{\frac{pq}{n}}$$

Standard Deviation of the Mann-Whitney *U* Distribution

$$\sigma_U = \sqrt{\frac{n_1 n_2 (n_1 + n_2 + 1)}{12}}$$

Standard Deviation of the Total Number of Runs (*r*)

$$\sigma_r = \sqrt{\frac{2n_1 n_2 (2n_1 n_2 - n_1 - n_2)}{(n_1 + n_2)^2 (n_1 + n_2 - 1)}}$$

Standard Deviation of the Wilcoxon *T* Distribution

$$\sigma_T = \sqrt{\frac{n(n + 1)(2n + 1)}{24}}$$

14

DECISION MAKING USING PRIOR INFORMATION

In recent years, in addition to lively developments in classical, or traditional, statistical inference, there have been parallel developments in theory and methodology concerned with the problem of decision making under conditions of uncertainty. This modern formulation has come to be known as **statistical decision theory** or **Bayesian decision theory**. The latter term is often used to emphasize the role of Bayes' theorem in this type of decision analysis. The two ways of referring to modern decision analysis can be used interchangeably and will be so used in this book.

14.1
THE IMPORTANCE OF
STATISTICAL DECISION THEORY

Statistical decision theory has become an important model for making rational selections among alternative courses of action when information is *incomplete* and *uncertain*. It is a prescriptive theory rather than a descriptive one. That is, it presents the principles and methods for making the best decisions under specified conditions, but it does not purport to describe how actual decisions are made in the real world.

14.2
STRUCTURE OF THE DECISION-MAKING PROBLEM

Managerial decision making has increased in complexity as the economy of the United States and the business units within it have grown larger and more intricate. However, Bayesian decision theory is based on the assumption that certain common characteristics of the decision problem can be discerned regardless of the type of decision (whether it involves long- or short-range consequences; whether it is in finance, production, marketing, or some other area; whether it is at a relatively high or low level of managerial responsibility). These characteristics constitute the formal description of the problem and provide the structure for a solution. The decision problem under study may be represented by a model comprising the following 5 elements:

The decision maker • The agent charged with the responsibility for making the decision, the **decision maker** is viewed as an entity and may be a single individual, a corporation, a government agency, and so on.

Alternative courses • The decision involves a selection among two or more alternative courses *of action* of action, referred to simply as **acts**. The problem is to choose the best of these alternative acts. Sometimes the decision maker must choose the best of alternative **strategies**, where each strategy is a decision rule indicating which act should be taken in response to a specific type of experimental or sample information.

Events • Occurrences that affect the achievement of the objectives, **events** are viewed as lying outside the control of the decision maker, who does not know for certain which event will occur. The events constitute a mutually exclusive and complete set of outcomes; that is, one and only one of them can occur. Events are also referred to as **states of nature** or **states of the world** or, simply, **outcomes**.

Payoff • A measure of net benefit to be received by the decision maker under particular circumstances, the payoffs are summarized in a **payoff table** or **payoff matrix**, which displays the consequences of each act selected and each event that occurs.

Uncertainty • The indefiniteness concerning which events or states of nature will occur, **uncertainty** is indicated in terms of probabilities assigned to events. One of the distinguishing characteristics of Bayesian decision theory is the assignment of personalistic, or subjective, probabilities as well as other types of probabilities.

The payoff table, expressed symbolically in general terms, is given in Table 14-1. We assume that there are n alternative acts, denoted $A_1, A_2, \ldots, A_n$. These different possible courses of action are listed as column headings in the table. There are m possible events or states of nature, denoted $\theta_1, \theta_2, \ldots, \theta_m$.

TABLE 14-1
The payoff table

Event	Act A_1	A_2	$\cdots$	A_n
θ_1	u_{11}	u_{12}	$\cdots$	u_{1n}
θ_2	u_{21}	u_{22}	$\cdots$	u_{2n}
$\vdots$	$\vdots$	$\vdots$	$\vdots$	$\vdots$
θ_m	u_{m1}	u_{m2}	$\cdots$	u_{mn}

The payoffs resulting from each combination of an act and an event are designated by the symbol u with appropriate subscripts. The letter u has been used because it is the first letter of the word *utility*.

> The net benefit, or payoff, of the selection of an act and the occurrence of a state of nature can be treated most generally in terms of the *utility* of this consequence to the decision maker.

Summary

How these utilities are determined is a technical matter that is discussed later in this chapter. In summary, the utility of selecting act A_1 and having event θ_1 occur is denoted u_{11}; the utility of selecting act A_2 and having event θ_1 occur is u_{12}, and so on. Note that the first subscript in these utilities indicates the event that prevails and the second subscript denotes the act chosen. A convenient general notation is the symbol u_{ij}, which denotes the utility of selecting act A_j if subsequently event θ_i occurs. The rows of a table (or matrix) are commonly denoted by the letter i (where i can take on values $1, 2, \ldots, m$) and the columns are denoted by j (where j can take on values $1, 2, \ldots, n$).

If the event that will occur (for example, θ_3) were known with certainty beforehand, then the decision maker could simply look along row θ_3 in the payoff table and select the act that yields the greatest payoff. However, in the real world, since the states of nature lie beyond the control of the decision maker, he or she ordinarily does not know with certainty which specific event will occur. The choice of the best course of action in the face of this uncertainty is the crux of the decision maker's problem.

14.3
AN ILLUSTRATIVE EXAMPLE: THE INVENTOR'S PROBLEM

To illustrate the ideas discussed in the preceding section, we will use a simplified business decision problem.[1]

---

[1] This problem will be continued in later sections to exemplify other principles.

TABLE 14-2
Payoff table for the inventor's problem (units of $10,000 profit)

Event	A_1 Inventor Manufactures Device	A_2 Inventor Sells Patent Rights
θ_1: Strong sales	$80	$40
θ_2: Average sales	20	7
θ_3: Weak sales	−5	1

Inventor's problem

> An inventor has patented a new device, and a bank is willing to lend the money to manufacture the device. Preliminary investigation establishes a suitable planning period of 5 years for the comparison of payoffs from this invention. According to the inventor's analysis, profits of $800,000 can be anticipated over the next 5 years if sales are strong; if sales are average, the inventor can expect to make $200,000; and if sales are weak, the inventor expects to lose $50,000. Nationwide Enterprises, Inc., has offered to purchase the patent rights. Based on the royalty arrangement, the inventor estimates that selling the patent rights may well bring a net profit of $400,000 if sales are strong, $70,000 if sales are average, and $10,000 if sales are weak.

The payoff table for the inventor's problem is given in Table 14-2.

The acts   In this problem, the alternative acts, denoted A_1 and A_2, respectively, are for the inventor to manufacture the device or to sell the patent rights. The *The events* events or states of nature (denoted θ_1, θ_2, and θ_3, respectively) are strong sales, average sales, and weak sales for the 5-year planning period. The payoffs are in terms of the net profits that would accrue to the inventor under each act– *The payoffs* event combination. To keep the numbers simple in this problem, the payoffs have been stated in units of $10,000; hence, a net profit of $800,000 has been recorded as $80, a net loss of $50,000 has been entered as −$5, and so on.[2]

The types of events used in the inventor's problem are, of course, simplified. Generally, an unlimited number of possible events could occur in the future relating to such matters as the customers, technological change, competitors, and so on, which lie beyond the decision maker's control. All of these states of nature may affect the potential payoffs of the alternative decisions to be made. However, in order to cut way through the maze of complexities

---

[2] It is good practice in the comparison of economic alternatives to compare the present values of discounted cash flows or equivalent annual rates of return. Both of these methods take into account the time value of money; that is, the fact that a dollar received today is worth more than a dollar received in some future period. These are conceptually the types of monetary payoff values that should appear in the payoff table. This point is amply discussed in standard texts dealing with economy studies or investment analysis. To avoid a lengthy tangential discussion, we will not elaborate on the point here.

involved, and to construct a manageable framework of analysis for the problem, we can think of the variable "demand" as the resultant of all of these other underlying factors.

In the inventor's problem, 3 different levels of demand (strong, average, and weak) have been distinguished. It is helpful in this regard to think of demand as a variable.

Demand as a variable

> In the inventor's problem, demand is a discrete random variable that can take on 3 possible values.

We could have considered demand as a discrete random variable taking on any finite or infinite number of values. (For example, it could have been stated in numbers of units demanded or in hundreds of thousands of units demanded.) Demand can also be treated as a continuous rather than a discrete variable. The conceptual framework of the solution to the decision problem remains the same, but the required mathematics differs somewhat from the case in which the events are stated in the form of a discrete variable.

14.4
CRITERIA OF CHOICE

Assuming that the inventor in our illustrative problem has carried out the thinking, experiments, data collection, and so on, required to construct the pay-off matrix (Table 14-2), how should he or she now compare the alternative acts? Neither act is preferable to the other under all states of nature. For example, if event θ_1 occurs, that is, if sales are strong, the inventor would be better off to manufacture the device (act A_1), realizing a profit of \$800,000, compared with selling the patent rights (acts A_2), which would yield a profit of only \$400,000. On the other hand, if even θ_3 occurs, and sales are weak, the preferable course of action would be to sell the patent rights, thereby earning a profit of \$10,000 compared with a loss of \$50,000. If the inventor knew with *certainty* which event was going to occur, the decision procedure would be simple: merely look along the row represented by that event and select the act that yields the highest payoff. However, the *uncertainty* with regard to which state of nature will prevail makes the decision problem an interesting one.

Maximin Criterion

Several different criteria for selecting the best act have been suggested. One of the earliest suggestions, made by mathematical statistician Abraham Wald,[3] is known as the *maximin criterion*.

---

[3] Abraham Wald, *Statistical Decision Functions* (New York: Johny Wiley & Sons, 1950).

Maximin criterion

> Under the **maximin criterion** method, the decision maker assumes that once a course of action has been chosen, nature might be malevolent and might select the state of nature that minimizes the decision maker's payoff. The decision maker chooses the act that maximizes the payoff under the most pessimistic assumption concerning nature's activity.

In other words, Wald suggested that selecting the "best of the worst" is a reasonable form of protection. By this criterion, if the inventor chose act A_1, nature would cause event θ_3 to occur and the payoff would be a loss of $50,000. If the decision maker chose A_2, nature would again cause θ_3 to occur, since that would yield the worst payoff—in this case, a profit of $10,000. Comparing these worst, or minimum, payoffs, we have

Event	A_1	A_2
Minimum payoffs (units of $10,000)	−5	1

The decision maker now must—in the face of this sort of perverse nature—select the act that yields the greatest minimum payoff, namely, act A_2. That is, the inventor should sell the patent rights, for which the minimum payoff is $10,000. Thus, the proposed decision procedure is to choose the act that yields the *maxi*mum of the *mini*mum payoffs—hence, the term **maximin**.

Obviously, the maximin is a pessimistic type of criterion. It is not reasonable to suppose that the executive would or should make decisions in this way. By following this decision procedure, the executive would always be concentrating on the worst things that could happen. In most situations, the maximin criterion would freeze the decision maker into complete inaction and would imply that it would be best to go out of business entirely. For example, let us consider an inventory stocking problem, in which the events are possible levels of demand, the acts are possible stocking levels (that is, the numbers of items to be stocked), and the payoffs are in terms of profits. If no items are stocked, the payoffs will be zero for every level of demand. For each of the other numbers of items stocked, we can assume that for some levels of demand, losses will occur. Since the worst that can happen if no items are stocked is that no profit will be made, and under all other courses of action the possibility of a loss exists, the maximin criterion would require the firm to carry no stock or, in effect, go out of business.

Such a procedure is not necessarily irrational, and it might be consistent with some people's attitudes toward risk. However, the person who is willing to take some risks would regard such an arbitrary decision rule as completely unacceptable. A number of other decision criteria have been suggested by various writers, but, to avoid a lengthy digression, they will not be discussed here.

It seems reasonable to argue that a decision maker should take into account the probabilities of occurrence of the different possible states of nature.

As an extreme example, if the state of nature that results in the minimum payoff for a given act has only one chance in a million of occurring, it would seem unwise to concentrate on the possibility of this occurrence. The decision procedures we will focus on include the probabilities of states of nature as an important part of the problem.

Expected Profit under Uncertainty

In a real decision-making situation, we may suppose that a decision maker would have some idea of the likelihood of occurrence of the various states of nature and that this knowledge would help in choosing a course of action. For example, in our illustrative problem, if the inventor felt confident that sales would be strong, he or she would move toward manufacturing the device, since the payoff under that act would exceed that of selling the patent rights. By the same reasoning, if the inventor were confident that sales would be weak, he or she would be influenced to sell the patent rights.

If many possible events and many possible courses of action exist, the problem becomes complex, and the decision maker clearly needs some orderly method of processing all the relevant information. Such a systematic procedure is provided by the computation of the *expected* monetary value of each course of action and the selection of the act that yields the highest of these expected values. As we shall see, this procedure yields reasonable results in a wide class of decision problems. Furthermore, we will see how this method can be adjusted for the computation of expected utilities rather than expected monetary values in cases where the maximization of expected monetary values is not an appropriate criterion of choice.

We now return to the inventor's problem to illustrate the calculations for decision making by maximization of the expected monetary value criterion. In this case, the maximization consists of selecting the act that yields the largest expected profit. Let us assume that the inventor carries out the following probability assignment procedure. On the basis of extensive investigation of past experience with similar devices, and on the basis of interviews with experts, the inventor concludes that the odds are 50:50 that sales will be average (that is, that the event we previously designed as θ_2 will occur). Furthermore, the inventor concludes that it is somewhat less likely that sales will be strong (event θ_1), than that they will be weak (event θ_3). On this basis, the inventor assigns the following subjective probability distribution to the events in question:

Calculations for decision making

Event	Probability
θ_1: Strong sales	0.2
θ_2: Average sales	0.5
θ_3: Weak sales	0.3
	1.0

TABLE 14-3
Inventor's expected profits (units of $10,000 profit)

Event	Act A_1: Inventor Manufactures Device			Act A_2: Inventor Sells Patent Rights		
	Probability	Profit	Weighted Profit	Probability	Profit	Weighted Profit
θ_1: Strong sales	0.2	$80	$16.0	0.2	$40	$ 8.0
θ_2: Average sales	0.5	20	10.0	0.5	7	3.5
θ_3: Weak sales	0.3	−5	−1.5	0.3	1	0.3
	1.0		$24.5	1.0		$11.8

Expected profit

= 24.5 (ten thousands of dollars)

= $245,000

Expected profit

= 11.8 (ten thousands of dollars)

= $118,000

To determine the basis for choice between the inventor's manufacturing the device (act A_1) and selling the patent rights (act A_2), we compute the expected profit for each of these courses of action. These calculations are shown in Table 14-3. As indicated in that table, profit is treated as a variable that takes on different values depending on which event occurs. We compute its expected value in the usual way, according to equation 3.11. The "expected value of an act" is the weighted average of the payoffs under that act, where the weights are the probabilities of the various events that can occur.

We see from Table 14-3 that the inventor's expected profit in manufacturing the device is $245,000, whereas the expected profit in selling the patent rights is only $118,000. To maximize the expected profit, our inventor will select A_1 and will manufacture the device rather than sell the patent.

It is useful to have a brief term to refer to the expected benefit of choosing the optimal act under conditions of uncertainty. We shall refer to the expected value of the monetary payoff of the best act as the **expected profit under uncertainty**. Hence, in the foregoing problem, the expected profit under uncertainty is $245,000.

Summary   We can summarize the method of calculating the expected profit under uncertainty as follows:

1. Calculate the expected profit for each act as the weighted average of the profits under that act, where the weights are the probabilities of the various events that can occur.

2. The expected profit under uncertainty is the maximum of the expected profits calculated in step 1.

Expected Opportunity Loss

A useful concept in the analysis of decisions under uncertainty is that of opportunity loss.

> An **opportunity loss** is the loss incurred because of failure to take the best possible action. Opportunity losses are calculated separately for each event that might occur.

Given the occurrence of a specific event, we can determine the best possible act. For a given event, the opportunity loss of an act is the difference between the payoff of that act and the payoff for the best act that could have been selected. For example, in the inventor's problem, if event θ_1 (strong sales) occurs, the best act is A_1, for which the payoff is $80 (in units of $10,000). The opportunity loss of that act is $80 - $80 = $0. The payoff for act A_2 is $40. The opportunity loss of act A_2 is the amount by which the payoff of the best act, $80, exceeds the $40 payoff of act A_2, which is $80 - $40 = $40.

It is convenient to asterisk the payoff of the best act for each event in the original payoff table in order to denote that opportunity losses are measured from these figures. Both the original payoff table and the opportunity loss table are given in Table 14-4 for the inventor's problem.

We can now proceed with the calculation of expected opportunity loss in a manner completely analogous to the calculation of expected profits. Again, we use the probabilities of events as weights and determine the weighted average opportunity loss for each act. Our goal is to select the act that yields the *minimum* expected opportunity loss. The calculation of the expected opportunity losses for the two acts in the inventor's problem is given in Table 14-5. The symbol EOL represents **expected opportunity loss**. Hence, $EOL(A_1)$ and $EOL(A_2)$ denote the expected opportunity losses of acts A_1 and A_2, respectively. The inventor's EOL in manufacturing the device is $18,000, and in selling the patent rights, the EOL is $145,000. If the inventor selects the act that minimizes the EOL, he or she will choose A_1, that is, to manufacture the device. This is the same act selected under the criterion of maximizing expected profit.

Expected opportunity loss (EOL)

> It can be proved that the best act according to the criterion of maximizing expected profit is also best if the decision maker follows the criterion of minimizing expected opportunity loss.

TABLE 14-4
Payoff table and opportunity loss table for the inventor's problem (units of $10,000)

	Payoff Table Acts		Opportunity Loss Table Acts	
Event	A_1	A_2	A_1	A_2
θ_1: Strong sales	$80*	$40	$0	$40
θ_2: Average sales	20*	7	0	13
θ_3: Weak sales	−5	1*	6	0

TABLE 14-5
Expected opportunity losses for the inventor's problem (units of $10,000)

Event	Act A_1: Inventor Manufactures Device			Act A_2: Inventor Sells Patent Rights		
	Probability	Opportunity Loss	Weighted Opportunity Loss	Probability	Opportunity Loss	Weighted Opportunity Loss
θ_1: Strong sales	0.2	$0	$ 0	0.2	$40	$ 8.0
θ_2: Average sales	0.5	0	0	0.5	13	6.5
θ_3: Weak sales	0.3	6	1.8	0.3	0	0
	1.0		$1.8	1.0		$14.5

$$\text{EOL}(A_1) = 1.8 \text{ (ten thousands of dollars)}$$
$$= \$18,000$$

$$\text{EOL}(A_2) = 14.5 \text{ (ten thousands of dollars)}$$
$$= \$145,000$$

The relationship between the maximum expected profit and the minimum expected opportunity loss will be examined later. Note that opportunity losses are not losses in the accountant's sense of profit and loss, because as we have seen, they occur even when only profits of different actions are compared for a given state. They represent opportunities foregone rather than monetary losses incurred.

Minimax Opportunity Loss

We noted earlier that various criteria of choice have been suggested for the decision problem. One that has been advanced in terms of opportunity losses is that of *minimax opportunity loss*.

> Under the **minimax opportunity loss** method, the decision maker selects the act that minimizes the worst possible opportunity loss that can be incurred among the various acts.

As with the maximin criterion for payoffs, the minimax criterion for opportunity loss takes a pessimistic view toward which states of nature will occur. Once the opportunity loss table has been prepared, as in Table 14-4, the decision maker determines for each act the largest opportunity loss that can be incurred. For example, in the inventor's problem, for act A_2, it is $400,000. The decision maker then chooses that act for which these worst possible losses are the least, that is, the act that *mini*mizes the *maxi*mum losses; hence, the term **minimax**. In the inventor's problem, the decision maker would choose act A_1, since the maximum possible opportunity loss ($60,000) under this course of action is less than the corresponding worst loss under act A_2 ($400,000). This criterion is also sometimes referred to as "minimax regret," where the opportunity losses are viewed as measures of regret for taking less than the best courses of action.[4]

[4] See L. J. Savage, "The Theory of Statistical Decision," *Journal of the American Statistical Association*, vol. 46, 1951, pp. 55–57.

In our illustrative problem, the minimax opportunity loss act, A_1, happens to be the same decision as would be made under the criterion of maximizing expected profit. However, this is not always the case. The minimax loss criterion, like the maximin payoff rule, singles out for each course of action the worst consequence that can befall the decision maker and then attempts to minimize this damage. As was true for the maximin payoff criterion, the minimax loss viewpoint yields results in many instances which imply that a business executive who faces risky ventures should simply go out of business.

Throughout the remainder of this book, we will use the Bayesian decision theory criterion of maximizing expected profit or its equivalent, minimizing expected opportunity loss.[5]

EXERCISES 14.4

1. Assume the following possible states of nature apply to the price of Smith Pharmaceutical stock:
   a. It will finish higher today than yesterday.
   b. It will finish the same as yesterday.
   c. It will finish lower today than yesterday.
   What is wrong with assessing prior probabilities of 0.4, 0.3, and 0.2, respectively?

2. During its first week of operation, the Net Set Tennis Shop sold 20 wooden rackets, 15 metal rackets, and five fiberglass rackets. Based solely on this past knowledge, what prior probability distribution would you formulate for the type of racket to be sold?

3. As a company manager, you must decide to invest $15,000 in either a cost reduction program or a new advertising campaign. You know that the cost reduction program will increase the profit-to-sales ratio from the current 11% to 12%. The advertising campaign, if successful, is expected to increase the current sales of $2 million by 13%. The probability that the campaign will be successful is 0.8. Which is the better course of action?

4. Lucky Leo has been offered a chance to play a card game with Shifty Sam using a deck of bridge cards. If Leo draws an ace, Sam will pay him $10,000. It will cost Leo $800 to try his luck. Should Leo play the card game?

5. A small town has only two automobile dealers, and the profit that each dealer makes depends on the pricing decisions of the other dealer. Realizing this, Dealer A wishes to set her price lower than Dealer B's during

---

[5] We shall indicate in section 14.8 that the criterion is actually the maximization of expected utility. In the case of a decision maker who is indifferent to risk—that is, one who has a linear utility function—expected utility is maximized by maximizing expected profit.

their common year-end close-out sales. There are only five prices that Dealer A and Dealer B can set. From past experience, Dealer A calculates what her profit would be under each of the pricing combinations that could occur. In the table below, these data are presented along with Dealer A's probability assessment of B's actions. What is Dealer A's optimal pricing decision and her expected profit?

Dealer A's profit (thousands of dollars)

Probability Assessment	Dealer B's Price	Dealer A's Price				
		1	2	3	4	5
0.1	1	2	0	−1	−1	−1
0.2	2	4	5	0	−1	−1
0.3	3	5	6	5	0	−1
0.2	4	6	7	6	6	0
0.2	5	7	7	7	6	6

6. An operations research team must decide whether to incorporate the predictions of the 10 leading investment advice newsletters in the information system the team is building. The cost of including the predictions is $4,900 per year. The research team estimates that in 20 decisions to be made in a year, the added information would result in a new decision only once. However, the decision change would result in an average saving of $80,000. Should the team include the newsletters in its information system?

7. A firm must decide whether to manufacture a new calculator. If demand is high, the firm will make a 10% profit on its investment. If demand is low, it will make only 2%. Instead of manufacturing the calculator, the firm could invest in government securities at a risk-free rate of 8%. If there is a 50% chance that demand will be high, should the firm manufacture the new calculator?

8. A presidential candidate has the opportunity to take a well-publicized view on a domestic spending bill currently before Congress. If the candidate speaks out in favor of the bill and the bill is passed, he could increase the number of his supporters by 12%. On the other hand, if he speaks out for the bill and the bill is defeated, he could lose 3% of his supporters. If he fails to take a stand on the issue, the number of his supporters will remain the same. What probability must his political advisors assign to the event "the bill will pass Congress" in order to make the candidate statistically indifferent regarding his course of action?

9. A businesswoman with $50,000 to invest must decide among three portfolios prepared by her financial advisor. The portfolios are characterized as high risk, average risk, and low risk, and their returns are dependent on the general economic situation. Assume that the only two possible

states of business conditions are "good" and "recession" and that the probability of a recession is 30%. Given the following payoff matrix (measured in thousands of dollars), which portfolio would she choose if she wanted to maximize expected profit? If she wanted to minimize expected opportunity loss?

State of Nature	Portfolio		
	High Risk	Average Risk	Low Risk
Good	10	4	2
Recession	−15	−2	1

10. A manufacturer of children's toys must determine its advertising budget for the coming year. Realizing that sales are dependent on the nation's economic situation, this manufacturer constructed the payoff table presented below. The table shows the profits (in millions of dollars) that the manufacturer expects with high, average, and low advertising costs in the event of a severe recession, a mild recession, a weak recovery, and a strong recovery.

State of Nature	Advertising Costs		
	High	Average	Low
Severe recession	−10	−6	0
Mild recession	−5	−1	1
Weak recovery	4	3	2
Strong recovery	12	9	4

a. Construct the corresponding opportunity loss table.
b. Assume that economists predict a 20% chance of a severe recession, a 25% chance of a mild recession, a 40% chance of a mild recovery, and a 15% chance of a strong recovery. Which advertising budget should the manufacturer choose if it wishes to minimize expected opportunity loss?

11. In the first week of operation of a bookstore, the owner sells 20 books from the "nonfiction" section, 100 books from the "classical fiction" section, 175 books from the "modern fiction" section, and 35 books from the "sports" section. What prior probability distribution would you formulate for the types of books sold?

12. A company will either invest $450,000 in 14% commercial paper or purchase $450,000 worth of new equipment. If it purchases new equipment, the company will earn 25% if business conditions are good and 8% if business conditions are bad. What must the probability of good business conditions be for the two investments to be equally attractive?

13. A manufacturing company must decide whether to sell its product in its present form or to modify it and sell it at a higher price. In its present form, the product can be sold with certainty for $50. Cost of modification is $10 and, if modified, there is a probability of 0.3 that the product can be sold for $75, a probability of 0.5 that it can be sold for $63, and a probability of 0.2 that it can be sold for $55. What should the company decide? (The cost of production of the unmodified product is $30.)

14. The owner of an electronics products store wishes to determine the number of television sets that must be kept in stock to maximize expected profits. Each set is bought for $200 and sold for $450. Unsold sets are later returned to the manufacturer for $100. The owner estimates that the demand for the sets has the following probability distribution:

Demand	0	1	2	3	4	5
Probability	0.10	0.10	0.15	0.35	0.20	0.10

How many sets should the owner stock?

14.5
EXPECTED VALUE OF PERFECT INFORMATION

Thus far in our discussion, we have considered situations in which the decision maker chooses among alternative courses of action on the basis of *prior information* without attempting to gather further information before making the decision. In other words, the probabilities used in computing the expected value of each act, as shown in Table 14-3, are **prior probabilities**.

Prior probabilities

> **Prior probabilities** are probabilities established prior to obtaining additional information through sampling.

Prior analysis

> The procedure of calculating expected value of each act based on these prior probabilities and selecting the optimal act is referred to in Bayesian decision theory as **prior analysis**.

In Chapter 15 we consider how courses of action may be compared after these prior probabilities are revised on the basis of sample information, experimental data, or information resulting from tests of any sort. However, the analysis we have carried out so far provides a yardstick for measuring the value

of perfect information concerning which events will occur. This yardstick will   *A yardstick for*
be referred to as the **expected value of perfect information**.   *measuring*

- To determine this value, we calculate the *expected profit with perfect information.*
- Then, we subtract the *expected profit under uncertainty*, (the calculation we previously examined) to find the *expected value of perfect information.*

These concepts will be explained in terms of the inventor's problem. We begin with the idea of expected profit with perfect information.

> The calculation of the *expected profit* of acting *with perfect information* is based on the expected payoff if the decision maker has access to a perfect predictor.

Expected profit with
perfect information

It is assumed that if this perfect predictor forecasts that a particular event will occur, then indeed that event will occur. The expected payoff under these conditions for the inventor's problem is given in Table 14-6.

To understand the meaning of this calculation, it is necessary to adopt a long-run relative frequency point of view. If the forecaster says the event "strong sales" will prevail, the decision maker can look along that row in the payoff table and select the act that yields the highest profit. In the case of strong sales, the best act is A_1, which yields a profit of $800,000. Hence, the figure $80 is entered under the profit column in Table 14-6. The same procedure is used to obtain the payoffs for each of the other possible events. The probabilities shown in the next column are the original probability assignments to the three states of nature. From a relative frequency viewpoint, these probabilities are now interpreted as the proportion of times the perfect predictor would forecast the occurrences of the given states of nature if the present situation were faced

---

TABLE 14-6
Calculation of expected profit with perfect information for the inventor's problem (units of $10,000 profit)

Predicted Event	Profit	Probability	Weighted Profit
θ_1: Strong sales	$80	0.2	$16.0
θ_2: Average sales	20	0.5	10.0
θ_3: Weak sales	1	0.3	0.3
			$26.3

Expected profit with perfect information

$\qquad = \$26.3$ (ten thousands of dollars)

$\qquad = \$263,000$

repeatedly. Each time the predictor makes a forecast, the decision maker selects the optimal payoff.

> The expected profit with perfect information is calculated (as in Table 14-6) by weighting the best payoffs by the probabilities and totaling the products.

The expected profit with perfect information in the inventor's problem is $263,000. This figure can be interpreted as the average profit the inventor could realize if he were faced with this decision problem repeatedly under identical conditions, and if he always took the best action after receiving the perfect indicator's forecast. Expected profit with perfect information has sometimes been called the "expected profit under certainty," but this term is somewhat misleading. The inventor is not *certain* to earn any one profit figure. The expected profit with perfect information should be interpreted as indicated in this discussion.

Expected value of perfect information (EVPI)

> The expected value of perfect information, abbreviated EVPI, is defined as *the expected profit with perfect information* minus *the expected profit under uncertainty.*

The interpretation of the EVPI is clear from its calculation (shown in Table 14-7). In the inventor's problem, the expected payoff of selecting the optimal act under conditions of uncertainty is $245,000 (see Table 14-3). On the other hand, if the perfect predictor were available and the inventor acted according to those predictions, the expected payoff would be $263,000 (see Table 14-6). The difference of $18,000 represents the increase in expected profit attributable to the use of the perfect forecaster. Hence, the expected value of perfect information may be interpreted as the most the inventor should be willing to pay to get perfect information on the sales level for the device.

The expected opportunity loss of selecting the optimal act under conditions of uncertainty in the inventor's problem was shown earlier to be $18,000 (see Table 14-5). That is, this figure represented the minimum value among the expected opportunity losses associated with each act. As shown in Table 14-7, this figure is equal to the expected value of perfect information. It can be

---

TABLE 14-7
Calculation of expected value of perfect information for the inventor's problem

Expected profit with perfect information	$263,000
Less: Expected profit under uncertainty	245,000
Expected value of perfect information (EVPI)	$ 18,000

EVPI = EOL of the optimal act under uncertainty = $18,000

mathematically proved that this equality holds in general. Another term used for the expected opportunity loss of the optimal act under uncertainty is the **cost of uncertainty**. This term highlights the "cost" attached to decision making under conditions of uncertainty. Expected profit would be larger if a perfect predictor were available and this uncertainty were removed. Hence, this cost of uncertainty is also equal to the expected value of perfect information. In summary, the following three quantities are equivalent:

Cost of uncertainty

- Expected value of perfect information (EVPI)
- Expected opportunity loss of the optimal act under uncertainty (EOL)
- Cost of uncertainty

14.6
REPRESENTATION BY A DECISION DIAGRAM

It is useful to represent the structure of a decision problem under uncertainty by a **decision tree diagram**, also called a **decision diagram** or, briefly, a **tree**. The diagram depicts the problem in terms of a series of choices made in alternating order by the decision maker and by "chance." Forks at which the decision maker is in control of choice are referred to as **decision forks** (represented by a square); those at which chance is in control are called **chance forks** (represented by a circle). Forks may also be referred to as **branching points** or **junctures**.

Decision tree diagram

A simplified decision diagram for the inventor's problem is given in Figure 14-1(a). After explaining this skeleton version, we will insert additional information to obtain a completed diagram. As we can see from Figure 14-1(a), the first choice is the decision maker's at branching point 1. He can follow either branch A_1 or branch A_2; that is, he can choose either act A_1 or A_2. Assuming that he follows path A_1, he comes to another juncture, branching point 2, which is a chance fork. Chance now determines whether the event that will occur is θ_1, θ_2, or θ_3. If chance takes him down the θ_1 path, the terminal payoff is $800,000; the corresponding payoffs are indicated for the other paths. An analogous interpretation holds if he chooses to follow branch A_2. Thus, the decision diagram depicts the basic structure of the decision problem in schematic form. In Figure 14-1(b), additional information is superimposed on the diagram to represent the analysis and solution to the problem.

The decision analysis process represented by Figure 14-1(b) (and other decision diagrams to be considered later) is known as **backward induction**. We imagine ourselves as located at the right-hand side of the tree diagram, where the monetary payoffs are. Let us consider first the upper three paths denoted θ_1, θ_2, and θ_3. Below these symbols, in parentheses, we enter the respective probability assignments (0.2, 0.5, and 0.3) as given in Table 14-3. These represent the probabilities assigned by chance to following these three paths, after the decision maker has selected act A_1. Moving back to the chance fork from which these three paths emanate, we can calculate the expected monetary value

Backward induction

FIGURE 14-1

(a) Simplified decision
diagram for inventor's
problem

(b) Decision diagram
for inventor's problem
(payoffs are in units of
$10,000 profit)

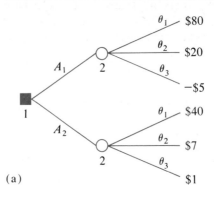

(a)

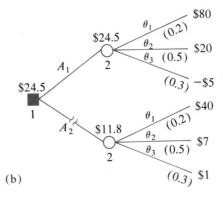

(b)

of being located at that fork. This expected monetary value is $24.5 (in units of $10,000, as are the other obvious corresponding numbers) and is calculated in the usual way:

$$\$24.5 = (0.2)(\$80) + (0.5)(\$20) + (0.3)(-\$5)$$

This figure is entered at the upper chance fork. It represents the value of standing at that fork after choosing act A_1, as chance is about to select one of the three paths. The analogous figure entered at the lower chance fork is $11.8. Therefore imagining ourselves as being transferred back to branching point 1, where the square represents a fork at which the decision maker can make a choice, we have the alternatives of selecting act A_1 or act A_2. Each of these acts leads us down a path at the end of which is a risky option whose expected profit has been indicated. Since following path A_1 yields a higher expected payoff than path A_2, we block off A_2 as a nonoptimal course of action. This is indicated on the diagram by the two wavy lines. Hence, A_1 is the optimal course of action, and it has the indicated expected payoff of $24.5. This expected payoff of the best act, $24.5, is shown above branching point 1 in Figure 14-1(b).

Thus, the decision tree diagram reproduces in compact schematic form the analysis given in Table 14-3. An analogous diagram would be constructed in terms of opportunity losses to reproduce the analysis of Table 14-5.

EXERCISES 14.6

1. Explain the meaning of *expected value of perfect information.*

2. Explain the difference between expected opportunity loss and expected value of perfect information.

3. Given an opportunity loss table, can you compute the corresponding pay-off table? Explain why or why not.

4. A New York City cab driver must decide whether to work a Friday night. If the demand for her services is good, she will make a $100 profit; if it is only fair she will make a $50 profit; and if demand is poor she will break even. Compute the appropriate opportunity loss table.

5. A farmer is trying to decide how many acres of corn to plant this year. He realizes that his profit will depend on the weather conditions. He identifies the following four weather types that he feels could occur: W_1, W_2, W_3, and W_4 (with W_1 being most conducive to a high corn yield per acre and W_4 being least conducive). This farmer has studied Bayesian decision theory and plans to use it in his analysis. He constructs the payoff matrix below in units of $10,000.

Weather Condition	Number of Acres to Plant			
	A_1 500	A_2 1,000	A_3 2,000	A_4 3,000
W_1	100	120	200	250
W_2	85	90	110	100
W_3	70	60	65	75
W_4	60	55	70	65

The farmer assigns the following prior probabilities to the possible weather conditions:

Weather Condition	Probability
W_1	0.2
W_2	0.4
W_3	0.3
W_4	0.1

Compute the EVPI by two methods.

6. The following is a store owner's payoff table in units of $1,000:

	Action			
Demand	A_1	A_2	A_3	A_4
S_1: Above average	20	17	18	13
S_2: Average	10	14	14	12
S_3: Below average	4	7	5	10

where A_1 = keep store open weekdays, evenings, and Saturdays
A_2 = keep store open weekdays and Wednesday evenings
A_3 = keep store open weekdays and Saturdays
A_4 = keep store open only weekdays

The prior probability distribution of demand is as follows:

S_i	$P(S_i)$
S_1	0.5
S_2	0.3
S_3	0.2

a. Find the expected profit with perfect information.
b. Find the expected profit under uncertainty.
c. How much should the store owner pay for information that yields the true state of nature?

7. George Frey has the opportunity to invest in the stock of a company that is developing an "electronic language translator." If the stock "takes off" as his financial advisor assures him it will, Frey will earn $60,000 next year. If the company dissolves, Frey could lose his initial investment of $30,000. After much research and consultation with others, Frey assesses the probability of the company's failing as $\frac{1}{3}$. Should he invest in the company? What is the expected value of perfect information? (Ignore the time value of money.)

8. Suppose you are considering taking some debatable deductions on your tax return. You will save $3,500 in taxes if you take the deductions and the IRS does not audit your return. If the IRS does audit your return and is not satisfied with your explanation, you will have to pay a penalty of $2,000 in addition to the tax payment of $3,500. If the rest of your return is correct and would not be challenged in an audit, the payoff table would be as follows:

State of Nature	Take Deductions	Do Note Take Deductions
IRS audits	− $2,000	$0
IRS does not audit	+ $3,500	0

Because of the complicated nature of your tax return, you estimate a 60% chance that the IRS will audit it. Calculate the expected value of perfect information.

9. Ted Jasper is one of two boat salespeople in a small resort town. Jasper realizes that his profit will depend a great deal on the prices his competitor, Roberta Bennett, charges. Jasper knows that Bennett is having a sale on a specific sailboat model in the coming week. He also knows that Bennett will set the price at either $1,800 or $2,000. Jasper decides that he too will price at either $1,800 or $2,000. From past experience with price and demand relationships, Jasper constructs the profit payoff table shown below.

Competitor's Action Price at	Jasper's Action	
	Price at $1,800	Price at $2,000
$1,800	$ 6,000	$ 4,000
2,000	18,000	20,000

a. Jasper feels that there is a 20% chance that Bennett will set the price at $1,800. What is the expected profit with perfect information?
b. What is the expected profit under uncertainty?
c. Assume there is some legitimate way that Jasper could obtain information on the price Bennett will charge. How much should Jasper be willing to pay for this information?

10. You are the president of a small but growing firm, and you are approached by a minicomputer sales representative, who tries to convince you to use a computer in your business. Specifically, he claims that the the first-year profits alone would increase by $20,000. Skeptical, you realize that the value of the computer depends largely on how well your employees adjust to it. Your subjective assessment of payoffs, shown below, reflects your feelings.

Employee Adjustment	Payoff
Good	$20,000
Fair	10,000
Poor	−2,000

a. From your knowledge of your employees, you determine the probabilities of the three levels of adjustment as follows: good, 0.3; fair, 0.3; and poor, 0.4. Should you buy the computer?
b. How much would you be willing to pay a perfect predictor to forecast the level of employee adjustment?

11. In exercise 4, the cab driver, after much consideration, feels that the probabilities that demand will be good and fair are 0.3 and 0.4, respectively. What is the expected opportunity loss for each action? What is the optimal decision?

12. A brewer who currently packages beer in old-style bottles is debating whether to change the packaging of the beer for next year. The choices are A_1, an easy-open aluminum can; A_2, a lift-top can; A_3, a wide-mouth screw-top bottle; or A_4, the same old-style bottle. Profits resulting from each move depend on what the brewer's competitor does for the next year. The payoff matrix and prior probabilities, measured in $10,000 units, are as follows:

Prior Probability	Competitor Uses	Act			
		A_1	A_2	A_3	A_4
0.4	Old-style bottles	16	16	15	18
0.3	Easy-open cans	14	13	12	10
0.2	Lift-top cans	8	11	10	8
0.1	Screw-top bottles	7	8	10	7

   a. Find the expected opportunity loss for each act.
   b. Determine the EVPI.
   c. Determine the optimal decision.

14.7
THE ASSESSMENT OF PROBABILITY DISTRIBUTIONS

We indicated earlier that it is reasonable to assume that a decision maker would have some idea of the probabilities of the occurrences of the various states of nature that are relevant to the decision to be made. In this section, we discuss how such probabilities may be assessed.[6]

> If the random variable representing states of nature is discrete and there are only a small number of possible outcomes, then the decision maker will probably be able to assign probabilities directly to each possible outcome.

For example, if the decision maker must assess the probability that a particular manufactured article is defective, he or she may be able to assign a probability

---

[6] For a comprehensive discussion of the philosophy and practice of the assessment of subjective probabilities, see C. S. Spetzler and C. S. Staël Von Holstein, "Probability Encoding in Decision Analysis," *Management Science*, vol. 22, no. 3 (November 1975).

of, say, 0.1 that the article is defective; hence, the complementary probability would be 0.9 that the article is not defective. If there were three possible outcomes—such as seriously defective, moderately defective, and not defective—again the decision maker can assign a probability to each of these three possible outcomes. As an example in a quite different context, the decision maker might have to assign probabilities to the following events: (1) the U.S. Congress *will* pass the bill on education currently before it or (2) the U.S. Congress *will not* pass the bill. In all of these cases, the decision maker may use a combination of past empirical information and a subjective evaluation of the effects of additional knowledge in making the probability assignments.

On the other hand, the random variable of interest may take on a large number of possible values. For example, if the states of nature are values for a company's sales next year for a certain product, the relevant values to be considered by the decision maker may range from, say, 100,000 to 500,000 units. It would be rather meaningless to attempt to assess a probability for each of the 400,001 possible outcomes for sales, that is, for 100,000, 100,001, . . . , 500,000.

> If the random variable has a large number of possible values, the decision maker should treat the random variable as continuous and should set up a cumulative distribution function by making probability assessments for a number of selected ranges of the random variable.

A distinction can be made between the following two situations:

- The decision maker directly establishes a subjective cumulative probability distribution without formal processing of data.
- There is a small quantity of past data, and the decision maker formally processes this information to set up a cumulative probability distribution.

We now discuss how these probability distributions may be constructed.

Direct Subjective Assessment

We first illustrate a situation in which a decision maker establishes a subjective cumulative probability distribution without formal processing of data. A decision maker, forecasting a company's sales of a certain product for next year, feels that sales may range from 100,000 to 500,000 units and wishes to establish a subjective cumulative probability distribution without explicitly using any data.

The basis of the procedure is to focus attention at a few key points, or *Finding the first three* **fractiles**, in the distribution. For example, the 0.50 fractile is a value such that *fractile values* the decision maker believes the probability is $\frac{1}{2}$ that the random variable is equal to or less than that value. (Note that the 0.50 fractile is simply another name for the median.) It is useful in this context to think in terms of hypothetical gambles. For example, one gamble might pay $100 if the random variable in question is less than or equal to some value selected by the decision maker, and a second gamble might pay $100 if the random variable turns out to be more than that value. The 0.50 fractile is the point at which the decision maker is indifferent to the choice between the two gambles. Let us assume that after some serious reflection on these gambles, the decision maker selects 350,000 units as the median or 0.50 fractile value.

We continue with the subjective assessment process. The decision maker should now select the 0.25 and 0.75 fractiles. For example, the 0.25 fractile is a figure such that the probability is $\frac{1}{4}$ that the value of the random variable will lie below that figure and $\frac{3}{4}$ that the value will lie above it. The analogous interpretation applies for the 0.75 fractile. We return to choices among gambles to determine the 0.25 fractile value. The decision maker might be asked to assume that sales next year will be less than 350,000 units and should then be presented with a pair of gambles: The first wager will pay $100 if sales fall between a value that the decision maker chooses and the 0.50 fractile; the second will pay $100 if sales are equal to or less than the chosen figure. The dividing value of the random variable determined by this "indifference point" between the two gambles is the 0.25 fractile. Obviously, if the decision maker selects a tentative dividing value but then finds one gamble more attractive than the other, the 0.25 fractile value has not been determined. The assessment procedure should continue until the indifference point has been found. Let us assume that the decision maker chooses 250,000 as the 0.25 fractile. The 0.75 fractile should then be determined by a similar procedure; let us assume that the decision maker determines 400,000 as the 0.75 fractile value.

Three selected points (the 0.25, 0.50, and 0.75 fractiles) have now been found in the cumulative distribution function.

> The usual procedure at this juncture is to determine two more values at the extremes of the distribution and to use the resulting five points as a basis for sketching the function.

Selecting the extreme values   The extreme values often used are either the 0.01 and 0.99 or the 0.001 and 0.999 fractiles. In the present illustration, let us assume that the decision maker, when queried about the range of 100,000 to 500,000 units for next year's sales, stated that these were not absolute limits, that it was possible for next year's sales to fall outside them. However, after some further hard thinking, the decision maker selects these two values as the 0.01 and 0.99 fractiles. That is, the subjectively assigned odds are 99:1 that sales will exceed 100,000 (corresponding to a probability of 0.99). Similarly, the subjective odds are 99:1 that next year's sales will not exceed 500,000 units. Actually, these extreme fractiles are difficult to assess. Empirical experiments have demonstrated that people tend to make the distributions too tight; that is, the 0.01 (or 0.001) and 0.99 (or 0.999) fractiles are generally not low enough and not high enough, respectively. In formulating these assessments, we must make a conscientious effort to spread the extreme values sufficiently.

To monitor further the probability assessments, the decision maker should carry out some tests for consistency of judgments. For example, since the 0.25 and 0.75 fractiles in this example are 250,000 and 400,000, the decision maker is asserting a 0.50 chance that next year's sales will fall within the range of these two figures (known, incidentally, as the **interquartile range**). Hence these probability assessments imply that it is equally likely that next year's sales will fall inside this range or outside this range. Perhaps, on reflection, the decision maker will revise one or both of the two fractiles. After a few introspective checks of this sort, the five selected points should be plotted and a smooth curve drawn through them. Figure 14-2 shows a cumulative probability distribution that could be drawn through the 0.01, 0.25, 0.50, 0.75, and 0.99 fractiles referred to in our illustration. The curve is the typical S-shaped form obtained for cumulative probability distributions for continuous random variables.

Testing for consistency of judgments

Sketching the function

Even after the graph has been drawn, the decision maker should continue to check or monitor the probability assignments. If the curve rises too slowly or too steeply in a certain portion of the random variable's range, the graph (and consequently the appropriate fractile values) should be adjusted. The reassessment procedure should continue until the decision maker feels that the curve appropriately describes the probability distribution.

Checking the probability assignments

The direct subjective assessment of probability distributions is appropriate for the following situations.

- when there is little or no past historical data on which to base the construction of the function

- when there *is* a substantial amount of past data but the decision maker is unwilling to use these data formally in the establishment of a probability distribution

Discarding past data

The latter situation might arise if the decision maker decides that the factors that gave rise to the past historical data have changed so greatly that these

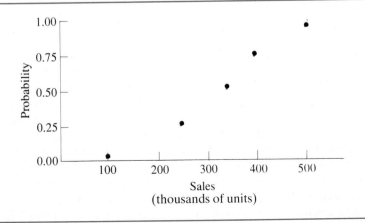

FIGURE 14-2
A cumulative probability distribution sketched through five selected fractiles

data are not a reliable guide for assessing future probabilities. For example, suppose a company that purchases one of its raw materials from a supplier has a large quantity of past data on the percentages of defective materials in past shipments from the supplier. Assuming no important changes in the supplier's manufacturing process and in policies governing the production of the product, the purchasing company may feel justified in using the past relative frequencies of occurrence of percentage defectives per shipment to assess a probability distribution for the percentage of defectives that might occur on the *next* shipment. On the other hand, if the supplier has made an important change in its method of manufacture or in the work force that produces the product, the purchaser may feel that the past relative frequencies of occurrence are no longer relevant for assessing the probability distribution. The purchaser may then appropriately make a direct subjective assessment of the probability distribution.

Use of Past Data

Large quantity of data

Small quantity of data

We turn now to the first situation listed at the beginning of this section, namely that there is a *small* quantity of data and the decision maker formally processes these data to construct a cumulative probability distribution for the relevant states of nature. If there were a *large* quantity of relevant historical data in the form of relative frequencies of occurrence, the decision maker could simply use these figures to set up a discrete probability distribution. If only a *small* quantity of relevant data exists, how should these data be processed to construct the desired probability distribution? We will discuss here a systematic method for dealing with this type of situation.

Let us assume that a retail establishment asks you, a consultant, to prepare a probabilistic forecast of the number of telephone orders that would be received next week for a certain product. The product has been sold for the past 20 weeks, and Table 14-8(a) gives a record of the number of weeks in which stated numbers of telephone orders were received. If you set up a relative frequency distribution of this discrete random variable, you would obtain the function shown in Table 14-8(b).

Interpreting the relative frequency distribution of Table 14-8(b) as a probability distribution, you find yourself making some rather odd statements. For example, starting at the beginning of the distribution, you would state that receiving 21 orders has a probability of 0.05, receiving 22 and 23 orders has a probability of 0, receiving 24 orders has a probability of 0.10, and so on. However, you really do not believe that it is impossible for 22 or 23 telephone orders to occur. You would recognize that these numbers of orders probably did not occur in the small sample of only 20 weeks of experience because of chance sampling fluctuations. If you work in terms of a cumulative probability distribution rather than the probability mass function of Table 14-8(b), you can develop a more reasonable interpretation of the data. The cumulative relative frequencies of occurrence for the telephone order data are shown in Table 14-9. These cumulative frequencies represent the proportion of weeks in which

TABLE 14-8

(a) Numbers of weeks in which specified numbers of telephone orders were received

(b) Relative frequency of occurrence of number of telephone orders

Number of Telephone Orders	Number of Weeks That Number of Orders Was Received	Number of Telephone Orders	Relative Frequency of Occurrence
21	1	21	0.05
22	0		
23	0	24	0.10
24	2	25	0.10
25	2	26	0.15
26	3	27	0.15
27	3	28	0.05
28	1		
29	0	31	0.05
30	0	32	0.05
31	1	33	0.05
32	1	34	0.05
33	1		
34	1	37	0.05
35	0	38	0.05
36	0	39	0.05
37	1		
38	1	42	0.05
39	1		
40	0		
41	0		
42	1		

TABLE 14-9

Cumulative relative frequency of occurrence of numbers of telephone orders

Number of Telephone Orders	Proportion of Weeks in Which the Specified Number or Fewer Orders Were Received	Number of Telephone Orders	Proportion of Weeks in Which the Specified Number or Fewer Orders Were Received
21	0.05	32	0.70
22	0.05	33	0.75
23	0.05	34	0.80
24	0.15	35	0.80
25	0.25	36	0.80
26	0.40	37	0.85
27	0.55	38	0.90
28	0.60	39	0.95
29	0.60	40	0.95
30	0.60	41	0.95
31	0.65	42	1.00

the indicated numbers of telephone orders or *fewer* were received. Interpreted as a probability, a cumulative frequency may be thought of as the probability that the random variable "number of telephone orders" is less than or equal to the specified value.

A graph of the cumulative frequency distribution of Table 14-9 is shown in Figure 14-3. In the present example, the raw data are ungrouped, and they are graphed as a step function in the figure. The upper point plotted on each vertical line in the graph is the cumulative frequency corresponding to the value of the random variable plotted on the horizontal axis. As with the original frequency distribution shown in Table-8(b), the irregularities of the step function graph of the cumulative relative frequency distribution in Figure 14-3 are doubtless the result of chance sampling variations. In an attempt to smooth out these sampling variations, you try to fit a smooth curve to the graph. Such a smooth curve is fitted to the step function data in Figure 14-3. In fitting this curve, you should draw through the *centers* of the flat portions ("flats") of the step function to show a representative cumulative probability for the *range* of values covered by these flats. The range of possible values shown on the graph should be wider than the analogous range observed in the small sample of data. Thus, for example, although the original data had a lowest value of 21 and a highest value of 42, the horizontal axis of Figure 14-3 ranges from 10 to 60 telephone orders. The lowest and highest values should reflect the decision maker's best judgment concerning the range of possible outcomes. In fact, the entire smooth cumulative probability distribution should represent the decision maker's judgment concerning the form of the curve (guided, of course, by the rough form suggested by the step function data).

FIGURE 14-3

Continuous cumulative probability distribution fitted to the step function graph of cumulative relative frequency distribution in Table 14-8(b)

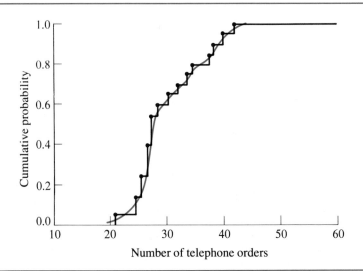

Summary

In the preceding subsections, we outlined the following ways of constructing probability distributions for states of nature depending on the type and quantity of data available.

- If there are no historical data available, or if the decision maker prefers not to process the available data because of doubts concerning the constancy of the mechanisms producing these data, he or she may proceed to a direct subjective assessment of the cumulative probability distribution. As we have seen, by using selected fractiles, the decision maker can construct a continuous cumulative probability function.[7]

- When a relatively *small* quantity of available data exists, we have noted how a cumulative relative frequency step function could be graphed and how a continuous cumulative probability distribution could be fitted to that function.

If there are *large* quantities of relative frequency data available concerning states of nature, and if we are confident about the constancy of the causal systems producing the data, we can simply set up a discrete probability distribution from those data.

We discussed how to construct continuous cumulative probability distributions for events or states of nature in decision-making problems in which there are no data or relatively small amounts of data. Continuous probability distributions are difficult to deal with in the analysis of many practical decision problems because they assign probability densities to every possible value of the states of nature.[8]

Bracket Medians

In the earlier sections of this chapter, we discussed the structure of the decision making problem and decision diagram representation in terms of *discrete* probability distributions for states of nature.

The **bracket median** technique is a method of approximating a continuous cumulative probability distribution by a discrete distribution. This technique enables us to carry out decision analysis for states of nature represented by

---

[7] See R. S. Schlaifer, *Analysis of Decisions Under Uncertainty* (New York: McGraw-Hill, 1969) for a comprehensive discussion of the assessment of probability distributions based on the quantity and type of data available.

[8] More advanced texts in statistical decision theory deal with methods of analysis for continuous probability distributions whose formulas can be given algebraically (for example, the normal, gamma, and beta distributions).

FIGURE 14-4

The representation of a continuous cumulative probability distribution by bracket medians

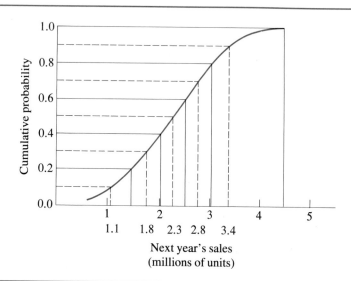

discrete random variables as outlined in this and subsequent chapters. As an illustration of the bracket median technique, we consider the continuous cumulative probability distribution for next year's sales of a certain product, shown in Figure 14-4. The basic rationale of the method is to break up the probability distribution into groups having equal probabilities. Thus, in Figure 14-4, the vertical scale of cumulative probabilities has five equal divisions, 0–0.2, 0.2–0.4, and so on. This procedure correspondingly divides the random variable sales into five equally likely groups, each having a probability of 0.2.

Our next step is to determine a representative value, or **bracket median**, for each of the five groups of the random variable and assign a 0.2 probability to each of these representative values. These bracket medians are determined by bisecting each of the probability groups along the vertical axis, drawing the broken lines shown at 0.1, 0.3, 0.5, 0.7, and 0.9 over to the curve. We then read down on the horizontal axis to determine the bracket medians. In Figure 14-4, these bracket medians are at 1.1, 1.8, 2.3, 2.8, and 3.4 millions of units of sales. The resulting discrete probability distribution is given in Table 14-10.

In this illustration, five bracket medians were used. Of course, we could obtain greater accuracy by using larger numbers of groups. For example, if we used 10 groups, we would assign probabilities of 0.1 to the equally likely groups, and so on. Computer programs are available to carry out the bracket median procedure for large numbers of equally probable groupings. The resulting discrete probability distributions can be used in a decision analysis in the same way that the probability distribution of sales was used in the example given earlier in this chapter.

TABLE 14-10
Approximation of the continuous probability distribution of Figure 14-4 by a discrete distribution using bracket medians

Bracket Median Next Year's Sales (millions of units)	Probability
1.1	0.2
1.8	0.2
2.3	0.2
2.8	0.2
3.4	0.2
	1.0

14.8
DECISION MAKING BASED ON EXPECTED UTILITY

In the decision analysis discussed up to this point, the criterion of choice was the maximization of expected monetary value. This criterion can be interpreted as a test of preference that selects the optimal act, that is, the act that yields the greatest long-run average profit. In a decision problem such as our example involving the inventor's choices, the optimal act is the one that would result in the largest long-run average profit if the same decision had to be made repeatedly under identical environmental conditions. In general, in such decision-making situations, the observed average payoff approaches the theoretical expected payoff as the number of repetitions increases. Gamblers, baseball managers, insurance companies, and others who engage in what is colloquially called "playing the percentages," are often characterized as using the aforementioned criterion.

However, many of the most important personal and business decisions are made under unique sets of conditions, and on some of these occasions it may not be realistic to think in terms of many repetitions of the same decision situation. Indeed, in the business world, many of management's most important decisions are unique, high-risk, high-stake choice situations, whereas the less important, routine, repetitive decisions are ones customarily delegated to subordinates. Therefore, it is useful to have an apparatus for dealing with one-time decision making. **Utility theory**, which we discuss in this section, provides such an apparatus, as well as a logical method for repetitive decision making.

One-time decision making

Whether an individual, a corporation, or other entity would be willing to make decisions on the basis of the expected monetary value criterion depends on the decision maker's attitude toward risk situations. Several simple choice

TABLE 14-11
Alternative courses of action with different expected monetary payoffs

Certainty Equivalent		Gamble
A_1: Receive $0 for certain (That is, you are certain to incur neither a gain nor a loss.)	or	A_2: Receive $0.60 with probability $\frac{1}{2}$ and lose $0.40 with probability $\frac{1}{2}$
B_1: Receive $0 for certain (That is, you are certain to incur neither a gain nor a loss.)	or	B_2: Receive $60,000 with probability $\frac{1}{2}$ and lose $40,000 with probability $\frac{1}{2}$
C_1: Receive a $1 million gift for certain	or	C_2: Receive $2.1 million with probability $\frac{1}{2}$ and receive $0 with probability $\frac{1}{2}$

situations are presented in Table 14-11 to illustrate that in choosing between two alternative acts we might select the one with the lower expected value. We might make a choice for lower expected value because we feel that the risk of gain is too great—that the increase in expected gain of the act with greater expected monetary value does not sufficiently reward us for the additional risk involved. Table 14-11 gives three choice situations for alternative acts grouped in pairs. For each pair, a decision must be made between the two alternatives.

The illustrative choices are to be made once and only once. That is, the decision experiment is not to be repeated. We shall assume for simplicity that all monetary payoffs are tax free. Suppose you choose acts A_2, B_1, and C_1. In the choice between A_1 and A_2, you might argue as follows: "The expected value of act A_1 is $0; the expected value of A_2 is $E(A_2) = (\frac{1}{2})(\$0.60) + (\frac{1}{2})(-\$0.40) = \$0.10$. A_2 has the higher expected value, and since I can sustain the loss of $0.40 with equanimity, I am willing to accept the risk involved in the selection of this course of action."

A useful way of viewing the choice between A_1 and A_2 is to think of A_2 as an option in which a fair coin is tossed. If it lands "heads" you recieve a payment of $0.60. If it lands "tails" you must pay $0.40. If you select act A_1, you are choosing to receive $0 for certain rather than to play the game involved in flipping the coin. Hence, you neither lose nor gain anything. The $0 figure *Certainty equivalent* is referred to as the **certainty equivalent** of the indicated gamble.

On the other hand, you might very well choose act B_1 rather than B_2, even though the respective expected values are

$$E(B_1) = \$0$$

$$E(B_2) = \left(\frac{1}{2}\right)(\$60,000) + \left(\frac{1}{2}\right)(-\$40,000) = \$10,000$$

In this case, you might reason that even though act B_2 has the higher expected monetary value, a calamity of no mean proportions would occur if the coin landed tails, and you incurred a loss (say, a debt) of $40,000. Your present asset level might make such a loss intolerable. Hence, you would refuse to play the game. If you look at the difference between the two choices just discussed (A_1 versus A_2 and B_1 versus B_2) you will note that the only difference is in the amounts of the gains and losses. In A_2 we had the payoffs \$0.60 and $-\$0.40$. In B_2 the decimal point has been moved five places to the right for each of these numbers, making the monetary gains and losses much larger than in A_2. In all other respects, the wording of the choice between A_1 and A_2 and between B_1 and B_2 is the same. Nevertheless, as we shall note after the ensuing discussion of the choice between acts C_1 and C_2, it is not necessarily irrational to select act A_2 over A_1, where A_1 has the greater expected monetary value, and B_1 over B_2, where B_1 has the smaller expected monetary value.

In the choice between acts C_1 and C_2 you would probably select act C_1, which has the lower expected monetary value. That is, most people would doubtless prefer a certain gift of \$1 million to a 50:50 chance at \$2.1 million and \$0, for which the expected payoff is

$$E(C_2) = \left(\frac{1}{2}\right)(\$2,100,000) + \left(\frac{1}{2}\right)(\$0) = \$1,050,000$$

In this case, you might argue that you would much prefer to have the \$1 million for certain, and go home to contemplate your good fortune in peace, than to play a game where on the flip of a coin you might receive nothing at all. You might also feel that there are relatively few things that you could do with \$2.1 million that you could not accomplish with \$1 million. Hence, the increase in satisfaction to be derived even from winning on the toss of the coin in C_2 might not convince you to take the risk involved compared with the "sure thing" of \$1 million in the selection of act C_1.

From the above discussion, we may conclude that it is reasonable to depart sometimes from the criterion of maximizing expected monetary values in making choices in risk situations. We cannot specify how a person *should* choose among alternative courses of action involving monetary payoffs, given only the type of information contained in Table 14-11. Our decisions will clearly depend upon our *attitudes toward risk*, which in turn will depend on a combination of factors such as level of assets, liking or distaste for gambling, and psychoemotional constitution.

Attitudes toward risk

> Large and small corporations do (and should) have different attitudes toward risk.

If we single out the factor of level of assets, for example, a large corporation with a substantial asset level may undertake certain risky ventures that a smaller corporation with fewer assets would avoid. In the case of the large corporation, an outcome of a loss of a certain number of dollars might represent an unfortunate occurrence but as a practical matter would not materially change the nature of operation of the business. In the case of the small corporation, a loss of the same magnitude might constitute a catastrophe and might require the liquidation of the business. In comparing a venturesome management of a small company with a highly conservative management of a large company, however, the attitudes toward risky ventures might well be found to be the reverse of the norm just indicated.

We can summarize the problem concerning decision making in problems involving payoffs that depend on risky outcomes as follows:

> When monetary payoffs are inappropriate as a measuring device, it may be appropriate to substitute a set of values that reflects the decision maker's attitude toward risk.

A clever approach to this problem has been furnished by John Von Neumann and Oskar Morgenstern, who developed the so-called Von Neumann and Morgenstern utility measure.[9] In the next section, we consider how these utilities may be derived and the procedures for using them in decision analysis.

Construction of Utility Functions

We have seen that in certain risk situations we might prefer one course of action to another even though the act selected has a lower expected monetary value. In the language of decision theory, we prefer the selected act because it possesses greater expected **utility** than does the act that is not selected. The procedure used to establish the utility function of a decision maker requires a series of choices. In each choice, the decision maker opts for certainty or for uncertainty. With certainty, one receives an amount of money denoted M (for "money"). With uncertainty, one gambles on receiving either an amount M_1 (with probability p) or an amount M_2 (with probability $1 - p$).

> The question the decision maker must answer is, "What probability p for consequence M_1 would make me *indifferent* to receiving M for certain and to participating in the gamble involving the receipt of either M_1 with probability p or M_2 with probability $1 - p$?"

This probability assessment provides the assignment of a utility index to the monetary value M. The data obtained from the series of questions posed to the decision maker result in a set of **money-utility pairs** that can be plotted on a graph and that constitutes the decision maker's utility function for money.

---

[9] The term *utility* as used by Von Neumann and Morgenstern and as used in this text differs from the economist's use of the same word. In traditional economics, utility refers to the inherent satisfaction delivered by a commodity and is measured in terms of psychic gains and losses. On the other hand, Von Neumann and Morgenstern conceived of utility as a measure of value used in the assessment of situations involving risk, which provides a basis for choice making. The two concepts can give rise to widely differing numerical measures of utility.

We will illustrate the procedure for constructing an individual's utility function by returning to one of the examples given in Table 14-11. *Constructing a utility function*

Suppose we ask the decision maker to choose between the following:

B_1: Receive $0 for certain

B_2: Receive $60,000 with probability $\frac{1}{2}$
and lose $40,000 with probability $\frac{1}{2}$

Suppose the decision maker chooses to receive $0 for certain. Our task then is to find out what probability for the receipt of the $60,000 would make the decision maker indifferent to the choice between the gamble and the certain receipt of $0. This will enable us to determine the utility assigned to $0. The first step is the arbitrary assignment of "utilities" to monetary consequences in the gamble, for example,

$$U(\$60,000) = 1$$

$$U(-\$40,000) = 0$$

where the symbol U denotes "utility" and $U(\$60,000) = 1$ is read "the utility of $60,000 is equal to one."

> The assignment of zero and one as the utilities of the lowest and highest outcomes of the gamble is entirely arbitrary. Any numbers can be assigned as long as the utility assigned to the higher monetary outcome is *greater than* that assigned to the lower outcome.

Thus, the utility scale has an arbitrary zero point, just like the $0°$ mark in temperature, which corresponds to different conditions depending on whether the Celsius or Fahrenheit scale is used.

The expected utility of the indicated gamble is

$$E[U(B_2)] = \frac{1}{2}[U(\$60,000)] + \frac{1}{2}[U(-\$40,000)] = \frac{1}{2}(1) + \frac{1}{2}(0) = \frac{1}{2}$$

Therefore, since the decision maker prefers $0 for certain to this gamble, it follows that the utility assigned to $0 is greater than $\frac{1}{2}$, or $U(\$0) > \frac{1}{2}$. In order to aid the decision maker in deciding how much greater than $\frac{1}{2}$ this utility is, we introduce the concept of a hypothetical lottery for calibrating the decision maker's utility assessment.

Let us assume that we have a box with 100 balls in it, 50 black and 50 white. The balls are identical in all other respects. Furthermore, we assume that if a ball is drawn at random from the box and its color is black, the decision maker receives a payoff of $60,000. On the other hand, if the ball is white, the payoff is $-\$40,000$. We now have constructed a gamble denoted B_2.

The question is this: If we retain the total number of balls in the calibrating box at 100 but vary the composition in terms of the numbers of black and white balls, how many black balls would be required for a decision maker to be indifferent between receiving $0 for certain and participating in the gamble? With 50 black balls in the box, the decision maker prefers $0 for certain. With 100 black balls (and no white balls), the decision maker would obviously prefer the gamble, since it would result in a payoff of $60,000 with certainty. For some number of black balls between 50 and 100, the decision maker should be indifferent, that is, at the threshold beyond which the gamble would become preferable to the certainty of the $0 payoff. Suppose we begin replacing white balls with black balls, and for some time the decision maker is still unwilling to participate in the gamble. Finally, when there are 70 black balls and 30 white balls, the decision maker announces that the point of indifference has been reached. We now can calculate the utility assigned to $0 as follows:

$$U(\$0) = 0.70[U(\$60{,}000)] + 0.30[U(-\$40{,}000)]$$

$$= 0.70(1) + 0.30(0) = 0.70$$

This utility calculation is a particular case of the general relationship

(14.1) $$U(M) = pU(M_1) + (1 - p)U(M_2)$$

where M is an amount of money received for certain and M_1 and M_2 are component prizes received in a gamble with respective probabilities p and $1 - p$.

We have now determined three money-utility ordered pairs: $(-\$40{,}000, 0)$, $(\$60{,}000, 1)$, and $(\$0, 0.70)$. The first figure in each ordered pair represents a monetary payoff in dollars and the second figure represents the utility index assigned to this amount. The utility figures for other monetary payoffs between $-\$40{,}000$ and $\$60{,}000$ can be assessed in exactly the same way as for $0, assuming that the patience of our long-suffering decision maker holds out. A relatively small number of points could be determined and the rest of the function interpolated. Suppose the utility function shown in Figure 14-5 results

FIGURE 14-5

A utility function

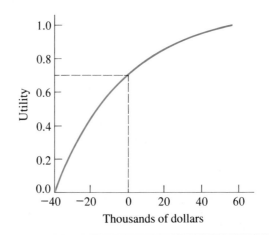

from the indifference probabilities assigned by the decision maker in the set of gambles proposed. The one point whose determination was illustrated ($0, 0.70) is depicted on the graph. This utility function can now be used to evaluate risk alternatives that might be presented to the decision maker. The expected utility of an alternative can be calculated by reading off the utility figure corresponding to each monetary outcome and then weighting these utilities by the probabilities that pertain to the outcomes. In other words, the utility figures can now be used by the decision maker in place of the original monetary values for calculation of expected utilities. For a person with this type of utility function, calculation of expected monetary values is clearly an inadequate guide for decision making.

The preceding discussion indicated one particular method of constructing a decision maker's utility function. That procedure was explained in terms of a basic gamble involving a maximum payoff of $60,000 and a minimum payoff of −$40,000 to which respective utility values of one and zero were assigned. Other points on the utility curve were obtained by determining the probabilities required for the receipt of the $60,000 to make the decision maker indifferent between the gamble and the certainty equivalent of specified dollar amounts ($0 was illustrated).

Another somewhat more practical method for assessing utility functions will now be discussed.

This method may be referred to as the **five-point procedure**,[10] because in its simplest form, five points on the utility curve are determined, namely, the points pertaining to utility values 1.00, 0.75, 0.50, 0.25, and 0.

Five-point procedure of assessing utility functions

Since it is helpful in understanding this method to have a particular example in mind, suppose you are considering two alternative investment proposals and wish to formalize your attitude toward risk by determining your utility function.

● You must first determine the **criterion** by which you should investigate your attitude toward risk.

A logical criterion would be your net asset position at the future time at which the consequences of the investment can be evaluated.[11]

● Then you should estimate the best and worst possible consequences in terms of your criterion.

---

[10] See R. S. Schlaifer, *Analysis of Decisions Under Uncertainty* (New York: McGraw-Hill 1969) for a detailed discussion of this method of assessing points on a utility curve.
[11] An alternative measure would be the *present value* of the net asset position at the future horizon data.

Let us assume, for example, that under the best possible outcome of the two investment proposals, you believe your net asset position two years hence (including the results of the proposed investments) would be $2.5 million; the worst possible outcome would be a net asset position of $0.

● You should then choose a pair of **reference consequences** for your criterion of net assets.

The range of the reference consequences should be sufficiently wide to include the best and worst possible outcomes. Suppose you feel that reference consequence limits of $3 million and $0 are wide enough to include your possible net asset positions.

● We then arbitrarily assign utility values of one and zero to $3 million and to $0, respectively.

As in our previous example, we plot these as the two extreme points on our utility curve, shown in Figure 14-6.

● We then obtain your certainty equivalents for three 50:50 gambles in the range between the two reference consequences.

Suppose that after considerable thought and discussion, you decide that you would take $800,000 for certain in exchange for 50:50 chances at the extreme outcomes of $3 million and $0. Then the expected utility of the gamble would be

$$U(\$800,000) = \frac{1}{2}\left[U(\$3,000,000)\right] + \frac{1}{2}\left[U(\$0)\right]$$

$$= \frac{1}{2}(1) + \frac{1}{2}(0) = 0.50$$

FIGURE 14-6

A utility function constructed by the five-point procedure

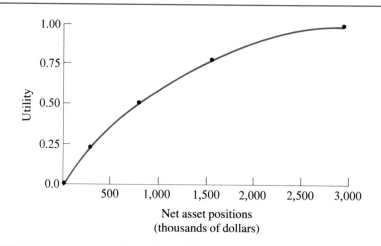

- We could then plot the point ($800,000, 0.50) on the graph.
- Now we can determine intermediate points between this 0.50 utility value and the two reference consequences.

Suppose that after suitable thought, you choose $1.6 million as your certainty equivalent for a 50:50 gamble at consequences of $3 million and $800,000. We now have

$$U(\$1,600,000) = \frac{1}{2}[U(\$3,000,000)] + \frac{1}{2}[U(\$800,000)]$$

$$= \frac{1}{2}(1) + \frac{1}{2}(0.50) = 0.75$$

- This yields the point ($1,600,000, 0.75) that can be plotted on the graph.
- Finally, let us assume that in a 50:50 gamble between $800,000 and $0, you choose a certainty equivalent of $300,000.

Then we have

$$U(\$300,000) = \frac{1}{2}[U(\$800,000)] + \frac{1}{2}[U(\$0)]$$

$$= \frac{1}{2}(0.50) + \frac{1}{2}(0) = 0.25$$

We have now determined the following five points:

Utility	1.0	0.75	0.50	0.25	0
Net asset positions	$3,000,000	$1,600,000	$800,000	$300,000	$0

The smooth curve drawn through these five points is shown in Figure 14-6.

> In a real-world situation, when the five points are plotted they may not lie along a smooth curve. When presented with this evidence, the decision maker should resolve inconsistencies.

Furthermore, additional gambles should be posed for other intermediate points as consistency checks. The purpose of this checking procedure is to make sure that the utility curve appropriately represents the decision maker's attitude toward risk for all outcomes in the range of the two reference consequences.

Characteristics and Types of Utility Functions

The utility functions depicted in Figures 14-5 and 14-6 rise consistently from the lower left to the upper right side of the chart. That is, the utility curves have positive slopes. This general characteristic of utility functions simply implies

FIGURE 14-7
Various types of utility
functions

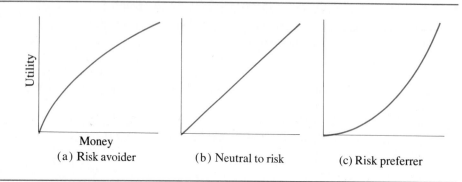

(a) Risk avoider   (b) Neutral to risk   (c) Risk preferrer

that people ordinarily attach greater utility to a large sum of money than to a small sum.[12] Economists have noted this psychological trait in traditional demand theory and have referred to it as a *positive marginal utility for money.* The concave downward shape shown in Figure 14-6 illustrates the utility curve of an individual who has a diminishing marginal utility for money, although the marginal utility is always positive. This type of utility curve is characteristic of a risk avoider, as indicated in Figure 14-7(a). A risk avoider would prefer a small but *certain* monetary gain to a gamble that involves either a large but unlikely gain or a large and likely (not unlikely) loss. The linear function in Figure 14-7(b) depicts the behavior of a person who is neutral or indifferent to risk. For such a person every increase of, say, $1,000 has an associated constant increase in utility. This type of individual would use the criterion of **maximizing expected monetary value** in decision making, because this would also maximize expected utility.

Risk avoider

Indifferent to risk

Risk preferrer

Figure 14-7(c) shows the utility curve for a risk preferrer. This type of person willingly accepts gambles that have a smaller expected monetary value than an alternative payoff received with certainty. For such an individual, the attractiveness of a possible large payoff in the gamble tends to outweigh the fact that the probability of such a payoff may indeed be very small.

Empirical research suggests that most individuals have utility functions in which for small changes in money amounts the slope does not change much. Over these ranges of money outcomes, the utility function may be considered approximately linear and constant in slope. However, in considering courses of action in which one of the consequences is adverse or in which one of the payoffs is large, individuals can be expected to depart from the maximization of expected monetary values as a guide to decision making.

[12] The infinite variety of human behavior is demonstrated by the fact that conduct that runs counter to this generalization is occasionally observed. A newspaper reported that an heir to a fortune of $30 million committed suicide at the age of 23, after indicating in a letter that his wealth prevented him from living a normal life.

> For many business decisions in which the monetary consequences represent only a small fraction of the total assets of the business unit, the use of maximization of expected monetary payoff constitutes a reasonable approximation to the decision-making criterion of maximization of expected utility.

In other words, in such cases, the utility function may often be treated as approximately linear over the range of monetary payoffs considered.

Assumptions Underlying Utility Theory

The utility measure we have discussed was derived by evoking the decision maker's preferences between sums of money obtainable with certainty and lotteries or gambles involving a set of basic alternative monetary outcomes. This procedure entails a number of assumptions.

- It is assumed that an individual, when faced with the types of choices discussed, can determine three things: whether act A_1 is preferable to act A_2, whether A_2 is preferred to A_1, or whether these acts A_1 and A_2 are regarded indifferently. If A_1 is preferred to A_2, then the utility assigned to A_1 should exceed the utility assigned to A_2; if A_2 is preferred, the utility assigned to A_2 should be greater.

- Another behavioral assumption is that an individual who prefers A_1 to A_2 and A_2 to A_3, will also prefer A_1 to A_3. This is referred to as the **principle of transitivity**. The assumption extends also to indifference relationships. That is, the decision maker who is indifferent to the choice between A_1 and A_2 and between A_2 and A_3 should also be indifferent to the choice between A_1 and A_3. *The principle of transitivity*

- Furthermore, it is assumed that if one is indifferent to the choice between a payoff or consequence of an act that replaces another consequence, one should also be indifferent to the choice between the old and new acts. This is often referred to as the **principle of substitution.** *The principle of substitution*

- Finally, it is assumed that the utility function is **bounded,** which means that utility cannot increase or decrease without limit. As a practical matter, this simply means that the range of possible monetary values is limited. For example, at the lower end the range may be limited by a bankruptcy condition.

We may argue that human beings do not always exhbit the type of consistency implied by these assumptions. However, the point is that in the construction of our utility function if it is observed that we are behaving inconsistently and these incongruities are indicated to us and if we are "reasonable" or "rational," we should adjust our choices accordingly. If we insist on being irrational and refuse to adjust the choices that violate the underlying assumptions of utility theory, then a utility function cannot be constructed for us and we cannot use maximization of expected utility as a criterion of rationality in choice making. It is important to keep in mind that the theory discussed here does not purport to describe the way people actually *do* behave in the real world, but rather specifies how they *should* behave if their decisions are to be consistent with their own expressed judgments as to preferences among consequences. Indeed, we may argue that since human beings are fallible and do make mistakes, it is useful to have prescriptive procedures that police their behavior and provide ways in which the behavior can be improved.

A Brief Note on Scales

The Von Neumann–Morgenstern utility scales are examples of *interval scales*.

> **Interval scales** have a constant unit of measurement but an arbitrary zero point. Differences between scale values can be expressed as multiples of one another, but individual values cannot be so expressed.

In decision making using utility measures, if a different zero point and a different scale are selected, the same choices will be made. A constant can be added to each utility value, and each utility value can be multiplied by a constant, without changing the properties of the utility function. Thus, if a is any constant, b is a positive constant, and x is an amount of money,

$$U_2(x) = a + bU_1(x)$$

and $U_2(x)$ is as legitimate a measure of utility as $U_1(x)$.

EXAMPLE 14-1

The familiar scales for temperature are examples of interval scales. We cannot say that 100°C is twice as hot as 50°C. The corresponding Fahrenheit measures would not exhibit a ratio of 2:1. On the other hand, we can say that the intervals or differences between 100°C and 50°C and between 75°C and 50°C are in a 2:1 ratio. Using the relationship $F = \frac{9}{5}C + 32°$, we have

$$C = 100°; \quad F = (\tfrac{9}{5})(100°) + 32° = 212°$$

$$C = 75°; \quad F = (\tfrac{9}{5})(75°) + 32° = 167°$$

$$C = 50°; \quad F = (\tfrac{9}{5})(50°) + 32° = 122°$$

The difference between 100° and 50° = 50°; between 75° and 50° = 25°. The ratio of 50° to 25° is 2:1.

The difference between 212° and 122° = 90°; between 167° and 122° = 45°. The ratio of 90° to 45° is 2:1.

1. a. The expected monetary return of the decision to buy life insurance is negative. Thus, it is irrational to buy life insurance. Do you agree or disagree? Explain.

    b. If a large company does not carry automobile insurance, why do you think this is so?

2. You have a choice of investing in risk-free government securities that pay interest equal to 10 units of utility (utiles) or investing in Rerox stock. With a probability of 0.4, Rerox will yield gains equal to 45 utiles; with a probability of 0.6, the stock will cause a loss of 15 utiles.

    a. What is the expected utility of the prospect "buy the stock"?

    b. Should you buy Rerox stock or the government securities?

3. The IRS has audited your last year's income tax return and has sent you a bill for $500 in back taxes. You now have the choice of paying the bill or disputing the audit. If you dispute it, you must pay an accountant $50 to prepare your case. After preliminary talks with your accountant, you feel your chances of winning the dispute are 0.05.

    a. Should you dispute the case based on monetary expectations?

    b. Assume that large losses of money are disastrous to you as a struggling student. This is reflected in your utility function, which indicates that $U(-\$50)$ is -4 units of utility, $U(-\$500)$ is -425 units of utility, and $U(-\$550)$ is -440 units of utility. Based on expected utility, what is your best course of action?

4. Drillwell Oil Company must make a decision concerning an option on a parcel of land. If the company takes the option, it can drill with a 100% interest or with a 50% interest (that is, it splits all costs and profits with another firm). It costs $1,000,000 to drill a well and $1,000,000 to operate a producing well until it is dry. The oil is worth $25 per barrel. Assume that the well is either dry or produces 200,000 or 400,000 barrels of oil. The firm assesses the probability of the outcomes as 0.8, 0.1, and 0.1, respectively.

    a. Based on expected monetary return, what is the company's best action?

    b. Suppose Drillwell's management has the following utility function:

Millions of Dollars	Units of Utility
-1.0	-30
-0.5	-15
1.5	45
3.0	90
4.0	120
8.0	240

What should the company decide based on expected utility?

c. Based on your answer in part (b), is Drillwell's management risk averse? Explain.

5. An investor who is considering buying a franchised furniture business estimates that the business will yield either a loss of $50,000 or a profit of $100,000, $200,000, or $500,000 per year with probabilities 0.5, 0.2, 0.1, and 0.2, respectively. The investor's utility function is found to be the following.

Dollars	Units of Utility	Dollars	Units of Utility	Dollars	Units of Utility
−50,000	−40	25,000	1.2	150,000	7.5
−37,500	−10	50,000	2.5	200,000	10.0
−25,000	−4.2	75,000	5.8	250,000	12.5
−12,500	−2	100,000	6.0	375,000	26.0
0	0	125,000	6.2	500,000	40.0

a. Graph the utility function and interpret the shape of the curve.
b. Based on expected utility, should the investor buy the business?
c. Suppose the investor can buy the whole franchise, three-fourths, one-half, or one-fourth of it (a one-fourth interest means that the investor receives one-fourth of all profits and pays one-fourth of all losses). What is the best investment decision?

6. An individual with a net worth of $100,000 contemplates investing in a risky venture with two equiprobable outcomes: It can either triple his net worth or bankrupt him completely. His preference for wealth, W (net worth), can be represented by the utility function

$$U(W) = \sqrt{W}$$

How much should he be willing to pay at most if he is to maximize
a. Expected monetary return?
b. Expected utility?

KEY TERMS

Backward Induction   the method used in solving a decision problem under uncertainty by means of a decision tree diagram.

Certainty Equivalent   the figure that a decision maker would be indifferent to receiving for certain as compared to participating in a particular gamble.

Expected Profit under Uncertainty   the maximum of the expected profits of all acts analyzed.

Expected Profit with Perfect Information   expected payoff if the decision maker has access to a perfect predictor.

Expected Value of Perfect Information   the expected profit with perfect information minus the expected profit under uncertainty (equivalent to the expected opportunity loss of the optimal act under uncertainty and the cost of uncertainty).

Interval Scale   a scale that has a constant unit of measurement and an arbitrary zero point.

Maximin Criterion   a decision procedure that chooses an act that yields the maximum of the minimum payoffs.

Opportunity Loss   the loss incurred because of failure to take the best possible action.

Prior Probabilities   probabilities established before obtaining additional information through sampling, experimentation, or surveys.

Utility Function   a function that incorporates attitude toward risk in decision making under uncertainty.

15

DECISION MAKING
WITH POSTERIOR PROBABILITIES

The discussion in Chapter 14 may be referred to as **prior analysis**, that is, decision making in which expected payoffs of acts are computed on the basis of *prior probabilities*. In this chapter, we discuss **posterior analysis**, in which expected payoffs are calculated on the basis of *revisions of prior probabilities*.

15.1
POSTERIOR PROBABILITIES

Bayes' theorem is utilized to accomplish revision of prior probabilities, which then become **posterior probabilities**. In this context, *prior* and *posterior* are relative terms. For example, subjective prior probabilities may be revised to incorporate the additional evidence of a particular sample. The revised probabilities then constitute posterior probabilities. If these probabilities are in turn revised on the basis of another sample or some experimental evidence, they represent prior probabilities relative to the new sample information, and the revised probabilities are posteriors.

As in Chapters 14 through 17, for simplicity of presentation, we have carried out decision analyses in terms of **monetary payoffs** rather than *utility figures*. As noted in section 14.8, this use of monetary values constitutes an assumption that the utility function is approximately linear over the range of monetary payoffs considered. Of course, when this assumption is not valid, the

Selecting terms of analyses

733

analyses should be performed with appropriate utility values substituted for monetary payoffs.

Reducing the cost of uncertainty   The basic purpose of attempting to incorporate more evidence through sampling is to reduce the expected cost of uncertainty. If the expected cost of uncertainty (or the expected opportunity loss of the optimal act) is high, then it will ordinarily be wise to engage in sampling. Sampling in this context includes statistical sampling, experimentation, testing, and any other methods used to acquire additional information.

The general method of incorporating sample evidence into the decision-making process can be illustrated in terms of two types of sample information, where

- The reliability of the sample information is specified.
- The sample size is specified.

We will illustrate the first type in terms of the inventor's problem in Chapter 14. Then we will consider an acceptance sampling problem in which the additional evidence is based on a sample of a given size.

15.2
POSTERIOR ANALYSIS: A SPECIFIED RELIABILITY ILLUSTRATION

Reconsidering the inventor's problem   In the problem discussed in Chapter 14, suppose the inventor decided not to rely solely on prior probabilities concerning the demand for the new device, but to have a market research organization conduct a sample survey of potential consumers to gather additional evidence for the probable level of sales for the product. Let us assume that the survey can result in three sample outcomes, denoted x_1, x_2, and x_3, corresponding to the three states of nature, sales levels θ_1, θ_2, and θ_3. Specifically, the possible results may be

x_1: Sample indicates strong sales

x_2: Sample indicates average sales

x_3: Sample indicates weak sales

The survey is conducted, and the sample indicates an average level of sales, that is, x_2 is observed. Assume that on the basis of previous surveys of this type, the market research organization can assess the reliability of the sample evidence in the following terms: In the past, when the actual sales level after a new device was placed on the market was average, sample surveys properly indicated an average level of demand 80% of the time. However, when the actual level was strong sales, about 10% of the sample surveys incorrectly demand as average; and when the actual level was weak sales, about 20% of the

TABLE 15-1
Computation of posterior probabilities in the inventor's problem for the sample
indication of an average level of sales

Event θ_i	$P(\theta_i)$	Conditional Probability $P(x_2\|\theta_i)$	Joint Probability $P(\theta_i)P(x_2\|\theta_i)$	Posterior Probability $P(\theta_i\|x_2)$
θ_1: Strong sales	0.2	0.1	0.02	0.042
θ_2: Average sales	0.5	0.8	0.40	0.833
θ_3: Weak sales	0.3	0.2	0.06	0.125
	1.0		0.48	1.000

sample surveys gave an indication of average sales. These relative frequencies represent conditional probabilities that the sample evidence indicates "average sales," given the three possible actual events concerning sales level. They can be symbolized as follows:

$$P(x_2|\theta_1) = 0.1$$

$$P(x_2|\theta_2) = 0.8$$

$$P(x_2|\theta_3) = 0.2$$

The revision by means of Bayes' theorem of the prior probabilities assigned to the three sales levels on the basis of the observed sample evidence x_2 (average sales) is given in Table 15-1. In terms of equation 2.17 for Bayes' theorem, x_2 plays the role of B, the sample observation; θ_i replaces A_i, the possible events, or states of nature. After the joint probabilities are calculated, they are divided by their total (in this case, 0.48) to yield posterior, or revised, probabilities for the possible events. We note the effect of the weighting given to the sample evidence by Bayes' theorem in the revision of the prior probabilities by comparing the posterior probabilities with the corresponding "priors" in Table 15-1. With a sample indication of average sales, the 0.5 prior probability of the event "average sales" was revised upward to 0.833. Correspondingly, the probabilities of events "strong sales" and "weak sales" declined from 0.2 to 0.042 and from 0.3 to 0.125, respectively.

Decision Making after the Observation of Sample Evidence

The revised probabilities calculated in Table 15-1 can now be used to compute the **posterior expected profits** of the inventor's alternative courses of action. In Table 14-3, expected payoffs were computed based on the subjective prior probabilities assigned to the possible events. We can now refer to these as **prior expected profits**. The calculation of the posterior expected profits (using the

TABLE 15-2
Calculation of posterior or expected profits in the inventor's problem using revised probabilities of events (units of $10,000)

| | Act A_1: Inventor Manufactures Device | | | Act A_2: Inventor Sells Patent Rights | | |
| | Probability | | Weighted | Probability | | Weighted |
Event	$P_1(\theta_i)$	Profit	Profit	$P_1(\theta_i)$	Profit	Profit
θ_2: Strong sales	0.042	$80	3.360	0.042	$40	1.680
θ_2: Average sales	0.833	20	16.660	0.833	7	5.831
θ_3: Weak sales	0.125	-5	-0.625	0.125	1	0.125
	1.000		19.395	1.000		7.636

Posterior expected profit A_1	Posterior expected profit A_2
= $19.395 (ten thousands of dollars)	= $7.636 (ten thousands of dollars)
= $193,950	= $76,360

revised, or posterior, probabilities as weights) is displayed in Table 15-2. By convention, we denote prior probabilities as $P_0(\theta_i)$ and posterior probabilities as $P_1(\theta_i)$. That is, the subscript zero is used to denote prior probabilities and the subscript one to signify posterior probabilities. A decision diagram of this posterior analysis is given in Figure 15-1. Note that the decision tree is essentially the same as the one for the prior analysis given in Figure 14-1(b), except that posterior probabilities have been substituted for prior probabilities.

Since the posterior expected profit of act A_1 exceeds that of A_2, the better of the two courses of action remains that of the inventor manufacturing the device himself. However, after the sample indication of "average sales," the expected profit of act A_1 has decreased from $245,000 based on the prior probabilities to $193,950 based on the revised probabilities. Moreover, the difference in the expected profits of the two acts has narrowed somewhat. The $245,000 and $193,950 figures are, respectively, the **prior expected profit under uncertainty** and the **posterior expected profit under uncertainty**.

> The optimal course of action under a posterior analysis may be different from that of the prior analysis.

In the present example, if the sample indication had been "weak sales," with appropriate conditional probabilities, the posterior expected profit of A_2 may have exceeded that of A_1. (Assume some figures and demonstrate this point.)

Insight can be gained into the cost of uncertainty and the value of obtaining additional information by calculating the **posterior expected value of perfect information**, which is simply the expected payoff, using posterior probabilities, of decision making in conjunction with a perfect predictor. We can now refer to the EVPI calculated in Chapter 14 (Table 14-7) as the **prior EVPI**. The prior EVPI of $18,000 means that the decision maker should have been willing to pay up to $18,000 for perfect information to eliminate his uncertainty concerning states of nature. Since no sample could be expected to yield

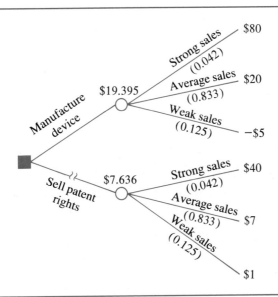

FIGURE 15-1

Decision diagram for posterior analysis of inventor's problem to manufacture or sell patent rights to a product (payoffs are in units of $10,000)

perfect information, the decision maker cannot yet determine the worth of obtaining additional information through sampling. Expected value of sample information is discussed in Chapter 16. However, the prior EVPI of $18,000 sets an upper limit to the worth of obtaining perfect information and eliminating uncertainty concerning events. After the decision maker obtains additional information through sampling, he can calculate the **posterior EVPI**. The change in EVPI may be used to evaluate the decision to be made and the worth of attempting to obtain further information.

The posterior EVPI is computed to be $7,500 in Table 15-3(a). Analogously to prior analysis, the posterior EVPI is calculated by subtracting the posterior expected profit under uncertainty from the posterior expected profit with perfect information. The alternative determination of the posterior EVPI as the expected opportunity loss of the optimal act using posterior probabilities is given in Table 15-3(b). The only difference between this calculation and the similar calculation in Table 14-5 for the prior expected opportunity loss for act A_1 is the substitution of posterior probabilities for the prior probabilities.

In summary, the EPVI has been reduced from $18,000 to $7,500 by the information obtained from the sample. Whereas the decision maker should have been willing to pay up to $18,000 for a perfect predictor prior to having the sample information, the expected value of perfect information is only $7,500 after the sample indication of "average sales." In other words, the decision maker has reduced his cost of uncertainty; the availability of a perfect forecaster is not as valuable as it was prior to the sample survey. Although this problem resulted in a decrease from the prior EPVI to the posterior EVPI, there might very well have been an increase. This could occur if there were a marked difference between the posterior and prior probability distributions,

TABLE 15-3

(a) Calculation of the posterior expected value of perfect information for the inventor's problem (units of $10,000)

Event	Profit	Posterior Probability	Weighted Profit
θ_1: Strong sales	$80	0.042	3.360
θ_2: Average sales	20	0.833	16.660
θ_3: Weak sales	1	0.125	0.125
		1.000	20.145

Posterior expected profit
with perfect information

$= \$20.145$ (ten thousands of dollars)

$= \$201,450$

Posterior expected profit with perfect information	$201,450
Less: Posterior expected profit under uncertainty	193,950
Posterior EVPI	$ 7,500

(b) Posterior expected opportunity loss of the optimal act for the inventor's problem (units of $10,000)

Act A_1: Inventor Manufactures Device

Event	Probability	Opportunity Loss	Weighted Opportunity Loss
θ_1: Strong sales	0.042	0	0
θ_2: Average sales	0.833	0	0
θ_3: Weak sales	0.125	6	0.75
			0.75

Posterior EOL of the optimal act

$=$ Posterior EVPI

$= \$0.75$ (ten thousands of dollars)

$= \$7,500$

and a reversal took place in the optimal act after the incorporation of sample information.

> An increase in the EVPI after inclusion of knowledge gained from sampling can be interpreted to mean that the additional evidence has increased the doubt concerning the decision.

Of course, the determination of the prior and posterior EVPIs by alternative methods is not necessary in practice, but the computations have been shown here to indicate the relationships involved.

EXERCISES 15.2

1. From the given information, fill in the blanks and interpret the data. X is the result of a survey of 1,000 construction companies.

State of Nature	$P_0(\theta)$	$P(X\|\theta)$	$P_0(\theta)P(X\|\theta)$	$P_1(\theta\|X)$
θ_1: Housing starts will increase next year	0.8	0.4		
θ_2: Housing starts will stay the same level or will decline		0.6		

2. A foreign exchange broker has ascertained the following probabilities for the spot rate of the Dutch guilder one month from a certain date:

θ	$P_0(\theta)$
$\theta_1 = \$0.50$	0.10
$\theta_2 = 0.51$	0.20
$\theta_3 = 0.52$	0.30
$\theta_4 = 0.53$	0.25
$\theta_5 = 0.54$	0.15

A sample observation X with the following properties occurs:

$$P(X|\theta_1) = 0.20 \qquad P(X|\theta_2) = 0.10 \qquad P(X|\theta_3) = 0.50$$

$$P(X|\theta_4) = 0.90 \qquad P(X|\theta_5) = 0.80$$

What are the revised probabilities?

3. Given an uncertain economic forecast and pending negotiations with the unions regarding possible concessions, Jeneral Motors' (JM) estimate of worldwide car sales for 1991 at the end of 1990 was as follow:

State of Nature	Probability $P_0(\theta)$
θ_1: High sales (more than five million cars)	0.2
θ_2: Mediocre sales (four million-five million cars)	0.6
θ_3: Poor sales (less than four million cars)	0.2

By February 1991, dealers were cutting back inventories. JM estimated that the probabilities of this situation given the expectations of high, mediocre, and poor sales were, respectively, 0.05, 0.20, and 0.95. Compute the revised probabilities of the three levels of sales.

4. Using prior probabilities, a decision maker determines that the best course of action is A_1 and that the EVPI is $1,000. After a sample is drawn, the best act is still A_1, and the revised EVPI is $400. The cost of sampling is $100. Can you conclude that the actual value of the sample information to the decision maker was $500? Explain your answer.

5. You are given the following results of an experiment where θ_1 is "product is superior to competitor's" and θ_2 is "product is as good as or inferior to competitor's."

State of Nature	$P(X|\theta)$	$P_1(\theta|X)$
θ_1	0.50	0.25
θ_2	0.50	0.75

Calculate the prior probability distribution.

6. Assume two possible actions to take, A_1 and A_2, and two states of nature, θ_1 and θ_2. A_1 is preferable if θ_1 is true, and A_2 is preferred if θ_2 is true. If the prior probabilities are $P(\theta_1) = 0.3$ and $P(\theta_2) = 0.7$ and you obtain a sample observation X such that $P(X|\theta_1) = 0.2$ and $P(X|\theta_2) = 0.9$, can you conclude that A_2 is the better act? Explain your answer.

7. Zuperior Oil Company uses three-dimensional scanning equipment to carry out seismic tests, resulting in a superior wildcat success ratio $P(\theta_1|X)$ relative to the industry. The symbol X indicates favorable results. The probability that the seismic tests show favorable indications if oil actually exists, $P(X|\theta_1)$, is 0.55 for a given project in the Williston Basin. The probability that the tests show favorable results when no oil exists for the same project, $P(X|\theta_2)$, is 0.10. Management's prior probability assessment of the project's success and payoffs were as follows:

State of Nature	Probability $P_0(\theta)$	Payoff
θ_1: Oil exists	0.5	$5 million
θ_2: No oil exists	0.5	−2 million

Given that the seismic tests are favorable, compute the revised probabilities of the two events and the posterior expected payoffs.

8. A bond broker is considering a sale of her inventory holdings of New York City obligations. Her profit on such a transaction would be $5,000 if undertaken immediately. If she waits six months before selling, her profit will depend on the direction of change in interest rates during that period. The potential profits and the trader's subjective probabilities for each possible event are shown:

State of Nature	Probability $P_0(\theta_i)$	Profit
θ_1: Higher interest rates	0.20	−$15,000
θ_2: No change	0.35	−5,000
θ_3: Lower interest rates	0.45	28,000

What action should she take?

Now suppose the latest business statistics show that a large number of companies are planning to raise capital through debt during the next six months. The probabilities of this situation, given expectations of lower rates, no change, and higher rates, are 0.80, 0.20, and 0.10, respectively. Compute the revised probabilities of the three events and the posterior expected profits of the two possible actions.

9. Kolgate faces two alternative courses of action for a major expansion project, and two possible states of nature will affect the payoffs. The payoff matrix in thousands of dollars and initial probabilities are as follows:

Act	State of Nature θ_1	θ_2
A_1	10	100
A_2	30	20
$P(\theta_1) = 0.4$	$P(\theta_2) = 0.6$	

To aid in the decision, Kolgate is considering buying a relevant data base from a consulting service. The reliability of the information, $P(X\,|\,\theta_i)$, is as follows:

State of Nature	$P(X\,|\,\theta_i)$
θ_1	0.9
θ_2	0.2

What are the revised probabilities, the best action, and the revised EVPI?

15.3
POSTERIOR ANALYSIS: AN ACCEPTANCE SAMPLING ILLUSTRATION

As the second illustration of posterior analysis, we consider a problem in acceptance sampling of a manufactured product. Let us assume that the Renny Corporation inspects incoming lots of articles produced by a supplier in order to determine whether to accept or reject these lots. In the past, incoming lots from this supplier have contained 10%, 20%, or 30% defective articles. On a relative frequency basis, lots with these percentages of defectives have occurred 50%, 30%, and 20% of the time, respectively. The Renny Corporation feels justified in using these past percentages as prior probabilities for a lot just delivered by the supplier. Renny Corporation draws a simple random sample of 10 units with replacement from the incoming lot, and two defectives are found.

The Renny Corporation has found from past experience that it should accept lots that have 10% defectives and should reject those that have 20% or 30% defectives. That is, the costs of rework make it uneconomical to accept lots with more than 10% defectives. On the basis of a careful analysis of past

TABLE 15-4

Payoff matrix showing opportunity losses for accepting and rejecting lots with specified proportions of defectives

Event p (lot proportion of defectives)	Act	
	A_1 Reject	A_2 Accept
0.10	$200	$ 0
0.20	0	100
0.30	0	200

costs, the Renny Corporation constructed the payoff matrix in terms of opportunity losses shown in Table 15-4. The two possible courses of action are act A_1, to reject the incoming lot, and act A_2, to accept the incoming lot. The three states of nature are the possible proportions of defectives in the lot—namely, 0.10, 0.20, and 0.30. (For convenience, we consider the defectiveness of the lots in terms of decimals rather than percentages.) The proportion of defectives in the lot is denoted p and will be treated as a discrete random variable that can take on only the three given values. Of course, it is rather unrealistic to assume that an incoming lot can be characterized only by a 0.10, 0.20, or 0.30 proportion of defectives. However, for convenience of exposition, we make that assumption here. The same general principles would hold if more realistic assumptions were made, as, for example, that the proportion of defectives could take on values at intervals of one percentage point—namely, 0.00, 0.01, . . . , 1.00—or that a continuous probability distribution for proportion of defectives was approximated by the bracket median method described in section 14.7. We now proceed to apply some of the principles of decision analysis we have learned.

Suppose the Renny Corporation had to take action concerning acceptance of the present lot before drawing the sample of ten units from this lot. Assuming that the firm is willing to make its decision on the basis of prior information, what action should be taken? In the absence of additional information, the company could reasonably use the past relative frequencies of lot proportion of defectives as prior probability assignments. The prior analysis of the company's two courses of action based on expected opportunity losses is given in Table 15-5 and is depicted in Figure 15-2.[1] As shown in the table, when the past relative frequencies are used as prior probabilities, the expected opportunity loss of rejecting the lot is $100 and that of accepting the lot is $70. Hence,

---

[1] The notation $P_0(p)$ is used in Table 15-5 for the prior probability distribution of the random variable p, and $P_1(p)$ is used in Table 15-6 for the corresponding posterior probability distribution. Moreover, the random variable is referred to by lower-case p in this and subsequent illustrations. This is a departure from the convention of using capital letters to denote random variables and lower-case letters to represent the values taken on by the random variable. This is done to avoid confusion because of the common use of the capital P to mean "probability."

TABLE 15-5
Prior expected opportunity losses for the Renny Corporation problem

	Act A_1: Reject the Lot			Act A_2: Accept the Lot		
Event p	Prior Probability $P_0(p)$	Opportunity Loss	Weighted Opportunity Loss	Prior Probability $P_0(p)$	Opportunity Loss	Weighted Opportunity Loss
0.10	0.50	$200	$100	0.50	$ 0	$ 0
0.20	0.30	0	0	0.30	100	30
0.30	0.20	0	0	0.20	200	40
	1.00		$100	1.00		$70

$$EOL(A_1) = \$100 \qquad\qquad EOL(A_2) = \$70$$
$$\text{Prior EVPI} = \$70$$

the optimal act is A_2, to accept the lot. The $70 figure is also the prior EVPI, since it represents the expected opportunity loss of the optimal act using the prior probability distribution.

We turn now to the posterior analysis. Assume that the Renny Corporation draws the simple random sample of 10 units with replacement from the incoming lot and observes two defectives. If we take this sample evidence into account, what is the optimal course of action?

Following the same general procedure as in the inventor's problem, we use the sample evidence to revise the prior probabilities assigned to the possible lot proportions of defectives. The application of Bayes' theorem for this purpose

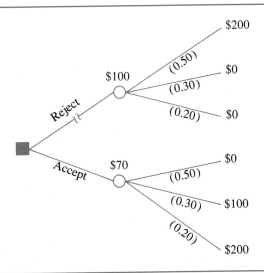

FIGURE 15-2
Decision diagram for prior analysis of the Renny Corporation problem of whether to accept or reject a lot produced by a supplier

TABLE 15-6

Computation of posterior probabilities for the Renny Corporation problem incorporating evidence based on a sample of size 10

Event p	Prior Probability $P_0(p)$	Conditional Probability $P(X = 2 \mid n = 10, p)$	Joint Probability $P_0(p)P(X = 2 \mid n = 10, p)$	Posterior Probability $P_1(p)$
0.10	0.50	0.1937	0.09685	0.4136
0.20	0.30	0.3020	0.09060	0.3869
0.30	0.20	0.2335	0.04670	0.1995
	1.00		0.23415	1.0000

is shown in Table 15-6. The conditional probabilities shown in the third column of Table 15-6 are often referred to as "likelihoods." That is, they represent the likelihoods of obtaining two defectives in 10 units in a simple random sample drawn with replacement from the assumed incoming lots. When the basic random variable is the parameter p of a Bernoulli process, as in this problem, the likelihoods of the observed "number of successes" in the sample are computed by the binomial distribution. The likelihood figures in Table 15-6 were obtained from Table A-2 of Appendix A. The notation $P(X = 2 \mid n = 10, p)$ means "the probability that the random variable 'number of successes' is equal to 2 in 10 trials of a Bernoulli process whose parameter is p." We will use this type of symbolism in this and other problems for likelihood calculations. With the evidence of two defectives in a sample of 10 units, or 20% defectives in the sample, the prior probability that the lot contains 20% defectives is revised upward from 0.30 to 0.3869, as indicated in Table 15-6. Correspondingly, the probabilities that the lots contain 10% or 30% defectives are revised downward.

Expected payoffs of the two acts can now be recomputed using the posterior probabilities. These computations are shown in Table 15-7 and are dis-

TABLE 15-7

Posterior expected opportunity losses for the Renny Corporation problem

	Act A_1: Reject the Lot			Act A_2: Accept the Lot		
Event p	Posterior Probability $P_1(p)$	Opportunity Loss	Weighted Opportunity Loss	Posterior Probability $P_1(p)$	Opportunity Loss	Weighted Opportunity Loss
0.10	0.4136	$200	$82.72	0.4136	$ 0	$ 0
0.20	0.3869	0	0	0.3869	100	38.69
0.30	0.1995	0	0	0.1995	200	39.90
	1.0000		$82.72	1.0000		$78.59

$$\text{EOL}(A_1) = \$82.72$$

$$\text{EOL}(A_2) = \$78.59$$
$$\text{Posterior EVPI} = \$78.59$$

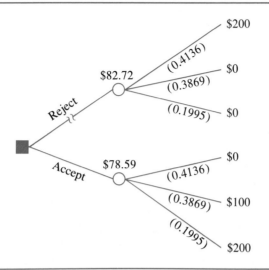

FIGURE 15-3
Decision diagram for posterior analysis of the Renny Corporation problem of whether to accept or reject a lot produced by a supplier

played in Figure 15-3. The optimal act is still A_2, accept the lot. However, the posterior expected opportunity losses of the two acts are much closer together than were the prior ones. Furthermore, the posterior expected opportunity loss of the optimal act, A_2, is $78.59, which represents an increase from the prior expected opportunity loss of the optimal act, $70. In other words, the posterior EVPI now exceeds the prior EVPI, which indicates that the value of having a perfect predictor available has increased.

Effect of Sample Size

We can use this acceptance sampling problem to illustrate an important point in Bayesian decision analysis, namely, the effect of sample size on the posterior probability distribution. Suppose that instead of a simple random sample size of 10 units being drawn from the incoming lot, a similar sample of 100 units was drawn. Furthermore, let us assume that 20 defectives were found in the sample; in other words, the fraction of defectives in this larger sample is 0.20, just as it was in the smaller sample. The computation of posterior probabilities using Bayes' theorem and the information from the sample of size 100 is given in Table 15-8. The conditional probabilities in Table 15-8 were obtained from a table of binomial probabilities that includes $n = 100$.

   Although 0.30 was the prior probability assigned to the state of nature that the incoming lot proportion of defectives was 0.20, in the last column of Table 15-8, the revised probability is 0.9336. Hence, because of the implicit weight given to the sample evidence by Bayes' theorem, it is much more probable that the defective lot fraction is 0.20 according to the posterior distribution than according to the prior distribution. Furthermore, comparing the

TABLE 15-8
Computation of posterior probabilities for the Renny Corporation problem incorporating evidence based on a sample of size 100

Event p	Prior Probability $P_0(p)$	Conditional Probability $P(X = 20 \mid n = 100, p)$	Joint Probability $P_0(p)P(X = 20 \mid n = 100, p)$	Posterior Probability $P_1(p)$
0.10	0.50	0.0012	0.00060	0.0188
0.20	0.30	0.0993	0.02979	0.9336
0.30	0.20	0.0076	0.00152	0.0476
	1.00		0.03191	1.0000

posterior probability distributions in Tables 15-6 and 15-8, a much higher probability (0.9336) is assigned to the event $p = 0.20$ after 20% defectives have been observed in a sample of size 100 than when that percentage of defectives is found in a sample of 10 units (0.3869).

> A generalization of this result is that as sample size increases, the posterior distribution of the random variable "proportion of defectives" is influenced more by the sample evidence and less by the prior distribution.

Prior and Posterior Means

Let us consider a somewhat more formal explanation of the relationship between the prior distribution, the sample evidence, and the posterior distribution. This explanation can be given in terms of the change that take place between the mean of the prior distribution and the mean of the corresponding posterior distribution. For brevity, we will refer to the mean of a prior distribution of a random variable representing states of nature as a *prior mean* or *prior expected value* and the corresponding mean of a posterior distribution as a *posterior mean* or *posterior expected value*. The prior mean is obtained by the usual method for computing the mean of any probability distribution. The calculation of the prior mean for the acceptance sampling problem is given in Table 15-9. Thus, the prior mean is 0.17 defective articles. The notation $E_0(p)$ is used for the prior mean, the letter E denoting "expected value" and the subscript 0 denoting "prior distribution." Analogously, the mean of the posterior distribution is denoted $E_1(p)$. The corresponding computations of the posterior means for the cases in which the posterior distributions reflect sample evidence of two defectives in a sample of 10 units and 20 defectives in a sample of 100 units are given in Table 15-10. The prior mean is the expected proportion of defectives in the supplier's lot based on the Renny Corporation's prior assessment, that is, before the use of sample information. The posterior mean is the expected proportion of defectives in the supplier's lot based on the Renny Corporation's posterior assessment, that is, after incorporation of the sample information.

TABLE 15-9
Calculation of the prior mean for the defective proportion in the Renny Corporation problem

Event p	Prior Probability $P_0(p)$	$pP_0(p)$
0.10	0.50	0.05
0.20	0.30	0.06
0.30	0.20	0.06
	1.00	0.17

Prior mean $= E_0(p) = 0.17$ defectives

Rounding off the results obtained in Table 15-10, we observe posterior means of 0.179 defectives based on the sample evidence of two defectives in a sample of 10 units, and 0.203 based on sample evidence of 20 defectives in a sample of 100. Hence, in the case of the smaller sample size, the posterior mean lies closer to the prior mean of 0.17 defectives than to the sample evidence of 0.20 defectives. On the other hand, when the larger sample is employed, the posterior mean falls closer to the sample evidence of 0.20 defectives than to the mean of the prior distribution.[2] This empirical finding agrees with our previous statement that as the sample size increases, the posterior distribution is

TABLE 15-10
Calculation of posterior means for the proportion of defectives in the Renny Corporation problem

Event p	Posterior Distribution Incorporating Sample Evidence $X = 2, n = 10$		Posterior Distribution Incorporating Sample Evidence $X = 20, n = 100$	
	Posterior Probability $P_1(p)$	$pP_1(p)$	Posterior Probability $P_1(p)$	$pP_1(p)$
0.10	0.4136	0.04136	0.0188	0.00188
0.20	0.3869	0.07738	0.9336	0.18672
0.30	0.1995	0.05985	0.0476	0.01428
	1.0000	0.17859	1.0000	0.20288

Posterior mean $= E_1(p) = 0.17859$ Posterior mean $= E_1(p) = 0.20288$

[2] The posterior mean of 0.203 exceeds both the value of the prior mean, 0.17, and the sample value, 0.20 defectives. An examination of the results for $p = 0.10$ and $p = 0.30$ in Table 15-8 gives the reason. The likelihood of 20 defectives in a sample of 100 for $p = 0.30$ is more than six times the similar likelihood for $p = 0.10$. Despite a lower prior probability for $p = 0.30$ than for $p = 0.10$, the joint probability and, therefore, the posterior probability in the case of 0.30 far exceed those for $p = 0.10$. The net effect is to pull $E_1(p)$ somewhat closer to 0.30 than to 0.10.

TABLE 15-11
Posterior expected opportunity losses for the Renny Corporation problem: the posterior probabilities incorporate sample evidence $X = 20$, $n = 100$

Event p	Act A_1: Reject the Lot			Act A_2: Accept the Lot		
	Posterior Probability $P_1(p)$	Opportunity Loss	Weighted Opportunity Loss	Posterior Probability $P_1(p)$	Opportunity Loss	Weighted Opportunity Loss
0.10	0.0188	$200	$3.76	0.0188	$ 0	$ 0
0.20	0.9336	0	0	0.9336	100	93.36
0.30	0.0476	0	0	0.0476	200	9.52
	1.0000		$3.76	1.0000		$102.88

$$\text{EOL}(A_1) = \$3.76 \qquad\qquad \text{EOL}(A_2) = \$102.88$$
$$\text{Posterior EVPI} = \$3.76$$

progressively more influenced by the sample evidence and less by the prior distribution.

It is instructive to determine the optimal act using posterior probabilities that incorporate the evidence of 20 defectives in a sample of size 100. Table 15-11 gives the posterior expected opportunity losses based on posterior probabilities. The posterior expected opportunity losses of act A_1 and act A_2 are further apart than in any of the preceding cases. The optimal act is A_1, reject the lot, with a posterior expected opportunity loss of only $3.76 compared with the corresponding figure of $102.88 for A_2. Comparing this decision in favor of act A_1 with the analogous choice of act A_2 based on prior expected opportunity losses illustrates a reversal of decision that takes place because of sample evidence. The low $3.76 figure, which can be interpreted as the posterior EVPI, indicates that after observing 20 defectives in a sample of 100 units, the Renny Corporation would not be wise to spend much money accumulating additional evidence before making its decision concerning acceptance or rejection of the lot.

EXERCISES 15.3

Note: For ease of computations in these exercises, assume sampling with replacement (binomial sampling distributions) where appropriate.

1. Let p be the true percentage of business executives who would subscribe to a new business magazine. Assume that p can take on the following values with the respective prior probabilities:

p	$P_0(p)$
0.10	0.80
0.20	0.20

A simple random sample of 20 business executives is asked whether they would subscribe to the magazine. What are the revised probabilities if

a. One executive would subscribe?

b. Two executives would subscribe?

c. Three executives would subscribe?

2. The personnel manager of Kelly Manufacturing Company is conducting an employee turnover study. He believes that employee turnover follows a Poisson process. Moreover, he thinks that the average number of employees leaving per month, μ, can assume the following values and respective prior probabilities:

μ	$P_0(\mu)$
6	0.7
7	0.3

If, in a particular month, seven workers leave Kelly's employ, what are the revised probabilities?

3. A buyer for a large department store is considering whether to buy a lot of hats offered to her by a manufacturer. Let p be the proportion of hats considered unacceptable because of flaws, and assume the following information:

p	$P_0(p)$	Opportunity Loss of Accepting Lot
0.05	0.4	$ 0
0.10	0.2	250
0.15	0.2	500
0.20	0.2	750

In a sample of 20 hats, two are unacceptable. Calculate the expected opportunity loss of the action "accept the lot" if

a. action is taken before sampling.

b. action is taken after sampling.

4. A new course is being offered at a college. Let p be the proportion of students who fail the midterm examination of the new course. The states of nature and their respective prior probabilities are as follows:

p	$P_0(p)$
0.05	0.3
0.10	0.4
0.15	0.2
0.20	0.1

The college draws a random sample of 20 students enrolled in the course. What are the revised probabilities if

a.   one student fails the midterm examination?

b.   three students fail the midterm examination?

5.   The government is trying to decide whether to build a reservoir in Dodd County or in Todd County. Dodd County has a population of 800,000, and Todd has a population of 1,200,000. The cost of building the reservoir is the same in each location. The immediate expected benefit of the reservoir is computed as $15 per person in either county immediately *after* the reservoir is built. However, building the reservoir will displace people, and the loss due to displacement is estimated at $60 per person. The reservoir in Dodd would displace 20,000 people, 10–20% of whom would leave Dodd. In Todd the reservoir would displace 80,000 people, 5–15% of whom would move out of the county. From economic studies and comparisons with similar previous reservoir sites, the prior probabilities of the proportions of people who would move are estimated for each county as follows:

θ	Dodd $P_0(\theta)$	Todd $P_0(\theta)$
0.05	0.4	0.0
0.10	0.3	0.2
0.15	0.3	0.6
0.20	0.0	0.2

a.   Write a mathematical expression for the benefit derived from each reservoir, letting X represent the proportion of displaced people who would move out of the county.

b.   Based on prior beliefs, where should the reservoir be built?

c.   In each county, the government selects a random sample of 20 persons, who are asked whether they would move out of the county if displaced. Four people in Dodd and six in Todd said that they would move. Taking this new information into consideration, where should the reservoir be built?

6.   A survey was taken by the Doody's Investor Service to determine the percentage of large corporations that use debt only as a last resort for financing, in order to keep their financial ratings as high as possible. The Investor Service's subjective probabilities associated with the possible percentage values, before considering the survey's results, were as follows:

p	$P_0(p)$
0.10	0.30
0.20	0.35
0.30	0.45

a. If the survey included 20 companies and 20% (four companies) maintained the conservative viewpoint on debt described above, what would be the revised probabilities?

b. If the survey included 100 companies and indicated the same percentage, what would be the revised probabilities?

7. Calculate the prior mean for the percentage of companies following the conservative financing practice described in exercise 6. Calculate the posterior mean for parts (a) and (b).

KEY TERMS

Posterior Expected Profit under Uncertainty   the maximum of expected profits after incorporating the most recent sample or experimental evidence.

Posterior Expected Value of Perfect Information   the expected payoff, using posterior probabilities, of decision making in conjunction with a perfect predictor.

Posterior Mean (expected value)   the expected value of the event distribution calculated with posterior probabilities.

16

DEVISING OPTIMAL STRATEGIES
PRIOR TO SAMPLING

In Chapter 14, we discussed **prior analysis**, a method for decision making prior to the incorporation of additional information obtained through sampling or experimentation. In Chapter 15, we studied **posterior analysis**, a corresponding method for decision making after additional information has been obtained through sampling or experimentation. In posterior analysis, we draw a sample and choose the best act, taking into account both the sample data and prior probabilities of the events that affect payoffs.

16.1
PREPOSTERIOR ANALYSIS

In this chapter, we will consider **preposterior analysis**, which determines whether it is worthwhile to collect sample or experimental data at all.

> If data collection *is* found to be worthwhile, preposterior analysis further specifies (1) the best courses of action for each possible type of sample or experimental outcome and (2) how large a sample to take.

This type of analysis leads to a more complex but more interesting view of decision making than those we have considered so far.

Terminal decision
In both prior and posterior analysis, a final decision is made among the alternative courses of action based on the information at hand. A decision that makes a final disposition of the choice of a best act is referred to in Bayesian analysis as a **terminal decision**. The act itself is referred to as a **terminal act**.

> In many business decision-making situations, the wise course of action is not to choose a terminal act, but rather to delay making a terminal decision in order to obtain further information.

Preposterior analysis, in addition to establishing whether additional information should be obtained (and, if so, how much), also delineates the optimal decision rule to employ based on the possible types of evidence that can be produced by the additional information if it is gathered. An obvious difficulty in this type of decision procedure is that the specific outcome of a sample (or additional information) is unknown prior to taking the sample. Yet, the decision whether sample information will be worthwhile must be made prior to drawing the sample. Because of the cost associated with acquiring additional knowledge through a sampling process, it pays to take a sample only if the anticipated worth of this sample information exceeds its cost. The methods discussed in this chapter provide a procedure for determining the **expected value of sample information**. We will begin by considering a simple problem involving a single-stage sample of fixed size. In section 16.2, we discuss two examples of preposterior analysis.

16.2
EXAMPLES OF PREPOSTERIOR ANALYSIS

A preposterior analysis, as the name implies, is an investigation that must be carried out *before* sample information is obtained and therefore *prior to* the availability of posterior probabilities based on a particular sample outcome. However, this type of analysis takes into account all possible sample results and computes the expected worth (or expected opportunity loss) of a strategy that assumes that the best acts are selected depending on the type of sample information observed. We will illustrate preposterior analysis by an oversimplified example in order to convey the basic principles of the procedure without getting bogged down in computational detail.

The A. B. Westerhoff Company, a firm that manufactures consumer products, considered marketing a new product it had developed. However, the company wanted to appraise the advisability of engaging a market research firm to help determine whether sufficient consumer demand existed to warrant placing the product on the market. The market research firm offered to conduct a nation-wide survey of consumers to obtain an appropriate indication of the market for the product. The fee for the survey was $15,000. The A. B. Westerhoff Company also wished to carry out an analysis that would specify whether it was better to act now (on the basis of prior betting odds as to the success or failure of the product and estimated payoffs) or to engage the market research firm and then act on the basis of the survey indication.

The A. B. Westerhoff Company problem

The company analyzed the situation as follows. The two available actions were

a_1: Market the product

a_2: Do not market the product

The states of nature, or possible events, were defined as

θ_1: Successful product

θ_2: Unsuccessful product

Although only two states of nature for success of the product are used in this example, many states could be employed to indicate degrees of success. Similarly more than two courses of action might be considered.

The company decided to view the problem in terms of opportunity losses of incorrect action. Based on appropriate estimates, the opportunity loss matrix (in thousands of dollars) shown in Table 16-1 was constructed.

The company, on the basis of its past experience with products of this type, assessed the odds that the product would be a failure at 3:1. That is, the

TABLE 16-1
Payoff table showing opportunity losses for the A. B. Westerhoff Company problem (thousands of dollars)

Event	Market a_1	Do Not Market a_2
θ_1: Successful product	$ 0	$200
θ_2: Unsuccessful product	160	0

TABLE 16-2
Calculation of prior expected opportunity losses for the A. B. Westerhoff Company problem (thousands of dollars)

Event	a_1: Market the Product			a_2: Do Not Market the Product		
	Probability $P(\theta_i)$	Opportunity Loss	Weighted Opportunity Loss	Probability $P(\theta_i)$	Opportunity Loss	Weighted Opportunity Loss
θ_1: Successful product	0.25	$ 0	$ 0	0.25	$200	$50
θ_2: Unsuccessful product	0.75	160	120	0.75	0	0
	1.00		$120	1.00		$50

EOL(a_1) = $120 (thousands of dollars)

EOL(a_2) = $50 (thousands of dollars)

EOL of the optimal act
= EOL(a_2) = $50 (thousands of dollars)

company assigned prior probabilities to the success and failure of the product as follows:

$$P(\theta_1) = P(\text{successful product}) = \frac{1}{4} = 0.25$$

$$P(\theta_2) = P(\text{unsuccessful product}) = \frac{3}{4} = 0.75$$

The prior analysis in Table 16-2 was conducted to determine the optimal course of action if no additional information was obtained. As shown in Table 16-2, the optimal course of action was a_2 (do not market). The expected opportunity loss for this course of action was $50,000. A decision diagram of this prior analysis is given in Figure 16-1.

The A. B. Westerhoff Company then turned its attention to the problem of analyzing its expected opportunity losses if its action was based on the survey results obtained by the market research firm. It requested the market research firm to indicate the nature and reliability of the consumer survey's

FIGURE 16-1
Decision diagram for action by the A. B. Westerhoff Company if no survey is conducted (all losses are in thousands of dollars)

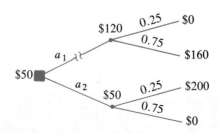

results. The market research firm replied that the survey would yield one of the following three types of indications:

$$X_1: \text{Favorable}$$

$$X_2: \text{Intermediate}$$

$$X_3: \text{Unfavorable}$$

That is, an X_1 indication meant an observed level of consumer demand in the survey that was favorable to the success of the product, an X_3 result was unfavorable, and an X_2 indication meant a situation falling between levels X_1 and X_3 and classified as "intermediate." Although we are using the verbal classifications "favorable," "intermediate," and "unfavorable" here for simplicity of reference, each of these indications will represent a specific numerical range of demand. We will explain later how this type of sample is used in decision making.

As a description of the anticipated *reliability* of the survey indications, the market research firm supplied the array of conditional probabilities shown in Table 16-3. The entries in this table are values of $P(X_j|\theta_i)$ based on past relative frequencies in similar types of surveys. That is, they represent the conditional probabilities of each type of sample evidence given that the product actually was successful or unsuccessful. For example, the $P(X_1|\theta_1) = 0.72$ entry in the upper left-hand corner of the table means the probability is 72% that the survey will yield a favorable indication, given a successful product. The subscript j is used for the sample indication, denoted X_j, analogous to the use of the subscript i for the states of nature, denoted θ_i. It may not always be feasible to use past relative frequencies as the conditional probability assignments as we are doing here. The use of past relative frequencies as probabilities for future events always involves the assumption of a continuation of the same conditions as existed when the relative frequencies were established. As we have seen in other examples, probability functions such as the binomial distribution are often appropriate for computing these "likelihoods" or conditional probabilities of sample results.

The A. B. Westerhoff Company decided to carry out a preposterior analysis to determine whether to engage the market research firm to conduct the

Anticipated reliability of survey indications

TABLE 16-3
Conditional probabilities of 3 types of survey evidence for the A.B. Westerhoff Company problem of deciding whether to conduct a survey

| Event | Conditional Probability $P(X_j|\theta_i)$ | | | |
	X_1	X_2	X_3	Total
θ_1: Successful product	0.72	0.16	0.12	1.00
θ_2: Unsuccessful product	0.08	0.12	0.80	1.00

TABLE 16-4
Calculation of the joint probability distribution of survey evidence and events
for the A. B. Westerhoff Company problem of deciding whether to conduct a survey

Event θ_i	Prior Probability $P(\theta_i)$	Conditional Probability $P(X_j \mid \theta_i)$			Joint Probability $P(X_j \text{ and } \theta_i)$			Total
		X_1	X_2	X_3	X_1	X_2	X_3	
θ_1	0.25	0.72	0.16	0.12	0.18	0.04	0.03	0.25
θ_2	0.75	0.08	0.12	0.80	0.06	0.09	0.60	0.75
Total	1.00				0.24	0.13	0.63	1.00

survey. For this purpose, it was necessary to compare (1) the expected opportunity loss of purchasing the survey and then selecting a terminal act to (2) the
expected opportunity loss of terminal action without the survey. The latter
expected loss figure was obtained from the prior analysis. As an intermediate
step, the company computed the joint probability distribution of the sample
evidence X_j and the events θ_i, as shown in Table 16-4. These joint probabilities
were obtained by multiplying the prior (marginal) probabilities by the appropriate conditional probabilities. For example, 0.18, the upper left entry in the
joint probability distribution in Table 16-4, was obtained by multiplying 0.25
by 0.72. In symbols, $P(X_1 \text{ and } \theta_1) = P(\theta_1)P(X_1 \mid \theta_1)$, and so on.

Interesting points about   It is useful to consider the following interesting points about the joint
joint probability distribution   probability distribution to understand the subsequent analysis.

- As in any joint bivariate probability distribution, the totals in the margins
  of the table are marginal probabilities.
- The row totals are the prior probabilities $P(\theta_1) = 0.25$ and $P(\theta_2) = 0.75$.
- The column totals are the marginal probabilities of the survey evidence;
  that is, $P(X_1) = 0.24$, $P(X_2) = 0.13$, and $P(X_3) = 0.63$.
- Conditional probabilities of events, given the survey evidence—that is,
  probabilities of the form $P(\theta_i \mid X_j)$—can be computed by dividing the joint
  probabilities by the appropriate column totals, $P(X_j)$.

For example, the probability of a successful product, given a favorable survey
indication, is

$$P(\theta_1 \mid X_1) = \frac{P(\theta_1 \text{ and } X_1)}{P(X_1)} = \frac{0.18}{0.24}$$

and so on. These probabilities can be viewed as revised or posterior probability
assignments to the events θ_1 and θ_2, given the survey evidence X_1, X_2, and
X_3. The calculation of these posterior probabilities represents an application

of Bayes' theorem (equations 2.12 and 2.16), as shown in the following method of symbolizing the probability $P(\theta_1|X_1)$ just calculated:

$$P(\theta_1|X_1) = \frac{P(\theta_1 \text{ and } X_1)}{P(X_1)} = \frac{P(\theta_1)P(X_1|\theta_1)}{P(\theta_1)P(X_1|\theta_1) + P(\theta_2)P(X_1|\theta_2)}$$

That is, the joint probability $P(\theta_1 \text{ and } X_1) = 0.18$ was calculated in Table 16-4 by multiplying the marginal probability $P(\theta_1) = 0.25$ by the conditional probability $P(X_1|\theta_1) = 0.72$; the marginal probability $P(X_1) = 0.24$ was obtained by adding the joint probability $P(\theta_1)P(X_1|\theta_1) = 0.18$ and $P(\theta_2)P(X_1|\theta_2) = 0.06$.

The A. B. Westerhoff Company proceeded with its analysis and constructed the decision diagram shown in Figure 16-2. All monetary figures are in thousands of dollars. Returning to the beginning of the problem, the first choice is to purchase the survey or not to purchase it. Therefore, starting at node (a) at the left and following the "no survey" branch of the tree, we move along to node (b). From node (b) to the right is the decision tree depicted in Figure 16-1 for the prior analysis with no survey. As indicated in that figure, the $50 entry at node (b) is the expected opportunity loss of choosing the optimal terminal act without conducting the survey.

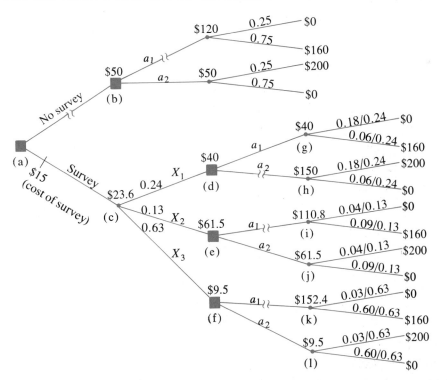

FIGURE 16-2

Decision diagram for preposterior analysis for the A. B. Westerhoff Company problem of deciding whether to conduct a survey (payoffs are in thousands of dollars)

Expected Value of Sample Information

On the other hand, suppose that at the outset [at node (a)], the decision is to conduct the survey, and we move down to point (c). The results of the survey then determine which branch to follow. The three branches emanating from node (c), which represent X_1, X_2, and X_3 types of survey information, are marked with their respective marginal probabilities, 0.24, 0.13, and 0.63. Suppose type X_1 information (that is, a "favorable" indication) were observed. Moving ahead to point (d), we can choose either act a_1 or act a_2, that is, market or not market the product. If act a_1 is selected, we move to node (g); if act a_2 is selected, we move to node (h). At these points, posterior or revised probability questions must be answered. For example, if act a_1 is chosen, the probabilities shown on the two branches stemming from node (g), $\frac{0.18}{0.24}$ and $\frac{0.06}{0.24}$, are the conditional (posterior) probabilities, given type X_1 information, that the product is successful or unsuccessful, or in symbols, $P(\theta_1|X_1)$ and $P(\theta_2|X_1)$. Thus, they represent revised probabilities of these two events after observation of a particular type of sample evidence. These conditional probabilities are calculated from Table 16-4, as indicated earlier, by dividing joint probabilities by the appropriate marginal probabilities. Looking forward from node (g) and applying the posterior probabilities ($\frac{0.18}{0.24}$ and $\frac{0.06}{0.24}$) as weights to the losses attached to the events "successful product" and "unsuccessful product," we obtain an expected opportunity loss payoff of $40 for act a_1. This figure is entered at node (g). Comparing it with the corresponding figure of $150 for a_2, we block off act a_2 as being nonoptimal. Therefore, $40 is carried down to node (d), representing the payoff for the optimal act upon observation of type X_1 information. Similar calculations yield $61.5 and $9.5 at nodes (e) and (f) for X_2 and X_3 types of information. Weighting these three payoffs by the marginal probabilities of X_1, X_2, and X_3 indications (0.24, 0.13, and 0.63, respectively), we obtain a loss of $23.6 as the expected payoff of conducting the survey and taking optimal action after the observation of the sample evidence.

Comparing the $23.6 figure with the $50 obtained under the "no survey" option, we see that it would be worthwhile to pay up to $50 − $23.6 = 26.4 (thousands of dollars) for the survey.

The difference represented by the 26.4 (thousands of dollars) is referred to as the **expected value of sample information,** denoted EVSI.

Expected value of sample information (EVSI)

> Hence, if we are considering a choice between immediate terminal action without obtaining sample information and a decision to sample and then select a terminal act, EVSI is the *expected amount by which the terminal opportunity loss is reduced by the information to be derived from the sample.*

This EVSI is a gross figure, since it has not taken into account the cost of obtaining the survey information.

> To calculate the **expected net gain of sample information**, which we denote as ENGS, we subtract the cost of obtaining the sample information from the expected value of this sample information.

Expected net gain of sample information (ENGS)

In general,

(16.1)   ENGS = EVSI − Cost of sample information

and in this problem,

ENGS = \$26.4 − \$15 = 11.4 (thousands of dollars)

In conclusion, since the expected value of sample information was \$26,400 and the cost of the survey was \$15,000, the ENGS was \$11,400. Therefore, the A. B. Westerhoff Company decided that it was worthwhile to engage the market research firm to conduct the survey.

It is important to recognize that the EVSI and ENGS computations have been made with respect to the particular prior probability distribution used in the analysis. If different prior probabilities were used, the survey would have had correspondingly different EVSI and ENGS values. Sensitivity analysis, discussed in section 16.5, can test how sensitive the alternative actions are to the size of the prior probabilities.

Some Considerations in Preposterior Analysis

A few points concerning preposterior analysis arise from consideration of the foregoing example. First of all, we note that in this problem, the optimal action if no survey were conducted was a_2, not to market the product. On the other hand, as seen from Figure 16-2, if the survey were carried out and a favorable indication of demand, X_1, were obtained, the best course of action would be a_1, to market the product. The fact that for at least one of the survey outcomes the decision maker may possibly change the selected act gives the survey some value. Clearly, if a decision maker's course of action cannot be modified regardless of the experimental outcome, then the experiment is without value.

Second, the calculations in the preceding problem were carried out in terms of opportunity losses. If the payoffs had been expressed in terms of profits, the obvious equivalent analysis would have been required. For example, instead of the EVSI being the expected amount by which the terminal opportunity loss is reduced by the information to be derived from the sample, it would be the *expected amount by which the terminal profit is increased* by this information. Hence, expected profit of terminal action without sampling and expected profit of choosing a terminal action after sampling would be calculated. The first quantity would be subtracted from the second to yield the same EVSI figure obtained in the opportunity loss analysis.

Third, in this problem, the terms "sample" and "survey" were used interchangeably. Actually, even if the survey had represented a complete enumeration, the same analysis could have been carried out. Thus, the term "sample" in this context is used in a general sense to include any sample from size one up to a complete census. Indeed, the sample outcomes X_1, X_2, and X_3 need not have arisen from a statistical sampling procedure but could represent any set of experimental outcomes. Therefore, a preposterior analysis may be viewed as yielding the "expected value of experimental information" rather than the "expected value of sample information." However, since the latter term is conventionally used in Bayesian decision analysis, we have resisted the temptation to adopt a new term.

Finally, another generalization suggests itself. Only one survey of a fixed size was considered in this problem. Many surveys of different types and different sizes could have been considered. The preposterior expected opportunity loss would then be calculated for each of these different surveys. The minimum of these figures would be subtracted from the prior expected opportunity loss to yield the EVSI. In order to obtain the ENGS, the total expected opportunity loss for each survey is calculated by adding the cost of the survey to the corresponding preposterior expected opportunity loss. Then these figures are subtracted from the prior expected loss. The maximum of these differences is the ENGS, since it represents the expected net gain of the survey with the lowest total expected opportunity loss. Theoretically, the number of possible surveys or experiments might be considered infinite, but obviously there would be a delimitation at the outset based on factors of practicability or feasibility. On the other hand, the use of computers increases considerably the number of possible alternatives that can practically be compared.

EXERCISES 16.2

1. Define and compare EVSI and ENGS.
2. Let θ_1 be the state of nature "a recession will occur during the upcoming year" and θ_2, "a recession will not occur during the upcoming year." The chairman of the Economics Department at a prestigious midwestern university will predict either X, "a recession will occur" or Y, "a recession will not occur." Calculate the missing entries of the following table:

State of Nature	$P(\theta)$	$P(X\|\theta)$	$P(Y\|\theta)$	$P(\theta)P(X\|\theta)$	$P(\theta)P(Y\|\theta)$	$P(\theta\|X)$	$P(\theta\|Y)$
θ_1	0.3	0.6					
θ_2	0.7	0.3					

3. An oil drilling firm is contemplating the following two states of nature concerning a particular drill site: θ_1, "oil will be found" and θ_2, "oil will not be found." The firm assigns a prior probability of 55% that oil will be found. A study of the site is now being conducted, which will indicate whether oil will be found or not. In the past, the persons conducting the study have been correct 75% of the time. Find the posterior probabilities for each of these indications.

4. You are given the decision tree diagram (see figure) with payoffs in terms of opportunity losses.
    a. Fill in the missing entries.
    b. Interpret each part of the tree.
    c. Find the EVPI before sampling.
    d. Find the EVSI.

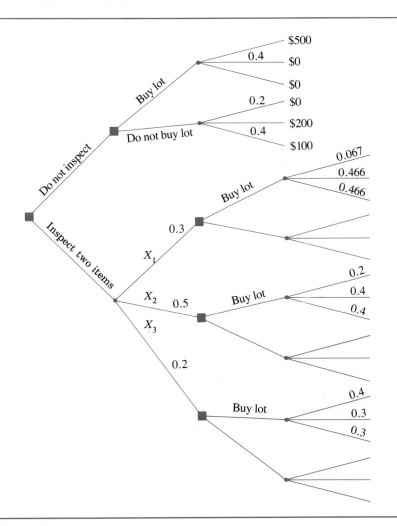

Decision tree diagram for exercise 4

5. A motel located near the site of a soon-to-be-opened world's fair is contemplating the construction of some temporary extra rooms. The rooms are of no value after the fair is closed, since the present size of the motel is more than adequate for normal demand. Let θ_1 be "demand for the temporary rooms is at least enough to cover the cost of building them," and θ_2 be "demand for the temporary rooms is not sufficient to warrant construction." A survey can be taken that would yield two possible indications, X_1 or X_2. The following is known:

State of Nature	$P(\theta)$	$P(X_1\|\theta)$		Opportunity Loss Table (units of $10,000) Build	Do No Build
θ_1	0.5	0.60	θ_1	0	17
θ_2	0.5	0.45	θ_2	12	0

a. Find the following:
   (1) the optimal act before sampling.
   (2) the optimal act if X_1 occurs.
   (3) the optimal act if X_2 occurs.
   (4) $P(X_1)$ and $P(X_2)$.
   (5) the EVSI.
b. Should the survey be taken?

6. A mutual fund is contemplating the sale of a certain common stock. A study can be made that will yield two possible results, X_1 or X_2. You are given the following information:

State of Nature	$P(\theta)$	$P(X_1\|\theta)$	$P(X_2\|\theta)$
θ_1: Stock price up	0.60	0.75	0.25
θ_2: Stock price the same or down	0.40	0.60	0.40

Decision	Opportunity Loss Table (units of $10,000) θ_1	θ_2
Sell	10	0
Do not sell	0	10

a. Find the following:
   (1) the optimal act before sampling.
   (2) the optimal act if X_1 results.
   (3) the optimal act if X_2 results.
   (4) $P(X_1)$ and $P(X_2)$.
   (5) the EVSI.
b. If the cost of the study is $500, what is the expected net gain from the study?

7. A cookie manufacturer recently developed a new snack food called Crunchie Munchies. The cookie manufacturer is unfamiliar with the market for snack foods and decides to have a study conducted before committing itself to the market. Three possible states of nature can occur: θ_1, "demand will be high," θ_2, "demand will be average," and θ_3, "demand will be low." Respective prior probabilities of occurrence are 0.58, 0.25, and 0.17. The manufacturer developed the following opportunity loss table, stated in thousands of dollars:

Decision	Opportunity Loss Table		
	θ_1	θ_2	θ_3
Manufacture Crunchie Munchies	0	0	600
Do not manufacture Crunchie Munchies	400	200	0

The test market study, conducted at a cost of $15,000, will result in one of three possible outcomes, X_1, X_2, or X_3. The probabilities of these outcomes are 0.6, 0.3, and 0.1, respectively. The posterior probabilities of θ_i, given the respective outcomes, are:

State of Nature	$P(\theta_i|X_1)$	$P(\theta_i|X_2)$	$P(\theta_i|X_3)$
θ_1	0.8	0.3	0.1
θ_2	0.1	0.5	0.4
θ_3	0.1	0.2	0.5
Total	1.0	1.0	1.0

Draw a decision tree diagram and make all the necessary entries on the branches dealing with outcome X_1.

8. The Superior Stereo Company recently decided to manufacture a new stereo model. As part of the marketing decision, a price for the new model must be established. The price set by the marketing department must be consistent with the public's perception of the model's worth. If the company's price is too high, consumers will not buy the stereo, and if the company's price is too low, it also will lose money. The opportunity loss table for the different pricing decisions is shown:

Pricing Decision	Public's Perception of Stereo's Worth		
	High Priced	Medium Priced	Low Priced
High	$ 0	$100,000	$500,000
Medium	100,000	0	300,000
Low	200,000	100,000	0

The marketing vice-president feels the probability that the public will consider the stereo to be medium priced is 0.6; low priced, 0.2. A forecasting model has been developed that has the following reliability [that is, $P(\text{indication}|\text{state of nature})$]:

| | Public's Perception of Stereo's Price | | |
Model Indication	High Priced	Medium Priced	Low Priced
High	0.80	0.10	0.10
Medium	0.10	0.85	0.05
Low	0.10	0.05	0.85

What is the expected value of information from the forecasting model for the new stereo?

9. The owner of an apartment building has recently read about a new energy-saving device. This device claims to save one-half of the consumer's electric bill. The owner defines three possible states of nature: S_1, "device will be a success," S_2, "device will have limited success," and S_3, "device will not be a success," with respective prior probabilities of occurrence of 0.60, 0.25, and 0.15. She also has constructed the following opportunity loss table, in thousands of dollars:

| | Opportunity Loss Table | | |
Decision	S_1	S_2	S_3
Buy device	0	0	60
Do not buy device	50	20	0

A private consultant has conducted an investigation costing $15,000 for which there are three outcomes—X, Y, and Z. The probabilities of these outcomes are 0.5, 0.3, and 0.2, respectively. The posterior probabilities of S_i, given the respective outcomes, are

State of Nature	$P(S_i|X)$	$P(S_i|Y)$	$P(S_i|Z)$
S_1	0.6	0.8	0.3
S_2	0.2	0.1	0.6
S_3	0.2	0.1	0.1

Draw a decision tree diagram for the owner and make all the necessary entries on the branches dealing with outcome X.

10. In exercise 5, suppose that the prior probabilities, $P(\theta_1)$ and $P(\theta_2)$, were instead 0.4 and 0.6, respectively. Answer the questions in exercise 5 using these values for the prior probabilities.

11. The manager of a bus company that operates between two large cities is debating whether to add an early morning route to the company's current schedule. Company management has calculated that 25 people must use this route each day for the company to break even. Management feels that there is a 30% chance that fewer than 25 people will use the route. The payoff table (over a one-month period) is as follows, stated in thousands of dollars:

Decision	Fewer than 25 People Use Route	25 People or More Use Route
Add route	−3	5
Do not add route	0	0

A private consulting firm offers to conduct a study to obtain a better estimate of the demand for his early morning route. The firm states that the study would have the following reliability.

| Study Indicates | State of Nature | |
	Fewer than 25 People Use Route	25 People or More Use Route
Fewer than 25 people	0.9	0.2
25 or more people	0.1	0.8

What is the maximum amount that the bus company should be willing to pay for the study?

12. Refer to exercise 6. Suppose the mutual fund felt that the prior probability the common stock would increase was 0.70 instead of 0.60. What would be the expected net gain from the study? Compare it to the expected net gain if the prior probability is 0.60.

13. A mail-order house with a fixed "market" of 200,000 people is deciding whether to sell a new line of goods. If more than 25% of its customers will purchase the new line, the company should market it, and if fewer than 25% will purchase the line, the company should not market it. For simplicity, assume that either 20% or 30% will purchase the new line. The manager of the firm believes the probability that only 20% of the market will buy is 0.7. The payoff table (in units of $10,000) is as follows:

Decision	20% Will Purchase	30% Will Purchase
Market new line	−2	5
Do not market new line	0	0

a. One person is selected at random from the 200,000 and asked whether he will purchase from the new line. What is the expected value of this information?

b. Suppose two persons are selected at random from the 200,000 and asked whether they will purchase goods from the new line. What is the expected value of this information?

16.3
EXTENSIVE-FORM AND NORMAL-FORM ANALYSES

The type of preposterior investigation carried out in the A. B. Westerhoff Company problem is known as **extensive-form analysis**. It is perhaps easiest to characterize this type of analysis in terms of a decision tree diagram, such as Figure 16-2. In that diagram, a prior analysis is given in the upper part of the tree with the resulting prior expected opportunity loss figure entered at node (b). An extensive-form analysis is given in the lower part of the tree with the resulting preposterior expected opportunity loss entered at node (c).

For purposes of comparison, we summarize the procedures for prior, posterior, and extensive-form preposterior analysis. This comparison assumes that experimental (sample) information is collected in a single-stage procedure as in the A. B. Westerhoff Company problem. Each type of analysis may be thought of as starting at the right side of a decision tree diagram and then proceeding inward by the process referred to in Chapter 14 as backward induction. See Table 16-5 for a comparison of the three types of analysis.

Concept of a Strategy

If many experiments are possible, rather than one as assumed in Table 16-5, the extensive-form analysis also specifies which of the possible experiments should be carried out. Moreover, the extensive-form analysis supplies a decision rule that selects an optimal act for each possible outcome of the chosen experiment. For example, in Figure 16-2, this decision rule was as follows:

$$X_1 \rightarrow a_1$$

$$X_2 \rightarrow a_2$$

$$X_3 \rightarrow a_2$$

where $X_1 \rightarrow a_1$ means that if experimental outcome X_1 is observed, select act a_1, and so on. Such a decision rule is referred to as a **strategy**.

> Mathematically, a strategy can be defined as a function in which an act is assigned to each possible experimental outcome.

TABLE 16-5
The process of backward induction

Prior Analysis	Posterior Analysis	Preposterior Analysis—Extensive-Form
1. Sketch a tree and depict states of nature at the right tips.		
2. Assign payoffs to each of the states for each possible action.		
3. Assign prior probabilities to each state of nature.	Assign posterior probabilities to each state of nature based on a specific outcome of the experimental information.	a. Assign marginal probabilities to each possible experimental (sample) outcome. b. Assign posterior probabilities to each state of nature given specific experimental outcomes. c. For each experimental outcome, carry out steps 4 and 5.
4. Calculate expected terminal payoffs for each act.		
5. Select the act with the highest expected terminal payoff.		
6.		For each type of experimental information, weight the expected terminal payoffs of the best act by the marginal probability of occurrence of that type of information, and add these products to yield an overall preposterior expected payoff. The difference between the preposterior expected payoff and a prior expected payoff is the EVSI.

We now turn to an alternative procedure to extensive-form analysis, known as **normal-form analysis**, in which the problem begins with a listing of all possible strategies that might be employed. Normal-form analysis then makes an explicit comparison of the worth of all of these strategies to arrive at the same optimal strategy as was derived in extensive-form analysis.

Normal-Form Analysis

In the preceding section, extensive-form analysis was applied to the A. B. Westerhoff Company problem. In order to indicate the nature of the normal-form procedure and the relationship between extensive and normal forms of analysis, we will now solve the same problem by the latter method. Then some factors relating to the choice between the two procedures will be given.

TABLE 16-6
A listing of all possible strategies in the A. B. Westerhoff Company problem
of deciding whether to conduct a survey

Sample Outcome	Strategy							
	s_1	s_2	s_3	s_4	s_5	s_6	s_7	s_8
X_1	a_1	a_1	a_1	a_1	a_2	a_2	a_2	a_2
X_2	a_1	a_1	a_2	a_2	a_1	a_1	a_2	a_2
X_3	a_1	a_2	a_1	a_2	a_1	a_2	a_1	a_2

All possible strategies or decision rules for the A. B. Westerhoff Company problem are listed in Table 16-6. The strategies (denoted $s_1, s_2, \ldots, s_8$) indicate the acts taken in response to each sample outcome. Hence, for example, strategy s_3 is

$$X_1 \rightarrow a_1$$

$$X_2 \rightarrow a_2$$

$$X_3 \rightarrow a_1$$

and s_4 is the one previously described. In this problem, there are 8 possible strategies, 3 sample outcomes (X_1, X_2, and X_3), and 2 possible acts (a_1 and a_2). In general, the number of possible strategies is given by n^r, where n denotes the number of acts and r is the number of sample outcomes. Hence, in this case, there are $2^3 = 8$ possible strategies.

Clearly, some of the strategies in Table 16-6 are not very sensible. For example, strategies s_1 and s_8 select the same act regardless of the experimental outcome (survey indication). Strategy s_5 selects act a_2, not to market the product, if the "favorable" survey indication X_1 is obtained, but perversely it would market the product if the intermediate or "unfavorable" indications X_2 or X_3 are obtained. On the other hand, strategies s_2 and s_4 seem to be quite logical and would probably be the only ones a reasonable person would seriously consider on an intuitive basis if he or she contemplated using the experimental information at all.

Continuing with the normal-form analysis, we now compute the expected payoff of each possible strategy. The method consists of calculating for each strategy the expected opportunity loss conditional on the occurrence of each state of nature.

> These conditional expected losses are referred to in Bayesian decision analysis as **risk**.

We will use that term or its equivalent, **conditional expected losses**. The weighted average, or expected value, of these risks, using prior probabilities of states as weights, yields the **expected opportunity loss**, or **expected risk**, of the strategy.

TABLE 16-7
Calculation of risks, or conditional expected opportunity losses, for strategies
s_2 and s_4 ($000)

State of Nature	Strategy s_2						
	Opportunity Loss		Probability of Action $s_2(a_1, a_1, a_2)$			Risk (conditional expected loss) $R(s_2\mid\theta_i)$	
	a_1	a_2	a_1	a_1	a_2		
θ_1	$ 0	$200	0.72	0.16	0.12	$24.00	
θ_2	160	0	0.08	0.12	0.80	32.00	

$$R(s_2\mid\theta_1) = 0.72(\$0) + 0.16(\$0) + 0.12(\$200) = \$24.00$$
$$R(s_2\mid\theta_2) = 0.08(\$160) + 0.12(\$160) + 0.80(\$0) = \$32.00$$

State of Nature	Strategy s_4						
	Opportunity Loss		Probability of Action $s_4(a_1, a_2, a_2)$			Risk (conditional expected loss) $R(s_4\mid\theta_i)$	
	a_1	a_2	a_1	a_2	a_2		
θ_1	$ 0	$200	0.72	0.16	0.12	$56.00	
θ_2	160	0	0.08	0.12	0.80	12.80	

$$R(s_4\mid\theta_1) = 0.72(\$0) + 0.16(\$200) + 0.12(\$200) = \$56.00$$
$$R(s_4\mid\theta_2) = 0.08(\$160) + 0.12(\$0) + 0.80(\$0) = \$12.80$$

Table 16-7 shows the calculation of the risks associated with strategies s_2 and s_4. For example, let us consider the calculations for strategy s_2. The left-hand portion of the table is the payoff table in terms of opportunity losses. In the next section are **probabilities of action**, which are probabilities of taking actions specified by the given strategy based on the observations of the possible types of sample evidence. A convenient notation for designating a strategy is given in the column heading of this section of the table. The notation $s_2(a_1, a_1, a_2)$ means strategy s_2 consists in taking act a_1 if the sample indication X_1 is observed; again selecting a_1 if X_2 is observed; and choosing a_2 if X_3 is observed. That is, the first element within the parentheses denotes the action to be taken upon observing the first sample indication, the second element specifies the action to be taken upon observing the second sample indication, and so on.

The risk, or expected opportunity loss, given the occurrence of a specific state of nature, is calculated by multiplying the probabilities of action by the respective losses incurred if these actions are taken and summing the products. Thus, the risk, or expected opportunity loss, associated with the use of strategy s_2, given that state of nature θ_1 occurs—denoted $R(s_2\mid\theta_1)$—is calculated to be $24.00, as shown in the top half of Table 16-7. Correspondingly, the expected opportunity loss of strategy s_2, conditional on the occurrence of θ_2—denoted $R(s_2|\theta_2)$—is calculated to be $32.00. The analogous calculation of risks for strategy s_4 is given in the bottom half of Table 16-7.

We may also think of these risks as simply *calculations for each state of nature of the loss due to taking the wrong act times the probability that the wrong act will be taken.* For example, let us consider $R(s_2|\theta_1)$, the conditional expected loss of strategy s_2, given that θ_1 occurs. Now, if θ_1 occurs, the correct (best) course of action is a_1; the incorrect act is a_2. The $24.00 figure for $R(s_2|\theta_1)$ is merely $200, the loss of taking act a_2, times 0.12, the total probability of selecting a_2 under the strategy s_2, given the occurrence of θ_1. Similarly, the $32.00 figure for $R(s_2|\theta_2)$ is equal to $160, the loss of taking act a_1, times 0.20, the total probability of selecting a_1 under strategy s_2, conditional on the occurrence of θ_2. The risk calculations of Table 16-7 are shown in Table 16-8, utilizing this alternative conception.

A summary of the risks, or conditional expected opportunity losses, for all eight strategies is given in Table 16-9. Let us review the interpetation of these figures, using strategy s_4 as an example. In a relative frequency sense (that is, in a large number of identical situations of successful new products, θ_1), if survey evidence X_1, X_2, and X_3 occur with the specified probabilities and if strategy s_4 is employed, the average opportunity loss per product would be $56.00. A similar interpretation holds for unsuccessful products, θ_2, with an average loss of $12.80. Now, we can weight these average losses by the prior probabilities

---

TABLE 16-8
Alternative calculation of risks, or conditional expected opportunity losses, for strategies s_2 and s_4 ($000)

| State of Nature | Strategy $s_2(a_1, a_1, a_2)$ | | |
	Opportunity Loss of Wrong Act	Probability of Wrong Act Given θ_i	Conditional Expected Opportunity Loss $R(s_2|\theta_i)$
θ_1	$200	0.12	$24.00
θ_2	160	0.20	32.00

| State of Nature | Strategy $s_4(a_1, a_2, a_2)$ | | |
	Opportunity Loss of Wrong Act	Probability of Wrong Act Given θ_i	Conditional Expected Opportunity Loss $R(s_4|\theta_i)$
θ_1	$200	0.28	$56.00
θ_2	160	0.08	12.80

TABLE 16-9

Risks, or conditional expected opportunity losses, for the 8 strategies in the A. B. Westerhoff Company problem ($000)

State of Nature	Strategy							
	s_1	s_2	s_3	s_4	s_5	s_6	s_7	s_8
θ_1	$ 0	$24.00	$ 32.00	$56.00	$144.00	$168.00	$176.00	$200.0
θ_2	160.00	32.00	140.80	12.80	147.20	19.20	128.00	0

of successful and unsuccessful products to obtain an overall expected opportunity loss for strategy s_4. The prior probabilities were given as $P(\theta_1) = 0.25$ and $P(\theta_2) = 0.75$. Since θ_1 occurs with probability 0.25 and the conditional expected loss if θ_1 occurs is $56.00, and since θ_2 occurs with probability 0.75 and the conditional expected loss if θ_2 occurs is $12.80, the expected opportunity loss of using strategy s_4 is

$$\text{EOL}(s_4) = (0.25)(\$56.00) + (0.75)(\$12.80) = \$23.60$$

In symbols, we have

$$\text{EOL}(s_4) = P(\theta_1)R(s_4|\theta_1) + P(\theta_2)R(s_4|\theta_2)$$

In general form, the expected opportunity loss of the kth strategy s_k is

$$(16.2) \quad \text{EOL}(s_k) = P(\theta_1)R(s_k|\theta_1) + P(\theta_2)R(s_k|\theta_2) + \cdots + P(\theta_m)R(s_k|\theta_m)$$

$$= \sum_{i=1}^{m} P(\theta_i)R(s_k|\theta_i)$$

The decision rule for which this expected opportunity loss is a minimum is known as the **Bayes' strategy**. The expected opportunity losses of the 8 strategies in the A. B. Westerhoff Company problem as calculated from equation 16.2 are given in Table 16-10. The optimal strategy, or the one with the lowest expected opportunity loss, is s_4. An asterisk has been placed beside the

TABLE 16-10

Expected opportunity losses of the 8 strategies in the A. B. Westerhoff Company problem ($000)

Strategy	s_1	s_2	s_3	s_4	s_5	s_6	s_7	s_8
EOL ($)	120.00	30.00	113.60	23.60*	146.40	56.40	140.00	50.00

* The minimum loss figure

$23.60 expected opportunity loss associated with decision rule s_4 to indicate that it is the minimum loss figure.

Summary   In summary, using normal-form analysis, the optimal strategy in this problem is $s_4(a_1, a_2, a_2)$; this strategy has an expected opportunity loss of $23.60 (thousands of dollars). Referring to Figure 16-2, we see that this is exactly the same solution arrived at by extensive-form analysis. In that figure, the $23.6 is shown at node (c) and the optimal strategy (a_1, a_2, a_2) is determined by noting for the survey outcomes X_1, X_2, and X_3 the forks that have not been blocked off.

We can now summarize the steps involved in a normal form-analysis, in which sample evidence is obtained from a single sample (experiment) as in the foregoing example.

Normal-form analysis   1. List all possible strategies in terms of actions to be taken upon observation of sample outcomes.

2. Calculate the conditional expected opportunity loss (risk) for each state of nature. The probabilities used in this calculation are conditional probabilities of sample outcomes, given states of nature.

3. Compute the (unconditional) expected opportunity loss of each strategy by weighting the conditional expected opportunity losses by the prior probabilities of states of nature.

4. Select the strategy that has the minimum expected opportunity loss.

This summary of normal-form analysis has been given in terms of a single experiment. If more than one experiment is conducted, the decision maker should carry out steps (1) through (4) and then select the experiment that yields the lowest expected opportunity loss. Furthermore, the summary has been expressed in terms of opportunity losses. If payoffs of utility or profits were used, the same procedures would be followed except that the decision maker would maximize expected utility or profit rather than minimize expected opportunity loss.

16.4
COMPARISON OF EXTENSIVE-FORM
AND NORMAL-FORM ANALYSES

As we have seen, the extensive and normal forms of analysis are equivalent approaches. In both procedures, an expected opportunity loss (or expected profit) is calculated before experimental results are observed. This expected payoff anticipates the selection of optimal acts after the observation of experimental outcomes. However, the two types of analysis differ in the way in which the various components of the procedures are performed. These differences give rise to advantages and disadvantages in the two forms of analysis.

We say in the A. B. Westerhoff Company example that the extensive-form solution can be calculated more rapidly. This follows from the fact that

> *in the extensive-form approach, it is not necessary to carry out expected loss calculations for every possible rule.*

Because nonoptimal courses of action posterior to the observation of sample evidence are blocked off, only the expected loss for the optimal strategy need be determined. In some problems, the number of decision rules or strategies that must be evaluated in a normal-form analysis may be large. For example, in the problem discussed in this chapter, the number of possible strategies was $2^3 = 8$, because there were two acts and three experimental outcomes. If there had been 3 acts ($n = 3$) and 4 experimental outcomes ($r = 4$), normal-form calculations would be required for $n^r = 3^4 = 81$ strategies.

On the other hand, the normal form of analysis may appeal to a decision maker who feels uneasy about making the subjective probability assessments involved in preposterior analysis. In many problems, the conditional probabilities of sample outcomes, given states of nature, may be based on relative frequencies, as in the A. B. Westerhoff Company case, or on an appropriate probability distribution, as in a problem discussed later in this chapter. However, the prior probabilities $P(\theta_i)$, will most likely represent subjective or judgmental assignments in most problems. In normal-form analysis, these prior probabilities are applied as the last step in the calculations. Therefore, it is possible to proceed all the way to this point without introducing subjective probability assessments. We can then judge how sensitive the choice of the optimal strategy is to the prior probability assignments. That is, we can determine by how much the magnitudes of the prior probabilities may be permitted to change without a shift occurring in the optimal decision rule to be employed. This procedure is discussed in the next section.

EXERCISES 16.4

1. Using the information in exercise 4 of section 16.2 (page 763), construct a table showing all possible strategies. Select the optimal strategy if the two items are inspected.

2. From the data in exercise 7 of section 16.2 (page 765), the following strategy table has been constructed. The two alternative courses of action are

   a_1: Manufacture Crunchie Munchies

   a_2: Do not manufacture Crunchie Munchies

Fill in those strategies that are missing and select the optimal strategy.

Sample Outcome	s_1	s_2	s_3	s_4	s_5	s_6	s_7	s_8
X	a_1	a_1	a_1	a_1	a_2	a_2	a_2	a_2
Y		a_1	a_2	a_2		a_2	a_1	a_1
Z		a_2	a_1	a_2		a_1	a_2	a_1

3. Exercise 8 of section 16.2 (page 765) contains 27 strategies. Assume that the optimal strategy is one of the following three:

Model Indication	s_1	s_2	s_3
High	M	H	H
Medium	M	M	M
Low	L	M	L

The letters L, M, and H mean low, medium, and high market pricing plans used, respectively. Compute conditional expected losses and expected opportunity losses to determine which of these three strategies is best.

4. The following table pertains to exercise 11 of section 16.2 (page 767). The two courses of action are

a_1: Add route

a_2: Do not add route

Fill in the missing strategies. Compute the conditional expected payoffs as well as the expected payoffs for strategies s_1 and s_4. Which strategy is optimal?

Sample Outcome	s_1	s_2	s_3	s_4
Fewer than 25 people	a_1	a_1	a_2	a_2
25 people or more		a_1	a_1	a_2

5. A certain decision situation has three states of nature, θ_1, θ_2, and θ_3; two actions, a_1 and a_2; and three forecast outcomes, X, Y, and Z. The opportunity loss table is

State of Nature	a_1	a_2
θ_1	0	1,300
θ_2	300	0
θ_3	800	0

A forecasting model has been developed that has the following reliability:

State of Nature	Model Indication		
	X	Y	Z
θ_1	0.60	0.20	0.20
θ_2	0.10	0.50	0.40
θ_3	0.10	0.10	0.80

Of the eight strategies, only the following two are plausible: $s_3(a_1, a_2, a_2)$ and $s_5(a_1, a_1, a_2)$. Use normal-form analysis to determine the optimal strategy. In computing the conditional expected losses for the two strategies, multiply the loss of taking the wrong act by the probability of taking the wrong act. The prior probabilities for θ_1, θ_2, and θ_3 are 0.35, 0.30, and 0.35, respectively.

16.5
SENSITIVITY ANALYSIS

In the new product decision problem discussed in this chapter, only two decision rules appeared reasonable on an intuitive basis, namely, s_2 and s_4. We saw in Table 16-10 that these strategies yielded the two lowest expected opportunity losses, with figures of \$30.00 and \$23.60, respectively, for s_2 and s_4. However, we can observe in Table 16-7 that although the conditional expected loss for s_2 is lower than s_4 if θ_1 occurs (\$24.00 versus \$56.00, respectively), the opposite is true if θ_2 occurs (\$32.00 versus \$12.80, respectively). Because these conditional expected losses were weighted by prior probabilities of states of nature, $P(\theta_1)$ and $P(\theta_2)$, clearly the choice between strategies s_2 and s_4 is dependent on the magnitudes of these prior probabilities. We can test how *sensitive* the choice between decision rules s_2 and s_4 is to the size of these prior probabilities.

> The **sensitivity analysis** tests how sensitive the solution to a decision problem is to changes in the data for the variables of the problem.

In the present problem, a sensitivity test can be accomplished by solving for a *break-even value for the prior probability of one of the two states of nature* such that the expected opportunity losses of the two strategies will be equal. If the prior probability rises above this break-even value, s_2 is the optimal act rather than s_4. We illustrate this procedure in the new product decision problem. As indicated earlier, the (unconditional) expected opportunity losses of the two strategies in question were computed as follows:

$$\text{EOL}(s_2) = (0.25)(\$24.00) + (0.75)(\$32.00) = \$30.00$$

$$\text{EOL}(s_4) = (0.25)(\$56.00) + (0.75)(\$12.80) = \$23.60$$

If we now substitute p for the 0.25 value of $P(\theta_1)$, and $1 - p$ for the 0.75 figure of $P(\theta_2)$, we can solve for the value of p for which $\text{EOL}(s_2) = \text{EOL}(s_4)$. Making this substitution and equation the resulting expressions for expected opportunity loss, we have

$$p(\$24.00) + (1 - p)(\$32.00) = p(\$56.00) + (1 - p)(\$12.80)$$

Carrying out the multiplications, we obtain

$$\$24.00p + \$32.00 - \$32.00p = \$56.00p + \$12.80 - \$12.80p$$

Collecting terms, we find the break-even value of p.

$$\$51.20p = \$19.20$$

$$p = \frac{19.20}{51.20} = 0.375$$

Summary   In summary, we conclude that if $p = P(\theta_1) = 0.375$, then the expected opportunity losses of strategies s_2 and s_4 would be equal, and we would have no preference between them. In the A. B. Westerhoff Company new product decision problem, $p = P(\theta_1) = 0.25$. Hence, we observe that this subjective prior probability assignment could have varied up to a value of 0.375 and s_4 would still have been a better strategy than s_2. However, for p values in excess of 0.375, s_2 has a lower expected opportunity loss and is therefore the better rule.

Of course, break-even values of p could be determined between s_4 and other strategies as well. In the case of a strategy such as s_5, whose conditional expected values given θ_1 and θ_2 are each higher than the corresponding figures for s_4 ($\$144.00$ versus $\$56.00$ and $\$147.20$ versus $\$12.80$, respectively), it is impossible for the weighted average or expected opportunity loss to be less than that of s_4, regardless of the weights p and $1 - p$. In this situation, the strategy s_4 is said to *dominate* s_5. As a general definition, a strategy, say s_1, is a **dominating strategy** with respect to s_2 if

Dominating strategy $$R(\theta|s_1) \leqslant R(\theta|s_2) \qquad \text{for all values of } \theta$$

and

$$R(\theta|s_1) < R(\theta|s_2) \qquad \text{for at least one value of } \theta$$

In words, s_1 is said to dominate s_2 if the conditional expected losses (risks) for s_1 are *equal to or less than* the corresponding figures for s_2 for every state of nature *and* if the conditional expected loss for s_1 is *less than* the corresponding figure for s_2 for at least one value of θ.

The preceding discussion illustrated the use of sensitivity analysis to determine the effects of variations in prior probabilities on the selection of a best act. We could also determine the effects of changes in payoff matrix entries on our choice of a best act.

> In general, in any decision analysis problem, it is useful to test how sensitive the solution is to changes in all of the important variables of the problem.

In problems involving numerous states of nature, experimental outcomes, and courses of action, the analyses may require many calculations. The use of computers in such situations is usually a practical necessity.

16.6
AN ACCEPTANCE SAMPLING EXAMPLE

As another illustration of preposterior analysis, we will return to the problem of acceptance sampling of a manufactured product discussed in section 15.2. Since the problem was posed in posterior analysis, we must make some changes to convert it to a problem in preposterior analysis. We assume, as previously, that the Renny Corporation inspects incoming lots of articles produced by a supplier in order to determine whether to accept or reject these lots. We also retain the assumptions that in the past, incoming lots from this supplier have contained 10%, 20%, or 30% defectives and that lots with these percentages of defectives have occurred 50%, 30%, and 20% of the time, respectively. The same payoff matrix as shown in Table 15-4 is assumed. Table 15-5 showed the calculation of prior expected opportunity losses for the two acts,

$$a_1: \text{Reject the incoming lot}^1$$

$$a_2: \text{Accept the incoming lot}$$

if action were taken without sampling. For convenience, this information on events, prior probabilities, payoffs, and prior expected opportunity losses is summarized in Table 16-11. The preferable action without sampling is act a_2, "accept the incoming lot," which has a prior expected opportunity loss of $70 as opposed to act a_1, "reject the incoming lot," which has a prior expected opportunity loss of $100.

We would like to know whether it is better to make the decision concerning acceptance or rejection of the incoming lot without sampling or to draw and inspect a random sample of items from the incoming shipment and then make the decision. We recognize this problem as one of preposterior analysis, specifically one that requires an evaluation of the *expected value of sample information* (EVSI) and the *expected net gain of sampling* (ENGS). We consider drawing a sample of two articles without replacement and inspecting these two articles at a cost of $5. Hence, our problem is to choose between a terminal decision without sampling and a decision to examine a fixed-size sample of two articles and then select a terminal act.

Deciding whether or not to sample

---

[1] The alternative actions have been denoted here by lower-case rather than capital letters to maintain consistency with the notation we have used in specifying strategies.

TABLE 16-11
Basic data for acceptance sampling problem for the Renny Corporation

Event p (lot proportion of defectives)	Prior Probability $P_0(p)$	Opportunity Loss	
		a_1 Reject	a_2 Accept
0.10	0.50	$200	$ 0
0.20	0.30	0	100
0.30	0.20	0	200
	1.00		

Prior $\text{EOL}(a_1) = (0.50)(\$200) + (0.30)(\$0) + (0.20)(\$0) = \$100$

Prior $\text{EOL}(a_2) = (0.50)(\$0) + (0.30)(\$100) + (0.20)(\$200) = \$70$

Of course, a sample of only two articles may be entirely too small to be realistic in many cases. On the other hand, in the case of certain manufactured complex assemblies of large unit cost where the testing procedure is destructive (for example, in missile testing or in space probing where the test vehicle is not retrievable), it may only be feasible to test a small number of items. However, the principles illustrated in this problem are perfectly general. The assumption of a larger sample size would merely increase the computational burden. In section 16.7, we will consider appropriate decision theory procedures for determining an optimal sample size.

As in the preceding example in this section, we will solve the problem in two different ways, first through the use of extensive-form analysis, and then by normal-form analysis.

Extensive-form Analysis

The decision diagram for the preposterior analysis of the Renny Corporation problem is given in Figure 16-3. As indicated in the legend, the alternative actions in this problem have been denoted

$$a_1: \text{Reject the lot}$$

$$a_2: \text{Accept the lot}$$

The possible sample outcomes based on a random sample of two articles drawn from the incoming lot are denoted

$$X = 0 \text{ defectives}$$

$$X = 1 \text{ defective}$$

$$X = 2 \text{ defectives}$$

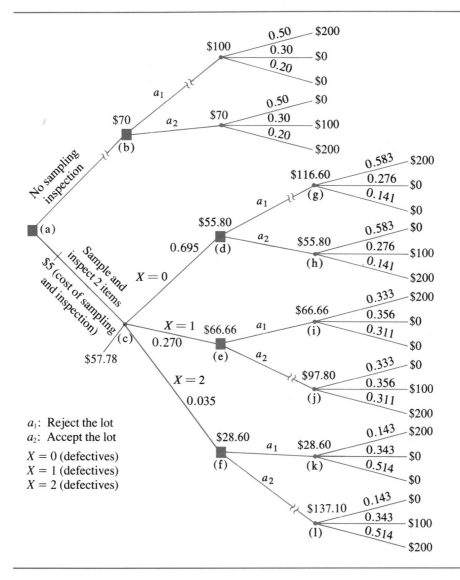

FIGURE 16-3
Decision diagram for
preposterior analysis of the lot
acceptance decision for the
Renny Corporation problem

At node (a), the two basic choices are shown: Not to engage in sampling *Two choices* inspection, or to sample and inspect two articles from the incoming lot. The prior analysis for the "no sampling" choice is shown in the upper portion of the tree. As previously observed in Table 16-11, the prior expected opportunity loss of rejection, a_1, is \$100 and that of acceptance, a_2, is \$70. Hence, the non-optimal act a_1 has been blocked off on the branch emanating from node (b) and the \$70 figure for a_2 has been entered at node (b). This \$70 figure is the expected opportunity loss of choosing the optimal act without sampling.

The alternative choice   The alternative choice is to sample and inspect two items prior to making the terminal decision of acceptance or rejection of the lot. To carry out the calculations for this part of the analysis, we begin at the right tips of the tree and proceed inward by backward induction. The expected values of $116.60 and $55.80 shown at nodes (g) and (h), for example, are the expected opportunity losses of acts a_1 and a_2, respectively, after observing no defectives ($X = 0$) in the sample of two articles. The probabilities used to calculate these expected values are the conditional (posterior) probabilities, given $X = 0$, that the lot proportion of defectives, p, is 0.10, 0.20, and 0.30, respectively. In symbols,

$$P(p = 0.10 | X = 0) = 0.583$$

$$P(p = 0.20 | X = 0) = 0.276$$

$$P(p = 0.30 | X = 0) = 0.141$$

The calculation of these posterior probabilities, shown in Table 16-14, will be explained shortly. Since acceptance of the lot (act a_2) has a lower expected loss than does rejection (act a_1), given that no defectives have been observed in the sample, the lower figure, $55.80, is entered at node (d). Furthermore, act a_1 is blocked off as nonoptimal. Corresponding figures of $66.66 and $28.60 for expected losses of optimal acts after observing one and two defectives are shown at nodes (e) and (f), respectively. Rejection, act a_1, is shown to be optimal after observation of either one or two defectives. Weighting the three payoffs shown at nodes (d), (e), and (f) by the marginal probabilities of observing zero, one, and two defectives (0.695, 0.270, and 0.035, respectively), we find a figure of $57.78 as the expected opportunity loss of sampling and inspecting two articles and taking terminal action after observing the sample evidence. Before proceeding with the calculation of EVSI and ENGS, let us consider the method of calculating the marginal and posterior probabilities mentioned in this paragraph.

Obtaining the joint   In order to calculate the marginal probabilities of the sample evidence,
probability distribution $P(X = 0)$, $P(X = 1)$, and $P(X = 2)$, and the posterior probabilities of the form
of the sample evidence $P(p | X)$, we must obtain the joint probability distribution of the sample evidence
and events   and events, that is, $P(X$ and $p)$, for $X = 0$, 1, and 2, and $p = 0.10$, 0.20, and 0.30. As in the A. B. Westerhoff Company problem, these joint probabilities are obtained by multiplying prior probabilities of events—in this case, $P_0(p)$— by the appropriate conditional probabilities of sample evidence, given events— in this case, $P(X | p)$. In the A. B. Westerhoff Company problem, we assumed that these P(sample outcomes|events) values were based on past relative frequencies of occurrence in surveys previously conducted by a market research firm. In the present problem, we must calculate these $P(X | p)$ values by using an appropriate probability distribution. Let us assume that the incoming lot from which the sample of two articles is drawn is large relative to the sample size. Then, as indicated in section 3.4, we can assume that even if the sample is selected without replacement, the drawings of the articles may be considered trials of a Bernoulli process, and the binomial distribution may be used to calculate probabilities of sample outcomes. That is, we may assume that since

TABLE 16-12
Conditional probabilities of specified numbers of defectives in a sample of 2 articles from an incoming lot in the Renny Corporation problem

$P(X=0\|p=0.10)=(0.90)^2$	$=0.81$		$P(X=0\|p=0.20)=(0.80)^2$	$=0.64$	
$P(X=1\|p=0.10)=2(0.90)(0.10)$	$=0.18$		$P(X=1\|p=0.20)=2(0.80)(0.20)$	$=0.32$	
$P(X=2\|p=0.10)=(0.10)^2$	$=0.01$		$P(X=2\|p=0.20)=(0.20)^2$	$=0.04$	
	$\overline{1.00}$			$\overline{1.00}$	

$$P(X=0\|p=0.30)=(0.70)^2 = 0.49$$
$$P(X=1\|p=0.30)=2(0.70)(0.30)=0.42$$
$$P(X=2\|p=0.30)=(0.30)^2 = 0.09$$
$$\overline{1.00}$$

there is so little change in the population (lot) because of the drawing of the first article, the probability of obtaining a defective item on the second draw is the same as on the first draw. If the incoming lot is not large relative to the sample size—that is, if it is not at least 10 times the sample size—then the hypergeometric distribution is appropriate. The $P(X\|p)$ values for the present problem (that is, the conditional probabilities of observing zero, one, or two defectives, given lot proportion defectives of 0.10, 0.20, and 0.30 calculated by the use of the binomial distribution) are shown in Table 16-12. As indicated in the table, the probabilities of zero, one, and two defectives in a sample of size 2 from a lot that contains 0.10 defectives are given by the respective terms of the binomial distribution whose parameters are $n=2$, $p=0.10$, and corresponding calculations provide the probabilities for the cases in which $p=0.20$ and 0.30.

Table 16-13 shows the computation of the joint probability distribution $P(X \text{ and } p)$. This joint distribution is obtained by multiplying the prior probabilities, $P_0(p)$, by the conditional probabilities, $P(X\|p)$, derived in Table 16-12. For example, 0.405, the upper left entry in the joint probability distribution in

Computation of the joint probability distribution P(X and p)

TABLE 16-13
Calculation of the joint probability distribution of sample outcomes and events in the Renny Corporation problem

Event p (lot proportion of defectives)	Prior Probability $P_0(p)$	Conditional Probability $P(X\|p)$			Joint Probability $P(X \text{ and } p)$			
		$X=0$	$X=1$	$X=2$	$X=0$	$X=1$	$X=2$	Total
0.10	0.50	0.81	0.18	0.01	0.405	0.090	0.005	0.50
0.20	0.30	0.64	0.32	0.04	0.192	0.096	0.012	0.30
0.30	0.20	0.49	0.42	0.09	0.098	0.084	0.018	0.20
					$\overline{0.695}$	$\overline{0.270}$	$\overline{0.035}$	$\overline{1.00}$

Table 16-13, was obtained by multiplying 0.50 by 0.81. In symbols, $P(X = 0$ and $p = 0.10) = P_0(p = 0.10)P(X = 0|p = 0.10)$. We can now obtain the marginal probabilities of the sample outcomes of zero, one, and two defectives. The column totals in the joint probability distribution are $P(X = 0) = 0.695$, $P(X = 1) = 0.270$, and $P(X = 2) = 0.035$, respectively. These marginal probabilities are entered in the decision diagram in Figure 16-3 on the branches emanating from node (c).

The calculations of posterior, or revised, probabilities of lot proportion of defectives, given the sample outcomes $X = 0$, 1, and 2, are shown in Table 16-14. These posterior probabilities were obtained by dividing the joint probabilities by the appropriate column totals, $P(X)$. For example, the probability that the incoming lot contains 0.10 defectives, given that no defectives were observed in the sample of two articles, is

$$P(p = 0.10|X = 0) = \frac{P(X = 0 \text{ and } p = 0.10)}{P(X = 0)} = \frac{0.405}{0.695} = 0.583$$

As previously indicated in connection with Table 16-4, the calculation of these posterior probabilities represents an application of Bayes' theorem. The posterior probabilities are shown in Figure 16-3 on the branches stemming from nodes (g), (h), (i), (j), (k), and (l).

We can now complete the extensive-form analysis. We noted earlier that the $70 at node (b) in Figure 16-3 represents the expected loss of terminal action without sampling. Correspondingly, the $57.78 at node (c) is the expected loss of sampling and inspecting two articles and then taking terminal action. Hence, the *expected value of sample information* is

$$\text{EVSI} = \$70 - \$57.78 = \$12.22$$

The expected amount by which the terminal opportunity loss of action without sampling is reduced by sampling two articles and taking action after inspection of the sample is $12.22. Since the sampling and inspection of two articles costs $5.00, the *expected net gain of sample information* is

$$\text{ENGS} = \$12.22 - \$5.00 = \$7.22$$

TABLE 16-14

Calculation of the posterior probabilities of lot proportion of defectives in the Renny Corporation problem

$P(p = 0.10|X = 0) = \dfrac{0.405}{0.695} = 0.583$	$P(p = 0.10|X = 1) = \dfrac{0.090}{0.270} = 0.333$	$P(p = 0.10|X = 2) = \dfrac{0.005}{0.035} = 0.143$
$P(p = 0.20|X = 0) = \dfrac{0.192}{0.695} = 0.276$	$P(p = 0.20|X = 1) = \dfrac{0.096}{0.270} = 0.356$	$P(p = 0.20|X = 2) = \dfrac{0.012}{0.035} = 0.343$
$P(p = 0.30|X = 0) = \dfrac{0.098}{0.695} = \dfrac{0.141}{1.000}$	$P(p = 0.30|X = 1) = \dfrac{0.084}{0.270} = \dfrac{0.311}{1.000}$	$P(p = 0.30|X = 2) = \dfrac{0.018}{0.035} = \dfrac{0.514}{1.000}$

Therefore, since the expected net gain of sample information is $7.22 for terminal action after sampling and inspecting two articles compared with terminal action without sampling, the Renny Corporation should follow the former course of action.

Normal-form Analysis

We now turn to the normal-form analysis of the Renny Corporation problem, which we have just solved by extensive-form procedures. As in the A. B. Westerhoff problem, we begin the normal-form analysis by listing all possible strategies, or decision rules. These are shown in Table 16-15. We can see on an intuitive basis that some of the strategies are illogical and would therefore tend to have relatively large expected opportunity losses associated with them. For example, strategy $s_2(a_1, a_1, a_2)$ rejects the incoming lot if the sample of two articles yields zero or one defective, but accepts the lot if two defectives are observed. Strategy $s_4(a_1, a_2, a_2)$ also employs perverse logic, since it rejects the lot on an observation of zero defectives but accepts it if one or two defectives are observed. Strategy $s_1(a_1, a_1, a_1)$ rejects the lot, and $s_8(a_2, a_2, a_2)$ accepts the lot, regardless of the sample observation. The two most logical decision rules appear to be strategy $s_5(a_2, a_1, a_1)$, which accepts the lot if zero defectives are observed in the sample and rejects the lot if one or two defectives are observed, and strategy $s_7(a_2, a_2, a_1)$, which accepts the lot if zero or one defective is observed and rejects the lot if two defectives are found.

    The next step is to calculate for each strategy the risks or conditional expected losses associated with the occurrence of each event (lot proportion of defectives). We will illustrate these calculations for strategies s_5 and s_7 by computing, for each proportion of defectives, the *loss* of taking the wrong act times the *probability* of taking the wrong act. These risks, denoted $R(s_5|p)$ and $R(s_7|p)$, are shown in Table 16-16. For example, let us consider strategy $s_5(a_2, a_1, a_1)$. Referring back to the original payoff matrix in Table 16-11, we recall that the correct course of action if $p = 0.10$ is to accept, and if $p = 0.20$ or 0.30, to reject

Normal-form Analysis

List all strategies

Select most-logical strategies

Calculate risks for each strategy

TABLE 16-15
A listing of all possible strategies in the Renny Corporation problem

Sample Outcome (number of defectives)	Strategy							
	s_1	s_2	s_3	s_4	s_5	s_6	s_7	s_8
$X = 0$	a_1	a_1	a_1	a_1	a_2	a_2	a_2	a_2
$X = 1$	a_1	a_1	a_2	a_2	a_1	a_1	a_2	a_2
$X = 2$	a_1	a_2	a_1	a_2	a_1	a_2	a_1	a_2

a_1: Reject the lot
a_2: Accept the lot

TABLE 16-16
Calculation of risks, or conditional expected opportunity losses, for strategies s_5 and s_7

Event p (lot proportion of defectives)	Strategy $s_5(a_2, a_1, a_1)$			Strategy $s_7(a_2, a_2, a_1)$		
	Opportunity Loss of Wrong Act	Probability of Wrong Act Given p	Risk (conditional expected loss) $R(s_5\|p)$	Opportunity Loss of Wrong Act	Probability of Wrong Act Given p	Risk (conditional expected loss) $R(s_7\|p)$
0.10	$200	0.19	$38	$200	0.01	$ 2
0.20	100	0.64	64	100	0.96	96
0.30	200	0.49	98	200	0.91	182

the lot. Strategy $s_5(a_2, a_1, a_1)$ accepts the lot on the observation of no defectives and rejects it otherwise. From Table 16-13 we find a 0.81 conditional probability of observing no defectives, given that the lot proportion of defectives is $p = 0.10$. Hence, with strategy s_5, the probability of making the wrong decision, given a lot proportion of defectives $p = 0.10$, is $1 - 0.81 = 0.19$. The loss associated with rejection if $p = 0.10$ is $200 (Table 16-11). Therefore, as shown in Table 16-16, the risk of strategy s_5, given $p = 0.10$, is $R(s_5|p = 0.10) = \$200 \times 0.19 = \$38$. Analogous calculations produce the other risks shown in Table 16-16. A summary of the risks for all eight strategies is presented in Table 16-17.

We can now calculate the expected loss of each strategy using equation 16.2. Adapting the notation of that equation to that used in the present problem, we obtain as the expected opportunity loss of the kth strategy

$$(16.3) \qquad \text{EOL}(s_k) = \sum_p P_0(p)R(s_k|p)$$

That is, for each strategy, we weight the conditional expected losses associated with each lot proportion of defectives (Table 16-17) by the prior probabilities

TABLE 16-17
Risks, or conditional expected opportunity losses, for the 8 strategies in the Renny Corporation problem

Event p (lot proportion of defectives)	Strategy							
	s_1	s_2	s_3	s_4	s_5	s_6	s_7	s_8
0.10	$200	$198	$164	$162	$38	$ 36	$ 2	$ 0
0.20	0	4	32	36	64	68	96	100
0.30	0	18	84	102	98	116	182	200

TABLE 16-18
Expected opportunity losses of the 8 strategies in the Renny Corporation problem

Strategy	s_1	s_2	s_3	s_4	s_5	s_6	s_7	s_8
EOL ($)	100.00	103.80	108.40	112.20	57.80*	61.60	66.20	70.00

* The minimum loss figure

of such incoming lots (Table 16-13). Hence, the expected opportunity losses (expected risks) of strategies s_5 and s_7 are

$$EOL(s_5) = (0.50)(\$38) + (0.30)(\$64) + (0.20)(\$98) = \$57.80$$

$$EOL(s_7) = (0.50)(\$2) + (0.30)(\$96) + (0.20)(\$182) = \$66.20$$

The expected opportunity losses of all eight strategies are given in Table 16-18. The optimal strategy is seen to be $s_5(a_2, a_1, a_1)$, since it has the minimum expected opportunity loss of $57.80. The strategy, which accepts the lot if no defectives are observed in the sample and rejects the lot if one or two defectives are found, is the same as the one found in the extensive-form analysis depicted in the decision diagram in Figure 16-3. The minor difference between the expected opportunity losses of the optimal strategy—$57.78 in the extensive-form analysis and $57.80 in the normal-form analysis—is attributable to rounding of decimal places.

EXERCISES 16.6

1. Trivia Press, Inc., estimates that the gain from a successful novel would be $8 million; the loss from an unsuccessful novel would be $4 million. The publisher sends a copy of the book to 10 critics for their opinions. The results are slightly unfavorable: Six out of 10 critics dislike the book. If the book should be a failure, the probability that a critic would dislike it is 0.8, and if the book should be a success, there is a 50:50 chance the critic would like it. At what value of prior probability of success would the decision concerning publication change?

2. A certain manufacturing process produces lots of 500 units each. In each lot, either 10% or 30% of the 500 items are defective. The quality control department, which inspects each lot before shipment, knows that 80% of the lots produced in the past contained 10% defectives. If accepted for shipment, the lot produces a profit of $500, but if rejected, the lot is sold for scrap at cost. If a 30% defective lot is sent out, the firm estimates its loss in goodwill, and hence in future orders, at $1,500. It costs $3 to test an item, but the test is not destructive (that is, the item is still good after the test and can be sold). What is the expected net gain from testing 3

items drawn at random and what is the best decision rule? Use a binomial probability distribution to compute conditional probabilities.

3. You are given the following information:

State of Nature	s_1	s_2
θ_1	50	70
θ_2	100	40

At what value of the prior probability of θ_1 would you be indifferent between the 2 strategies?

16.7
OPTIMAL SAMPLE SIZE

In the preceding acceptance sampling problem, we indicated how the expected value of sample information can be derived prior to the actual drawing of a sample. This EVSI figure was obtained by subtracting the expected opportunity loss of the best terminal act without sampling from the expected loss of the optimal strategy with sampling. More precisely, the latter value is the expected opportunity loss of a decision to sample and then take optimal terminal action after observation of the sample outcome. In that problem, we assumed a fixed sample size of two articles. However, as previously indicated (page 780), the method of analysis presented is a general one, and the only practical effect of an assumption of a larger sample size would have been an increase in the computational burden. Nevertheless, can an *optimal sample size* be derived in a problem such as the one presented? The answer is yes, and the general method for obtaining such an optimal value follows.

Obtaining an optimal sample size   As might be suspected intuitively, an increase in sample size brings about an increase in the EVSI. However, an increase in sample size also results in an increase in the cost of sampling. (The term "cost of sampling" here means the total cost of sampling and inspection.) Therefore, we would like to find the sample size for which the difference between the EVSI and the cost of sampling is the largest—this is the **optimal sample size**. An equivalent and more convenient approach is to minimize *total loss*, where the total loss associated with any sample size n is defined as

Minimizing total loss   (16.4)

$$\text{Total loss} = \frac{\text{Cost of}}{\text{sampling}} + \frac{\text{Expected opportunity loss}}{\text{of the optimal strategy}}$$

Let us consider how the quantities in equation 16.4 might be calculated. The cost of sampling would ordinarily not be difficult to calculate. In many instances, this cost may be entirely variable, that is, proportional to the number of articles sampled. In that case, the cost of sampling would be equal to

(16.5)
$$C = vn$$

Cost of sampling
when cost is
proportional to
size of sample

where C = cost of sampling
   v = cost of sampling each unit
   n = number of units in the sample

In the Renny Corporation example, where the cost of sampling was $5.00 for two articles, if costs were entirely variable, the cost of sampling each unit would be $2.50. Hence, v = $2.50 and n = 2. The cost of sampling for a sample of size 10 would be C = $2.50 × 10 = $25.00, and so on.

   In certain situations, a portion of the total cost of sampling might be fixed and the remaining part variable. In that case, the cost of sampling would be

Cost of sampling
with fixed and
variable costs

(16.6)
$$C = f + vn$$

where C = cost of sampling
   f = fixed cost
   v = cost of sampling each unit
   n = number of units in the sample

Thus, for example, in a case in which the fixed cost of sampling is $10 and the variable cost per unit is $2, the cost of sampling 20 units would be

$$C = \$10 + (\$2)(20) = \$50$$

Other formulas can be derived for more complex situations.

   Turning now to the other term in equation 16.4 for **total loss**, namely, the **expected opportunity loss of the optimal strategy**, we first introduce more convenient notation. Let us assume we are dealing with an acceptance sampling problem in which the sample size is 10. Although in the normal-form analysis of the Renny Corporation problem we considered every possible strategy for the acceptance of the incoming lot, it can be shown that the only strategies worth considering as potential optimal decision rules are those that accept the incoming lot if a certain number of defectives or less are observed and reject the lot otherwise. It is conventional to refer to this critical number as "the acceptance number," denoted c. Hence, in the present problem, in which the sample size is 10, the possible values for the acceptance number c are 0, 1, 2, ..., 10. For example, if c = 0, the lot is accepted if no defectives are observed in a sample of size n = 10 and rejected otherwise. If c = 1, the lot is accepted if *one or fewer* defectives are observed and rejected otherwise. If c = 2, the lot is accepted if *two or fewer* defectives are observed and rejected otherwise, and so on. Therefore, we can characterize the optimal strategy for each sample size by two figures: c, the acceptance number, and n, the sample size. For example, the c, n pair, denoted (c, n), for an optimal strategy with an acceptance number c = 2 and sample size n = 10 would be (2, 10). In the Renny Corporation problem, the c, n pair was (0, 2) for the optimal strategy s_5.

More convenient
notation

Theoretically, we could calculate for every possible sample size the expected opportunity loss of the optimal strategy (c, n). Adding this figure to the cost of sampling, we would obtain the total loss associated with each sample size. We would then select as the optimal sample size the one that yielded the *minimum total loss*.

At first, this might seem to be an impossible procedure, because if the population is infinite, the sample size could conceivably take on any positive integral value. However, n must be a finite value because of the cost of sampling involved. The expected value of sample information (EVSI) cannot exceed the expected value of perfect information (EVPI). As we saw in section 14.5, EVPI is the expected opportunity loss of the optimal act prior to sampling. That is, it is the expected loss of the best terminal action without sampling. It would never be worthwhile to take a sample so large that the cost of sampling exceeded the EVPI. Hence, n for the optimal sample size will be a finite number.

Binomial sampling
The acceptance sampling problem we have been discussing can be described as involving **binomial sampling**. That is, the conditional probabilities of sample outcomes of the form $P(X|p)$ were calculated in the Renny Corporation problem by the binomial probability distribution. In this type of problem, the optimal strategies for the various sample sizes must be calculated by the methods we have indicated, which may be characterized as trial-and-error procedures. That is, no simple general formula enables us to derive the optimal (c, n) pairs for every problem. Therefore, a considerable amount of calculation may be involved to determine the optimal sample size. For sufficiently large and important problems, the use of computers may represent the only practical method of carrying out the computations.

16.8
GENERAL COMMENTS

Preposterior analysis
In this chapter, we have discussed decision-making procedures for the selection of optimal strategies prior to obtaining experimental data or sample information. The method used, preposterior analysis, anticipates the adoption of best actions after the observation of the experimental or sample results. The general principles of this type of analysis have been discussed using two examples.

- The first illustration involved the introduction of a new product. Experimental evidence in the form of sample survey results was obtainable for revising prior probabilities of occurrence of the states of nature "successful product" and "unsuccessful product."

- The second example involved a decision concerning acceptance or rejection of an incoming lot. Here, information in the form of results of a random sample drawn from a lot was obtainable for revising prior probabilities of the states of nature "lot proportion of defectives."

In the first example, an empirical joint frequency distribution on the success or failure of past new products and survey results provided information for the calculation of probabilities required for the problem solution. In the second example, the decision depended on the proportion of defectives, denoted p. The observable sample data were represented by the random variable "number of defectives," denoted X, in a sample drawn from the incoming lot. This random variable was taken to be binomially distributed.

    In both of these problems, the states of nature and the experimental or sample evidence were discrete random variables. Another class of problems comprises situations in which the states of nature and sample evidence are represented by continuous random variables. For example, the decision may depend on a parameter μ, the mean of a population, and the observed sample evidence may be a sample mean $\bar{X}$. The mathematics required for the solution of this class of problems is outside the scope of this book. However, the same basic principles discussed in this chapter for extensive-form and normal-form analysis are applicable, regardless of whether the random variables representing states of nature and experimental outcomes are discrete or continuous.

States of nature

KEY TERMS

Dominating Strategy   a strategy s_1 dominates strategy s_2 if the risks $R(s_1|\theta_i)$ for s_1 are equal to or less than the corresponding figures for s_2 for every state of nature and the risk for s_1 is less than the corresponding figure for s_2 for at least one value of θ.

Expected Net Gain of Sample Information (ENGS)   the expected value of sample information minus the cost of obtaining the sample information.

Expected Risk (expected opportunity loss)   the expected value of conditional expected losses using prior probabilities of states as weights.

Expected Value of Sample Information (EVSI)   the difference between the expected payoff with sample information and the expected payoff without sample information.

Extensive-Form Analysis   preposterior analysis carried out in the form of a decision diagram.

Normal-Form Analysis   preposterior analysis carried out by a comparison of the net payoffs of all possible strategies.

Optimal Sample Size   a sample size that yields the minimum total loss.

Preposterior Analysis   an analysis that specifies the optimal decision rule based on the possible types of evidence that can be produced by additional information if it is collected.

Risk   conditional expected loss in a normal-form decision analysis.

Sensitivity Analysis   a test to determine how sensitive the solution to a decision problem is to changes in the data for the variables of the problem.

Terminal Decision   a decision that makes a final disposition of the choice of a best act.

KEY FORMULAS

Expected Opportunity Loss of s_k

$$\text{EOL}(s_k) = \sum_{i=1}^{m} P(\theta_i)R(s_k|\theta_i)$$

Dominating Strategy
s_1 dominates s_2 if
$R(\theta|s_1) \leqslant R(\theta|s_2)$ for all values of θ
$R(\theta|s_1) < R(\theta|s_2)$ for at least one value of θ

17

COMPARISON OF CLASSICAL
AND BAYESIAN STATISTICS

Topics in classical statistical inference and Bayesian decision theory were discussed in earlier chapters. In this chapter, we compare some aspects of classical statistics and Bayesian statistics.

17.1
COMPARING CLASSICAL AND BAYESIAN
METHODS

Classical statistics is a broad term that includes the two main topics of classical statistical inference (hypothesis testing and confidence interval estimation) as well as other topics, such as classical regression analysis discussed in Chapters 9 and 10. **Bayesian statistics** is also a broad term that analogously may be thought of as including Bayesian decision theory, Bayesian estimation, and other topics such as Bayesian regression analysis.

Although the terminologies of classical and Bayesian statistics differ, there are many similarities in the structure of the problems they address and in their methods of analysis. However, important differences, particularly in their methods of analysis, are a matter of continuing discussion and study.

To compare these two important types of statistical analysis, we consider in section 17.2 an illustrative problem that presents a comparison of classical hypothesis-testing methods and Bayesian decision theory; in section 17.3, we compare classical and Bayesian estimation procedures. In conclusion, section 17.4 presents some general comments on both the common ground and the differences between these two schools of thought.

Testing the Null Hypothesis against the Alternative Hypothesis

Classical approach

To introduce the comparison, let us consider a standard hypothesis-testing problem. Suppose we wish to test the null hypothesis, H_0: $p \leqslant p_0$ (where p_0 is a known or hypothesized population proportion), against the alternative hypothesis, H_1: $p > p_0$. For example, we might test the hypothesis that p (the proportion of defectives in a shipment of a manufactured product) is less than or equal to 0.03 against the alternative hypothesis that $p > 0.03$. Using classical hypothesis-testing methods, we could design a decision rule, which would tell us whether to accept or to reject the null hypothesis on the basis of a random sample drawn from the shipment. We would fix α, the desired maximum probability of making a Type I error, and (using the power curve) we could determine the risks of making Type II errors for values of p for which the alternative hypothesis H_1 is true. Table 17-1 summarizes the relationship between actions concerning these hypotheses and the truth or falsity of the hypotheses. For convenience, the table is given in terms of the null hypothesis H_0. However, it is understood that when H_0 is true, H_1 is false and when H_0 is false, H_1 is true. We refer to the truth or falsity of H_0 as the prevailing "state of nature." As indicated in the column headings of the table, the symbols a_1 and a_2 denote the actions "accept H_0" and "reject H_0," respectively.

We see that this hypothesis-testing problem includes three components.

1. states of nature representing the truth or falsity of the null hypothesis
2. actions a_1 and a_2, which accept or reject the null hypothesis
3. sample or experimental data, which—when examined in the light of a decision rule—lead to one of the actions a_1 or a_2

Bayesian approach

Let us rephrase the example in terms of Bayesian decision theory. We are dealing with a two-action problem involving acts a_1 and a_2, where the states of nature are the possible values of the proportion of defectives p. Although p

---

TABLE 17-1

Relationships between actions concerning a null hypothesis and the truth or falsity of the hypothesis

State of Nature	Action Concerning the Null Hypothesis	
	a_1: Accept H_0	a_2: Reject H_0
H_0 is true	No error	Type I error
H_0 is false	Type II error	No error

TABLE 17-2
Payoff table in terms of opportunity losses for the two-action problem to accept or reject a shipment

State of Nature	Act	
	a_1	a_2
θ_1	0	$L(a_1\|\theta_1)$
θ_2	$L(a_1\|\theta_2)$	0

varies along a continuum, and may be considered a continuous random variable, we assume for comparative purposes that only two states of nature are distinguished, namely, $\theta_1: p \leqslant 0.03$ and $\theta_2: p > 0.03$. Hence, θ_1 and θ_2 correspond to truth and falsity of H_0, respectively, in the classical hypothesis-testing problem. Finally, a random sample can be drawn from the shipment, and the observed sample or experimental data can be used to help choose the better action of a_1 and a_2. Therefore, the same three components of the decision-theory problem are present as were discussed in the hypothesis-testing problem.

1. states of nature
2. alternative actions
3. experimental data that aid in the choice of actions

Furthermore, Table 17-2, a payoff table for this problem in terms of opportunity losses, is similar to Table 17-1. The symbols $L(a_2|\theta_1)$ and $L(a_1|\theta_2)$ denote the opportunity loss of action a_2 given that θ_1 is the true state of nature and a_1 given that θ_2 is the true state of nature. The zeros in the other two cells of the table indicate no opportunity loss when the correct action is taken for the specified states of nature. Actually, payoffs would ordinarily be treated as a function of p and would vary with p. However, we have assumed in this discussion that only two states of nature are distinguished.

Difference in approach

Now, let us consider the difference in the two approaches. In hypothesis testing, the choice of α, the significance level, establishes the decision rule and is the overriding feature in the choice between alternative actions. In symbols, $\alpha = P(a_2|H_0$ is true). That is, α is the conditional probability of rejecting the null hypothesis given that it is true. Hence, a major criterion of choice among actions in hypothesis testing is the relative frequency of occurrence of this type of error. But how is α chosen? In many applications, conventional significance levels such as 0.05 and 0.01 are used uncritically with little or no thought given to underlying considerations. However, it would be unfair to criticize a methodological approach simply because there are misuses of it.

Classical statistics

In classical statistics, the investigator must consider the relative seriousness of both Type I and Type II errors in establishing alternative hypotheses and significance levels at which these hypotheses are to be tested.

Also, the investigator is aided by prior knowledge concerning the likelihood that H_0 and H_1 are true. For example, in the problem just discussed, why did the investigator not set up the null hypothesis as, say, $H_0: p \leqslant 0.001$ or $H_0: p \leqslant 0.60$? In this particular acceptance sampling problem, we may know that two hypotheses such as these would be utterly ridiculous because of the extremely low and extremely high proportion of defectives implied. We may be virtually certain that the first of these hypotheses is false and that the second is true. Consequently, it would not be useful to set up the hypotheses in these forms.

Prior knowledge concerning the likelihood of truth of the competing hypotheses also helps the investigator in establishing the significance level. Hence, if it is considered likely that the null hypothesis is true, we will tend to set α at a low figure in order to maintain a low probability of erroneously rejecting that hypothesis.

Bayesian decision theory

> Advocates of Bayesian decision theory criticize the classical hypothesis-testing procedures for informality and for excessive reliance on unaided intuition and judgment.

The Bayesians argue that the structure of their decision theory represents a logical extension of classical hypothesis testing, since it explicitly provides for the assignment of prior probability distributions to states of nature and incorporates losses into the formal structure of the problem. These decision theorists contend that losses should be considered in a classical hypothesis testing in evaluating the relative seriousness of Type I and Type II errors. But how can losses be considered if no explicit loss function is formulated?

The following acceptance sampling problem compares the Bayesian approach with classical hypothesis testing. The problem demonstrates that non-optimal decisions may be made if tests of hypotheses are conducted in the usual manner (establishing decision rules of rejecting or failing to reject hypotheses at preset levels of significance).

17.2
A COMPARATIVE PROBLEM

Let us assume a situation in which a company inspects incoming lots of articles produced by a particular supplier. Acceptance sampling inspection is carried out to decide whether to accept or reject these incoming lots by selecting a random sample of n articles from each lot. As in previous problems of this type, we make the simplifying assumption that only a few levels of proportion of defectives are possible, in this case, 0.02, 0.05, and 0.08. On the basis of an analysis of past costs, the company constructed the payoff table in terms of opportunity losses depicted in Table 17-3. From long experience, the firm has

TABLE 17-3
Payoff matrix showing opportunity losses for actions of acceptance and rejection

State of Nature (p = lot proportion of defectives)	Prior Probability	Act	
		a_1 Reject	a_2 Accept
0.02	0.50	$200	$ 0
0.05	0.25	0	300
0.08	0.25	0	500
	1.00		

determined that lots containing 0.02 defectives are "good" and should be accepted. Hence, as indicated in Table 17-3, "accept" is the best act in the case of a 0.02 defective lot, and the opportunity loss is $0. On the other hand, "reject" is the optimal act for lots containing 0.05 and 0.08 defectives, and correspondingly the opportunity loss is $0 for such correct action.

On the basis of past performance, it has been determined that half of this supplier's lots are 2% defective, a fourth are 5% defective, and a fourth are 8% defective. In the absence of any further information, these past relative frequencies are adopted as the prior probabilities that such lots will be submitted by the supplier for acceptance or rejection.

To compare the approaches of Bayesian decision theory and traditional *Comparison procedure* hypothesis testing, we will first carry out a study of possible single sampling plans by extensive- and normal-form preposterior analyses. The results of these analyses will determine the optimal sampling plan or strategy. Then a hypothesis-testing solution will be given, and a comparison will be made of the two approaches.

A decision tree diagram is given in Figure 17-1, beginning with the decision *Sampling plan* to sample and inspect n items. We move to branch point (b), where the results of the sampling inspection determine which branch to follow. The possible results of sampling have been classified into three categories. The number of defectives, denoted X, may have been less than or equal to some number c_1, where $c_1 < n$. It may have been greater than c_1 but less than or equal to c_2, where $c_1 < c_2 < n$. Finally, the number of defectives may have been greater than c_2. These three types of results, for purposes of brevity, are referred to as type L (low), type M (middle), and type H (high) information, respectively. In Table 17-4, a joint frequency distribution is given for sample results and states of nature. We will assume that these frequencies were derived from a large number of past observations and therefore may represent probability in the relative frequency sense.[1] For example, in 0.42 of the lots inspected in the past,

---

[1] Alternatively, the conditional probabilities of sample results (likelihoods) derivable from this table may be thought of as having been calculated from an appropriate probability distribution such as the binomial or hypergeometric distribution. The basic methodological discussion remains unchanged.

FIGURE 17-1

Decision tree diagram for the
acceptance sampling problem

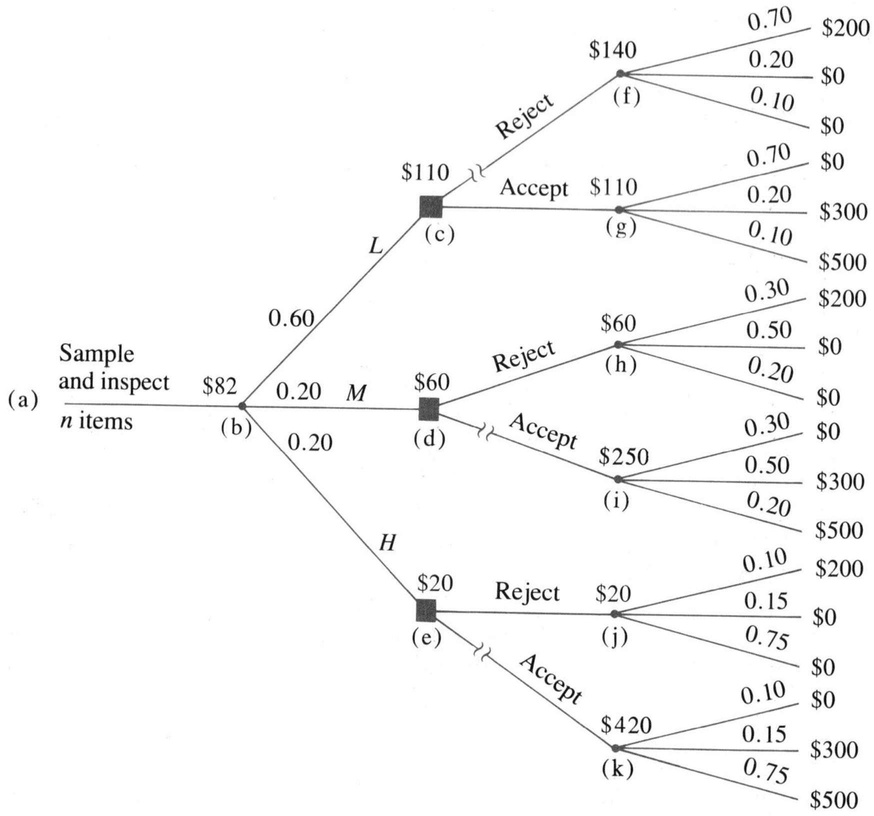

the number of defectives observed was c_1 or less, and the lots contained 0.02
defectives. In terms of marginal frequencies, type L information $(X \leqslant c_1)$ was
observed in 0.60 of the lots, type M $(c_1 < X \leqslant c_2)$ was observed in 0.20 of the
lots, and type H $(X > c_2)$ was observed in 0.20 of the lots.

Returning to the decision tree, we find the three branches representing L,
M, and H types of information emanating from node (b) marked with their respective probabilities, 0.60, 0.20, and 0.20. We will give a brief explanation of

Extensive-form analysis   the extensive-form analysis, using type L information as an example. If type
L information is observed, we move to branch point (c), where we can either
accept or reject the lot. If we reject, we move to node (f), if we accept, to node
(g). The probabilities shown on the three branches stemming from (f)—that is,
0.70, 0.20, and 0.10—are the posterior probabilities, given type L information,
that the lots contain 0.02, 0.05, and 0.08 defectives, respectively. The calculation
of these posterior probabilities by Bayes' theorem is given in Table 17-5. These

TABLE 17-4

Joint frequency distribution of sample results and states of nature

State of Nature $(p =$ lot proportion of defectives)	Sample Results			
	Type L $X \leqslant c_1$	Type M $c_1 < X \leqslant c_2$	Type H $X > c_2$	Total
0.02	0.42	0.06	0.02	0.50
0.05	0.12	0.10	0.03	0.25
0.08	0.06	0.04	0.15	0.25
	0.60	0.20	0.20	1.00

probabilities can also be derived from Table 17-4 by dividing joint probabilities by the appropriate marginal probabilities, for example, $0.70 = (\frac{0.42}{0.60})$.

We now use the standard backward induction technique (see section 14.6) to obtain the expected opportunity loss of the optimal strategy. Looking forward from node (f) and using the posterior probabilities 0.70, 0.20, and 0.10 as weights attached to the three states of nature (0.02, 0.05, and 0.08 defective lots), we obtain an expected opportunity loss of $140 for the act "reject." Comparing this figure with the corresponding $110 figure for "accept," we block off the action "reject" as nonoptimal. Therefore, $110 is entered at node (c), representing the payoff for the optimal act upon observing type L $(X \leqslant c_1)$ information. Similar calculations yield $60 and $20 at (d) and (e) for types M and H information. Weighting these three payoffs by the marginal probabilities of obtaining types L, M, and H information, we obtain a loss of $82 as the expected payoff of sampling and inspecting n items. The cost of sampling and inspection would then have to be added if, for example, we wished to make a comparison with the expected loss of terminal action without sampling. However, for our comparison of the Bayesian decision theory approach with hypothesis testing, we will focus attention on the $82 figure, which has been entered at node (b).

Backward induction technique

TABLE 17-5

Calculation of posterior probabilities of states of nature

State of Nature $(p =$ lot proportion of defectives)	Type L Information $(X \leqslant c_1)$			
	Prior Probability $P_0(p)$	Conditional Probability $P(L|p)$	Joint Probability $P_0(p)P(L|p)$	Posterior Probability $P_1(p)$
0.02	0.50	0.84	0.42	0.70
0.50	0.25	0.58	0.12	0.20
0.08	0.25	0.24	0.06	0.10
			0.60	1.00

TABLE 17-6
Possible decision rules based on information derived from single samples of size n

Sample Information	Strategy							
	s_1	s_2	s_3	s_4	s_5	s_6	s_7	s_8
Type L $(X \leqslant c_1)$	R	A	A	A	A	R	R	R
Type M $(c_1 < X \leqslant c_2)$	R	A	A	R	R	A	R	A
Type H $(X > c_2)$	R	A	R	R	A	A	A	R

We note, in summary, that $82 expected loss of the optimal strategy, which accepts the lot if type L information is observed and rejects it otherwise.

Normal-form analysis   We turn now to normal-form analysis, in which all possible decision rules or strategies will be considered as a means of commenting on traditional hypothesis-testing procedures. The eight possible strategies implicit in the decision tree diagram in Figure 17-1 are enumerated in Table 17-6. (An R denotes reject; an A denotes accept.) For example, strategy s_3 means accept the lot if type L or type M information is observed, that is, if c_2 or fewer defectives are found. Strategy s_4 signifies acceptance if c_1 or fewer defectives are observed. Thus, a choice between strategies s_3 and s_4 means (in acceptance sampling terms) a selection between a single sampling plan with an acceptance number of c_2 and a plan with an acceptance number of c_1. (The conclusion of the extensive-form analysis was that s_4 is the optimal strategy, that is, s_4 has the minimum expected opportunity loss.)

Decision rules may not appear to be logical   As in previous problems, certain decision rules do not make much sense. For example, strategy s_1 would reject the lot and strategy s_2 would accept the lot, regardless of the information revealed by the sample. Strategy s_6 would reject the lot if a small number of defectives $(X \leqslant c_1)$ were observed in the sample, but would accept the lot for larger numbers of defectives $(X > c_1)$. The only strategies that appear to be at all logical are s_3 and s_4.

Hypothesis-testing procedure   Now, let us suppose this problem had been approached from the standpoint of a hypothesis-testing procedure. The two alternative hypotheses would be

$$H_0: p = 0.02$$

$$H_1: p = 0.05 \quad \text{or} \quad 0.08$$

Acceptance or rejection of the null hypothesis H_0 would mean acceptance or rejection of the lot, respectively. As indicated earlier, the company conducting the acceptance sampling wishes to accept lots that are 0.02 defective and to reject otherwise. Hence, the rejection of a good lot (one that contains 0.02 defectives) constitutes a Type I error. Let us assume that the company decides to test the null hypothesis at a 0.05 significance level. That is, the company specifies that it wants to reject lots containing 0.02 defectives no more than 5% of the time. We will examine what this selection of $\alpha = 0.05$ implies concerning the choice of a decision rule.

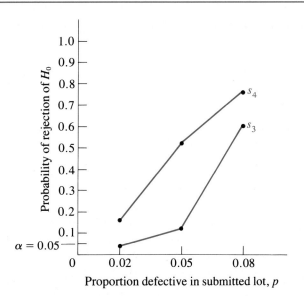

FIGURE 17-2

Power curves for strategies s_3 and s_4

Power curves may be plotted for each of the strategies or decision rules given in Table 17-6. However, we will show only the ones for s_3 and s_4, since none of the other strategies are worthy of further consideration. The power curves are depicted in Figure 17-2.   *Power curves*

Let us consider how the power curves are plotted by taking the points for $p = 0.02$ as an example. Strategy s_3 accepts H_0 (accepts the lot) if type L or type M information is observed. From Table 17-4, we find that the conditional probability of observing type L or type M information, given $p = 0.02$, is $\left(\frac{0.42}{0.50}\right) + \left(\frac{0.06}{0.50}\right) = 0.96$. Therefore, the probability that H_0 will be accepted, given $p = 0.02$, is 0.96. The probability that H_0 will be rejected, given $p = 0.02$, is $1 - 0.96 = 0.04$. Symbolically, for strategy s_3, $P(\text{Rejection of } H_0 | p = 0.02) = 0.04$. Analogously, we find that for strategy s_4, $P(\text{Rejection of } H_0 | p = 0.02) = 1 - \left(\frac{0.42}{0.50}\right) = 0.16$.

Now, if we impose the condition of a 0.05 significance level, that is, lots containing 0.02 defectives should be rejected no more than 5% of the time, we find that strategy s_3 meets this criterion but strategy s_4 does not! Therefore, traditional hypothesis-testing procedures would require the use of strategy s_3, which has been shown to be nonoptimal. Looking at Figure 17-1, we can see why this is so. Under strategy s_3, if type M information is observed, the lot must be accepted, incurring an expected loss of $250, whereas under strategy s_4, if type M information is observed, the lot is rejected, with a loss of only $60. In summary, the expected opportunity losses of the two strategies are

$$\text{EOL}(s_3) = (0.60)(\$110) + (0.20)(\$250) + (0.20)(\$20) = \$120$$

$$\text{EOL}(s_4) = (0.60)(\$110) + (0.20)(\$60) + (0.20)(\$20) = \$82$$

A major criticism

The major criticism of traditional hypothesis-testing procedures implied by this example is that too much burden is placed on significance levels as a means of deciding between alternative acts.

> The inclusion of economic costs, or more generally, opportunity losses is not a standard procedure in the decision-making process.

Another illustrative problem is given in the next section, which more thoroughly contrasts the two sets of procedures. This is followed by a general comparative discussion.

EXERCISES 17.2

1. A marketing researcher for Standard Foods Company wishes to determine whether to accept or reject a null hypothesis regarding the proportion of buyers of Instant Standard Foods Coffee that will switch to a new instant coffee, Minim. He believes the proportion that will switch, denoted p, is equal to 0.20. The alternative hypotheses are

$$H_0: p = 0.20$$

$$H_1: p = 0.25 \text{ or } 0.30$$

He takes a random sample of current Standard Foods Coffee buyers who were contacted three months later to see whether they switched. The following joint frequency distribution is obtained for the sample proportion $\bar{p}$ of switches from Standard Foods Coffee to Minim.

State of Nature	Sample Results $\bar{p} \leqslant c_1$	$c_1 < \bar{p} \leqslant c_2$	$\bar{p} > c_2$	Total
$p = 0.20$	0.40	0.06	0.02	0.48
$p = 0.25$	0.08	0.22	0.03	0.33
$p = 0.30$	0.04	0.04	0.11	0.19
Total	0.52	0.32	0.16	1.00

Of the possible strategies, the only ones that make sense are $s_3(A, R, R)$ and $s_4(A, A, R)$, where A means accept H_0 and R means reject H_0. As an example, $s_3(A, R, R)$ means accept H_0 if $\bar{p} \leqslant c_1$, but reject H_0 if $\bar{p} > c_1$. Construct power curves for both of these strategies. If the probability of a Type I error has been set at 0.05, which strategy satisfies the requirement?

2. Referring to exercise 1, suppose that the decision maker's payoffs, including the cost of sampling, are as follows:

State of Nature (p = proportion of buyers that will switch)	Accept Null Hypothesis	Reject Null Hypothesis	Prior Probability
0.20	$5,000	−$1,000	0.48
0.25	2,500	0	0.33
0.30	−2,000	2,000	0.19

a. Using the joint frequency distribution provided in exercise 1, calculate the posterior probabilities for p, given each of the sample results.

b. Construct the marketing researcher's decision tree diagram. Enter the appropriate probabilities onto the branches and place the payoffs along the respective end positions.

3. The Stalwart Appliance Center regularly inspects incoming shipments of small appliances from its supplier to regulate the acceptance of defective items. From an analysis of past costs and past performance, the company has developed the following payoff table showing prior probabilities and opportunity losses:

State of Nature (p = lot proportion of defectives)	Prior Probability	Act A_1 Reject	Act A_2 Accept
0.01	0.45	$400	$ 0
0.02	0.35	0	700
0.05	0.20	0	900

The company has set two limits, x_1 and x_2, for classification of the results of an inspection into one of three categories:

$$\text{Type } L: X \leqslant x_1$$

$$\text{Type } M: x_1 < X \leqslant x_2$$

$$\text{Type } H: X > x_2$$

where X is the number of defectives in a sample of n items. From a large number of past samples, the following results are obtained:

State of Nature (p = lot proportion of defectives)	Type L $X \leqslant x_1$	Type M $x_1 < X \leqslant x_2$	Type H $X > x_2$	Total
0.01	0.30	0.11	0.04	0.45
0.02	0.15	0.13	0.07	0.35
0.05	0.05	0.06	0.09	0.20
Total	0.50	0.30	0.20	1.00

a. Draw a decision tree for this problem and make all necessary entries. What is the optimal decision (reject or accept shipment) if type M information is obtained from a sample?

b. Construct power curves for the two strategies $s_3(A_2, A_2, A_1)$ and $s_4(A_2, A_1, A_1)$. If the probability of a Type I error has been set at 0.10, which strategy satisfies the requirement? The alternative hypotheses are

$$H_0: p = 0.01$$

$$H_1: p = 0.02 \text{ or } 0.05$$

17.3
CLASSICAL AND BAYESIAN ESTIMATION

In the preceding section, a comparison was made between hypothesis-testing procedures in classical statistical inference and the corresponding approaches in Bayesian decision theory. In this section, a comparison is made between the estimation techniques in the two approaches.

In Chapter 6, a brief description was given of classical point estimation techniques, that is, methods in which a population parameter value is estimated by a single statistic computed from the observations in a sample. For example, the mean of a sample may be used as the best single estimate of a population mean. In most practical problems, it is not sufficient to have merely a point estimate. If we were given two different point estimates of a population parameter and no further information, we could not distinguish the degree of reliability of these estimates. Yet one estimate might be based on a sample of size 10,000 and the other on a sample of size 10. Clearly, these estimates differ greatly in reliability. As we have seen, traditional statistics handles the problem of indicating reliability by the use of the confidence interval procedure. In this section, we will compare this classical technique to the corresponding Bayesian approach. However, before making this comparison, we comment briefly on point estimation techniques in the two approaches.

Point Estimation

In section 6.2, criteria of goodness of estimation were discussed. We have become familiar with point estimators such as the observed sample proportion of successes $\bar{p}$ in a Bernoulli process, which is used as the estimator of the population parameter p, and the observed sample mean $\bar{x}$ in a process described by the normal distribution, which is used as the estimator of the population mean μ.

Bayesian decision theory   Bayesian decision theory takes a different approach to the problem of point estimation. It views estimation as a straightforward problem of decision making.

> The estimator is the decision rule, the estimate is the action, and the possible values that the population parameter can assume are the states of nature.

For example, the sample mean $\bar{x}$ might be the estimator (decision rule), 10.6 might be the estimate (action), and the possible values that the population mean μ can assume are the parameter values (states of nature). In this formulation, the unknown population parameter is treated as a random variable.

To clarify the method, we will introduce some notation. Let θ be the true value of the parameter we want to estimate and $\hat{\theta}$ the estimator or action. A loss is involved if the value of $\hat{\theta}$ differs from θ, and the amount of the loss is some function of the difference between $\hat{\theta}$ and θ. Two possible loss functions might be

(17.1)
$$L(\hat{\theta}|\theta) = |\hat{\theta} - \theta|$$
Linear loss function

and

(17.2)
$$L(\hat{\theta}|\theta) = (\hat{\theta} - \theta)^2$$
Quadratic loss function

where $L(\hat{\theta}|\theta)$ is the loss involved in estimating (taking action) $\hat{\theta}$ when the parameter value (state of nature) is θ.

Somewhat more generally, the loss functions 17.1 and 17.2 may be written as

(17.3)
$$L(\hat{\theta}|\theta) = k(\theta)|\hat{\theta} - \theta|$$

and

(17.4)
$$L(\hat{\theta}|\theta) = k(\theta)(\hat{\theta} - \theta)^2$$

respectively, where $k(\theta)$ is a constant for a particular value of θ. This constant may be in money units, utility units, and so on. For simplicity, in the ensuing discussion, we will assume $k(\theta) = 1$ unit of utility, sometimes referred to as a **utile**. Therefore, we are dealing with functions of the form of equations 17.1 and 17.2, and the losses are given in units of utility.

Expression 17.1 is referred to as a **linear loss function**; expression 17.2 is a **quadratic loss function** (or **squared error loss function**). The nature of these functions can be illustrated by simple examples. Assume that the true value of the parameter θ is 10. Consider the losses involved if we estimate this parameter incorrectly as $\hat{\theta} = 11$ and $\hat{\theta} = 12$. For these two estimates, the respective linear loss functions 17.1 are

$$L(11|10) = |11 - 10| = 1$$

and

$$L(12|10) = |12 - 10| = 2$$

On the other hand, the quadratic loss function, 17.2 is equal to

$$L(11|10) = (11 - 10)^2 = 1$$

and

$$L(12|10) = (12 - 10)^2 = 4$$

In other words, in the linear case the loss in overestimating by two units is *twice* as much as in overestimating by one unit. In the quadratic case, the loss in overestimating by two units is *four* times as much as in overestimating by one unit. Note that in both functions, an underestimate of a given size, say two units, is as serious as an overestimate of the same size. Such loss functions are said to be **symmetrical**.

The ideas of these two loss functions were referred to earlier in section 1.17 in somewhat different forms. There, we were concerned with guessing the value of an observation selected at random from a frequency distribution. The penalty of an incorrect guess or estimate was referred to as the "cost of error." The "cost" corresponds to "loss" in the present discussion. We pointed out that if the cost of error varies directly with the size of error regardless of sign (the linear loss function 17.1), the median is the "best guess," since it minimizes average absolute deviations. On the other hand, if the cost of error varies according to the square of the error (the quadratic loss function 17.2), the mean should be the estimated value, since the average of the squared deviations around it is less than around any other figure. Note that least squares methods of estimation in classical statistics assume a quadratic loss function, since they obtain estimates for which the average squared error is minimized. Whether or not this is an appropriate loss function for the particular problem involved is rarely investigated.

Bayesian method of point estimation   The Bayesian method of point estimation begins with setting up whatever loss function appears to be appropriate. Then these losses are used in the standard decision procedure. Risks (conditional expected losses) are computed for each decision rule, or estimator. Prior probabilities are assigned to states of nature, or parameter values. Expected risks are computed for each decision rule. Then the estimator for which the expected risk is the least is the one chosen.

No Bayesian point estimators will be derived here, but one result is of particular interest. If the parameter p of a Bernoulli process is estimated using a squared error loss function, and a uniform or rectangular (continuous) prior probability distribution for p is assumed (that is, all values between zero and one are assumed to be equally likely), then the Bayesian estimator of p, denoted $\hat{p}$, is

(17.5)
$$\hat{p} = \frac{X + 1}{n + 2}$$

where X = the number of successes
         n = the number of trials

This value of $\hat{p}$ is also the mean or expected value of the posterior distribution of p if the prior distribution of p is assumed to be rectangular (and continuous) and the sample evidence is an observation of X successes in n trials. The estimate of p that we used earlier was simply the observed proportion of successes X/n. If the sample size n is large, these two estimates are approximately equal. Furthermore, there are other prior probability distributions besides the rectangular (uniform) distribution for which the mean of the posterior distribution will have a difference of this magnitude when compared with an estimator derived by classical methods. This brings out a very interesting point.

> From the Bayesian point of view, the standard use of the X/n estimate in such situations carries with it assumptions concerning the nature of the prior distribution of p.

The Bayesian decision analyst would argue that some of these prior distributions are unreasonable in the context of particular problems. For example, a rectangular prior distribution implies that all values of the parameter (in its admissible range) are equally likely. Such an implication may be quite unrealistic based on the analyst's prior knowledge.

Interval Estimation

We turn now to a consideration of interval estimation in classical and Bayesian statistics. We have seen that in estimating a population parameter in classical confidence interval estimation, an interval is set up on the basis of a sample of n observations and a so-called confidence coefficient is associated with this interval. Suppose, for example, we wanted to make a confidence interval estimate of p, the proportion of all customers on an importer's mailing list who would purchase special jars of cocktail onions if advertisements were sent to them, and let us assume that we want to base this estimate on the proportion $\bar{p}$ who purchased the jars in a simple random sample of 100 drawn from the list. We could establish a 95% confidence interval around $\bar{p}$ for the estimation of p in the usual way. Let us review the interpretation of this confidence interval. According to the classical school, it is definitely *incorrect* to say that the probability is 95% that the parameter is included in the interval. The population parameter is a particular value and therefore cannot be considered a random variable. Indeed, in all of classical statistics, it is forbidden to make conditional probability statements about a *population parameter* given the value of a sample statistic, such as $P(p|\bar{p})$. Permissible statements concern conditional probabilities of sample statistics given the value of a population parameter. For example, in a problem involving a Bernoulli process, we could compute probabilities of the type $P(\bar{p}|p)$.

Classical approach

Returning to the importer's problem, from the classical viewpoint, the confidence interval estimate of p cannot be interpreted as a probability statement about the proportion of all customers on the mailing list who would purchase the product. Since the interval is considered to be the random variable, the confidence coefficient refers to the concept that 95% of the intervals so constructed would bracket or include the true value of the population parameter. Thus, on a relative frequency basis, 95% of the statements made on the basis of such intervals would be correct. Furthermore, in keeping with the classical viewpoint, only the evidence of this particular sample can be used in establishing the confidence interval. Prior knowledge of any sort is not a part of the estimation procedure. Finally, just as in hypothesis testing, the use of the sample observations must be determined prior to the examination of these observations.

Bayesian approach   The Bayesian approach to this general problem contrasts sharply with the classical procedure. The Bayesian argues that if the value of the population parameter is unknown, then it can and should be treated as a random variable. In a setting such as the importer's problem, we would view the population parameter p as a random variable affecting a decision that must be made. Hence, we would be willing to compute conditional probabilities of the type $P(p|\bar{p})$. Furthermore, we would state that these conditional probabilities are relevant to the decision maker, rather than those of the form $P(\bar{p}|p)$. For example, in problems similar to that of the importer's, we might be interested in the probability that at least a certain proportion of the population would purchase the product based on the sample evidence. We would not be interested in the reverse conditional probability concerning a proportion in the sample given some postulated value for the population. The Bayesian decision analyst would argue that the confidence interval information is not particularly relevant. The decision maker is not interested in the proportion of correct statements that would be made in the long run, but rather in making a correct decision in this particular case.

The Bayesian approach also maintains that the analyst should not be restricted to the evidence of the particular sample that has been drawn but that the sample evidence should be incorporated with prior information through the use of Bayes' theorem to produce a posterior probability distribution. This leads to the Bayesian approach to the problem that classical inference solves by confidence interval estimation. The Bayesian procedure begins with the assignment of a prior probability distribution to the parameter being estimated. Then a sample is drawn, and the sample evidence is used to revise the prior probability distribution. The revision generates a posterior probability distribution. Then statements such as the following can be made: The probability is 0.90 that p lies between 0.04 and 0.06. The probability is 0.95 that the value of p is 0.07 or less, and so on. We may have a large number of possible values for p if that random variable is discrete, or we may have a probability density function over p if the random variable is continuous. The principle remains the same. If the prior distribution was a subjective probability distribution, then the posterior probabilities similarly represent revised degrees of belief or betting odds.

An interesting result occurs that is analogous to a relationship indicated earlier between classical and Bayesian point estimation when a rectangular prior distribution was assumed.

> If a rectangular distribution is assumed for the random variable p, and if the sample size is large, then there is a close coincidence between the statements made under the classical and Bayesian schools of thought.

Specifically, for example, the posterior probability that p lies in a 0.95 confidence interval is approximately 0.95. A roughly rectangular or uniform prior distribution is often referred to by Bayesian decision theorists as a "diffuse" or "gentle" prior distribution. Such a distribution implies roughly equal likelihood of occurrence of all values of the random variable in its admissible range. This type of distribution is thought of as an appropriate subjective prior distribution when the decision maker has virtually no knowledge of the value of the parameter being estimated. Doubtless, such states of almost complete lack of knowledge about parameter values are rare. Hence, Bayesian decision theorists argue that the uncritical use of confidence interval estimates may imply unreasonable assumptions about the investigator's prior knowledge concerning the parameter being estimated.

17.4
SOME REMARKS ON CLASSICAL AND BAYESIAN STATISTICS

As might be surmised from the material in this chapter, some controversy exists between adherents of the classical, or orthodox, school of statistics and advocates of the Bayesian viewpoint. In this section, we will comment on some of the areas of common ground and some of the points of difference between the two schools of thought.

● Despite differences in terminology, both schools conceptualize a problem of decision making consisting of states of nature and actions that must be taken in the light of sample or experimental evidence about these states of nature.

Common ground

● Both schools use conditional probabilities of sample outcomes, given states of nature (population parameters) for the decision process. These conditional probabilities provide the error characteristic curve from which the classicist chooses the decision rule. Informally, we should take into account the relative seriousness of Type I and Type II errors by considering the entire error characteristic curve, but since losses virtually always vary with population parameter values, it is not clear how we can actually do this.

● The Bayesian approach supplements or completes the classical analysis by formally providing a loss function that specifies the seriousness of errors in selecting acts and by assigning prior probabilities to states of nature on either an objective or a subjective basis. However, serious measurement problems are clearly present both in the establishment of loss functions and prior probability distribution.

Points of difference

● Some classicists have affirmed that hypothesis testing is not a decision problem, but rather one of drawing conclusions or inferences. However, other classical adherents have specifically formulated hypothesis testing as an action problem. In any event, it is not always clear whether a problem is one of inference or decision making.

● An important area of disagreement between non-Bayesian and Bayesian analysts is the matter of subjective prior probability distributions. The non-Bayesians argue that the only legitimate types of probabilities are "objective" or relative frequency of occurrence probabilities. They find it difficult to accept the idea that subjective or personalistic probabilities should be processed together with the relative frequencies, as in the Bayesian's use of Bayes' theorem, to arrive at posterior probabilities. The Bayesian argues that in actual decision making we do exactly that type of analysis. We have prior betting odds on events that influence the payoffs of our actions. On the observation of sample or experimental information, we revise these prior betting odds. This argument centers on "descriptive" behavior, that is, a purported description of how people actually behave. However, the Bayesian goes further, saying that Bayesian procedures are "normative" or "prescriptive," that is, they specify how a reasonable person *should* choose among alternatives to be consistent with one's own evaluations of payoffs and degrees of belief attached to uncertain events. The Bayesian also argues that if we rigidly maintain that only objective probabilities have meaning, we prevent ourselves from handling some of the most important uncertainties involved in problems of decision making. This latter point is surely a cogent one, particularly in areas such as business and economic decision making.

● The problem of how to assign prior probabilities is troublesome, even to convinced Bayesians, and is a subject of ongoing research. There are unresolved problems involved in determining whether all events should be considered equally likely under ignorance, how to pose questions to a decision maker to derive that individual's distribution of betting odds, or more generally, how best to quantify judgments about uncertainty.

● The Bayesian turns the tables on the orthodox school, which makes an accusation of excessive subjectivity, and directs a similar charge against classical statistics. The choices of hypotheses to test, probability distributions to use, significance and confidence levels, and what data to collect in order to obtain a relative frequency distribution are all inextricably interwoven with subjective judgments.

The preceding indication of some of the disagreements between the classical and Bayesian schools tends to emphasize a polarization of points of view. However, the fact is that even within each of these schools there are philosophical and methodological disagreements as well. These diversities of viewpoint between and within schools of thought make statistical analysis for decision making a lively and growing field.

Bayesian Interval Estimation   a method in which interval estimates of a population parameter are derived from the posterior probability distribution of that parameter.

Bayesian Point Estimation   a method of estimation in which the estimator with the minimum expected risk is selected.

Bayesian Statistics   an area that deals primarily with topics such as Bayesian decision theory, Bayesian estimation, and Bayesian regression analysis.

Classical Statistics   an area that deals primarily with topics such as statistical inference and regression analysis.

Loss Function   a function in terms of the difference between an estimate and the value of the parameter being estimated.

Linear Loss Function

$$L(\hat{\theta}|\theta) = k(\theta)|\hat{\theta} - \theta|$$

Quadratic Loss Function

$$L(\hat{\theta}|\theta) = k(\theta)(\hat{\theta} - \theta)^2$$

Cumulative (handwritten)

Appendix _A_
STATISTICAL TABLES

h = # OF TRIALS (handwritten)
c = OUTCOMES POSSIBLE (handwritten)

P = PROBABILITY of SUCCESS ON "ONE" GIVEN TRIAL. (handwritten)

TABLE A-1
Selected values of the binomial cumulative distribution function

$$F(c) = P(X \leqslant c) = \sum_{x=0}^{c} \binom{n}{x} (1-p)^{n-x} p^x$$

LESS THAN (handwritten)

Example   If $p = 0.20$, $n = 7$, $c = 2$, then $F(2) = P(X \leqslant 2) = 0.8520$.

n	c	0.05	0.10	0.15	0.20	0.25	p 0.30	0.35	0.40	0.45	0.50
2	0	0.9025	0.8100	0.7225	0.6400	0.5625	0.4900	0.4225	0.3600	0.3025	0.2500
	1	0.9975	0.9900	0.9775	0.9600	0.9375	0.9100	0.8775	0.8400	0.7975	0.7500
3	0	0.8574	0.7290	0.6141	0.5120	0.4219	0.3430	0.2746	0.2160	0.1664	0.1250
	1	0.9928	0.9720	0.9392	0.8960	0.8438	0.7840	0.7182	0.6480	0.5748	0.5000
	2	0.9999	0.9990	0.9966	0.9920	0.9844	0.9730	0.9571	0.9360	0.9089	0.8750
4	0	0.8145	0.6561	0.5220	0.4096	0.3164	0.2401	0.1785	0.1296	0.0915	0.0625
	1	0.9860	0.9477	0.8905	0.8192	0.7383	0.6517	0.5630	0.4752	0.3910	0.3125
	2	0.9995	0.9963	0.9880	0.9728	0.9492	0.9163	0.8735	0.8208	0.7585	0.6875
	3	1.0000	0.9999	0.9995	0.9984	0.9961	0.9919	0.9850	0.9744	0.9590	0.9375
5	0	0.7738	0.5905	0.4437	0.3277	0.2373	0.1681	0.1160	0.0778	0.0503	0.0312
	1	0.9774	0.9185	0.8352	0.7373	0.6328	0.5282	0.4284	0.3370	0.2562	0.1875
	2	0.9988	0.9914	0.9734	0.9421	0.8965	0.8369	0.7648	0.6826	0.5931	0.5000
	3	1.0000	0.9995	0.9978	0.9933	0.9844	0.9692	0.9460	0.9130	0.8688	0.8125
	4	1.0000	1.0000	0.9999	0.9997	0.9990	0.9976	0.9947	0.9898	0.9815	0.9688

Source: From Irwin Miller and John E. Freund, _Probability and Statistics for Engineers_, Second Edition, © 1977, pp. 477–481. Reprinted by permission of Prentice-Hall, Inc., Englewood Cliffs, NJ.

TABLE A-1   (continued)

n	c	0.05	0.10	0.15	0.20	0.25	p 0.30	0.35	0.40	0.45	0.50
6	0	0.7351	0.5314	0.3771	0.2621	0.1780	0.1176	0.0754	0.0467	0.0277	0.0156
	1	0.9672	0.8857	0.7765	0.6554	0.5339	0.4202	0.3191	0.2333	0.1636	0.1094
	2	0.9978	0.9842	0.9527	0.9011	0.8306	0.7443	0.6471	0.5443	0.4415	0.3438
	3	0.9999	0.9987	0.9941	0.9830	0.9624	0.9295	0.8826	0.8208	0.7447	0.6562
	4	1.0000	0.9999	0.9996	0.9984	0.9954	0.9891	0.9777	0.9590	0.9308	0.8906
	5	1.0000	1.0000	1.0000	0.9999	0.9998	0.9993	0.9982	0.9959	0.9917	0.9844
7	0	0.6983	0.4783	0.3206	0.2097	0.1335	0.0824	0.0490	0.0280	0.0152	0.0078
	1	0.9556	0.8503	0.7166	0.5767	0.4449	0.3294	0.2338	0.1586	0.1024	0.0625
	2	0.9962	0.9743	0.9262	0.8520	0.7564	0.6471	0.5323	0.4199	0.3164	0.2266
	3	0.9998	0.9973	0.9879	0.9667	0.9294	0.8740	0.8002	0.7102	0.6083	0.5000
	4	1.0000	0.9998	0.9988	0.9953	0.9871	0.9712	0.9444	0.9037	0.8471	0.7734
	5	1.0000	1.0000	0.9999	0.9996	0.9987	0.9962	0.9910	0.9812	0.9643	0.9375
	6	1.0000	1.0000	1.0000	1.0000	0.9999	0.9998	0.9994	0.9984	0.9963	0.9922
8	0	0.6634	0.4305	0.2725	0.1678	0.1001	0.0576	0.0319	0.0168	0.0084	0.0039
	1	0.9428	0.8131	0.6572	0.5033	0.3671	0.2553	0.1691	0.1064	0.0632	0.0352
	2	0.9942	0.9619	0.8948	0.7969	0.6785	0.5518	0.4278	0.3154	0.2201	0.1445
	3	0.9996	0.9950	0.9786	0.9437	0.8862	0.8059	0.7064	0.5941	0.4770	0.3633
	4	1.0000	0.9996	0.9971	0.9896	0.9727	0.9420	0.8939	0.8263	0.7396	0.6367
	5	1.0000	1.0000	0.9998	0.9988	0.9958	0.9887	0.9747	0.9502	0.9115	0.8555
	6	1.0000	1.0000	1.0000	0.9999	0.9996	0.9987	0.9964	0.9915	0.9819	0.9648
	7	1.0000	1.0000	1.0000	1.0000	1.0000	0.9999	0.9998	0.9993	0.9983	0.9961
9	0	0.6302	0.3874	0.2316	0.1342	0.0751	0.0404	0.0207	0.0101	0.0046	0.0020
	1	0.9288	0.7748	0.5995	0.4362	0.3003	0.1960	0.1211	0.0705	0.0385	0.0195
	2	0.9916	0.9470	0.8591	0.7382	0.6007	0.4628	0.3373	0.2318	0.1495	0.0898
	3	0.9994	0.9917	0.9661	0.9144	0.8343	0.7297	0.6089	0.4826	0.3614	0.2539
	4	1.0000	0.9991	0.9944	0.9804	0.9511	0.9012	0.8283	0.7334	0.6214	0.5000
	5	1.0000	0.9999	0.9994	0.9969	0.9900	0.9747	0.9464	0.9006	0.8342	0.7461
	6	1.0000	1.0000	1.0000	0.9997	0.9987	0.9957	0.9888	0.9750	0.9502	0.9102
	7	1.0000	1.0000	1.0000	1.0000	0.9999	0.9996	0.9986	0.9962	0.9909	0.9805
	8	1.0000	1.0000	1.0000	1.0000	1.0000	1.0000	0.9999	0.9997	0.9992	0.9980
10	0	0.5987	0.3487	0.1969	0.1074	0.0563	0.0282	0.0135	0.0060	0.0025	0.0010
	1	0.9139	0.7361	0.5443	0.3758	0.2440	0.1493	0.0860	0.0464	0.0232	0.0107
	2	0.9885	0.9298	0.8202	0.6778	0.5256	0.3828	0.2616	0.1673	0.0996	0.0547
	3	0.9990	0.9872	0.9500	0.8791	0.7759	0.6496	0.5138	0.3823	0.2660	0.1719
	4	0.9999	0.9984	0.9901	0.9672	0.9219	0.8497	0.7515	0.6331	0.5044	0.3770
	5	1.0000	0.9999	0.9986	0.9936	0.9803	0.9527	0.9051	0.8338	0.7384	0.6230
	6	1.0000	1.0000	0.9999	0.9991	0.9965	0.9894	0.9740	0.9452	0.8980	0.8281
	7	1.0000	1.0000	1.0000	0.9999	0.9996	0.9984	0.9952	0.9877	0.9726	0.9453
	8	1.0000	1.0000	1.0000	1.0000	1.0000	0.9999	0.9995	0.9983	0.9955	0.9893
	9	1.0000	1.0000	1.0000	1.0000	1.0000	1.0000	1.0000	0.9999	0.9997	0.9990

TABLE A-1   (continued)

n	c	0.05	0.10	0.15	0.20	0.25	p 0.30	0.35	0.40	0.45	0.50
11	0	0.5688	0.3138	0.1673	0.0859	0.0422	0.0198	0.0088	0.0036	0.0014	0.0005
	1	0.8981	0.6974	0.4922	0.3221	0.1971	0.1130	0.0606	0.0302	0.0139	0.0059
	2	0.9848	0.9104	0.7788	0.6174	0.4552	0.3127	0.2001	0.1189	0.0652	0.0327
	3	0.9984	0.9815	0.9306	0.8389	0.7133	0.5696	0.4256	0.2963	0.1911	0.1133
	4	0.9999	0.9972	0.9841	0.9496	0.8854	0.7897	0.6683	0.5328	0.3971	0.2744
	5	1.0000	0.9997	0.9973	0.9883	0.9657	0.9218	0.8513	0.7535	0.6331	0.5000
	6	1.0000	1.0000	0.9997	0.9980	0.9924	0.9784	0.9499	0.9006	0.8262	0.7256
	7	1.0000	1.0000	1.0000	0.9998	0.9988	0.9957	0.9878	0.9707	0.9390	0.8867
	8	1.0000	1.0000	1.0000	1.0000	0.9999	0.9994	0.9980	0.9941	0.9852	0.9673
	9	1.0000	1.0000	1.0000	1.0000	1.0000	1.0000	0.9998	0.9993	0.9978	0.9941
	10	1.0000	1.0000	1.0000	1.0000	1.0000	1.0000	1.0000	1.0000	0.9998	0.9995
12	0	0.5404	0.2824	0.1422	0.0687	0.0317	0.0138	0.0057	0.0022	0.0008	0.0002
	1	0.8816	0.6590	0.4435	0.2749	0.1584	0.0850	0.0424	0.0196	0.0083	0.0032
	2	0.9804	0.8891	0.7358	0.5583	0.3907	0.2528	0.1513	0.0834	0.0421	0.0193
	3	0.9978	0.9744	0.9078	0.7946	0.6488	0.4925	0.3467	0.2253	0.1345	0.0730
	4	0.9998	0.9957	0.9761	0.9274	0.8424	0.7237	0.5833	0.4382	0.3044	0.1938
	5	1.0000	0.9995	0.9954	0.9806	0.9456	0.8822	0.7873	0.6652	0.5269	0.3872
	6	1.0000	0.9999	0.9993	0.9961	0.9857	0.9614	0.9154	0.8418	0.7393	0.6128
	7	1.0000	1.0000	0.9999	0.9994	0.9972	0.9905	0.9745	0.9427	0.8883	0.8062
	8	1.0000	1.0000	1.0000	0.9999	0.9996	0.9983	0.9944	0.9847	0.9644	0.9270
	9	1.0000	1.0000	1.0000	1.0000	1.0000	0.9998	0.9992	0.9972	0.9921	0.9807
	10	1.0000	1.0000	1.0000	1.0000	1.0000	1.0000	0.9999	0.9997	0.9989	0.9968
	11	1.0000	1.0000	1.0000	1.0000	1.0000	1.0000	1.0000	1.0000	0.9999	0.9998
13	0	0.5133	0.2542	0.1209	0.0550	0.0238	0.0097	0.0037	0.0013	0.0004	0.0001
	1	0.8646	0.6213	0.3983	0.2336	0.1267	0.0637	0.0296	0.0126	0.0049	0.0017
	2	0.9755	0.8661	0.6920	0.5017	0.3326	0.2025	0.1132	0.0579	0.0269	0.0112
	3	0.9969	0.9658	0.8820	0.7473	0.5843	0.4206	0.2783	0.1686	0.0929	0.0461
	4	0.9997	0.9935	0.9658	0.9009	0.7940	0.6543	0.5005	0.3530	0.2279	0.1334
	5	1.0000	0.9991	0.9925	0.9700	0.9198	0.8346	0.7159	0.5744	0.4268	0.2905
	6	1.0000	0.9999	0.9987	0.9930	0.9757	0.9376	0.8705	0.7712	0.6437	0.5000
	7	1.0000	1.0000	0.9998	0.9988	0.9944	0.9818	0.9538	0.9023	0.8212	0.7095
	8	1.0000	1.0000	1.0000	0.9998	0.9990	0.9960	0.9874	0.9679	0.9302	0.8666
	9	1.0000	1.0000	1.0000	1.0000	0.9999	0.9993	0.9975	0.9922	0.9797	0.9539
	10	1.0000	1.0000	1.0000	1.0000	1.0000	0.9999	0.9997	0.9987	0.9959	0.9888
	11	1.0000	1.0000	1.0000	1.0000	1.0000	1.0000	1.0000	0.9999	0.9995	0.9983
	12	1.0000	1.0000	1.0000	1.0000	1.0000	1.0000	1.0000	1.0000	1.0000	0.9999
14	0	0.4877	0.2288	0.1028	0.0440	0.0178	0.0068	0.0024	0.0008	0.0002	0.0001
	1	0.8470	0.5846	0.3567	0.1979	0.1010	0.0475	0.0205	0.0081	0.0029	0.0009
	2	0.9699	0.8416	0.6479	0.4481	0.2811	0.1608	0.0839	0.0398	0.0170	0.0065
	3	0.9958	0.9559	0.8535	0.6982	0.5213	0.3552	0.2205	0.1243	0.0632	0.0287
	4	0.9996	0.9908	0.9533	0.8702	0.7415	0.5842	0.4227	0.2793	0.1672	0.0898

TABLE A-1   (continued)

n	c	0.05	0.10	0.15	0.20	0.25	*p* 0.30	0.35	0.40	0.45	0.50
	5	1.0000	0.9985	0.9885	0.9561	0.8883	0.7805	0.6405	0.4859	0.3373	0.2120
	6	1.0000	0.9998	0.9978	0.9884	0.9617	0.9067	0.8164	0.6925	0.5461	0.3953
	7	1.0000	1.0000	0.9997	0.9976	0.9897	0.9685	0.9247	0.8499	0.7414	0.6047
	8	1.0000	1.0000	1.0000	0.9996	0.9978	0.9917	0.9757	0.9417	0.8811	0.7880
	9	1.0000	1.0000	1.0000	1.0000	0.9997	0.9983	0.9940	0.9825	0.9574	0.9102
	10	1.0000	1.0000	1.0000	1.0000	1.0000	0.9998	0.9989	0.9961	0.9886	0.9713
	11	1.0000	1.0000	1.0000	1.0000	1.0000	1.0000	0.9999	0.9994	0.9978	0.9935
	12	1.0000	1.0000	1.0000	1.0000	1.0000	1.0000	1.0000	0.9999	0.9997	0.9991
	13	1.0000	1.0000	1.0000	1.0000	1.0000	1.0000	1.0000	1.0000	1.0000	0.9999
15	0	0.4633	0.2059	0.0874	0.0352	0.0134	0.0047	0.0016	0.0005	0.0001	0.0000
	1	0.8290	0.5490	0.3186	0.1671	0.0802	0.0353	0.0142	0.0052	0.0017	0.0005
	2	0.9638	0.8159	0.6042	0.3980	0.2361	0.1268	0.0617	0.0271	0.0107	0.0037
	3	0.9945	0.9444	0.8227	0.6482	0.4613	0.2969	0.1727	0.0905	0.0424	0.0176
	4	0.9994	0.9873	0.9383	0.8358	0.6865	0.5155	0.3519	0.2173	0.1204	0.0592
	5	0.9999	0.9978	0.9832	0.9389	0.8516	0.7216	0.5643	0.4032	0.2608	0.1509
	6	1.0000	0.9997	0.9964	0.9819	0.9434	0.8689	0.7548	0.6098	0.4522	0.3036
	7	1.0000	1.0000	0.9996	0.9958	0.9827	0.9500	0.8868	0.7869	0.6535	0.5000
	8	1.0000	1.0000	0.9999	0.9992	0.9958	0.9848	0.9578	0.9050	0.8182	0.6964
	9	1.0000	1.0000	1.0000	0.9999	0.9992	0.9963	0.9876	0.9662	0.9231	0.8491
	10	1.0000	1.0000	1.0000	1.0000	0.9999	0.9993	0.9972	0.9907	0.9745	0.9408
	11	1.0000	1.0000	1.0000	1.0000	1.0000	0.9999	0.9995	0.9981	0.9937	0.9824
	12	1.0000	1.0000	1.0000	1.0000	1.0000	1.0000	0.9999	0.9997	0.9989	0.9963
	13	1.0000	1.0000	1.0000	1.0000	1.0000	1.0000	1.0000	1.0000	0.9999	0.9995
	14	1.0000	1.0000	1.0000	1.0000	1.0000	1.0000	1.0000	1.0000	1.0000	1.0000
16	0	0.4401	0.1853	0.0743	0.0281	0.0100	0.0033	0.0010	0.0003	0.0001	0.0000
	1	0.8108	0.5147	0.2839	0.1407	0.0635	0.0261	0.0098	0.0033	0.0010	0.0003
	2	0.9571	0.7892	0.5614	0.3518	0.1971	0.0994	0.0451	0.0183	0.0066	0.0021
	3	0.9930	0.9316	0.7899	0.5981	0.4050	0.2459	0.1339	0.0651	0.0281	0.0106
	4	0.9991	0.9830	0.9209	0.7982	0.6302	0.4499	0.2892	0.1666	0.0853	0.0384
	5	0.9999	0.9967	0.9765	0.9183	0.8103	0.6598	0.4900	0.3288	0.1976	0.1051
	6	1.0000	0.9995	0.9944	0.9733	0.9204	0.8247	0.6881	0.5272	0.3660	0.2272
	7	1.0000	0.9999	0.9989	0.9930	0.9729	0.9256	0.8406	0.7161	0.5629	0.4018
	8	1.0000	1.0000	0.9998	0.9985	0.9925	0.9743	0.9329	0.8577	0.7441	0.5982
	9	1.0000	1.0000	1.0000	0.9998	0.9984	0.9929	0.9771	0.9417	0.8759	0.7728
	10	1.0000	1.0000	1.0000	1.0000	0.9997	0.9984	0.9938	0.9809	0.9514	0.8949
	11	1.0000	1.0000	1.0000	1.0000	1.0000	0.9997	0.9987	0.9951	0.9851	0.9616
	12	1.0000	1.0000	1.0000	1.0000	1.0000	1.0000	0.9998	0.9991	0.9965	0.9894
	13	1.0000	1.0000	1.0000	1.0000	1.0000	1.0000	1.0000	0.9999	0.9994	0.9979
	14	1.0000	1.0000	1.0000	1.0000	1.0000	1.0000	1.0000	1.0000	1.0000	0.9997
	15	1.0000	1.0000	1.0000	1.0000	1.0000	1.0000	1.0000	1.0000	1.0000	1.0000

TABLE A-1   (continued)

n	c	0.05	0.10	0.15	0.20	0.25	p 0.30	0.35	0.40	0.45	0.50
17	0	0.4181	0.1668	0.0631	0.0225	0.0075	0.0023	0.0007	0.0002	0.0000	0.0000
	1	0.7922	0.4818	0.2525	0.1182	0.0501	0.0193	0.0067	0.0021	0.0006	0.0001
	2	0.9497	0.7618	0.5198	0.3096	0.1637	0.0774	0.0327	0.0123	0.0041	0.0012
	3	0.9912	0.9174	0.7556	0.5489	0.3530	0.2019	0.1028	0.0464	0.0184	0.0064
	4	0.9988	0.9779	0.9013	0.7582	0.5739	0.3887	0.2348	0.1260	0.0596	0.0245
	5	0.9999	0.9953	0.9681	0.8943	0.7653	0.5968	0.4197	0.2639	0.1471	0.0717
	6	1.0000	0.9992	0.9917	0.9623	0.8929	0.7752	0.6188	0.4478	0.2902	0.1662
	7	1.0000	0.9999	0.9983	0.9891	0.9598	0.8954	0.7872	0.6405	0.4743	0.3145
	8	1.0000	1.0000	0.9997	0.9974	0.9876	0.9597	0.9006	0.8011	0.6626	0.5000
	9	1.0000	1.0000	1.0000	0.9995	0.9969	0.9873	0.9617	0.9081	0.8166	0.6855
	10	1.0000	1.0000	1.0000	0.9999	0.9994	0.9968	0.9880	0.9652	0.9174	0.8338
	11	1.0000	1.0000	1.0000	1.0000	0.9999	0.9993	0.9970	0.9894	0.9699	0.9283
	12	1.0000	1.0000	1.0000	1.0000	1.0000	0.9999	0.9994	0.9975	0.9914	0.9755
	13	1.0000	1.0000	1.0000	1.0000	1.0000	1.0000	0.9999	0.9995	0.9981	0.9936
	14	1.0000	1.0000	1.0000	1.0000	1.0000	1.0000	1.0000	0.9999	0.9997	0.9988
	15	1.0000	1.0000	1.0000	1.0000	1.0000	1.0000	1.0000	1.0000	1.0000	0.9999
	16	1.0000	1.0000	1.0000	1.0000	1.0000	1.0000	1.0000	1.0000	1.0000	1.0000
18	0	0.3972	0.1501	0.0536	0.0180	0.0056	0.0016	0.0004	0.0001	0.0000	0.0000
	1	0.7735	0.4503	0.2241	0.0991	0.0395	0.0142	0.0046	0.0013	0.0003	0.0001
	2	0.9419	0.7338	0.4797	0.2713	0.1353	0.0600	0.0236	0.0082	0.0025	0.0007
	3	0.9891	0.9018	0.7202	0.5010	0.3057	0.1646	0.0783	0.0328	0.0120	0.0038
	4	0.9985	0.9718	0.8794	0.7164	0.5187	0.3327	0.1886	0.0942	0.0411	0.0154
	5	0.9998	0.9936	0.9581	0.8671	0.7175	0.5344	0.3550	0.2088	0.1077	0.0481
	6	1.0000	0.9988	0.9882	0.9487	0.8610	0.7217	0.5491	0.3743	0.2258	0.1189
	7	1.0000	0.9998	0.9973	0.9837	0.9431	0.8593	0.7283	0.5634	0.3915	0.2403
	8	1.0000	1.0000	0.9995	0.9957	0.9807	0.9404	0.8609	0.7368	0.5778	0.4073
	9	1.0000	1.0000	0.9999	0.9991	0.9946	0.9790	0.9403	0.8653	0.7473	0.5927
	10	1.0000	1.0000	1.0000	0.9998	0.9988	0.9939	0.9788	0.9424	0.8720	0.7597
	11	1.0000	1.0000	1.0000	1.0000	0.9998	0.9986	0.9938	0.9797	0.9463	0.8811
	12	1.0000	1.0000	1.0000	1.0000	1.0000	0.9997	0.9986	0.9942	0.9817	0.9519
	13	1.0000	1.0000	1.0000	1.0000	1.0000	1.0000	0.9997	0.9987	0.9951	0.9846
	14	1.0000	1.0000	1.0000	1.0000	1.0000	1.0000	1.0000	0.9998	0.9990	0.9962
	15	1.0000	1.0000	1.0000	1.0000	1.0000	1.0000	1.0000	1.0000	0.9999	0.9993
	16	1.0000	1.0000	1.0000	1.0000	1.0000	1.0000	1.0000	1.0000	1.0000	0.9999
19	0	0.3774	0.1351	0.0456	0.0144	0.0042	0.0011	0.0003	0.0001	0.0000	0.0000
	1	0.7547	0.4203	0.1985	0.0829	0.0310	0.0104	0.0031	0.0008	0.0002	0.0000
	2	0.9335	0.7054	0.4413	0.2369	0.1113	0.0462	0.0170	0.0055	0.0015	0.0004
	3	0.9868	0.8850	0.6841	0.4551	0.2630	0.1332	0.0591	0.0230	0.0077	0.0022
	4	0.9980	0.9648	0.8556	0.6733	0.4654	0.2822	0.1500	0.0696	0.0280	0.0096

TABLE A-1 (continued)

n	c	0.05	0.10	0.15	0.20	0.25	0.30	0.35	0.40	0.45	0.50
						p					
	5	0.9998	0.9914	0.9463	0.8369	0.6678	0.4739	0.2968	0.1629	0.0777	0.0318
	6	1.0000	0.9983	0.9837	0.9324	0.8251	0.6655	0.4812	0.3081	0.1727	0.0835
	7	1.0000	0.9997	0.9959	0.9767	0.9225	0.8180	0.6656	0.4878	0.3169	0.1796
	8	1.0000	1.0000	0.9992	0.9933	0.9713	0.9161	0.8145	0.6675	0.4940	0.3238
	9	1.0000	1.0000	0.9999	0.9984	0.9911	0.9674	0.9125	0.8139	0.6710	0.5000
	10	1.0000	1.0000	1.0000	0.9997	0.9977	0.9895	0.9653	0.9115	0.8159	0.6762
	11	1.0000	1.0000	1.0000	1.0000	0.9995	0.9972	0.9886	0.9648	0.9129	0.8204
	12	1.0000	1.0000	1.0000	1.0000	0.9999	0.9994	0.9969	0.9884	0.9658	0.9165
	13	1.0000	1.0000	1.0000	1.0000	1.0000	0.9999	0.9993	0.9969	0.9891	0.9682
	14	1.0000	1.0000	1.0000	1.0000	1.0000	1.0000	0.9999	0.9994	0.9972	0.9904
	15	1.0000	1.0000	1.0000	1.0000	1.0000	1.0000	1.0000	0.9999	0.9995	0.9978
	16	1.0000	1.0000	1.0000	1.0000	1.0000	1.0000	1.0000	1.0000	0.9999	0.9996
	17	1.0000	1.0000	1.0000	1.0000	1.0000	1.0000	1.0000	1.0000	1.0000	1.0000
20	0	0.3585	0.1216	0.0388	0.0115	0.0032	0.0008	0.0002	0.0000	0.0000	0.0000
	1	0.7358	0.3917	0.1756	0.0692	0.0243	0.0076	0.0021	0.0005	0.0001	0.0000
	2	0.9245	0.6769	0.4049	0.2061	0.0913	0.0355	0.0121	0.0036	0.0009	0.0002
	3	0.9841	0.8670	0.6477	0.4114	0.2252	0.1071	0.0444	0.0160	0.0049	0.0013
	4	0.9974	0.9568	0.8298	0.6296	0.4148	0.2375	0.1182	0.0510	0.0189	0.0059
	5	0.9997	0.9887	0.9327	0.8042	0.6172	0.4164	0.2454	0.1256	0.0553	0.0207
	6	1.0000	0.9976	0.9781	0.9133	0.7858	0.6080	0.4166	0.2500	0.1299	0.0577
	7	1.0000	0.9996	0.9941	0.9679	0.8982	0.7723	0.6010	0.4159	0.2520	0.1316
	8	1.0000	0.9999	0.9987	0.9900	0.9591	0.8867	0.7624	0.5956	0.4143	0.2517
	9	1.0000	1.0000	0.9998	0.9974	0.9861	0.9520	0.8782	0.7553	0.5914	0.4119
	10	1.0000	1.0000	1.0000	0.9994	0.9961	0.9829	0.9468	0.8725	0.7507	0.5881
	11	1.0000	1.0000	1.0000	0.9999	0.9991	0.9949	0.9804	0.9435	0.8692	0.7483
	12	1.0000	1.0000	1.0000	1.0000	0.9998	0.9987	0.9940	0.9790	0.9420	0.8684
	13	1.0000	1.0000	1.0000	1.0000	1.0000	0.9997	0.9985	0.9935	0.9786	0.9423
	14	1.0000	1.0000	1.0000	1.0000	1.0000	1.0000	0.9997	0.9984	0.9936	0.9793
	15	1.0000	1.0000	1.0000	1.0000	1.0000	1.0000	1.0000	0.9997	0.9985	0.9941
	16	1.0000	1.0000	1.0000	1.0000	1.0000	1.0000	1.0000	1.0000	0.9997	0.9987
	17	1.0000	1.0000	1.0000	1.0000	1.0000	1.0000	1.0000	1.0000	1.0000	0.9998
	18	1.0000	1.0000	1.0000	1.0000	1.0000	1.0000	1.0000	1.0000	1.0000	1.0000

TABLE A-2
Selected values of the binomial probability distribution

$$P(x) = \binom{n}{x}(1 - p)^{n-x}p^x$$

Example If $p = 0.15$, $n = 4$, and $x = 3$, then $P(3) = 0.0115$. When $p > 0.5$, the value of $P(x)$ for a given n, x, and p is obtained by finding the tabular entry for the given n, with $n - x$ in place of the given x and $1 - p$ in place of the given p.

n	x	0.05	0.10	0.15	0.20	0.25	0.30	0.35	0.40	0.45	0.50
1	0	0.9500	0.9000	0.8500	0.8000	0.7500	0.7000	0.6500	0.6000	0.5500	0.5000
	1	0.0500	0.1000	0.1500	0.2000	0.2500	0.3000	0.3500	0.4000	0.4500	0.5000
2	0	0.9025	0.8100	0.7225	0.6400	0.5625	0.4900	0.4225	0.3600	0.3025	0.2500
	1	0.0950	0.1800	0.2550	0.3200	0.3750	0.4200	0.4550	0.4800	0.4950	0.5000
	2	0.0025	0.0100	0.0225	0.0400	0.0625	0.0900	0.1225	0.1600	0.2025	0.2500
3	0	0.8574	0.7290	0.6141	0.5120	0.4219	0.3430	0.2746	0.2160	0.1664	0.1250
	1	0.1354	0.2430	0.3251	0.3840	0.4219	0.4410	0.4436	0.4320	0.4084	0.3750
	2	0.0071	0.0270	0.0574	0.0960	0.1406	0.1890	0.2389	0.2880	0.3341	0.3750
	3	0.0001	0.0010	0.0034	0.0080	0.0156	0.0270	0.0429	0.0640	0.0911	0.1250
4	0	0.8145	0.6561	0.5220	0.4096	0.3164	0.2401	0.1785	0.1296	0.0915	0.0625
	1	0.1715	0.2916	0.3685	0.4096	0.4219	0.4116	0.3845	0.3456	0.2995	0.2500
	2	0.0135	0.0486	0.0975	0.1536	0.2109	0.2646	0.3105	0.3456	0.3675	0.3750
	3	0.0005	0.0036	0.0115	0.0256	0.0469	0.0756	0.1115	0.1536	0.2005	0.2500
	4	0.0000	0.0001	0.0005	0.0016	0.0039	0.0081	0.0150	0.0256	0.0410	0.0625
5	0	0.7738	0.5905	0.4437	0.3277	0.2373	0.1681	0.1160	0.0778	0.0503	0.0312
	1	0.2036	0.3280	0.3915	0.4096	0.3955	0.3602	0.3124	0.2592	0.2059	0.1562
	2	0.0214	0.0729	0.1382	0.2048	0.2637	0.3087	0.3364	0.3456	0.3369	0.3125
	3	0.0011	0.0081	0.0244	0.0512	0.0879	0.1323	0.1811	0.2304	0.2757	0.3125
	4	0.0000	0.0004	0.0022	0.0064	0.0146	0.0284	0.0488	0.0768	0.1128	0.1562
	5	0.0000	0.0000	0.0001	0.0003	0.0010	0.0024	0.0053	0.0102	0.0185	0.0312
6	0	0.7351	0.5314	0.3771	0.2621	0.1780	0.1176	0.0754	0.0467	0.0277	0.0156
	1	0.2321	0.3543	0.3993	0.3932	0.3560	0.3025	0.2437	0.1866	0.1359	0.0938
	2	0.0305	0.0984	0.1762	0.2458	0.2966	0.3241	0.3280	0.3110	0.2780	0.2344
	3	0.0021	0.0146	0.0415	0.0819	0.1318	0.1852	0.2355	0.2765	0.3032	0.3125
	4	0.0001	0.0012	0.0055	0.0154	0.0330	0.0595	0.0951	0.1382	0.1861	0.2344
	5	0.0000	0.0001	0.0004	0.0015	0.0044	0.0102	0.0205	0.0369	0.0609	0.0938
	6	0.0000	0.0000	0.0000	0.0001	0.0002	0.0007	0.0018	0.0041	0.0083	0.0156
7	0	0.6983	0.4783	0.3206	0.2097	0.1335	0.0824	0.0490	0.0280	0.0152	0.0078
	1	0.2573	0.3720	0.3960	0.3670	0.3115	0.2471	0.1848	0.1306	0.0872	0.0547
	2	0.0406	0.1240	0.2097	0.2753	0.3115	0.3177	0.2985	0.2613	0.2140	0.1641
	3	0.0036	0.0230	0.0617	0.1147	0.1730	0.2269	0.2679	0.2903	0.2918	0.2734
	4	0.0002	0.0026	0.0109	0.0287	0.0577	0.0972	0.1442	0.1935	0.2388	0.2734

TABLE A-2   (continued)

n	x	0.05	0.10	0.15	0.20	0.25	p 0.30	0.35	0.40	0.45	0.50
	5	0.0000	0.0002	0.0012	0.0043	0.0115	0.0250	0.0466	0.0774	0.1172	0.1641
	6	0.0000	0.0000	0.0001	0.0004	0.0013	0.0036	0.0084	0.0172	0.0320	0.0547
	7	0.0000	0.0000	0.0000	0.0000	0.0001	0.0002	0.0006	0.0016	0.0037	0.0078
8	0	0.6634	0.4305	0.2725	0.1678	0.1001	0.0576	0.0319	0.0168	0.0084	0.0039
	1	0.2793	0.3826	0.3847	0.3355	0.2670	0.1977	0.1373	0.0896	0.0548	0.0312
	2	0.0515	0.1488	0.2376	0.2936	0.3115	0.2965	0.2587	0.2090	0.1569	0.1094
	3	0.0054	0.0331	0.0839	0.1468	0.2076	0.2541	0.2786	0.2787	0.2568	0.2188
	4	0.0004	0.0046	0.0185	0.0459	0.0865	0.1361	0.1875	0.2322	0.2627	0.2734
	5	0.0000	0.0004	0.0026	0.0092	0.0231	0.0467	0.0808	0.1239	0.1719	0.2188
	6	0.0000	0.0000	0.0002	0.0011	0.0038	0.0100	0.0217	0.0413	0.0703	0.1094
	7	0.0000	0.0000	0.0000	0.0001	0.0004	0.0012	0.0033	0.0079	0.0164	0.0312
	8	0.0000	0.0000	0.0000	0.0000	0.0000	0.0001	0.0002	0.0007	0.0017	0.0039
9	0	0.6302	0.3874	0.2316	0.1342	0.0751	0.0404	0.0207	0.0101	0.0046	0.0020
	1	0.2985	0.3874	0.3679	0.3020	0.2253	0.1556	0.1004	0.0605	0.0339	0.0176
	2	0.0629	0.1722	0.2597	0.3020	0.3003	0.2668	0.2162	0.1612	0.1110	0.0703
	3	0.0077	0.0446	0.1069	0.1762	0.2336	0.2668	0.2716	0.2508	0.2119	0.1641
	4	0.0006	0.0074	0.0283	0.0661	0.1168	0.1715	0.2194	0.2508	0.2600	0.2461
	5	0.0000	0.0008	0.0050	0.0165	0.0389	0.0735	0.1181	0.1672	0.2128	0.2461
	6	0.0000	0.0001	0.0006	0.0028	0.0087	0.0210	0.0424	0.0743	0.1160	0.1641
	7	0.0000	0.0000	0.0000	0.0003	0.0012	0.0039	0.0098	0.0212	0.0407	0.0703
	8	0.0000	0.0000	0.0000	0.0000	0.0001	0.0004	0.0013	0.0035	0.0083	0.0176
	9	0.0000	0.0000	0.0000	0.0000	0.0000	0.0000	0.0001	0.0003	0.0008	0.0020
10	0	0.5987	0.3487	0.1969	0.1074	0.0563	0.0282	0.0135	0.0060	0.0025	0.0010
	1	0.3151	0.3874	0.3474	0.2684	0.1877	0.1211	0.0725	0.0403	0.0207	0.0098
	2	0.0746	0.1937	0.2759	0.3020	0.2816	0.2335	0.1757	0.1209	0.0763	0.0439
	3	0.0105	0.0574	0.1298	0.2013	0.2503	0.2668	0.2522	0.2150	0.1665	0.1172
	4	0.0010	0.0112	0.0401	0.0881	0.1460	0.2001	0.2377	0.2508	0.2384	0.2051
	5	0.0001	0.0015	0.0085	0.0264	0.0584	0.1029	0.1536	0.2007	0.2340	0.2461
	6	0.0000	0.0001	0.0012	0.0055	0.0162	0.0368	0.0689	0.1115	0.1596	0.2051
	7	0.0000	0.0000	0.0001	0.0008	0.0031	0.0090	0.0212	0.0425	0.0746	0.1172
	8	0.0000	0.0000	0.0000	0.0001	0.0004	0.0014	0.0043	0.0106	0.0229	0.0439
	9	0.0000	0.0000	0.0000	0.0000	0.0000	0.0001	0.0005	0.0016	0.0042	0.0098
	10	0.0000	0.0000	0.0000	0.0000	0.0000	0.0000	0.0000	0.0001	0.0003	0.0010
11	0	0.5688	0.3138	0.1673	0.0859	0.0422	0.0198	0.0088	0.0036	0.0014	0.0005
	1	0.3293	0.3835	0.3248	0.2362	0.1549	0.0932	0.0518	0.0266	0.0125	0.0054
	2	0.0867	0.2131	0.2866	0.2953	0.2581	0.1998	0.1395	0.0887	0.0513	0.0269
	3	0.0137	0.0710	0.1517	0.2215	0.2581	0.2568	0.2254	0.1774	0.1259	0.0806
	4	0.0014	0.0158	0.0536	0.1107	0.1721	0.2201	0.2428	0.2365	0.2060	0.1611

TABLE A-2   (continued)

n	x	0.05	0.10	0.15	0.20	0.25	*p* 0.30	0.35	0.40	0.45	0.50
	5	0.0001	0.0025	0.0132	0.0388	0.0803	0.1321	0.1830	0.2207	0.2360	0.2256
	6	0.0000	0.0003	0.0023	0.0097	0.0268	0.0566	0.0985	0.1471	0.1931	0.2256
	7	0.0000	0.0000	0.0003	0.0017	0.0064	0.0173	0.0379	0.0701	0.1128	0.1611
	8	0.0000	0.0000	0.0000	0.0002	0.0011	0.0037	0.0102	0.0234	0.0462	0.0806
	9	0.0000	0.0000	0.0000	0.0000	0.0001	0.0005	0.0018	0.0052	0.0126	0.0269
	10	0.0000	0.0000	0.0000	0.0000	0.0000	0.0000	0.0002	0.0007	0.0021	0.0054
	11	0.0000	0.0000	0.0000	0.0000	0.0000	0.0000	0.0000	0.0000	0.0002	0.0005
12	0	0.5404	0.2824	0.1422	0.0687	0.0317	0.0138	0.0057	0.0022	0.0008	0.0002
	1	0.3413	0.3766	0.3012	0.2062	0.1267	0.0712	0.0368	0.0174	0.0075	0.0029
	2	0.0988	0.2301	0.2924	0.2835	0.2323	0.1678	0.1088	0.0639	0.0339	0.0161
	3	0.0173	0.0852	0.1720	0.2362	0.2581	0.2397	0.1954	0.1419	0.0923	0.0537
	4	0.0021	0.0213	0.0683	0.1329	0.1936	0.2311	0.2367	0.2128	0.1700	0.1208
	5	0.0002	0.0038	0.0193	0.0532	0.1032	0.1585	0.2039	0.2270	0.2225	0.1934
	6	0.0000	0.0005	0.0040	0.0155	0.0401	0.0792	0.1281	0.1766	0.2124	0.2256
	7	0.0000	0.0000	0.0006	0.0033	0.0115	0.0291	0.0591	0.1009	0.1489	0.1934
	8	0.0000	0.0000	0.0001	0.0005	0.0024	0.0078	0.0199	0.0420	0.0762	0.1208
	9	0.0000	0.0000	0.0000	0.0001	0.0004	0.0015	0.0048	0.0125	0.0277	0.0537
	10	0.0000	0.0000	0.0000	0.0000	0.0000	0.0002	0.0008	0.0025	0.0068	0.0161
	11	0.0000	0.0000	0.0000	0.0000	0.0000	0.0000	0.0001	0.0003	0.0010	0.0029
	12	0.0000	0.0000	0.0000	0.0000	0.0000	0.0000	0.0000	0.0000	0.0001	0.0002
13	0	0.5133	0.2542	0.1209	0.0550	0.0238	0.0097	0.0037	0.0013	0.0004	0.0001
	1	0.3512	0.3672	0.2774	0.1787	0.1029	0.0540	0.0259	0.0113	0.0045	0.0016
	2	0.1109	0.2448	0.2937	0.2680	0.2059	0.1388	0.0836	0.0453	0.0220	0.0095
	3	0.0214	0.0997	0.1900	0.2457	0.2517	0.2181	0.1651	0.1107	0.0660	0.0349
	4	0.0028	0.0277	0.0838	0.1535	0.2097	0.2337	0.2222	0.1845	0.1350	0.0873
	5	0.0003	0.0055	0.0266	0.0691	0.1258	0.1803	0.2154	0.2214	0.1989	0.1571
	6	0.0000	0.0008	0.0063	0.0230	0.0559	0.1030	0.1546	0.1968	0.2169	0.2095
	7	0.0000	0.0001	0.0011	0.0058	0.0186	0.0442	0.0833	0.1312	0.1775	0.2095
	8	0.0000	0.0000	0.0001	0.0011	0.0047	0.0142	0.0336	0.0656	0.1089	0.1571
	9	0.0000	0.0000	0.0000	0.0001	0.0009	0.0034	0.0101	0.0243	0.0495	0.0873
	10	0.0000	0.0000	0.0000	0.0000	0.0001	0.0006	0.0022	0.0065	0.0162	0.0349
	11	0.0000	0.0000	0.0000	0.0000	0.0000	0.0001	0.0003	0.0012	0.0036	0.0095
	12	0.0000	0.0000	0.0000	0.0000	0.0000	0.0000	0.0000	0.0001	0.0005	0.0016
	13	0.0000	0.0000	0.0000	0.0000	0.0000	0.0000	0.0000	0.0000	0.0000	0.0001
14	0	0.4877	0.2288	0.1028	0.0440	0.0178	0.0068	0.0024	0.0008	0.0002	0.0001
	1	0.3593	0.3559	0.2539	0.1539	0.0832	0.0407	0.0181	0.0073	0.0027	0.0009
	2	0.1229	0.2570	0.2912	0.2501	0.1802	0.1134	0.0634	0.0317	0.0141	0.0056
	3	0.0259	0.1142	0.2056	0.2501	0.2402	0.1943	0.1366	0.0845	0.0462	0.0222
	4	0.0037	0.0349	0.0998	0.1720	0.2202	0.2290	0.2022	0.1549	0.1040	0.0611

TABLE A-2   (continued)

n	x	0.05	0.10	0.15	0.20	0.25	0.30	0.35	0.40	0.45	0.50
							p				
	5	0.0004	0.0078	0.0352	0.0860	0.1468	0.1963	0.2178	0.2066	0.1701	0.1222
	6	0.0000	0.0013	0.0093	0.0322	0.0734	0.1262	0.1759	0.2066	0.2088	0.1833
	7	0.0000	0.0002	0.0019	0.0092	0.0280	0.0618	0.1082	0.1574	0.1952	0.2095
	8	0.0000	0.0000	0.0003	0.0020	0.0082	0.0232	0.0510	0.0918	0.1398	0.1833
	9	0.0000	0.0000	0.0000	0.0003	0.0018	0.0066	0.0183	0.0408	0.0762	0.1222
	10	0.0000	0.0000	0.0000	0.0000	0.0003	0.0014	0.0049	0.0136	0.0312	0.0611
	11	0.0000	0.0000	0.0000	0.0000	0.0000	0.0002	0.0010	0.0033	0.0093	0.0222
	12	0.0000	0.0000	0.0000	0.0000	0.0000	0.0000	0.0001	0.0005	0.0019	0.0056
	13	0.0000	0.0000	0.0000	0.0000	0.0000	0.0000	0.0000	0.0001	0.0002	0.0009
	14	0.0000	0.0000	0.0000	0.0000	0.0000	0.0000	0.0000	0.0000	0.0000	0.0001
15	0	0.4633	0.2059	0.0874	0.0352	0.0134	0.0047	0.0016	0.0005	0.0001	0.0000
	1	0.3658	0.3432	0.2312	0.1319	0.0668	0.0305	0.0126	0.0047	0.0016	0.0005
	2	0.1348	0.2669	0.2856	0.2309	0.1559	0.0916	0.0476	0.0219	0.0090	0.0032
	3	0.0307	0.1285	0.2184	0.2501	0.2252	0.1700	0.1110	0.0634	0.0318	0.0139
	4	0.0049	0.0428	0.1156	0.1876	0.2252	0.2186	0.1792	0.1268	0.0780	0.0417
	5	0.0006	0.0105	0.0449	0.1032	0.1651	0.2061	0.2123	0.1859	0.1404	0.0916
	6	0.0000	0.0019	0.0132	0.0430	0.0917	0.1472	0.1906	0.2066	0.1914	0.1527
	7	0.0000	0.0003	0.0030	0.0138	0.0393	0.0811	0.1319	0.1771	0.2013	0.1964
	8	0.0000	0.0000	0.0005	0.0035	0.0131	0.0348	0.0710	0.1181	0.1647	0.1964
	9	0.0000	0.0000	0.0001	0.0007	0.0034	0.0116	0.0298	0.0612	0.1048	0.1527
	10	0.0000	0.0000	0.0000	0.0001	0.0007	0.0030	0.0096	0.0245	0.0515	0.0916
	11	0.0000	0.0000	0.0000	0.0000	0.0001	0.0006	0.0024	0.0074	0.0191	0.0417
	12	0.0000	0.0000	0.0000	0.0000	0.0000	0.0001	0.0004	0.0016	0.0052	0.0139
	13	0.0000	0.0000	0.0000	0.0000	0.0000	0.0000	0.0001	0.0003	0.0010	0.0032
	14	0.0000	0.0000	0.0000	0.0000	0.0000	0.0000	0.0000	0.0000	0.0001	0.0005
	15	0.0000	0.0000	0.0000	0.0000	0.0000	0.0000	0.0000	0.0000	0.0000	0.0000
16	0	0.4401	0.1853	0.0743	0.0281	0.0100	0.0033	0.0010	0.0003	0.0001	0.0000
	1	0.3706	0.3294	0.2097	0.1126	0.0535	0.0228	0.0087	0.0030	0.0009	0.0002
	2	0.1463	0.2745	0.2775	0.2111	0.1336	0.0732	0.0353	0.0150	0.0056	0.0018
	3	0.0359	0.1423	0.2285	0.2463	0.2079	0.1465	0.0888	0.0468	0.0215	0.0085
	4	0.0061	0.0514	0.1311	0.2001	0.2252	0.2040	0.1553	0.1014	0.0572	0.0278
	5	0.0008	0.0137	0.0555	0.1201	0.1802	0.2099	0.2008	0.1623	0.1123	0.0667
	6	0.0001	0.0028	0.0180	0.0550	0.1101	0.1649	0.1982	0.1983	0.1684	0.1222
	7	0.0000	0.0004	0.0045	0.0197	0.0524	0.1010	0.1524	0.1889	0.1969	0.1746
	8	0.0000	0.0001	0.0009	0.0055	0.0197	0.0487	0.0923	0.1417	0.1812	0.1964
	9	0.0000	0.0000	0.0001	0.0012	0.0058	0.0185	0.0442	0.0840	0.1318	0.1746
	10	0.0000	0.0000	0.0000	0.0002	0.0014	0.0056	0.0167	0.0392	0.0755	0.1222
	11	0.0000	0.0000	0.0000	0.0000	0.0002	0.0013	0.0049	0.0142	0.0337	0.0667
	12	0.0000	0.0000	0.0000	0.0000	0.0000	0.0002	0.0011	0.0040	0.0115	0.0278
	13	0.0000	0.0000	0.0000	0.0000	0.0000	0.0000	0.0002	0.0008	0.0029	0.0085
	14	0.0000	0.0000	0.0000	0.0000	0.0000	0.0000	0.0000	0.0001	0.0005	0.0018

TABLE A-2   (continued)

n	x	0.05	0.10	0.15	0.20	0.25	*p* 0.30	0.35	0.40	0.45	0.50
	15	0.0000	0.0000	0.0000	0.0000	0.0000	0.0000	0.0000	0.0000	0.0001	0.0002
	16	0.0000	0.0000	0.0000	0.0000	0.0000	0.0000	0.0000	0.0000	0.0000	0.0000
17	0	0.4181	0.1668	0.0631	0.0225	0.0075	0.0023	0.0007	0.0002	0.0000	0.0000
	1	0.3741	0.3150	0.1893	0.0957	0.0426	0.0169	0.0060	0.0019	0.0005	0.0001
	2	0.1575	0.2800	0.2673	0.1914	0.1136	0.0581	0.0260	0.0102	0.0035	0.0010
	3	0.0415	0.1556	0.2359	0.2393	0.1893	0.1245	0.0701	0.0341	0.0144	0.0052
	4	0.0076	0.0605	0.1457	0.2093	0.2209	0.1868	0.1320	0.0796	0.0411	0.0182
	5	0.0010	0.0175	0.0668	0.1361	0.1914	0.2081	0.1849	0.1379	0.0875	0.0472
	6	0.0001	0.0039	0.0236	0.0680	0.1276	0.1784	0.1991	0.1839	0.1432	0.0944
	7	0.0000	0.0007	0.0065	0.0267	0.0668	0.1201	0.1685	0.1927	0.1841	0.1484
	8	0.0000	0.0001	0.0014	0.0084	0.0279	0.0644	0.1134	0.1606	0.1883	0.1855
	9	0.0000	0.0000	0.0003	0.0021	0.0093	0.0276	0.0611	0.1070	0.1540	0.1855
	10	0.0000	0.0000	0.0000	0.0004	0.0025	0.0095	0.0263	0.0571	0.1008	0.1484
	11	0.0000	0.0000	0.0000	0.0001	0.0005	0.0026	0.0090	0.0242	0.0525	0.0944
	12	0.0000	0.0000	0.0000	0.0000	0.0001	0.0006	0.0024	0.0081	0.0215	0.0472
	13	0.0000	0.0000	0.0000	0.0000	0.0000	0.0001	0.0005	0.0021	0.0068	0.0182
	14	0.0000	0.0000	0.0000	0.0000	0.0000	0.0000	0.0001	0.0004	0.0016	0.0052
	15	0.0000	0.0000	0.0000	0.0000	0.0000	0.0000	0.0000	0.0001	0.0003	0.0010
	16	0.0000	0.0000	0.0000	0.0000	0.0000	0.0000	0.0000	0.0000	0.0000	0.0001
	17	0.0000	0.0000	0.0000	0.0000	0.0000	0.0000	0.0000	0.0000	0.0000	0.0000
18	0	0.3972	0.1501	0.0536	0.0180	0.0056	0.0016	0.0004	0.0001	0.0000	0.0000
	1	0.3763	0.3002	0.1704	0.0811	0.0338	0.0126	0.0042	0.0012	0.0003	0.0001
	2	0.1683	0.2835	0.2556	0.1723	0.0958	0.0458	0.0190	0.0069	0.0022	0.0006
	3	0.0473	0.1680	0.2406	0.2297	0.1704	0.1046	0.0547	0.0246	0.0095	0.0031
	4	0.0093	0.0700	0.1592	0.2153	0.2130	0.1681	0.1104	0.0614	0.0291	0.0117
	5	0.0014	0.0218	0.0787	0.1507	0.1988	0.2017	0.1664	0.1146	0.0666	0.0327
	6	0.0002	0.0052	0.0301	0.0816	0.1436	0.1873	0.1941	0.1655	0.1181	0.0708
	7	0.0000	0.0010	0.0091	0.0350	0.0820	0.1376	0.1792	0.1892	0.1657	0.1214
	8	0.0000	0.0002	0.0022	0.0120	0.0376	0.0811	0.1327	0.1734	0.1864	0.1669
	9	0.0000	0.0000	0.0004	0.0033	0.0139	0.0386	0.0794	0.1284	0.1694	0.1855
	10	0.0000	0.0000	0.0001	0.0008	0.0042	0.0149	0.0385	0.0771	0.1248	0.1669
	11	0.0000	0.0000	0.0000	0.0001	0.0010	0.0046	0.0151	0.0374	0.0742	0.1214
	12	0.0000	0.0000	0.0000	0.0000	0.0002	0.0012	0.0047	0.0145	0.0354	0.0708
	13	0.0000	0.0000	0.0000	0.0000	0.0000	0.0002	0.0012	0.0045	0.0134	0.0327
	14	0.0000	0.0000	0.0000	0.0000	0.0000	0.0000	0.0002	0.0011	0.0039	0.0117
	15	0.0000	0.0000	0.0000	0.0000	0.0000	0.0000	0.0000	0.0002	0.0009	0.0031
	16	0.0000	0.0000	0.0000	0.0000	0.0000	0.0000	0.0000	0.0000	0.0001	0.0006
	17	0.0000	0.0000	0.0000	0.0000	0.0000	0.0000	0.0000	0.0000	0.0000	0.0001
	18	0.0000	0.0000	0.0000	0.0000	0.0000	0.0000	0.0000	0.0000	0.0000	0.0000

TABLE A-2   (continued)

						p					
n	*x*	0.05	0.10	0.15	0.20	0.25	0.30	0.35	0.40	0.45	0.50
19	0	0.3774	0.1351	0.0456	0.0144	0.0042	0.0011	0.0003	0.0001	0.0000	0.0000
	1	0.3774	0.2852	0.1529	0.0685	0.0268	0.0093	0.0029	0.0008	0.0002	0.0000
	2	0.1787	0.2852	0.2428	0.1540	0.0803	0.0358	0.0138	0.0046	0.0013	0.0003
	3	0.0533	0.1796	0.2428	0.2182	0.1517	0.0869	0.0422	0.0175	0.0062	0.0018
	4	0.0112	0.0798	0.1714	0.2182	0.2023	0.1491	0.0909	0.0467	0.0203	0.0074
	5	0.0018	0.0266	0.0907	0.1636	0.2023	0.1916	0.1468	0.0933	0.0497	0.0222
	6	0.0002	0.0069	0.0374	0.0955	0.1574	0.1916	0.1844	0.1451	0.0949	0.0518
	7	0.0000	0.0014	0.0122	0.0443	0.0974	0.1525	0.1844	0.1797	0.1443	0.0961
	8	0.0000	0.0002	0.0032	0.0166	0.0487	0.0981	0.1489	0.1797	0.1771	0.1442
	9	0.0000	0.0000	0.0007	0.0051	0.0198	0.0514	0.0980	0.1464	0.1771	0.1762
	10	0.0000	0.0000	0.0001	0.0013	0.0066	0.0220	0.0528	0.0976	0.1449	0.1762
	11	0.0000	0.0000	0.0000	0.0003	0.0018	0.0077	0.0233	0.0532	0.0970	0.1442
	12	0.0000	0.0000	0.0000	0.0000	0.0004	0.0022	0.0083	0.0237	0.0529	0.0961
	13	0.0000	0.0000	0.0000	0.0000	0.0001	0.0005	0.0024	0.0085	0.0233	0.0518
	14	0.0000	0.0000	0.0000	0.0000	0.0000	0.0001	0.0006	0.0024	0.0082	0.0222
	15	0.0000	0.0000	0.0000	0.0000	0.0000	0.0000	0.0001	0.0005	0.0022	0.0074
	16	0.0000	0.0000	0.0000	0.0000	0.0000	0.0000	0.0000	0.0001	0.0005	0.0018
	17	0.0000	0.0000	0.0000	0.0000	0.0000	0.0000	0.0000	0.0000	0.0001	0.0003
	18	0.0000	0.0000	0.0000	0.0000	0.0000	0.0000	0.0000	0.0000	0.0000	0.0000
	19	0.0000	0.0000	0.0000	0.0000	0.0000	0.0000	0.0000	0.0000	0.0000	0.0000
20	0	0.3585	0.1216	0.0388	0.0115	0.0032	0.0008	0.0002	0.0000	0.0000	0.0000
	1	0.3774	0.2702	0.1368	0.0576	0.0211	0.0068	0.0020	0.0005	0.0001	0.0000
	2	0.1887	0.2852	0.2293	0.1369	0.0669	0.0278	0.0100	0.0031	0.0008	0.0002
	3	0.0596	0.1901	0.2428	0.2054	0.1339	0.0716	0.0323	0.0123	0.0040	0.0011
	4	0.0133	0.0898	0.1821	0.2182	0.1897	0.1304	0.0738	0.0350	0.0139	0.0046
	5	0.0022	0.0319	0.1028	0.1746	0.2023	0.1789	0.1272	0.0746	0.0365	0.0148
	6	0.0003	0.0089	0.0454	0.1091	0.1686	0.1916	0.1712	0.1244	0.0746	0.0370
	7	0.0000	0.0020	0.0160	0.0545	0.1124	0.1643	0.1844	0.1659	0.1221	0.0739
	8	0.0000	0.0004	0.0046	0.0222	0.0609	0.1144	0.1614	0.1797	0.1623	0.1201
	9	0.0000	0.0001	0.0011	0.0074	0.0271	0.0654	0.1158	0.1597	0.1771	0.1602
	10	0.0000	0.0000	0.0002	0.0020	0.0099	0.0308	0.0686	0.1171	0.1593	0.1762
	11	0.0000	0.0000	0.0000	0.0005	0.0030	0.0120	0.0336	0.0710	0.1185	0.1602
	12	0.0000	0.0000	0.0000	0.0001	0.0008	0.0039	0.0136	0.0355	0.0727	0.1201
	13	0.0000	0.0000	0.0000	0.0000	0.0002	0.0010	0.0045	0.0146	0.0366	0.0739
	14	0.0000	0.0000	0.0000	0.0000	0.0000	0.0002	0.0012	0.0049	0.0150	0.0370
	15	0.0000	0.0000	0.0000	0.0000	0.0000	0.0000	0.0003	0.0013	0.0049	0.0148
	16	0.0000	0.0000	0.0000	0.0000	0.0000	0.0000	0.0000	0.0003	0.0013	0.0046
	17	0.0000	0.0000	0.0000	0.0000	0.0000	0.0000	0.0000	0.0000	0.0002	0.0011
	18	0.0000	0.0000	0.0000	0.0000	0.0000	0.0000	0.0000	0.0000	0.0000	0.0002
	19	0.0000	0.0000	0.0000	0.0000	0.0000	0.0000	0.0000	0.0000	0.0000	0.0000
	20	0.0000	0.0000	0.0000	0.0000	0.0000	0.0000	0.0000	0.0000	0.0000	0.0000

Selected values of the Poisson cumulative distribution

POISSON *(handwritten)*

$$F(c) = P(X \leqslant c) = \sum_{x=0}^{c} \frac{\mu^x e^{-\mu}}{x!}$$

μ = MEAN OF DISTRIBUTION (handwritten)
x = NUMBER OF SUCCESS (handwritten)

Example If $\mu = 1.00$, then $F(2) = P(X \leqslant 2) = 0.920$.

NOTE: TABLE IS CUMULATIVE (handwritten)

EX: 3 FISH (handwritten)

.647 - .423 = (handwritten)

PROBABILITY OF CATCHING (handwritten)

μ	c = 0	1	2	3	4	5	6	7	8	9
0.02	0.980	1.000								
0.04	0.961	0.999	1.000							
0.06	0.942	0.998	1.000							
0.08	0.923	0.997	1.000							
0.10	0.905	0.995	1.000							
0.15	0.861	0.990	0.999	1.000						
0.20	0.819	0.982	0.999	1.000						
0.25	0.779	0.974	0.998	1.000						
0.30	0.741	0.963	0.996	1.000						
0.35	0.705	0.951	0.994	1.000						
0.40	0.670	0.938	0.992	0.999	1.000					
0.45	0.638	0.925	0.989	0.999	1.000					
0.50	0.607	0.910	0.986	0.998	1.000					
0.55	0.577	0.894	0.982	0.998	1.000					
0.60	0.549	0.878	0.977	0.997	1.000					
0.65	0.522	0.861	0.972	0.996	0.999	1.000				
0.70	0.497	0.844	0.966	0.994	0.999	1.000				
0.75	0.472	0.827	0.959	0.993	0.999	1.000				
0.80	0.449	0.809	0.953	0.991	0.999	1.000				
0.85	0.427	0.791	0.945	0.989	0.998	1.000				
0.90	0.407	0.772	0.937	0.987	0.998	1.000				
0.95	0.387	0.754	0.929	0.984	0.997	1.000				
1.00	0.368	0.736	0.920	0.981	0.996	0.999	1.000			
1.10	0.333	0.699	0.900	0.974	0.995	0.999	1.000			
1.20	0.301	0.663	0.879	0.966	0.992	0.998	1.000			
1.30	0.273	0.627	0.857	0.957	0.989	0.998	1.000			
1.40	0.247	0.592	0.833	0.946	0.986	0.997	0.999	1.000		
1.50	0.223	0.558	0.809	0.934	0.981	0.996	0.999	1.000		
1.60	0.202	0.525	0.783	0.921	0.976	0.994	0.999	1.000		
1.70	0.183	0.493	0.757	0.907	0.970	0.992	0.998	1.000		
1.80	0.165	0.463	0.731	0.891	0.964	0.990	0.997	0.999	1.000	
1.90	0.150	0.434	0.704	0.875	0.956	0.987	0.997	0.999	1.000	
2.00	0.135	0.406	0.677	0.857	0.947	0.983	0.995	0.999	1.000	
2.20	0.111	0.355	0.623	0.819	0.928	0.975	0.993	0.998	1.000	
2.40	0.091	0.308	0.570	0.779	0.904	0.964	0.988	0.997	0.999	1.000
2.60	0.074	0.267	0.518	0.736	0.877	0.951	0.983	0.995	0.999	1.000
2.80	0.061	0.231	0.469	0.692	0.848	0.935	0.976	0.992	0.998	0.999
3.00	0.050	0.199	0.423	0.647	0.815	0.916	0.966	0.988	0.996	0.999

Source: From Eugene L. Grant, *Statistical Quality Control*, Copyright 1964 by McGraw-Hill Book Company. Used with permission of McGraw-Hill Book Company.

TABLE A-3   (continued)

μ \ c	0	1	2	3	4	5	6	7	8	9
3.20	0.041	0.171	0.380	0.603	0.781	0.895	0.955	0.983	0.994	0.998
3.40	0.033	0.147	0.340	0.558	0.744	0.871	0.942	0.977	0.992	0.997
3.60	0.027	0.126	0.303	0.515	0.706	0.844	0.927	0.969	0.988	0.996
3.80	0.022	0.107	0.269	0.473	0.668	0.816	0.909	0.960	0.984	0.994
4.00	0.018	0.092	0.238	0.433	0.629	0.785	0.889	0.949	0.979	0.992
4.20	0.015	0.078	0.210	0.395	0.590	0.753	0.867	0.936	0.972	0.989
4.40	0.012	0.066	0.185	0.359	0.551	0.720	0.844	0.921	0.964	0.985
4.60	0.010	0.056	0.163	0.326	0.513	0.686	0.818	0.905	0.955	0.980
4.80	0.008	0.048	0.143	0.294	0.476	0.651	0.791	0.887	0.944	0.975
5.00	0.007	0.040	0.125	0.265	0.440	0.616	0.762	0.867	0.932	0.968
5.20	0.006	0.034	0.109	0.238	0.406	0.581	0.732	0.845	0.918	0.960
5.40	0.005	0.029	0.095	0.213	0.373	0.546	0.702	0.822	0.903	0.951
5.60	0.004	0.024	0.082	0.191	0.342	0.512	0.670	0.797	0.886	0.941
5.80	0.003	0.021	0.072	0.170	0.313	0.478	0.638	0.771	0.867	0.929
6.00	0.002	0.017	0.062	0.151	0.285	0.446	0.606	0.744	0.847	0.916

μ	10	11	12	13	14	15	16
2.80	1.000						
3.00	1.000						
3.20	1.000						
3.40	0.999	1.000					
3.60	0.999	1.000					
3.80	0.998	0.999	1.000				
4.00	0.997	0.999	1.000				
4.20	0.996	0.999	1.000				
4.40	0.994	0.998	0.999	1.000			
4.60	0.992	0.997	0.999	1.000			
4.80	0.990	0.996	0.999	1.000			
5.00	0.986	0.995	0.998	0.999	1.000		
5.20	9.982	0.993	0.997	0.999	1.000		
5.40	0.977	0.990	0.996	0.999	1.000		
5.60	0.972	0.988	0.995	0.998	0.999	1.000	
5.80	0.965	0.984	0.993	0.997	0.999	1.000	
6.00	0.957	0.980	0.991	0.996	0.999	0.999	1.000

μ \ c	0	1	2	3	4	5	6	7	8	9
6.20	0.002	0.015	0.054	0.134	0.259	0.414	0.574	0.716	0.826	0.902
6.40	0.002	0.012	0.046	0.119	0.235	0.384	0.542	0.687	0.803	0.886
6.60	0.001	0.010	0.040	0.105	0.213	0.355	0.511	0.658	0.780	0.869
6.80	0.001	0.009	0.034	0.093	0.192	0.327	0.480	0.628	0.755	0.850
7.00	0.001	0.007	0.030	0.082	0.173	0.301	0.450	0.599	0.729	0.830

TABLE A-3   (continued)

c / μ	0	1	2	3	4	5	6	7	8	9
7.20	0.001	0.006	0.025	0.072	0.156	0.276	0.420	0.569	0.703	0.810
7.40	0.001	0.005	0.022	0.063	0.140	0.253	0.392	0.539	0.676	0.788
7.60	0.001	0.004	0.019	0.055	0.125	0.231	0.365	0.510	0.648	0.765
7.80	0.000	0.004	0.016	0.048	0.112	0.210	0.338	0.481	0.620	0.741
8.00	0.000	0.003	0.014	0.042	0.100	0.191	0.313	0.453	0.593	0.717
8.50	0.000	0.002	0.009	0.030	0.074	0.150	0.256	0.386	0.523	0.653
9.00	0.000	0.001	0.006	0.021	0.055	0.116	0.207	0.324	0.456	0.587
9.50	0.000	0.001	0.004	0.015	0.040	0.089	0.165	0.269	0.392	0.522
10.00	0.000	0.000	0.003	0.010	0.029	0.067	0.130	0.220	0.333	0.458

μ	10	11	12	13	14	15	16	17	18	19
6.20	0.949	0.975	0.989	0.995	0.998	0.999	1.000			
6.40	0.939	0.969	0.986	0.994	0.997	0.999	1.000			
6.60	0.927	0.963	0.982	0.992	0.997	0.999	0.999	1.000		
6.80	0.915	0.955	0.978	0.990	0.996	0.998	0.999	1.000		
7.00	0.901	0.947	0.973	0.987	0.994	0.998	0.999	1.000		
7.20	0.887	0.937	0.967	0.984	0.993	0.997	0.999	0.999	1.000	
7.40	0.871	0.926	0.961	0.980	0.991	0.996	0.998	0.999	1.000	
7.60	0.854	0.915	0.954	0.976	0.989	0.995	0.998	0.999	1.000	
7.80	0.835	0.902	0.945	0.971	0.986	0.993	0.997	0.999	1.000	
8.00	0.816	0.888	0.936	0.966	0.983	0.992	0.996	0.998	0.999	1.000
8.50	0.763	0.849	0.909	0.949	0.973	0.986	0.993	0.997	0.999	0.999
9.00	0.706	0.803	0.876	0.926	0.959	0.978	0.989	0.995	0.998	0.999
9.50	0.645	0.752	0.836	0.898	0.940	0.967	0.982	0.991	0.996	0.998
10.00	0.583	0.697	0.792	0.864	0.917	0.951	0.973	0.986	0.993	0.997

μ	20	21	22
8.50	1.000		
9.00	1.000		
9.50	0.999	1.000	
10.00	0.998	0.999	1.000

c / μ	0	1	2	3	4	5	6	7	8	9
10.50	0.000	0.000	0.002	0.007	0.021	0.050	0.102	0.179	0.279	0.397
11.00	0.000	0.000	0.001	0.005	0.015	0.038	0.079	0.143	0.232	0.341
11.50	0.000	0.000	0.001	0.003	0.011	0.028	0.060	0.114	0.191	0.289
12.00	0.000	0.000	0.001	0.002	0.008	0.020	0.046	0.090	0.155	0.242
12.50	0.000	0.000	0.000	0.002	0.005	0.015	0.035	0.070	0.125	0.201

TABLE A-3   (continued)

c / μ	0	1	2	3	4	5	6	7	8	9
13.00	0.000	0.000	0.000	0.001	0.004	0.011	0.026	0.054	0.100	0.166
13.50	0.000	0.000	0.000	0.001	0.003	0.008	0.019	0.041	0.079	0.135
14.00	0.000	0.000	0.000	0.000	0.002	0.006	0.014	0.032	0.062	0.109
14.50	0.000	0.000	0.000	0.000	0.001	0.004	0.010	0.024	0.048	0.088
15.00	0.000	0.000	0.000	0.000	0.001	0.003	0.008	0.018	0.037	0.070

	10	11	12	13	14	15	16	17	18	19
10.50	0.521	0.639	0.742	0.825	0.888	0.932	0.960	0.978	0.988	0.994
11.00	0.460	0.579	0.689	0.781	0.854	0.907	0.944	0.968	0.982	0.991
11.50	0.402	0.520	0.633	0.733	0.815	0.878	0.924	0.954	0.974	0.986
12.00	0.347	0.462	0.576	0.682	0.772	0.844	0.899	0.937	0.963	0.979
12.50	0.297	0.406	0.519	0.628	0.725	0.806	0.869	0.916	0.948	0.969
13.00	0.252	0.353	0.463	0.573	0.675	0.764	0.835	0.890	0.930	0.957
13.50	0.211	0.304	0.409	0.518	0.623	0.718	0.798	0.861	0.908	0.942
14.00	0.176	0.260	0.358	0.464	0.570	0.669	0.756	0.827	0.883	0.923
14.50	0.145	0.220	0.311	0.413	0.518	0.619	0.711	0.790	0.853	0.901
15.00	0.118	0.185	0.268	0.363	0.466	0.568	0.664	0.749	0.819	0.875

	20	21	22	23	24	25	26	27	28	29
10.50	0.997	0.999	0.999	1.000						
11.00	0.995	0.998	0.999	1.000						
11.50	0.992	0.996	0.998	0.999	1.000					
12.00	0.988	0.994	0.997	0.999	0.999	1.000				
12.50	0.983	0.991	0.995	0.998	0.999	0.999	1.000			
13.00	0.975	0.986	0.992	0.996	0.998	0.999	1.000			
13.50	0.965	0.980	0.989	0.994	0.997	0.998	0.999	1.000		
14.00	0.952	0.971	0.983	0.991	0.995	0.997	0.999	0.999	1.000	
14.50	0.936	0.960	0.976	0.986	0.992	0.996	0.998	0.999	0.999	1.000
15.00	0.917	0.947	0.967	0.981	0.989	0.994	0.997	0.998	0.999	1.000

c / μ	4	5	6	7	8	9	10	11	12	13
16.00	0.000	0.001	0.004	0.010	0.022	0.043	0.077	0.127	0.193	0.275
17.00	0.000	0.001	0.002	0.005	0.013	0.026	0.049	0.085	0.135	0.201
18.00	0.000	0.000	0.001	0.003	0.007	0.015	0.030	0.055	0.092	0.143
19.00	0.000	0.000	0.001	0.002	0.004	0.009	0.018	0.035	0.061	0.098
20.00	0.000	0.000	0.000	0.001	0.002	0.005	0.011	0.021	0.039	0.066
21.00	0.000	0.000	0.000	0.000	0.001	0.003	0.006	0.013	0.025	0.043
22.00	0.000	0.000	0.000	0.000	0.001	0.002	0.004	0.008	0.015	0.028
23.00	0.000	0.000	0.000	0.000	0.000	0.001	0.002	0.004	0.009	0.017
24.00	0.000	0.000	0.000	0.000	0.000	0.000	0.001	0.003	0.005	0.011
25.00	0.000	0.000	0.000	0.000	0.000	0.000	0.001	0.001	0.003	0.006

TABLE A-3   (continued)

	14	15	16	17	18	19	20	21	22	23
16.00	0.368	0.467	0.566	0.659	0.742	0.812	0.868	0.911	0.942	0.963
17.00	0.281	0.371	0.468	0.564	0.655	0.736	0.805	0.861	0.905	0.937
18.00	0.208	0.287	0.375	0.469	0.562	0.651	0.731	0.799	0.855	0.899
19.00	0.150	0.215	0.292	0.378	0.469	0.561	0.647	0.725	0.793	0.849
20.00	0.105	0.157	0.221	0.297	0.381	0.470	0.559	0.644	0.721	0.787
21.00	0.072	0.111	0.163	0.227	0.302	0.384	0.471	0.558	0.640	0.716
22.00	0.048	0.077	0.117	0.169	0.232	0.306	0.387	0.472	0.556	0.637
23.00	0.031	0.052	0.082	0.123	0.175	0.238	0.310	0.389	0.472	0.555
24.00	0.020	0.034	0.056	0.087	0.128	0.180	0.243	0.314	0.392	0.473
25.00	0.012	0.022	0.038	0.060	0.092	0.134	0.185	0.247	0.318	0.394

	24	25	26	27	28	29	30	31	32	33
16.00	0.978	0.987	0.993	0.996	0.998	0.999	0.999	1.000		
17.00	0.959	0.975	0.985	0.991	0.995	0.997	0.999	0.999	1.000	
18.00	0.932	0.955	0.972	0.983	0.990	0.994	0.997	0.998	0.999	1.000
19.00	0.893	0.927	0.951	0.969	0.980	0.988	0.993	0.996	0.998	0.999
20.00	0.843	0.888	0.922	0.948	0.966	0.978	0.987	0.992	0.995	0.997
21.00	0.782	0.838	0.883	0.917	0.944	0.963	0.976	0.985	0.991	0.994
22.00	0.712	0.777	0.832	0.877	0.913	0.940	0.959	0.973	0.983	0.989
23.00	0.635	0.708	0.772	0.827	0.873	0.908	0.936	0.956	0.971	0.981
24.00	0.554	0.632	0.704	0.768	0.823	0.868	0.904	0.932	0.953	0.969
25.00	0.473	0.553	0.629	0.700	0.763	0.818	0.863	0.900	0.929	0.950

	34	35	36	37	38	39	40	41	42	43
19.00	0.999	1.000								
20.00	0.999	0.999	1.000							
21.00	0.997	0.998	0.999	0.999	1.000					
22.00	0.994	0.996	0.998	0.999	0.999	1.000				
23.00	0.988	0.993	0.996	0.997	0.999	0.999	1.000			
24.00	0.979	0.987	0.992	0.995	0.997	0.998	0.999	0.999	1.000	
25.00	0.966	0.978	0.985	0.991	0.994	0.997	0.998	0.999	0.999	1.000

TABLE A-4
Four-place common logarithms

N	0	1	2	3	4	5	6	7	8	9	1	2	3	4	5	6	7	8	9
											Proportional Parts								
10	0000	0043	0086	0128	0170	0212	0253	0294	0334	0374	4	8	12	17	21	25	29	33	37
11	0414	0453	0492	0531	0569	0607	0645	0682	0719	0755	4	8	11	15	19	23	26	30	34
12	0792	0828	0864	0899	0934	0969	1004	1038	1072	1106	3	7	10	14	17	21	24	28	31
13	1139	1173	1206	1239	1271	1303	1335	1367	1399	1430	3	6	10	13	16	19	23	26	29
14	1461	1492	1523	1553	1584	1614	1644	1673	1703	1732	3	6	9	12	15	18	21	24	27
15	1761	1790	1818	1847	1875	1903	1931	1959	1987	2014	3	6	8	11	14	17	20	22	25
16	2041	2068	2095	2122	2148	2175	2201	2227	2253	2279	3	5	8	11	13	16	18	21	24
17	2304	2330	2355	2380	2405	2430	2455	2480	2504	2529	2	5	7	10	12	15	17	20	22
18	2553	2577	2601	2625	2648	2672	2695	2718	2742	2765	2	5	7	9	12	14	16	19	21
19	2788	2810	2833	2856	2878	2900	2923	2945	2967	2989	2	4	7	9	11	13	16	18	20
20	3010	3032	3054	3075	3096	3118	3139	3160	3181	3201	2	4	6	8	11	13	15	17	19
21	3222	3243	3263	3284	3304	3324	3345	3365	3385	3404	2	4	6	8	10	12	14	16	18
22	3424	3444	3464	3483	3502	3522	3541	3560	3579	3598	2	4	6	8	10	12	14	15	17
23	3617	3636	3655	3674	3692	3711	3729	3747	3766	3784	2	4	6	7	9	11	13	15	17
24	3802	3820	3838	3856	3874	3892	3909	3927	3945	3962	2	4	5	7	9	11	12	14	16
25	3979	3997	4014	4031	4048	4065	4082	4099	4116	4133	2	3	5	7	9	10	12	14	15
26	4150	4166	4183	4200	4216	4232	4249	4265	4281	4298	2	3	5	7	8	10	11	13	15
27	4314	4330	4346	4362	4378	4393	4409	4425	4440	4456	2	3	5	6	8	9	11	13	14
28	4472	4487	4502	4518	4533	4548	4564	4579	4594	4609	2	3	5	6	8	9	11	12	14
29	4624	4639	4654	4669	4683	4698	4713	4728	4742	4757	1	3	4	6	7	9	10	12	13
30	4771	4786	4800	4814	4829	4843	4857	4871	4886	4900	1	3	4	6	7	9	10	11	13
31	4914	4928	4942	4955	4969	4983	4997	5011	5024	5038	1	3	4	6	7	8	10	11	12
32	5051	5065	5079	5092	5105	5119	5132	5145	5159	5172	1	3	4	5	7	8	9	11	12
33	5185	5198	5211	5224	5237	5250	5263	5276	5289	5302	1	3	4	5	6	8	9	10	12
34	5315	5328	5340	5353	5366	5378	5391	5403	5416	5428	1	3	4	5	6	8	9	10	11
35	5441	5453	5465	5478	5490	5502	5514	5527	5539	5551	1	2	4	5	6	7	9	10	11
36	5563	5575	5587	5599	5611	5623	5635	5647	5658	5670	1	2	4	5	6	7	8	10	11
37	5682	5694	5705	5717	5729	5740	5752	5763	5775	5786	1	2	3	5	6	7	8	9	10
38	5798	5809	5821	5832	5843	5855	5866	5877	5888	5899	1	2	3	5	6	7	8	9	10
39	5911	5922	5933	5944	5955	5966	5977	5988	5999	6010	1	2	3	4	5	7	8	9	10
40	6021	6031	6042	6053	6064	6075	6085	6096	6107	6117	1	2	3	4	5	6	8	9	10
41	6128	6138	6149	6160	6170	6180	6191	6201	6212	6222	1	2	3	4	5	6	7	8	9
42	6232	6243	6253	6263	6274	6284	6294	6304	6314	6325	1	2	3	4	5	6	7	8	9
43	6335	6345	6355	6365	6375	6385	6395	6405	6415	6425	1	2	3	4	5	6	7	8	9
44	6435	6444	6454	6464	6474	6484	6493	6503	6513	6522	1	2	3	4	5	6	7	8	9
45	6532	6542	6551	6561	6571	6580	6590	6599	6609	6618	1	2	3	4	5	6	7	8	9
46	6628	6637	6646	6656	6665	6675	6684	6693	6702	6712	1	2	3	4	5	6	7	7	8
47	6721	6730	6739	6749	6758	6767	6776	6785	6794	6803	1	2	3	4	5	5	6	7	8
48	6812	6821	6830	6839	6848	6857	6866	6875	6884	6893	1	2	3	4	4	5	6	7	8
49	6902	6911	6920	6928	6937	6946	6955	6964	6972	6981	1	2	3	4	4	5	6	7	8
50	6990	6998	7007	7016	7024	7033	7042	7050	7059	7067	1	2	3	3	4	5	6	7	8
51	7076	7084	7093	7101	7110	7118	7126	7135	7143	7152	1	2	3	3	4	5	6	7	8
52	7160	7168	7177	7185	7193	7202	7210	7218	7226	7235	1	2	2	3	4	5	6	7	7
53	7243	7251	7259	7267	7275	7284	7292	7300	7308	7316	1	2	2	3	4	5	6	6	7
54	7324	7332	7340	7348	7356	7364	7372	7380	7388	7396	1	2	2	3	4	5	6	6	7
N	0	1	2	3	4	5	6	7	8	9	1	2	3	4	5	6	7	8	9

N	0	1	2	3	4	5	6	7	8	9	Proportional Parts								
											1	2	3	4	5	6	7	8	9
55	7404	7412	7419	7427	7435	7443	7451	7459	7466	7474	1	2	2	3	4	5	5	6	7
56	7482	7490	7497	7505	7513	7520	7528	7536	7543	7551	1	2	2	3	4	5	5	6	7
57	7559	7566	7574	7582	7589	7597	7604	7612	7619	7627	1	2	2	3	4	5	5	6	7
58	7634	7642	7649	7657	7664	7672	7679	7686	7694	7701	1	1	2	3	4	4	5	6	7
59	7709	7716	7723	7731	7738	7745	7752	7760	7767	7774	1	1	2	3	4	4	5	6	7
60	7782	7789	7796	7803	7810	7818	7825	7823	7839	7846	1	1	2	3	4	4	5	6	6
61	7853	7860	7868	7875	7882	7889	7896	7903	7910	7917	1	1	2	3	4	4	5	6	6
62	7924	7931	7938	7945	7952	7959	7966	7973	7980	7987	1	1	2	3	3	4	5	6	6
63	7993	8000	8007	8014	8021	8028	8035	8041	8048	8055	1	1	2	3	3	4	5	5	6
64	8062	8069	8075	8082	8089	8096	8102	8109	8116	8122	1	1	2	3	3	4	5	5	6
65	8129	8136	8142	8149	8156	8162	8169	8176	8182	8189	1	1	2	3	3	4	5	5	6
66	8195	8202	8209	8215	8222	8228	8235	8241	8248	8254	1	1	2	3	3	4	5	5	6
67	8261	8267	8274	8280	8287	8293	8299	8306	8312	8319	1	1	2	3	3	4	5	5	6
68	8325	8331	8338	8344	8351	8357	8363	8370	8376	8382	1	1	2	3	3	4	4	5	6
69	8388	8395	8401	8407	8414	8420	8426	8432	8439	8445	1	1	2	2	3	4	4	5	6
70	8451	8457	8463	8470	8476	8482	8488	8494	8500	8506	1	1	2	2	3	4	4	5	6
71	8513	8519	8525	8531	8537	8543	8549	8555	8561	8567	1	1	2	2	3	4	4	5	5
72	8573	8579	8585	8591	8597	8603	8609	8615	8621	8627	1	1	2	2	3	4	4	5	5
73	8633	8639	8645	8651	8657	8663	8669	8675	8681	8686	1	1	2	2	3	4	4	5	5
74	8692	8698	8704	8710	8716	8722	8727	8733	8739	8745	1	1	2	2	3	4	4	5	5
75	8751	8756	8762	8768	8774	8779	8785	8791	8797	8802	1	1	2	2	3	3	4	5	5
76	8808	8814	8820	8825	8831	8837	8842	8848	8854	8859	1	1	2	2	3	3	4	5	5
77	8865	8871	8876	8882	8887	8893	8899	8904	8910	8915	1	1	2	2	3	3	4	4	5
78	8921	8927	8932	8938	8943	8949	8954	8960	8965	8971	1	1	2	2	3	3	4	4	5
79	8976	8982	8987	8993	8998	9004	9009	9015	9020	9025	1	1	2	2	3	3	4	4	5
80	9031	9036	9042	9047	9053	9058	9063	9069	9074	9079	1	1	2	2	3	3	4	4	5
81	9085	9090	9096	9101	9106	9112	9117	9122	9128	9133	1	1	2	2	3	3	4	4	5
82	9138	9143	9149	9154	9159	9165	9170	9175	9180	9186	1	1	2	2	3	3	4	4	5
83	9191	9196	9201	9206	9212	9217	9222	9227	9232	9238	1	1	2	2	3	3	4	4	5
84	9243	9248	9253	9258	9263	9269	9274	9279	9284	9289	1	1	2	2	3	3	4	4	5
85	9294	9299	9304	9309	9315	9320	9325	9330	9335	9340	1	1	2	2	3	3	4	4	5
86	9345	9350	9355	9360	9365	9370	9375	9380	9385	9390	1	1	2	2	3	3	4	4	5
87	9395	9400	9405	9410	9415	9420	9425	9430	9435	9440	0	1	1	2	2	3	3	4	4
88	9445	9450	9455	9460	9465	9469	9474	9479	9484	9489	0	1	1	2	2	3	3	4	4
89	9494	9499	9504	9509	9513	9518	9523	9528	9533	9538	0	1	1	2	2	3	3	4	4
90	9542	9547	9552	9557	9562	9566	9571	9576	9581	9586	0	1	1	2	2	3	3	4	4
91	9590	9595	9600	9605	9609	9614	9619	9624	9628	9633	0	1	1	2	2	3	3	4	4
92	9638	9643	9647	9652	9657	9661	9666	9671	9675	9680	0	1	1	2	2	3	3	4	4
93	9685	9689	9694	9699	9703	9708	9713	9717	9722	9727	0	1	1	2	2	3	3	4	4
94	9731	9736	9741	9745	9750	9754	9759	9763	9768	9773	0	1	1	2	2	3	3	4	4
95	9777	9782	9786	9791	9795	9800	9805	9809	9814	9818	0	1	1	2	2	3	3	4	4
96	9823	9827	9832	9836	9841	9845	9850	9854	9859	9863	0	1	1	2	2	3	3	4	4
97	9868	9872	9877	9881	9886	9890	9894	9899	9903	9908	0	1	1	2	2	3	3	4	4
98	9912	9917	9921	9926	9930	9934	9939	9943	9948	9952	0	1	1	2	2	3	3	4	4
99	9956	9961	9965	9969	9974	9978	9983	9987	9991	9996	0	1	1	2	2	3	3	3	4
N	0	1	2	3	4	5	6	7	8	9	1	2	3	4	5	6	7	8	9

Areas under the standard normal probability distribution
between the mean and successive values of z

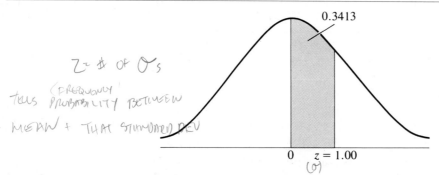

0.3413

$z =$ # of σs

TELLS (FREQUENCY
PROBABILITY BETWEEN

MEAN + THAT STANDARD DEV

0   z = 1.00
      (σ)

Example   If z = 1.00, then the area between the mean and this value of z is 0.3413.

z	0.00	0.01	0.02	0.03	0.04	0.05	0.06	0.07	0.08	0.09
0.0	0.0000	0.0040	0.0080	0.0120	0.0160	0.0199	0.0239	0.0279	0.0319	0.0359
0.1	0.0398	0.0438	0.0478	0.0517	0.0557	0.0596	0.0636	0.0675	0.0714	0.0753
0.2	0.0793	0.0832	0.0871	0.0910	0.0948	0.0987	0.1026	0.1064	0.1103	0.1141
0.3	0.1179	0.1217	0.1255	0.1293	0.1331	0.1368	0.1406	0.1443	0.1480	0.1517
0.4	0.1554	0.1591	0.1628	0.1664	0.1700	0.1736	0.1772	0.1808	0.1844	0.1879
0.5	0.1915	0.1950	0.1985	0.2019	0.2054	0.2088	0.2123	0.2157	0.2190	0.2224
0.6	0.2257	0.2291	0.2324	0.2357	0.2389	0.2422	0.2454	0.2486	0.2518	0.2549
0.7	0.2580	0.2612	0.2642	0.2673	0.2704	0.2734	0.2764	0.2794	0.2823	0.2852
0.8	0.2881	0.2910	0.2939	0.2967	0.2995	0.3023	0.3051	0.3078	0.3106	0.3133
0.9	0.3159	0.3186	0.3212	0.3238	0.3264	0.3289	0.3315	0.3340	0.3365	0.3389
1.0	0.3413	0.3438	0.3461	0.3485	0.3508	0.3531	0.3554	0.3577	0.3599	0.3621
1.1	0.3643	0.3665	0.3686	0.3708	0.3729	0.3749	0.3770	0.3790	0.3810	0.3830
1.2	0.3849	0.3869	0.3888	0.3907	0.3925	0.3944	0.3962	0.3980	0.3997	0.4015
1.3	0.4032	0.4049	0.4066	0.4082	0.4099	0.4115	0.4131	0.4147	0.4162	0.4177
1.4	0.4192	0.4207	0.4222	0.4236	0.4251	0.4265	0.4279	0.4292	0.4306	0.4319
1.5	0.4332	0.4345	0.4357	0.4370	0.4382	0.4394	0.4406	0.4418	0.4429	0.4441
1.6	0.4452	0.4463	0.4474	0.4484	0.4495	0.4505	0.4515	0.4525	0.4535	0.4545
1.7	0.4554	0.4564	0.4573	0.4582	0.4591	0.4599	0.4608	0.4616	0.4625	0.4633
1.8	0.4641	0.4649	0.4656	0.4664	0.4671	0.4678	0.4686	0.4693	0.4699	0.4706
1.9	0.4713	0.4719	0.4726	0.4732	0.4738	0.4744	0.4750	0.4756	0.4761	0.4767
2.0	0.4772	0.4778	0.4783	0.4788	0.4793	0.4798	0.4803	0.4808	0.4812	0.4817
2.1	0.4821	0.4826	0.4830	0.4834	0.4838	0.4842	0.4846	0.4850	0.4854	0.4857
2.2	0.4861	0.4864	0.4868	0.4871	0.4875	0.4878	0.4881	0.4884	0.4887	0.4890
2.3	0.4893	0.4896	0.4898	0.4901	0.4904	0.4906	0.4909	0.4911	0.4913	0.4916
2.4	0.4918	0.4920	0.4922	0.4925	0.4927	0.4929	0.4931	0.4932	0.4934	0.4936
2.5	0.4938	0.4940	0.4941	0.4943	0.4945	0.4946	0.4948	0.4949	0.4951	0.4952
2.6	0.4953	0.4955	0.4956	0.4957	0.4959	0.4960	0.4961	0.4962	0.4963	0.4964
2.7	0.4965	0.4966	0.4967	0.4968	0.4969	0.4970	0.4971	0.4972	0.4973	0.4974
2.8	0.4974	0.4975	0.4976	0.4977	0.4977	0.4978	0.4979	0.4979	0.4980	0.4981
2.9	0.4981	0.4982	0.4982	0.4983	0.4984	0.4984	0.4985	0.4985	0.4986	0.4986
3.0	0.49865	0.4987	0.4987	0.4988	0.4988	0.4989	0.4989	0.4989	0.4990	0.4990
4.0	0.49997									

TABLE A-6
Student's t distribution

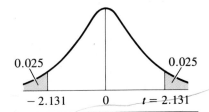

Example For 15 degrees of freedom, the t value that corresponds to an area of 0.05 in both tails combined is 2.131.

0.025   0.025

−2.131   0 $t = 2.131$

Degrees of Freedom	Area in Both Tails Combined			
	0.10	0.05	0.02	0.01
1	6.314	12.706	31.821	63.657
2	2.920	4.303	6.965	9.925
3	2.353	3.182	4.541	5.841
4	2.132	2.776	3.747	4.604
5	2.015	2.571	3.365	4.032
6	1.943	2.447	3.143	3.707
7	1.895	2.365	2.998	3.499
8	1.860	2.306	2.896	3.355
9	1.833	2.262	2.821	3.250
10	1.812	2.228	2.764	3.169
11	1.796	2.201	2.718	3.106
12	1.782	2.179	2.681	3.055
13	1.771	2.160	2.650	3.012
14	1.761	2.145	2.624	2.977
15	1.753	2.131	2.602	2.947
16	1.746	2.120	2.583	2.921
17	1.740	2.110	2.567	2.898
18	1.734	2.101	2.552	2.878
19	1.729	2.093	2.539	2.861
20	1.725	2.086	2.528	2.845
21	1.721	2.080	2.518	2.831
22	1.717	2.074	2.508	2.819
23	1.714	2.069	2.500	2.807
24	1.711	2.064	2.492	2.797
25	1.708	2.060	2.485	2.787
26	1.706	2.056	2.479	2.779
27	1.703	2.052	2.473	2.771
28	1.701	2.048	2.467	2.763
29	1.699	2.045	2.462	2.756
30	1.697	2.042	2.457	2.750
40	1.684	2.021	2.423	2.704
60	1.671	2.000	2.390	2.660
120	1.658	1.980	2.358	2.617
Normal Distribution	1.645	1.960	2.326	2.576

Source: From Table III of Fisher and Yates: *Statistical Tables for Biological, Agricultural and Medical Research*, published by Longman Group, Ltd., London (1974) 6th edition (previously published by Oliver and Boyd, Ltd., Edinburgh), and by permission of the authors and publishers.

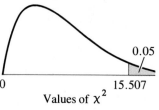

Values of χ^2

All one-tailed tests—one-way.

Example In a chi-square distribution with $v = 8$ degrees of freedom, the area to the right of a chi-square value of 15.507 is 0.05.

Degrees of Freedom	Area in Right Tail				
v	0.20	0.10	0.05	0.02	0.01
1	1.642	2.706	3.841	5.412	6.635
2	3.219	4.605	5.991	7.824	9.210
3	4.642	6.251	7.815	9.837	11.345
4	5.989	7.779	9.488	11.668	13.277
5	7.289	9.236	11.070	13.388	15.086
6	8.558	10.645	12.592	15.033	16.812
7	9.803	12.017	14.067	16.622	18.475
8	11.030	13.362	15.507	18.168	20.090
9	12.242	14.684	16.919	19.679	21.666
10	13.442	15.987	18.307	21.161	23.209
11	14.631	17.275	19.675	22.618	24.725
12	15.812	18.549	21.026	24.054	26.217
13	16.985	19.812	22.362	25.472	27.688
14	18.151	21.064	23.685	26.873	29.141
15	19.311	22.307	24.996	28.259	30.578
16	20.465	23.542	26.296	29.633	32.000
17	21.615	24.769	27.587	30.995	33.409
18	22.760	25.989	28.869	32.346	34.805
19	23.900	27.204	30.144	33.687	36.191
20	25.038	28.412	31.410	35.020	37.566
21	26.171	29.615	32.671	36.343	38.932
22	27.301	30.813	33.924	37.659	40.289
23	28.429	32.007	35.172	38.968	41.638
24	29.553	33.196	36.415	40.270	42.980
25	30.675	34.382	37.652	41.566	44.314
26	31.795	35.563	38.885	42.856	45.642
27	32.912	36.741	40.113	44.140	46.963
28	34.027	37.916	41.337	45.419	48.278
29	35.139	39.087	42.557	46.693	49.588
30	36.250	40.256	43.773	47.962	50.892

Source: From Table IV of Fisher and Yates: *Statistical Tables for Biological, Agricultural and Medical Research*, published by Longman Group, Ltd., London (1974) 6th edition (previously published by Oliver and Boyd, Ltd., Edinburgh), and by permission of the authors and publishers.

TABLE A-8
F distribution

Example In an *F* distribution with $v_1 = 5$ and $v_2 = 6$ degrees of freedom, the area to the right of an *F* value of 4.39 is 0.05. The value on the *F* scale to the right of which lies 0.05 of the area is in lightface type. The value on the *F* scale to the right of which lies 0.01 of the area is in boldface type. For the numerator, v_1 = number of degrees of freedom; v_2 = number of degrees of freedom for the denominator.

Positively Skewed

0.05

4.39

Values of *F*

Each cell shows the 0.05 value (lightface) / 0.01 value (boldface).

v_2 \ v_1	1	2	3	4	5	6	7	8	9	10	11	12	14	16	20	24	30	40	50	75	100	200	500	∞
1	161 / 4,052	200 / 4,999	216 / 5,403	225 / 5,625	230 / 5,764	234 / 5,859	237 / 5,928	239 / 5,981	241 / 6,022	242 / 6,056	243 / 6,082	244 / 6,106	245 / 6,142	246 / 6,169	248 / 6,208	249 / 6,234	250 / 6,261	251 / 6,286	252 / 6,302	253 / 6,323	253 / 6,334	254 / 6,352	254 / 6,361	254 / 6,366
2	18.51 / 98.49	19.00 / 99.00	19.16 / 99.17	19.25 / 99.25	19.30 / 99.30	19.33 / 99.33	19.36 / 99.36	19.37 / 99.37	19.38 / 99.39	19.39 / 99.40	19.40 / 99.41	19.41 / 99.42	19.42 / 99.43	19.43 / 99.44	19.44 / 99.45	19.45 / 99.46	19.46 / 99.47	19.47 / 99.48	19.47 / 98.48	19.48 / 99.49	19.49 / 99.49	19.49 / 99.49	19.50 / 99.50	19.50 / 99.50
3	10.13 / 34.12	9.55 / 30.82	9.28 / 29.46	9.12 / 28.71	9.01 / 28.24	8.94 / 27.91	8.88 / 27.67	8.84 / 27.49	8.81 / 27.34	8.78 / 27.23	8.76 / 27.13	8.74 / 27.05	8.71 / 26.92	8.69 / 26.83	8.66 / 26.69	8.64 / 26.60	8.62 / 26.50	8.60 / 26.41	8.58 / 26.35	8.57 / 26.27	8.56 / 26.23	8.54 / 26.18	8.54 / 26.14	8.53 / 26.12
4	7.71 / 21.20	6.94 / 18.00	6.59 / 16.69	6.39 / 15.98	6.26 / 15.52	6.16 / 15.21	6.09 / 14.98	6.04 / 14.80	6.00 / 14.66	5.96 / 14.54	5.93 / 14.45	5.91 / 14.37	5.87 / 14.24	5.84 / 14.15	5.80 / 14.02	5.77 / 13.93	5.74 / 13.83	5.71 / 13.74	5.70 / 13.69	5.68 / 13.61	5.66 / 13.57	5.65 / 13.52	5.64 / 13.48	5.63 / 13.46
5	6.61 / 16.26	5.79 / 13.27	5.41 / 12.06	5.19 / 11.39	5.05 / 10.97	4.95 / 10.67	4.88 / 10.45	4.82 / 10.29	4.78 / 10.15	4.74 / 10.05	4.70 / 9.96	4.68 / 9.89	4.64 / 9.77	4.60 / 9.68	4.56 / 9.55	4.53 / 9.47	4.50 / 9.38	4.46 / 9.29	4.44 / 9.24	4.42 / 9.17	4.40 / 9.13	4.38 / 9.07	4.37 / 9.04	4.36 / 9.02
6	5.99 / 13.74	5.14 / 10.92	4.76 / 9.78	4.53 / 9.15	4.39 / 8.75	4.28 / 8.47	4.21 / 8.26	4.15 / 8.10	4.10 / 7.98	4.06 / 7.87	4.03 / 7.79	4.00 / 7.72	3.96 / 7.60	3.92 / 7.52	3.87 / 7.39	3.84 / 7.31	3.81 / 7.23	3.77 / 7.14	3.75 / 7.09	3.72 / 7.02	3.71 / 6.99	3.69 / 6.94	3.68 / 6.90	3.67 / 6.88
7	5.59 / 12.25	4.74 / 9.55	4.35 / 8.45	4.12 / 7.85	3.97 / 7.46	3.87 / 7.19	3.79 / 7.00	3.73 / 6.84	3.68 / 6.71	3.63 / 6.62	3.60 / 6.54	3.57 / 6.47	3.52 / 6.35	3.49 / 6.27	3.44 / 6.15	3.41 / 6.07	3.38 / 5.98	3.34 / 5.90	3.32 / 5.85	3.29 / 5.78	3.28 / 5.75	3.25 / 5.70	3.24 / 5.67	3.23 / 5.65
8	5.32 / 11.26	4.46 / 8.65	4.07 / 7.59	3.84 / 7.01	3.69 / 6.63	3.58 / 6.37	3.50 / 6.19	3.44 / 6.03	3.39 / 5.91	3.34 / 5.82	3.31 / 5.74	3.28 / 5.67	3.23 / 5.56	3.20 / 5.48	3.15 / 5.36	3.12 / 5.28	3.08 / 5.20	3.05 / 5.11	3.03 / 5.06	3.00 / 5.00	2.98 / 4.96	2.96 / 4.91	2.94 / 4.88	2.93 / 4.86
9	5.12 / 10.56	4.26 / 8.02	3.86 / 6.99	3.63 / 6.42	3.48 / 6.06	3.37 / 5.80	3.29 / 5.62	3.23 / 5.47	3.18 / 5.35	3.13 / 5.26	3.10 / 5.18	3.07 / 5.11	3.02 / 5.00	2.98 / 4.92	2.93 / 4.80	2.90 / 4.73	2.86 / 4.64	2.82 / 4.56	2.80 / 4.51	2.77 / 4.45	2.76 / 4.41	2.73 / 4.36	2.72 / 4.33	2.71 / 4.31
10	4.96 / 10.04	4.10 / 7.56	3.71 / 6.55	3.48 / 5.99	3.33 / 5.64	3.22 / 5.39	3.14 / 5.21	3.07 / 5.06	3.02 / 4.95	2.97 / 4.85	2.94 / 4.78	2.91 / 4.71	2.86 / 4.60	2.82 / 4.52	2.77 / 4.41	2.74 / 4.33	2.70 / 4.25	2.67 / 4.17	2.64 / 4.12	2.61 / 4.05	2.59 / 4.01	2.56 / 3.96	2.55 / 3.93	2.54 / 3.91
11	4.84 / 9.65	3.98 / 7.20	3.59 / 6.22	3.36 / 5.67	3.20 / 5.32	3.09 / 5.07	3.01 / 4.88	2.95 / 4.74	2.90 / 4.63	2.86 / 4.54	2.82 / 4.46	2.79 / 4.40	2.74 / 4.29	2.70 / 4.21	2.65 / 4.10	2.61 / 4.02	2.57 / 3.94	2.53 / 3.86	2.50 / 3.80	2.47 / 3.74	2.45 / 3.70	2.42 / 3.66	2.41 / 3.62	2.40 / 3.60
12	4.75 / 9.33	3.88 / 6.93	3.49 / 5.95	3.26 / 5.41	3.11 / 5.06	3.00 / 4.82	2.92 / 4.65	2.85 / 4.50	2.80 / 4.39	2.76 / 4.30	2.72 / 4.22	2.69 / 4.16	2.64 / 4.05	2.60 / 3.98	2.54 / 3.86	2.50 / 3.78	2.46 / 3.70	2.42 / 3.61	2.40 / 3.56	2.36 / 3.49	2.35 / 3.46	2.32 / 3.41	2.31 / 3.38	2.30 / 3.36

numerator

denominator

$F > F_c \Rightarrow F_c > F_t$ reject H_0

Source: From George W. Snedecor and William G. Cochran, *Statistical Methods*, Seventh Edition, © 1980 by the Iowa State University Press, Ames, IA. 50010. Reprinted by permission.

TABLE A-8 (continued)

Each cell lists the upper 5% (light) value over the upper 1% (bold) value.

v_2 \ v_1	1	2	3	4	5	6	7	8	9	10	11	12	14	16	20	24	30	40	50	75	100	200	500	∞
13	4.67 / 9.07	3.80 / 6.70	3.41 / 5.74	3.18 / 5.20	3.02 / 4.86	2.92 / 4.62	2.84 / 4.44	2.77 / 4.30	2.72 / 4.19	2.67 / 4.10	2.63 / 4.02	2.60 / 3.96	2.55 / 3.85	2.51 / 3.78	2.46 / 3.67	2.42 / 3.59	2.38 / 3.51	2.34 / 3.42	2.32 / 3.37	2.28 / 3.30	2.26 / 3.27	2.24 / 3.21	2.22 / 3.18	2.21 / 3.16
14	4.60 / 8.86	3.74 / 6.51	3.34 / 5.56	3.11 / 5.03	2.96 / 4.69	2.85 / 4.46	2.77 / 4.28	2.70 / 4.14	2.65 / 4.03	2.60 / 3.94	2.56 / 3.86	2.53 / 3.80	2.48 / 3.70	2.44 / 3.62	2.39 / 3.51	2.35 / 3.43	2.31 / 3.34	2.27 / 3.26	2.24 / 3.21	2.21 / 3.14	2.19 / 3.11	2.16 / 3.06	2.14 / 3.02	2.13 / 3.00
15	4.54 / 8.68	3.68 / 6.36	3.29 / 5.42	3.06 / 4.89	2.90 / 4.56	2.79 / 4.32	2.70 / 4.14	2.64 / 4.00	2.59 / 3.89	2.55 / 3.80	2.51 / 3.73	2.48 / 3.67	2.43 / 3.56	2.39 / 3.48	2.33 / 3.36	2.29 / 3.29	2.25 / 3.20	2.21 / 3.12	2.18 / 3.07	2.15 / 3.00	2.12 / 2.97	2.10 / 2.92	2.08 / 2.89	2.07 / 2.87
16	4.49 / 8.53	3.63 / 6.23	3.24 / 5.29	3.01 / 4.77	2.85 / 4.44	2.74 / 4.20	2.66 / 4.03	2.59 / 3.89	2.54 / 3.78	2.49 / 3.69	2.45 / 3.61	2.42 / 3.55	2.37 / 3.45	2.33 / 3.37	2.28 / 3.25	2.24 / 3.18	2.20 / 3.10	2.16 / 3.01	2.13 / 2.96	2.09 / 2.98	2.07 / 2.86	2.04 / 2.80	2.02 / 2.77	2.01 / 2.75
17	4.45 / 8.40	3.59 / 6.11	3.20 / 5.18	2.96 / 4.67	2.81 / 4.34	2.70 / 4.10	2.62 / 3.93	2.55 / 3.79	2.50 / 3.68	2.45 / 3.59	2.41 / 3.52	2.38 / 3.45	2.33 / 3.35	2.29 / 3.27	2.23 / 3.16	2.19 / 3.08	2.15 / 3.00	2.11 / 2.92	2.08 / 2.86	2.04 / 2.79	2.02 / 2.76	1.99 / 2.70	1.97 / 2.67	1.96 / 2.65
18	4.41 / 8.28	3.55 / 6.01	3.16 / 5.09	2.93 / 4.58	2.77 / 4.25	2.66 / 4.01	2.58 / 3.85	2.51 / 3.71	2.46 / 3.60	2.41 / 3.51	2.37 / 3.44	2.34 / 3.37	2.29 / 3.27	2.25 / 3.19	2.19 / 3.07	2.15 / 3.00	2.11 / 2.91	2.07 / 2.83	2.04 / 2.78	2.00 / 2.71	1.98 / 2.68	1.95 / 2.62	1.93 / 2.59	1.92 / 2.57
19	4.38 / 8.18	3.52 / 5.93	3.13 / 5.01	2.90 / 4.50	2.74 / 4.17	2.63 / 3.94	2.55 / 3.77	2.48 / 3.63	2.43 / 3.52	2.38 / 3.43	2.34 / 3.36	2.31 / 3.30	2.26 / 3.19	2.21 / 3.12	2.15 / 3.00	2.11 / 2.92	2.07 / 2.84	2.02 / 2.76	2.00 / 2.70	1.96 / 2.63	1.94 / 2.60	1.91 / 2.54	1.90 / 2.51	1.88 / 2.49
20	4.35 / 8.10	3.49 / 5.85	3.10 / 4.94	2.87 / 4.43	2.71 / 4.10	2.60 / 3.87	2.52 / 3.71	2.45 / 3.56	2.40 / 3.45	2.35 / 3.37	2.31 / 3.30	2.28 / 3.23	2.23 / 3.13	2.18 / 3.05	2.12 / 2.94	2.08 / 2.86	2.04 / 2.77	1.99 / 2.69	1.96 / 2.63	1.92 / 2.56	1.90 / 2.53	1.87 / 2.47	1.85 / 2.44	1.84 / 2.42
21	4.32 / 8.02	3.47 / 5.78	3.07 / 4.87	2.84 / 4.37	2.68 / 4.04	2.57 / 3.81	2.49 / 3.65	2.42 / 3.51	2.37 / 3.40	2.32 / 3.31	2.28 / 3.24	2.25 / 3.17	2.20 / 3.07	2.15 / 2.99	2.09 / 2.88	2.05 / 2.80	2.00 / 2.72	1.96 / 2.63	1.93 / 2.58	1.89 / 2.51	1.87 / 2.47	1.84 / 2.42	1.82 / 2.38	1.81 / 2.36
22	4.30 / 7.94	3.44 / 5.72	3.05 / 4.82	2.82 / 4.31	2.66 / 3.99	2.55 / 3.76	2.47 / 3.59	2.40 / 3.45	2.35 / 3.35	2.30 / 3.26	2.26 / 3.18	2.23 / 3.12	2.18 / 3.02	2.13 / 2.94	2.07 / 2.83	2.03 / 2.75	1.98 / 2.67	1.93 / 2.58	1.91 / 2.53	1.87 / 2.46	1.84 / 2.42	1.81 / 2.37	1.80 / 2.33	1.78 / 2.31
23	4.28 / 7.88	3.42 / 5.66	3.03 / 4.76	2.80 / 4.26	2.64 / 3.94	2.53 / 3.71	2.45 / 3.54	2.38 / 3.41	2.32 / 3.30	2.28 / 3.21	2.24 / 3.14	2.20 / 3.07	2.14 / 2.97	2.10 / 2.89	2.04 / 2.78	2.00 / 2.70	1.96 / 2.62	1.91 / 2.53	1.88 / 2.48	1.84 / 2.41	1.82 / 2.37	1.79 / 2.32	1.77 / 2.28	1.76 / 2.26
24	4.26 / 7.82	3.40 / 5.61	3.01 / 4.72	2.78 / 4.22	2.62 / 3.90	2.51 / 3.67	2.43 / 3.50	2.36 / 3.36	2.30 / 3.25	2.26 / 3.17	2.22 / 3.09	2.18 / 3.03	2.13 / 2.93	2.09 / 2.85	2.02 / 2.74	1.98 / 2.66	1.94 / 2.58	1.89 / 2.49	1.86 / 2.44	1.82 / 2.36	1.80 / 2.33	1.76 / 2.27	1.74 / 2.23	1.73 / 2.21
25	4.24 / 7.77	3.38 / 5.57	2.99 / 4.68	2.76 / 4.18	2.60 / 3.86	2.49 / 3.63	2.41 / 3.46	2.34 / 3.32	2.28 / 3.21	2.24 / 3.13	2.20 / 3.05	2.16 / 2.99	2.11 / 2.89	2.06 / 2.81	2.00 / 2.70	1.96 / 2.62	1.92 / 2.54	1.87 / 2.45	1.84 / 2.40	1.80 / 2.32	1.77 / 2.29	1.74 / 2.23	1.72 / 2.19	1.71 / 2.17
26	4.22 / 7.72	3.37 / 5.53	2.98 / 4.64	2.74 / 4.14	2.59 / 3.82	2.47 / 3.59	2.39 / 3.42	2.32 / 3.29	2.27 / 3.17	2.22 / 3.09	2.18 / 3.02	2.15 / 2.96	2.10 / 2.86	2.05 / 2.77	1.99 / 2.66	1.95 / 2.58	1.90 / 2.50	1.85 / 2.41	1.82 / 2.36	1.78 / 2.28	1.76 / 2.25	1.72 / 2.19	1.70 / 2.15	1.69 / 2.13

TABLE A-8 (continued)

ν_2	1	2	3	4	5	6	7	8	9	10	11	12	14	16	20	24	30	40	50	75	100	200	500	∞
27	4.21 **7.68**	3.35 **5.49**	2.96 **4.60**	2.73 **4.11**	2.57 **3.79**	2.46 **3.56**	2.37 **3.39**	2.30 **3.26**	2.25 **3.14**	2.20 **3.06**	2.16 **2.98**	2.13 **2.93**	2.08 **2.83**	2.03 **2.74**	1.97 **2.63**	1.93 **2.55**	1.88 **2.47**	1.84 **2.38**	1.80 **2.33**	1.76 **2.25**	1.74 **2.21**	1.71 **2.16**	1.68 **2.12**	1.67 **2.10**
28	4.20 **7.64**	3.34 **5.45**	2.95 **4.57**	2.71 **4.07**	2.56 **3.76**	2.44 **3.53**	2.36 **3.36**	2.29 **3.23**	2.24 **3.11**	2.19 **3.03**	2.15 **2.95**	2.12 **2.90**	2.06 **2.80**	2.02 **2.71**	1.96 **2.60**	1.91 **2.52**	1.87 **2.44**	1.81 **2.35**	1.78 **2.30**	1.75 **2.22**	1.72 **2.18**	1.69 **2.13**	1.67 **2.09**	1.65 **2.06**
29	4.18 **7.60**	3.33 **5.42**	2.93 **4.54**	2.70 **4.04**	2.54 **3.73**	2.43 **3.50**	2.35 **3.33**	2.28 **3.20**	2.22 **3.08**	2.18 **3.00**	2.14 **2.92**	2.10 **2.87**	2.05 **2.77**	2.00 **2.68**	1.94 **2.57**	1.90 **2.49**	1.85 **2.41**	1.80 **2.32**	1.77 **2.27**	1.73 **2.19**	1.71 **2.15**	1.68 **2.10**	1.65 **2.06**	1.64 **2.03**
30	4.17 **7.56**	3.32 **5.39**	2.92 **4.51**	2.69 **4.02**	2.53 **3.70**	2.42 **3.47**	2.34 **3.30**	2.27 **3.17**	2.21 **3.06**	2.16 **2.98**	2.12 **2.90**	2.09 **2.84**	2.04 **2.74**	1.99 **2.66**	1.93 **2.55**	1.89 **2.47**	1.84 **2.38**	1.79 **2.29**	1.76 **2.24**	1.72 **2.16**	1.69 **2.13**	1.66 **2.07**	1.64 **2.03**	1.62 **2.01**
32	4.15 **7.50**	3.30 **5.34**	2.90 **4.46**	2.67 **3.97**	2.51 **3.66**	2.40 **3.42**	2.32 **3.25**	2.25 **3.12**	2.19 **3.01**	2.14 **2.94**	2.10 **2.86**	2.07 **2.80**	2.02 **2.70**	1.97 **2.62**	1.91 **2.51**	1.86 **2.42**	1.82 **2.34**	1.76 **2.25**	1.74 **2.20**	1.69 **2.12**	1.67 **2.08**	1.64 **2.02**	1.61 **1.98**	1.59 **1.96**
34	4.13 **7.44**	3.28 **5.29**	2.88 **4.42**	2.65 **3.93**	2.49 **3.61**	2.38 **3.38**	2.30 **3.21**	2.23 **3.08**	2.17 **2.97**	2.12 **2.89**	2.08 **2.82**	2.05 **2.76**	2.00 **2.66**	1.95 **2.58**	1.89 **2.47**	1.84 **2.38**	1.80 **2.30**	1.74 **2.21**	1.71 **2.15**	1.67 **2.08**	1.64 **2.04**	1.61 **1.98**	1.59 **1.94**	1.57 **1.91**
36	4.11 **7.39**	3.26 **5.25**	2.86 **4.38**	2.63 **3.89**	2.48 **3.58**	2.36 **3.35**	2.28 **3.18**	2.21 **3.04**	2.15 **2.94**	2.10 **2.86**	2.06 **2.78**	2.03 **2.72**	1.98 **2.62**	1.93 **2.54**	1.87 **2.43**	1.82 **2.35**	1.78 **2.26**	1.72 **2.17**	1.69 **2.12**	1.65 **2.04**	1.62 **2.00**	1.59 **1.94**	1.56 **1.90**	1.55 **1.87**
38	4.10 **7.35**	3.25 **5.21**	2.85 **4.34**	2.62 **3.86**	2.46 **3.54**	2.35 **3.32**	2.26 **3.15**	2.19 **3.02**	2.14 **2.91**	2.09 **2.82**	2.05 **2.75**	2.02 **2.69**	1.96 **2.59**	1.92 **2.51**	1.85 **2.40**	1.80 **2.32**	1.76 **2.22**	1.71 **2.14**	1.67 **2.08**	1.63 **2.00**	1.60 **1.97**	1.57 **1.90**	1.54 **1.86**	1.53 **1.84**
40	4.08 **7.31**	3.23 **5.18**	2.84 **4.31**	2.61 **3.83**	2.45 **3.51**	2.34 **3.29**	2.25 **3.12**	2.18 **2.99**	2.12 **2.88**	2.07 **2.80**	2.04 **2.73**	2.00 **2.66**	1.95 **2.56**	1.90 **2.49**	1.84 **2.37**	1.79 **2.29**	1.74 **2.20**	1.69 **2.11**	1.66 **2.05**	1.61 **1.97**	1.59 **1.94**	1.55 **1.88**	1.53 **1.84**	1.51 **1.81**
42	4.07 **7.27**	3.22 **5.15**	2.83 **4.29**	2.59 **3.80**	2.44 **3.49**	2.32 **3.26**	2.24 **3.10**	2.17 **2.96**	2.11 **2.86**	2.06 **2.77**	2.02 **2.70**	1.99 **2.64**	1.94 **2.54**	1.89 **2.46**	1.82 **2.35**	1.78 **2.26**	1.73 **2.17**	1.68 **2.08**	1.64 **2.02**	1.60 **1.94**	1.57 **1.91**	1.54 **1.85**	1.51 **1.80**	1.49 **1.78**
44	4.06 **7.24**	3.21 **5.12**	2.82 **4.26**	2.58 **3.78**	2.43 **3.46**	2.31 **3.24**	2.23 **3.07**	2.16 **2.94**	2.10 **2.84**	2.05 **2.75**	2.01 **2.68**	1.98 **2.62**	1.92 **2.52**	1.88 **2.44**	1.81 **2.32**	1.76 **2.24**	1.72 **2.15**	1.66 **2.06**	1.63 **2.00**	1.58 **1.92**	1.56 **1.88**	1.52 **1.82**	1.50 **1.78**	1.48 **1.75**
46	4.05 **7.21**	3.20 **5.10**	2.81 **4.24**	2.57 **3.76**	2.42 **3.44**	2.30 **3.22**	2.22 **3.05**	2.14 **2.92**	2.09 **2.82**	2.04 **2.73**	2.00 **2.66**	1.97 **2.60**	1.91 **2.50**	1.87 **2.42**	1.80 **2.30**	1.75 **2.22**	1.71 **2.13**	1.65 **2.04**	1.62 **1.98**	1.57 **1.90**	1.54 **1.86**	1.51 **1.80**	1.48 **1.76**	1.46 **1.72**
48	4.04 **7.19**	3.19 **5.08**	2.80 **4.22**	2.56 **3.74**	2.41 **3.42**	2.30 **3.20**	2.21 **3.04**	2.14 **2.90**	2.08 **2.80**	2.03 **2.71**	1.99 **2.64**	1.96 **2.58**	1.90 **2.48**	1.86 **2.40**	1.79 **2.28**	1.74 **2.20**	1.70 **2.11**	1.64 **2.02**	1.61 **1.96**	1.56 **1.88**	1.53 **1.84**	1.50 **1.78**	1.47 **1.73**	1.45 **1.70**

TABLE A-8   (continued)

v_2 \ v_1	1	2	3	4	5	6	7	8	9	10	11	12	14	16	20	24	30	40	50	75	100	200	500	∞
50	4.03 / **7.17**	3.18 / **5.06**	2.79 / **4.20**	2.56 / **3.72**	2.40 / **3.41**	2.29 / **3.18**	2.20 / **3.02**	2.13 / **2.88**	2.07 / **2.78**	2.02 / **2.70**	1.98 / **2.62**	1.95 / **2.56**	1.90 / **2.46**	1.85 / **2.39**	1.78 / **2.26**	1.74 / **2.18**	1.69 / **2.10**	1.63 / **2.00**	1.60 / **1.94**	1.55 / **1.86**	1.52 / **1.82**	1.48 / **1.76**	1.46 / **1.71**	1.44 / **1.68**
55	4.02 / **7.12**	3.17 / **5.01**	2.78 / **4.16**	2.54 / **3.68**	2.38 / **3.37**	2.27 / **3.15**	2.18 / **2.98**	2.11 / **2.85**	2.05 / **2.75**	2.00 / **2.66**	1.97 / **2.59**	1.93 / **2.53**	1.88 / **2.43**	1.83 / **2.35**	1.76 / **2.23**	1.72 / **2.15**	1.67 / **2.06**	1.61 / **1.96**	1.58 / **1.90**	1.52 / **1.82**	1.50 / **1.78**	1.46 / **1.71**	1.43 / **1.66**	1.41 / **1.64**
60	4.00 / **7.08**	3.15 / **4.98**	2.76 / **4.13**	2.52 / **3.65**	2.37 / **3.34**	2.25 / **3.12**	2.17 / **2.95**	2.10 / **2.82**	2.04 / **2.72**	1.99 / **2.63**	1.95 / **2.56**	1.92 / **2.50**	1.86 / **2.40**	1.81 / **2.32**	1.75 / **2.20**	1.70 / **2.12**	1.65 / **2.03**	1.59 / **1.93**	1.56 / **1.87**	1.50 / **1.79**	1.48 / **1.74**	1.44 / **1.68**	1.41 / **1.63**	1.39 / **1.60**
65	3.99 / **7.04**	3.14 / **4.95**	2.75 / **4.10**	2.51 / **3.62**	2.36 / **3.31**	2.24 / **3.09**	2.15 / **2.93**	2.08 / **2.79**	2.02 / **2.70**	1.98 / **2.61**	1.94 / **2.54**	1.90 / **2.47**	1.85 / **2.37**	1.80 / **2.30**	1.73 / **2.18**	1.68 / **2.09**	1.63 / **2.00**	1.57 / **1.90**	1.54 / **1.84**	1.49 / **1.76**	1.46 / **1.71**	1.42 / **1.64**	1.39 / **1.60**	1.37 / **1.56**
70	3.98 / **7.01**	3.13 / **4.92**	2.74 / **4.08**	2.50 / **3.60**	2.35 / **3.29**	2.23 / **3.07**	2.14 / **2.91**	2.07 / **2.77**	2.01 / **2.67**	1.97 / **2.59**	1.93 / **2.51**	1.89 / **2.45**	1.84 / **2.35**	1.79 / **2.28**	1.72 / **2.15**	1.67 / **2.07**	1.62 / **1.98**	1.56 / **1.88**	1.53 / **1.82**	1.47 / **1.74**	1.45 / **1.69**	1.40 / **1.62**	1.37 / **1.56**	1.35 / **1.53**
80	3.96 / **6.96**	3.11 / **4.88**	2.72 / **4.04**	2.48 / **3.56**	2.33 / **3.25**	2.21 / **3.04**	2.12 / **2.87**	2.05 / **2.74**	1.99 / **2.64**	1.95 / **2.55**	1.91 / **2.48**	1.88 / **2.41**	1.82 / **2.32**	1.77 / **2.24**	1.70 / **2.11**	1.65 / **2.03**	1.60 / **1.94**	1.54 / **1.84**	1.51 / **1.78**	1.45 / **1.70**	1.42 / **1.65**	1.38 / **1.57**	1.35 / **1.52**	1.32 / **1.49**
100	3.94 / **6.90**	3.09 / **4.82**	2.70 / **3.98**	2.46 / **3.51**	2.30 / **3.20**	2.19 / **2.99**	2.10 / **2.82**	2.03 / **2.69**	1.97 / **2.59**	1.92 / **2.51**	1.88 / **2.43**	1.85 / **2.36**	1.79 / **2.26**	1.75 / **2.19**	1.68 / **2.06**	1.63 / **1.98**	1.57 / **1.89**	1.51 / **1.79**	1.48 / **1.73**	1.42 / **1.64**	1.39 / **1.59**	1.34 / **1.51**	1.30 / **1.46**	1.28 / **1.43**
125	3.92 / **6.84**	3.07 / **4.78**	2.68 / **3.94**	2.44 / **3.47**	2.29 / **3.17**	2.17 / **2.95**	2.08 / **2.79**	2.01 / **2.65**	1.95 / **2.56**	1.90 / **2.47**	1.86 / **2.40**	1.83 / **2.33**	1.77 / **2.23**	1.72 / **2.15**	1.65 / **2.03**	1.60 / **1.94**	1.55 / **1.85**	1.49 / **1.75**	1.45 / **1.68**	1.39 / **1.59**	1.36 / **1.54**	1.31 / **1.46**	1.27 / **1.40**	1.25 / **1.37**
150	3.91 / **6.81**	3.06 / **4.75**	2.67 / **3.91**	2.43 / **3.44**	2.27 / **3.14**	2.16 / **2.92**	2.07 / **2.76**	2.00 / **2.62**	1.94 / **2.53**	1.89 / **2.44**	1.85 / **2.37**	1.82 / **2.30**	1.76 / **2.20**	1.71 / **2.12**	1.64 / **2.00**	1.59 / **1.91**	1.54 / **1.83**	1.47 / **1.72**	1.44 / **1.66**	1.37 / **1.56**	1.34 / **1.51**	1.29 / **1.43**	1.25 / **1.37**	1.22 / **1.33**
200	3.89 / **6.76**	3.04 / **4.71**	2.65 / **3.88**	2.41 / **3.41**	2.26 / **3.11**	2.14 / **2.90**	2.05 / **2.73**	1.98 / **2.60**	1.92 / **2.50**	1.87 / **2.41**	1.83 / **2.34**	1.80 / **2.28**	1.74 / **2.17**	1.69 / **2.09**	1.62 / **1.97**	1.57 / **1.88**	1.52 / **1.79**	1.45 / **1.69**	1.42 / **1.62**	1.35 / **1.53**	1.32 / **1.48**	1.26 / **1.39**	1.22 / **1.33**	1.19 / **1.28**
400	3.86 / **6.70**	3.02 / **4.66**	2.62 / **3.83**	2.39 / **3.36**	2.23 / **3.06**	2.12 / **2.85**	2.03 / **2.69**	1.96 / **2.55**	1.90 / **2.46**	1.85 / **2.37**	1.81 / **2.29**	1.78 / **2.23**	1.72 / **2.12**	1.67 / **2.04**	1.60 / **1.92**	1.54 / **1.84**	1.49 / **1.74**	1.42 / **1.64**	1.38 / **1.57**	1.32 / **1.47**	1.28 / **1.42**	1.22 / **1.32**	1.16 / **1.24**	1.13 / **1.19**
1,000	3.85 / **6.66**	3.00 / **4.62**	2.61 / **3.80**	2.38 / **3.34**	2.22 / **3.04**	2.10 / **2.82**	2.02 / **2.66**	1.95 / **2.53**	1.89 / **2.43**	1.84 / **2.34**	1.80 / **2.26**	1.76 / **2.20**	1.70 / **2.09**	1.65 / **2.01**	1.58 / **1.89**	1.53 / **1.81**	1.47 / **1.71**	1.41 / **1.61**	1.36 / **1.54**	1.30 / **1.44**	1.26 / **1.38**	1.19 / **1.28**	1.13 / **1.19**	1.08 / **1.11**
∞	3.84 / **6.63**	2.99 / **4.60**	2.60 / **3.78**	2.37 / **3.32**	2.21 / **3.02**	2.09 / **2.80**	2.01 / **2.64**	1.94 / **2.51**	1.88 / **2.41**	1.83 / **2.32**	1.79 / **2.24**	1.75 / **2.18**	1.69 / **2.07**	1.64 / **1.99**	1.57 / **1.87**	1.52 / **1.79**	1.46 / **1.69**	1.40 / **1.59**	1.35 / **1.52**	1.28 / **1.41**	1.24 / **1.36**	1.17 / **1.25**	1.11 / **1.15**	1.00 / **1.00**

TABLE A-9
Critical values of *T* in the Wilcoxon matched-pairs signed-ranks test
Critical values of *T* at various levels of probability

The symbol *T* denotes the smaller sum of ranks associated with differences that are all of the same sign. For any given *N* (number of ranked differences), the obtained *T* is significant at a given level if it is equal to or *less than* the value shown in the table.

	Level of Significance for One-tailed Test					Level of Significance for One-tailed Test			
	0.05	0.025	0.01	0.005		0.05	0.025	0.01	0.005
	Level of Significance for Two-tailed Test					Level of Significance for Two-tailed Test			
N	0.10	0.05	0.02	0.01	*N*	0.10	0.05	0.02	0.01
5	0	—	—	—	28	130	116	101	91
6	2	0	—	—	29	140	126	110	100
7	3	2	0	—	30	151	137	120	109
8	5	3	1	0	31	163	147	130	118
9	8	5	3	1	32	175	159	140	128
10	10	8	5	3	33	187	170	151	138
11	13	10	7	5	34	200	182	162	148
12	17	13	9	7	35	213	195	173	159
13	21	17	12	9	36	227	208	185	171
14	25	21	15	12	37	241	221	198	182
15	30	25	19	15	38	256	235	211	194
16	35	29	23	19	39	271	249	224	207
17	41	34	27	23	40	286	264	238	220
18	47	40	32	27	41	302	279	252	233
19	53	46	37	32	42	319	294	266	247
20	60	52	43	37	43	336	310	281	261
21	67	58	49	42	44	353	327	296	276
22	75	65	55	48	45	371	343	312	291
23	83	73	62	54	46	389	361	328	307
24	91	81	69	61	47	407	378	345	322
25	100	89	76	68	48	426	396	362	339
26	110	98	84	75	49	446	415	379	355
27	119	107	92	83	50	466	434	397	373

(Slight discrepancies will be found between the critical values appearing in the table above and in Table 2 of the 1964 revision of F. Wilcoxon and R.A. Wilcox, *Some Rapid Approximate Statistical Procedures*, New York, Lederle Laboratories, 1964. The disparity reflects the latter's policy of selecting the critical value nearest a given significance level, occasionally overstepping that level. For example, for *N* = 8, the probability of a *T* of three equals 0.0390 (two-tail), and the probability of a *T* of four equals 0.0546 (two-tail). Wilcoxon and Wilcox select a *T* of four as the critical value at the 0.05 level of significance (two-tail), whereas Table A-9 reflects a more conservative policy by setting a *T* of three as the critical value at this level.)

Source: From Frank Wilcoxon and Roberta A. Wilcox, *Some Rapid Approximate Statistical Procedures*. Revised 1964 by Lederle Laboratories, Pearl River, NY. Reproduced with the permission of the American Cyanamid Company.

TABLE A-10
Table of exponential functions

x	e^x	e^{-x}	x	e^x	e^{-x}
0.00	1.000	1.000	3.00	20.086	0.050
0.10	1.105	0.905	3.10	22.198	0.045
0.20	1.221	0.819	3.20	24.533	0.041
0.30	1.350	0.741	3.30	27.113	0.037
0.40	1.492	0.670	3.40	29.964	0.033
0.50	1.649	0.607	3.50	33.115	0.030
0.60	1.822	0.549	3.60	36.598	0.027
0.70	2.014	0.497	3.70	40.447	0.025
0.80	2.226	0.449	3.80	44.701	0.022
0.90	2.460	0.407	3.90	49.402	0.020
1.00	2.718	0.368	4.00	54.598	0.018
1.10	3.004	0.333	4.10	60.340	0.017
1.20	3.320	0.301	4.20	66.686	0.015
1.30	3.669	0.273	4.30	73.700	0.014
1.40	4.055	0.247	4.40	81.451	0.012
1.50	4.482	0.223	4.50	90.017	0.011
1.60	4.953	0.202	4.60	99.484	0.010
1.70	5.474	0.183	4.70	109.95	0.009
1.80	6.050	0.165	4.80	121.51	0.008
1.90	6.686	0.150	4.90	134.29	0.007
2.00	7.389	0.135	5.00	148.41	0.007
2.10	8.166	0.122	5.10	164.02	0.006
2.20	9.025	0.111	5.20	181.27	0.006
2.30	9.974	0.100	5.30	200.34	0.005
2.40	11.023	0.091	5.40	221.41	0.005
2.50	12.182	0.082	5.50	244.69	0.004
2.60	13.464	0.074	5.60	270.43	0.004
2.70	14.880	0.067	5.70	298.87	0.003
2.80	16.445	0.061	5.80	330.30	0.003
2.90	18.174	0.055	5.90	365.04	0.003
3.00	20.086	0.050	6.00	403.43	0.002

TABLE A-11

Values of d_L and d_U for the Durbin–Watson Test for $\alpha = 0.05$

n = number of observations
k = number of independent variables

n	$k=1$ d_L	d_U	$k=2$ d_L	d_U	$k=3$ d_L	d_U	$k=4$ d_L	d_U	$k=5$ d_L	d_U
15	1.08	1.36	0.95	1.54	0.82	1.75	0.69	1.97	0.56	2.21
16	1.10	1.37	0.98	1.54	0.86	1.73	0.74	1.93	0.62	2.15
17	1.13	1.38	1.02	1.54	0.90	1.71	0.78	1.90	0.67	2.10
18	1.16	1.39	1.05	1.53	0.93	1.69	0.82	1.87	0.71	2.06
19	1.18	1.40	1.08	1.53	0.97	1.68	0.86	1.85	0.75	2.02
20	1.20	1.41	1.10	1.54	1.00	1.68	0.90	1.83	0.79	1.99
21	1.22	1.42	1.13	1.54	1.03	1.67	0.93	1.81	0.83	1.96
22	1.24	1.43	1.15	1.54	1.05	1.66	0.96	1.80	0.86	1.94
23	1.26	1.44	1.17	1.54	1.08	1.66	0.99	1.79	0.90	1.92
24	1.27	1.45	1.19	1.55	1.10	1.66	1.01	1.78	0.93	1.90
25	1.29	1.45	1.21	1.55	1.12	1.66	1.04	1.77	0.95	1.89
26	1.30	1.46	1.22	1.55	1.14	1.65	1.06	1.76	0.98	1.88
27	1.32	1.47	1.24	1.56	1.16	1.65	1.08	1.76	1.01	1.86
28	1.33	1.48	1.26	1.56	1.18	1.65	1.10	1.75	1.03	1.85
29	1.34	1.48	1.27	1.56	1.20	1.65	1.12	1.74	1.05	1.84
30	1.35	1.49	1.28	1.57	1.21	1.65	1.14	1.74	1.07	1.83
31	1.36	1.50	1.30	1.57	1.23	1.65	1.16	1.74	1.09	1.83
32	1.37	1.50	1.31	1.57	1.24	1.65	1.18	1.73	1.11	1.82
33	1.38	1.51	1.32	1.58	1.26	1.65	1.19	1.73	1.13	1.81
34	1.39	1.51	1.33	1.58	1.27	1.65	1.21	1.73	1.15	1.81
35	1.40	1.52	1.34	1.58	1.28	1.65	1.22	1.73	1.16	1.80
36	1.41	1.52	1.35	1.59	1.29	1.65	1.24	1.73	1.18	1.80
37	1.42	1.53	1.36	1.59	1.31	1.66	1.25	1.72	1.19	1.80
38	1.43	1.54	1.37	1.59	1.32	1.66	1.26	1.72	1.21	1.79
39	1.43	1.54	1.38	1.60	1.33	1.66	1.27	1.72	1.22	1.79
40	1.44	1.54	1.39	1.60	1.34	1.66	1.29	1.72	1.23	1.79
45	1.48	1.57	1.43	1.62	1.38	1.67	1.34	1.72	1.29	1.78
50	1.50	1.59	1.46	1.63	1.42	1.67	1.38	1.72	1.34	1.77
55	1.53	1.60	1.49	1.64	1.45	1.68	1.41	1.72	1.38	1.77
60	1.55	1.62	1.51	1.65	1.48	1.69	1.44	1.73	1.41	1.77
65	1.57	1.63	1.54	1.66	1.50	1.70	1.47	1.73	1.44	1.77
70	1.58	1.64	1.55	1.67	1.52	1.70	1.49	1.74	1.46	1.77
75	1.60	1.65	1.57	1.68	1.54	1.71	1.51	1.74	1.49	1.77
80	1.61	1.66	1.59	1.69	1.56	1.72	1.53	1.74	1.51	1.77
85	1.62	1.67	1.60	1.70	1.57	1.72	1.55	1.75	1.52	1.77
90	1.63	1.68	1.61	1.70	1.59	1.73	1.57	1.75	1.54	1.78
95	1.64	1.69	1.62	1.71	1.60	1.73	1.58	1.75	1.56	1.78
100	1.65	1.69	1.63	1.72	1.61	1.74	1.59	1.76	1.57	1.78

Source: From J. Durbin and G.S. Watson, "Testing for Serial Correlation in Least Squares Regression," *Biometrika*, 38 June 1951. Reproduced by permission of the Biometrika Trustees.

TABLE A-11   (continued)
Values of d_L and d_U for the Durbin–Watson Test for $\alpha = 0.01$

	$k = 1$		$k = 2$		$k = 3$		$k = 4$		$k = 5$	
n	d_L	d_U	d_L	d_U	d_L	d_U	d_L	d_U	d_L	d_U
15	0.81	1.07	0.70	1.25	0.59	1.46	0.49	1.70	0.39	1.96
16	0.84	1.09	0.74	1.25	0.63	1.44	0.53	1.66	0.44	1.90
17	0.87	1.10	0.77	1.25	0.67	1.43	0.57	1.63	0.48	1.85
18	0.90	1.12	0.80	1.26	0.71	1.42	0.61	1.60	0.52	1.80
19	0.93	1.13	0.83	1.26	0.74	1.41	0.65	1.58	0.56	1.77
20	0.95	1.15	0.86	1.27	0.77	1.41	0.68	1.57	0.60	1.74
21	0.97	1.16	0.89	1.27	0.80	1.41	0.72	1.55	0.63	1.71
22	1.00	1.17	0.91	1.28	0.83	1.40	0.75	1.54	0.66	1.69
23	1.02	1.19	0.94	1.29	0.86	1.40	0.77	1.53	0.70	1.67
24	1.04	1.20	0.96	1.30	0.88	1.41	0.80	1.53	0.72	1.66
25	1.05	1.21	0.98	1.30	0.90	1.41	0.83	1.52	0.75	1.65
26	1.07	1.22	1.00	1.31	0.93	1.41	0.85	1.52	0.78	1.64
27	1.09	1.23	1.02	1.32	0.95	1.41	0.88	1.51	0.81	1.63
28	1.10	1.24	1.04	1.32	0.97	1.41	0.90	1.51	0.83	1.62
29	1.12	1.25	1.05	1.33	0.99	1.42	0.92	1.51	0.85	1.61
30	1.13	1.26	1.07	1.34	1.01	1.42	0.94	1.51	0.88	1.61
31	1.15	1.27	1.08	1.34	1.02	1.42	0.96	1.51	0.90	1.60
32	1.16	1.28	1.10	1.35	1.04	1.43	0.98	1.51	0.92	1.60
33	1.17	1.29	1.11	1.36	1.05	1.43	1.00	1.51	0.94	1.59
34	1.18	1.30	1.13	1.36	1.07	1.43	1.01	1.51	0.95	1.59
35	1.19	1.31	1.14	1.37	1.08	1.44	1.03	1.51	0.97	1.59
36	1.21	1.32	1.15	1.38	1.10	1.44	1.04	1.51	0.99	1.59
37	1.22	1.32	1.16	1.38	1.11	1.45	1.06	1.51	1.00	1.59
38	1.23	1.33	1.18	1.39	1.12	1.45	1.07	1.52	1.02	1.58
39	1.24	1.34	1.19	1.39	1.14	1.45	1.09	1.52	1.03	1.58
40	1.25	1.34	1.20	1.40	1.15	1.46	1.10	1.52	1.05	1.58
45	1.29	1.38	1.24	1.42	1.20	1.48	1.16	1.53	1.11	1.58
50	1.32	1.40	1.28	1.45	1.24	1.49	1.20	1.54	1.16	1.59
55	1.36	1.43	1.32	1.47	1.28	1.51	1.25	1.55	1.21	1.59
60	1.38	1.45	1.35	1.48	1.32	1.52	1.28	1.56	1.25	1.60
65	1.41	1.47	1.38	1.50	1.35	1.53	1.31	1.57	1.28	1.61
70	1.43	1.49	1.40	1.52	1.37	1.55	1.34	1.58	1.31	1.61
75	1.45	1.50	1.42	1.53	1.39	1.56	1.37	1.59	1.34	1.62
80	1.47	1.52	1.44	1.54	1.42	1.57	1.39	1.60	1.36	1.62
85	1.48	1.53	1.46	1.55	1.43	1.58	1.41	1.60	1.39	1.63
90	1.50	1.54	1.47	1.56	1.45	1.59	1.43	1.61	1.41	1.64
95	1.51	1.55	1.49	1.57	1.47	1.60	1.45	1.62	1.42	1.64
100	1.52	1.56	1.50	1.58	1.48	1.60	1.46	1.63	1.44	1.65

SYMBOLS, SUBSCRIPTS, AND SUMMATIONS

In statistics, **symbols** such as X, Y, and Z are used to represent different sets of data. Hence, if we have data for five families, we might let

Symbols

$$X = \text{family income}$$

$$Y = \text{family clothing expenditures}$$

$$Z = \text{family savings}$$

Subscripts are used to represent individual observations within these sets of data. Thus, X_i represents the income of the ith family, where i takes on the values 1, 2, 3, 4, and 5. In this notation X_1, X_2, X_3, X_4, and X_5 stand for the incomes of the first family, the second family, and so on. The data are arranged in some order, such as by size of income, the order in which the data were gathered, or any other way suitable to the purposes or convenience of the investigator. The subscript i is a variable used to index the individual data observations. Therefore, X_i, Y_i, and Z_i represent the income, clothing expenditures, and savings of the ith family. For example, X_2 represents the income of the second family, Y_2 clothing expenditures of the second (same) family, and Z_5 the savings of the fifth family.

Subscripts

Now, let us suppose that we have data for two different samples, say the net worths of 100 corporations and the test scores of 20 students. To refer to individual observations in these samples, we can let X_i denote the net worth of the ith corporation, where i assumes values from 1 to 100. (This latter idea is indicated by the notation $i = 1, 2, 3, \ldots, 100$.) We can also let Y_j denote the test score of the jth student, where $j = 1, 2, 3, \ldots, 20$. The different subscript letters make it clear that different samples are involved. Letters such as X, Y, and Z generally represent the different variables or types of measurements involved, whereas subscripts such as i, j, k, and l designate individual observations.

Summations   We now turn to the method of expressing **summations** of sets of data. Suppose we want to add a set of four observations, denoted X_1, X_2, X_3, and X_4. A convenient way of designating this addition is

$$\sum_{i=1}^{4} X_i = X_1 + X_2 + X_3 + X_4$$

where the symbol Σ (Greek capital "sigma") means "the sum of." Hence, the symbol

$$\sum_{i=1}^{4} X_i$$

is read "the sum of the X_i's, i going from 1 to 4." For example, if $X_1 = 3$, $X_2 = 1$, $X_3 = 10$, and $X_4 = 5$,

$$\sum_{i=1}^{4} X_i = 3 + 1 + 10 + 5 = 19$$

In general, if there are n observations, we write

$$\sum_{i=1}^{n} X_i = X_1 + X_2 + \cdots + X_n$$

EXAMPLE B-1

Let $X_1 = -2$, $X_2 = 3$, and $X_3 = 5$. Find

a. $\displaystyle\sum_{i=1}^{3} X_i$ b. $\displaystyle\sum_{j=1}^{3} X_j^2$ c. $\displaystyle\sum_{j=1}^{3} (2X_j + 3)$

Solutions

a. $\displaystyle\sum_{i=1}^{3} X_i = X_1 + X_2 + X_3 = -2 + 3 + 5 = 6$

b. $\displaystyle\sum_{j=1}^{3} X_j^2 = X_1^2 + X_2^2 + X_3^2 = (-2)^2 + (3)^2 + (5)^2 = 38$

c. $\displaystyle\sum_{j=1}^{3} (2X_j + 3) = (2X_1 + 3) + (2X_2 + 3) + (2X_3 + 3)$

$= (-4 + 3) + (6 + 3) + (10 + 3) = -1 + 9 + 13 = 21$

EXAMPLE B-2

Prove

a. $\displaystyle\sum_{i=1}^{n} aX_i = a \sum_{i=1}^{n} X_i$ b. $\displaystyle\sum_{i=1}^{n} a = na$ c. $\displaystyle\sum_{i=1}^{n} (X_i + Y_i) = \sum_{i=1}^{n} X_i + \sum_{i=1}^{n} Y_i$

where a is a constant.

a. $\displaystyle\sum_{i=1}^{n} aX_i = aX_1 + aX_2 + \cdots + aX_n = a(X_1 + X_2 + \cdots + X_n) = a\sum_{i=1}^{n} X_i$ *Solutions*

b. $\displaystyle\sum_{i=1}^{n} a = a\sum_{i=1}^{n} 1 = a\underbrace{(1 + 1 + \cdots + 1)}_{n \text{ terms}} = na$

c. $\displaystyle\sum_{i=1}^{n} (X_i + Y_i) = X_1 + Y_1 + X_2 + Y_2 + \cdots + X_n + Y_n$

$$= (X_1 + X_2 + \cdots + X_n) + (Y_1 + Y_2 + \cdots + Y_n) = \sum_{i=1}^{n} X_i + \sum_{i=1}^{n} Y_i$$

These three summation properties are listed as rules 1, 2, and 3 at the end of this appendix.

Double summations are used to indicate summations of more than one variable, where different subscript indexes are involved. For example, the symbol   *Double summations*

$$\sum_{j=1}^{3} \sum_{i=1}^{2} X_i Y_j$$

means "the sum of the products of X_i and Y_j where $i = 1, 2$ and $j = 1, 2, 3$." Thus, we can write

$$\sum_{j=1}^{3} \sum_{i=1}^{2} X_i Y_j = X_1 Y_1 + X_2 Y_1 + X_1 Y_2 + X_2 Y_2 + X_1 Y_3 + X_2 Y_3$$

Simplified Summation Notations

In this text, simplified summation notations are often used in which subscripts are eliminated. For example, ΣX, ΣX^2, and ΣY^2 are used instead of

$$\sum_{i=1}^{n} X_i \qquad \sum_{i=1}^{n} X_i^2 \qquad \text{and} \qquad \sum_{i=1}^{n} Y_i^2$$

Also in this text, subscripts have ordinarily been dropped in the case of probability distributions. The statement that the sum of the probabilities is equal to one is given by

$$\sum_{i=1}^{3} f(x_i) = 1$$

In this textbook, we use the customary simplified notation, as shown in these two examples of a discrete probability distribution.

Standard Notation		Simplified Notation	
x_i	$f(x_i)$	x	$f(x)$
$x_1 = 0$	0.2	0	0.2
$x_2 = 1$	0.3	1	0.3
$x_3 = 2$	0.5	2	0.5
	1.0		1.0

The corresponding summation statement is $\sum_x f(x) = 1$ where $\sum_x$ means "sum over all values of x." The notation is also further simplified by writing

$$\sum f(x) = 1$$

Summation Properties

Rule 1

$$\sum_{i=1}^{n} aX_i = a \sum_{i=1}^{n} X_i$$

Rule 2

$$\sum_{i=1}^{n} a = \underbrace{a + a + \cdots + a}_{n \text{ terms}} = na$$

Rule 3

$$\sum_{i=1}^{n} (X_i + Y_i) = \sum_{i=1}^{n} X_i + \sum_{i=1}^{n} Y_i$$

Appendix C

PROPERTIES OF EXPECTED VALUES AND VARIANCES

In keeping with notational conventions used in this text, a and b represent constants, whereas X represents a random variable. The symbols $E(X)$ and $\text{VAR}(X)$ denote the expected value and variance of the random variable X, and the symbols $E(X_1 + X_2 + \cdots + X_n)$ and $\text{VAR}(X_1 + X_2 + \cdots + X_n)$ denote the expected value and variance of the sum of the random variables $X_1, X_2, \ldots, X_n$, and so forth.

Rule 1 declares that the expected value of a constant is equal to that constant.

$$E(a) = a$$

Rule 1

Rule 2 states that the expected value of a constant times a random variable is equal to the constant times the expected value of the random variable.

$$E(bX) = bE(X)$$

Rule 2

Rule 3 combines rules 1 and 2.

$$E(a + bX) = a + bE(X)$$

Rule 3

A brief proof for rule 3 illustrates a general method of proofs for expected values.

Let X denote a discrete random variable that takes on values $x_1, x_2, \ldots,$ $x_i, \ldots, x_n$ with probabilities $f(x_1), f(x_2), \ldots, f(x_i), \ldots, f(x_n)$. Then, using the definition of an expected value given in equation 3.11 in Chapter 3, we have

$$E(a + bX) = \sum_{i=1}^{n} (a + bx_i)f(x_i) = \sum_{i=1}^{n} af(x_i) + \sum_{i=1}^{n} bx_i f(x_i)$$

$$= a \sum_{i=1}^{n} f(x_i) + b \sum_{i=1}^{n} x_i f(x_i)$$

$$= a(1) + bE(X) = a + bE(X)$$

Rule 4 says the expected value of a sum equals the sum of the expected values.

Rule 4
$$E(X_1 + X_2 + \cdots + X_n) = E(X_1) + E(X_2) + \cdots + E(X_n)$$

where $X_1, X_2, \ldots, X_n$ are random variables. The X_i's are not restricted in any way; that is, they may be either independent or dependent.

Expressing this rule in somewhat different symbols, we have

$$E\left[\sum_{i=1}^{n} (X_i) \right] = \sum_{i=1}^{n} [E(X_i)]$$

Treating the expected value and summation symbols as operators (that is, as defining specific operations on the X_i's), we have the result that the summation sign and expected value symbol are interchangeable operators.

Rule 5 states that the variance of a constant is equal to zero.

Rule 5
$$\text{VAR}(a) = 0$$

Rule 6 says the variance of a constant times a random variable is equal to the constant squared times the variance of the random variable.

Rule 6
$$\text{VAR}(bX) = b^2 \text{VAR}(X)$$

Rule 7 combines rules 5 and 6.

Rule 7
$$\text{VAR}(a + bX) = b^2 \text{VAR}(X)$$

As in the case of rule 3 for the expected value, a simple application of the definition of a variance yields the desired result. The proof is left to the reader as an exercise.

Rule 8 states that for n independent random variables, the variance of a sum is equal to the sum of the variances.

Rule 8
$$\text{VAR}(X_1 + X_2 + \cdots + X_n) = \text{VAR}(X_1) + \text{VAR}(X_2) + \cdots + \text{VAR}(X_n)$$

where $X_1, X_2, \ldots, X_n$ are independent random variables; that is, every pair of X_i's is independent.

Thus, if the X_i's are independent, the variance of a sum is equal to the sum of the variances.

Expressing rule 8 in summation terminology, we obtain

$$\text{VAR}\left[\sum_{i=1}^{n} (X_i)\right] = \sum_{i=1}^{n} [\text{VAR}(X_i)]$$

The variance and summation symbols are interchangeable operators if the X_i's are independent.

Rule 9 is derived by applying rules 7 and 8.

$$\text{VAR}(a_1 X_1 + a_2 X_2) = a_1^2 \text{VAR}(X_1) + a_2^2 \text{VAR}(X_2) \qquad \text{\textit{Rule 9}}$$

if X_1 and X_2 are independent.

Special cases of rule 9 are given as rules 10 and 11. In **rule 10**, $a_1 = +1$ and $a_2 = +1$.

$$\text{VAR}(X_1 + X_2) = \text{VAR}(X_1) + \text{VAR}(X_2) \qquad \text{\textit{Rule 10}}$$

if X_1 and X_2 are independent. In **rule 11**, $a_1 = +1$ but $a_2 = -1$.

$$\text{VAR}(X_1 - X_2) = \text{VAR}(X_1) + \text{VAR}(X_2) \qquad \text{\textit{Rule 11}}$$

if X_1 and X_2 are independent.

Rule 12 says that the variance of a sample mean is equal to the population variance divided by the sample size.

$$\text{VAR}(\bar{X}) = \frac{\sigma^2}{n} \qquad \text{\textit{Rule 12}}$$

where X is a random variable, μ and σ are its mean and standard deviation, respectively, and $\bar{X}$ is the arithmetic mean in a sample of n independent observations of X. If $X_1, X_2, \ldots, X_n$ denote the n observations, then

$$\bar{X} = \frac{1}{n} \sum_{i=1}^{n} X_i$$

Proof: Rule 12 may be proved in a few steps.

1. $$\text{VAR}(\bar{X}) = \text{VAR}\left(\frac{1}{n}\sum_{i=1}^{n} X_i\right) = \frac{1}{n^2} \text{VAR}\left(\sum_{i=1}^{n} X_i\right) \qquad \text{(by rule 6)}$$

2. $$= \frac{1}{n^2} \sum_{i=1}^{n} [\text{VAR}(X_i)] \qquad \text{(by rule 8)}$$

But since every X_i has the same probability distribution as X, then $\text{VAR}(X_i) = \text{VAR}(X)$ for each i. Hence,

3. $$\sum_{i=1}^{n} [\text{VAR}(X_i)] = n\sigma^2$$

Substituting step 3 into step 2 gives

4. $$\text{VAR}(\bar{X}) = \left(\frac{1}{n^2}\right)(n\sigma^2) = \frac{\sigma^2}{n}$$

which completes the proof.

Let us express rule 12 in the language and symbolism of sampling theory. If a simple random sample of size n is drawn from an infinite population (or a finite population with replacement) with standard deviation σ, the variance of the sample mean is given by

5.
$$\sigma_{\bar{X}}^2 = \frac{\sigma^2}{n}$$

Rule 13 states that the expected value of the sample mean is equal to the population mean where the same conditions prevail as in rule 12; X is a random variable and μ and σ are its mean and standard deviation.

Rule 13
$$E(\bar{X}) = \mu$$

Proof: Rule 13 is easily proved as follows:

1.
$$E(\bar{X}) = E\left(\frac{1}{n} \sum_{i=1}^{n} X_i\right)$$

$$= \frac{1}{n}\left[E\left(\sum_{i=1}^{n} X_i\right)\right] \qquad \text{(by rule 2)}$$

2.
$$= \frac{1}{n}\left[\sum_{i=1}^{n} E(X_i)\right] \qquad \text{(by rule 4)}$$

But since every X_i has the same probability distribution as X, then $E(X_i) = E(X)$ for each i. Hence,

3.
$$\sum_{i=1}^{n} E(X_i) = nE(X) = n\mu$$

Substituting step 3 into step 2 gives

4.
$$E(\bar{X}) = \left(\frac{1}{n}\right)(n\mu) = \mu$$

As in rule 12, **let us express this result in terms of sampling theory.** If a simple random sample of size n is drawn from an infinite population (or a finite population with replacement) with mean μ, the expected value (arithmetic mean) of the sample mean is given by

5.
$$E(\bar{X}) = \mu_{\bar{x}} = \mu$$

The expected value of the sample variance defined with divisor $n - 1$ is equal to the population variance.

This result is expressed in symbols in **rule 14**.

Rule 14
$$E(s^2) = E\left[\frac{\Sigma(X_i - \bar{X})^2}{n - 1}\right] = \sigma^2$$

where the same conditions prevail as in rules 12 and 13.

SHORTCUT FORMULAS TO USE WHEN CLASS INTERVALS ARE EQUAL IN SIZE

Shortcut Calculation of the Mean

A shortcut calculation known as the step-deviation method is useful when class intervals in a frequency distribution are of equal size. The method results in simpler arithmetic than the direct definitional formula, particularly if the class intervals and frequencies involve a large number of digits.

The **step-deviation method** of computing the mean involves three basic steps:

1. Selection of an assumed (or arbitrary) mean.

2. Calculation of an average deviation from this assumed mean.

3. Addition of this average deviation as a correction factor to the assumed mean to obtain the true mean. This correction factor is positive if the assumed mean lies below the true mean, negative if above.

Step-deviation method

To accomplish step 2, a midpoint of a class (preferably near the center of the distribution) is taken as the assumed mean. Then deviations of the midpoints of the other classes are taken from the assumed mean in class interval units. These deviations are denoted d. After the d values are averaged, the result is multiplied by the size of the class interval to return to the units of the original data.

TABLE D-1

Calculation of the arithmetic mean for grouped data by the step-deviation method for bottle weights data

Weight (in ounces)	Number of bottles f	d	fd
14.0 and under 14.5	8	-4	-32
14.5 and under 15.0	10	-3	-30
15.0 and under 15.5	15	-2	-30
15.5 and under 16.0	18	-1	-18
16.0 and under 16.5	22	0	0
16.5 and under 17.0	14	1	14
17.0 and under 17.5	8	2	16
17.5 and under 18.0	5	3	15
	100		-65

$$\bar{X}_a = 16.25$$

$$\bar{X} = \bar{X}_a + \left(\frac{\Sigma fd}{n}\right)(i) = 16.25 + \left(\frac{-65}{100}\right)(0.5) = 15.93 \text{ ounces}$$

The formula for the step-deviation method is given in equation D.1.

Step-deviation method for the arithmetic mean (grouped data)   **(D.1)**

$$\bar{X} = \bar{X}_a + \left(\frac{\Sigma fd}{n}\right)(i)$$

where $\bar{X}$ = the arithmetic mean
$\bar{X}_a$ = the assumed arithmetic mean
f = frequencies
d = deviations of midpoints from the assumed mean in class interval units
n = the number of observations
i = the size of a class interval

In Table D-1, the assumed arithmetic mean is 16.25. The values in the d column indicate the number of class intervals below or above the one in which the assumed mean is taken.

Shortcut Calculation of the Standard Deviation

Just as in the case of the arithmetic mean, the step-deviation method of calculating the standard deviation is useful when class intervals in a frequency distri-

TABLE D-2
Calculation of the standard deviation for grouped data by the step-deviation
method for bottle weights data

Weight (in ounces)	Number of Bottles f	d	fd	fd^2
14.0 and under 14.5	8	-4	-32	128
14.5 and under 15.0	10	-3	-30	90
15.0 and under 15.5	15	-2	-30	60
15.5 and under 16.0	18	-1	-18	18
16.0 and under 16.5	22	0	0	0
16.5 and under 17.0	14	1	14	14
17.0 and under 17.5	8	2	16	32
17.5 and under 18.0	5	3	15	45
	100		-65	387

$$\bar{X} = 16.25$$

$$s = (i)\sqrt{\frac{\Sigma fd^2 - \dfrac{(\Sigma fd)^2}{n}}{n-1}} = (0.5)\sqrt{\frac{387 - \dfrac{(-65)^2}{100}}{100-1}} = (0.5)(1.866)$$

$$= 0.933 \text{ ounce}$$

bution are the same size. The saving in computational effort accomplished by
the use of the step-deviation method is illustrated for the bottle weights data
given in Table D-1.

As with the calculation for the arithmetic mean, the procedure involves
taking deviations of midpoints of classes from an assumed mean and stating
them in class interval units. Only one additional column of values, fd^2, is
required to compute the standard deviation by the step-deviation method compared with the corresponding arithmetic mean computation given in Table D-1.
The formula for the step-deviation method is given in equation D.2. All of the
symbols have the same meaning as in equation D.1 for the arithmetic mean
and the computation in Table D-1. The standard deviation is equal to 0.933
ounce.

(D.2)
$$s = (i)\sqrt{\frac{\Sigma fd^2 - \dfrac{(\Sigma fd)^2}{n}}{n-1}}$$

*Step-deviation method
for the sample
standard deviation
(grouped data)*

The use of equation D.2 is illustrated in Table D-2. The same assumed mean,
$\bar{X}_a = 16.25$ ounces, was used in Table D-2 as in the calculation of the arithmetic
mean given in Table D-1.

Appendix **E**

DATA BANK FOR COMPUTER EXERCISES

These data are found in the Minitab file 'databank . mtw' and are used in the Minitab exercises for Chapter 7.

Family	Annual Food Expenditures ($000)	Annual Income ($000)	Family Size	Age of the Highest Income Earner (in years)	Home Owner (0) or Renter (1)
1	5.2	28	3	32	1
2	5.1	26	3	28	1
3	5.6	32	2	25	0
4	4.6	24	1	43	1
5	11.3	54	4	50	0
6	8.1	59	2	55	0
7	7.8	44	3	32	0
8	5.8	30	2	28	1
9	5.1	40	1	37	0
10	18.0	82	6	54	0
11	4.9	42	3	30	1
12	11.8	58	4	31	0
13	5.2	28	1	28	1
14	4.8	20	5	48	0
15	7.9	42	3	42	0
16	6.4	47	1	32	1
17	20.0	112	6	60	0
18	13.7	85	5	47	0
19	5.1	31	2	33	0

Family	Annual Food Expenditures ($000)	Annual Income ($000)	Family Size	Age of the Highest Income Earner (in years)	Home Owner (0) or Renter (1)
20	2.9	26	2	29	1
21	3.3	18	1	26	1
22	6.8	30	1	45	1
23	4.9	31	4	43	1
24	10.2	62	3	30	0
25	10.7	54	4	55	0
26	4.8	29	3	33	1
27	5.0	30	2	29	1
28	4.1	34	1	26	0
29	4.4	30	1	25	1
30	4.3	25	1	31	1
31	10.3	57	6	48	0
32	9.8	52	4	55	0
33	7.3	47	3	31	1
34	5.1	36	1	32	1
35	3.3	19	4	29	1
36	4.6	28	4	49	0
37	2.8	17	2	43	1
38	12.4	47	5	33	1
39	8.0	49	3	35	0
40	13.8	87	3	63	0
41	10.4	72	2	34	0
42	2.5	12	1	23	1
43	4.3	28	2	27	1
44	3.3	16	1	25	1
45	3.4	19	1	28	1
46	8.7	45	4	30	0
47	4.2	27	2	51	0
48	15.3	75	5	45	1
49	11.1	63	5	47	0
50	5.6	32	3	28	1
51	5.9	35	3	29	1
52	8.7	46	4	32	0
53	5.4	29	1	25	1
54	4.8	24	1	27	1
55	10.8	58	7	46	0
56	8.9	48	5	48	0
57	8.0	51	4	52	0
58	6.3	38	3	36	1
59	2.9	17	1	29	1
60	5.1	26	2	32	0
61	3.2	15	1	24	1
62	4.1	21	1	28	1
63	7.5	50	2	42	0
64	13.1	78	3	58	0

Family	Annual Food Expenditures ($000)	Annual Income ($000)	Family Size	Age of the Highest Income Earner (in years)	Home Owner (0) or Renter (1)
65	5.5	27	1	68	0
66	5.1	31	2	33	1
67	12.5	73	2	43	0
68	4.5	29	3	38	1
69	3.2	20	1	31	1
70	9.8	60	4	35	0
71	9.7	51	5	51	0
72	5.3	33	3	29	1
73	12.2	58	4	52	0
74	6.8	53	3	30	1
75	6.2	51	1	33	1
76	6.4	32	2	29	1
77	8.1	41	3	34	0
78	15.6	63	4	52	0
79	3.1	19	1	27	1
80	10.4	62	5	64	0
81	5.8	28	2	29	1
82	3.4	12	1	25	1
83	8.6	41	3	46	0
84	6.3	41	2	33	1
85	9.3	50	3	36	1
86	8.4	51	3	39	0
87	8.1	45	2	28	1
88	12.1	60	6	54	0
89	2.6	18	1	37	1
90	2.4	17	1	35	1
91	8.8	43	3	44	0
92	9.6	50	3	31	0
93	5.4	31	2	34	0
94	4.4	30	2	30	1
95	12.1	62	4	49	0
96	8.3	52	2	36	1
97	8.4	53	3	28	0
98	14.8	72	5	55	0
99	9.3	50	3	48	0
100	5.2	46	1	32	1

GLOSSARY OF SYMBOLS

Numbers in parentheses refer to the section in which the symbol is introduced.

General Mathematical Symbols	Meaning	Probability and Statistical Symbols	Meaning
$\lvert d \rvert$	Absolute value of d (13.3)	A	Y intercept in a two-variable linear regression model. The value of a in a sample regression equation is an estimator of A (9.1)
e	A constant equal to 2.71828 ...; the base of the Naperian, or natural, logarithm system (3.7)		
$\log X$	Logarithm of X to the base 10 (9.8)	$\bar{A}$	Complement of event A (9.1)
$n!$	n factorial, or $(n)(n-1)\cdots(2)(1)$ (2.4)	$A_1, A_2, \ldots, A_n$	Alternative acts (14.2)
		A_t	Actual figure for time period t in exponential smoothing (11.6)
π	A constant equal to 3.1416 ... (5.3)	a	Y intercept calculated from a sample of observations; computed value of Y when $X = 0$ in a two-variable regression equation; computed value of Y when values of all independent variables are 0 in a multiple regression equation (9.4)
Σ	Sum of (see Appendix B) (1.9)		
$\displaystyle\sum_{i=1}^{n}$	Sum of the terms that follow, from $i = 1$ to $i = n$ (see Appendix B) (1.9)		
$\displaystyle\sum_{x}$ or $\sum_x$	Sum of the terms that follow, for all values x takes on (3.1)	a, b	Constants in a straight-line trend equation (11.4)
$\displaystyle\sum_{j}\sum_{i}$	Summation of terms that follow, first over all values of i, then over all values of j (8.3)	a, b, c	Constants in a parabolic trend equation (11.4)
$x \leqslant a$	x is less than or equal to a (3.1)	α	Probability of a Type I error, or of rejecting H_0 when it is true; the significance level (7.2)
$x \geqslant a$	x is greater than or equal to a (3.1)		

B	Regression coefficient in a two-variable linear regression model. The value of b in a sample regression equation is an estimator of B (9.1)	F	F ratio; the ratio of the between-treatment variance to the within-treatment variance (8.3)
b	Regression coefficient calculated from a sample of observations; slope of the regression line in a sample two-variable regression equation (9.4)	F	F ratio; the ratio of explained variance to unexplained variance (10.6)
b_1, b_2	Sample net regression coefficients; the coefficients of independent variables X_1 and X_2 (10.1)	$F(v_1, v_2)$	F ratio with v_1 and v_2 degrees of freedom for the numerator and denominator, respectively (8.3)
β	Probability of a Type II error, or of accepting H_0 when it is false (7.2)	f	Number of observations (frequency) in a class interval of a frequency distribution (1.8)
β_1, β_2	Beta coefficients in a multiple regression equation; β_1 measures the number of standard deviations of change in $\hat{Y}$ for each change of one standard deviation in X_1, when X_2 is held constant (10.11)	$f = \dfrac{n}{N}$	Sampling fraction (5.5)
		f_o	Observed frequency in a χ^2 test (8.1)
C	Correction term in shortcut computation of an analysis of variance (8.3)	f_t	Theoretical, or expected, frequency in χ^2 test (8.1)
C	Effect of the cyclical factor in time-series analysis (11.4)	F_t	Forecast for time period t in exponential smoothing (11.6)
$\chi^2 = \sum \dfrac{(f_o - f_t)^2}{f_t}$	χ^2 statistic in a test of independence or goodness of fit (8.1)	$F(x) = P(X \leqslant x)$	Cumulative probability that random variable X is less than or equal to x (3.1)
CV	Coefficient of variation (1.16)	f_{Md}	Frequency of the class containing the median (1.11)
c	Number of columns in an arrangement of data to which analysis of variance is applied (8.3)	f_p	Frequencies in classes preceding the one containing the median (1.11)
		$f(x) = P(X = x)$	Probability that random variable X is equal to the value x (3.1)
D	Tolerated sampling error in a determination of sample size (6.5)	$f(Y\|X)$	Conditional probability distribution of Y given X in the linear regression model (9.1)
d	Durbin–Watson statistic; a statistic used in tests for autocorrelation in time series (10.9)	G	Geometric mean (1.13)
		H	Value of the highest observation (1.2)
$\bar{d}$	Mean difference of pairs of observations made on the same individuals or objects (7.6)	H_0	Null hypothesis; basic hypothesis being tested (7.1)
$d = X - Y$	Difference between ranks of paired observations in rank correlation (13.7)	H_1	Alternative hypothesis; rejection of H_0 implies tentative acceptance of H_1 (7.1)
ENGS	Expected net gain of sample information (16.2)	I	Effect of the irregular factor in time-series analysis (11.4)
$\mathrm{EOL}(A_j)$	Expected opportunity loss of act A_j (14.4)	i	Size of the class interval (1.2)
EVPI	Expected value of perfect information (14.5)	K	Kruskal–Wallis test statistic in a one-factor analysis of variance by ranks (13.6)
EVSI	Expected value of sample information (16.2)	k	Number of classes in a frequency distribution (1.2)
$E(X)$	Expected value of the random variable X, that is, the expected value of the probability distribution of X (3.10)	L	Value of the lowest observation (1.2)
		$L(a_1\|\theta_1)$	Opportunity loss of act a_1 given state of nature θ_1 (17.1)
		L_{Md}	Lower limit of class containing the median (1.11)
		$L(\hat{\theta}\|\theta)$	Loss involved in estimating $\hat{\theta}$ when the parameter value is θ (17.3)

MA	Moving average figures in seasonal variation analysis (11.5)	P_0	Price in a base period in an index number formula (12.2)
Md	Median (1.10)	$P_0(p)$	Prior probability distribution of random variable p (15.2)
m	Midpoint of a class interval of a frequency distribution (1.8)	$P_1(p)$	Posterior probability distribution of random variable p (15.2)
μ	Arithmetic mean of a population (1.8)	$P(X = x \mid Y = y)$	Conditional probability that random variable X is equal to the value x given that random variable Y is equal to y (3.13)
μ	Arithmetic mean of a probability distribution (3.10)		
$\mu_{n\bar{p}}$	Mean of sampling distribution of number of occurrences (5.5)	$P(X = x \text{ and } Y = y)$	Joint probability that X takes on the value x and Y takes on the value y (3.13)
$\mu_{\bar{p}}$	Mean of sampling distribution of a proportion (5.6)	p	Probability of a success on a given trial (binomial distribution); also used as population proportion of successes (3.4)
$\mu_{\bar{p}_1 - \bar{p}_2}$	Mean of sampling distribution of the difference between two proportions (6.3)		
μ_r	Mean of sampling distribution of the number of runs r (13.5)	$\bar{p} = \dfrac{x}{n}$	Proportion of successes in a sample of size n (5.6)
μ_T	Mean of sampling distribution of T in the Wilcoxon matched-pairs signed rank test (13.3)	$\hat{p}$	Weighted mean of two sample proportions (7.3)
μ_U	Mean of sampling distribution of U in the Mann–Whitney U test (rank sum test) (13.4)	Q_f	A fixed set of weights in an index number formula (12.2)
		Q_n	Quantity in a nonbase period in an index number formula (12.2)
$\mu_{\bar{x}_1 - \bar{x}_2}$	Mean of sampling distribution of the difference between two sample means (6.3)	Q_0	Quantity in a base period in an index number formula (12.2)
$\mu_{Y.X}$	Mean of conditional probability distribution of Y given X in the linear regression model; the population value that corresponds to the $\hat{Y}$ value computed from sample observations (9.1)	$q = 1 - p$	Probability of failure on a given trial (binomial distribution); also used as population proportion of failures (3.4)
		R_1, R_2	Sum of ranks of the items in samples 1 and 2, respectively, in the Mann–Whitney U test (rank sum test) (13.4)
N	Number of observations in a population (1.8)	$R(s_1 \mid \theta_1)$	Risk or expected opportunity loss associated with the use of strategy s_1, given that state of nature θ_1 occurs (16.3)
n	Number of observations in a sample (1.8)		
$\dbinom{n}{x}$	Number of combinations of n objects taken x at a time (2.4)	$R^2_{Y.12\ldots(k-1)}$	Sample coefficient of multiple determination for a regression equation involving $k - 1$ independent variables $X_1, X_2, \ldots, X_{k-1}$ and dependent variable Y (10.4)
$_nP_x$	Number of permutations of n objects taken x at a time (2.4)		
ν	*nu*: Number of degrees of freedom (6.4)	r	Number of rows in an arrangement of data to which analysis of variance is applied (8.3)
$P(A)$	Probability of event A (2.1)	r	Sample correlation coefficient (9.6)
$P(A_1 \text{ or } A_2)$	Probability of the occurrence of at least one of events A_1 and A_2 (2.2)	r^2	Sample coefficient of determination (9.6)
$P(A_1 \text{ and } A_2)$	Joint probability of events A_1 and A_2 (2.2)	r^2_c	Corrected or adjusted sample coefficient of determination (9.6)
$P(B_1 \mid A_1)$	Conditional probability of event B_1 given A_1 (2.2)	r_j	Number of observations in the jth column (8.3)
P_n	Price in a nonbase period in an index number formula (12.2)	r_r	Rank correlation coefficient (13.7)

r_{12}	Correlation coefficient for variables X_1 and X_2 (10.5)		two-variable regression line. An estimator of $\sigma_{Y.X}$ (9.5)
ρ	rho: Population correlation coefficient (9.6)	σ	Standard deviation of a population (1.15)
ρ^2	Population coefficient of determination (9.6)	σ	Standard deviation of a probability distribution (3.11)
S	Sample space (2.1)	σ^2	Variance of a population (1.15)
SI	Seasonal index (11.5)	σ^2	Variance of a probability distribution (3.11)
SSA	Between-treatment sum of squares (8.3)	$\sigma_{n\bar{p}}$	Standard deviation of sampling distribution of number of occurrences (5.5)
SSE	Within-treatment (error) sum of squares (8.3)		
SST	Total sum of squares (8.3)	$\sigma_{\bar{p}}$	Standard error of a proportion, that is, standard deviation of sampling distribution of a proportion (5.6)
$S^2_{Y.12\ldots(k-1)}$	Sample variance around a regression equation involving $k-1$ independent variables $X_1, X_2, \ldots, X_{k-1}$ and dependent variable Y (10.3)		
		$\sigma_{\bar{p}_1-\bar{p}_2}$	Standard error of the difference between two proportions (7.3)
s	Standard deviation of a sample (1.15)	σ_r	Standard error of sampling distribution of the number of runs r (13.5)
s^2	Variance of a sample (1.15)		
$s_1, s_2, \ldots$	Strategies (16.3)	σ_T	Standard deviation of sampling distribution of T in the Wilcoxon matched-pairs signed rank test (13.3)
s_b	Standard error of regression coefficient b (9.7)		
s_d	Standard deviation of differences between pairs of observations made on the same individuals or objects (7.6)	σ_U	Standard error of U statistic in the Mann–Whitney U test (rank sum test) (13.4)
		$\sigma_{\bar{x}}$	Standard error of the mean, that is, standard deviation of sampling distribution of the mean (5.4)
$s_{\bar{d}}$	Standard error of $\bar{d}$, the mean difference of pairs of observations made on the same individuals or objects (7.6)		
		$\sigma_{\bar{x}_1-\bar{x}_2}$	Standard error of the difference between two means (6.3)
$s_{\bar{p}}$	Estimated standard error of a proportion (6.3)	$\sigma_{Y.X}$	Standard deviation of conditional probability distribution of Y given X in the linear regression model; the population value that corresponds to the $s_{Y.X}$ value computed from sample observations (9.1)
s_{IND}	Standard error of forecast in a two-variable regression analysis. Used to establish prediction intervals for individual Y values		
$s_{\bar{p}_1-\bar{p}_2}$	Estimated or approximate standard error of the difference between two proportions (6.3)	T	Grand total of all observations in a shortcut computation of an analysis of variance (8.3)
$s_{\bar{x}_1-\bar{x}_2}$	Estimated or approximate standard error of the difference between two means (6.3)	T	Effect of the trend factor in time-series analysis (11.4)
$s_{Y.}$	Standard deviation of a sample of Y values (9.5)	T	Wilcoxon's T statistic; the smaller of two ranked sums (13.3)
$s_{\hat{Y}}$	Estimated standard error of the conditional mean in a two-variable regression analysis. Used to establish confidence intervals for a conditional mean (9.5)	T_j	Total of the observations in the jth column (8.3)
		$t = \dfrac{\bar{x} - \mu}{s/\sqrt{n}}$	The t statistic, distributed according to the Student t distribution with v degrees of freedom (6.4)
$s_{Y.X}$	Standard error of estimate. Measures scatter of observed values of Y around the corresponding $\hat{Y}$ values on a	θ	Population parameter (6.2)
		$\hat{\theta}$	Estimator of θ (6.2)
		$\theta_1, \theta_2, \ldots, \theta_m$	States of nature (14.2)

U — A measure of the difference between the ranked observations of two samples in the Mann-Whitney U test (rank sum test)   (13.4)

$U(x)$ — Utility of monetary payoff x (14.8)

u_{ij} — Utility of selecting act A_j when state of nature θ_i occurs   (14.2)

$VAR(X)$ — Variance of the random variable X, that is, the variance of the probability distribution of X (3.11)

w — Weight applied to an observation (1.9)

X — Value of an observation   (1.8)

$\bar{\bar{X}}$ — Grand (arithmetic) mean of all observations   (8.3)

X_{ij} — Value of an observation in the ith row and jth column   (8.3)

$\bar{X}_j$ — Arithmetic mean of the jth column of observations   (8.3)

$\bar{X}_w$ — Weighted arithmetic mean   (1.9)

$\bar{X}, \bar{x}$ — Arithmetic mean of a sample   (1.8)

x — Number of successes in a sample of size n (5.6)

$\hat{Y}$ — Computed value of Y in a sample regression equation   (9.4)

$\hat{Y}_t$ — Computed trend value for the time-series variable Y (11.4)

$z = \dfrac{x - \mu}{\sigma}$ — Standard score; deviation of value of an observation from the arithmetic mean of a distribution expressed in multiples of the standard deviation (5.3)

BIBLIOGRAPHY

Statistics for the General Reader

Anderson, A. *Interpreting Data*. England: Chapman and Hall, 1988.

Bailey, M. *Reducing Risks to Life*. Washington, DC: American Enterprise Institute, 1980.

Bickel, P.J., and O'Connell, J.W. "Sex Bias in Graduate Admissions: Data from Berkeley," *Science*, Vol. 187 (February 1975), pp. 398–404.

Brightman, H.J. *Statistics in Plain English*. Cincinnati: South-Western, 1985.

Campbell, S.K. *Flaws and Fallacies in Statistical Thinking*. Englewood Cliffs, NJ: Prentice-Hall, 1974.

Fairley, W.B., and Mosteller, F., eds. *Statistics and Public Policy*. Reading, MA: Addison-Wesley, 1977.

Federer, W.T. *Statistics and Society*. New York: Marcel Dekker, 1973.

Folks, J.L. *Ideas of Statistics*. New York: John Wiley, 1981.

Gilbert, J.P., McPeek, B., and Mosteller, F. "Statistics and Ethics in Surgery and Anesthesia," *Science*, Vol. 198 (November 1977), pp. 684–89.

Hamburg, M. *Basic Statistics: A Modern Approach*, 3rd ed. San Diego, CA: Harcourt Brace Jovanovich, 1985.

Hooke, R. *How to Tell the Liars from the Statisticians*. New York: Dekker, 1983.

Huff, D. *How to Lie with Statistics*. New York: W.W. Norton, 1954.

"Is Vitamin C Really Good for Colds?" *Consumer Reports*, Vol. 41, No. 2 (February 1976), pp. 68–70.

Kemp, K.W. *Dice, Data and Decisions: Introductory Statistics*. New York: Halstead, 1984.

Larson, R.J., and Stroup, D.F. *Statistics in the Real World*. New York: Macmillan, 1976.

Levinson, H.C. *Chance, Luck, and Statistics.* New York: Dover Publications, 1963.

McNeil, B.J., Weichselbaum, R., and Pareker, S.G. "Fallacy of the Five-Year Survival in Lung Cancer," *New England Journal of Medicine*, Vol. 229 (December 1978), pp. 1397–401.

Moore, D.S. *Statistics: Concepts and Controversies.* San Francisco: Freeman, 1979.

Moroney, M.J. *Facts from Figures.* New York: Penguin Books, 1956.

Mosteller, F., Pieters, R.S., Kruskal, W.H., Rising, G.R., Link, R.F., Carlson, R., and Zelinka, M. *Statistics by Example.* Reading, MA: Addison-Wesley, 1973.

Odell, J.W. *Basic Statistics: An Introduction to Problem Solving with Your Personal Computer.* Blue Ridge Summit, PA: TAB Books, Inc., 1984.

Reichman, W.J. *Use and Abuse of Statistics.* New York: Oxford University Press, 1962.

Tanur, J.M., et al., eds. *Statistics: A Guide to the Unknown.* San Francisco: Holden-Day, 1977. Also revised as three paperbacks: (1) *Statistics: A Guide to Business and Economics;* (2) *Statistics: A Guide to Biological and Health Sciences;* (3) *Statistics: A Guide to Political and Social Issues.*

Tufte, E.R. *Data Analysis for Politics and Policy.* Englewood Cliffs, NJ: Prentice-Hall, 1974.

U.S. Surgeon-General, *Smoking and Health: A Report of the Surgeon-General.* Washington, DC: U.S. Department of Health, Education and Welfare, 1979.

Wallis, W.A., and Roberts, H.V. *The Nature of Statistics.* New York: The Free Press, 1965.

Probability

Aldrich, J.H., and Nelson, F.D. *Linear Probability, Logit and Probit Models.* Beverly Hills, CA: Sage Publicatons, Inc. 1984.

Clarke, B.A., and Disney, Ralph L. *Probability and Random Processes: A First Course with Applications.* New York: John Wiley & Sons, 1985.

Feller, W. *An Introduction to Probability Theory and Its Applications*, 3rd ed. Vol. I. New York: John Wiley & Sons, 1968.

Freund, J.E. *Introduction to Probability.* Encino, CA: Dickenson Publishing, 1973.

Galambos, J. *Introductory Probability Theory.* New York: Dekker, 1984.

Goldberg, S. *Probability: An Introduction.* Englewood Cliffs, NJ: Prentice-Hall, 1960.

Hacking, I. *The Emergence of Probability: A Philosophical Study of Early Ideas about Probability, Induction and Statistical Inference.* New York: Cambridge University Press, 1984.

Hodges, J.L., and Lehman, E.L. *Basic Concepts of Probability and Statistics.* San Francisco: Holden-Day, 1964.

Hodges, J.L., and Lehman, E.L. *Elements of Finite Probability.* San Francisco: Holden-Day, 1965.

Hoel, P.G., Port, S.C., and Stone, C.J. *Introduction to Probability Theory.* Boston: Houghton Mifflin, 1972.

Hogg, R.V., and Tanis, E.A. *Probability and Statistical Inference.* New York: Macmillan, 1988.

Ingram, O., Glaser, L.J., and Derman, C. *Probability Models and Applications.* New York: Macmillan, 1980.

Jeffreys, H. *Theory of Probability*, 3rd ed. New York: Oxford University Press, 1983.

Malcolm, J.G. "Practical Application of Bayes' Formulas." Annual Reliability and Maintainability Symposium. IEEE: 1983, pp. 180–86.

Mosteller, F., Rourke, R., and Thomas, G., Jr. *Probability and Statistics.* Reading, MA: Addison-Wesley, 1961.

Newman, R.W., and White, R.M. *Reference Anthropometry of Army Men.* Report No. 180. Lawrence, MA: Environmental Climatic Research Laboratory.

Parzen, E. *Modern Probability Theory and Its Applications.* New York: John Wiley & Sons, 1960.

Ross, S. *A First Course in Probability*, 3rd ed. New York: Macmillan, 1984.

Regression and Correlation Analysis

Berry, W., and Feldman, S. *Multiple Regression in Practice.* Beverly Hills, CA: Sage, 1985.

Breiman, L., and Freedman, D. "How Many Variables Should Be Entered in a Regression Equation?" *Journal of the American Statistical Association* Vol. 78 (March 1983), pp. 131–36.

Chatterjee, S., and Price, B. *Regression Analysis by Example.* New York: John Wiley & Sons, 1978.

Cohen, J., and Cohen, P. *Applied Multiple Regression: Correlation Analysis for the Behavioral Sciences*, 2nd ed. Hillsdale, NJ: L. Erlbaum Associates, 1983.

Draper, N.R., and Smith, H. *Applied Regression Analysis*, 2nd ed. New York: John Wiley & Sons, 1981.

Edwards, A.L. *Multiple Regression Analysis and the Analysis of Variance and Covariance*, 2nd ed. New York: W.H. Freeman, 1985.

Ezekiel, M., and Fox, K.A. *Methods of Correlation and Regression Analysis*, 3rd ed. New York: John Wiley & Sons, 1959.

Johnston, J. *Econometric Methods*, 2nd ed. New York: McGraw-Hill, 1971.

Mendenhall, W., and Sincich, T. *A Second Course in Business Statistics: Regression Analysis*, 2nd ed. San Francisco: Dellen, 1986.

Morrison, Donald F. *Applied Linear Statistical Methods.* Englewood Cliffs, NJ: Prentice-Hall, 1983.

Mosteller, F., and Tukey, J.W. *Data Analysis and Regression: A Second Course in Statistics.* Reading, MA: Addison-Wesley, 1977.

Neter, J., and Kutner, M.H. *Applied Linear Regression Analysis.* Homewood, IL: Richard D. Irwin, 1983.

Neter, J., and Wasserman, W. *Applied Linear Statistical Models.* Homewood, IL: Richard D. Irwin, 1974.

Ryan, T.A., Joiner, B.L., and Ryan, B.F. *Minitab Student Handbook.* N. Scituate, MA: Duxbury Press, 1976.

SAS Institute, Inc. *SAS User's Guide, 1979 Edition.* Cary, NC: 1979.

Schroeder, L.D. *Understanding Regression Analysis.* Beverly Hills, CA: Sage, 1986.

Snedecor, G.W., and Cochran, W.G. *Statistical Methods*, 8th ed. Iowa State University Press, 1989.

SPSS* (Statistical Package for the Social Sciences) User's Guide. New York: McGraw-Hill, 1983.

Weisberg, S. *Applied Linear Regression*, 2nd ed. New York: John Wiley & Sons, 1985.

Williams, E.J. *Regression Analysis.* New York: John Wiley & Sons, 1959.

Wonnacott, R.J., and Wonnacott, T.H. *Econometrics*, 2nd ed. New York: John Wiley & Sons, 1979.

Wonnacott, T.H., and Wonnacott, R.J. *Regression: A Second Course in Statistics.* New York: John Wiley & Sons, 1981.

Younger, M. *First Course in Linear Regression*, 2nd ed. Boston: PWS Publishers, 1985.

Times-Series Analysis and Forecasting

Abraham, B., and Ledholter, J. *Statistical Methods for Forecasting.* New York: Wiley, 1983.

Bowerman, B.L., and O'Connell, R.T. *Forecasting and Time Series.* Boston: Duxbury Press, 1979.

Box, G.E.P., and Jenkins, G.M. *Time-Series Analysis: Forecasting and Control*, 2nd ed. San Francisco: Holden-Day, 1977.

Brown, Robert G. *Smoothing, Forecasting, and Prediction of Discrete Time Series.* Englewood Cliffs, NJ: Prentice-Hall, 1963.

Chambers, J.C., Mullick, S.K., and Smith, D.D. *An Executive's Guide to Forecasting.* New York: John Wiley & Sons, 1974.

Chatfield, C. *Analysis of Time Series: An Introduction*, 3rd ed. New York: Methuen, Inc., 1984.

Cryer. *Time-Series Analysis with Minitab*. Boston: PWS Publishers, 1986.

Daniels, L.M. *Business Forecasting for the 1980s—and Beyond*. Baker Library, Reference List No. 31. Boston: Harvard Business School, 1980.

Gilchrist, W. *Statistical Forecasting*. New York: John Wiley & Sons, 1976.

Granger, C.W.J. *Forecasting in Business and Economics*. New York: Academic Press, 1980.

Makridakis, S. *The Forecasting Accuracy of Major Time Series Methods*. New York: John Wiley & Sons, 1984.

Makridakis, S., Wheelwright, S.C., and McGee, V.E. *Forecasting: Methods and Applications*, 2nd ed. New York: John Wiley & Sons, 1983.

Milne, T.E. *Business Forecasting—A Managerial Approach*. London: Longman Group, Ltd., 1975.

Nelson, C.R. *Applied Time-Series Analysis for Managerial Forecasting*. San Francisco: Holden-Day, 1973.

1980 Supplement to Economic Indicators. Washington, DC: U.S. Government Printing Office, 1980.

Roberts, H.V. *Time Series and Forecasting with IDA*. New York: McGraw-Hill, 1984.

Sullivan, W.G., and Claycombe, W.W. *Fundamentals of Forecasting*. Reston, VA: Reston Publishing, 1977.

Wheelwright, S.C., and Makridakis, S. *Forecasting Methods for Management*, 3rd ed. New York: John Wiley & Sons, 1980.

Woods, D., and Fildes, R. *Forecasting for Business: Methods and Applications*. New York: Longman Group, Ltd., 1976.

Index Numbers

Allen, R.E.D. *Index Numbers in Theory and Practice*. Chicago: Aldine Publishing, 1975.

Fisher, I. *Making of Index Numbers*, 3rd reprint. New York: Kelly, 1927.

Banerjee, K.S. *Cost of Living Index Numbers: Practice, Precision, and Theory*. New York: Dekker, 1975.

Mudgett, Bruce D. *Index Numbers*. New York: John Wiley & Sons, 1951.

U.S. Department of Labor, Bureau of Labor Statistics. *The Consumer Price Index: Concepts and Content over the Years*, Bulletin 2134-2 (April 1984).

U.S. Department of Labor, Bureau of Labor Statistics. *CPI Issues*, Report 593 (February 1980).

U.S. Department of Labor, Bureau of Labor Statistics. *Monthly Labor Review*, Vol. 104, No. 9 (September 1981), pp. 20–25.

Analysis of Variance and Design of Experiments

Anderson, V.L., and McLean, R.A. *Design of Experiments: A Realistic Approach.* New York: Dekker, 1974.

Cochran, G., and Cox, M. *Experimental Designs,* 2nd ed. New York: John Wiley & Sons, 1957.

Cox, R. *Planning of Experiments.* New York: John Wiley & Sons, 1958.

Finney, D.J. *Experimental Design and Its Statistical Basis.* Chicago: University of Chicago Press, 1955.

Guenther, W.C. *Analysis of Variance.* Englewood Cliffs, NJ: Prentice-Hall, 1964.

Fisher, R.A. *The Design of Experiments.* New York: Hafner, 1974.

Hicks, C.R. *Fundamental Concepts in the Design of Experiments,* 3rd ed. New York: Holt, Rinehart & Winston, 1982.

Mendenhall, W. *An Introduction to Linear Models and the Design and Analysis of Experiments.* Belmont, CA: Wadsworth, 1968.

Mendenhall, W., and Sincich, T. *A Second Course in Business Statistics: Regression Analysis,* 2nd ed. San Francisco: Dellen, 1986.

Morrison, Donald F. *Applied Linear Statistical Methods.* Englewood Cliffs, NJ: Prentice-Hall, 1983.

Neter, J., and Wasserman, W. *Applied Linear Statistical Models.* Homewood, IL: Richard D. Irwin, 1974.

Scheffé, H. *The Analysis of Variance.* New York: Wiley, 1959.

Wolach, A.H. *Basic Analysis of Variance Programs for Microcomputers.* Belmont, CA: Wadsworth Publishers, 1983.

Wolter, K.M. *Introduction to Variance Estimation.* New York: Springer-Verlag, 1985.

Nonparametric Statistics

Bradley, J.V. *Distribution-Free Statistical Tests.* Englewood Cliffs, NJ: Prentice-Hall, 1968.

Conover, W.J. *Practical Nonparametric Statistics,* 2nd ed. New York: John Wiley & Sons, 1980.

Gibbons, J.D. *Nonparametric Statistical Inference.* New York: Dekker, 1985.

Hajek, J. *Nonparametric Statistics.* San Francisco: Holden-Day, 1969.

Hollander, M., and Wolfe, D.A. *Nonparametric Statistical Methods.* New York: Wiley, 1973.

Krishnaiah, P.R. and Sen, P.K., ed. *Nonparametric Methods.* New York: Elsevier, 1985.

Marascuilo, L., and McSweeney, M. *Nonparametric and Distribution-Free Methods for the Social Sciences.* Monterey, CA: Brooks/Cole, 1977.

Noether, G.E. *Introduction to Statistics: A Nonparametric Approach*, 2nd ed. Boston: Houghton Mifflin, 1976.

Puri, M.L. and Sen, P.K. *Nonparametric Methods in General Linear Models.* New York: John Wiley & Sons, 1985.

Siegel, S. *Nonparametric Statistics* New York: McGraw-Hill, 1965.

Decision Analysis

Aitchison, J. *Choice Against Chance.* Reading, MA: Addison-Wesley, 1970.

Baird, B.F. *Introduction to Decision Analysis.* Boston: Duxbury Press, 1978.

Brown, R.V., Kahr, A.S., and Peterson, C. *Decision Analysis for the Manager.* New York: Holt, Rinehart & Winston, 1974.

Bunn, D.W. *Applied Decision and Analysis.* New York: McGraw-Hill, 1984.

Chernoff, H., and Moses, L. *Elementary Decision Theory.* New York: Dover Publications, 1987.

Edwards, W., and Tversky, A., eds. *Decision Making: Selected Readings.* Harmondsworth, England: Penguin Books, 1967.

Forester, J. *Statistical Selection of Business Strategies.* Homewood, IL: Richard D. Irwin, 1968.

Gupta, M.M., and Sanchez, E., eds. *Approximate Reasoning in Decision Analysis.* New York: Elsevier, 1983.

Harvard Business Review. Statistical Decision Series, Parts I–IV. Boston: Harvard University Press, 1951–1970.

Howard, R.A. "The Decision to Seed Hurricanes," *Science*, Vol. 176 (June 1972), pp. 1191–1201.

Keeney, Ralph L. "Decision Analysis: An Overview," *Journal of Operations Research* 30 (September/October 1982), pp. 803–38.

Keeney, R.L., and Raiffa, H. *Decisions with Multiple Objectives: Preferences and Value Tradeoffs.* New York: John Wiley & Sons, 1976.

Kostolansky, J., and Bailey, A.D. *Decision Analysis, Including Modeling and Information Systems.* Baton Rouge, LA: Malibu Publications, 1984.

Lindley, D.V. *Introduction to Probability and Statistics from a Bayesian Viewpoint.* Part 2, "Inference." New York: Cambridge University Press, 1965.

Lindley, D.V. *Making Decisions.* London: Wiley-Interscience, 1971.

Luce, R.D., and Raiffa, H. *Games and Decisions.* New York: John Wiley & Sons, 1957.

Morris, W.T. *Management Science: A Bayesian Introduction.* Englewood Cliffs, NJ: Prentice-Hall, 1968.

Pessemier, E.A. *New Product Decisions—An Analytic Approach.* New York: McGraw-Hill, 1966.

Pratt, J., Raiffa, H., and Schlaifer, R. *Introduction to Statistical Decision Theory.* New York: McGraw-Hill, 1965.

Raiffa, H. *Decision Analysis: Introductory Lectures on Choices Under Uncertainty*, New York: Random, 1986.

Raiffa, H., and Schlaifer, R. *Applied Statistical Decision Theory.* Cambridge, MA: Division of Research, Graduate School of Business Administration, Harvard University, 1961.

Risk Analysis—Proceedings of the United States Army Operations Research Symposium, May 15–18, 1972. Washington, DC: Office of the Chief of Research and Development, Department of the Army, 1972.

Schlaifer, R. *Probability and Statistics for Business Decisions.* New York McGraw-Hill, 1959.

Vatter, R.A., Bradley, S.P., Frey, S.C., Jr., and Jackson, B.B. *Quantitative Methods in Management: Text and Cases.* Homewood, IL: Richard D. Irwin, 1978.

Winkler, R.L. *An Introduction to Bayesian Inference and Decision.* New York: Holt, Rinehart & Winston, 1972.

Statistical Tables

Beyer, W.H. *Handbook of Tables for Probability and Statistics*, 2nd ed. Cleveland: The Chemical Rubber Company, 1968.

Burington, R.S., and May, D.C. *Handbook of Probability and Statistics with Tables*, 2nd ed. New York: McGraw-Hill, 1970.

Fisher, R.A., and Yates, F. *Statistical Tables for Biological, Agricultural, and Medical Research*, 6th ed. London: Longman Group, Ltd., 1978.

Hald, A. *Statistical Tables and Formulas.* New York: John Wiley & Sons, 1952.

Kres, H., and Wadsack, P., trans. *Statistical Tables for MultivariateAnalysis: A Handbook with References to Applications.* New York: Springer-Verlag, 1983.

Lindley, D.V., and Scott, W.F. *New Cambridge Elementary Statistical Tables.* New York: Cambridge University Press, 1984.

National Bureau of Standards. *Tables of the Binomial Probability Distribution.* Applied Mathematical Series 6. Washington, DC: U.S. Department of Commerce, 1950.

Odeh, E. *Tables for Tests, Confidence Limits and Plans Based on Proportions.* New York: Dekker, 1983.

Odeh, E. and Owen, D., eds. *Tables for Normal Tolerance Limits, Sampling Plans, and Screening.* New York: Dekker, 1980.

Owen, D. *Handbook of Statistical Tables.* Reading, MA: Addison-Wesley, 1962.

Pearson, E.S., and Hartley, H.O. *Biometrika Tables for Statisticians*, 2nd ed. Cambridge, England: Cambridge University Press, 1962.

RAND Corporation. *A Million Random Digits with 100,000 Normal Deviates.* New York: The Free Press of Glencoe, 1955.

Romig, H.G. *50–100 Binomial Tables.* New York: John Wiley & Sons, 1953.

White, J., and Yeats, A. *Tables for Statisticians.* New York: University of Queensland Press, 1984.

Zehna, P.W., and Barr, D.R. *Tables of the Common Probability Distributions.* Monterey, CA: U.S. Naval Postgraduate School, 1970.

Dictionary of Statistical Terms

Dudewicz, E.J., and Koo, J.O. *The Complete Categorized Guide to Statistical Selection and Ranking Procedures.* New York: American Sciences Press, 1982.

Freund, J., and Williams, F. *Dictionary/Outline of Basic Statistics.* New York: McGraw-Hill, 1966.

Kendall, M.G., and Buckland, W.R. *A Dictionary of Statistical Terms,* 4th ed. New York: Wiley, 1986.

Omuircheartaigh, C., and Francis, D.P. *Statistics: A Dictionary of Terms and Ideas.* New York: Hippocrene Books, Inc., 1983.

Tietjen, G.L. *A Topical Dictionary of Statistics.* New York: Routledge Chapman and Hall, 1985.

Statistical Computer Software Handbooks

Dixon, W.J., ed. BMDP, *Statistical Software Manual.* Los Angeles: University of California Press, 1985.

Klecka, William R., Norman H. Nie, and C. Hadlai Hull. *SPSS Primer.* New York: McGraw Hill, 1975.

Miller, Robert B. *Minitab Handbook for Business and Economics.* Boston: PWS-Kent Publishing, 1988.

Ryan, Barbara F., Brian L. Joiner, and Thomas A. Ryan. *Minitab Handbook,* 2nd ed. Boston: Duxbury Press, 1985.

SAS (Statistical Analysis System) User's Guide: Statistics, 1985 Edition. Cary, NC: SAS Institute, Inc., 1985.

SPSS (Statistical Package for the Social Sciences) User's Guide, 2nd ed. Chicago: SPSS Inc., 1986.

Trower. *Using Minitab for Introductory Statistical Analysis.* Merrill, 1989.

SOLUTIONS TO EVEN-NUMBERED EXERCISES

2. a. Using seven classes, we have the following distribution for donations received at Union-Path offices:

Donations (thousands of dollars)	Midpoints	Frequency
20.0 and under 55.0	37.5	1
55.0 and under 90.0	72.5	3
90.0 and under 125.0	107.5	6
125.0 and under 160.0	142.5	8
160.0 and under 195.0	177.5	4
195.0 and under 230.0	212.5	2
230.0 and under 265.0	247.5	1
Total		25

b. That distribution yields the following histogram and "less than" ogive, respectively.

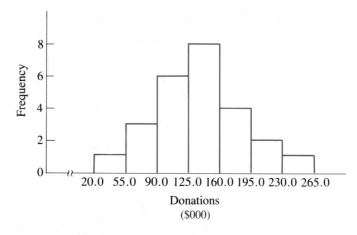

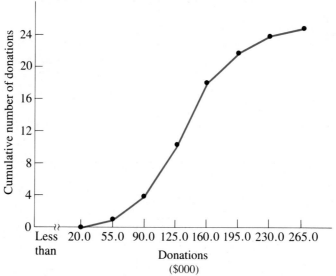

4. a. The class intervals are not mutually exclusive; the class limits are unclear and overlapping.

   b. There is a gap in values between classes; thus, the classes are not exhaustive.

6. a and b. Number of boxes sold during the one-week period:

Number of Boxes	Total	Number of Stores Philadelphia	Camden
500 and under 1,000	3	1	2
1,000 and under 1,500	4	1	3
1,500 and under 2,000	8	2	6
2,000 and under 2,500	5	3	2
2,500 and under 3,000	10	9	1
3,000 and under 3,500	6	6	0
Total	36	22	14

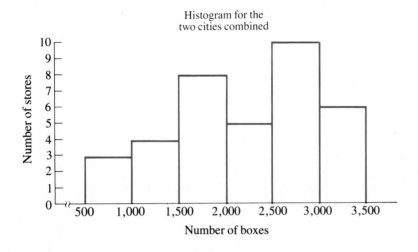

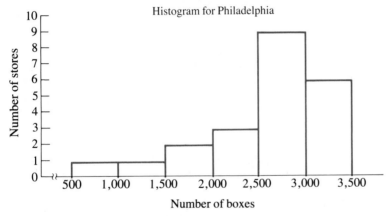

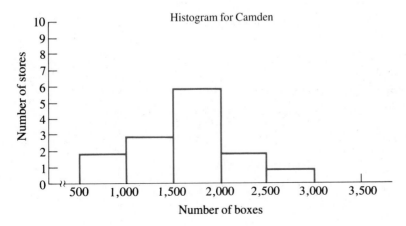

c. The answer in part (b) is preferable, because a bimodal distribution results if the data for the two market cities are merged.

Exercises 1.9

2. a.

x	f	fx
2.5%	34	85%
7.5	46	345
12.5	38	475
17.5	22	385
22.5	10	225
	150	1515%

$$\bar{X} = \frac{1515\%}{150} = 10.1\%$$

4. No, this is not enough information for us to make a decision. Because we do not know the proportions of the ingredients in the two spreads, we cannot say that Brandex has a lower *overall* Oxy-toxin content . For example, if the proportions are as follows, we could conclude that Brandex actually has a higher Oxy-toxin level:

	Proportion		Weighted Oxy-toxin Level (%)	
Ingredient	Brandex	Leading Spread	Brandex	Leading Spread
Moonflower oil	0.4	0.1	0.48	0.18
Processed butter-glop	0.3	0.4	0.18	0.32
Flavoring	0.1	0.4	0.03	0.12
Coloring	0.2	0.1	0.10	0.06
Total	1.0	1.0	0.79	0.68

From this table, we see an overall level of 0.79% for Brandex compared with an overall level of 0.68% for the leading spread. Therefore, we need to know the proportions of ingredients before we can make a decision. We also need to know about any ingredients that have not been advertised and the levels of Oxy-toxin in those ingredients as well.

6. a. Weighted average of pollutant level in the three tanks is

$$\bar{X} = \frac{(4.5)(261,432) + (15.0)(118,300) + (21.3)(287,456)}{667,188}$$

$$= 13.6 \text{ parts per 10,000}$$

b. Unweighted average is

$$\bar{X} = \frac{4.5 + 15.0 + 21.3}{3} = 13.6 \text{ parts per 10,000}$$

c. Yes, the answer in part (b) shows an unexpected correspondence to that in part (a).

Exercises 1.12

2. Agree, since in all three cases it is quite logical to assume that there would be extreme values at the upper end of the scale and that such values would tend to make the arithmetic mean larger than the median.

4. a.

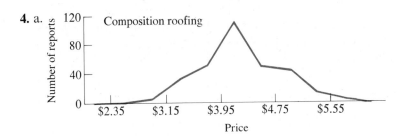

b. Composition roofing: Median class is $3.95 and under $4.35; modal class is $3.95 and under $4.35. Douglas and Inland Firs: Median class is $136 − $145; modal class is $146 − $155.

c.

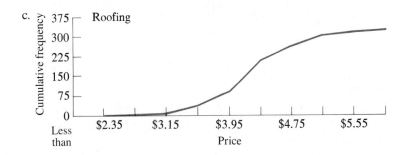

d. The distribution of composition roofing prices is nearly symmetrical, and the distribution of fir prices is skewed to the left.

6. a. $\bar{X} = \dfrac{\$129,850}{168} = \$772.92$

b. No. This ratio gives the exact mean, whereas part (a) gives only a close approximation, since a frequency table was used.

c. Median class is "$750 and under $800"; median observation is the $(168 + 1)/2 = 84.5$th observation; median $= \$750 + [(84 - 66)/36](\$50) = \$775$.

d. Slightly skewed to the left.

e. No, since Σfx is $129,850$. This figure is an *estimate* of the total payroll and is close to the true figure of $129,832.

f. Yes, the mean here seems typical, since there is little skewness to distort it.

Exercises 1.13

2. a. $1 + r = \sqrt[6]{\dfrac{2000,000}{4000,000}}$

$\qquad r = -.1092 = -10.92\%$ per yr.

4. a. $(\$4000)(1.09)^n = \$8000$

$$n = \frac{.3010}{.0374} = 8.05$$

Hence rounding off to the nearest whole number of years above 8.05, we find that 9 years are required.

b. $(\$4,000)(1.09)^{4.5}$ c. $(\$4,000)(1.09)^{18}$

antilog $3.7704 = \$5,890$ antilog $4.2753 = \$18,850$

Compound interest has an exponential growth pattern. As the base becomes larger, the increase becomes greater.

2. a. Tiny Tot: arithmetic mean = $\bar{X}$ = \$1.10; standard deviation = s = $\sqrt{1.22/(10-1)}$ = \$0.37. Gigantic Game: arithmetic mean = $\bar{X}$ = \$7.80; standard deviation = s = $\sqrt{6.28/(10-1)}$ = \$0.84. Gigantic Game Corporation earnings per share showed greater absolute variation.

   b. Tiny Tot: coefficient of variation = $s/\bar{X}$ = \$0.37/\$1.10 = 33.6%. Gigantic Game: coefficient of variation = $s/\bar{X}$ = \$0.84/\$7.80 = 10.77%. Tiny Tot earnings per share showed relatively greater variation than Gigantic Game, according to the coefficient of variation.

4. a. $\bar{X} = \dfrac{165}{5} = 33$ years

   $s = \sqrt{\dfrac{974}{4}} = 15.6$ years

   b. If a constant c is added to or subtracted from all items in a distribution, the mean is increased or decreased by that constant c.
   If a constant c is added to or subtracted from all items in a distribution, the standard deviation is unchanged.

CHAPTER 2

2. 0 = fall
   1 = same
   2 = rise

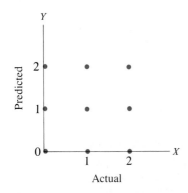

4. {(022), (202), (220), (122), (212), (221), (222)}

6.

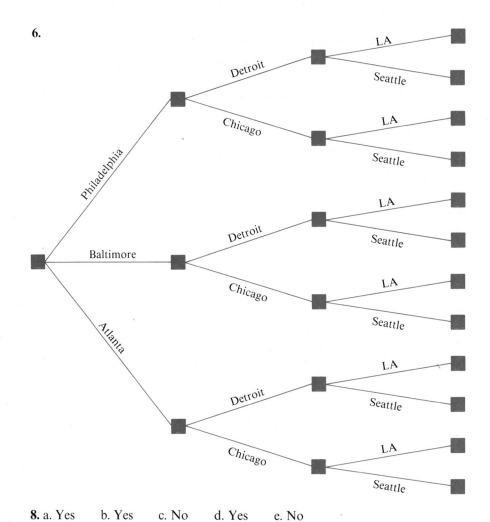

8. a. Yes   b. Yes   c. No   d. Yes   e. No

Exercises 2.2

2. $1 - (0.02) - (0.04) = 0.94$

4. $P(\text{meets quota}) = 1 - P(\text{does not meet quota})$
$$= 1 - 0.2 = 0.8$$
odds $= 4:1$

6. a. $\{(SSS), (SSN), (SNS), (NSS), (SNN), (NSN), (NNS), (NNN)\}$ b. 1/8
c. $P(\text{do not vote unanimously})$
$$= 1 - P(\text{all vote to strike}) - P(\text{all vote to work})$$
$$= 1 - (1/2 \times 1/2 \times 1/2) - (1/2 \times 1/2 \times 1/2) = 6/8$$

8. a. $(0.8 \times 0.6 \times 0.5) = 0.24$ b. $(0.2 \times 0.4 \times 0.5) = 0.04$
c. Assume statistical independence

10. a.

Opinion	Size		
	Large	Small	Subtotal
Favor	100	150	250
Not in favor	200	50	250
Totals	300	200	500

b. $50/500 = 0.1$ c. $250/500 = 0.5$ d. $50/200 = 0.25$

e. P(large corporation and favors reform) $\neq$ P(large corporation) $\times$ P(favors reform)

Therefore, size of corporation and opinion of tax reform are not statistically independent.

Exercises 2.3

2. $\dfrac{(0.75)(0.07)}{(0.75)(0.07) + (0.25)(0.04)} = 0.84$

4. Let θ = state of nature

X = sample indicates high demand

State of Nature θ	$P(\theta)$	$P(x\mid\theta)$	Joint Probabilities	$P(\theta\mid x)$
High	0.55	0.80	$(0.55)(0.80) = 0.44$	0.71
Average	0.25	0.60	$(0.25)(0.60) = 0.15$	0.24
Low	0.20	0.15	$(0.20)(0.15) = 0.03$	0.05
Total	1.00		0.62	1.00

Exercises 2.4

2. a. $10^3 = 1,000$ b. $(26)(10^2) = 2,600$ c. $(36)^3 = 46,656$

4. $4! = 4 \times 3 \times 2 \times 1 = 24$

6. $\binom{8}{3}\binom{6}{3} = (56)(20) = 1120$

a. $\binom{8}{6} = 28$ b. $\binom{6}{6} = 1$

8. a. $\binom{5}{3} = 10$ b. $(5)(4)(3) = 60$ c. $\dfrac{(3)(2)(1)}{(5)(4)(3)} = \dfrac{6}{60} = \dfrac{1}{10},$ or $\dfrac{1}{\binom{5}{3}} = \dfrac{1}{10}$

10. a. $\dfrac{\binom{8}{1}\binom{8}{6}}{\binom{16}{7}} = \dfrac{(8)(28)}{11,440} = \dfrac{224}{11,440} = 0.0196$

b. $\dfrac{\binom{8}{1}\binom{7}{5}}{\binom{15}{6}} = \dfrac{(8)(21)}{5005} = \dfrac{168}{5005} = 0.0336$

c. $\dfrac{\binom{8}{6}}{\binom{15}{6}} = \dfrac{28}{5005} = 0.0056$

d. $\dfrac{\binom{9}{1}\binom{9}{6}}{\binom{18}{7}} = \dfrac{(9)(84)}{31,824} = \dfrac{756}{31,824} = 0.0238$

CHAPTER 3

Exercises 3.1

2. a. $P(X = 1) = f(1) = \dfrac{1}{14}(1)^2 = \dfrac{1}{14}$

$P(X = 2) = f(2) = \dfrac{1}{14}(2)^2 = \dfrac{4}{14}$

$P(X = 3) = f(3) = \dfrac{1}{14}(3)^2 = \dfrac{9}{14}$

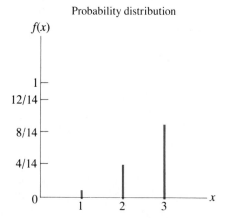

Probability distribution

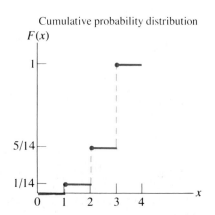

Cumulative probability distribution

b. Probability distribution

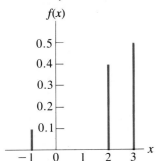

Cumulative probability distribution

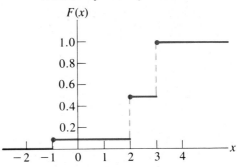

c. $\sum\limits_{x} f(x) = 1$ and $f(x) \geqslant 0$ for all x in both functions.

4. a. Elements of the sample space: $\bar{K}\bar{K}\bar{K}$, $K\bar{K}\bar{K}$, $\bar{K}K\bar{K}$, $\bar{K}\bar{K}K$, $KK\bar{K}$, $K\bar{K}K$, $\bar{K}KK$, KKK

b. $\binom{3}{0} = \dfrac{3!}{0!3!} = 1$

$\binom{3}{1} = \dfrac{3!}{1!2!} = 3$

$\binom{3}{2} = \dfrac{3!}{2!1!} = 3$

$\binom{3}{3} = \dfrac{3!}{3!0!} = 1$

$P(X = 0) = f(0) = 1(0.65)^3(0.35)^0 = 0.275$
$P(X = 1) = f(1) = 3(0.65)^2(0.35)^1 = 0.444$
$P(X = 2) = f(2) = 3(0.65)^1(0.35)^2 = 0.239$
$P(X = 3) = f(3) = 1(0.65)^0(0.35)^3 = \underline{0.042}$
$\overline{1.000}$

c. $f(x)$

Number of Kool Kola customers

6. $f(1) = \dfrac{1}{25}$; $f(2) = \dfrac{3}{25}$; $f(3) = \dfrac{5}{25} = \dfrac{1}{5}$; $f(4) = \dfrac{7}{25}$; $f(5) = \dfrac{9}{25}$

Exercises 3.3

2. a. $f(x) = 1/96$, $x = 1, \ldots, 96$
   b. 70/96   **c.** 40/96   **d.** 41/96   **e.** 25/96

Exercises 3.4

Although tables can be used, analytical solutions are given here. For example in exercise 1(a), from Table A-2, $P(3) = 0.2787$ for $n = 8$ and $p = 0.40$.

2. $\dbinom{4}{3}(0.85)^1(0.15)^3 = 0.0115$

Assuming independent trials means that the probability of being awarded a contract is not affected by previous biddings.

4. $\dbinom{10}{10}\left(\dfrac{16}{34}\right)^{10} = 0.0005$ The feat is highly unlikely.

6. $\displaystyle\sum_{x=2}^{10} \dbinom{10}{x}(0.95)^{10-x}(0.05)^x = 0.0861$

The probability is less than 10%; therefore, she should contract with Forte Iron.

8. $\dbinom{4}{2}\left(\dfrac{1}{3}\right)^2\left(\dfrac{2}{3}\right)^2 + \dbinom{4}{3}\left(\dfrac{1}{3}\right)^1\left(\dfrac{2}{3}\right)^3 = 0.6913$

10. $\dbinom{5}{0}\left(\dfrac{5}{6}\right)^5\left(\dfrac{1}{6}\right)^0 = 3{,}125/7{,}776$

$\dbinom{5}{5}\left(\dfrac{5}{6}\right)^0\left(\dfrac{1}{6}\right)^5 = 1/7{,}776$

12.

x	$p = 0.30$ $f(x)$	$p = 0.50$ $f(x)$	$p = 0.70$ $f(x)$
0	0.1681	0.0312	0.0024
1	0.3601	0.1563	0.0284
2	0.3087	0.3125	0.1323
3	0.1323	0.3125	0.3087
4	0.0284	0.1563	0.3601
5	0.0024	0.0312	0.1681

The binomial distribution becomes increasingly skewed to the left for all values of $p < 0.50$ and skewed to the right for all values of $p > 0.50$ (see figure).

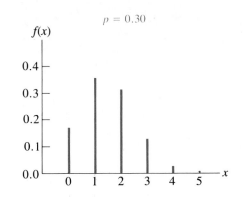

$p = 0.30$

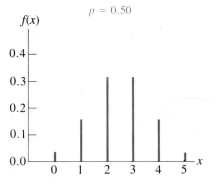

$p = 0.50$

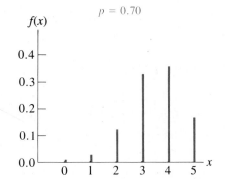

$p = 0.70$

14. a. $\binom{16}{1}(0.90)^{15}(0.10)^1 = 0.3294$

b. $\binom{16}{1}(0.80)^{15}(0.20)^1 = 0.1126$

c. $\binom{16}{1}(0.75)^{15}(0.25)^1 = 0.0535$

Exercises 3.5

2. $\dfrac{15!}{3!4!6!2!}(0.40)^3(0.30)^4(0.20)^6(0.10)^2 = 0.002$

Since this probability is low, there is probably something wrong in the distribution of errors.

Exercises 3.6

2. $\dfrac{\binom{13}{3}\binom{13}{4}\binom{13}{5}\binom{13}{1}}{\binom{52}{13}}$

4. Probability of appointing Mr. Hedge:

$$\frac{\binom{7}{0}\binom{5}{3}}{\binom{12}{3}} = \frac{10}{220} = 0.045$$

Probability if only a majority vote were needed:

$$\frac{\binom{7}{1}\binom{5}{2}}{\binom{12}{3}} + \frac{\binom{7}{0}\binom{5}{3}}{\binom{12}{3}} = \frac{70 + 10}{220} = 0.364$$

6. a. $\dfrac{\binom{2}{1}\binom{3}{1}\binom{1}{1}}{\binom{6}{3}} = \dfrac{6}{20} = 0.30$ b. $\dfrac{\binom{1}{1}\binom{5}{2}}{\binom{6}{3}} = \dfrac{10}{20} = 0.50$

   c. $\dfrac{\binom{2}{2}\binom{4}{1}}{\binom{6}{3}} = \dfrac{4}{20} = 0.20$

Exercises 3.7

2. $f(0) = \dfrac{6^0 e^{-6}}{0!} = 0.002$

4. a. $f(12) = \dfrac{12^{12} e^{-12}}{12!} = 0.114$ b. $1 - \sum\limits_{x=0}^{10} \dfrac{12^x e^{-12}}{x!} = 1 - 0.347 = 0.653$

   c. $\sum\limits_{x=0}^{5} \dfrac{12^x e^{-12}}{x!} = 0.020$

6. a. The probability that all computers will be working is

$$\binom{5}{5}(0.05)^0(0.95)^5 = 0.774$$

b. The probability that at least three computers will not be working is

$$1 - \sum\limits_{x=0}^{2} \binom{5}{x}(0.95)^{5-x}(0.05)^x = 1 - 0.9988 = 0.0012$$

Exercises 3.13

2. $E(\text{gain}) = (\$315)(0.97) + (-\$9,685)(0.03) = \$15$
No. On any single policy, the company will either earn the premium of $315 or pay out $9,685 ($10,000 − $315). The $15 represents the average return per policy if an infinite number of policies are issued.

4. a. $R(0) = -\$3,000$

$R(1) = \$5,000(1) - \$1,000(2) = \$3,000$

$R(2) = \$5,000(2) - \$1,000(1) = \$9,000$

$R(3) = \$5,000(3) = \$15,000$

$R(4) = \$5,000(3) = \$15,000$

b. $E(3) = \dfrac{10}{20}(-\$3,000) + \dfrac{2}{20}(\$3,000) + \dfrac{4}{20}(\$9,000)$

$\qquad + \dfrac{3}{20}(\$15,000) + \dfrac{1}{20}(\$15,000)$

$\quad = \$3,600$

$\text{VAR}(3) = \dfrac{10}{20}(-\$3,000 - \$3,600)^2 + \dfrac{2}{20}(\$3,000 - \$3,600)^2$

$\qquad + \dfrac{4}{20}(\$9,000 - \$3,600)^2 + \dfrac{3}{20}(\$15,000 - \$3,600)^2$

$\qquad + \dfrac{1}{20}(\$15,000 - \$3,600)^2$

$\quad = 53,640,000$

$\sigma(3) = \$7,324$

c. Assuming that Heavy Machineries, Inc., desires to maximize expected monetary returns, the optimal number of pieces of equipment to be sent can be obtained by comparing the expected returns for all decisions (that is, to send 1, 2, 3, or 4 pieces). The optimal decision is associated with the highest return.

6. $\mu = $ mean without fifth outlet $= \mu_1 + \mu_2 + \mu_3 + \mu_4$

$\quad = \$6,500 + \$6,200 + \$6,000 + \$5,800 = \$24,500$

$\mu = $ mean with fifth outlet $= \mu + \mu_5 = \$24,500 + 8,000$

$\quad = \$32,500$

The variance of total sales will increase by 15,000 with the acquisition of the fifth outlet; that variance will amount to 47,300 per month.

8. a. $E(T) = (0)(0.02) + 1(0.04) + 2(0.12) + 3(0.20) + 4(0.30) + 5(0.20) + 6(0.12)$

$\quad = 3.8$ copies per week

$\text{VAR}(T) = (0 - 3.8)^2(0.02) + (1 - 3.8)^2(0.04) + (2 - 3.8)^2(0.12)$

$\qquad + (3 - 3.8)^2(0.20) + (4 - 3.8)^2(0.30) + (5 - 3.8)^2(0.20)$

$\qquad + (6 - 3.8)^2(0.12)$

$\quad = 2.0$

$E(S) = 0(0.10) + 1(0.10) + 2(0.10) + 3(0.15) + 4(0.20) + 5(0.15) + 6(0.20)$

$\quad = 3.5$ copies per week

$\text{VAR}(S) = (0 - 3.5)^2(0.10) + (1 - 3.5)^2(0.10) + (2 - 3.5)^2(0.10)$

$\qquad + (3 - 3.5)^2(0.15) + (4 - 3.5)^2(0.20) + (5 - 3.5)^2(0.15)$

$\qquad + (6 - 3.5)^2(0.20)$

$\quad = 3.75$

$E(T) > E(S)$ and $\text{VAR}(T) < \text{VAR}(S)$

b. $\text{Profit}(T) = (0.40)(3.8) - (0.15)(6) = \$0.62$
$\text{Profit}(S) = (0.60)(3.5) - (0.30)(6) + (0.05)(2.5)$
$= \$0.425$

c. *TV Times.*

10. Let $X_1, X_2, \ldots, X_{12}$ represent the net profit per month for the 12 stores. Then $X_1 + X_2 + \cdots + X_{12} = $ total net profit per month for the 12 stores combined.

$$E(X_1 + X_2 + \cdots + X_{12}) = E(X_1) + E(X_2) + \cdots + E(X_{12})$$
$$= 12(\$6500) = \$78,000$$

$$\text{VAR}(X_1 + X_2 + \cdots + X_{12}) = \text{VAR}(X_1) + \text{VAR}(X_2) + \cdots + \text{VAR}(X_{12})$$
$$= 12(30,000) = 360,000$$

$$\sigma(X_1 + X_2 + \cdots + X_{12}) = \$600$$

12. $\mu = E(X) = \Sigma x f(x) = 0.0500$; $\text{VAR}(X) = \Sigma(x - \mu)^2 f(x) = 0.000592$; $\sigma(X) = 0.0243$; expected percentage of users affected: 5%; standard deviation: 2.43%.

14. $E(\text{total time}) = E(\text{I} + \text{II} + \text{III} + \text{IV}) = E(\text{I}) + E(\text{II}) + E(\text{III}) + E(\text{IV})$
$= 5 + 14 + 8 + 3 = 30$ weeks
$\text{VAR}(\text{total time}) = \text{VAR}(\text{I} + \text{II} + \text{III} + \text{IV})$
$= \text{VAR}(\text{I}) + \text{VAR}(\text{II}) + \text{VAR}(\text{III}) + \text{VAR}(\text{IV})$
$= 4 + 25 + 9 + 1 = 39$
$\sigma(\text{total time}) \quad = \sqrt{39} = 6.24$ weeks

Exercises 3.15

2. Yes; $P(\text{for}|\text{skilled}) = 275/600 \neq P(\text{for}) = 500/1,400$. A higher proportion of skilled employees than unskilled employees is in favor of the labor proposal.

4. a.

| | \multicolumn{4}{c}{x} | | |
y	0	1	2	3
0	0	2/56	4/56	6/56
1	1/56	3/56	5/56	7/56
2	4/56	6/56	8/56	10/56

b.

x	$f(x)$	y	$g(y)$
0	5/56	0	12/56
1	11/56	1	16/56
2	17/56	2	28/56
3	23/56		1
	1		

c.

y	$g(y|0)$	$g(y|1)$	$g(y|2)$	$g(y|3)$
0	0	2/11	4/17	6/23
1	1/5	3/11	5/17	7/23
2	4/5	6/11	8/17	10/23

d. No. $f(x)g(y) \neq f(x, y)$

CHAPTER 4

Exercises 4.2

See the discussion in section 4.2 for the answer to question 2.

Exercises 4.4

2. See the discussion in section 4.4

4. a. The control group would consist of athletes who perform the same exercises, but who are not given the new heat treatment.
b. The control group would be a sample of children who are given the information, but without visual aids.
c. No control group is needed to test this assertion. A large and ongoing sample to find any marathon runners who have suffered a heart attack (which would disprove the hypothesis) is needed.

6. The director used those employees who had attended the previous year's party as the universe. The director's sample was biased because employees who thought spouses should be included probably had a lower attendance rate than other employees—such as singles or married persons who did not prefer the inclusion of spouses. The subgroup of employees who attended the Christmas party last year undoubtedly differed from the "all other employees" group with respect to preferences for an "employees only" party.

8. Yes, probably both sampling and systematic errors are present. Sampling error would exist in a random sample of 10,000 compared with the population of 200,000 subscribers. Systematic error is probably introduced because of the difference between the attitudes of the 1,500 respondents and the corresponding attitudes of the total sample of 10,000.

Exercises 4.5

2. a. The universe sampled was all of those parents with children in Lower Fenwick public schools. However, a simple random sample of pupils did not constitute a simple random sample of the parent population. Parents with more than one child in the public schools had a higher probability of inclusion in the sample than did parents with only one child in these schools.
b. See part (a).
c. We would not approve the universe studied. If the school board wished to ascertain voter opinion, it should have sampled the population of voters in Lower Fenwick, many of whom may not have children in the public schools. A relevant sampling frame would have been a current list of registered voters in Lower Fenwick.

CHAPTER 5

Exercises 5.1

2. a.

Possible Samples	Sample Mean ($\bar{X}$)	Probability $P(\bar{X})$
(4, 4)	4.0	1/15
(4, 5)	4.5	1/15
(4, 6)(5, 5)	5.0	2/15
(4, 7)(5, 6)	5.5	2/15
(4, 8)(5, 7)(6, 6)	6.0	3/15
(5, 8)(6, 7)	6.5	2/15
(6, 8)(7, 7)	7.0	2/15
(7, 8)	7.5	1/15
(8, 8)	8.0	1/15
		1

b. $E(\bar{X}) = (4.0)(1/15) + (4.5)(1/15) + \cdots + (8.0)(1/15) = 6$

$$\mu = \frac{4 + 5 + 6 + 7 + 8}{5} = 6$$

The sample mean is an unbiased estimator of the population mean in simple random sampling, with or without replacement.

Exercises 5.3

2. a. 0.5000   **b.** 0.0062   **c.** 0.8664
   d. 0.6170   **e.** 0.0228   **f.** 0.8413

4. a. 0.50   **b.** $P = 2(0.4772) = 0.9544$
   c. z of 0.350 is 1.04

$$-1.04 = \frac{x - 450}{75}$$

$$x = 372$$

6. a. $P\left(\dfrac{2.0 - 3}{1} < z < \dfrac{4.0 - 3}{1}\right) = 0.6826$

b. $P\left(z < \dfrac{2.8 - 3}{1}\right) = 0.4207$

$$P\left(\frac{4.5 - 6}{2} < z < \frac{6.5 - 6}{2}\right) = 0.3721$$

$$P\{(X < 2.8) \text{ and } (4.5 < Y < 6.5)\} = (0.4207)(0.3721)$$
$$= 0.1565$$

c. $P\left(z < \dfrac{2.8 - 3}{1}\right) P\left(z > \dfrac{3.0 - 6.0}{2}\right) = (0.4207)(0.9332)$

$\qquad\qquad\qquad\qquad\qquad\qquad\quad = 0.3926$

d. $P\{(X < 2.8) \text{ or } (Y > 3.0)\}$
$\qquad = P(X < 2.8) + P(Y > 3.0) - P[(X < 2.8) \text{ and } (Y > 3.0)]$
$\qquad = (0.4207) + (0.9332) - (0.3926)$
$\qquad = 0.9613$

e. $2P\left(0 \leqslant z \leqslant \dfrac{X - 3}{1}\right) = 0.51$

$\quad P\left(0 \leqslant z \leqslant \dfrac{X - 3}{1}\right) = 0.255$

$\quad \dfrac{(X - 3)}{1} = 0.69 \text{ (from Table A-5)}$

$\qquad X = 3.69$

Hence, the middle 51% of the scores lie between 2.31 and 3.69.

8. a. $P\left(z < \dfrac{21.5 - 18}{3}\right) = P(z < 1.17) = 0.8790$

b. $P\left(z \leqslant \dfrac{16.5 - 18}{3}\right) = P(z \leqslant -0.5) = 0.3085$

c. $P\left(z \leqslant \dfrac{x - 18}{3}\right) = 0.0668 \qquad z = -1.5$

$\quad$ Therefore, $x = 18 - 1.5(3) = 13.5$ minutes.

5.4

Exercises 5.5

2. a. $P\left(z > \dfrac{30{,}000 - 32{,}000}{1{,}500}\right) = 0.9082$

b. $P\left(z > \dfrac{35{,}000 - 32{,}000}{1{,}500}\right) = 0.0228$

c. $P\left(z > \dfrac{30{,}000 - 32{,}000}{1{,}500/\sqrt{4}}\right) = 0.9962$

4. a. Disagree. The Central Limit Theorem applies to the distribution of the sample means, not to the population distribution.
b. Agree.
c. Disagree. The Central Limit Theorem applies to the approach of the sampling distribution of the mean to normality as n becomes large.

6. $P\left(z \geqslant \dfrac{30 - 28}{\sqrt{81/100}}\right) = 0.0132$

No, because of the Central Limit Theorem.

8. Each week: $4500 \pm 2(450)$, or $3600-5400$

Each month: $4500 \pm 2(450/\sqrt{4})$, or $4050-4950$

10. Use the Central Limit Theorem

$z_0 = \$200/(2{,}000/\sqrt{100}) = 1$

$P(|\bar{x} - \mu| > \$200) = P(|z| > 1)$

$= 0.3174$

Exercises 5.6

2. a. p **b.** $f(\bar{p}|p = 0.90)$ **c.** $f(\bar{p}|p = 0.5)$

p	$f(\bar{p}\|p = 0.90)$	$f(\bar{p}\|p = 0.5)$
0.00	0	0.0156
0.17	0.0001	0.0938
0.33	0.0012	0.2344
0.50	0.0146	0.3124
0.67	0.0984	0.2344
0.83	0.3543	0.0938
1.00	0.5314	0.0156

4. $\mu_{\bar{p}} = p = 0.25$

$\sigma_{\bar{p}} = \sqrt{\dfrac{pq}{n}} = \sqrt{\dfrac{(0.25)(0.75)}{20}} = 0.0968$

Exercises 5.7

2. a. $P\left(\dfrac{5.5 - 6}{\sqrt{(0.15)(0.85)(40)}} \leqslant z \leqslant \dfrac{6.5 - 6}{\sqrt{(0.15)(0.85)(40)}}\right)$

$= P(-0.22 \leqslant z \leqslant 0.22) = 0.1742$

b. $P(z > 0.22) = 0.4129$

c. $P(z < -0.22) = 0.4129$

4. From Alaska: $P\left(z > \dfrac{90 - 100}{\sqrt{(0.25)(0.75)(400)}}\right) = P(z > -1.15) = 0.8749$

From Texas: $P\left(z > \dfrac{90 - 80}{\sqrt{(0.2)(0.8)(400)}}\right) = P(z > 1.25) = 0.1056$

6. a. Using the binomial distribution, $p = 0.15$, $n = 10$, $x = 1$; $P(X = 1) = 0.3474$

b. $P\left(\dfrac{56 - 68}{12} \leqslant z \leqslant \dfrac{92 - 68}{12}\right) = P(-1 \leqslant z \leqslant 2) = 0.8185$

c. $P\left(\dfrac{70 - 68}{12/\sqrt{49}} \leqslant z \leqslant \dfrac{74 - 68}{12/\sqrt{49}}\right) = P(1.17 \leqslant z \leqslant 3.5)$

$= 0.8779$

d. $P\left(z > \dfrac{x - 68}{12}\right) = 0.1357;\ z = 1.10$

$x = 68 + 1.10(12) = 81.2$

8. $z_0 = \dfrac{0.75 - 0.80}{\sqrt{\dfrac{(0.8)(0.2)}{100}}} = -1.25$

$P(\bar{p} \geqslant 0.75) = P(z \geqslant -1.25) = 0.8944$

10. a. $\mu_{\bar{x}} = \$98{,}900,\ \sigma_{\bar{x}} = \dfrac{\$4210}{\sqrt{25}}\sqrt{\dfrac{100 - 25}{100 - 1}} = \$732.87$

b. $\mu_{\bar{x}} = \$98{,}900,\ \sigma_{\bar{x}} = \dfrac{\$4210}{\sqrt{50}}\sqrt{\dfrac{100 - 50}{100 - 1}} = \$423.12$

c. $\mu_{\bar{x}} = \$98{,}900,\ \sigma_{\bar{x}} = 0$, since only one sample is possible.

Review Exercises for Chapters 1 through 5

2. a. $P(X \geqslant 1.2) = P\left(z \geqslant \dfrac{1.2 - 1.06}{.3140}\right) = P(z \geqslant .45) = .5 - .1736 = .3264$

Empirical probability $= 12/21 = .57$. Not a good approximation.

b. Mode $= 1.2$

Median $= 1.2$ (value of 10.5th item)

4. a. $\dfrac{\dbinom{3}{2}\dbinom{17}{0}}{\dbinom{20}{2}} = \dfrac{3}{\dfrac{20!}{18!2!}} = \left(\dfrac{3}{20}\right)\left(\dfrac{2}{19}\right) = 0.016$ or $\dfrac{\dbinom{2}{2}\dbinom{18}{1}}{\dbinom{20}{1}} = 0.016$

b. Poisson distribution

$\mu = np = (50)(0.016) = 0.80$

$P(X \geqslant 1) = 1 - F(0) = 1 - 0.449 = 0.551$ (Table A-3 for $\mu = 0.8$)

or by binomial distribution

$P(X \geqslant 1) = 1 - (0.984)^{50} = 0.554$

c. $E(X) = (-\$1)(0.984) + (\$50)(0.016) = -\$0.184$

6. a. $z_0 = \dfrac{0 - 5}{9} = -0.56$

$P(X \leqslant 0) = P(z < -0.56) = 0.5000 - 0.2123 = 0.2877 = 0.29$

Using the Binomial Distribution

$p = 0.29$ (probability of early arrival)

$q = 0.71$

$n = 5$

$P(X = 3) = \dbinom{5}{3}(.71)^2(.29)^3 = 0.12$

b. $\sigma_{\bar{x}} = \dfrac{\sigma}{\sqrt{n}} = \dfrac{9}{\sqrt{9}} = 3$

$z_0 = \dfrac{4 - 5}{3} = -1/3 = -0.33$

$P(X \geqslant 4) = P(z > -0.33) = 0.1293 + 0.5000 = 0.6293$

8. a. Assume symmetrical distributions. Therefore, mean equals mode. C has lowest mean productivity. Point C (mean) is lowest (farthest left on X-axis).

b. Employee B has the highest σ. Dispersion around the mean is greatest.

c. Employee A has the lowest coefficient of variation (CV). $CV = \dfrac{\sigma}{\mu}$, A has the smallest numerator (σ) and the largest mean (μ). Both factors lead to a smaller CV.

10. a. $\bar{X} = \dfrac{20}{8} = 2\frac{1}{2}$

$$s^2 = \dfrac{\begin{array}{c}(2 - 2.5)^2 + (3 - 2.5)^2 + (1 - 2.5)^2 + (4 - 2.5)^2 + (2 - 2.5)^2 \\ + (5 - 2.5)^2 + (3 - 2.5)^2 + (0 - 2.5)^2\end{array}}{7}$$

$s^2 = 18/7 = 2.57$

$s = \sqrt{2.57} = 1.6$ people

b. Let μ be estimated by $\bar{X} = 2.5$

From Table A-3, interpolate between 2.4 and 2.6

$P(X \geqslant 4) = 1 - P(X \leqslant 3) = 1 - 0.757 = 0.243$

12. a. $Md = 5$ (value of the $100\frac{1}{2}$th item)

b. $\bar{X} = \dfrac{(1)(6) + (2)(22) + (3)(30) + (4)(35) + (5)(62) + (6)(45)}{200} = 4.3$

$\bar{X} \pm s = 4.3 \pm \sqrt{2} = 4.3 \pm 1.4 = 2.9$ to $5.7 = 3, 4, 5$

$30 + 35 + 62 = 127$ items. Therefore, $127/200 = 63.5\%$ of the employees fall in the interval.

c. $P\begin{pmatrix}\text{at least one} \\ \text{of 5 records} \\ \text{scored a "1"}\end{pmatrix} = 1 - \dfrac{\dbinom{6}{0}\dbinom{22}{5}}{\dbinom{28}{5}}$

d. *Binomial*

$q = .80$ responders

$p = .20$ non responders $P(X \leqslant 2$ nonresponders$) = 0.6778$ (Table A-1)

$n = 10$

14. a. $\bar{X} = \dfrac{0(73) + 1(74) + 2(37) + 3(12) + 4(4)}{200} = 1$

$$s^2 = \dfrac{(0 - 1)^2(73) + (1 - 1)^2(74) + (2 - 1)^2(37) + (3 - 1)^2(12) + (4 - 1)^2(4)}{199}$$

$= .9749$

b. *Poisson*, because $\bar{X} = s^2$ and $\mu = \sigma^2$ in Poisson

No. of units	0	1	2	3	4
Actual	73	74	37	12	4
Theoretical:					
Poisson $\mu = 1$	73.6	73.6	36.8	12.2	3

c. Poisson $\mu = 1$ $F(4) - F(1) = .996 - .736 = .26$

Estimate for number of days $= 250(.26) = 65$

16. a.

x	$f(x)$	$xf(x)$	y	$f(y)$	$yf(y)$
0	0.5	0	0	0.45	0
1	0.2	0.2	2	0.15	0.3
2	0.3	0.6	4	0.40	1.6
		$\overline{0.8}$			$\overline{1.9}$

b. If $X = 1$ we have $P(Y = 0 | X = 1) = \frac{1}{4}$; $P(Y = 2 | X = 1) = \frac{1}{4}$; and $P(Y = 4 | X = 1) = \frac{1}{2}$ $E(Y | X = 1) = 0(\frac{1}{4}) + 2(\frac{1}{4}) + 4(\frac{1}{2}) = 2\frac{1}{2}$ (million dollars). If proposal 1 is 1 million dollars then the average net profit from investment proposal 2 is $2\frac{1}{2}$ million dollars.

18. a. *Poisson Distribution* (Table A-3)

$\mu = 2$; $P(X = 2) = 0.677 - 0.406 = 0.271$

$\mu = 4$; $P(X = 3) = 0.433 - 0.238 = 0.195$

$P(\text{pattern: 2, 2, 3}) = (0.271)(0.271)(0.195) = 0.0143$

b. *Poisson Distribution* (Table A-3)

$\mu = 6$ calls/2 hour period

$P(4 \leqslant X \leqslant 6) = F(6) - F(3) = 0.606 - 0.151 = 0.455$

c. *Poisson Distribution* (Table A-3)

$\mu = 4$ calls

$P(X \geqslant 3) = 1 - F(2) = 1 - 0.238 = 0.762$

Binomial Distribution

$$P(X = 4 \text{ successes}) = \binom{5}{4}(.24)^1(.76)^4 = 0.4003$$

Alternatively, let $p = 0.75$, $n = 5$

Since $P(4 \text{ successes}) = P(1 \text{ failure})$, then using $p = 0.25$, $n = 5$ and Table A-2 $P(X = 1) = 0.3955 \approx 0.40$

CHAPTER 6

Exercises 6.2

2. See the discussion in section 6.2.

4. Part (c) is best since (a) is a single point estimate which can be either right or wrong. As for (b), despite the perfect forecast, the range of estimate is too wide and, hence, useless.

Exercises 6.3

2. a. $5.2 \pm 2.00(0.9/\sqrt{36})$, or $4.90 - 5.50$ flavors

b. In repeated samples randomly drawn from the population, 95.5% of the intervals as drawn in part (a) would include the population mean.

4. $50 \pm 2.17(15/\sqrt{400})$, or $48.37 - 51.63$ minutes

6. No. From the viewpoint of classical statistics, it is incorrect to make probability statements about population parameters.

8. $1.9 \pm 1.65(1.0/\sqrt{100})$, or $1.735 - 2.065$

10. a. $150 \pm 2(40/\sqrt{100}) = 150 \pm 8$, or $142 - 158$ gallons

b. Accept a lower level of confidence or increase the sample size.

12. a. $232 \pm 2.00(40/\sqrt{400})$, or from 228 to 236 employees

b. In repeated samples randomly drawn from the sample population, 95.5% of the intervals constructed as in (a) would include the true population mean.

14. a. $0.60 \pm 1.65\sqrt{(0.6)(0.4)/400}$, or from 0.560 to 0.640 preferred sales help.

b. $(0.30 - 0.27) \pm (1.96)\sqrt{\dfrac{(0.30)(0.70)}{100} + \dfrac{(0.27)(0.73)}{300}}$, or from -0.073 to 0.133. In all possible pairs of samples drawn from the same population, 95% of the confidence intervals established by this method would contain the actual difference between the population proportions.

Exercises 6.4

2. a. $73 \pm 3.355(11.18/\sqrt{9}) = 73 \pm 12.50$, or $60.50 - 85.50$

b. We assume in the derivation of the t distribution that the underlying population is normally distributed. This assumption would be violated by a highly skewed population distribution. Hence, the range set up in part (a) would not have exactly a 99% confidence coefficient associated with it.

4. a. $\$.60 \pm 2.861(\$.15/\sqrt{20}) = \$.60 \pm \$.096$, or $\$.504 - \$.696$

b. For a 90% confidence coefficient

$\$.60 \pm 1.729(\$.15/\sqrt{20}) = \$.60 \pm \$.058$, or $\$.542 - \$.658$

For a 95% confidence coefficient

$\$.60 \pm 2.093(\$.15/\sqrt{20}) = \$.60 \pm 0.07$, or $\$.53 - \$.67$

6. a. $51 \pm 3.355(16.10/\sqrt{9}) = 51 \pm 18.02$, or from 32.98 minutes to 69.02 minutes.

b. It is assumed in the derivation of the t distribution that the underlying population is normally distributed, and this assumption would be violated by a highly skewed population distribution. Hence, the range set up in part (a) would not have exactly a 99% confidence coefficient associated with it.

2. a. $0.56 \pm (2.33)\sqrt{(0.56)(0.44)/100}$, or $0.444 - 0.676$ of the time

b. $2.33\sigma_{\bar{p}} = 0.03$

$$\frac{0.03}{2.33} = \sqrt{\frac{(0.5)(0.5)}{n}}$$

$n = 1,508.0$

Therefore, the necessary sample size is 1,508.

4. a. Of all possible samples he could draw, 95.5% will yield a correct interval estimate, whereas 4.5% of all possible samples will lead to an incorrect interval estimate.

b. $2(55,840/\sqrt{n}) = 5,000$; $n \approx 499$

6. The cost of taking a larger sample might be uneconomical when compared with the return on the sample information.

8. $2(\$.05/\sqrt{n}) = \$0.005$

$n = 400$

10. a. 95.5% of all the possible samples he could draw will yield a correct interval estimate, whereas 4.5% of all possible samples will lead to an incorrect interval estimate.

b. $2(\$0.50/\sqrt{n}) = \$.05$; $n = 400$

CHAPTER 7

2. The statement is incorrect, since there are two types of errors that can occur, Type I and Type II. If α (the probability of a Type I error) is made extremely small, the Type II error probability (with a fixed sample size) will tend to become quite large. The two errors, for a fixed sample size, move inversely to one another. As an extreme case, for example, if we always accept H_0 regardless of test results, the α error is 0, but the β error is equal to 1.

4. Power is defined as the probability of rejecting H_0 when H_1 is the true state of nature. Power $= 1 - \beta$.

6. a. $H_0: \mu \leqslant 250$

$H_1: \mu > 250$

$\alpha = 0.01$

Critical value $= 250 + 2.33(40/\sqrt{200}) = 257$

Decision Rule

1. Reject H_0 if $\bar{x} > 257$ and widen the bridge
2. Do not reject H_0 otherwise

b. $z = \dfrac{(255 - 250)}{40/\sqrt{200}} = 1.77$

p value $= P(z > 1.77) = 0.5000 - 0.4616 = 0.0384$

8. a. $H_0: \mu \leqslant 0.012$

$H_1: \mu > 0.012$

Critical value $= 0.012 + 2.33(0.002/\sqrt{50})$

$\qquad\qquad = 0.0127$

Decision Rule
  1. Reject H_0 if $\bar{x} > 0.0127$
  2. Do not reject H_0 otherwise

Since $\bar{x} = 0.015 > 0.0127$, we reject H_0. We would conclude that the average mortality rate in these cities is above the national rate.

b. $z = \dfrac{0.015 - 0.012}{0.002/\sqrt{50}} = 10.6$

p value $= 0.0000$.

10. $H_0: \mu = 8.0\%$

$H_1: \mu \neq 8.0\%$

Critical values $= 8\% \pm 1.96(0.3\%/\sqrt{200})$, or 7.9584% and 8.0416%

Decision Rule
  1. Reject H_0 if $7.9584\% > \bar{x}$ or $\bar{x} > 8.0416\%$
  2. Do not reject H_0 if $7.9584\% < \bar{x} < 8.0416\%$

a. No. We would not be willing to conclude that a change had taken place. Because $\bar{x} = 8.03\%$ does not fall in the rejection region, we cannot reject $H_0: \mu = 8.0\%$

b. $P(\text{Type II error}) = P(-3.00 < z < 1.02) = 0.8448$

c. The probability is 0.8448 of concluding that the arithmetic mean interest rate has not changed from 8.0%, when in fact two years later it was 8.02%. Note: This is a tricky problem. Students are tempted to treat it as a case of hypothesis testing concerning a proportion rather than a mean.

12. Probability of Type I error $= P(\bar{p} \leqslant 0.25 \,|\, p = 0.30)$

$$= P\left(z \leqslant \dfrac{0.25 - 0.30}{\sqrt{(0.3)(0.7)/300}}\right)$$

$$= P(z \leqslant -1.89) = 0.0294$$

The Type I error here is an incorrect decision to discontinue the program.

14. $H_0: p \geqslant 0.06$

$H_1: p < 0.06$

$\alpha = 0.05$

Critical value $= 0.06 - 1.65\sqrt{(0.06)(0.94)/1,000} = 0.048$

Decision Rule
1. Reject H_0 if $\bar{p} < 0.048$
2. Do not reject H_0 otherwise

Reject H_0, since $0.038 < 0.048$ and decide that the system is effective.

16. a. $H_0: p = 0.35$
$H_1: p \neq 0.35$
$\alpha = 0.02$
Critical values $= 0.35 \pm (2.33)\sqrt{(0.35)(0.65)/600}$
$\qquad\qquad\quad = 0.35 \pm 0.045$ or 0.305 and 0.395

.228

24.495

Decision Rule
1. Reject H_0 if $\bar{p} < 0.305$ or $\bar{p} > 0.395$
2. Do not reject H_0 if $0.305 \leqslant \bar{p} \leqslant 0.395$

Reject H_0, because $\bar{p} = (175/600) = 0.292 < 0.305$. The J.T. McClay Company should alter its marketing campaign.

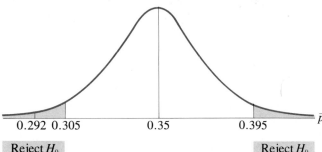

| 0.292 0.305 | 0.35 | 0.395 | $\bar{p}$ |

Reject H_0 Reject H_0

b. $H_0: p \geqslant 0.35$
$H_1: p < 0.35$
$\alpha = 0.02$
Critical value $= 0.35 - 2.05\sqrt{(0.35)(0.65)/600}$
$\qquad\qquad\quad = 0.35 - 0.040$
$\qquad\qquad\quad = 0.31$

Decision Rule
1. Reject H_0 if $\bar{p} < 0.31$
2. Do not reject H_0 if $\bar{p} \geqslant 0.31$

Reject H_0, since $\bar{p} = (175/600) = 0.292 < 0.31$.

Exercises 7.3

2. $H_0: \mu_1 = \mu_2$ (The two countries have the same mean height.)
$H_1: \mu_1 \neq \mu_2$ (The two countries have different mean heights.)
$\alpha = 0.05$
Critical value $= 0 \pm (1.96)\sqrt{(3)^2/2{,}000 + (2)^2/1{,}500}$
$\qquad\qquad\quad = 0 \pm 0.166$

Decision Rule
1. Reject H_0 if $\bar{x}_1 - \bar{x}_2 < -0.166$ or $\bar{x}_1 - \bar{x}_2 > 0.166$
2. Do not reject H_0 if $-0.166 \leqslant \bar{x}_1 - \bar{x}_2 \leqslant 0.166$

Reject H_0, since $69.1 - 68.3 = 0.8 > 0.166$.

4. Because a census was conducted (and assuming the data are accurate), we can conclude that the average balance due on customers' accounts at Bombergers exceeds that due on the accounts at Zimbels. Testing a hypothesis to draw an inference about the population means would be a waste. We already know that the population mean of $56 is greater than the population mean of $38.

6. $H_0: p_A = p_B$
$H_1: p_A \neq p_B$
$\alpha = 0.01$
$\bar{p} = (72 + 84)/(100 + 100) = 0.78$
Critical value $= 0 \pm (2.57)\sqrt{(0.78)(0.22)(1/100 + 1/100)}$
$= 0 \pm 0.15$

Decision Rule
1. Reject H_0 if $|\bar{p}_A - \bar{p}_B| > 0.15$
2. Do not reject H_0 if $|\bar{p}_A - \bar{p}_B| \leqslant 0.15$

Do not reject H_0, since $|0.72 - 0.84| < 0.15$. Therefore, we cannot conclude that the two groups differed with respect to whether they would purchase the cookies tasted.

8. a. $H_0: p_T = p_{PE}$ T = traders
     $H_1: p_T \neq p_{PE}$ PE = professors of economics
   Assume that $\alpha = 0.05$

$$\bar{p} = \frac{135 + 154}{225 + 200} = \frac{289}{425} = 0.68$$

Critical value $= 0 \pm (1.96)\sqrt{(0.68)(0.32)(\frac{1}{200} + \frac{1}{225})} = 0 \pm 1.96(0.045) = 0 \pm 0.088$

Decision Rule
1. Reject H_0 if $|\bar{p}_T - \bar{p}_{PE}| > 0.088$
2. Do not reject H_0 if $|\bar{p}_T - \bar{p}_{PE}| \leqslant 0.088$

$$\bar{p}_T = \frac{135}{225} = 0.60, \qquad \bar{p}_{PE} = \frac{154}{200} = 0.77$$

Decision: Reject H_0, since $|\bar{p}_T - \bar{p}_{PE}| = |0.6 - 0.77| = 0.17 > 0.088$

b. The error of concluding that there is a difference in the proportions of foreign exchange traders and professors of economics who believe that flexible exchange rates will enhance international monetary stability, when in fact the proportions are the same.

10. $H_0: p_1 = p_2$
$H_1: p_1 \neq p_2$
$\alpha = 0.05$
$\bar{p} = 398/900 = 0.442$
Critical value:

$$0 \pm (1.96) \sqrt{(0.442)(0.558)\left(\frac{1}{400} + \frac{1}{500}\right)} = \pm 0.065$$

Decision Rule
1. Reject H_0 if $|\bar{p}_1 - \bar{p}_2| > 0.065$
2. Do not reject H_0 if $|\bar{p}_1 - \bar{p}_2| \leqslant 0.065$

Decision: Do not reject H_0, since $|0.47 - 0.42| < 0.065$

Exercises 7.4

2. a. $\hat{s}^2 = \dfrac{13(0.25)^2 + 9(0.10)^2}{14 + 10 - 2} = 0.041$

$s_{\bar{x}_1 - \bar{x}_2} = \sqrt{0.041} \sqrt{1/14 + 1/10} = 0.084$

$t = \dfrac{\$1.22 - \$1.48}{0.084} = -3.095$

For 22 degrees of freedom and $\alpha = 0.10$, the critical value of t is -1.717. Since $-3.095 < -1.717$, we conclude that the difference between the sample averages is statistically significant.

b. $H_0: \mu_1 - \mu_2 = 0$
$H_1: \mu_1 - \mu_2 \neq 0$

Decision Rule
1. Reject H_0 if $t < -1.717$ or $t > 1.717$
2. Do not reject H_0 if $-1.717 \leqslant t \leqslant 1.717$

4.

Brand A	Brand B
$\bar{x}_A = 22$	$\bar{x}_B = 20$
$s_A = 4.77$	$s_B = 3.42$
$n_A = 15$	$n_B = 15$

$H_0: \mu_A = \mu_B$

$H_1: \mu_A \neq \mu_B$

For 28 degrees of freedom and $\alpha = 0.05$, the critical value of t is 2.048.

Decision Rule
1. Reject H_0 if $t < -2.048$ or $t > 2.048$
2. Do not reject H_0 if $-2.048 \leqslant t \leqslant 2.048$

$$s_A = \sqrt{318/14} = 4.77$$
$$s_B = \sqrt{164/14} = 3.42$$
$$\hat{s}^2 = \frac{14(4.77)^2 + 14(3.42)^2}{15 + 15 - 2} = 17.22$$
$$s_{\bar{X}_A - \bar{X}_B} = \sqrt{17.22(\sqrt{1/15 + 1/15})} = 1.52$$
$$t = (22 - 20)/1.52 = 1.32$$

Since $-2.048 \leqslant 1.32 \leqslant 2.048$, we cannot conclude that there was a statistically significant difference between the sample means.

6. $H_0: \mu_1 - \mu_2 = 0$
$H_1: \mu_1 - \mu_2 \neq 0$
$\alpha = 0.02$

$$\hat{s}^2 = \frac{(14)(0.05)^2 + (12)(0.03)^2}{15 + 13 - 2} = 0.001762$$

$$s_{\bar{x}_1 - \bar{x}_2} = \sqrt{0.001762} \left(\sqrt{\frac{1}{15} + \frac{1}{13}} \right) = 0.016$$

Critical values $= 0 \pm (2.479)(0.016)$
$$= \pm 0.040$$

Decision Rule
1. Retain H_0 if $-0.040 \leqslant \bar{x}_1 - \bar{x}_2 \leqslant 0.040$
2. Otherwise, reject H_0.

Since $0.02 < 0.040$, we retain H_0. The assistant cannot conclude that there is a significant difference between the sample average success rates of the two investment banking firms.

Exercises 7.5

2. $n = \left[\dfrac{(1.65 + 2.33)\$9}{\$30 - \$28)} \right]^2 \approx 321$

Decision Rule
1. If $\bar{x} < \$29.17$, reject $H_0: \mu = \$30$
2. If $\bar{x} \geqslant \$29.17$, retain $H_0: \mu = \$30$

Exercises 7.6

2. Let μ_A and μ_B be the mean population comprehension ratings after and before the course, respectively.
$H_0: \mu_A = \mu_B$
$H_1: \mu_A > \mu_B$
$s_{\bar{d}} = 0.455/\sqrt{13} = 0.126$
The critical value of t is 1.782 for $v = 12$ in a one-tailed test.

Decision Rule
   1. Reject H_0 if $t > 1.782$
   2. Do not reject H_0 otherwise

Since $t = 0.40/0.126 = 3.175 > 1.782$, reject H_0 and conclude that the Blair Good Reading Comprehension Course is effective in increasing reading comprehension ratings.

CHAPTER 8

Exercises 8.1

2. H_0: The distribution is Poisson with $\mu = 2.6$
   H_1: The distribution is not Poisson with $\mu = 2.6$

(1) Number of Plants	(2) Number of Countries Observed f_o	(3) Column 1 × Column 2	(4) $f(x)$	(5) Expected Number of Countries f_t
0	6	0	0.074	7.4
1	7	7	0.193	19.3
2	40	80	0.251	25.1
3	24	72	0.218	21.8
4	14	56	0.141	14.1
5 or more	9	45	0.123	12.3
Total	100	260	1.000	100.0

Mean $= 260/100 = 2.6$
   $\chi^2 = 18.195$
    $v = 4$
$\chi^2_{0.01} = 13.277$
Hence, reject H_0 and conclude that the Poisson distribution with $\mu = 2.6$ is not a good fit.

4. H_0: The distribution is Poisson with $\mu = 1.5$
   H_1: The distribution is not Poisson with $\mu = 1.5$

(1) Ratings	(2) Number of Companies f_o	(3) (Col. 1 × Col. 2)	(4) $f(x)$	(5) Expected Number of Companies f_t
0	10	0	0.223	11.15
1	18	18	0.335	16.75
2	13	26	0.251	12.55
3	6 ⎤	18	0.125	6.25 ⎤
4	2 ⎬ 9	8	0.047	2.35 ⎬ 9.35
5	1 ⎦	5	0.015	0.75 ⎦
Total	50	75	0.996	49.80

$$\text{Mean} = 75/50 = 1.5$$
$$\chi^2 = 0.249$$
$$v = 2$$
$$\chi^2_{0.05} = 5.991$$

Retain H_0 and conclude that the evidence is consistent with the hypothesis that the distribution of ratings is Poisson with $\mu = 1.5$.

6. H_0: The department store's credit customers are drawn from a population that has an equal distribution among the six classes

H_1: The department store's credit customers are not drawn from a population that has an equal distribution among the six classes

Classification	f_o	f_t
A	46	50
B	42	50
C	56	50
D	64	50
E	48	50
F	44	50
Total	300	300

$$\chi^2 = 7.04$$
$$v = 5$$
$$\chi^2_{0.05} = 11.070$$

Retain H_0 at $\alpha = 0.05$. We cannot reject the null hypothesis.

8. H_0: The bank's employees with MBA degrees are equally distributed among the four schools

H_1: The bank's employees with MBA degrees are not equally distributed among the four schools

School	f_o	f_t
A	4	6
B	9	6
C	8	6
D	3	6
Total	24	24

$$\chi^2 = 4.333$$
$$v = 3$$
$$\chi^2_{0.05} = 7.815$$

Hence, retain H_0 at $\alpha = 0.05$. We cannot reject the null hypothesis.

Exercises 8.3

2. H_0: Preferences for ice creams are independent of buyer's sex

H_1: Preferences for ice creams are not independent of buyer's sex

a. $\chi^2 = 7.50$
$v = (2-1)(3-1) = 2$
$\chi^2_{0.05} = 5.991$
Hence, reject H_0 at $\alpha = 0.05$.
b. $\chi^2 = 10.50$
$v = (2-1)(4-1) = 3$
$\chi^2_{0.01} = 11.345$
Hence, retain H_0 at $\alpha = 0.01$.

4. H_0: Impact of advertisement is independent of nature of advertisement
H_1: Impact of advertisement is not independent of nature of advertisement

f_o	f_t
75	49.5
25	50.5
23	48.5
75	49.5
198	198.0

$\chi^2 = 52.555$
$v = (2-1)(2-1) = 1$
$\chi^2_{0.01} = 6.635$
Reject H_0 at $\alpha = 0.01$ and conclude that impact of advertisement is not independent of nature of advertisement.

6. H_0: The proportions hired were the same for the three categories of applicants
H_1: The proportions hired were not the same for the three categories of applicants

f_o	f_t
76	60
23	29
26	36
164	180
93	87
118	108
500	500

$\chi^2 = 11.048$
$v = 2$
$\chi^2_{0.05} = 5.991$
Hence, reject H_0 at $\alpha = 0.05$ and conclude that the proportions hired were not the same for the three categories of applicants.

8. H_0: Share prices are independent of declared dividends.
H_1: Share price are not independent of declared dividends.

f_0	f_t
6	6.0
12	12.0
2	2.0
14	13.8
28	27.6
4	4.6
10	10.2
20	20.4
4	3.4
100	100.0

$$\chi^2 = 0.205$$
$$df = (3 - 1)(3 - 1) = 4$$
$$\chi^2_{0.01} = 13.277$$

Since $0.205 < 13.277$, we retain H_0 and conclude that share price is independent of declared dividends.

Exercises 8.3

2. H_0: The collectors are equally successful in collecting overdue accounts
H_1: The collectors are not equally successful in collecting overdue accounts

Analysis of Variance Table

Source of Variation	Sum of Squares	Degrees of Freedom	Mean Square
Between collectors	430	2	215
Within collectors	580	12	48.33
Total	1,010	14	

$$F(2, 12) = 215/48.33 = 4.45$$

$F_{0.05}(2, 12) = 3.88$. Since $4.45 > 3.88$, reject H_0 and conclude that the collectors are not equally successful in collecting overdue accounts.

4. H_0: The average ratios of expected return to standard deviation do not differ among the three industries
H_1: The average ratios of expected return to standard deviation differ among the three industries

Analysis of Variance Table

Source of Variation	Sum of Squares	Degrees of Freedom	Mean Square
Between industries	0.507	2	0.254
Within industries	0.222	6	0.037
Total	0.729	8	

$$F(2, 6) = 0.254/0.037 = 6.86$$
$$F_{0.05}(2, 6) = 5.14$$

Since $6.86 > 5.14$, reject H_0. Conclude that the average ratios of expected return to standard deviation differ among the three industries.

6.

Two-factor Analysis of Variance Table

Source of Variation	Sum of Squares	Degrees of Freedom	Mean Square
Relative prices	20.016	3	6.672
National income	0.487	2	0.244
Error	6.446	6	1.074
Total	26.949	11	

(1) H_0: There is no difference in national trade surpluses (deficits) among the three levels of national income.

H_1: There is a difference in the national trade surpluses (deficits) among the three levels of national income.

$F(2, 6) = 0.244/1.074 = 0.227$

$F_{0.05}(2, 6) = 5.14$

Since $0.227 < 5.14$, retain H_0 and conclude that national trade surpluses (deficits) do not differ among the three levels of national income.

(2) H_0: There is no difference in national trade surpluses (deficits) at the four levels of relative prices

H_1: There is a difference in the national trade surpluses (deficits) at the four levels of relative prices

$F(3, 6) = 6.672/1.074 = 6.21$

$F_{0.05}(3, 6) = 4.76$

Since $6.21 > 4.76$, reject H_0 and conclude that national trade surpluses (deficits) differ at the four levels of relative prices.

8.

Two-factor Analysis of Variance Table

Source of Variation	Sum of Squares	Degrees of Freedom	Mean Square	F
Operators	11.19	3	3.73	2.66
Machines	5.69	3	1.90	1.36
Error	12.56	9	1.40	
Total	29.44	15		

H_0: The operators do not differ with respect to average output per minute

H_1: The operators differ with respect to average output per minute

$F(3, 9) = 3.73/1.40 = 2.66$

$F_{0.01}(3, 9) = 6.99$

Since $2.66 < 6.99$, retain H_0.

H_0: The machines do not differ with respect to average output per minute

H_1: The machines differ with respect to average output per minute

$F(3, 9) = 1.9/1.4 = 1.36$

$F_{0.01}(3, 9) = 6.99$

Since $1.36 < 6.99$, retain H_0.

10. H_0: No difference in performance exists among the three funds.
H_1: Performances differ among the three funds.

Analysis of Variance Table

Source of Variation	Sum of Squares	Degrees of Freedom	Mean Square
Between funds	21,332.87	2	10,666.44
Within each fund	2,667.0	27	98.78
Total	22,483.74	29	

$F(2, 27) = 107.98$
$F_{0.05}(2, 27) = 3.35$
Since $107.98 > 3.35$, reject H_0, and conclude that there are real differences in performance among the three funds.

Review Exercises for Chapters 6 through 8

2. a. H_0: $p \leqslant 0.5$ $\bar{p} = \dfrac{60}{100} = 0.60$

H_1: $p > 0.5$
$\alpha = 0.01$

critical value $= 0.50 + 2.33 \sqrt{\dfrac{(0.5)(0.5)}{100}} = 0.6165$

Since $0.60 < 0.6165$, we retain H_0. The union cannot be fairly certain that the proportion of favorable union members exceeds 0.5.

b. $P(\text{Type II error}) = P(\text{Retain } H_0 | p = 0.64)$

$$P(\bar{p} \leqslant .6165 | p = 0.64) = P\left(z \leqslant \dfrac{0.6165 - .64}{\sqrt{(0.64)(0.36)/100}} \right)$$

$P(z \leqslant 0.49) = 0.3121$

4. a. H_0: $p \geqslant 0.20$
H_1: $p < 0.20$
Critical value $= 68/400 = 0.17$

$$\text{Critical } z = \dfrac{0.17 - 0.20}{\sqrt{\dfrac{(0.20)(0.80)}{400}}} = -1.50$$

$\alpha = 0.5000 - 0.4332 = 0.0668$
Nature of the Type I error: incorrectly discontinuing marketing of the product, that is, discontinuing marketing when $p \geqslant 0.20$.

b. $z = \dfrac{0.17 - 0.15}{\sqrt{\dfrac{(0.15)(0.85)}{400}}} = 1.12$

$\beta = P(z > 1.12) = 0.5000 - 0.3686 = 0.1314$

6. a. $H_0: \mu \leqslant 5$ $\alpha = 0.05$
   $H_1: \mu > 5$ $v = 5 - 1 = 4$

Critical value $= 5 + 2.132\left(\dfrac{0.6}{\sqrt{5}}\right) = 5.572$

Since $\bar{x} = 6 > 5.572$, reject H_0. This implies that it is safe to hire high school graduates.

b. $H_0: \mu_3 - \mu_1 = 0$ $\alpha = 0.05$
   $H_1: \mu_3 - \mu_1 \neq 0$ $v = (5 - 1) + (5 - 1) = 8$

$$\hat{s}^2 = \frac{4(0.25) + 4(0.36)}{4 + 4} = 0.305$$

$$t = \frac{\bar{x}_3 - \bar{x}_1}{s_{\bar{x}_3 - \bar{x}_1}} = \frac{8 - 6}{(\sqrt{0.305})\left(\sqrt{\dfrac{1}{5} + \dfrac{1}{5}}\right)} = 5.73$$

$t_{0.05,8} = 2.306$

Since $5.73 > 2.306$, reject H_0, and conclude that there is a difference in performance of high school graduates and MBAs.

c.

Source of Variation	Sum of Squares	Degrees of Freedom	Mean Square
Between Groups	10	2	5
Within Groups	5	12	5/12
Total	15	14	

Note:
$s_1^2 = \Sigma(x_1 - \bar{x}_1)^2/(n_1 - 1)$
$\Sigma(X_1 - \bar{X}_1)^2 = (n_1 - 1)s_1^2$
$SSA = 5(6 - 7)^2 + 5(7 - 7)^2 + 5(8 - 7)^2 = 10$
$SSE = 4(0.36) + 4(0.64) + 4(0.25) = 5$

$$F(2, 12) = \frac{5}{5/12} = 12$$

$F_{0.05}(2, 12) = 3.88$

Since $12 > 3.88$, reject H_0. Conclude that there are differences in performance among the three groups.

8. a. $\bar{p} = \dfrac{150}{250} = 0.6$

Confidence interval

$$0.6 \pm 1.645\sqrt{\frac{(0.6)(0.4)}{250}} = 0.549 \text{ to } 0.651$$

b. $n \geqslant (2.33)^2(1/4)/(0.01)^2 = 13{,}572.25$
   Cost $= 13{,}573(\$2.50) = \$33{,}931$

10. a. $H_0: p \leqslant 0.50$
   $H_1: p > 0.50$
   $\alpha = 0.05$

$$\text{Critical value} = .50 + 1.65 \sqrt{\frac{(0.50)(0.50)}{30}} = 0.65$$

Since $0.67 > 0.65$, reject H_0. Conclude that the data support the claim.

b. $\dfrac{0.65 - 0.60}{\sqrt{\dfrac{(.6)(.4)}{30}}} = 0.56$

$P(\text{Retain } H_0 | p = 0.60) = P(\bar{p} < 0.65 | p = 0.60) = P(z < 0.56) = 0.7123$

12. a.

Source of Variation	Sum of Squares	Degrees of Freedom	Mean Square
Between industries	1,340	2	670
Within industries	2,530	27	93.70
Total	3,870	29	

$SSA = 10[(21 - 14)^2 + (16 - 14)^2 + (5 - 14)^2] = 1,340$
$SSE = 9[(87.78) + (98.89) + (94.44)] = 2,530$
$F(2, 27) = 670/93.7 = 7.15$
$F_{0.05}(2, 27) = 3.35$
Since $7.15 > 3.35$, conclude that the mean yields of the three industries differ.

b. See figure.

Figure for exercise 12(b)

Observed

	A	B	
I	2	8	10
II	5	5	10
III	8	2	10
	15	15	30

Expected

	A	B	
I	5	5	10
II	5	5	10
III	5	5	10
	15	15	30

$$\chi^2 = \frac{(2 - 5)^2}{5} + \frac{(8 - 5)^2}{5} + \frac{(5 - 5)^2}{5} + \frac{(5 - 5)^2}{5} + \frac{(8 - 5)^2}{5} + \frac{(2 - 5)^2}{5}$$
$$= 7.2$$

$\chi^2_{0.05,2} = 5.99$
Since $7.2 > 5.99$, conclude that there is dependence between industry classification and percentage yield.

14. a. $H_0: \mu_1 - \mu_2 = 0 \qquad \alpha = 0.01$
$H_1: \mu_1 - \mu_2 \neq 0$
$s_{\bar{x}_1 - \bar{x}_2} = \sqrt{5^2/100 + 12^2/100} = 1.3$
Critical $\bar{x}_1 - \bar{x}_2 = 0 \pm 2.58(1.3) = \pm 3.354$

Since $19.0 - 24.5 = -5.5 < -3.354$, reject H_0 and conclude that there is a difference in average coffee consumption in the two regions.

b. True, because of symmetry in the power curve.

c. False. The size of the critical region decreases.

d. False. $\alpha + \beta \neq 1$. However, for a fixed sample size, increasing α does bring about decreases in probabilities of Type II errors.

16. a. $10 \pm 2.17(4.5/\sqrt{64}) = 8.78$ to 11.22 minutes

b. $n = (2.57)^2(6)^2/3^2 = 26.4 \approx 27$

c. $10 + z\sqrt{20.25}/\sqrt{64} = 10.721$ $z = 1.282$ 80% confidence interval

18. a. $H_0: \mu = 200$ $\alpha = 0.05$

$H_1: \mu < 200$

$\sigma_{\bar{x}} = \$42/\sqrt{9} = \$14$

Critical $\bar{x} = \$200 - (1.65)(\$14) = \$176.90$

Decision Rule

1. If $\bar{x} < 176.90$, reject H_0
2. If $\bar{x} \geqslant 176.90$, retain H_0

Conclusion: Since $\$188 > \$176.90$, retain H_0. Therefore, we cannot reject the hypothesis that the population of additional stores has a monthly sales mean of $200.

b. $H_0: \mu = \$200$ $\alpha = 0.05$

$H_1: \mu < \$200$

Critical t value for $v = 9 - 1 = 8$, one-tailed test $= -1.86$.

Decision Rule

1. If $t < -1.86$, reject H_0
2. If $t \geqslant -1.86$, retain H_0

$$t = \frac{\$188 - \$200}{\$\sqrt{350}/\sqrt{9}} = -1.92$$

Conclusion: Since $-1.92 < -1.86$, reject H_0. Therefore, we reject the hypothesis that the population of additional stores has a monthly sales mean of $200.

c. $z_0 = (\$176.90 - \$162.90)/\$14 = 1$

$\beta = P(z > 1) = 0.5000 - 0.3413 = 0.16$

20. a. $H_0: p_1 = 0.60, p_2 = 20, p_3 = 0.20$

$H_1:$ Not so

$\chi^2 = (110 - 120)^2/120 + (40 - 40)^2/40 + (50 - 40)^2/40 = 3.33$

$\chi^2_{0.01, 2} = 9.21$

Since $3.33 < 9.21$, retain H_0 and conclude that the percentages observed in the sample of 200 offers are not inconsistent with the claimed percentages.

b. $H_0:$ Acceptances and rejections of offers are independent of type of option.

$H_1:$ Acceptances and rejections of offers are not independent of type of option.

$$\chi^2 = \frac{(61-66)^2}{66} + \frac{(49-44)^2}{44} + \frac{(25-24)^2}{24} + \frac{(15-16)^2}{16}$$

$$+ \frac{(34-30)^2}{30} + \frac{(16-20)^2}{20} = 2.38$$

$$v = (2-1)(3-1) = 2$$

$$\chi^2_{0.05,2} = 5.99$$

Since $2.38 < 5.99$, retain H_0. We cannot conclude that the type of option affects the acceptance or rejection of offers.

22. a. $H_0: \mu_{ML} = \mu_{NP}$ $\alpha = 0.01$
    $H_1: \mu_{ML} \neq \mu_{NP}$
    $$\hat{s}^2 = [6(0.081) + 10(0.072)]/(6+10) = 0.075375$$
    $$s_{\bar{x}_{ML} - \bar{x}_{NP}} = \sqrt{0.075375}\sqrt{1/7 + 1/11} = 0.1327$$
    $$t = (3.7 - 3.4)/0.1327 = 2.26$$
    $$t_{0.01,16} = 2.921$$
    Since $2.26 < 2.921$, retain H_0. Therefore, the result supports the claim of virtually identical weight.

    b. $H_0: \mu_{NP} \geqslant \mu_{ML}$
       $\mu_{NP} < \mu_{ML}$
    $t = 2.26$
    $t_{0.05,16}$ in a one-tailed test $= 1.746$
    Since $2.26 > 1.746$, reject H_0

    c. $3.7 + t\sqrt{0.081}/\sqrt{7} = 4.038$
    $t = 3.143$
    Since $v = 6$, we find this t value to pertain to a 98% confidence level.

CHAPTER 9

Exercises 9.4

2. a. $Y =$ Merit rating of employees after two years of service
   $X =$ Aptitude test score
   $$b = \frac{5,759 - 10(80.7)(7)}{66,707 - 10(80.7)^2} = 0.0695$$
   $$a = 7 - 0.0695(80.7) = 1.3914$$
   $$\hat{Y} = 1.3914 + 0.0695X$$
   $\hat{Y}$: 6.604, 5.839, 7.785, 6.951, 6.673, 5.422, 8.063, 7.577, 8.202, 6.882

   b. The regression coefficient $b = 0.0695$ means that for two employees whose aptitude scores differ by one point, the estimated difference in their merit ratings is 0.0695 points.

   c. $\hat{Y} = 1.3914 + (0.0695)(90) = 7.646$

4. a. $$b = \frac{4,470 - (45)(14.16)(5.2)}{10,124 - (45)(14.16)^2} = 1.05$$
   $$a = 5.2 - (1.05)(14.16) = -9.668$$
   $$\hat{Y} = -9.668 + 1.05X$$

b. $\hat{Y} = -9.668 + 1.05(12) = 2.932$ or 3 defective articles

6. a. $\hat{Y} =$ return on average investment (percent)

$X =$ market share (percent)

$$b = \frac{2,940.30 - [(5)(25.40)(18.84)]}{4,371.00 - [(5)(25.40)^2]}$$

$$= \frac{547.62}{1,145.20} = 0.48$$

$a = 18.84 - [(0.48)(25.40)] = 6.65$

$\hat{Y} = 6.65 + 0.48X$

$\hat{Y}$: 8.57%; 23.93%; 18.65%; 13.85%; 29.21%

b. A regression coefficient of $b = 0.48$ means that for two products with market-share levels that differ by 1 percentage point, the estimated difference in the return on average investment for these two products is 0.48 percentage points.

c. $\hat{Y} = 6.65 + [(0.48)(30)] = 21.05\%$

Exercise 9.5

2. a. $Y =$ earnings (in millions of dollars) in 1990

$X =$ research and development expenditures (in millions of dollars) in 1989

$$b = \frac{10,186.5 - (15)(6.2)(73)}{895 - (15)(6.2)^2} = 10.671$$

$a = 73 - (10.671)(6.2) = 6.840$

$Y_c = 6.840 + 10.671X$

b. $\hat{Y}_c = 6.840 + 10.671(10) = 113.55$

c. $s_{Y.X} = \sqrt{\dfrac{122,825 - (6.84)(1095) - (10.671)(10,186.5)}{13}}$

$= \sqrt{510.389} = 22.592$ (millions of dollars)

The standard error of estimate measures the scatter of the observed values of Y around the corresponding computed $\hat{Y}$ values on the regression line.

d. $s_{\hat{Y}} = s_{Y.X} \sqrt{\dfrac{1}{15} + \dfrac{(10 - 6.2)^2}{895 - (93)^2/15}} = (22.592)\sqrt{0.1121} = 7.564$

The estimated interval is $113.55 \pm (2.160)(7.564) = 113.55 \pm 16.338$, or from \$97.212 million to \$129.888 million

4. a. $Y =$ First year sales (millions of dollars)

$X =$ Customer awareness (percent)

$b = [39,815 - (14)(40)(55)]/[28,750 - (14)(40)^2] = 1.42$

$a = 55 - (1.42)(40) = -1.80$

$\hat{Y} = -1.80 + 1.42X$

b. $s_{Y.X} = \sqrt{[56,476 - (-1.8)(770) - 1.42(39.815)]/12}$

$= \sqrt{110.392} = 10.507$ (millions of dollars)

See footnote 2 of Chapter 9

c. $s_{\hat{y}} = s_{Y.X}\sqrt{\dfrac{1}{14} + \dfrac{(35-40)^2}{28{,}750 - (560)^2/14}}$

$= (10.507)\sqrt{0.0754} = 2.885$ (millions of dollars)

The estimated interval is $47.9 \pm (2.179)(2.885) = 47.9 \pm 6.286$ or from $41.614 million to $54.186 million.

d. $s_{IND} = s_{Y.X}\sqrt{1 + \dfrac{1}{14} + \dfrac{(35-40)^2}{28{,}750 - (560)^2/14}}$

$= (10.507)\sqrt{1.7054} = 10.896$

The estimated interval is $47.9 \pm (2.179)(10.896) = 47.9 \pm 23.742$, or from $24.158 million to $71.642 million.

e. The interval estimated in part (d) is much wider than that estimated in part (c). Individual values of Y cannot be predicted with as much precision as a conditional mean, because s_{IND} was used in part (d).

Exercises 9.8

2. a. $r_c^2 = 1 - \dfrac{s_{Y.X}^2}{s_Y^2} = 1 - \dfrac{0.04}{1.00} = 0.96$

$r_c = \sqrt{0.96} = 0.980$

The use of r values gives the impression of a stronger relationship than is actually present. Moreover, the use of r^2 is convenient because it can be interpreted as a percentage of variability explained by the regression equation.

b. $s_b = \dfrac{s_{Y.X}}{\sqrt{\Sigma(X - \bar{X})^2}} = \dfrac{0.2}{\sqrt{0.816}} = 0.221$

$t = \dfrac{(b - B)}{s_b} = \dfrac{4.0 - 0}{0.221} = 18.10$, with 60 degrees of freedom

Since $18.10 > 2.660$, we reject at the 1% level of significance the null hypothesis that the population regression coefficient is zero.

4. a. For two students whose studying time differed by one hour, the one who studied more would score an estimated 7.6 points higher.

$s_b = \dfrac{5.4}{\sqrt{196}} = 0.386$

$t = \dfrac{7.6 - 0}{0.386} = 19.689$, with 60 degrees of freedom

Since $19.689 > 2.66$, we conclude that the estimated coefficient is significantly different from zero at the 1% level of significance.

b. $r_c^2 = 1 - \dfrac{(5.4)^2}{(8.2)^2} = 0.566$

c. $s_{IND} = (5.4)\sqrt{1 + \dfrac{1}{62} + \dfrac{(1)^2}{196}} = (5.4)\sqrt{1.02123} = 5.457$

An estimated 95% prediction interval for the test score for a student who studied nine hours would be $74.8 \pm (2)(5.457)$, or from about 63.9 to 85.7. Since this student's score is considerably below the lower limit of this interval, we may conclude that he is a poor performer.

6. a. $r_c^2 = 1 - \dfrac{8}{25} = 0.68$

b. The regression coefficient -0.01 indicates an inverse relationship between checking-account balance and bad checks written on the account. For every $100 increase in the checking-account balance, one *fewer* bad check is expected to be written on the account.

c. $t = \dfrac{0.82}{\sqrt{0.32/198}} = 20.4$

The critical t value at 5% for $df = 198$ is approximately 1.96. Since 20.4 is far in excess of 1.96, we reject H_0 and conclude that a linear relationship exists for all checking account balances and the number of bad checks written.

CHAPTER 10

Exercises 10.13

2. a. The b_1 coefficient (5.8) indicates that if a firm spends $1 million more on research and development than another firm, and the two firms spend the same amount on advertising, then the estimated annual sales for the first firm will be $5.8 million more than the second firm's annual sales.

b. $H_0: B_1 = 0$
$H_1: B_1 = 0$

$t_1 = \dfrac{b_1 - 0}{s_{b_1}} = \dfrac{5.8}{1.20} = 4.83$, with 19 degrees of freedom

Since $4.83 > 2.861$, we reject H_0 at the 1% level of significance and conclude that b_1 is significantly different from zero.

$H_0: B_2 = 0$
$H_1: B_2 \neq 0$

$t_2 = \dfrac{b_2 - 0}{s_{b_2}} = \dfrac{4.2}{1.31} = 3.21$, with 19 degrees of freedom

Since $3.21 > 2.861$, we reject H_0 and conclude that b_2 is significantly different from zero at the 1% level of significance.

$H_0: B_3 = 0$
$H_1: B_3 \neq 0$

$t_3 = \dfrac{b_3 - 0}{s_{b_3}} = \dfrac{7.4}{1.56} = 4.74$, with 19 degrees of freedom

Since $4.74 > 2.861$, we reject H_0 and conclude that b_1 is significantly different from zero at the 1% level of significance,

c. If multicollinearity exists, that is, if the independent variables are highly correlated, the estimated regression coefficients for these variables tend to be unreliable. Separating the individual influences of the variables becomes extremely difficult.

d. $R_c^2 = 1 - \dfrac{S_{Y.12}^2}{s_Y^2} = 1 - \dfrac{(12.4)^2}{(25)^2} = 0.754$

e. $\hat{Y} = -2.3 + 5.8(6) + 4.2(10) + 7.4(7) = 126.3$ (millions of dollars)

f. $\beta_1 = b_1 \dfrac{s_{X_1}}{s_Y} = (5.8)\dfrac{(15)}{25} = 3.48$

$\beta_2 = b_2 \dfrac{s_{X_2}}{s_Y} = (4.2)\dfrac{(10)}{25} = 1.68$

$\beta_3 = b_3 \dfrac{s_{X_3}}{s_Y} = (7.4)\dfrac{(5)}{25} = 1.48$

β_1 measures the number of standard deviations by which $\hat{Y}$ changes with each change of one standard deviation in X_1.

4. a. The value $b_2 = 0.8$ indicates that if an option's expiration date is one month farther away than another option, and the underlying stocks have the same price volatility, then the estimated value of the first option exceeds that of the second by $0.80.

b. The reliability of the individual net regression coefficients is particularly important if the purpose of the analysis is to measure the separate effects of the independent variables on the dependent variable.

c. $S_{Y.12} = \sqrt{\dfrac{\Sigma(Y - \hat{Y})^2}{n - k}} = \sqrt{\dfrac{236}{40}} = \$2.429$

d. $R_c^2 = 1 - \dfrac{S_{Y.12}^2}{s_Y^2} = 1 - \dfrac{(2.429)^2}{17.643} = 0.666$

$s_Y^2 = \dfrac{\Sigma(Y - \bar{Y})^2}{n - 1} = \dfrac{741}{42} = 17.643$

e. $\hat{Y} = 3.0 + (0.3)(20) + (0.8)(6) = \$13.80$

f. H_0: The regression is not significant
   H_1: The regression is significant

$F = \dfrac{\Sigma(\hat{Y} - \bar{Y})^2/(k - 1)}{\Sigma(Y - \hat{Y})^2/(n - k)} = \dfrac{622/2}{236/40} = 52.71$

Since $52.71 > 5.18$, we reject the null hypothesis that the regression is not significant at the 1% level of significance.

6. a. $t_1 = \dfrac{b_1 - 0}{s_{b_1}} = \dfrac{-0.25120}{0.03411} = -7.364$, with 30 degrees of freedom

Since $-7.364 < -2.750$, we reject the null hypothesis and conclude that b_1 differs significantly from zero at the 1% level of significance.

$$t_2 = \frac{b_2 - 0}{s_{b_2}} = \frac{-0.03015}{0.00712} = -4.235, \text{ with 30 degrees of freedom}$$

Since $-4.235 < -2.750$, we reject H_0 and conclude that b_2 is significantly different from zero.

b. $S_{Y.12}^2 = \dfrac{2.8712}{30} = 0.09571$

$s_Y^2 = \dfrac{6.4224}{32} = 0.2007$

$R_c^2 = 1 - \dfrac{s_{Y.12}^2}{s_Y^2} = 1 - \dfrac{0.09571}{0.2007} = 0.523$

$R_c = \sqrt{R_c^2} = 0.723$

R_c^2 is easier to interpret, because it is a percentage figure and R_c is not.

c. $F = \dfrac{3.5512/2}{2.8712/30} = 18.553$, with (2,30) degrees of freedom

Since $18.553 > 5.39$, we reject H_0 and conclude that the regression is significant at the 1% level of significance.

$$F = \frac{\Sigma(\hat{Y} - \bar{Y})^2/(k-1)}{\Sigma(Y - \hat{Y})^2/(n-k)}$$

The F statistic compares the mean square due to regression (explained variance) to the residual mean square (unexplained variance).

Review Exercises for Chapter 10

2. a. Selected computer output:

Variate correlations (r values)

	GPA	V-SAT	RANK / C-SIZE
GPA	1.000	0.614	−0.483
V-SAT		1.000	0.023
RANK			1.000
C-SIZE			

Slope coefficients and standard errors

	COEFF.	ST. ERROR	t
Y-INTERCEPT	−1.1660702	0.97660667	−1.194
V-SAT	0.087295002	0.02071934	4.21321
RANK / C-SIZE	−2.8664136	0.85566342	−3.34993

$\hat{Y} = -1.1661 + 0.0873(\text{V-SAT}) - 2.8664(\text{RANK/C-SIZE})$

b. The interpretations of slope coefficients are similar to those in exercise 1.

$v = n - k = 20 - 3 = 17$

$t_{0.05,17} = |2.110|$

Therefore, the slope coefficients for V-SAT and RANK/C-SIZE are significant at the 5% level.

The signs of both slope coefficients are reasonable.

Analysis of variance table

	SUM OF SQUARES	DEGREES OF FREEDOM	MEAN SUM OF SQUARES	F
REGRESSION	7.4504564	2	3.7252282	14.1682
ERROR	4.4697986	17	0.26292933	
TOTAL	11.9202550	19		

Since $F = 14.1682 > F_{0.05;2,17} = 3.59$, reject the null hypothesis of no relationship between the dependent variable and the independent variables. $R^2 = 0.625025$, uncorrected for degrees of freedom

c. The revised model that includes only two independent variables, V-SAT and RANK/C-SIZE, explains more of the variation in freshman GPA ($R^2 = 0.625$) than does the first model ($R^2 = 0.549$). V-SAT is the one variable most highly correlated with freshman grade point averages. Note that since the revised model only accounts for 62.5% of the variation in freshman grade point averages, other factors not included in the regression equation have a substantial effect on freshman grade point averages.

CHAPTER 11

Exercises 11.4

2. a.

```
MTB > BRIEF 3
MTB > REGRESS C3 ON 1 PREDICTOR C2, C4 C5

THE REGRESSION EQUATION IS
GSERVCON = 1246 + 41.2 TIME

PREDICTOR          COEF        STDEV      T-RATIO          P
CONSTANT         1245.90       12.43       100.21      0.000
TIME               41.181       1.460       28.20      0.000

S = 22.02    R-SQ = 98.5%    R-SQ (ADJ) = 98.4%
```

b.

```
MTB > PRINT C1 C2 C3 C5 C6

ROW        YEAR         TIME        GSERVCON          TREND         %TREND
  1        1975           1          1286.4        1287.08         99.947
  2        1976           2          1324.4        1328.26         99.709
  3        1977           3          1368.7        1369.44         99.946
  4        1978           4          1426.9        1410.62        101.154
  5        1979           5          1478.6        1451.80        101.846
  6        1980           6          1511.1        1492.99        101.213
  7        1981           7          1533.4        1534.17         99.950
  8        1982           8          1547.5        1575.35         98.232
  9        1983           9          1585.5        1616.53         98.081
 10        1984          10          1625.2        1657.71         98.039
 11        1985          11          1684.3        1698.89         99.141
 12        1986          12          1738.1        1740.07         99.887
 13        1987          13          1801.1        1781.25        101.114
 14        1988          14          1855.4        1822.43        101.809
```

The original annual series contains trend, cyclical, and irregular factors.
The percentage of trend series contains only cyclical and irregular factors,
since the division by trend eliminates that factor.

c. See figure: Straight-line trend fitted to GNP–services in 1982 constant
dollars in the United States, 1975–1988.

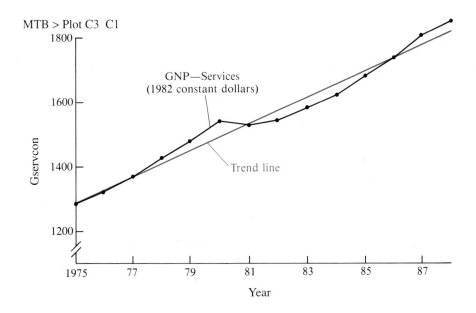

MTB > Plot C3 C1

4.

ROW	YEAR	TIME	PURCHASES
1	1980	1	246.9
2	1981	2	259.6
3	1982	3	272.7
4	1983	4	275.1
5	1984	5	290.8
6	1985	6	326.0
7	1986	7	333.4
8	1987	8	339.0
9	1988	9	326.1

```
MTB > REGRESS C3 1 C2;
SUBC> PREDICT C4.

THE REGRESSION EQUATION IS
PURCHASE = 236 + 12.1 TIME
```

PREDICTOR	COEF	STDEV	T-RATIO	P
CONSTANT	236.014	8.581	27.50	0.000
TIME	12.122	1.525	7.95	0.000

```
S = 11.81    R-SQ = 90.0%    R-SQ(ADJ) = 88.6%
```

PROJECTED FIGURE	
YEAR	(BILLIONS OF 1982 DOLLARS)
1990	357.23
1991	369.35
1992	381.47
1993	393.60
1994	405.72
1995	417.84

6. a. The labor force of this county is increasing by decreasing amounts and at a decreasing percentage rate.

b. $\hat{Y}_t = 202.53 + 6.85(16) - 0.52(16)^2 = 179.01$ (thousands) The deviation of 25,990 from the computed trend figure of 179,010 does not necessarily imply a poor fit. Possibly, the fit is adequate and the deviation represents an unusually strong cyclical influence.

8. a.

```
MTB > PRINT C1-C3

ROW          YEAR          TIME          EXPENSE
  1          1975             1           235.2
  2          1976             2           254.9
  3          1977             3           273.2
  4          1978             4           301.3
  5          1979             5           327.7
  6          1980             6           363.2
  7          1981             7           391.4
  8          1982             8           414.3
  9          1983             9           440.2
 10          1984            10           475.9
 11          1985            11           516.7
 12          1986            12           561.9
 13          1987            13           602.8
 14          1988            14           646.6

MTB > NAME C4 'TREND1'  NAME C5 'TREND2'
MTB > REGRESS C3 1 C2 C6 C4

THE REGRESSION EQUATION IS
EXPENSE = 179 + 31.4 TIME

PREDICTOR          COEF          STDEV          T-RATIO          P
CONSTANT        179.344          9.023           19.88        0.000
TIME             31.376          1.060           29.61        0.000

S = 15.98    R-SQ = 98.6%    R-SQ(ADJ) = 98.5%
```

b.

```
MTB > NAME C7 'TIMETIME'
MTB > LET C7=C2**2
MTB > REGRESS C3 2 C2 C7 C6 C5

THE REGRESSION EQUATION IS
EXPENSE = 218 + 16.8 TIME + 0.971 TIMETIME

PREDICTOR          COEF          STDEV          T-RATIO          P
CONSTANT        218.193          5.014           43.52        0.000
TIME             16.808          1.538           10.93        0.000
TIMETIME         0.97122         0.09972          9.74        0.000

S = 5.381    R-SQ = 99.9%    R-SQ(ADJ) = 99.8%
```

c.

```
MTB > NAME C4 '%TREND1' C5 '%TREND2'

MTB > LET C4 = 100*(C3/C4)

MTB > LET C5 = 100*(C3/C5)

MTB > PRINT C1 C2 C3 C4 C5
```

ROW	YEAR	TIME	EXPENSE	%TREND1	%TREND2
1	1975	1	235.2	111.617	99.673
2	1976	2	254.9	105.289	99.690
3	1977	3	273.2	99.901	98.501
4	1978	4	301.3	98.836	100.112
5	1979	5	327.7	97.465	100.364
6	1980	6	363.2	98.803	102.598
7	1981	7	391.4	98.101	102.077
8	1982	8	414.3	96.270	99.876
9	1983	9	440.2	95.337	98.230
10	1984	10	475.9	96.511	98.450
11	1985	11	516.7	98.517	99.252
12	1986	12	561.9	101.087	100.386
13	1987	13	602.8	102.651	100.328
14	1988	14	646.6	104.525	100.426

The second-degree trend line appears to lie closer to the actual data.

Exercises 11.5

2. a.

Quarter	Y/SI (thousands of gallons)
Spring	28,333
Summer	25,500
Fall	24,500
Winter	25,200

The deseasonalized sales volumes are quite close to 25,000,000 gallons except for the spring quarter of 1990. Cyclical and random fluctuations are quite large for the spring quarter of 1990.

b. Quarterly sales volume for 1991 = 1.15 × 25,000,000 gallons per quarter = 28,750,000 gallons per quarter.

Quarter	$\hat{Y}_t \times SI$ Sales Forecast (gallons)
Spring	25,875,000
Summer	35,937,500
Fall	30,187,500
Winter	23,000,000

4. a. (1) June
(2) May
b. April
c. In the multiplicative model of time-series analysis, the original monthly figures consist of the product of the trend level, the cyclical relative (combined effect of cyclical and irregular factor), and the seasonal index. In symbols,

$$Y = \hat{Y}_t \left(\frac{Y/SI}{\hat{Y}_t} \right)(SI)$$

For March 1990:
$$\hat{Y}_t = 641.04 + 0.96(63) = 701.5$$

$$\frac{Y}{SI} = \frac{570}{0.86} = 662.8$$

$$\frac{Y/SI}{\hat{Y}_t} = \frac{662.8}{701.5} = 0.945$$

$$SI = 0.86$$

The equation $570 = (701.5)(0.945)(0.86)$ represents the breakdown of the March 1990 figure into its components.

　　If an additive model were used—that is, $Y = T + (C + I) + S$—then the breakdown would be as follows:
$$\hat{Y}_t = 701.5$$
Since $Y/SI = 662.8$, the effect of the seasonal factor is $570.0 - 662.8 = -92.8$. The effects of the cyclical and irregular factors are measured by the absolute cyclical residual
$$Y/SI - \hat{Y}_t = 662.8 - 701.5 = -38.7$$
In summary, the breakdown of components of the March 1990 figure using the additive model is as follows:

Trend	701.5
Seasonal	− 92.8
	608.7
Cyclical and Irregular	− 38.7
Original	570.0

6. a. $\dfrac{Y}{SI} = \dfrac{\$2.5}{1.10} = \$2.27$ million

$$\hat{Y}_t = 1.85 + 0.16(3) + 0.03(3)^2$$
$$= \$2.60 \text{ million}$$

Relative cyclical residual $= \dfrac{2.27 - 2.6}{2.60}(100) = -12.7\%$

b. A relative cyclical residual of -12.7% indicates that the seasonally adjusted figure is 12.7% below the trend figure because of cyclical and irregular factors.

c. For December 1991:
$$\hat{Y}_t = 1.85 + 0.16(4) + 0.03(4)^2$$
$$= 2.97$$
$$Y = (2.97)(1.10) = \$3.27 \text{ million}$$

8. a. (3)   b. (2)   c. (4)   d. (2)   e. (1)   f. (1)   g. (3)

Exercises 11.6

2.

Month	Actual Demand	Next Forecast $w = 0.1$	Next Forecast $w = 0.6$
1	$10 = A_1$	$10 = F_2$	$10 = F_2$
2	$10 = A_2$	$10 = F_3$	$10 = F_3$
3	$10 = A_3$	$10 = F_4$	$10 = F_4$
4	$15 = A_4$	$10.5 = F_5$	$13 = F_5$
5	$15 = A_5$	$10.95 = F_6$	$14.2 = F_6$
6	$15 = A_6$	$11.355 = F_7$	$14.68 = F_7$
7	$15 = A_7$	$11.7195 = F_8$	$14.872 = F_8$
8	$15 = A_8$	$12.04755 = F_9$	$14.9488 = F_9$
9	$15 = A_9$	$12.342795 = F_{10}$	$14.97952 = F_{10}$
10	$15 = A_{10}$	$12.608516 = F_{11}$	$14.991808 = F_{11}$

4.

Year	Actual Corporate Bonds Interest Rate	Next Forecast $w = 0.5$
1	$4.35 = A_1$	$4.350 = F_2$
2	$4.33 = A_2$	$4.340 = F_3$
3	$4.26 = A_3$	$4.300 = F_4$
4	$4.40 = A_4$	$4.350 = F_5$
5	$4.49 = A_5$	$4.420 = F_6$
6	$5.13 = A_6$	$4.775 = F_7$
7	$5.51 = A_7$	$5.143 = F_8$
8	$6.18 = A_8$	$5.661 = F_9$
9	$7.03 = A_9$	$6.346 = F_{10}$
10	$8.04 = A_{10}$	$7.193 = F_{11}$
11	$7.39 = A_{11}$	$7.291 = F_{12}$
12	$7.21 = A_{12}$	$7.251 = F_{13}$
13	$7.44 = A_{13}$	$7.345 = F_{14}$
14	$8.57 = A_{14}$	$7.958 = F_{15}$
15	$8.83 = A_{15}$	$8.394 = F_{16}$
16	$8.43 = A_{16}$	$8.412 = F_{17}$
17	$8.02 = A_{17}$	$8.216 = F_{18}$
18	$8.73 = A_{18}$	$8.473 = F_{19}$
19	$9.63 = A_{19}$	$9.052 = F_{20}$

The simple exponential smoothing procedure used here is appropriate for series that are relatively stable and do not have pronounced trends.

The corporate bonds interest rate series has a significant upward trend movement; therefore, the exponential smoothing forecasts generally underestimate the actual yearly values.

CHAPTER 12

Exercises 12.2

2. a. For 1989 on a 1988 base:

$$\frac{\Sigma P_{89}}{\Sigma P_{88}} \cdot 100 = \frac{\$265}{\$230} \times 100 = 115.22$$

For 1990 on a 1988 base:

$$\frac{\Sigma P_{90}}{\Sigma P_{88}} \cdot 100 = \frac{\$270}{\$230} \times 100 = 117.39$$

This index is unduly influenced by high-priced stocks, which may be relatively unimportant in the consumption or purchasing pattern of the group to which the index pertains. Moreover, in some situations, the calculation of the index may be arbitrary depending on the quoted units for which the prices are stated.

b. For 1989 on a 1988 base:

$$\frac{\Sigma P_{89}Q_{88}}{\Sigma P_{88}Q_{88}} \cdot 100 = \frac{\$52,100}{\$46,300} \times 100 = 112.53$$

For 1990 on a 1988 base:

$$\frac{\Sigma P_{90}Q_{88}}{\Sigma P_{88}Q_{88}} \cdot 100 = \frac{\$54,550}{\$46,300} \times 100 = 117.82$$

By keeping quantities fixed as of the base period, the Laspeyres index assumes a frozen consumption or purchasing pattern.

c. For 1989 on a 1988 base:

$$\frac{\Sigma P_{89}Q_{89}}{\Sigma P_{88}Q_{89}} \cdot 100 = \frac{\$66,150}{\$56,300} \times 100 = 117.50$$

For 1990 on a 1988 base:

$$\frac{\Sigma P_{90}Q_{90}}{\Sigma P_{88}Q_{90}} \cdot 100 = \frac{\$48,850}{\$39,825} \times 100 = 122.66$$

The use of current year weights makes year-to-year comparisons of price changes impossible. Another practical disadvantage is the necessity of obtaining a new set of weights each period.

Exercises 12.3

2. a. For October 11 on an October 4 base:

$$\frac{\Sigma\left(\dfrac{P_{11}}{P_4}\right)(P_4 Q_4)}{\Sigma(P_4 Q_4)}(100) = \frac{\Sigma P_{11}Q_4}{\Sigma P_4 Q_4}(100)$$

$$= \frac{\$103,230}{\$99,700}(100) = 103.5$$

For October 18 on an October 4 base:

$$\frac{\Sigma\left(\dfrac{P_{18}}{P_4}\right)(P_4 Q_4)}{\Sigma(P_4 Q_4)}(100) = \frac{\Sigma P_{18}Q_4}{\Sigma P_4 Q_4}(100)$$

$$= \frac{98,720}{\$99,700}(100) = 99.0$$

For October 25 on an October 4 base:

$$\frac{\Sigma\left(\dfrac{P_{25}}{P_4}\right)(P_4 Q_4)}{\Sigma(P_4 Q_4)}(100) = \frac{\Sigma P_{25}Q_4}{\Sigma P_4 Q_4}(100)$$

$$= \frac{\$101,975}{\$99,700}(100) = 102.3$$

b. Same numerical results as in part (a).

Exercises 12.6

2.

Year	(1) Industry Sales (in 1985 constant thousands of dollars)	(2) Company Sales (in 1985 constant thousands of dollars)	(3) Market Share (%) Col 2 ÷ Col 1 × 100
1989	$11,025	$2,205	20%
1990	16,667	3,666	22
1991	19,644	4,910	25

4. Deflated GNP for 1987:

$$\frac{\$3,847.0 \text{ billion}}{1.177} = \$3,268.5 \text{ billion}$$

Deflated GNP for 1986:

$$\frac{\$3,721.7 \text{ billion}}{1.139} = \$3,267.5 \text{ billion}$$

"Real" GNP increased $\dfrac{3,268.5}{3,267.5} - 1 = 0.0\%$

CHAPTER 13

Exercises 13.2

2. H_0: $p = 0.50$
  H_1: $p \neq 0.50$
  $\alpha = 0.05$
  $\sigma_{\bar{p}} = \sqrt{(0.50)(0.50)/382} = 0.026$
  $\bar{p} = 260/382 = 0.68$
  $z = \dfrac{0.68 - 0.50}{0.026} = 6.92$

Since $6.92 > 1.96$, we reject the null hypothesis that the proportion of boxes with an excess of 100 crayons is 0.50.

4. H_0: $p \leqslant 0.50$
  H_1: $p > 0.50$
  $\alpha = 0.02$
  $\sigma_{\bar{p}} = \sqrt{(0.50)(0.50)/412} = 0.025$
  $\bar{p} = 252/412 = 0.61$
  $z = (0.61 - 0.50)/0.025 = 4.4$

Since $4.4 > 2.05$, we reject the null hypothesis that $p \leqslant 0.50$. Hence, we conclude that more than 50% of U.S. economists felt that there had been an increase in the influence of Federal Reserve stabilization policies over the past decade.

6. H_0: $p = 0.50$
  H_1: $p \neq 0.50$
  $\alpha = 0.05$

$$\sigma_{\bar{p}} = \sqrt{\frac{(0.50)(0.50)}{238}} = 0.0324$$

$$\bar{p} = \frac{135}{238} = 0.567$$

$$z = \frac{0.567 - 0.500}{0.0324} = 2.07$$

Since $z = 2.07 > 1.96$, we reject the hypothesis that equal proportions of students scored above and below 600.

Exercises 13.3

2. Complete table of values:

Sales Representative	X_1	X_2	$d = X_2 - X_1$	Rank of $\lvert d \rvert$	Signed Rank Rank (+)	Signed Rank Rank (−)
A	22	25	+3	4	4	
B	10	14	+4	5	5	
C	15	17	+2	3	3	
D	21	31	+10	9	9	
E	28	33	+5	6	6	
F	27	27	0			
G	33	32	−1	1.5		1.5
H	18	19	+1	1.5	1.5	
I	17	28	+11	10	10	
J	16	22	+6	7	7	
K	15	22	+7	8	8	
					53.5	1.5 = T

$n = 10$ $T = 1.5$ $T_{0.01} = 5$.
Since $T = 1.5 < T_{0.01} = 5$, we reject the null hypothesis of identical population distributions. Therefore, we conclude that the promotional campaign was accompanied by an increase in number of sales.

Exercises 13.4

In the solutions to section 13.4, the data are ranked from highest to lowest, with rank 1 assigned to the highest value.

2. $R_1 = 110$, $R_2 = 190$, $n_1 = 12$, $n_2 = 12$, $U = 112$, $\mu_U = 72$, $\alpha = 0.05$
$\sigma_U = \sqrt{(12)(12)(25)/12} = 17.32$
$z = (112 - 72)/17.32 = 2.31$
Since $2.31 > 1.96$, we reject the hypothesis of no difference between the average research and development expenditure levels of the two industries.

4. $R_1 = 143$, $R_2 = 263$, $n_1 = 13$, $n_2 = 15$, $U = 143$, $\mu_U = 97.5$, $\alpha = 0.05$
$\sigma_U = \sqrt{(13)(15)(29)/12} = 21.7$
$z = (143 - 97.5)/21.7 = 2.10$
Since $2.10 > 1.96$, we reject the hypothesis of no difference between the true average times for the two different methods.

6. $R_1 = 168$, $R_2 = 238$, $n_1 = 13$, $n_2 = 15$, $U = 118$, $\mu_U = 97.5$, $\alpha = 0.05$.
$\sigma_U = \sqrt{(13)(15)(29)/12} = 21.7$
$z = (118 - 97.5)/21.7 = 0.94$

Since $0.94 < 1.96$, we retain the hypothesis of no difference between the true average times for the two different methods.

2. $n_1 = 24$, $n_2 = 26$, $r = 20$

$$\mu_r = \frac{2(24)(26)}{24 + 26} + 1 = 25.96$$

$$\sigma_r = \sqrt{\frac{2(24)(26)[2(24)(26) - 24 - 26]}{(24 + 26)^2(24 + 26 - 1)}} = 3.49$$

$$z = \frac{20 - 25.96}{3.49} = -1.71$$

Since $|-1.71| < 1.96$ and $|-1.71| < 2.58$, we retain the hypothesis of randomness of runs above and below the mean 4.5 at both the 0.05 and 0.01 levels of significance.

4. $n_1 = 15$
$n_2 = 11$
$r = 16$

$$\mu_r = \frac{2(15)(11)}{15 + 11} + 1 = 13.69$$

$$\sigma_r = \sqrt{\frac{[2(15)(11)][(2)(15)(11) - 15 - 11]}{(15 + 11)^2(15 + 11 - 1)}} = 2.44$$

$$z = \frac{16 - 13.69}{2.44} = 0.95$$

Since $0.95 < 1.96$, we retain the null hypothesis of randomness of runs above and below the mean.

2. a.

Number of accounts collected:	20	25	26	30	31	32	35	36	38
Rank:	1	2	3	4	5	6	7	8	9
Number of accounts collected:	39	40	42	45	46	48	49	50	52
Rank:	10	11	12	13	14	15	16	17	18

$$v = 3 - 1 = 2 \quad \chi^2_{0.01} = 9.210$$

$$K = \frac{12}{18(18 + 1)}\left(\frac{36^2}{6} + \frac{62^2}{6} + \frac{73^2}{6}\right) - 3(19) = 4.222$$

Since $4.222 < 9.210$, we cannot reject the null hypothesis of no difference in effectiveness among the three collectors at the 1% level of significance.

b.

Analysis of Variance Table

(1) Source of Variation	(2) Sum of Squares	(3) Degrees of Freedom	(4) Mean Squares
Between collectors	366.33	2	183.17
Error	1,127.67	15	75.18
	1,494.00	17	

$F_{0.01}(2, 15) = 6.36$

Since $2.44 < 6.36$, we cannot reject the null hypothesis of no difference in average numbers of accounts removed from delinquency by the three collectors.

4. a.

Number of units sold:	30	36	38	43	47	52	55	59	...	82
Rank:	1	2	3	4	5	6	7	8	...	18
Number of units sold:	61	64	66	67	69	73	75	77	...	80
Rank:	9	10	11	12	13	14	15	16	...	17

$v = 3 - 1 = 2 \quad \chi^2_{0.01} = 9.210$

$$K = \frac{12}{18(18 + 1)}\left(\frac{31^2}{6} + \frac{68^2}{6} + \frac{72^2}{6}\right) - 3(18 + 1) = 5.977$$

Since $5.977 < 9.210$, we cannot reject the null hypothesis of no difference in effectiveness among the three advertisement designs at the 1% level of significance.

Analysis of Variance Table

(1) Source of Variation	(2) Sum of Squares	(3) Degrees of Freedom	(4) Mean Square
Between columns	1425	2	712.50
Between rows	2831	15	188.73
Total	4256	17	

$$F(2, 15) = \frac{712.50}{188.73} = 3.78$$

$F_{0.01}(2, 15) = 6.36$

Since $3.78 < 6.36$, we cannot reject the null hypothesis of no difference in average numbers of sales resulting from the three advertisements.

Exercises 13.7

2. $r_r = 1 - \dfrac{6(116)}{10(100 - 1)} = 0.30$

4. $r_r = 1 - \dfrac{6(62)}{12(12^2 - 1)} = 0.78$

$t = \dfrac{0.78}{\sqrt{(1 - 0.61)/(12 - 2)}} = 3.95$

Since $3.95 > 2.228$ (two-tailed test), we reject the null hypothesis of no rank correlation.

CHAPTER 14

Exercises 14.4

2. $P(\text{wooden racket}) = 0.50$; $P(\text{metal racket}) = 0.375$; $P(\text{fiberglass racket}) = 0.125$

4. Expected profit (playing game) $= (1/13)(\$10,000) + (12/13)(-\$800)$
$$= \$30.77$$
Therefore, on an expected monetary value basis, it pays to play the game.

6. The cost per year is $4,900. The expected gain from the information is $(0.05)(\$80,000) = \$4,000$. Since cost exceeds the expected gain do not include newsletters.

8. $0.12(x) + (-0.03)(1.00 - x) = 0$
$$x = 20\%$$
Therefore, when the chance of the bill passing is 20%, the candidate will be statistically indifferent.

10. a.   Opportunity Loss Table

	Advertising Costs		
State of Nature	High	Average	Low
Severe recession	10	6	0
Mild recession	6	2	0
Weak recovery	0	1	2
Strong recovery	0	3	8

b. $\text{EOL(high)} = 10(0.20) + 6(0.25) + 0(0.55) = 3.5$
$\text{EOL(average)} = 6(0.20) + 2(0.25) + 1(0.40) + 3(0.15)$
$$= 2.55$$
$\text{EOL(low)} = 0(0.20) + 0(0.25) + 2(0.40) + 8(0.15)$
$$= 2.00$$
The manufacturer should choose the low advertising cost budget if the company wishes to minimize the expected opportunity loss.

12. Return from 14% commercial paper: $63,000
$$[P(\text{good business condition})(\$112,500)]$$
$$+ [1 - P(\text{good business condition})]$$
$$\cdot (\$36,000) = \$63,000$$
$$P(\text{good business condition}) = 0.353$$

14.

Profit Table

Number Demanded	Probability	0	1	2	3	4	5
				Number of Television Sets in Stock			
0	0.10	0	−100	−200	−300	−400	−500
1	0.10	0	250	150	50	−50	−150
2	0.15	0	250	500	400	300	200
3	0.35	0	250	500	750	650	550
4	0.20	0	250	500	750	1,000	900
5	0.10	0	250	500	750	1,000	1,250
Expected values ($)		0	215	395	522.50	527.50	462.50

The optimal stock level is 4 television sets.
The expected profit is $527.50.

Exercises 14.6

2. The expected value of perfect information is the expected opportunity loss of the optimal act under uncertainty.

4.   Opportunity Loss Table

Demand	Work	Do not Work
Good	$0	$100
Fair	0	50
Poor	0	0

6. Money figures are in units of $1,000.
 a. Expected profit with perfect information
   $= (0.5)(20) + (0.3)(14) + (0.2)(10) = 16.2$
 b. Expected profits:

$$A_1 = (0.5)(20) + (0.3)(10) + (0.2)(4) = 13.8$$

$$A_2 = (0.5)(17) + (0.3)(14) + (0.2)(7) = 14.1$$

$$A_3 = (0.5)(18) + (0.3)(14) + (0.2)(5) = 14.2$$

$$A_4 = (0.5)(13) + (0.3)(12) + (0.2)(10) = 12.1$$

 The optimal act is A_3. Hence, the expected profit under uncertainty is 14.2.
 c. EVPI $= 16.2 − 14.2 = 2.0$

8. Under certainty:
 Expected return $= (0.60)(0) + (0.40)($3,500)$
   $= $1,400$
 Under uncertainty:
 Expected return(take deductions)
   $= (0.60)(−$2,000) + (0.40)(+$3,500) = 200
 Expected return(do not take deductions)
   $= (0.60)(0) + (0.40)(0) = 0$

EVPI = $1,400 − $200 = $1,200
Alternatively,
EOL(take deductions) = (0.60)($2,000) + (0.40)(0)
$$= \$1,200$$
EOL(do not take deductions)
$$= (0.60)(0) + (0.40)(\$3,500) = \$1,400$$
EVPI = $1,200

10. a. Expected gain(buy) = (0.3)($20,000) + (0.3)($10,000) + (0.4)(−$2,000)
$$= \$8,200$$
Since the expected gain is positive, you should buy the computer.
b. Expected gain with perfect information
$$= (0.3)(\$20,000) + (0.3)(\$10,000) + (0.4)(\$0) = \$9,000$$
Therefore, EVPI = $9,000 − $8,200 = $800

12. a. In $10,000 units,
$$EOL(A_1) = (0.4)(2) + (0.3)(0) + (0.2)(3) + (0.1)(3)$$
$$= 1.7$$
$$EOL(A_2) = (0.4)(2) + (0.3)(1) + (0.2)(0) + (0.1)(2)$$
$$= 1.3$$
$$EOL(A_3) = (0.4)(3) + (0.3)(2) + (0.2)(1) + (0.1)(0)$$
$$= 2.0$$
$$EOL(A_4) = (0.4)(0) + (0.3)(4) + (0.2)(3) + (0.1)(3)$$
$$= 2.1$$
b. EVPI = 1.3
c. Since A_2 has the lowest EOL, the optimal decision is A_2.

Exercises 14.8

2. a. Expected utility(buy the stock)
$$= (0.4)(45) + (0.6)(-15) = 9 \text{ units of utility}$$
b. Expected utility(buy government securities)
$$= 10 \text{ units of utility}$$
Therefore, you should buy the government securities.

4. a. Expected gain(100% interest)
$$= (0.8)(-\$1) + (0.1)(\$3) + (0.1)(\$8) = \$0.3 \text{ million}$$
Expected gain(50% interest)
$$= (0.8)(-\$0.5) + (0.1)(\$1.5) + (0.1)(\$4) = \$0.15 \text{ million}$$
Expected gain(don't drill) = 0
The best act is to drill with a 100% interest.
b. Expected utility(100% interest)
$$= (0.8)(-30) + (0.1)(90) + (0.1)(240) = 9 \text{ units of utility}$$
Expected utility(50% interest)
$$= (0.8)(-15) + (0.1)(45) + (0.1)(120) = 4.5 \text{ units of utility}$$
The best act is to drill with a 100% interest.
c. The fact that the decision using an expected utility criterion is the same
as that using an expected monetary value criterion is no basis to make

conclusions on risk aversion. However, in Drillwell's utility function, note that the utility figures are always 30 times the monetary figures, that is, monetary values and utilities are linearly related. Hence, we conclude that Drillwell's management is *risk neutral*.

6. a. Expected monetary return
$$= (0.5)(\$200,000) + (0.5)(-\$100,000) = \$50,000$$
He would be willing to pay at most \$50,000.

b. Expected utility of final wealth
$$= (0.5)(\sqrt{300,000}) + (0.5)(\sqrt{0}) = 273.9 \text{ units of utility}$$
$$U(\$100,000) = \sqrt{100,000} = 316.2$$
Since $316.2 > 273.9$, the individual will not pay anything to invest.

CHAPTER 15

Exercises 15.2

2.

State of Nature	$P_0(\theta)$	$P(X\|\theta)$	$P_0(\theta)P(X\|\theta)$	$P_1(\theta\|X)$
θ_1	0.10	0.20	0.020	0.037
θ_2	0.20	0.10	0.020	0.037
θ_3	0.30	0.50	0.150	0.280
θ_4	0.25	0.90	0.225	0.421
θ_5	0.15	0.80	0.120	0.224
			0.535	1.000

4. No, because after sampling the best act remains the same. The actual value of the sample information is zero.

6. No. Although both prior and sample information indicate that θ_2 is true, the relative sizes of payoffs determine whether act A_2 is better than A_1. The expected payoff of act A_1 can conceivable exceed that of act A_2.

8. Prior expected profit $= (-\$15,000)(0.20) + (-\$5,000)(0.35)$
$$+ (\$28,000)(0.45)$$
$$= \$7,850$$

Before considering the additional information, her best action is to wait six months.

State of Nature	$P_0(\theta)$	$P(X\|\theta)$	$P_0(\theta)P(X\|\theta)$	$P_1(\theta\|X)$
θ_1: Higher interest rates	0.20	0.10	0.02	0.044
θ_2: No change	0.35	0.20	0.07	0.156
θ_3: Lower interest rates	0.45	0.80	0.36	0.800
			0.45	1.000

Posterior expected profit $= (-\$15,000)(0.044) + (-\$5,000)(0.156)$
$$+ (\$28,000)(0.800)$$
$$= \$20,960$$

After considering the additional information, her best action is still to wait six months.

Exercises 15.3

2.

μ	$P_0(\mu)$	$P(X = 7|\mu)$	$P_0(\mu)P(X = 7|\mu)$	$P_1(\mu)$
6	0.7	0.138	0.0966	0.68
7	0.3	0.149	0.0447	0.32
			0.1413	1.00

4. a. $P(0.05|1) = 0.45$ $P(0.10|1) = 0.42$
$P(0.15|1) = 0.11$ $P(0.20|1) = 0.02$
b. $P(0.05|3) = 0.11$ $P(0.10|3) = 0.47$
$P(0.15|3) = 0.30$ $P(0.20|3) = 0.12$

6. a.

Complete Table of Values

Event p	Prior Probability $P_0(p)$	Conditional Probability $P(X = 4|n = 20, p)$	Joint Probability $P_0(p)P(X = 4|n = 20, p)$	Posterior Probability $P_1(p)$
0.10	0.20	0.0898	0.01796	0.1174
0.20	0.35	0.2182	0.07637	0.4991
0.30	0.45	0.1304	0.05868	0.3835
			0.15301	1.0000

b.

Complete Table of Values

Event p	Joint Probability $P_0(p)$	Conditional Probability $P(X = 20|n = 100, p)$	Joint Probability $P_0(p)P(X = 20|n = 100, p)$	Posterior Probability $P_1(p)$
0.10	0.20	0.0012	0.00024	0.00625
0.20	0.35	0.0993	0.03476	0.90474
0.30	0.45	0.0076	0.00342	0.08901
			0.03842	1.00000

CHAPTER 16

Exercises 16.2

2.

State of Nature	$P(\theta)$	$P(X|\theta)$	$P(Y|\theta)$	$P(\theta)P(X|\theta)$	$P(\theta)P(Y|\theta)$	$P(\theta|X)$	$P(\theta|Y)$
θ_1	0.3	0.6	0.4	0.18	0.12	0.46	0.20
θ_2	0.7	0.3	0.7	0.21	0.49	0.54	0.80
				0.39	0.61	1.00	1.00

4. a. and b. See figure.

Decision tree
diagram for
exercises 4

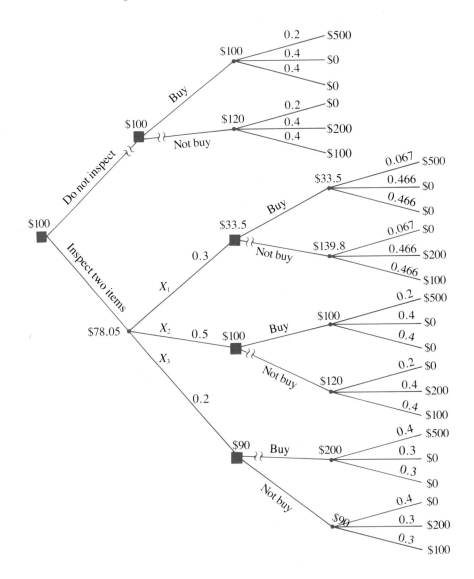

c. EVPI(before) = $100
d. EVSI = $100 − $78.05 = $21.95

6. In this exercise, money figures are units of $10,000.
  a. (1) EOL(sell) = (0.60)(10) = 6.0
     EOL(do not sell) = (0.40)(10) = 4.0
     Therefore, the optimal act is not to sell.

(2)

State of Nature	$P(\theta)$	$P(X_1\|\theta)$	$P(\theta)P(X_1\|\theta)$	$P(\theta\|X_1)$
θ_1	0.60	0.75	0.45	0.65
θ_2	0.40	0.60	0.24	0.35
			0.69	1.00

EOL(sell) = (0.65)(10) = 6.5
EOL(do not sell) = (0.35)(10) = 3.5
Therefore, the optimal act is not to sell.

(3)

State of Nature	$P(\theta)$	$P(X_2\|\theta)$	$P(\theta)P(X_2\|\theta)$	$P(\theta\|X_2)$
θ_1	0.60	0.25	0.15	0.48
θ_2	0.40	0.40	0.16	0.52
			0.31	1.00

EOL(sell) = (0.48)(10) = 4.8
EOL(do not sell) = (0.52)(10) = 5.2
Therefore, the optimal act is to sell.
(4) $P(X_1) = 0.69$
   $P(X_2) = 0.31$
(5) EVSI = $4.0 - [(0.69)(3.5) + (0.31)(4.8)] = 0.097$
b. ENGS = $970 - $500 = $470

8. EVSI = $8 - 2.80 = 5.20$ (tens of thousands of dollars)
EVSI = $52,000
See figure.

10. a. (1) EOL(build) = (0.6)(12) = 7.2
   EOL(do not build) = (0.4)(17) = 6.8
   Therefore, the optimal act is not to build.
   (2) $P(\theta_1|X_1) = 0.471$; $P(\theta_2|X_1) = 0.529$
   EOL(build) = (0.529)(12) = 6.348
   EOL(do not build) = (0.471)(17) = 8.007
   Therefore, the optimal act is to build.
   (3) $P(\theta_1|X_2) = 0.327$; $P(\theta_2|X_2) = 0.673$
   EOL(build) = (0.673)(12) = 8.075
   EOL(do not build) = (0.327)(17) = 5.559
   Therefore, the optimal act is not to build.
   (4) $P(X_1) = 0.51$
      $P(X_2) = 0.49$
   (5) EVSI = $6.8 - [(0.51)(6.348) + (0.49)(5.559)]$
      = 0.84 (ten thousands of dollars)
      = $8,400
b. Since the EVSI is $8,400, we should be willing to spend up to that amount for the survey.

Decision tree
diagram for
exercise 8

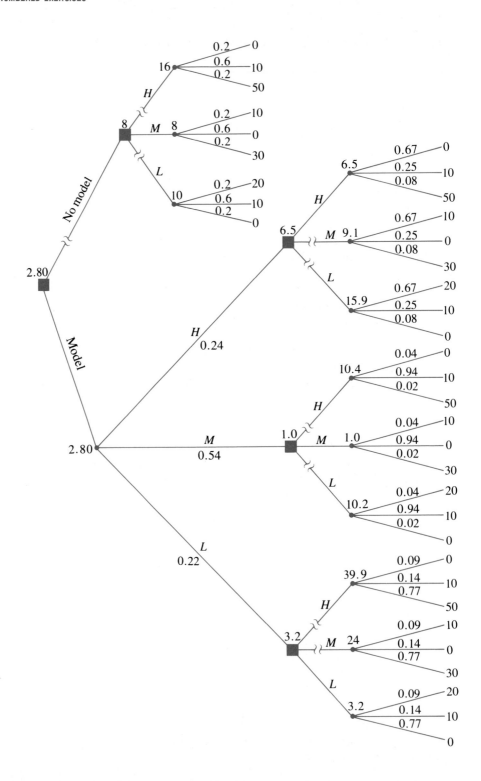

12. In this exercise, money figures are in tens of thousands of dollars.

State of Nature	$P(\theta)$	$P(X_1|\theta)$	$P(X_2|\theta)$	$P(\theta)P(X_1|\theta)$	$P(\theta)P(X_2|\theta)$	$P(\theta|X_1)$	$P(\theta|X_2)$
θ_1	0.70	0.75	0.25	0.525	0.175	0.745	0.593
θ_2	0.30	0.60	0.40	0.180	0.120	0.255	0.407
				$P(X_1) = \overline{0.705}$	$P(X_2) = \overline{0.295}$	$\overline{1.000}$	$\overline{1.000}$

a. (1) Without a study:
   EOL(sell) = (0.70)(10) = 7.0
   EOL(do not sell) = (0.30)(10) = 3.0
   Therefore, the optimal act is not to sell.
   (2) With information X_1 from the study:
   EOL(sell) = (0.745)(10) = 7.45
   EOL(do not sell) = (0.255)(10) = 2.55
   Therefore, the optimal act is not to sell.
   (3) With information X_2 from the study:
   EOL(sell) = (0.593)(10) = 5.93
   EOL(do not sell) = (0.407)(10) = 4.07
   Therefore, the optimal act is not to sell.
   (4) EVSI = 3.0 − [(0.705)(2.55) + (0.295)(4.07)]
         = $0
   (5) ENGS = $0 − $500 = −$500
b. In exercise 6, ENGS = $470. In this exercise, EVSI is equal to zero because the study always leads to the decision "do not sell," regardless of the study outcome. Since "do not sell" is the better action without a study and the study results cannot change this decision, sample information has no value.

Exercises 16.4

2. Sample

Outcome	s_1	s_2	s_3	s_4	s_5	s_6	s_7	s_8
X	a_1	a_1	a_1	a_1	a_2	a_2	a_2	a_2
Y	a_1	a_1	a_2	a_2	a_2	a_2	a_1	a_1
Z	a_1	a_2	a_1	a_2	a_2	a_1	a_2	a_1

a_1 denotes "manufacture Crunchie Munchies"
a_2 denotes "do not manufacture Crunchie Munchies"
s_2 is the optimal strategy

4.

Sample Outcome	s_1	s_2	s_3	s_4
Less than 25 people	a_1	a_1	a_2	a_2
25 people or more	a_2	a_1	a_2	a_1

Strategy $s_1(a_1, a_2)$

State of Nature	(thousands of dollars) a_1	a_2	Probability of Action $s_1(a_1, a_2)$ a_1	a_2	Conditional Expected Payoff
Less than 25 people	-3	0	0.1	0.8	-0.3
25 people or more	5	0	0.9	0.2	4.5

Strategy $s_4(a_2, a_1)$

State of Nature	(thousands of dollars) a_1	a_2	Probability of Action $s_4(a_2, a_1)$ a_2	a_1	Conditional Expected Payoff
Less than 25 people	-3	0	0.1	0.8	-2.4
25 people or more	5	0	0.9	0.2	1.0

Expected Payoffs of the Two Strategies
(Thousands of Dollars):

Strategy	Expected Payoffs
s_1	$(0.30)(-0.3) + (0.70)(4.5) = \quad 3.06$
s_4	$(0.30)(-2.4) + (0.70)(1.0) = -0.02$

Strategy s_1 is optimal.

Exercises 16.6

2. ENGS $= \$200 - \$167.14 - \$9.00 = \$23.86$
Decision Rule
If 0 defective, then ship
If 1 defective, then ship
If 2 defective, then scrap
If 3 defective, then scrap
See figure.

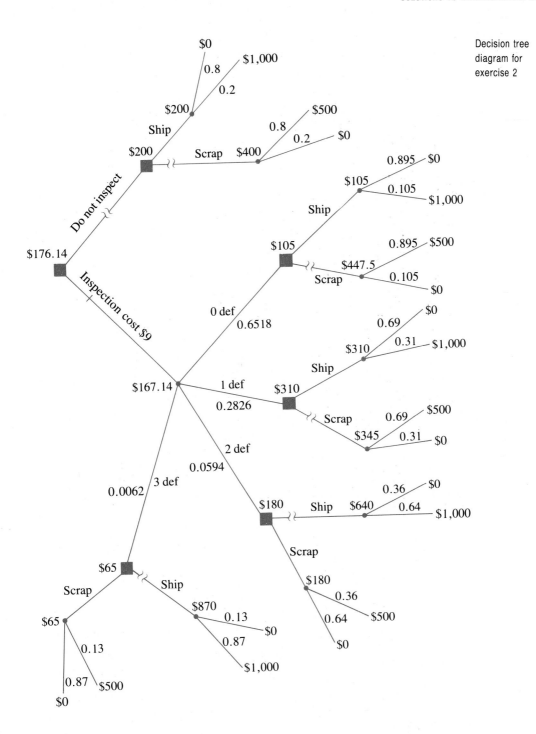

Decision tree
diagram for
exercise 2

CHAPTER 17

Exercises 17.2

2. a. $P(p = 0.20|\bar{p} \leqslant c_1) = 0.40/0.52 = 0.769$
$P(p = 0.20|c_1 < \bar{p} \leqslant c_2) = 0.06/0.32 = 0.187$
$P(p = 0.20|\bar{p} > c_2) = 0.02/0.16 = 0.125$
$P(p = 0.25|\bar{p} \leqslant c_1) = 0.08/0.52 = 0.154$
$P(p = 0.25|c_1 < \bar{p} \leqslant c_2) = 0.22/0.32 = 0.688$
$P(p = 0.25|\bar{p} > c_2) = 0.03/0.16 = 0.188$
$P(p = 0.30|\bar{p} \leqslant c_1) = 0.04/0.52 = 0.077$
$P(p = 0.30|c_1 < \bar{p} \leqslant c_2) = 0.04/0.32 = 0.125$
$P(p = 0.30|\bar{p} > c_2) = 0.11/0.16 = 0.687$
b. See figure.

Decision tree diagram for
exercise 2(b)

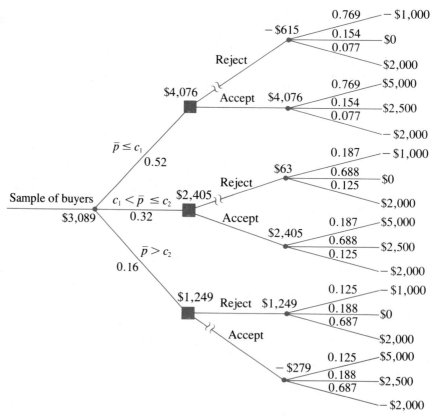

INDEX